Human Genetics

Concepts and Applications

Ricki Lewis

Genetic Counselor
CareNet Medical Group
Schenectady, New York

Adjunct Assistant Professor of
Medical Education
Alden March Bioethics Institute
Albany Medical College

Writer, Medscape Medical News
  Genetic Literacy Project

Blogger, Public Library of Science

HUMAN GENETICS

Published by McGraw Hill LLC, 2 Penn Plaza, New York, NY 10121. Copyright ©2021 by McGraw Hill LLC. All rights reserved. Printed in the United States of America. No part of this publication may be reproduced or distributed in any form or by any means, or stored in a database or retrieval system, without the prior written consent of McGraw Hill LLC, including, but not limited to, in any network or other electronic storage or transmission, or broadcast for distance learning.

Some ancillaries, including electronic and print components, may not be available to customers outside the United States.

This book is printed on acid-free paper.

1 2 3 4 5 6 7 8 9 LWI 24 23 22 21 20

ISBN 978-1-260-57046-5
MHID 1-260-57046-0

Cover Image: *Portraits-Rawpixel.com/Shutterstock, Rawpixel Ltd/Getty Images, Kivilcim Pinar/iStock/ Getty Images. World Map-vectorlight/Shutterstock. DNA-Olena_T/Getty Images.*

All credits appearing on page or at the end of the book are considered to be an extension of the copyright page.

The Internet addresses listed in the text were accurate at the time of publication. The inclusion of a website does not indicate an endorsement by the authors or McGraw Hill LLC, and McGraw Hill LLC does not guarantee the accuracy of the information presented at these sites.

mheducation.com/highered

About the Author

Courtesy of Larry Lewis

Ricki Lewis has built an eclectic career in communicating the excitement of genetics and genomics, combining skills as a geneticist and a journalist. She currently writes the popular weekly blog, DNA Science, at Public Library of Science and contributes frequent articles to Medscape Medical News and the Genetic Literacy Project.
Dr. Lewis has authored or coauthored several university-level textbooks and is the author of the narrative nonfiction book, *The Forever Fix: Gene Therapy and the Boy Who Saved It*, as well as an essay collection, a novel, and a short "basics" book on human genetics. She teaches an online course on "Genethics" for the Alden March Bioethics Institute of Albany Medical College and is a genetic counselor for a private medical practice. Her passion is rare genetic diseases; she writes often about affected families who are pioneering DNA-based treatments.

Dedicated to the families who live with genetic diseases, the health care providers who help them, and the researchers who develop new tests and treatments.

Brief Contents

PART 1
Introduction 1

CHAPTER 1
The Information in a Human Genome 1

CHAPTER 2
Cells 17

CHAPTER 3
Meiosis, Development, and Aging 43

PART 2
Transmission Genetics 71

CHAPTER 4
Single-Gene Inheritance 71

CHAPTER 5
Beyond Mendel's Laws 91

CHAPTER 6
Matters of Sex 111

CHAPTER 7
Complex Traits 130

CHAPTER 8
Genetics of Behavior 148

PART 3
DNA and Chromosomes 165

CHAPTER 9
DNA Structure and Replication 165

CHAPTER 10
Gene Action: From DNA to Protein 182

CHAPTER 11
Gene Expression and Epigenetics 202

CHAPTER 12
Gene Mutation 217

CHAPTER 13
Chromosomes 241

CHAPTER 14
Genomes 265

PART 4
Population Genetics 277

CHAPTER 15
Constant Allele Frequencies 277

CHAPTER 16
Changing Allele Frequencies 286

CHAPTER 17
Human Ancestry and Evolution 306

CHAPTER 18
The Genetics of Identity 328

PART 5
Immunity and Cancer 347

CHAPTER 19
Genetics of Immunity 347

CHAPTER 20
Cancer Genetics and Genomics 373

PART 6
Genetic Technology 395

CHAPTER 21
DNA Technologies 395

CHAPTER 22
Genetic Testing and Treatment 414

CHAPTER 23
Reproductive Technologies 433

Part 1: Tom Grill/Getty Images; Part 2: Courtesy, Jane Mervar; Part 3: Dinodia/age fotostock; Part 4: Portraits-Rawpixel.com/Shutterstock, Rawpixel Ltd/Getty Images, Kivilcim Pinar/iStock/Getty Images. World Map-vectorlight/Shutterstock. DNA-Olena_T/Getty Images; Part 5: CDC/C. Goldsmith, P. Feorino, E. L. Palmer, W. R. McManus; Part 6: Ingram Publishing/Alamy

Contents

About the Author iii
Preface ix
Applying Human Genetics xiv
The Human Touch xv
The Lewis Guided Learning System xvi

PART 1 Introduction 1

CHAPTER 1
The Information in a Human Genome 1

1.1 Introducing Genes and Genomes 2
1.2 Levels of Genetics and Genomics 3
1.3 Applications of Genetics and Genomics 8
1.4 A Global Perspective on Genomes 11

CHAPTER 2
Cells 17

2.1 Introducing Cells 18
2.2 Cell Components 18
2.3 Cell Division and Death 28
2.4 Stem Cells 33
2.5 The Human Microbiome 38

CHAPTER 3
Meiosis, Development, and Aging 43

3.1 The Reproductive System 44
3.2 Meiosis 45
3.3 Gametes Mature 49
3.4 Prenatal Development 53
3.5 Birth Defects 62
3.6 Maturation and Aging 64

PART 2 Transmission Genetics 71

CHAPTER 4
Single-Gene Inheritance 71

4.1 Following the Inheritance of One Gene 72
4.2 Single-Gene Inheritance Is Rare 76
4.3 Following the Inheritance of More Than One Gene 80
4.4 Pedigree Analysis 83

Chapter 1: Tom Grill/Getty Images; Chapter 2: milosljubicic/Shutterstock; Chapter 3: Courtesy The Progeria Research Foundation; Chapter 4: Courtesy, Jane Mervar

CHAPTER 5

Beyond Mendel's Laws 91

5.1 When Gene Expression Appears to Alter Mendelian Ratios 92
5.2 Mitochondrial Genes 98
5.3 Linkage 101

CHAPTER 6

Matters of Sex 111

6.1 Our Sexual Selves 112
6.2 Traits Inherited on Sex Chromosomes 116
6.3 Sex-Limited and Sex-Influenced Traits 121
6.4 X Inactivation 122
6.5 Parent-of-Origin Effects 124

CHAPTER 7

Complex Traits 130

7.1 Genes and the Environment Mold Traits 131
7.2 Polygenic Traits Are Continuously Varying 133
7.3 Methods to Investigate Complex Traits 135
7.4 A Closer Look: Body Weight 143

CHAPTER 8

Genetics of Behavior 148

8.1 Genes and Behavior 149
8.2 Sleep 150
8.3 Intelligence and Intellectual Disability 151
8.4 Drug Addiction 152
8.5 Mood Disorders 155
8.6 Schizophrenia 156
8.7 Autism 159

PART 3 DNA and Chromosomes 165

CHAPTER 9

DNA Structure and Replication 165

9.1 Experiments Identify and Describe the Genetic Material 166
9.2 DNA Structure 170
9.3 DNA Replication—Maintaining Genetic Information 174
9.4 Sequencing DNA 177

CHAPTER 10

Gene Action: From DNA to Protein 182

10.1 The Importance of Proteins 183
10.2 Transcription Copies the Information in DNA 184
10.3 Translation of a Protein 189
10.4 Processing a Protein 195

Chapter 5: Courtesy of Matthew Garcia ; Chapter 6: Shutterstock/Monkey Business Images; Chapter 7: anetta/Shutterstock; Chapter 8: Sami Sarkis/Photographer's Choice RF/Getty Images; Chapter 9: Dinodia/age fotostock; Chapter 10: NASA Photo/Alamy Stock Photo

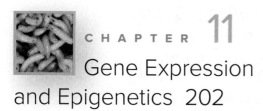

CHAPTER 11

Gene Expression and Epigenetics 202

- 11.1 Gene Expression Through Time and Tissue 203
- 11.2 Control of Gene Expression 206
- 11.3 Maximizing Genetic Information 210
- 11.4 Most of the Human Genome Does *Not* Encode Protein 212

CHAPTER 12

Gene Mutation 217

- 12.1 The Nature of Gene Variants 218
- 12.2 A Closer Look at Two Mutations 220
- 12.3 Allelic Diseases 223
- 12.4 Causes of Mutation 224
- 12.5 Types of Mutations 227

- 12.6 The Importance of Position 232
- 12.7 DNA Repair 233

CHAPTER 13

Chromosomes 241

- 13.1 Portrait of a Chromosome 242
- 13.2 Detecting Chromosomes 244
- 13.3 Atypical Chromosome Number 247
- 13.4 Atypical Chromosome Structure 254
- 13.5 Uniparental Disomy—A Double Dose from One Parent 260

CHAPTER 14

Genomes 265

- 14.1 From Genetics to Genomics 266
- 14.2 Analysis of Human Genome Content 271
- 14.3 A Genomic View Expands Knowledge 274

PART 4 Population Genetics 277

CHAPTER 15

Constant Allele Frequencies 277

- 15.1 Population Genetics Underlies Evolution 278
- 15.2 Constant Allele Frequencies 279
- 15.3 Applying Hardy-Weinberg Equilibrium 281

CHAPTER 16

Changing Allele Frequencies 286

- 16.1 Population Matters: Steel Syndrome in East Harlem 287

- 16.2 Nonrandom Mating 288
- 16.3 Migration 290
- 16.4 Genetic Drift 291
- 16.5 Mutation 294
- 16.6 Natural Selection 295
- 16.7 Eugenics 300

CHAPTER 17

Human Ancestry and Evolution 306

- 17.1 Human Origins 307
- 17.2 Methods to Study Molecular Evolution 315
- 17.3 The Peopling of the Planet 318
- 17.4 What Makes Us Human? 322

Chapter 11: The Natural History Museum/Alamy Stock Photo; Chapter 12: "The Blue People," copyright ©1982-2016 by Walt Spitzmiller. All Rights Reserved; Chapter 14: Olena_T/Getty Images; Chapter 15: SDI Productions/Getty Images; Chapter 16: MaraZe/Shutterstock; Chapter 17: Peter Johnson/Corbis/VCG/Getty Images

CHAPTER 18
The Genetics of Identity 328

18.1 Genetics as an Informational Science 329
18.2 Forensic DNA Typing 329
18.3 Genetic Genealogy 338
18.4 Forensic STRs Meet Genetic Genealogy SNPs 342

PART 5 Immunity and Cancer 347

CHAPTER 19
Genetics of Immunity 347

19.1 The Importance of Cell Surfaces 348
19.2 The Human Immune System 351
19.3 Abnormal Immunity 356
19.4 The Immune Response During Pregnancy 361
19.5 Altering Immunity 362
19.6 Using Genomics to Fight Infection 367

CHAPTER 20
Cancer Genetics and Genomics 373

20.1 Cancer Is an Abnormal Growth That Invades and Spreads 374
20.2 Cancer at the Cellular Level 377
20.3 Cancer Genes and Genomes 381
20.4 Diagnosing and Treating Cancer 388

PART 6 Genetic Technology 395

CHAPTER 21
DNA Technologies 395

21.1 Patenting DNA 396
21.2 Modifying DNA 397
21.3 Monitoring Gene Function 404
21.4 Gene Silencing and Genome Editing 406

CHAPTER 22
Genetic Testing and Treatment 414

22.1 Genetic Counseling 415
22.2 Genetic Testing 417
22.3 Treating Genetic Disease 423
22.4 CRISPR-Cas9 in Diagnosis and Treatment 426

CHAPTER 23
Reproductive Technologies 433

23.1 Savior Siblings and More 434
23.2 Infertility and Subfertility 435
23.3 Assisted Reproductive Technologies 438
23.4 Extra Embryos 444

Index I-1

Chapter 18: Andrew Brookes/Getty Images; Chapter 19: CDC/C. Goldsmith, P. Feorino, E. L. Palmer, W. R. McManus; Chapter 20: Source: Jonathan Bailey, National Human Genome Research Institute; Chapter 21: Ingram Publishing/Alamy; Chapter 22: Courtesy Glenn O'Neill; Chapter 23: MedicalRF.com

Preface

The Evolution of *Human Genetics: Concepts and Applications*

This textbook began as a table of contents scribbled on a cocktail napkin at a publisher's meeting circa 1990, when genetics was on the precipice of becoming genomics. Since then, in addition to clearly explaining the basics of inheritance, the book has chronicled the beginnings of genome sequencing, the first clinical trials of gene therapy, the gradual expansion and explosion of genetic testing, and now the entrance of DNA information into our everyday lives.

In the middle editions, focus shifted to human genetics as the science of our diversity and introduced consumer DNA testing. That has now detonated as millions of people spit into tubes and swipe their cheeks to learn about their health, traits, families, and ancestry. At the same time, DNA evidence in forensic investigations has opened up new concerns about genetic privacy.

In 2019, the author learned, out-of-the-blue thanks to consumer DNA testing, of several half-siblings, and with further information, that she was donor-conceived. The impact was stunning, but ultimately positive. As a result, this edition of Human Genetics: Concepts and Applications *has been revised and updated with a new sensitivity to the power of genetic information, including a new chapter, "The Genetics of Identity." It covers genetic genealogy, privacy, and forensic DNA typing.*

What Sets This Book Apart

The vivid narrative writing style reflects Dr. Lewis's eclectic experience as a health and science writer, professor, speaker, and genetic counselor. Cases, historical asides, and descriptions of new technologies propel the clear explanations of concepts and mechanisms. In this new edition, a broader coverage of topics, yet with more practical information, prepares the reader to evaluate media reports, interpret genetic test results, and question their health care providers.

Updates to this edition include

- Using DNA to find the Golden State Killer
- The right *not* to know genetic information
- Percent of the genome that relatives share
- Genetics and transgender identity
- An astronaut's altered gene expression
- Tracking RNA to pinpoint time of death
- Inbreeding in Charles Darwin's family
- DNA typing reunites families
- Engineering a plant-based burger
- CRISPR, RNAi, antisense applications

Pie charts ease understanding of big data:

- Parts of a human genome (figure 11.14)
- Gene variant classification (figure 12B)
- Five sources of evolutionary change (figure 16.2)
- Ethnicity estimates (figure 18.12a)

A minor reorganization moves Genomes (in past editions under Technology, near the book's end) to follow Chromosomes (chapter 13), now that it is no longer futuristic. The new chapter 18, The Genetics of Identity, fits perfectly in the population genetics unit. Figures that did not add to content or concepts have been dropped, as have been Technology Timelines and some Glimpse of History boxes. These changes emphasize the Bioethics boxes, as the science of genetics becomes more a part of everyday life. Some boxes in past editions are now part of the text.

The Lewis Guided Learning System

Each chapter begins with two views of the content. *Learning Outcomes* embedded in the table of contents guide the student in setting goals and mastering material, while *The Big Picture* encapsulates the chapter's theme. The chapter opener essay and figure grab attention, pulling the reader into a true narrative.

Content flows logically through three to five major sections per chapter. Each section concludes with *Key Concepts Questions*. Answers to the questions marked with an asterisk are provided to the instructor in the Instructor Resources in Connect®. Instructors can choose to share these with the students.

Clinical Connections and *Bioethics* boxed readings pepper the chapters, each including challenging *Questions for Discussion*. *A Glimpse of History* boxes provide context and perspective.

Each chapter ends with a *Study Guide*. It includes a *Summary*, defined *Key Terms*, and straight recall *Review Questions*.

Then follow *Applied Questions* that begin with one question that relates to the Chapter Opener, *Forensic Focus* questions, and *Case Studies and Research Results*. These types of inquiry challenge critical thinking and data evaluation skills.

Dynamic Art

Outstanding photographs and dimensional illustrations, vibrantly colored, are featured throughout *Human Genetics: Concepts and Applications*. Figure types include process figures with numbered steps, micro to macro representations, and the combination of art and photos to relate stylized drawings to real-life biological structures.

New to This Edition

The new edition embraces the broadening of human genetics from an academic and medical discipline to an informational science that can be highly personal, yet have societal impact. Changes to enhance learning include better contrast in figures, particularly for chromosomes; additional subheads; new figures, tables, and end-of-chapter questions; and key terms with definitions at the ends of chapters. Answers to even-numbered Key Concepts questions are available to instructors in the Instructor Resources in Connect®. Some *Glimpses of History* and all *Technology Timelines* have been cut.

Chapter 1 The Information in A Human Genome
- Types of discoveries from consumer DNA testing

Chapter 2 Cells
- Stem cell tourism

Chapter 4 Single-Gene Inheritance
- Monogenic versus complex traits
- The right *not* to know genetic information

Chapter 5 Beyond Mendel's Laws
- The complex inheritance of red hair

Chapter 6 Matters of Sex
- Transgender identity
- Brain cancer as a sex-influenced trait

Chapter 7 Complex Traits
- De-emphasis on European ancestry in examples
- Missing heritability indicating environmental component
- Polygenic risk scores

Chapter 9 DNA Structure and Replication
- Many small changes to improve clarity

Chapter 10 Gene Action: From DNA to Protein
- Altered gene expression in an astronaut
- A missing enzyme that knocks out nine others

Chapter 11 Gene Expression and Epigenetics
- Tracking RNA to pinpoint time of death
- Gene expression in the placenta

Chapter 13 Chromosomes
- Shifted focus to newer techniques

Chapter 14 Genomes (previously chapter 22)
- How a genomic view expands knowledge
- Sequencing the genomes of newborns

Chapter 15 Constant Allele Frequencies (formerly chapter 14)
- Tailoring genetic tests to population groups

Chapter 16 Changing Allele Frequencies (formerly chapter 15)
- Inbreeding in Charles Darwin's family
- One pie chart replaces several shape figures

Chapter 17 Human Ancestry and Evolution (formerly chapter 16)
- Interpreting Neanderthal DNA in consumer tests
- Migration through "vegetated corridors"

Chapter 18 Genetics of Identity
- New chapter, which blends tests and technologies used in ancestry and forensic testing
- DNA typing to reunite families
- Forensic DNA phenotyping to identify remains
- Limitations of forensic DNA evidence
- Capturing the Golden State Killer

Chapter 20 Cancer Genetics and Genomics (formerly chapter 18)
- Targeted treatments

Chapter 21 DNA Technologies (formerly chapter 19)
- Engineering a better veggie burger
- CRISPR patent battle
- Antisense and RNAi in agriculture and health care
- Editing the human germline

Chapter 22 Genetic Testing and Treatment (formerly chapter 20)
- Gene therapy vs. gene editing
- Pharmacogenetics and depression

Chapter 23 Reproductive Technologies (formerly chapter 21)
- Effect of pollution on female fertility
- Changes in sperm donation

New Figures

1.1 Consumer genetics
1.2 The cost of sequencing a human genome has plunged
1.7 A normal female karyotype
1.9 Percent of the genome shared with relatives
3.14 Oogenesis and spermatogenesis compared
3.24 The nuclear membrane of a dying cell in progeria
10A Multiple sulfatase deficiency
12.1 Somatic mutation puts patches on a golden retriever
12.10 A point mutation can alter a splice site
12.11 A pseudogene causes mutation
12B Gene variant classification
13A Trisomy vs. translocation in Down syndrome

15.1	Different populations have different frequencies of single-gene diseases	19.15	Herd immunity
15.2	Microevolution and macroevolution in speech	20.17	PD-L1 immune checkpoint inhibitor
15.7	Risk of inheriting CF	20.18	Chimeric antigen receptor T cell (CAR T) therapy
16.1	Inbreeding in the family of Charles Darwin	22.7	Comparison of gene therapy and gene editing

New Tables

16.2	The five sources of evolutionary change at the population level
16.3	Types of genetic drift
16A	Bili lights for Crigler-Najjar syndrome
16.4	A telltale mutation for PKU
17.17	Combinations of characteristics reflect ancestors from different populations
18.2-5	Applications of comparing 20 CODIS loci
18.8	Grandmothers of the Plaza de Mayo
18.9	Forensic DNA phenotyping
18.10	A maternal haplogroup
18.11	A paternal haplogroup
18.12	Ethnicity estimates
18.13	Centimorgans shared across chromosomal segments
18.14	Combining SNPs and STRs as a forensic tool

5.2	Shared Centimorgans Indicate Relationships
18.2	A Maternal Lineage
18.3	Selected Y Haplogroups
18.4	Distribution of 3,000 SNP Profiles Across 26 Reference Populations
19A	Human Organoids
20.2	Viruses That Can Cause Cancer
21.1	Key Biotechnology Inventions
21.3	Drugs Made in Transgenic Animals
21.4	Engineered Crops
22.5	Gene Therapies
23.1	Mutations That Interfere with Early Prenatal Development
23.2	Live Births Following IVF

New to Connect
Remote Proctoring & Browser-Locking Capabilities

New remote proctoring and browser-locking capabilities, hosted by Proctorio within Connect, provide control of the assessment environment by enabling security options and verifying the identity of the student.

Seamlessly integrated within Connect, these services allow instructors to control students' assessment experience by restricting browser activity, recording students' activity, and verifying students are doing their own work.

Instant and detailed reporting gives instructors an at-a-glance view of potential academic integrity concerns, thereby avoiding personal bias and supporting evidence-based claims.

Writing Assignment
Available within McGraw-Hill Connect® and McGraw-Hill Connect® Master, the Writing Assignment tool delivers a learning experience to help students improve their written communication skills and conceptual understanding. As an instructor you can assign, monitor, grade, and provide feedback on writing more efficiently and effectively.

ACKNOWLEDGMENTS

Human Genetics: Concepts and Applications, Thirteenth Edition, would not have been possible without the assistance of the editorial and production team: portfolio managers Ian Townsend and Andrew Urban, senior product developer Donna Nemmers, marketing manager Kelly Brown, project manager Kelly Hart, designer Beth Blech, content licensing specialist Beth Cray, and copyeditor Cheryl Ferguson. I am grateful to the reviewers of this textbook over the years, whose suggestions have contributed to its clarity and accuracy, and to readers of my weekly articles who have provided feedback on specific topics. As always, many thanks to my wonderful husband, Larry, for his support and encouragement.

13th Edition Reviewers

Justin Blau
New York University

Bruce Bowerman
University of Oregon

Chi-hua Chiu
Kent State University

Joseph Francis
The Master's University

Rivka Glaser
Stevenson University

Jaime Grace
Bradley University

Katie Hollis
William Rainey Harper College

Patricia Matthews
Grand Valley State University

Harper G. Mazock
Oakton Community College

Kiran Misra
Edinboro University

Geoffrey Mitchell
Wofford College

Robert Moss
Wofford College

Jennifer N. Ness
Bradley University

Wade A. Nichols
Illinois State University

Alexey G. Nikitin
Grand Valley State University

Sharon Rittman
North Carolina Central University

Christine Rushlow
New York University

Mark Sanders
University of California–Davis

Roxann Schroeder
Humboldt State University

Yunqiu Wang
University of Miami

Applying Human Genetics

Chapter Openers

1	Eve's Genome	1
2	Diagnosis of Rett Syndrome from a Tooth	17
3	Progress for Progeria	43
4	Juvenile Huntington Disease: The Cruel Mutation	71
5	The Complex Inheritance of Red Hair	91
6	China's One-Child Policy Created an Aged Population	111
7	The Complex Genetics of Athletics	130
8	Posttraumatic Stress Disorder: Who Is Predisposed?	148
9	Elephant Forensics	165
10	Altering Gene Expression in Space	182
11	Tracking RNA to Pinpoint Time of Death	202
12	The Blue People of Troublesome Creek	217
13	Shattered!	241
14	Sequencing the Genomes of the Deceased	265
15	Tailoring Genetic Tests to Population Groups	277
16	The Evolution of Lactose Tolerance	286
17	Indigenous Peoples	306
18	Postconviction DNA Testing	328
19	Mimicking a Mutation to Protect Against HIV	347
20	Treating Cancer by Location and Mutation	373
21	Engineering a Better Veggie Burger	395
22	Eliza Gets Gene Therapy	414
23	Choosing an Embryo	433

Matt Garcia

Source: CDC/Carl Flint; Armed Forces Institute of Pathology

A GLIMPSE OF HISTORY

3	Evolution of pregnancy testing	
4	Gregor Mendel	
5	The murdered Romanovs and mitochondrial DNA	
10	The RNA tie club	
12	Fava beans and G6PD deficiency	
12	The discovery of sickled cells	
13	Turner syndrome discovery	
16	Malaria in the United States	
17	Neanderthal discoveries	
18	Famous forensics cases	
19	Smallpox	
20	Chronic myelogenous leukemia	

The Human Touch

Clinical Connections

1.1	Genome Sequencing Ends a Child's "Diagnostic Odyssey"	11
2.1	Inborn Errors of Metabolism Affect the Major Biomolecules	20
2.2	A Girl with Giant Axons	28
3.1	When an Arm Is Really a Leg: Homeotic Mutations	58
4.1	"65 Roses": Progress in Treating Cystic Fibrosis	79
5.1	The Roots of Familial Alzheimer Disease	96
6.1	Colorblindness and Tetrachromacy	118
7.1	Many Genes Control Heart Health	132
8.1	Cannabis: Addictive Drug and Seizure Treatment	154
10.1	A Missing Enzyme Knocks out Nine Others: Multiple Sulfatase Deficiency	194
11.1	The Dutch Hunger Winter: One Route to Schizophrenia	209
12.1	Fragile X Mutations Affect Boys and Their Grandfathers	231
13.1	Distinguishing a Robertsonian Translocation from a Trisomy	257
16.1	The Founder Effect and "Plain" Populations	292
19.1	Viruses	349
20.1	A Liquid Biopsy Monitors Cancer Recurrence and Response to Treatment	391
22.1	The Rocky History of Gene Therapy	428
23.1	The Case of the Round-Headed Sperm	436

All rights reserved by Sally McBurney

Jane Ades/National Human Genome Research Institute

Bioethics

1	Genetic Privacy and Pandora's Box	12
2	Stem Cell Tourism	37
3	Why a Clone Is Not an Exact Duplicate	55
4	The Right Not to Know Genetic Information: Huntington Disease	87
5	Replacing Mitochondria	100
6	Probing a Genetic Basis of Transgender Identity	116
8	Finding Fault for Autism	161
9	Infidelity Testing	178
12	A Diagnostic Dilemma: What to Do with a VUS?	237
13	Will Trisomy 21 Down Syndrome Disappear?	252
14	Should We Create Genomes?	273
16	Designer Babies: Is Prenatal Genetic Testing Eugenic?	301
18	Should DNA Collected Today Be Used to Solve a Past Crime?	337
19	Replacement Parts	366
21	EPO: Built-in Blood Booster or Performance-Enhancing Drug?	401
22	Secondary Findings: Does DNA Sequencing Provide Too Much Information?	420
23	Removing and Using Gametes after Death	442

The Lewis Guided Learning System

In-Chapter Review Tools include summary tables and the end-of-chapter Study Guide, which has a summary of each section, Key Terms, and several sets of review questions, which are handy tools for reference and study. Boldfaced terms within the chapter and in the end-of-chapter Summary are defined in the Key Terms section of the Study Guide.

Learning Outcomes preview major chapter topics in an inquiry-based format according to numbered sections.

The Big Picture encapsulates chapter content at the start.

Chapter Openers vividly relate content to real life.

Key Concepts Questions follow each numbered section.

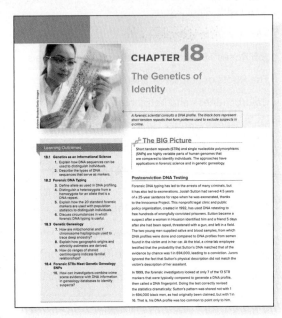

Clinical Connection boxes discuss how genetics and genomics impact health and health care and include questions for discussion.

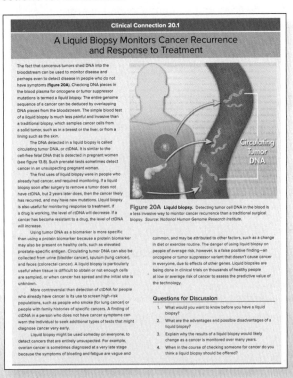

Bioethics boxes include Questions for Discussion.

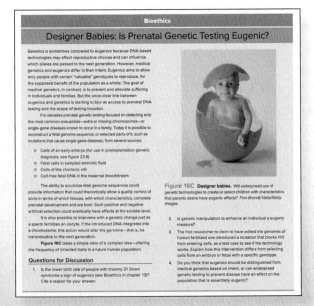

Each chapter ends with a point-by-point **Summary**.

Key Terms that are boldfaced within the chapter are defined in the Key Terms section.

Review Questions assess content knowledge.

Applied Questions help students develop problem-solving skills. The first question in this section relates back to the chapter opener.

Forensics Focus questions probe the use of genetic information in criminal investigations.

Case Studies and Research Results use stories based on accounts in medical and scientific journals; real clinical cases; posters and reports from professional meetings; interviews with researchers; and fiction to ask students to analyze data and predict results.

Study Guide

Summary

2.1 Introducing Cells

1. Cells are the basic units of life and comprise the human body. Inherited traits and illnesses can be understood at the cellular and molecular levels.
2. All cells share certain features, but they are also specialized because they express different subsets of genes.
3. Somatic (body) cells are **diploid**, and **germ cells** (sperm and egg cells) are **haploid**. Stem cells produce new cells.

2.2 Cell Components

4. A **prokaryotic cell** does not have a nucleus or other **organelles**. A **eukaryotic cell** has organelles, including a nucleus.
5. Cells consist primarily of water and several types of macromolecules: **carbohydrates, lipids, proteins,** and **nucleic acids**.
6. Organelles sequester related biochemical reactions, improving the efficiency of life functions and protecting the cell. The cell also consists of **cytoplasm** and chemicals.
7. The nucleus contains DNA and a **nucleolus**, a site of ribosome synthesis. **Ribosomes** provide scaffolds for protein synthesis; they exist free in the cytoplasm or complexed with the rough **endoplasmic reticulum (ER)**. In protein synthesis, **messenger RNA** (mRNA) carries the information held in a gene's DNA sequence and **transfer RNA** (tRNA) connects the mRNA to protein building blocks (amino acids).
8. In secretion, the rough ER is where protein synthesis begins. Smooth ER is the site of lipid synthesis, transport, and packaging; the **Golgi apparatus** packages secretions into **vesicles**, which exit through the **plasma membrane**. **Exosomes** ferry molecules between cells.
9. **Lysosomes** contain **enzymes** that dismantle debris through **autophagy**. **Endosomes** are vesicles derived from the plasma membrane that ferry materials to lysosomes. **Peroxisomes** house enzymes that catalyze a variety of chemical reactions.

Key Terms

anaphase (ăn′ə-fāz′) Stage of mitosis when the centromeres of replicated chromosomes part.
apoptosis (ăp″ō-tō′sĭs) A form of cell death that is a normal part of growth and development.
autophagy (au-toph′-a-gy) A process in which a cell dismantles its own debris.
carbohydrate (kar″bo-hī′-drāt) An organic compound that consists of carbon, hydrogen, and oxygen in a 1:2:1 ratio. A sugar or a starch.
cell cycle A cycle of events describing a cell's preparation for division and division itself.
cellular adhesion A series of interactions among the proteins that connect cells.
centriole (sĕn′-trē-ohl) A structure that organizes microtubules into the mitotic spindle.
centromere (sĕn′-trō-mîr) The largest constriction in a chromosome, at a specific site in each chromosome type.
centrosome (sĕn′-trō-sōm) A structure built of centrioles and proteins that organizes microtubules into a spindle during cell division.
chromatid (krō′ mə-tĭd) A single, very long DNA molecule and its associated proteins, forming a longitudinal half of a replicated chromosome.
cytokinesis (sī-tō-kin-ē′-sis) Division of the cytoplasm and its contents.
cytoplasm (sī′-tō-plăzm) Cellular contents other than organelles.
cytoskeleton (sī-tō-skĕl′ĭ-tn) A framework of protein tubules and rods that supports the cell and gives it a distinctive form.
diploid (dĭp′ loid) A cell containing two sets of chromosomes.
embryonic stem (ES) cell (ĕmbrē-ŏn′ĭk) A cell derived in laboratory culture from inner cell mass cells of very early embryos that can self-renew and produce some daughter cells that can differentiate as any cell type.
endoplasmic reticulum (ĕn′-dō-plaz-mĭk rĕtĭk′-u-lum) An organelle consisting of a labyrinth of membranous tubules on which proteins, lipids, and sugars are synthesized.
endosome A vesicle that buds inward from the plasma membrane.
enzyme (ĕnzīm) A protein that speeds the rate of a specific biochemical reaction.
eukaryotic cell (yōō-kar′ē-ŏt′ĭk sel) A complex cell containing organelles, including a nucleus.
exosome (x-ō-sōm) A vesicle that carries molecules from cell to cell.
G₀ phase An offshoot of the cell cycle in which the cell remains specialized but does not replicate its DNA or divide.
G₁ phase The stage of the cell cycle following mitosis in which the cell resumes synthesis of proteins, lipids, and carbohydrates.
G₂ phase The stage of the cell cycle following S phase but before mitosis, when certain proteins are synthesized.
germ cells Cells that give rise to sperm or eggs.
Golgi apparatus (gŏl′jē) An organelle, consisting of flattened, membranous sacs, that packages secretion components.
haploid (hap′ loid). A cell containing one set of chromosomes.
hormone (hor′ mōn) A biochemical produced in a gland and carried in the blood to a target organ, where it exerts an effect.
induced pluripotent stem (iPS) cells Somatic cells that are reprogrammed toward an alternative developmental fate by altering their gene expression.
intermediate filament A type of cytoskeletal component made of different proteins in different cell types.

Review Questions

1. Match each organelle to its function.

Organelle	Function
a. lysosome	1. lipid synthesis
b. rough ER	2. houses DNA
c. nucleus	3. energy extraction
d. smooth ER	4. dismantles debris
e. Golgi apparatus	5. detoxification
f. mitochondrion	6. protein synthesis
g. peroxisome	7. processes secretions

2. Describe a disease or symptoms that result when each of the following organelles or structures malfunctions.
   a. lysosome
   b. peroxisome
   c. mitochondrion
   d. cytoskeleton
   e. ion channel
3. What advantage does compartmentalization provide to a large and complex cell?
4. Distinguish between an exosome and an endosome.
5. Give two examples of how the plasma membrane functions as more than just a covering of the cell's insides.
6. List four types of controls on cell cycle rate.
7. How can all of a person's cells contain exactly the same genetic material, yet specialize as bone cells, nerve cells, muscle cells, and connective tissue cells?
8. Distinguish between
   a. a bacterial cell and a eukaryotic cell.

Applied Questions

1. The chapter opener describes diagnosing Rett syndrome, which affects neurons in the brain, from DNA extracted from a saved baby tooth. Explain why a mutation detected in one cell type can be used to help diagnose an inherited condition that affects a different cell type.
2. If you wanted to create a synthetic organelle to test new drugs for toxicity, which natural organelle's function would you try to replicate?
3. An inherited form of migraine is caused by a mutation in a gene (SCN1A) that encodes a sodium channel in neurons. What is a sodium channel, and in which cell structure is it located?
4. In a form of Parkinson disease, a protein called alpha synuclein accumulates because of impaired autophagy. Which organelle is implicated in this form of the disease?
5. Why wouldn't a cell in an embryo likely be in phase G₀?
6. Describe two ways to derive stem cells without using human embryos.
7. Two roommates regularly eat 100-calorie packages of cookies. After a semester of each roommate eating one package a night, and following similar diet and exercise plans otherwise, one roommate has gained 12 pounds, but the other's weight has stayed the same. If genetics does not account for the difference in weight gain, what other factor discussed in the chapter might?
8. A "telomere health DNA test" sold online promises to "track your cellular age based on your telomere length" so that you can "improve it." The company sells supplement packages to help in the quest for biological youth. What further information might be helpful in deciding whether to purchase this type of genetic test and the suggested supplements?

Forensics Focus

1. A man who owned a company that provides human tissues was sentenced to serve many years in prison for trafficking in body parts taken, without consent, from dismembered corpses from funeral homes. Thousands of parts from hundreds of bodies were used in surgical procedures, including hip replacements and dental implants. The most commonly used product was a bone paste. Many family members of the tissue sources testified at the trials. Explain why cells from bone tissue can be matched to blood or cheek lining cells from blood relatives.

Case Studies and Research Results

1. A boy with "congenital skin crease Kunze type" has rings of creases on his arms and legs, unusual facial features, intellectual disability, slow growth, and a cleft palate. He has a mutation in the gene TUBB, and his condition is termed a "tubulinopathy." What part of his cells is affected?
2. Members of a large, three-generation family in Italy do not notice burns, broken bones, or even red hot chili pepper powder injected under the skin. What might their abnormality be on a cellular level?
3. Researchers isolated stem cells from fat removed from people undergoing liposuction, a procedure to remove fat. The stem cells can give rise to muscle, fat, bone, and cartilage cells.
   a. Are the stem cells totipotent or pluripotent?
   b. Are these stem cells ES cells, iPS cells, or adult stem cells?

xvii

Dynamic Art Program

Multilevel Perspective

Illustrations depicting complex structures show microscopic and macroscopic views to help students see relationships among increasingly detailed drawings.

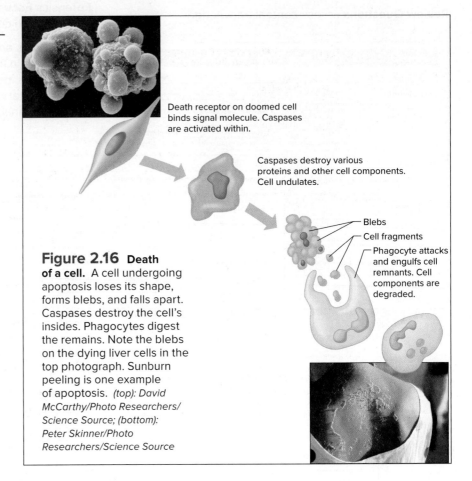

Figure 2.16 Death of a cell. A cell undergoing apoptosis loses its shape, forms blebs, and falls apart. Caspases destroy the cell's insides. Phagocytes digest the remains. Note the blebs on the dying liver cells in the top photograph. Sunburn peeling is one example of apoptosis. *(top): David McCarthy/Photo Researchers/Science Source; (bottom): Peter Skinner/Photo Researchers/Science Source*

New Technologies

CAR T cell therapy harnesses a personalized immune response against cancer cells.

Figure 20.18 Chimeric antigen receptor therapy arms a patient's T cells to attack cancer cells. (a) A blood sample is taken from a patient and the T cells are separated. (b) In laboratory glassware, the patient's T cells are exposed to a gene modified to include the genetic instructions for both a T-cell receptor and an antibody. (c) Millions of copies of the modified T cells are cultured in the dish and then (d) transferred to the patient. The modified cells circulate and produce the hybrid protein that consists of both the T-cell receptor (e) and the antibody. The receptor guides the T cell to a cancer cell, where the antibody part binds, which alerts the immune system to send perforins that shatter the cancer cell and granulysins that dismantle the cancer cell from within. *(c): 2017 Terese Winslow LLC U.S. Govt. has certain rights*

Practical Information

Consumer genetics reveals information that may be pertinent to a person's health, traits, family, and ancestry.

Figure 1.1 Consumer genetics. *DNA tests reveal several types of information and can bring surprises.*

Process Figures

Complex processes are broken down into a series of smaller steps that are easy to follow.

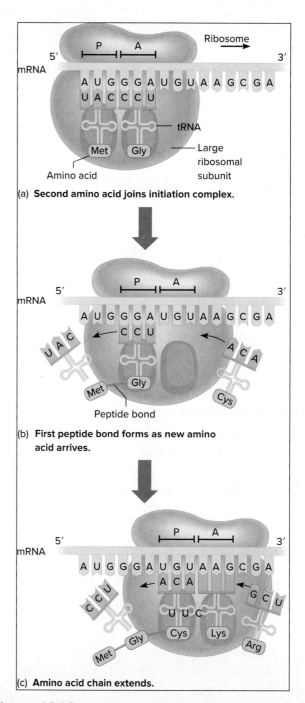

Figure 10.16 Building a polypeptide. (a) A large ribosomal subunit binds to the initiation complex, and a tRNA bearing a second amino acid (glycine, in this example) forms hydrogen bonds between its anticodon and the mRNA's second codon at the A site. The first amino acid, methionine, occupies the P site. **(b)** The methionine brought in by the first tRNA forms a peptide bond with the amino acid brought in by the second tRNA, and a third tRNA arrives, in this example carrying the amino acid cysteine, at the temporarily vacant A site. **(c)** A fourth and then fifth amino acid link to the growing polypeptide chain. The process continues until reaching a stop codon.

Instructors: Student Success Starts with You

Tools to enhance your unique voice

Want to build your own course? No problem. Prefer to use our turnkey, prebuilt course? Easy. Want to make changes throughout the semester? Sure. And you'll save time with Connect's auto-grading too.

65% Less Time Grading

Laptop: McGraw Hill; Woman/dog: George Doyle/Getty Images

Study made personal

Incorporate adaptive study resources like SmartBook® 2.0 into your course and help your students be better prepared in less time. Learn more about the powerful personalized learning experience available in SmartBook 2.0 at **www.mheducation.com/highered/connect/smartbook**

Affordable solutions, added value

Make technology work for you with LMS integration for single sign-on access, mobile access to the digital textbook, and reports to quickly show you how each of your students is doing. And with our Inclusive Access program you can provide all these tools at a discount to your students. Ask your McGraw Hill representative for more information.

Padlock: Jobalou/Getty Images

Solutions for your challenges

A product isn't a solution. Real solutions are affordable, reliable, and come with training and ongoing support when you need it and how you want it. Visit **www.supportateverystep.com** for videos and resources both you and your students can use throughout the semester.

Checkmark: Jobalou/Getty Images

Students: Get Learning that Fits You

Effective tools for efficient studying

Connect is designed to make you more productive with simple, flexible, intuitive tools that maximize your study time and meet your individual learning needs. Get learning that works for you with Connect.

Study anytime, anywhere

Download the free ReadAnywhere app and access your online eBook or SmartBook 2.0 assignments when it's convenient, even if you're offline. And since the app automatically syncs with your eBook and SmartBook 2.0 assignments in Connect, all of your work is available every time you open it. Find out more at **www.mheducation.com/readanywhere**

> "I really liked this app—it made it easy to study when you don't have your textbook in front of you."
>
> - Jordan Cunningham, Eastern Washington University

Calendar: owattaphotos/Getty Images

Everything you need in one place

Your Connect course has everything you need—whether reading on your digital eBook or completing assignments for class, Connect makes it easy to get your work done.

Learning for everyone

McGraw Hill works directly with Accessibility Services Departments and faculty to meet the learning needs of all students. Please contact your Accessibility Services Office and ask them to email accessibility@mheducation.com, or visit **www.mheducation.com/about/accessibility** for more information.

Top: Jenner Images/Getty Images, Left: Hero Images/Getty Images, Right: Hero Images/Getty Images

PART 1 Introduction

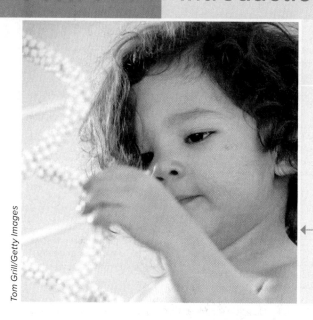

CHAPTER 1

The Information in a Human Genome

A child's genome holds information on her health, where she came from, and what she might experience and achieve—but the environment is very important too in guiding who she is and will become.

Learning Outcomes

1.1 Introducing Genes and Genomes
1. Explain what genetics is and what it is not.
2. Distinguish among gene, exome, and genome.
3. Define *bioethics*.

1.2 Levels of Genetics and Genomics
4. List the levels of genetics.
5. Explain how DNA is maintained and how it provides the information to construct a protein.
6. Explain how a mutation can cause a disease.
7. State the basis of genetic diversity.
8. Explain the relationship between DNA and chromosomes.
9. Distinguish between Mendelian and complex traits.
10. Explain how genetics underlies evolution.

1.3 Applications of Genetics and Genomics
11. List some practical uses of DNA information.
12. Distinguish between traditional breeding and genetically modifying organisms.

1.4 A Global Perspective on Genomes
13. Explain how investigating genomes extends beyond interest in ourselves.

 The BIG Picture

A human genome is a vast store of information encoded in the sequence of building blocks of the molecule deoxyribonucleic acid (DNA). Genetic information affects our health and traits, and holds clues to how we are biologically related to one another.

Eve's Genome

A baby is born. Drops of blood from her heel are placed into a small device that sends personal information into her electronic medical record. The device deciphers the entire sequence of DNA building blocks wound into the nucleus of a white blood cell. This is Eve's genome. Past, present, and future are encoded in nature's master informational molecule, deoxyribonucleic acid, or DNA—with room for environmental influences.

Eve's genome indicates overall good genetic health. She has a mild clotting disease that the nurse suspected when two gauze patches were needed to stop the bleeding from the heel stick. Two rare variants of the gene that causes cystic fibrosis (CF) mean that Eve is susceptible to certain respiratory infections and sensitive to irritants, but her parents knew that possibility from prenatal testing. Fortunately, the family lives in a rural area far from industrial pollution or wildfires, which will make it easier for Eve to avoid irritants such as smoke and dust.

The inherited traits that will emerge as Eve grows and develops range from interesting to important. Her hair will darken and curl, and variants of genes that contribute to bone development indicate that she'll have a small nose, broad forehead, and chiseled cheekbones. If she follows a healthy diet, she'll be as tall as her parents. On the serious side, Eve has inherited a mutation in a gene that greatly raises her risk of developing certain types of cancers. Her genes predict a healthy heart, but she might develop diabetes unless she exercises regularly and limits carbohydrates in her diet.

Many traits are difficult to predict because of environmental influences, including experiences. What will Eve's personality be like? How intelligent will she be? How will she react to stress? What will be her passions?

Genome sequencing also reveals clues to Eve's past, which is of special interest to her father, who was adopted. She has gene variants common among the Eastern European population of her mother's origin, and others that match people from Morocco. Is that her father's heritage? Eve is the beautiful consequence of a mix of her parents' genomes, receiving half of her genetic material from each.

Do you want to know the information in your genome?

1.1 Introducing Genes and Genomes

Genetics is the study of inherited traits, rooted in DNA, and their variations and transmission. The field, a type of life science, began more than a century ago with plant breeding experiments. Genetics evolved during the last century from a minor medical discipline dealing with rare diseases to forming the underpinnings of the emerging field of precision medicine, discussed in section 1.3. Precision medicine attempts to prevent and treat disease based on individual differences in gene variants, environmental exposures, and lifestyle factors such as diet and exercise.

In this new century, genetics has exploded into a powerful source of information about our identities that can reveal or revise what we know about our pasts, reach diagnoses and target health treatments in the present, and predict our medical futures. Human genetics touches forensics, bioethics, psychology, and even history. Consumer genetics enables anyone to learn about their DNA by sending samples of saliva or cheek cells to companies (**figure 1.1**) that return and store DNA data.

Genetics is not the same as traditional genealogy, which considers relationships but not traits. The newer area of **genetic genealogy** considers how people are related and where their ancestors lived, using and comparing information from DNA sequences and evidence such as documents, old photographs, maps, and family stories and memories.

Figure 1.1 Consumer genetics. DNA tests reveal several types of information and can bring surprises.

Heredity is the transmission of traits and biological information between generations. Inherited traits range from obvious physical characteristics, such as freckles and red hair, to many aspects of health, including disease and risk of developing disease. Talents, quirks, personality traits, and other difficult-to-define characteristics might appear to be inherited if they affect several family members, but might also reflect shared genetic and environmental influences. Attributing some traits to genetics, such as sense of humor or whether or not one votes, are oversimplifications. These connections are associations, not causes.

Like all sciences, genetics has its own vocabulary. Some technical terms and expressions may be familiar, but actually have precise scientific definitions. Conversely, the language of genetics sometimes enters casual conversation. "*It's in her DNA,*" for example, usually means an inborn trait, not a specific DNA sequence. The terms and concepts introduced in this chapter are explained and explored in detail in subsequent chapters. Boldfaced key terms are defined at the chapter's end.

Genes function as the units of heredity in that copies of genes are passed from one generation to the next. Genes are biochemical instructions that tell **cells**, the basic units of life, how to manufacture certain proteins. These proteins, in turn, impart or control the characteristics that create much of our individuality. A gene consists of the long molecule **deoxyribonucleic acid (DNA)**. The DNA transmits information in its sequence of four types of building blocks, which function like an alphabet. Most of a cell's DNA is in a structure called the **nucleus**.

The complete set of genetic instructions characteristic of an organism, including protein-encoding genes and other DNA sequences, constitutes a **genome**. Nearly all

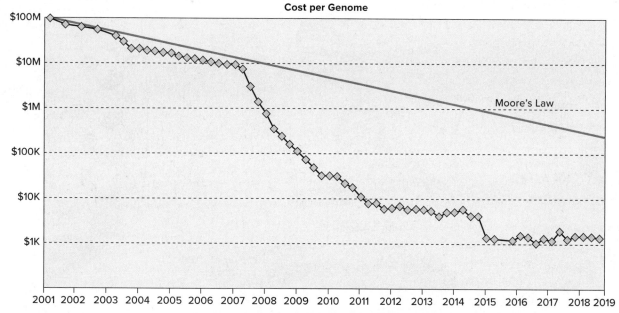

Figure 1.2 **The cost of sequencing a human genome has plunged.** Today people can be paid for use of their genetic information. *National Human Genome Research Institute*

of our cells contain two copies of the genome. Researchers published the deciphered sequences of the first human genomes in 2003, following a multi year, international effort. **Figure 1.2** shows how the cost of sequencing a human genome has plummeted since the first genomes were sequenced in the early 2000s. When the idea to sequence a human genome was proposed in 1990, the cost was estimated to be $1 billion. As technology steadily improved, the cost fell, drastically so in 2008, when a much more efficient method of sequencing genomes came into wide use. That is when the cost defied Moore's law, which is a term from business used to describe regular doubling of an event. Today, companies pay citizens for access to their genomes in doing research and developing products. Although millions of people have had their genomes sequenced, researchers are still analyzing what each of our genes does and how genes interact and respond to environmental stimuli.

Only about 1 percent of the 3.2 billion building blocks of our genomes specify proteins. This tiny slice of the genome, called the **exome**, is responsible for many aspects of health and our traits, including our differences.

The exome includes about 20,325 genes, accounting for about 85 percent of the genome known to contribute to genetic disease. Much of the rest of the genome controls how the body uses the genes of the exome. When genome sequencing was expensive, exome sequencing provided a shortcut to identify mutations that could explain a patient's mysterious symptoms and provide a diagnosis. Today sequencing genomes in search of a diagnosis is more economically feasible, as **Clinical Connection 1.1** discusses.

Analyzing and comparing genomes constitutes the field of **genomics**.

Genetics directly affects our lives and those of our relatives, including our descendants. Principles of genetics also may be relevant in politics, economics, sociology, anthropology, art, the law, and athletics. Using genetic technologies forces us to wrestle with concepts of benefit and risk, even tapping our deepest feelings about right and wrong. The field of **bioethics** addresses concerns that arise from the use of new genetic technologies, including tests, treatments, privacy, and discrimination. Essays throughout this book address bioethical issues.

Key Concepts Questions 1.1

a. Distinguish between genetics and heredity.
*b. What is the type of chemical that makes up a gene?
c. Distinguish among a gene, an exome, and a genome.
*d. Define *bioethics*.

1.2 Levels of Genetics and Genomics

Genetics considers the transmission of information at several levels. It begins with the molecular level and broadens through cells, tissues and organs, individuals, families, and finally to populations and the evolution of species.

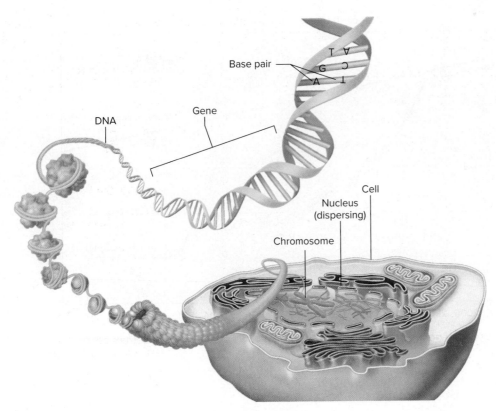

Figure 1.3 Levels of genetics. Genetics can be considered at several levels, from the sequences of DNA base pairs, to genes, to chromosomes, to genomes, to the more familiar individuals, families, and populations.

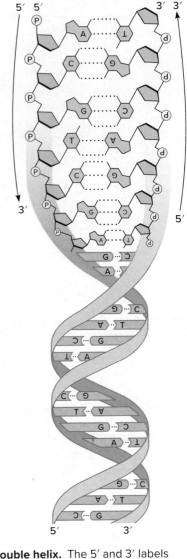

Figure 1.4 The DNA double helix. The 5′ and 3′ labels indicate the head-to-tail organization of the DNA double helix. A, C, T, and G are bases. S stands for sugar and P for phosphate. The green five-sided shapes represent the sugars. The dots that connect the bases into pairs represent hydrogen bonds.

Instructions and Information: DNA

A DNA molecule resembles a spiral staircase or double helix (**figure 1.3**). The "rails," or backbone, consist of alternating chemical groups (sugars and phosphates) and are the same in all DNA molecules. The two strands of the double helix are oriented in opposite directions, like two snakes biting each other's tails. The "steps" of the DNA double helix are pairs of the four types of building blocks, or **nitrogenous bases**: **adenine** (A) and **thymine** (T), which attract each other, and **cytosine** (C) and **guanine** (G), which attract each other (**figure 1.4**). The information that a DNA molecule imparts is in the sequences of A, T, C, and G.

The chemical structure of DNA gives the molecule two key abilities that are essential for being the basis of life: DNA can replicate itself when a cell divides and its information accessed to manufacture specific proteins.

In **DNA replication**, the chains of the double helix untwist and separate, and then each half builds a new partner chain from free DNA bases. In this way, the resulting "daughter" cells inherit identical copies of the genome during cell division. A and T recognize each other and attract, as do C and G. Chemical attractions called hydrogen bonds hold the bases of a pair together. Then **transcription** copies the sequence of part of one strand of a DNA molecule into a related type of molecule, messenger **ribonucleic acid (RNA)**. RNA molecules include a fifth type of base, **uracil** (U), in place of the T in DNA. Transcription is also called **gene expression**. In **translation**, each three RNA bases in a row attract another type of RNA that functions as a connector, bringing in a particular amino acid. The amino acids align and link like snap beads, forming a protein.

Proteins provide the traits associated with genes. The disease cystic fibrosis (CF) illustrates how a missing or abnormal protein causes the symptoms of an inherited disease (**figure 1.5**). In CF, the protein is the cystic fibrosis transmembrane conductance regulator (*CFTR*). The functioning protein works like a selective doorway in cells lining the airways and certain other body parts, thickening secretions when it doesn't work properly.

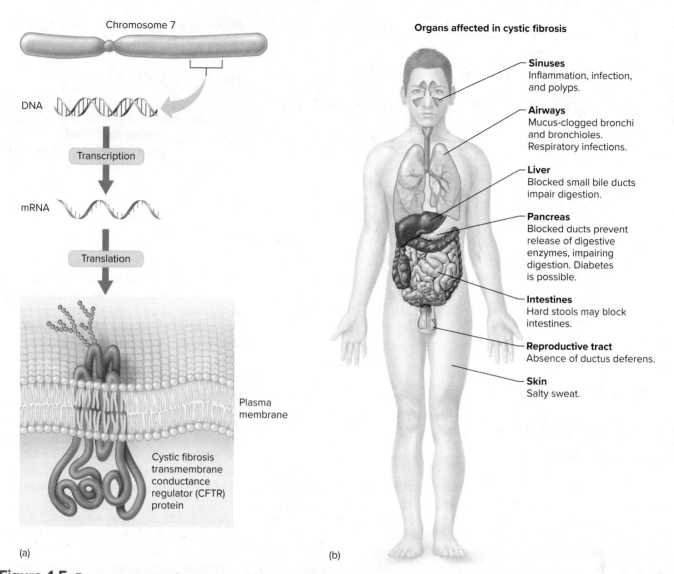

(a) (b)

Figure 1.5 From gene to protein to person. (a) The gene encoding the CFTR protein, causing cystic fibrosis when in a variant form (a mutation), is part of the seventh largest chromosome. CFTR normally folds into a channel that regulates the flow of salt components (ions) into and out of cells lining the respiratory tract, pancreas, intestines, and elsewhere. **(b)** Cystic fibrosis causes several symptoms.
Source: Data from "Reverse genetics and cystic fibrosis" by M. C. Iannuzi and F. S. Collins. *American Journal of Respiratory Cellular and Molecular Biology* 2:309–316 [1990].

A change in a gene is a **mutation**, and it can have an effect at the whole-person level, such as causing a disease. **Figure 1.6** depicts the effect of a mutation in *CFTR*. A change of a "C" in the DNA sequence at a specific location in the gene to a "T" inserts the amino acid aspartic acid rather than the amino acid glycine as the protein forms. The resulting protein cannot open to the cell's surface, removing channels for certain salt components, causing the symptoms described in figure 1.5. Difficulty breathing, impaired digestion, and other symptoms result (see Clinical Connection 4.1).

The same protein-encoding gene may vary slightly in DNA base sequence from person to person. These gene variants are called **alleles**. The changes in DNA sequence that distinguish alleles arise by mutation. (The word *mutation* is also used as a noun to refer to the changed gene.) Once a gene mutates, the change is passed on when the cell that contains it replicates its DNA and divides. If the change is in a sperm or egg cell that becomes a fertilized egg, it is passed to the next generation.

Some mutations cause disease, and others provide variation such as freckled skin. Mutations can also help. One rare mutation makes a person's cells unable to manufacture a surface protein that binds HIV. These people are resistant to HIV infection. Mutations that have no detectable effect because they do not change the encoded protein in a way that affects its function are sometimes called gene variants. They are a little like a minor spelling error that does not obscure the meaning of a sentence.

The DNA sequences of the human genome are grouped among 23 structures called **chromosomes**. When a cell is not

Figure 1.6 **A mutation can alter a protein, causing symptoms.** One type of mutation in the *CFTR* gene replaces one amino acid type (glycine) with another (aspartic acid) at a specific site, altering the encoded protein in a way that closes a type of ion channel that participates in secretion.

dividing, the chromosomes are in an unwound state within the nucleus. When a cell is dividing, the chromosomes wind up so tightly that they can be seen under a microscope when stained, appearing rod shaped. The DNA of a chromosome is continuous, but it includes hundreds of genes, plus other sequences.

A human **somatic cell** (non-sex cell) has 23 pairs of chromosomes, equaling two complete sets of genetic information (genomes). Twenty-two of these 23 pairs are **autosomes**, which do not differ between the sexes. The autosomes are numbered from 1 to 22, with 1 being the largest. The other two chromosomes, the X and the Y, are **sex chromosomes**.

In humans, a female has two X chromosomes and a male has one X and one Y chromosome. The Y chromosome bears genes that determine maleness. Charts called **karyotypes** display the chromosome pairs from largest to smallest (**figure 1.7**).

A trait caused predominantly by a single gene is termed Mendelian, named for Gregor Mendel, who discovered the patterns of trait transmission. Most characteristics are **complex traits**, which means that they are determined by one or more genes and environmental factors (**figure 1.8**). The more factors that contribute to a trait or illness—inherited or environmental—the more difficult it is to predict the risk of occurrence in a particular family member. The bone-thinning condition osteoporosis illustrates factors that can contribute to a disease. Mutations in several genes elevate osteoporosis risk by conferring susceptibility to fractures, but so do smoking, lack of weight-bearing exercise, and a calcium-poor diet.

The Body: Cells, Tissues, and Organs

A human body consists of approximately 30 trillion cells. All somatic cells except red blood cells contain two copies of the genome, but cells differ in appearance and activities because they use only some of their genes. Which genes a cell uses at any given time depends on environmental conditions inside and outside the body.

Like the internet, a genome contains a wealth of information, but only some of it is needed in a particular cell under particular circumstances. The use, or expression, of different subsets of genes to manufacture proteins drives the **differentiation**, or specialization, of distinctive cell types. An adipose cell is filled with fat, but not the contractile proteins of muscle cells. Both cell types, however, have two complete genomes. Groups of differentiated cells assemble and interact with each other and the nonliving materials that they secrete to form aggregates called **tissues**. Table 2.1 lists the four basic tissue types: connective tissues, epithelium (linings), muscle, and nervous. Altogether, they are composed of more than 290 types of cells.

Tissues intertwine and layer to form organs, which connect into organ systems. The stomach, for example, is a sac made of muscle that also has a lining of epithelial tissue, nervous tissue, and a supply of blood, which is a type of connective tissue. Many organs include rare, unspecialized **stem cells**.

Figure 1.7 **A normal female karyotype.** *Kateryna Kon/Shutterstock*

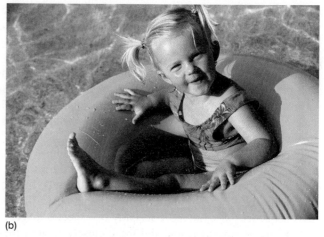

Figure 1.8 Mendelian versus complex traits. (a) Polydactyly—extra fingers and/or toes—is a Mendelian trait (single gene). **(b)** Hair color is complex (also called multifactorial), controlled by at least three genes plus environmental influences, such as the bleaching effects of sun exposure. *(a): Lester V. Bergman/Getty Images; (b): Steve Mason/Getty Images*

A stem cell can divide to yield another stem cell and a cell that differentiates. Stem cells provide a reserve supply of cells that enable an organ to grow and repair damage.

Relationships: From Individuals to Families

Two terms distinguish the alleles that are *present* in an individual from the alleles that are *expressed*. The **genotype** refers to the underlying DNA instructions (alleles present), whereas the **phenotype** is the visible trait, biochemical change, or effect on health (alleles expressed). Alleles are further distinguished by how many copies are necessary to affect the phenotype. A **dominant** allele has an effect when present in just one copy (on one chromosome), whereas a **recessive** allele must be present on both chromosomes of a pair to be expressed.

Individuals are genetically connected into families. A person has approximately half of his or her gene variants in common with each parent, sibling, and offspring, and one-quarter with each grandparent. First cousins share one-eighth of their gene variants. Charts called **pedigrees** depict the members of a family and indicate which individuals have particular inherited traits. **Figure 1.9** shows how the percent of the genome shared with a direct ancestor decreases as the generations go back and partners introduce new genomes. In this way, the proportion of the genome an individual shares with an ancestor halves at each generation.

The Bigger Picture: From Populations to Evolution

Above the family level of genetic organization is the population. In a strict biological sense, a population is a group of individuals that can have healthy offspring together. In a genetic sense, a population is a large collection of alleles, distinguished by their frequencies. People from a Swedish population, for example, would have a greater frequency of alleles that specify

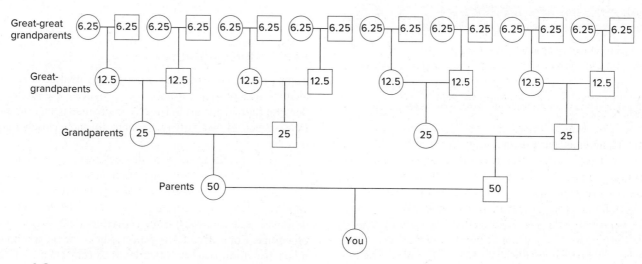

Figure 1.9 Percent of the genome shared with relatives. The percent of the DNA sequence that a person shares with ancestors halves at each generation.

light hair and skin than people from a population in Nigeria, who tend to have dark hair and skin. All the alleles in a population constitute the **gene pool**. (An individual does not have a gene pool.)

Population genetics is applied in health care, forensics, and other fields. It is also the basis of evolution, which is defined as changing allele frequencies in populations. Small-scale genetic changes in populations underlie the species distinctions we most often associate with evolution.

Comparing DNA sequences for individual genes, or the amino acid sequences of the proteins that the genes encode, can reveal how closely related different types of organisms are. The assumption is that the more similar the DNA sequences are, the more recently two species diverged from a shared ancestor, and the more closely related they are. This is a more plausible explanation than two species having evolved similar or identical gene sequences coincidentally. The same logic applies to family patterns of inherited traits. It is more likely that a brother and sister share approximately half of their gene variants because they have the same parents than that half of their genetic material is identical by chance.

More information is available in full genome sequences than in single genes. Humans, for example, share more than 98 percent of the DNA sequence with chimpanzees. Our genomes differ from theirs more in gene organization and in the number of copies of genes. Learning the functions of the human-specific genes may explain the differences between us and them—such as our sparse body hair and use of spoken language. Figure 17.16 highlights some of our distinctively human traits.

At the genome level, we are much more like each other genetically than are other mammals. Chimpanzees are more distinct from each other than we are! The most genetically diverse modern people are from Africa, where humanity arose. The gene variants among different modern ethnic groups include subsets of our ancestral African gene pool.

Key Concepts Questions 1.2

*a. List the levels of genetic information.
b. Explain how DNA carries and maintains information.
*c. List the nitrogenous bases that are part of DNA.
d. Explain how a mutation can cause a disease.
*e. Define *allele*.
*f. Name the two types of chromosomes.
g. Distinguish between Mendelian and complex traits.
*h. Define *gene expression*.
i. Distinguish between genotype and phenotype; dominant and recessive.
*j. Define *gene pool*.
k. Explain how comparing DNA sequences can clarify evolutionary relationships.

1.3 Applications of Genetics and Genomics

Genetics is impacting several areas of our lives, from health care choices, to what we eat, to unraveling our pasts and guiding our futures. "Citizen scientists" are discovering genetic information about themselves while helping researchers compile databases that will speed development of new diagnostic tests and treatments.

Thinking about genetics evokes fear, hope, anger, wonder, and despair, depending on context and circumstance. Following are a few uses of DNA information, then glimpses of applications of genetics and genomics that are explored more fully in subsequent chapters:

- Predicting shelf life of fruits and vegetables, detecting spoiled meat, identifying allergens, and indicating degree of fermentation in cheese
- Identifying victims of human trafficking at transportation centers by comparing the DNA of suspected victims to DNA from concerned relatives
- Detecting disease-causing mutations or abnormal chromosome numbers in a fetus from DNA in a pregnant woman's blood
- Identifying which of several pets produced feces, so a stool sample can be brought to a veterinarian to diagnose the sick animal
- Performing rapid diagnosis of an infectious disease on the battlefield
- Creating a tree of life depiction of how all species are related
- Selecting crops and show animals for breeding
- Choosing people to date
- Identifying criminals by comparing DNA from crime scenes, from a suspect, and to DNA from a suspect's relative who has uploaded information to a public genetic genealogy database
- Detecting tiny amounts of DNA in fur, feathers, or feces of rare or elusive species to sequence their genomes and learn more about them

Establishing Identity

Comparing DNA sequences among individuals can rule out identity, relationships, or ancestry, or indicate the probability that two individuals are related. The premise is that the more DNA sequences two individuals share, the more closely related they are.

DNA profiling refers to the techniques, statistical analyses, and machine learning approaches that are used to compare DNA sequences between and among individuals. It has varied applications. The term is most often used in the context of forensic science, which is the collecting of physical evidence of a crime. Comparing DNA collected at crime scenes to DNA in samples from suspects often leads to convictions, and also to reversing convictions erroneously made using other forms of evidence.

DNA profiling is also useful in identifying victims of natural disasters, such as violent storms and earthquakes. In happier circumstances, DNA profiles maintained in databases assist adopted individuals in locating blood relatives and children of sperm donors in finding their biological fathers and half-siblings. Chapter 18 explores the genetics of identity.

Another use of DNA profiling is to analyze food, because foods were once organisms, which have species-specific DNA sequences. Analyzing DNA sequences revealed horsemeat in meatballs sold at a restaurant chain, cheap fish sold as gourmet varieties, and worms in cans of sardines.

Illuminating History

DNA analysis is a time machine of sorts. It can connect past to present, from determining family relationships to establishing geographic origins of specific populations. DNA evidence sometimes confirms findings from anthropology and history and sometimes contradicts it.

DNA analysis confirmed that Thomas Jefferson had children with his slave Sally Hemings. The president was near Hemings 9 months before each of her seven children was born, and the children resembled him. Male descendants of Sally Hemings share an unusual Y chromosome sequence with the president's male relatives. His only son with his wife died in infancy, so researchers deduced the sequence of the president's Y chromosome from descendants of his uncle. Today the extended family holds reunions (**figure 1.10**).

DNA testing can provide views into past epidemics. For example, analysis of DNA in the mummy of the Egyptian king Tutankhamun, who died in 1323 B.C.E. at age 19, revealed DNA from the microorganism that causes malaria. The famous boy king likely died from complications of malaria following a leg fracture from weakened bones, rather than from intricate murder plots, a kick from a horse, or a fall from a chariot, as had been thought. His tomb included a cane and drugs, supporting the diagnosis based on DNA evidence.

Conservation Genetics

Combining analysis of genetic diversity with reproductive technologies creates a way to rebuild populations that are headed toward species extinction. This is the case for the northern white rhinoceros of Africa. The animals die after poachers remove their horns, which are displayed as decorations or ground up and used in medicines.

The last male northern white rhinoceros, named Sudan, died from an infection in 2018 at a nature conservancy in Kenya, leaving only Najin and her daughter Fatu, who are infertile. However, the genomes of cells from nine northern white rhinos stored at the San Diego Frozen Zoo reveal enough genetic diversity to suggest that researchers can bring back the species by borrowing from the genomes of the subspecies to the south. The southern rhino population is several thousand strong thanks to efforts to prevent poaching.

Researchers are working on three ways to bring back the animals:

- Treat defrosted northern white rhino cells with chemicals to make them differentiate as sperm and eggs, and then allow fertilization to occur in a lab dish. Transfer resulting embryos to the uteruses of surrogate southern rhinos.
- Transfer a nucleus from a frozen cell from a northern rhino to an egg from a southern rhino that has had its nucleus removed (this is cloning).
- Defrost northern rhino sperm and use them to fertilize southern white rhino eggs. Resulting offspring are then bred to reconstruct a northern white rhino population.

Precision Medicine

In several nations, people are volunteering to have their genomes sequenced so that researchers can learn more about health and disease. The DNA data are considered along with other types of information that can impact health, such as environmental exposures, exercise, diet, lifestyle factors, family histories, and the many microbes that live in and on the human body, collectively termed the **microbiome**. Evaluating genetic data is a large part of precision medicine, which is the tailoring of treatments to individuals.

On a smaller scale, a precision medicine approach consults DNA information to select drugs that are most likely to work and least likely to have side effects in a particular individual. This strategy, called **pharmacogenetics**, is already used to guide prescription of more than 150 drugs, according to the Food and Drug Administration. Some highly effective new drugs that collectively treat a variety of conditions, from cystic

Figure 1.10 **DNA reveals and clarifies history.** After DNA evidence showed that Thomas Jefferson likely fathered children with his slave Sally Hemings, confirming gossip of the time, descendants of both sides of the family met, and continue to do so. *Leslie Close/AP Photo*

fibrosis to cancers, are targeted to patients who have specific mutations in specific genes.

Healthy people can contribute to precision medicine by helping researchers identify gene variant combinations that are associated with, and therefore might contribute to, wellness and longevity. The "All of Us" initiative in the United States and efforts elsewhere, such as the United Kingdom's 100,000 Genomes Project, are involving the general population in collecting information that could improve health care for many. For example, imagine identifying a group of people who all have a mutation known to cause a specific disease, but some of them do not have the disease. Something in the environment or something the protected people are doing might explain their health, and the finding used to develop treatments.

Genetic Modification

Genetic modification means altering a gene or genome in a way that does not occur in nature, such as giving a carrot a gene from a green bean that isn't part of the carrot genome. Traditional agriculture and animal breeding do not result in "genetically modified organisms" (GMOs) because they select traits within one species.

In health care, GMOs in the form of bacteria bearing human genes have provided many drugs, such as insulin and clotting factors, since the 1970s. Foods are genetically modified to be more nutritious, easier to cultivate, or able to grow in the presence of herbicides and pesticides. In the United States, more than 93 percent of the crops of corn, soybeans, and cotton are GMOs and the public has been eating them for more than a quarter century (according to the U.S. Department of Agriculture). However, some GMO foods have failed. The "Flavr Savr" tomato, for example, was genetically modified to have a longer shelf life, but it had a terrible taste!

Genetically modified foods are not dangerous to human health, except if a person is allergic to a molecule in one species that appears in a species in which it wouldn't naturally be present. However, some people object to GM foods for ecological reasons. For example, wind can disperse GM pollen farther than intended. Another objection is that seeds from altered plants may be modified to germinate only under certain conditions or not at all, forcing small farmers and home gardeners to purchase new seeds every spring.

Genetically modified crops and drugs have been available for many years. They arise from **recombinant DNA technology**, which adds a gene from a different species. A newer technology, **genome editing**, is more specific and powerful. It can replace, remove, or add specific genes into the cells of any organism. Genome editing is being developed for use in agriculture and in health care. However, using it to edit the genomes of human gametes (sperm or eggs) or fertilized eggs is controversial because it would create a genetically modified human and create a change that can be passed to the next generation. Another controversial application of genome editing is to render pest populations extinct, such as mosquitoes that carry the microorganism that causes malaria.

The most talked-about genome-editing tool is CRISPR-Cas9, but similar technologies have been used since 2009. The techniques use enzymes that cut both strands of a DNA double helix at a selected site. The possibilities related to editing the genome have been discussed since the 1980s.

Exome Sequencing

Exome sequencing determines the order of the DNA bases of all parts of the genome that encode proteins. It is valuable after more conventional diagnostic tests, such as tests for single gene diseases and chromosome abnormalities, do not explain a person's symptoms. The information is compared to databases that list many gene variants (alleles) and their associations with specific phenotypes, such as diseases.

The information in an exome sequence can be lifesaving. For example, it showed that a 2-year-old boy who had severe eating difficulties and was very thin had a mutation known to cause Marfan syndrome, which usually causes long limbs and fingers but did not in this child. However, another symptom of Marfan syndrome is an enlarged aorta, the largest artery, and often this is not obvious until the aorta bursts and the person dies. After exome sequencing revealed a Marfan mutation, an ultrasound scan indeed showed a bulge in the wall of the aorta near the boy's heart, which could have resulted in his sudden death. Doctors successfully patched the bulging blood vessel.

Exome sequencing is valuable in identifying extremely rare diseases—swiftly. In the past, parents of children with unusual symptoms referred to the multiyear effort to find a physician who recognized the disease as a "diagnostic odyssey," in reference to Homer's epic poem about Greek hero Odysseus' 10-year journey home. Most diagnostic odysseys for genetic diseases took 5 years or longer. With exome as well as full genome sequencing, diagnosis can take just hours. **Clinical Connection 1.1** describes how genome sequencing led to the diagnosis of a rare genetic disease in a child.

Key Concepts Questions 1.3

*a. Define *DNA profiling*.

b. Explain how DNA testing can add to what we know about history.

*c. What is precision medicine?

d. Distinguish between traditional breeding and genetic modification.

*e. What is exome sequencing?

Clinical Connection 1.1

Genome Sequencing Ends a Child's "Diagnostic Odyssey"

Millie McWilliams was born September 2, 2005. At first, Millie seemed healthy, lifting her head and rolling over when most babies do. "But around 6 months, her head became shaky, like an infant's. Then she stopped saying 'dada'," recalled her mother Angela.

By her first birthday, Millie couldn't crawl or sit, and her head shaking had become a strange, constant swaying. She had bouts of irritability and vomiting and the peculiar habit of biting her hands and fingers. In genetic diseases, odd habits and certain facial features can be clues. None of the many tests, scans, and biopsies that Millie underwent led to a diagnosis.

By age 6, Millie had lost the ability to speak, was intellectually disabled, and confined to a wheelchair, able to crawl only a short distance. Today she requires intensive home-based therapies. But Millie can communicate with her parents. "She likes to look at what she wants, with an intense stare," said Angela. She loves country music and Beyoncé, and every once in awhile something funny happens and she'll break into a big smile.

Millie's pediatrician, Dr. Sarah Soden, suggested that genome sequencing, already being done at the medical center where Millie receives care, might explain the worsening symptoms. So the little girl and her parents had their genomes sequenced in December 2011. Dr. Soden's team identified a suspicious mutation, but the gene had never been linked to a childhood disease.

In February 2013, a medical journal published a report about four children with mutations in this gene who had symptoms strikingly like those of Millie. An answer had finally emerged: Millie has Bainbridge-Ropers syndrome. Even her facial structures—arched eyebrows, flared nostrils, and a high forehead—matched, as well as the hand-biting symptom.

Millie is missing two DNA bases in the gene *ASXL3*. DNA bases are "read" three at a time to indicate the amino acids in a protein, so missing two bases garbled the code, leading to tiny, nonfunctional proteins for that particular gene. Somehow the short proteins caused the symptoms. Because Angela and Millie's father, Earl, do not have the mutation, it originated in either the sperm or the egg that went on to become Millie.

Since Millie was diagnosed, more than 600 other people have been found to have Bainbridge-Ropers syndrome, and families have formed a support group and Facebook page. Although there is no treatment yet, the families are happy to have an answer, because sometimes parents blame themselves. Said Angela, "It was a relief to finally put a name on it and figure out what was actually going on with her, and then to understand that other families have this too. I've been able to read about her diagnosis and what other kids are going through."

Soon, newborn genome sequencing may be routine. A project called BabySeq analyzed genomes of an initial group of 159 newborns. The findings enabled researchers to predict that 15 of the children would develop diseases before adolescence that were not in the family or indicated with other tests; 8 of the children would have dangerous reactions to certain medications; and 3 would develop certain treatable diseases as adults. Will genome sequencing at birth become a medical crystal ball?

Questions for Discussion

1. Millie has a younger brother and an older sister. Why don't they have Bainbridge-Ropers syndrome?
2. Would exome sequencing have discovered Millie's mutation?
3. Do you think it is valuable to have a diagnosis of a condition that has no treatment? Why or why not?

1.4 A Global Perspective on Genomes

We share the planet with many thousands of other species. We aren't familiar with many of Earth's residents because we can't observe their habitats, or we can't grow them in laboratories. **Metagenomics** is a field that describes much of the invisible living world by sequencing all of the DNA in a habitat, such as soil, an insect's gut, garbage, or a volume of captured air over a city. A project that involves swabbing various surfaces in the New York City subway system is painting a portrait of an "urban microbiome." Metagenomics studies are showing how species interact, and may yield information useful in developing new drugs or energy sources.

Metagenomics researchers collect and sequence DNA, then consult databases of known genomes to imagine what the organisms might be like. The first metagenomics project described life in the Sargasso Sea. This 2-million-square-mile area off the coast of Bermuda was thought to lack life beneath its thick cover of seaweed, which is so abundant that Christopher Columbus thought he'd reached land when his ships came upon it. Boats have been lost in the Sargasso Sea, which includes the area known as the Bermuda Triangle. Researchers collected more than a billion DNA molecules from the depths, representing about 1,800 microbial species and including more than a million unknown genes.

One interesting subject of metagenomics is the human body, which is home to trillions of bacteria. Section 2.5 discusses the human microbiome, the collection of life within us.

Efforts are underway to limit "genetic prospecting," which is the collection of DNA from organisms unique to a nation that have potential industrial use, such as microorganisms that provide antibiotics. Under the Nagoya Protocol, databases would store digitized genome sequences of useful species, and the home nations would charge licensing fees for use of the DNA information. Funds raised would go toward preserving biodiversity in those nations, most of which are developing. An example of protection is the sweetener stevia, which comes from a plant native to Paraguay and Brazil. When food and beverage manufacturers extract and concentrate stevia on an industrial scale, the farmers in the native countries lose markets for their sweet leaves.

Genetics is a special branch of life science because it affects us intimately. Equal access to genetic tests and treatments, misuse of genetic information, and abuse of genetics to intentionally cause harm are compelling social issues that parallel scientific progress.

Genetics and genomics are spawning technologies that may vastly improve quality of life. But tests and treatments are costly and not widely available, especially at first. Not everyone can afford genetic tests or exome or genome sequencing. In an African nation where two of five children suffer from AIDS and many die from other infectious diseases, newborn screening for rare single-gene diseases may seem impractical. However, genetic diseases weaken people so that they become more susceptible to infectious diseases, which they can pass to others.

Recognizing that human genome information can ultimately benefit everyone, organizations such as the United Nations, World Health Organization, and World Bank are discussing how nations can share new diagnostic tests and therapeutics that arise from genome information about ourselves and the microbes that make us sick. Applications of genetic testing, however, are not all related to health. **Bioethics** discusses instances when genetic testing can affect privacy.

Key Concepts Questions 1.4

*a. Define *metagenomics*.

b. Name social issues that arise from technologies based on genetics.

Bioethics

Genetic Privacy and Pandora's Box

The field of bioethics began in the 1950s to address issues raised by medical experimentation during World War II. Bioethics initially centered on informed consent, paternalism, autonomy, allocation of scarce medical resources, justice, and definitions of life and death. Today, the field covers medical and biotechnologies and the dilemmas they present. Genetic testing is a key issue in bioethics because its informational nature affects privacy. That is, findings that apply to one person may also affect relatives. In just a few years, these issues have become more common, due to the increasing popularity of consumer genetic testing and the ease of sharing personal information (**figure 1A**). But consequences can be both positive and negative.

Genetic privacy issues arising from testing have evolved. Some of the first to learn about their genetic make-up were students in genetics classes and medical students having their exomes or genomes sequenced to explore how such information might be used in patient care. An early example of the dangers of required testing occurred just before the fall semester in 2010 at the University of California, Berkeley. Freshmen received, along with class schedules and dorm assignments, kits to send in DNA samples to test three genes that control three supposedly harmless traits. Participation was voluntary, and because the intent was to gather data, informed consent was not required. However, after genetics groups, bioethicists, policy analysts, and consumer groups protested, the Department of Public Health ruled that the tests provided personal medical information and should be conducted by licensed medical labs. Because this quintupled the cost, the

Figure 1A Genetic privacy. How can we protect DNA sequence data? *Jane Ades/National Human Genome Research Institute*

university changed the program to collect aggregate data, rather than individually identified results.

The three genetic tests were to detect lactose intolerance, alcohol metabolism, and folic acid metabolism. The alcohol test detects variants of a gene that cause a facial flush, nausea, and heart palpitations after drinking, particularly in East Asians—who made up a significant part of the freshman class. Certain mutations in this gene raise the risk of developing esophageal cancer, and so test results may be useful, but they could also encourage drinking.

Eight years after the failed Berkeley genetic testing program, the issue of genetic privacy arose in an unexpected manner: the use of a public DNA database to capture the Golden State Killer. He had killed at least a dozen people and raped more than 50, from 1976 to 1986 in certain towns in California. Forensic scientists thought to look for relatives of the suspect who might have DNA data with consumer genetics companies, which were becoming popular. But when the companies refused to provide data without a court order, police instead went to GEDmatch. This is a website where people voluntarily upload their DNA data to learn if they have possible relatives they hadn't known about. Investigators compared the DNA sequences that millions of people had uploaded to GEDmatch to crime scene DNA evidence and a suspect's DNA collected from a piece of trash. A hit to a distant male relative led directly, and quickly, to the suspect's arrest.

Soon, similar DNA matches solved other "cold cases." Meanwhile, genetic privacy issues were arising as more people took DNA-based ancestry tests. While some people delighted in finding out that they are distant relatives of Elvis Presley or Abraham Lincoln, others were learning family secrets, most centered around discovering "not parent expected," or NPE, status. People who thought that they were full siblings have found out that they are half-siblings, while people have discovered, many late in life, half-siblings that they never knew they had. The discoveries bring up questions about adoptions, multiple partners, rape, and sperm donors. Reactions to such news range from joy at finding new family members to anger at the circumstances and secrecy—or both. Chapter 18 explores the repercussions of unexpected DNA test results.

Genetic testing that yields surprises has been compared to opening Pandora's box, a Greek myth that describes unanticipated challenges arising from a curiosity-driven action. Genetic privacy is today a major focus of bioethics research.

Questions for Discussion

1. Why did people object to the genetic tests that students at Berkeley were asked to take in 2010?
2. Under what circumstances, if any, should a government require DNA testing?
3. What are the pros and cons of students in a class taking genetic tests?
4. Discuss different ways that people might react to learning that they are not related to certain family members in the way that they thought they were, or who learn that they have half-siblings.

Study Guide

Summary

1.1 Introducing Genes and Genomes

1. **Genetics** investigates inherited traits and their variation. **Genetic genealogy** examines how people are related. **Heredity** concerns the transmission of traits.
2. **Genes** are the instructions to manufacture proteins, which determine inherited traits. Genes are composed of **deoxyribonucleic acid (DNA)**.
3. A **genome** is a complete set of genetic information. A **cell**, the basic unit of life, contains two genomes. The **exome** is the part of the genome that encodes protein. **Genomics** compares and analyzes the functions of many genes.
4. **Bioethics** addresses issues and controversies that arise in applying medical technology and using genetic information.

1.2 Levels of Genetics and Genomics

5. The **nitrogenous base** sequences of genes encode proteins. The DNA bases are **adenine** (A), **guanine** (G), **cytosine** (C), and **thymine** (T). **Ribonucleic acid (RNA)** molecules have **uracil** (U) instead of thymine. **DNA replication** maintains the genetic information. **Transcription**, also called **gene expression**, copies DNA information into RNA, and **translation** uses the information in RNA to assemble amino acids into proteins.
6. A **mutation** is a change in a gene that can cause a disease if it alters the amino acid sequence of the specified protein. Mutation also refers to the process of change.
7. Variants of a gene are called **alleles**. They are inherited or arise by mutation. Alleles may differ slightly from one another, but encode the same protein.
8. **Chromosomes** consist of DNA and protein. When a cell is not dividing, the chromosomes are unwound and contained in a structure called the **nucleus**. A **somatic cell** in humans has 23 chromosome pairs. The 22 types of **autosomes** do not include genes that specify sex. The X and Y **sex chromosomes** bear genes that determine sex. **Karyotypes** are chromosome charts.
9. Single genes determine **Mendelian traits**. **Complex traits** reflect the influence of one or more genes and the environment.
10. Cells undergo **differentiation** by expressing subsets of genes. **Stem cells** divide to yield other stem cells and cells that differentiate. **Tissues** are groups of cells with a shared function.

11. The **phenotype** is the gene's expression. An allele combination constitutes the **genotype**. Alleles may be **dominant** (exerting an effect in a single copy) or **recessive** (requiring two copies for expression).
12. **Pedigrees** are diagrams used to study traits in families.
13. Genetic populations are defined by their collections of alleles, termed the **gene pool**. Genome comparisons among species reveal evolutionary relationships.

1.3 Applications of Genetics and Genomics

14. **DNA profiling** compares DNA sequences of individuals to rule out identities, relationships, or ancestry, and can reveal if two individuals are related.
15. The **microbiome** consists of all of the microorganisms that are on or in a human body.
16. Precision medicine considers DNA information along with other types of information to learn about health and to develop new treatments for diseases. **Pharmacogenetics** considers gene variants to predict whether a specific drug will be effective or cause side effects in an individual. Genetically modified organisms are altered in ways that may not occur in nature. **Genome editing** adds, removes, or replaces specific genes.

1.4 A Global Perspective on Genomes

17. In **metagenomics**, DNA collected from specific habitats, including the human body, is sequenced to learn more about microbiomes.

Key Terms

adenine (A) (ăd′en-ēn′) One of the four types of nitrogenous bases in DNA and RNA.

allele (ə-lēl′) An alternate (variant) form of a gene.

autosome (ôtə-sōm) A chromosome that does not have any genes that determine sex.

bioethics A field that addresses personal and societal issues that arise in applying medical technology and using genetic information.

cell The fundamental unit of life.

chromosome (krō′ mə-sōm′) A highly wound, continuous molecule of DNA and the proteins associated with it.

complex trait A phenotype caused by one or more genes and an environmental influence.

cytosine (C) (sī′-tō-sēn) One of the four types of nitrogenous bases in DNA and RNA.

deoxyribonucleic acid (DNA) (dē-ŏksē-rībō-nōō-klā′ĭk) The genetic material; the biochemical that forms genes.

differentiation The specialization of distinctive cell types through the use of subsets of genes to manufacture proteins.

DNA profiling A group of biotechnologies that detects differences among individuals used to identify them or rule out relationships. Includes numbers of repeats, single nucleotide polymorphism patterns, and gene and genome sequencing.

DNA replication Construction of a new DNA double helix using the parental strand nucleotide sequence as a template.

dominant A gene variant expressed when present in one copy.

exome (x-ōm) The approximately 1 percent of the genome that encodes proteins.

gene (jēn) A section of a DNA molecule whose sequence of building blocks instructs a cell to produce a particular protein.

gene expression The transcription of mRNA molecules from a gene.

gene pool All the alleles in a population.

genetic genealogy Comparing DNA sequences among individuals to determine relatedness and geographic origins.

genetics The study of inherited traits and their variation.

genome (jē′nōm) The complete set of genetic instructions in the cells of a type of organism.

genome editing Creating double-stranded breaks in the DNA double helix, enabling insertion or removal of a specific DNA sequence.

genomics (jē nōm ĭks) The field that analyzes and compares genomes.

genotype (jē n′ə- tīp) The allele combinations in an individual that cause traits or diseases.

guanine (G) (gwa′ nēn) One of the four types of nitrogenous bases in DNA and RNA.

heredity Transmission of inherited traits from generation to generation.

karyotype (kă′rē-ō-tīp) A size-order chart of chromosome pairs.

Mendelian trait A characteristic specified by expression of one gene, with a transmission pattern in accordance with Mendel's law of segregation.

metagenomics Sequencing all of the genomes present in a sample of a particular environment.

microbiome (mahy-kroh-bahy′-ohm) All of the organisms that live in and on another organism.

mutation (myōō-tā′shən) A change in a DNA sequence that affects the phenotype and is rare in a population.

nitrogenous bases (nī-trŏj′ə-nəs) The nitrogen-containing bases that are the information-carrying part of DNA or RNA.

nucleus (nōō-klēəs) A structure in a nondividing cell that houses DNA.

pedigree A chart of symbols connected by lines that depicts the genetic relationships of and transmission of inherited traits in a family.

pharmacogenetics (farm a kō jə-nĕt-ĭks) Testing for a gene variant that affects metabolism of a specific drug.

phenotype (fē′ nō-tīp) The expression of a gene in traits or symptoms.

recessive An allele that must be present on both chromosomes of a pair to be expressed.

recombinant DNA technology Adding a gene from one species to the genome of another, to program production of a desired protein or trait.

ribonucleic acid (RNA) (rī′bō-nōō-klē′ĭk) A nucleic acid whose bases are A, C, U, and G.

sex chromosome (krō′mə-sōōm) A chromosome including genes that specify sex.

somatic cell (sō-măt′ĭk sĕl) A non-sex cell, with 23 pairs of chromosomes in humans.

stem cell A cell that can divide to yield another stem cell (self-renew) and a cell that differentiates.

thymine (T) One of the four types of nitrogenous bases in DNA.

tissue Aggregate of cells with a shared function.

transcription Manufacturing complementary RNA from a gene on a strand of DNA.

translation Assembly of an amino acid chain according to the sequence of base triplets in a molecule of mRNA.

uracil (yōōr′ə-sĭl) One of the four types of nitrogenous bases in RNA.

Review Questions

1. Place the following terms in size order, from largest to smallest, based on the structures or concepts they represent.
   a. chromosome
   b. gene pool
   c. gene
   d. DNA
   e. genome

2. Provide an example of a bioethical issue.

3. Explain how DNA encodes information.

4. Distinguish between
   a. an autosome and a sex chromosome.
   b. genotype and phenotype.
   c. DNA and RNA.
   d. recessive and dominant traits.
   e. pedigrees and karyotypes.
   f. gene and genome.
   g. exome and genome.
   h. a Mendelian and a complex trait.

5. Define *allele*.

6. Explain how all cells in a person's body have the same genome, but are of hundreds of different types that look and function differently.

7. What are the two types of information that pedigrees used in genetics show?

8. Define *gene pool*.

9. List three uses of DNA sequencing or manipulation.

10. Explain how metagenomics analyzes genomes other than those from humans.

Applied Questions

1. What are risks and benefits of consumer genetics?

2. Why might people accept drugs that are manufactured in genetically modified organisms, but not "GM" foods?

3. Would you carry your exome or genome information in your smartphone? Cite a reason for your answer.

4. Describe a frivolous use and a serious use of DNA testing.

5. What do you want to learn, if anything, about your own genome?

Forensics Focus

1. It is springtime and Romeo, a fluffy black-and-white Maine Coon cat, bolts outdoors. Hours later he returns, a headless rodent in his mouth, drops it, and then consumes a can of seafood cat food and a few liver treats, then grabs some fried eggplant from the counter but spits it out. He licks his fur for awhile, swallowing wads of hair, then falls into a deep sleep. The next morning, Romeo's humans awaken to the unmistakable sounds of a cat vomiting. They poke around in the material to deduce what may be disturbing Romeo's digestion, but it is too slimy to tell. Describe how DNA testing might be used to discover the source of Romeo's discomfort.

2. Researchers in Canada published a report on 468 legal cases that cited "genetic predisposition" to explain physical injury or criminal behavior due to mental illness or substance use disorder. (In 86 cases the genetic information didn't harm the plaintiff; in 134 cases it did; and in 248 cases the legal significance of the genetic findings was unclear.) Under what circumstances, if any, do you think genetic information might be useful in a legal case, and from whose point of view?

3. Suggest how DNA testing might be used to reunite children with parents or other relatives from whom they have been separated during situations of civil unrest or war.

Case Studies and Research Results

1. Many large-scale genome sequencing projects focus on individuals who have similar diagnoses, such as adults with cardiovascular disease or children with developmental delay, autism, and intellectual disability. In contrast, the Genomic Postmortem Research Project is sequencing the genomes of 300 deceased individuals who had been patients at a large clinic that kept meticulous electronic medical records (see the opener to chapter 14). What type of information might be gleaned from this genome information?

2. An artist travels around New York City collecting hairs, fingernails, discarded paper cups, gum, cigarette butts, and other trash; then extracts the DNA and uses a computer program to construct faces from the genetic information. The artist then prints the results in three dimensions, producing sculptures that she displays in a gallery along with the original DNA-bearing evidence for each sculpture. The algorithm considers 50 inherited traits that affect facial features, including information on ancestry. What would you do if you wandered into a gallery and found a sculpture of yourself or someone you know? How private is the DNA on a wad of discarded chewing gum?

3. A study collected DNA from children at a state fair, to identify genes that contribute to normal health and development. If you were a parent of a child at the fair, what questions would you have asked before donating your child's DNA?

CHAPTER 2

Cells

One sign of Rett syndrome, a genetic condition that affects mostly girls, is "loss of purposeful hand use." Today it is widely recognized and diagnosed early, but in 2004, a case was diagnosed years after the child passed away based on analysis of DNA from a baby tooth that the mother had saved. The DNA in cells can hold clues to whole-body symptoms.

Learning Outcomes

2.1 Introducing Cells
1. Explain why it is important to know the cellular basis of a disease.
2. Distinguish somatic cells from germ cells, and diploid from haploid.

2.2 Cell Components
3. Explain how cells differentiate.
4. List the four major chemicals in cells.
5. Describe how organelles interact to secrete, use energy, and clean up debris.
6. Describe the structure and function of a biological membrane.
7. List the components of the cytoskeleton.

2.3 Cell Division and Death
8. Describe the main events of the cell cycle.
9. Indicate how chromosomes change and interact during mitosis.
10. Discuss control of the cell cycle.
11. List the events of apoptosis.

2.4 Stem Cells
12. List the characteristics of a stem cell and a progenitor cell.
13. Describe the three main sources of human stem cells.

2.5 The Human Microbiome
14. Describe the components of the human microbiome.
15. List factors that can alter the human microbiome.

The BIG Picture

Our bodies are built of trillions of cells that interact in complex ways. All cells in a body use the same genome, but have different structures and functions because they access different parts of the genome.

Diagnosis of Rett Syndrome from a Tooth

The genome is like a database containing clues to health even after death. A mother in Australia gave her daughter's saved baby tooth to researchers. When they extracted and analyzed the DNA, the researchers finally put a name to the disease that had caused the child's symptoms: Rett syndrome.

The girl had seemed healthy until 16 months, when she still wasn't walking, and she began to lose words she'd used frequently. By age 2 she could no longer manipulate small objects and developed repetitive hand-wringing movements. By age 3 seizures began. She had autism and became increasingly disabled; she died in 1991, undiagnosed.

In 2004, the mother read about Rett syndrome and recognized her daughter's symptoms. Fortunately, she had kept her daughter's baby teeth. Could researchers test DNA from a tooth for the mutation in the *MECP2* gene that causes Rett syndrome? She read that teeth contain a soup of cells in the pulp, and realized that each such cell had two copies of her deceased daughter's genome. People from the Australian Rett Syndrome Study introduced the mother to the researchers, who diagnosed the condition.

2.1 Introducing Cells

Our inherited traits, quirks, and illnesses arise from the activities of cells. Understanding cell function reveals how a healthy body works, and how it develops from one cell to trillions. Understanding what goes wrong in certain cells to cause symptoms can suggest ways to treat the condition, because we learn what to repair or replace. For example, genes tell cells how to make the proteins that align to form the contractile apparatus of muscles. In Duchenne muscular dystrophy, one type of muscle protein is missing, and as a result muscle cells collapse under forceful contraction. Certain muscles become weak in early childhood, and gradually the person becomes paralyzed. This form of muscular dystrophy is lethal by early adulthood.

Our bodies include more than 290 specialized, or differentiated, cell types that aggregate and interact to form the four basic tissue types: epithelial, connective, muscle, and nervous tissue (**table 2.1**). However, specialized cells have subtypes, and even cells that look alike may differ in their patterns of gene expression.

Most cells are somatic cells, also called body cells. Somatic cells have two copies of the genome and are said to be **diploid**. In contrast, the **germ cells**, which are sperm and egg cells, have one copy of the genome and are **haploid**. The meeting of sperm and egg restores the diploid state. Stem cells are diploid cells that divide to give rise to differentiated cells, and to other stem cells in a process called **self-renewal**. Stem cells enable a body to develop, grow, and repair damage.

Cells interact. They send, receive, and respond to information, as figure 8.1 depicts for two nerve cells communicating. Some cells aggregate with others of like function, forming tissues, which in turn interact to form organs and organ systems. Other cells move about the body. Cell numbers and the proportions of different cell types are important for development, growth, and healing. Staying healthy reflects a balance between cell division, which adds cells, and cell death, which takes them away.

Key Concepts Questions 2.1

a. Name the four types of tissues.

*b. How many copies of the genome does a diploid cell have?

*c. How many copies of the genome does a haploid cell have?

d. Distinguish a somatic cell from a germ cell.

Table 2.1	Tissue Types
Tissue	**Function/Location/Description**
Connective tissues	A variety of cell types and surrounding materials protect, support, bind to cells, and fill spaces throughout the body; include cartilage, bone, blood, and fat
Epithelium	Tight cell layers form linings that protect, secrete, absorb, and excrete
Muscle	Cells contract, providing movement
Nervous	Neurons transmit information as electrochemical impulses that coordinate movement and also sense and respond to environmental stimuli; neuroglia support and nourish neurons

2.2 Cell Components

All cells share features that enable them to perform the basic life functions of reproduction, growth, response to stimuli, and energy use. Specialized characteristics emerge as cells express different subsets of the thousands of protein-encoding genes.

All multicellular organisms, including other animals, fungi, and plants, have differentiated cells. Some single-celled organisms, such as the paramecium and amoeba, have cells as complex as our own. The most abundant organisms on the planet, however, are single celled, such as bacteria. These microorganisms have occupied Earth much longer than we have, and so in an evolutionary sense have been more successful than us.

Biologists recognize three basic types of cells that define three major "domains" of life: the Archaea, the Bacteria, and the Eukarya. A domain is a broader classification than a kingdom.

Members of the Archaea and Bacteria are single celled, but they differ from each other in the sequences of many of their genes and in the types of molecules in their membranes. Archaea and Bacteria are prokaryotes. A **prokaryotic cell** does not have a nucleus, the structure that contains DNA in the cells of other types of organisms. Members of the third domain of life, the Eukarya, also known as eukaryotes, include single-celled organisms that have nuclei and all multicellular organisms (**figure 2.1**). We are eukaryotes. A **eukaryotic cell** is also distinguished from a prokaryotic cell by structures called **organelles**, which perform specific functions. The cells of all three domains contain globular assemblies of RNA and protein called **ribosomes** that are essential for protein synthesis. The eukaryotes may have arisen from an ancient fusion of a bacterium with an archaean.

Chemical Constituents

Cells are composed of molecules. Some of the chemicals of life (biochemicals) are so large that they are called macromolecules.

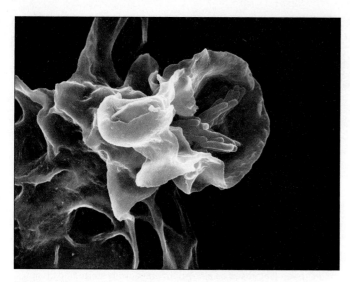

Figure 2.1 Eukaryotic and prokaryotic cells. The large purplish structure is a macrophage, part of the human immune system. Here it engulfs much smaller cells stained pink, which are the bacteria that cause tuberculosis. *SPL/Science Source*

The major macromolecules that make up cells and that cells use as fuel are **carbohydrates** (sugars and starches), **lipids** (fats and oils), **proteins**, and **nucleic acids** (DNA and RNA). Cells require vitamins and minerals in much smaller amounts.

Carbohydrates provide energy and contribute to cell structure. Lipids form the basis of some hormones, form membranes, provide insulation, and store energy. Proteins have many diverse functions in the human body. They enable blood to clot, form the contractile fibers of muscle cells, and form the bulk of the body's connective tissues. Antibodies that fight infection are proteins. **Enzymes** are especially important proteins because they facilitate, or catalyze, biochemical reactions so that they occur fast enough to sustain life. Most important to the study of genetics are the nucleic acids DNA and RNA, which translate information from past generations into specific collections of proteins that give a cell its characteristics.

Macromolecules can combine in cells, forming larger structures. For example, the membranes that surround cells and compartmentalize their interiors consist of double layers (bilayers) of lipids embedded with carbohydrates, proteins, and other lipids.

Life is based on the chemical principles that govern all matter; genetics is based on a highly organized subset of the chemical reactions of life. **Clinical Connection 2.1** describes diseases that affect these major types of biological molecules.

Organelles

A typical eukaryotic cell holds a thousand times the volume of a bacterial cell. To carry out the activities of life in such a large cell, organelles divide the labor by partitioning off certain areas or serving specific functions. The coordinated functioning of the organelles in a eukaryotic cell is much like the organization of departments in a big-box store, compared to the prokaryote-like simplicity of a small grocery store. In general, organelles keep related biochemicals and structures close enough to one another to interact efficiently. This eliminates the need to maintain a high concentration of a particular biochemical throughout the cell.

Organelles have a variety of functions. They enable a cell to retain as well as to use its genetic instructions to secrete substances, dismantle debris, and acquire energy. Saclike organelles sequester biochemicals that might harm other cellular constituents. Some organelles consist of membranes studded with enzymes embedded in the order in which they participate in the chemical reactions that produce a particular molecule. **Figure 2.2** depicts organelles.

The most prominent organelle of most cells is the nucleus. A layer called the nuclear envelope surrounds the nucleus. Biochemicals can exit or enter the nucleus through nuclear pores, which are rings of several types of proteins around an opening, like lined portholes in a ship's side (**figure 2.3**).

On the inner face of the nuclear membrane is a layer of fibrous material called the nuclear lamina. This layer is not just an inert covering, but turns off the expression of genes that contact it from within. The nuclear lamina also provides mechanical support and holds the nuclear pores in place. The opener to chapter 3 discusses a rare, accelerated aging disease that results from an abnormal nuclear lamina.

Inside the nucleus is an area that appears darkened under a microscope, called the **nucleolus** ("little nucleus"). Here, ribosomes are produced. The nucleus contains DNA complexed with many proteins to form chromosomes. Other proteins form fibers that fill out the nucleus, giving it a roughly spherical shape. RNA is abundant too, as are enzymes and other types of proteins that are required to synthesize RNA from DNA. The fluid in the nucleus, minus these contents, is called nucleoplasm.

The remainder of the cell—that is, everything but the nucleus, organelles, and the outer boundary, or **plasma membrane**—is **cytoplasm**. (The plasma membrane is also called the cell membrane.) Other cellular components include stored proteins, carbohydrates, and lipids; pigment molecules; and various other small chemicals. (The cytoplasm is called cytosol when these other parts are removed.) A closer look at three cellular functions—secretion, digestion inside cells, and energy production—reveals how organelles interact in ways that coordinate basic life functions and provide the characteristics of specialized cell types.

Secretion—The Eukaryotic Production Line

Secretion is the release of a substance from a cell. The process of secretion begins when the body sends a biochemical message to a cell to begin producing the substance. For example, when a newborn suckles the mother's breast, her brain releases hormones that signal cells in her breast to rapidly increase the production of the complex mixture that makes

Clinical Connection 2.1

Inborn Errors of Metabolism Affect the Major Biomolecules

Enzymes speed specific chemical reactions by lowering the "activation energy" required to start the reaction. They control a cell's production of all types of macromolecules. When an enzyme is not produced or cannot function, the chemical it acts on builds up and too little of the product of the reaction forms, like a block in a garden hose. The resulting buildups and breakdowns may cause symptoms. Genetic diseases that result from deficient or absent enzymes are called "inborn errors of metabolism." Following are examples of conditions that result from deficiencies or buildups of the major biological molecules.

Carbohydrates

The newborn yelled and pulled up her chubby legs in pain a few hours after each feeding, and had watery diarrhea. She had *lactase deficiency*, a lack of the enzyme lactase, which enables the digestive system to break down the carbohydrate lactose. Bacteria multiplied in the undigested lactose in the child's intestines, producing gas, cramps, and bloating. Switching to a soybean-based, lactose-free infant formula helped. A different, milder condition is lactose intolerance, common in adults (see the opening essay to chapter 16).

Lipids

A sudden sharp pain began in the man's arm and spread to his chest. At age 36, he was younger than most people who suffer heart attacks, but he had inherited a mutation that halved the number of protein receptors for cholesterol on his liver cells. Because cholesterol could not enter liver cells efficiently, it built up in his arteries, constricting blood flow in his heart and causing the mild heart attack. He inherited *familial hypercholesterolemia* (see figure 5.2). Taking a cholesterol-lowering drug and exercising regularly lowered his risk of future heart attacks.

Proteins

Newborn Tim slept most of the time, and he vomited so often that he hardly grew. A blood test that was part of a newborn screening panel done just a day after his birth revealed *maple syrup urine disease*, which causes urine to smell like maple syrup. Tim could not digest three types of amino acids (protein building blocks), which accumulated in his bloodstream. A diet very low in these amino acids controlled the symptoms.

Nucleic Acids

From birth, Troy's wet diapers held orange, sandlike particles, but otherwise he seemed healthy. By 6 months of age, urinating had become painful. A physician noted that Troy's writhing movements were involuntary rather than normal crawling. The orange particles in Troy's diaper indicated *Lesch-Nyhan syndrome*, caused by deficiency of the enzyme HGPRT. Troy's body could not recycle two of the four types of DNA building blocks (purines), instead converting them into uric acid, which crystallizes in urine. Later he developed severe intellectual disability, seizures, and aggressive and self-destructive behavior. By age 3, he uncontrollably bit his fingers and lips. His teeth were removed to keep him from harming himself, and he was kept in restraints. Life expectancy for the condition can reach 30 years.

Vitamins

Vitamins enable the body to use dietary carbohydrates, lipids, and proteins. Kaneesha inherited *biotinidase deficiency*, which greatly slows her body's use of the vitamin biotin. If Kaneesha hadn't been diagnosed as a newborn and quickly started on biotin supplements, she would have developed intellectual disability, seizures, skin rash, and loss of hearing, vision, and hair. Her slow growth, caused by her body's inability to extract energy from nutrients, would have eventually proved lethal.

Minerals

Ingrid, in her thirties, lived in an institution, unable to talk or walk. She grinned and drooled, but was alert and communicated using a picture board and an alphabet board, in the 1980s. When she was a healthy high-school senior, symptoms of *Wilson disease* began as her liver could no longer control the excess copper her digestive tract absorbed from food. The initial symptoms were stomachaches, headaches, and an inflamed liver (hepatitis). Then other changes began—loss of balance; a gravelly, slurred, low-pitched voice due to weak mouth muscles; and altered handwriting. The greenish rings around her irises, caused by copper buildup, indicated Wilson disease (**figure 2A**). A drug, penicillamine, enabled Ingrid to excrete excess copper in her urine. The treatment halted the course of the illness, saving her life, but she did not improve.

Figure 2A
Wilson disease. A greenish ring around the brownish iris is one sign of the copper buildup of Wilson disease. *SPL/Science Source*

Questions for Discussion

1. Explain how inherited diseases can affect molecules in the body other than proteins.
2. Explain how absence of or abnormal forms of enzymes can cause different diseases.
3. Describe how treatments for lactase deficiency, familial hypercholesterolemia, maple syrup urine disease, biotinidase deficiency, and Wilson disease help to alleviate symptoms.

Figure 2.2 Generalized animal cell. Organelles provide specialized functions for the cell. Most of these structures are transparent; colors are used here to distinguish them. Different cell types have different numbers of organelles. All cell types have a single nucleus, except for red blood cells, which expel their nuclei as they mature. *(top): David M. Phillips/Science Source; (left): Keith R. Porter/Science Source; (right): EM Research Services, Newcastle University*

up milk (**figure 2.4**). In response, information in certain genes is copied into molecules of **messenger RNA (mRNA)**, which then exit the nucleus (see steps 1 and 2 in figure 2.4). In the cytoplasm, the mRNAs, with the help of ribosomes and another type of RNA called **transfer RNA (tRNA)**, direct the manufacture of milk proteins. These include nutritive proteins

Chapter 2 *Cells* 21

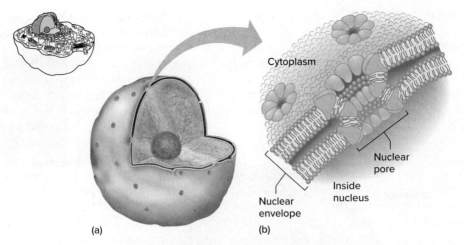

Figure 2.3 The nucleus is the genetic headquarters. (a) The largest structure in a typical human cell, the nucleus lies within two membrane layers that make up the nuclear envelope. **(b)** Nuclear pores allow specific molecules to move in and out of the nucleus through the envelope.

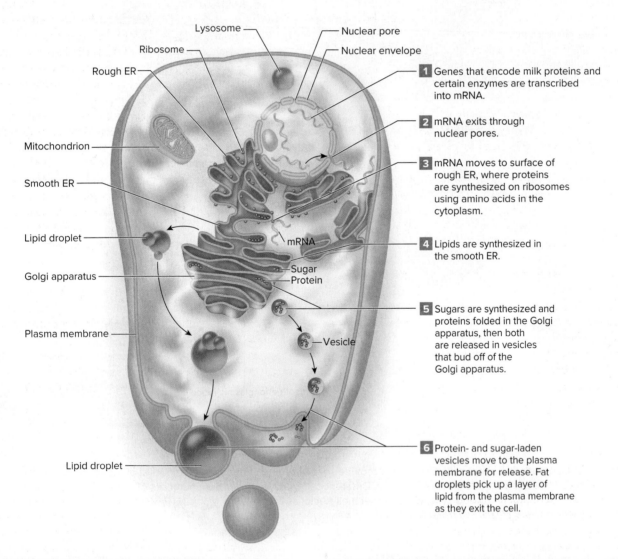

Figure 2.4 Secretion: making milk. Milk production and secretion illustrate organelle functions and interactions in a cell from a mammary gland: (1) through (6) indicate the order in which organelles participate in this process. Lipids are secreted in separate droplets from proteins and their attached sugars. This cell is highly simplified in terms of the relative sizes of structures.

called caseins, antibodies that protect against infection, and enzymes.

Most protein synthesis occurs on a maze of interconnected membranous tubules and sacs called the **endoplasmic reticulum (ER)** (see step 3 in figure 2.4). The ER winds from the nuclear envelope outward to the plasma membrane, forming a vast tubular network that transports molecules from one part of a cell to another. The ER nearest the nucleus, which is flattened and studded with ribosomes, is called rough ER, because the ribosomes make it appear fuzzy when viewed under an electron microscope. Protein synthesis begins on the rough ER when messenger RNA attaches to the ribosomes. Amino acids from the cytoplasm are then linked, following the instructions in the mRNA's sequence, to form particular proteins that will either exit the cell or become part of membranes along with lipids (step 3, figure 2.4). Proteins are also synthesized on ribosomes not associated with the ER. These proteins remain in the cytoplasm.

The ER acts as a quality control center for the cell. Its chemical environment enables the forming protein to start folding into the three-dimensional shape necessary for its specific function. Misfolded proteins are pulled out of the ER and degraded, much as an obviously defective toy might be pulled from an assembly line at a toy factory and discarded. Misfolded proteins that can cause disease are discussed further in chapter 10.

As the rough ER winds out toward the plasma membrane, the ribosomes become fewer and the tubules widen, forming a section called smooth ER. Here, lipids are made and added to the proteins arriving from the rough ER (step 4, figure 2.4). The lipids and proteins are transported until the tubules of the smooth ER narrow and end. Then the proteins exit the ER in membrane-bounded, saclike organelles called **vesicles**, which pinch off from the tubular endings of the membrane. Lipids are exported without a vesicle, because a vesicle is itself made of lipid.

A loaded vesicle takes its contents to the next stop in the secretory production line, a **Golgi apparatus** (step 5, figure 2.4), which looks like a stack of pancakes. This processing center is a column of four to six interconnected flat, membrane-enclosed sacs. Here, sugars, such as the milk sugar lactose, are made. Some sugars attach to proteins to form glycoproteins or to lipids to form glycolipids, which become parts of plasma membranes. Proteins finish folding in the Golgi apparatus and become active. Some cell types have just a few Golgi apparatuses, but those that secrete may have hundreds.

The components of complex secretions, such as milk, are temporarily stored in Golgi apparatuses. Droplets of proteins and sugars then bud off in vesicles that move outward to the plasma membrane, fleetingly becoming part of it until they are secreted to the cell's exterior. Lipids exit the plasma membrane directly, taking bits of it with them (step 6, figure 2.4).

In the breast, epithelial (lining) cells called lactocytes ("milk cells") align, forming ducts into which they secrete the components of milk. When the baby suckles, contractile cells squeeze the milk through the ducts and out of holes in the nipples.

A Golgi apparatus processes many types of proteins. These products are sorted when they exit the Golgi into different types of vesicles, a little like merchandise leaving a warehouse being packaged into different types of containers, depending on the destination. Molecular tags target vesicles to specific locations in the cell, or indicate that the contents are to be released to outside the cell.

A type of transport of molecules between cells, in contrast to within cells, uses vesicles called **exosomes** that bud from one cell and then travel to, merge with, and empty their contents into other cells. Exosomes are only 30 to 100 nanometers (billionths of a meter) in diameter. They may carry proteins, lipids, and RNA, and have been identified in many cell types. Exosomes remove debris, transport immune system molecules, and provide a vast communication network among cells.

Intracellular Digestion—Lysosomes and Peroxisomes

Just as clutter and garbage accumulate in an apartment, debris builds up in cells. Organelles called **lysosomes** ("bodies that cut") handle the garbage. Lysosomes are membrane-bounded sacs that contain enzymes that dismantle bacterial remnants, worn-out organelles, and other material such as excess cholesterol (**figure 2.5**). The enzymes also break down some digested nutrients into forms that the cell can use.

Lysosomes fuse with vesicles carrying debris from outside or within the cell, and the lysosomal enzymes then degrade the contents. The cell's disposing of its own trash is called **autophagy** ("eating self"). For example, a type of vesicle that forms from the plasma membrane, called an **endosome**, ferries extra low-density lipoprotein (LDL) cholesterol to lysosomes. A loaded lysosome moves toward the plasma membrane and fuses with it, releasing its digested contents to the outside. Lysosomes maintain the highly acidic environment that their enzymes require to function, without harming other cell parts that acids could destroy.

Cells differ in their number of lysosomes. The macrophage in figure 2.1 that is engulfing a bacterium has many lysosomes. Liver cells require many lysosomes to break down cholesterol, toxins, and drugs.

All lysosomes contain forty-three types of digestive enzymes, which must be maintained in balance. Absence or malfunction of an enzyme causes a "lysosomal storage disease," in which the molecule that the missing or abnormal enzyme normally degrades accumulates. The lysosomes swell, crowding other organelles and interfering with the cell's functions. In Tay-Sachs disease, for example, an enzyme is deficient that normally breaks down lipids in the cells that surround nerve cells. As the nervous system becomes buried in lipid, the infant begins to lose skills, such as sight, hearing, and mobility. Death is typically within 3 years. Even before birth, the lysosomes of affected cells swell. Lysosomal storage diseases affect about 10,000 people worldwide. (See table 22.4 and figure 22.4.)

Peroxisomes are sacs with single outer membranes that are studded with several types of proteins and that house enzymes that perform a variety of functions. The enzymes

Figure 2.5 Lysosomes are trash centers. Lysosomes fuse with vesicles or damaged organelles, activating the enzymes within to recycle the molecules. Lysosomal enzymes also dismantle bacterial remnants. These enzymes require a highly acidic environment to function. *Professors Pietro M. Motta & Tomonori Naguro/Science Source*

catalyze reactions that break down certain lipids and rare biochemicals, synthesize bile acids used in fat digestion, and detoxify compounds that result from exposure to oxygen free radicals (oxygen atoms that have unpaired electrons and damage macromolecules). Peroxisomes are large and abundant in liver and kidney cells, which handle toxins.

Adrenoleukodystrophy is an example of a genetic disease caused by an absent peroxisomal enzyme. A type of lipid called a very-long-chain fatty acid builds up in the brain and spinal cord. Early symptoms include low blood sugar, skin darkening, muscle weakness, visual loss, altered behavior and cognition, and irregular heartbeat. The patient eventually loses control over the limbs and usually dies within a few years. Providing a functional copy of the gene, called gene therapy, can halt progression of the disease or be given to newborns to prevent symptoms. (Chapter 22 discusses gene therapy.)

Energy Production—Mitochondria

Cells require continual energy to power the chemical reactions of life. Organelles called **mitochondria** provide energy by breaking the chemical bonds that hold together the nutrient molecules in food.

A mitochondrion has an outer membrane similar to those in the ER and Golgi apparatus and an inner membrane that forms folds called cristae (**figure 2.6**). These folds hold enzymes that catalyze the biochemical reactions that release energy from nutrient molecules. The freed energy is captured and stored in the bonds that hold together a molecule called adenosine triphosphate (ATP). In this way, ATP functions a little like an energy debit card.

A cell may have a few hundred to tens of thousands of mitochondria, depending on activity level. A typical liver cell has about 1,700 mitochondria, but a muscle cell, with its very high energy requirements, may have 10,000. A major symptom of diseases that affect mitochondria is fatigue. Mitochondria contain a small amount of DNA different in sequence from the DNA in the nucleus (see figure 5.8). Chapter 5 discusses mitochondrial inheritance and disease, and chapter 17 describes how mitochondrial genes provide insights into early human migrations.

Table 2.2 summarizes the structures and functions of organelles.

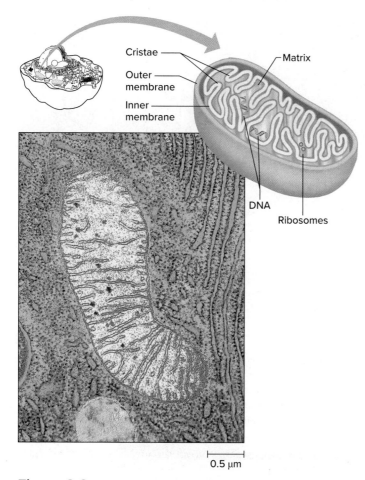

Figure 2.6 A mitochondrion extracts energy. Cristae, which are infoldings of the inner membrane of a mitochondrion, increase the available surface area containing enzymes for energy reactions. Mitochondria have their own DNA.
Bill Longcore/Science Source

Biological Membranes

Just as the character of a community is molded by the people who enter and leave it, the special characteristics of different cell types are shaped in part by the substances that enter and leave them. The plasma membrane completely surrounds the cell and monitors the movements of molecules in and out. How the chemicals that comprise the plasma membrane associate with each other determines which substances can enter or leave the cell.

Membranes also form the outer boundaries of mitochondria, lysosomes, peroxisomes, and vesicles, and completely comprise other organelles, such as the endoplasmic reticulum and Golgi apparatus. A cell's membranes are more than simple coverings, and the molecules embedded in them have specific functions.

Membrane Structure

A biological membrane has a distinctive structure. It is a double layer (bilayer) of molecules called phospholipids. A phospholipid is a fat molecule with attached phosphate groups. A phosphate group is a phosphorus atom bonded to four oxygen atoms. A phospholipid is depicted as a head with two parallel tails, shown in the inset to **figure 2.7**.

The tendency of lipid molecules to self-assemble into sheets makes the formation of biological membranes possible. The molecules form sheets because their ends react oppositely to water: The phosphate end of a phospholipid is attracted to water, and thus is hydrophilic ("water loving"); the other end, which consists of two chains of fatty acids, moves away from water, and is therefore hydrophobic ("water fearing"). Because of these forces, phospholipid molecules in water spontaneously form bilayers. Their hydrophilic surfaces are exposed to the watery exterior and interior of the cell. The hydrophobic surfaces face each other on the inside of the bilayer, away from water, and block entry and exit to most substances that dissolve in water. However, certain molecules can cross the membrane through proteins that form passageways, or when temporarily joined to a "carrier" protein.

Table 2.2 Structures and Functions of Organelles

Organelle	Structure	Function
Endoplasmic reticulum	Membrane network; rough ER has ribosomes, smooth ER does not	Site of protein synthesis and folding; lipid synthesis
Golgi apparatus	Stacks of membrane-enclosed sacs	Site where sugars are made and linked into starches or joined to lipids or proteins; proteins finish folding; secretions stored
Lysosome	Sac containing digestive enzymes	Degrades debris; recycles cell contents
Mitochondrion	Two membranes; inner membrane enzyme-studded	Releases energy from nutrients
Nucleus	Porous, double-membraned sac containing DNA	Separates DNA within cell
Peroxisome	Sac containing enzymes	Breaks down and detoxifies various molecules
Ribosome	Two associated globular subunits of RNA and protein	Scaffold and catalyst for protein synthesis
Vesicle	Membrane-bounded sac	Temporarily stores or transports substances

Some membrane proteins form channels for ions (atoms or molecules with electrical charge). "Channelopathies" are diseases that stem from faulty ion channels. For example, figure 1.6 depicts the ion channel that is abnormal in cystic fibrosis. Another type of channelopathy can cause either extreme pain in response to no apparent stimulus, or lack of pain. A famous case involved a boy who became a performer, walking on hot coals and stabbing himself to entertain crowds in Pakistan. He had a mutation in one of ten genes that encode proteins that form parts of channels that admit sodium ions into cells. Researchers are investigating these pain syndromes to develop new painkillers.

Proteins are embedded in the phospholipid bilayer of biological membranes. Some proteins traverse the entire structure, while others extend from one or both faces (figure 2.7). The phospholipid bilayer is oily, and some proteins move within it like ships on a sea.

The Plasma Membrane Enables Cell-to-Cell Communication

The proteins, glycoproteins, and glycolipids that extend from a plasma membrane create surface topographies that are important in a cell's functions and to its interactions with other cells. The surfaces of a person's cells indicate not only that they are part of that particular body, but also that they are part of a specific organ and type of tissue.

Many molecules that extend from the plasma membrane are receptors, which are structures that have indentations or other shapes that fit and hold specific molecules outside the cell. The molecule that binds to the receptor, called the ligand, may set into motion a cascade of chemical reactions inside the cell that carries out a particular activity, such as dividing.

In a cellular communication process called **signal transduction**, a series of molecules that are part of the plasma membrane form pathways that detect signals from outside the cell and transmit them inward, where yet other molecules orchestrate the cell's response. In a different process called **cellular adhesion**, the plasma membrane helps cells attach to certain other cells. These cell-to-cell connections are important in forming tissues.

Faulty signal transduction or cellular adhesion can harm health. For example, cancer is due to cells' failure to recognize or react to signals to cease dividing. Cancer cells also have abnormal cellular adhesion, which enables them to invade healthy tissue.

The Cytoskeleton

The **cytoskeleton** is a meshwork of protein rods and tubules that serves as the cell's architecture, positioning organelles and providing overall three-dimensional shapes. The proteins of the cytoskeleton are continually broken down and built up as a cell performs specific activities. Some cytoskeletal elements function as rails that transport cellular contents; other parts, called motor molecules, power the movement of organelles

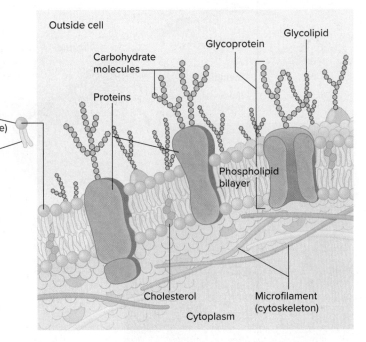

Figure 2.7 Anatomy of a plasma membrane. Mobile proteins are embedded throughout a phospholipid bilayer. Those proteins with attached carbohydrates are glycoproteins. Carbohydrates that jut from the membrane's outer face are glycolipids. The inset (left) shows a phospholipid molecule.

along these rails as they convert chemical energy into mechanical energy.

The cytoskeleton includes three major types of elements—**microtubules**, **microfilaments**, and **intermediate filaments** (**figure 2.8**). They are distinguished by protein type, diameter, and how they aggregate into larger structures. Other proteins connect these components, creating a framework that provides the cell's strength and ability to resist force and maintain shape.

Long, hollow microtubules provide many cellular movements. A microtubule is composed of pairs (dimers) of a protein, called tubulin, assembled into a hollow tube. Adding or removing tubulin molecules changes the length of the microtubule.

Cells contain both formed microtubules and individual tubulin molecules. When the cell requires microtubules to carry out a specific function—cell division, for example—free tubulin dimers self-assemble into more microtubules. After the cell divides, some of the microtubules fall apart into individual tubulin dimers, replenishing the cell's supply of building blocks. Cells perpetually build up and break down microtubules.

Microtubules maintain cellular organization and enable transport of substances within the cell. Abnormal microtubules cause several neurodegenerative diseases. For example, they may prevent neurotransmitters from reaching the synapses between neurons or between neurons and muscle cells.

Microtubules form hairlike structures called cilia, from the Latin meaning "cells' eyelashes" (**figure 2.9**). Cilia are of two types: motile cilia that move, and primary cilia that do not move but serve a sensory function.

Figure 2.8 The cytoskeleton is made of protein rods and tubules. The three major components of the cytoskeleton are microtubules, intermediate filaments, and microfilaments. Using special stains, the cytoskeletons in the cells in the inset appear orange with greenish fibers under the microscope. (The abbreviation nm stands for nanometer, which is a billionth of a meter.) *Photo: Dr. Gopal Murti/Science Source*

Figure 2.9 Cilia are built of microtubules. Motile cilia move secretions such as mucus on the cell surfaces of the respiratory tubes. In contrast, primary cilia do not move, functioning like antennae to sense changes. This figure shows only motile cilia.

Motile cilia have one more pair of microtubules than primary microtubules. Coordinated movement of motile cilia generates a wave that moves the cell or propels substances along its surface. The coordinated motion of many cilia moving a substance resembles "crowd surfing"—when a performer jumps into the audience and is then held aloft and passed back up to the stage. Similarly, motile cilia pass inhaled particles up and out of respiratory tubules and move egg cells in the female reproductive tract. Because motile cilia are so widespread, defects in them can cause multiple symptoms. One such ciliopathy—"sick cilia disease"—is Bardet-Biedl syndrome, which causes obesity, visual loss, diabetes, cognitive impairment, and extra fingers and/or toes. The several types of Bardet-Biedl syndrome affect both motile and primary cilia.

Many types of cells have primary cilia, which serve as antennae, sensing signals from outside cells and passing them to specific locations inside cells. Primary cilia sense light entering the eyes, urine leaving the kidney tubules, blood flowing from vessels in the heart, and pressure on cartilage. Although primary cilia do not move, they stimulate some cells to move, such as those that form organs in an embryo, and cells that help wounds to heal. Absence of primary cilia can harm health, such as in polycystic kidney disease. Cells may have many motile cilia but usually have only one cilium that does not move.

Microfilaments are long, thin rods composed of many molecules of the protein actin. They are solid and narrower than microtubules, enable cells to withstand stretching and compression, and help anchor one cell to another. Microfilaments provide many other functions in the cell through proteins that interact with actin. When any of these proteins is absent or abnormal, a genetic disease results. The actin part of the cytoskeleton is abnormal in Alzheimer disease and in some diseases of heart muscle.

Intermediate filaments have diameters intermediate between those of microtubules and microfilaments. Unlike those other components of the cytoskeleton, intermediate filaments are composed of different types of proteins in different cell types. However, all intermediate filaments consist of paired proteins entwined to form nested coiled rods. Intermediate filaments are scarce in many cell types but are abundant in nerve cells and skin cells. **Clinical Connection 2.2** describes how abnormal intermediate filaments affect several cell types in a girl who has a rare disease called giant axonal neuropathy.

Key Concepts Questions 2.2

*a. What does a eukaryotic cell have that a prokaryotic cell does not?

*b. List the chemical constituents of cells.

c. Explain the general functions of organelles.

d. Describe the organelles that interact as a cell secretes, degrades debris, and acquires energy.

*e. List components of the plasma membrane.

f. Define *cytoskeleton*.

Clinical Connection 2.2

A Girl with Giant Axons

Hannah Sames's cells do not make the protein gigaxonin, which normally breaks down intermediate filaments and recycles their components. In Hannah's hair cells (**figure 2B**), intermediate filaments made of keratin proteins build up, kinking the strands. In her neurons, different types of intermediate filament proteins accumulate, swelling the long nerve extensions (axons) that send messages to her muscles. Hannah is one of a few dozen people in the world known to have giant axonal neuropathy (GAN).

When Hannah was born on March 5, 2004, her parents and sisters were charmed that her hair was kinky while theirs was stick-straight. All seemed well until Hannah was 2 years, 5 months old, and her grandmother noticed her left arch rolling inward. Doctors thought she would outgrow it, but by her third birthday, both arches were involved and her gait had become awkward. Still physicians were not alarmed. Then Hannah's aunt showed phone video of Hannah walking to a physical therapist she worked with, who thought Hannah's gait resembled that of a child with muscular dystrophy.

With the idea of muscular dystrophy mentioned to the pediatrician, Hannah finally had genetic testing for the more common inherited childhood neurological conditions, but results were all normal. Then, at a visit with a pediatric neurologist, the doctor took one look at Hannah and reached for a large textbook high on a shelf, flipped through it, and showed Hannah's parents, Lori and Matt, a photo of a skinny boy with kinky hair and a high forehead and braces that went just below the knee—he had GAN, and he looked exactly like Hannah.

A test for the gigaxonin gene confirmed the diagnosis—Hannah had two identical "deletion" mutations—she did not have the gene at all. Lori and Matt were devastated when a genetic counselor told them that there were no treatments and no research on the extremely rare disease. The parents would not accept that, and within days had formed Hannah's Hope Foundation, which funded development of gene therapy (see section 22.3), the first into the spinal cord.

Figure 2B Hannah Sames has giant axonal neuropathy, which affects intermediate filaments. Her beautiful curls are one of the symptoms, but she prefers today to straighten her hair. *Courtesy, Lori Sames. Photo by Dr. Wendy Josephs*

Hannah received functional gigaxonin genes just after her twelfth birthday, in 2016. By that time, she was using a wheelchair and beginning to have trouble speaking and seeing. If all goes well, the gene therapy will halt the progression of the disease. Meanwhile, Hannah is working hard to strengthen her muscles. Gene therapy into the spinal cord may be developed to treat other, more common conditions, and perhaps spinal cord injuries.

Questions for Discussion

1. Why is the experimental gene therapy for GAN important, even though only a few people have this disease?
2. What is unusual about intermediate filaments, compared to microtubules and microfilaments?
3. Did the pediatric neurologist base his diagnosis of Hannah on her genotype or on her phenotype?

2.3 Cell Division and Death

In a human body, new cells form as old ones die, at different rates in different tissues. Growth, development, maintaining health, and healing from disease or injury require an intricate interplay between the rates of **mitosis** and **cytokinesis**, which divide the DNA and the rest of the cell, respectively, and **apoptosis**, a form of cell death.

An adult human body consists of about 30 trillion cells, and billions are replaced daily. Cells must die as part of normal development, molding organs much as a sculptor carefully removes clay to shape the desired object. Apoptosis carves fingers and toes, for example, from weblike structures that telescope out from an embryo's developing form (**figure 2.10**). *Apoptosis* is a Greek word for "leaves falling from a tree." It is a precise, genetically programmed sequence of events that is a normal part of development.

The Cell Cycle

Many cell divisions enable a fertilized egg to develop into a many-trillion-celled organism. A series of events called the **cell cycle** describes the sequence of activities as a cell prepares for and undergoes division.

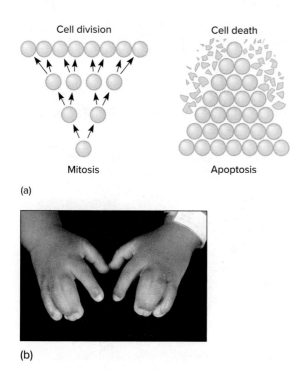

Figure 2.10 **Mitosis and apoptosis mold a body.** (a) Cell numbers increase from mitosis and decrease from apoptosis. (b) In the embryo, apoptosis normally carves fingers and toes from webbed structures. In syndactyly, apoptosis fails to carve digits, and webbing persists, as it does in these hands.
(b): Mediscan/Medical-on-Line/Alamy Stock Photo

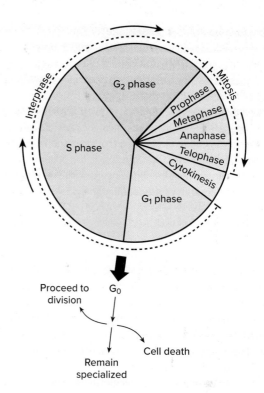

Figure 2.11 **The cell cycle.** The cell cycle is divided into interphase, when cellular components are replicated, and mitosis, when the cell distributes its contents into two daughter cells. Interphase is divided into G_1 and G_2, when the cell duplicates specific molecules and structures, and S phase, when it replicates DNA. Mitosis is divided into four stages plus cytokinesis, when the cells separate. G_0 is a "time-out" when a cell "decides" which course of action to follow.

Cell cycle rate varies in different tissues at different times. A cell lining the small intestine's inner wall may divide frequently, throughout life, whereas a neuron in the brain may never divide. A cell in the deepest skin layer may divide as long as a person lives, and even divide a few times after a person dies! Frequent mitosis enables the embryo and fetus to grow rapidly. By birth, the mitotic rate slows dramatically. Later, mitosis maintains the numbers and positions of specialized cells in tissues and organs.

The cell cycle is continual, but we describe it with stages. The two major stages are **interphase** (not dividing) and mitosis (dividing) (**figure 2.11**). In mitosis, a cell duplicates its chromosomes, then in cytokinesis it apportions one set of chromosomes, along with organelles, into each of two resulting cells, called daughter cells. Mitosis maintains the set of 23 chromosome pairs characteristic of a human somatic cell. Another form of cell division, **meiosis**, produces sperm or eggs, which have half the amount of genetic material that somatic cells do (or 23 single chromosomes, comprising one copy of the genome). Chapter 3 discusses meiosis in the context of development.

Interphase—A Time of Great Activity

During interphase a cell continues the basic biochemical functions of life, while also replicating its DNA and some organelles. Interphase is divided into two gap (G_1 and G_2) **phases** and one synthesis (**S**) **phase**. In addition, a cell can exit the cell cycle at G_1 to enter a quiet phase called G_0. A cell in G_0 is alive and maintains its specialized characteristics but does not replicate its DNA or divide. From G_0, a cell may also proceed to mitosis and divide, or die. Apoptosis may ensue if the cell's DNA is so damaged that cancer might result. G_0, then, is when a cell's fate is either decided or put on hold.

During G_1, which follows mitosis, the cell resumes synthesis of proteins, lipids, and carbohydrates. These molecules will contribute to building the extra plasma membrane required to surround the two new cells that form from the original one. G_1 is the period of the cell cycle that varies the most in duration among different cell types. Slowly dividing cells, such as those in the liver, may exit at G_1 and enter G_0, where they remain for years. In contrast, the rapidly dividing cells in bone marrow speed through G_1 in 16 to 24 hours. Cells of the early embryo may skip G_1 entirely.

During S phase, the cell replicates its entire genome. This happens simultaneously from several starting points, to get the enormous job done. After DNA replication, each chromosome consists of two copies of the genome joined at an area called the **centromere**. In most human cells, S phase takes 8 to 10 hours. Many proteins are also synthesized during this phase, including those that form the mitotic **spindle**, which will pull the chromosomes apart. Microtubules form structures called **centrioles** near the nucleus. Centriole microtubules, along with other proteins, are oriented at right angles to each other, forming paired, oblong structures called **centrosomes** that organize still other microtubules into the spindle.

G_2 occurs after the DNA has been replicated but before mitosis begins. More proteins are synthesized during this phase. Membranes are assembled from molecules made during G_1 and are stored as small, empty vesicles beneath the plasma membrane. These vesicles will merge with the plasma membrane to make enough of a boundary to enclose the two daughter cells.

Mitosis—The Cell Divides

DNA winds up tightly as mitosis begins, until the replicated chromosomes are condensed enough to be visible, when stained, under a microscope (see figure 9.13). The long strands of chromosomal material in replicated chromosomes are called **chromatids**, and if attached at a centromere, they are called sister chromatids. The space between sister chromatids is called a furrow (**figure 2.12**). At a certain point during mitosis, a replicated chromosome's centromere splits. This allows its sister chromatids to separate into two individual chromosomes. So a chromosome can be in an unreplicated form, in a replicated form consisting of two sister chromatids, or unwound and not visible when stained. Although the centromere of a replicated chromosome appears as a constriction, its DNA is replicated.

During **prophase**, the first stage of mitosis, DNA coils tightly. This coiling shortens and thickens the chromosomes, easing their separation (**figure 2.13**). Microtubules assemble from tubulin building blocks in the cytoplasm, forming the spindles. Toward the end of prophase, the nuclear membrane breaks down. The nucleolus is no longer visible.

Metaphase follows prophase. Chromosomes attach to the spindle at their centromeres and align along the center of the cell, which is called the equator. Metaphase chromosomes are under great tension, but they appear motionless because they are pulled with equal force on both sides, like a tug-of-war rope pulled taut.

Next, during **anaphase**, the plasma membrane indents at the center, where the metaphase chromosomes line up. A band of actin microfilaments forms on the inside face of the plasma membrane, constricting the cell down the middle. Then the centromeres part, which relieves the tension and releases one chromatid from each pair to move to opposite ends of the cell—like a

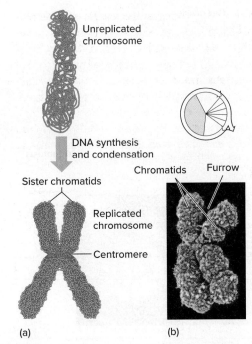

Figure 2.12 Replicated and unreplicated chromosomes. Chromosomes are replicated during S phase, before mitosis begins. Two genetically identical chromatids of a replicated chromosome join at the centromere **(a)**. The photomicrograph **(b)** shows a replicated human chromosome. An unreplicated chromosome and each sister chromatid are very long individual DNA molecules. *(b): SPL/Science Source*

Figure 2.13 Mitosis in a human cell. Replicated chromosomes separate and are distributed into two cells from one. In a separate process called cytokinesis, the cytoplasm and other cellular structures distribute and pinch off into two daughter cells. (Not all chromosome pairs are depicted.) *Photos: Ed Reschke*

Interphase
Chromosomes are uncondensed.

Prophase
Condensed chromosomes take up stain. The spindle assembles, centrioles appear, and the nuclear envelope breaks down.

tug-of-war rope breaking in the middle and the participants falling into two groups. Microtubule movements stretch the dividing cell. During the very brief anaphase, a cell fleetingly contains twice the normal number of chromosomes because each chromatid becomes an independently moving chromosome, but the cell has not yet physically divided.

In **telophase**, the final stage of mitosis, the cell looks like a dumbbell with a set of chromosomes at each end. The spindle falls apart, and nucleoli and the membranes around the nuclei re-form at each end of the elongated cell. Division of the genetic material is now complete. Next, during cytokinesis, organelles and macromolecules are distributed between the two daughter cells. Finally, the microfilament band contracts like a drawstring, separating the newly formed cells.

Control of the Cell Cycle

Control of mitosis is a daunting task. Quadrillions of mitoses occur in a lifetime, and not at random. When and where a somatic cell divides is crucial to health. Too little mitosis, and an injury goes unrepaired; too much, and an abnormal growth forms.

Groups of interacting proteins function at specific times in the cell cycle, called checkpoints, and they ensure that chromosomes are correctly replicated and apportioned into daughter cells (**figure 2.14**). A "DNA damage checkpoint" temporarily pauses the cell cycle while special proteins repair

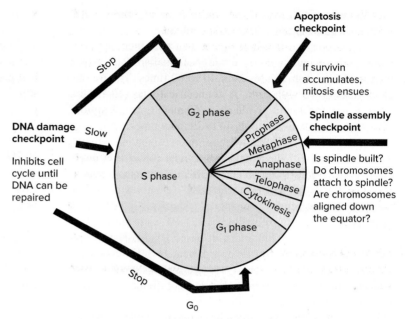

Figure 2.14 Cell cycle checkpoints. Checkpoints ensure that mitotic events occur in the correct sequence. Many types of cancer result from faulty checkpoints.

damaged DNA. An "apoptosis checkpoint" turns on as mitosis begins. During this checkpoint, proteins called survivins override signals telling the cell to die, so that mitosis (division) rather than apoptosis (death) occurs. Later during mitosis, the

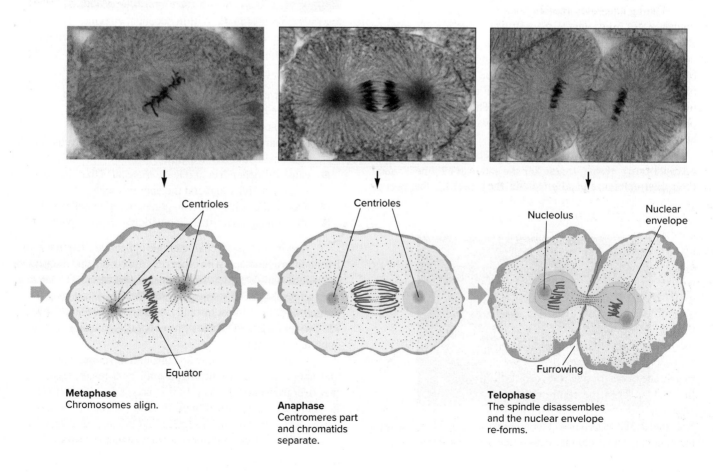

Chapter 2 Cells 31

"spindle assembly checkpoint" oversees construction of the spindle and the binding of chromosomes to it.

Cells obey an internal "clock" that tells them approximately how many times to divide. Mammalian cells grown (cultured) in a dish divide about 40 to 60 times. The mitotic clock ticks down with time. A connective tissue cell from a fetus, for example, will divide about 50 more times. A similar cell from an adult divides only 14 to 29 more times.

How can a cell "know" how many divisions remain? The answer lies in the chromosome tips, called **telomeres** (**figure 2.15**). Telomeres function like cellular fuses that burn down as pieces are lost from the ends. Telomeres consist of hundreds to thousands of repeats of a specific six DNA-base sequence (TTAGGG, see figure 13.2). At each mitosis, the telomeres lose 50 to 200 endmost bases, gradually shortening the chromosome. After approximately 50 divisions, a critical length of telomere DNA is lost, which signals mitosis to stop. The cell may remain alive but not divide again, or it may die.

Not all cells have shortening telomeres. In eggs and sperm, in cancer cells, and in a few types of normal cells that must continually supply new cells (such as bone marrow cells), an enzyme called telomerase keeps chromosome tips long. However, most cells do not produce telomerase, and their chromosomes gradually shrink.

The rate of telomere shortening provides a cellular "clock." It not only counts down a cell's remaining life span but may also sense environmental stimuli. Chronic stress, obesity, lack of exercise, and elevated blood sugar are associated with accelerated telomere shortening.

Factors from outside the cell can affect a cell's mitotic clock. Crowding can slow or halt mitosis. Normal cells growing in culture stop dividing when they form a one-cell-thick layer lining the container. This limitation to division is called contact inhibition. If the layer tears, the cells that border the tear grow and divide, filling in the gap. The cells stop dividing once the space is filled. Perhaps a similar mechanism in the body limits mitosis.

Hormones and growth factors are chemical signals that control the cell cycle from outside the cell. A **hormone** is made in a gland and transported in the bloodstream to another part of the body, where it exerts a specific effect. Hormones secreted in the brain, for example, signal the cells lining a young woman's uterus to build up each month by mitosis in preparation for possible pregnancy. A growth factor acts more locally. Epidermal growth factor, for example, stimulates cell division in the skin beneath a scab. Certain cancer drugs work by plugging growth factor receptors on cancer cells, blocking the signals to divide.

Two types of proteins, the cyclins and kinases, interact inside cells, activating the genes whose products carry out mitosis. The two types of proteins form pairs. Cyclin levels fluctuate regularly throughout the cell cycle, while kinase levels stay the same. A certain number of cyclin-kinase pairs turn on the genes that trigger mitosis. Then, as mitosis begins, enzymes degrade the cyclin. The cycle starts again as cyclin begins to build up during the next interphase. A class of highly effective cancer drugs target kinases, discussed in Clinical Connection 20.1.

Apoptosis

Apoptosis rapidly and neatly dismantles a cell into membrane-enclosed pieces that a phagocyte (a cell that engulfs and destroys another) can mop up. It is a little like packing the contents of a messy room into garbage bags, then disposing of it all. In contrast is necrosis, a form of cell death associated with inflammation and damage, rather than an orderly, contained destruction.

Like mitosis, apoptosis is a continuous process. It begins when a "death receptor" on the cell's plasma membrane receives a signal to die. Within seconds, enzymes called caspases are activated inside the doomed cell, stimulating each other and snipping apart various cell components. These killer enzymes:

- destroy enzymes that replicate and repair DNA;
- activate enzymes that cut DNA into similarly sized pieces;
- tear apart the cytoskeleton, collapsing the nucleus and condensing the DNA within;
- cause mitochondria to release molecules that increase caspase activity and end the energy supply;
- abolish the cell's ability to adhere to other cells; and
- attract phagocytes that dismantle the cell remnants.

A dying cell has a characteristic appearance (**figure 2.16**). It rounds up as contacts with other cells are cut off, and the plasma membrane undulates, forming bulges called blebs. The nucleus bursts, releasing DNA pieces, as mitochondria decompose. Then the cell shatters. Almost instantly, pieces of membrane encapsulate the cell fragments, which prevents inflammation. Within an hour, the cell is gone.

From the embryo onward through development, mitosis and apoptosis are synchronized and, as a result, tissue neither overgrows nor shrinks. In this way, a child's liver retains its shape as she grows into adulthood, yet enlarges. During early development, mitosis and apoptosis orchestrate the ebb and flow of cell numbers as new structures form. Later,

Figure 2.15 Telomeres. Fluorescent tags mark the telomeres in this human cell. *Health Protection Agency/Science Source*

these processes protect—mitosis produces new skin to heal a scraped knee; apoptosis peels away sunburnt skin cells that might otherwise become cancerous.

Cancer is a profound derangement of the balance between cell division and cell death. In cancer, mitosis occurs too frequently or too many times, or apoptosis happens too infrequently. This is why understanding the cell cycle has led to the development of many cancer drugs. Chapter 20 discusses cancer in detail.

Figure 2.16 Death of a cell. A cell undergoing apoptosis loses its shape, forms blebs, and falls apart. Caspases destroy the cell's insides. Phagocytes digest the remains. Note the blebs on the dying liver cells in the top photograph. Sunburn peeling is one example of apoptosis. *(top): David McCarthy/Photo Researchers/ Science Source; (bottom): Peter Skinner/Photo Researchers/Science Source*

Key Concepts Questions 2.3

a. Explain what can happen if the rates of mitosis and apoptosis are not balanced.

*b. List the phases of the cell cycle.

*c. List the stages of mitosis.

d. Describe what happens at each stage of mitosis.

*e. What are sister chromatids?

f. Explain how telomeres, crowding, hormones and growth factors, and cyclins and kinases control the cell cycle.

*g. List the events of apoptosis.

2.4 Stem Cells

A growing body must maintain the proportions and numbers of different cell types. Considering the composition of a human body by cell type, red blood cells are by far the most abundant. It is strange to realize that the cell types we are most aware of—fat and muscle cells—comprise only 0.2 percent or less of the total cell count. This is because the cells of fat and muscle tissue are much larger than red blood cells.

Bodies grow and heal thanks to cells that retain the ability to divide, generating new cells like themselves and cells that will specialize. Stem cells and **progenitor cells** renew tissues so that as the body grows or loses cells to apoptosis, injury, and disease, new cells are produced that take their places.

Cell Lineages

A stem cell divides by mitosis to yield either two daughter cells that are stem cells like itself, or one that is a stem cell and one that is a partially specialized progenitor cell (**figure 2.17**). The characteristic of self-renewal is what makes a stem cell a stem cell—its ability to continue the lineage of cells that can

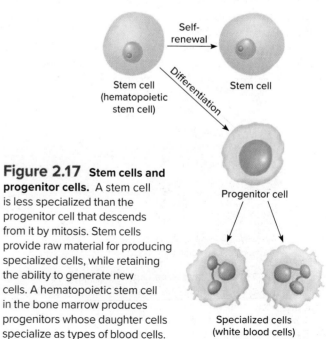

Figure 2.17 Stem cells and progenitor cells. A stem cell is less specialized than the progenitor cell that descends from it by mitosis. Stem cells provide raw material for producing specialized cells, while retaining the ability to generate new cells. A hematopoietic stem cell in the bone marrow produces progenitors whose daughter cells specialize as types of blood cells.

divide to give rise to another cell like itself. In contrast to a stem cell, a progenitor cell cannot self-renew, and its daughters specialize as any of a restricted number of cell types.

A fully differentiated cell, such as a mature blood cell, descends from a sequence of increasingly specialized progenitor cell intermediates, each one less like a stem cell and more like a blood cell. Our more than 290 differentiated cell types develop from sequences, called lineages, of stem, progenitor, and their increasingly differentiated daughter cells. **Figure 2.18** shows parts of a few cell lineages.

Stem cells and progenitor cells are described in terms of developmental potential—that is, according to the number

Figure 2.18 Pathways to cell specialization. All cells in the human body descend from stem cells, through the processes of mitosis and differentiation. The differentiated cells on the bottom are all connective tissues (blood, connective tissue, and bone), but the blood cells are more closely related to each other than they are to the other two cell types. On the upper right, the skin and sebaceous (oil) gland cells share a recent progenitor, and both share a more distant progenitor with neurons and supportive astrocytes. Imagine how complex the illustration would be if it depicted all 290-plus types of cells in a human body!

of possible fates of their daughter cells. A fertilized ovum is the ultimate stem cell. It is totipotent, which means that it can give rise to every cell type, including cells of the membranes that support the embryo. Other stem cells and progenitor cells are pluripotent: Their daughter cells have fewer possible fates. Some are multipotent: Their daughter cells have only a few developmental "choices." This is a little like a college freshman's consideration of many majors, compared to a junior's more narrowed focus in selecting courses.

As stem cell descendants specialize, they express some genes and ignore others. An immature bone cell forms from a progenitor cell by manufacturing mineral-binding proteins and enzymes. In contrast, an immature muscle cell forms from a muscle progenitor cell that accumulates contractile proteins. The bone cell does not produce muscle proteins, nor does the muscle cell produce bone proteins. All cells, however, synthesize proteins for basic "housekeeping" functions, such as energy acquisition and protein synthesis.

Many, if not all, of the organs in an adult human body have stem or progenitor cells. These cells can divide when injury or illness occurs and generate new cells to replace damaged ones. Stem cells in the adult may have been set aside in the embryo or fetus as repositories of future healing. Some stem cells, such as those from bone marrow, can travel to and replace damaged or dead cells elsewhere in the body, in response to signals that are released in injury or disease. Because every cell except red blood cells contains all of an individual's DNA, any cell type, given appropriate signals, can in theory become any other. This concept is the basis of some technologies that use stem cells.

Researchers study stem cells to learn more about biology and to develop treatments for a variety of diseases and injuries—not just inherited conditions. These cells come from both donors' and patients' own bodies, as **figure 2.19** illustrates. The cells can be mass produced in laboratory glassware, and if they originate from a patient's cell or nucleus, they are a genetic match.

Stem Cell Sources

There are three general sources of human stem cells: **embryonic stem (ES) cells**, **induced pluripotent stem (iPS) cells**, and "adult" stem cells (**table 2.3**).

Embryonic stem (ES) cells form in a laboratory dish from certain cells taken from a very early embryo, at a stage

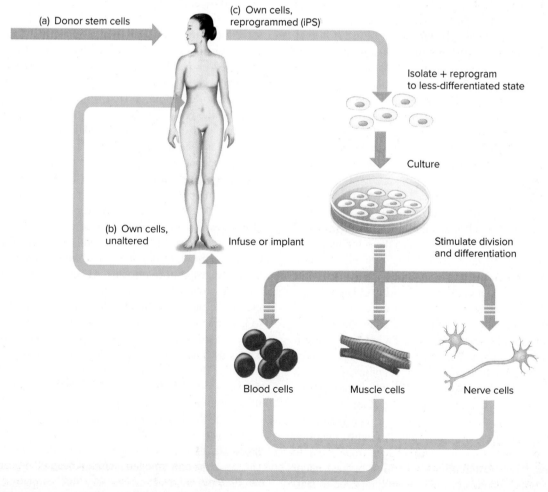

Figure 2.19 Using stem cells to heal. (a) Stem cells from donors (bone marrow or umbilical cord blood) are already in use. **(b)** A person's cells may be used, unaltered, to replace damaged tissue, such as bone marrow. **(c)** It is possible to "reprogram" a person's cells in culture, taking them back to a less specialized state and then nurturing them to differentiate as a needed cell type.

Table 2.3	Stem Cell Sources
Stem Cell Type	**Source**
Embryonic stem cell	Inner cell mass of very early embryo; somatic cell nuclear transfer into egg cell
Induced pluripotent stem cell	Genes or other chemicals reprogram somatic cell nucleus; no embryos required
"Adult" stem cell	Somatic cells that normally function as stem cells, from any stage of development from fertilized ovum through elderly

called the inner cell mass (ICM) (see figure 3.16). Some ICM cells, under certain conditions, become pluripotent and can self-renew—that is, they function as stem cells.

The ICM cells used to derive human ES cells can come from two sources. One is "leftover" embryos from fertility clinics that would otherwise be destroyed. The second source is a process called somatic cell nuclear transfer, in which a nucleus from a somatic cell (such as a skin fibroblast) is transferred to an egg cell that has had its own nucleus removed. An embryo develops and its ICM cells are used to make ES cells. Researchers can also derive ES cells from somatic cell nuclear transfer into egg cells directly, without using embryos. Somatic cell nuclear transfer is popularly called "cloning" because it copies the nucleus donor's genome. It was made famous by Dolly the sheep, whose original nucleus came from a mammary gland cell of a 6-year-old ewe. The age of the donor may be the reason why Dolly did not live as long as a sheep conceived the natural way. **Bioethics** in chapter 3 discusses "Why a Clone Is Not an Exact Duplicate."

Induced pluripotent stem (iPS) cells are somatic cells that are "reprogrammed" to differentiate into any of several cell types. Reprogramming a cell may take it back through developmental time to an ES cell-like state. Then the cell divides and gives rise to cells that specialize as different, desired cell types. Or, cells can be reprogrammed directly into another cell type. Reprogramming instructions may be in the form of DNA, RNA, or other chemical factors. Induced pluripotent stem cells do not require cells from an embryo, and they are a genetic match to the person of origin. For now, iPS cells are a research tool and will not be used to treat disease until researchers learn more about how they function in a human body. A common approach is to derive iPS cells from cells called fibroblasts in skin and reprogram them into a differentiated cell type of interest.

To return to the college major analogy, reprogramming a cell is like a senior in college deciding to change major. A French major wanting to become an engineer would have to start over, taking very different courses. But a biology major wanting to become a chemistry major would not need to start from scratch because many of the same courses apply to both majors. So it is for stem cells. Taking a skin cell from a man with heart disease and deriving a healthy heart muscle cell might require taking that initial cell back to an ES or iPS state, because these cells come from very different lineages. But turning a skeletal muscle cell into a smooth muscle cell requires fewer steps backward because the two cell types differentiate from the same type of progenitor cell.

A third source of stem cells are those that naturally are part of the body, called "adult" stem cells. They are more accurately called tissue-specific or somatic stem cells because they are found in the tissues of embryos, fetuses, and children, and not just in adult bodies. Adult stem cells self-renew, but most give rise only to a few types of specialized daughter cells. Researchers are still discovering niches of adult stem cells in the body, as well as organs that do not have them, such as the heart. Many potentially valuable adult stem cells are discarded as medical waste, such as those in fat, placenta, and brain tissue.

Stem Cell Applications

Stem cells are being used in four basic ways. In the first, drug discovery and development, stem cell cultures supply the human cells that are affected in a particular disease, which may be difficult or impossible to culture in a laboratory. Drugs are tested on these cells. Liver cells derived from stem cells are particularly useful in testing drugs for side effects, because the liver detoxifies many drugs. Using stem cells in drug development can minimize the need to experiment on animals and can eliminate candidate drugs with adverse effects before they are tested on people. Stem cells are also used to engineer bits of organs, called organoids, on which drugs can be tested. For example, an organoid derived from a piece of small intestine develops the velvety outpouchings of the lining through which nutrients and drugs are absorbed into the bloodstream. The organoid is used to study intestinal conditions.

A second application of stem cells growing in culture is to observe the earliest signs of a disease, which may begin long before symptoms are noted. The first disease for which human iPS cells were derived was amyotrophic lateral sclerosis (ALS), also known as Lou Gehrig's disease. In ALS, motor neurons that enable a person to move gradually fail. A few years after the first signs of weakness or stumbling, death comes from failure of the respiratory muscles. ALS had been difficult to study because motor neurons do not survive for long in laboratory glassware and do not divide. However, iPS cells reprogrammed to give rise to ALS motor neurons provide an endless source of the hard-to-culture cells from the abundant fibroblasts. The cells have practical uses, too. For example, drugs to treat muscular dystrophy can be tested on a patient's muscle cells derived from iPS cells, rather than from muscle cells removed using painful muscle biopsies.

The third application of stem cells is to create tissues and organs for use in implants, transplants, or infusions into the bloodstream. A patient's bone marrow stem cells, for example, can be removed, isolated, perhaps altered, grown, bathed in selected factors, and infused back into the patient, where they follow natural signals to damaged tissues. Bone marrow transplants have been done since the late 1950s. More recently, a child with the genetic disease junctional epidermolysis bullosa, in which the skin blisters and falls off, had his entire epidermis replaced with his own skin cells that had been given functional copies of the gene that is mutant in the condition. In health care, stem cell treatments like this experimental one are part of *regenerative medicine*. Researchers have created cell banks that store human iPS cells corresponding to different diseases.

The fourth application of stem cells became clear with the creation of iPS cells. It might be possible to introduce reprogramming proteins directly into the body to stimulate stem cells in their natural niches. That is, once we understand the signals, we might not need to introduce the cells.

The applications of stem cells seem limited only by our imaginations, but some companies take advantage of what the public does not know about the science. Bioethicists call this practice "stem cell tourism." **Bioethics** discusses stem cell banking and treatments.

Key Concepts Questions 2.4

a. Explain why different cell types are not present in the body in equal numbers.

*b. Define *self-renewal*.

c. Distinguish stem cells from progenitor cells.

*d. Define *cell lineage*.

e. Describe the functions of stem cells throughout life.

*f. Name the type of cell that derives from division of a stem cell and also can divide to give rise to a differentiated cell.

g. Describe how stem cells might be used in health care.

Bioethics

Stem Cell Tourism

The parents-to-be were excited by the company's promise: *Bank your baby's cord blood stem cells and benefit from breakthroughs. Be prepared for the unknowns in life.*

The website profiled children saved from certain diseases using stored umbilical cord blood. The statistics were persuasive: More than 70 diseases are currently treatable with cord blood transplants, and more than 10,000 procedures have already been done.

With testimonials like that, it is little wonder that parents collectively spend more than $100 million per year to store cord blood. The ads and statistics are accurate but misleading, because of what they *don't* say. Most people never actually use stored umbilical cord blood stem cells, not only because the treatable diseases are rare, but because cord blood stem cells are not very pluripotent, limiting their use. A compelling reason that stem cell banks are rarely used beyond initial storage is that for a person with an inherited disease, *healthy* stem cells are required—not the patient's, which could cause the disease all over again by adding the mutation in every introduced cell. The patient needs a well-matched donor, such as a healthy sibling.

Commercial cord blood banks may charge more than $1,000 for the initial collection plus an annual fee. However, not-for-profit banks may not charge fees. Donations of cord blood to these facilities are not to help the donors directly, but to help whomever can use the cells.

Online offerings of stem cell–based therapies compete with physician referrals to proven stem cell treatments available at medical centers. In the United States, hundreds of companies provide stem cells to hundreds of clinics. Globally, those numbers are in the thousands.

The International Society for Stem Cell Research maintains guidelines that enable consumers to distinguish treatments tested in clinical trials and with regulatory approval from interventions that are advertised but have not been rigorously investigated.

Governments, pharmaceutical companies, and foundations sponsor clinical trials and rarely ask participants to pay.

Crowdfunding has become a conduit of stem cell tourism, enabling patients or their families to seek help for treatments that may or may not have been adequately tested. One study evaluated 408 crowdfunding campaigns for stem cell treatments provided by 50 companies. Nearly half of the campaigns claimed the procedure's efficacy was "definitive or certain" and about a third were "optimistic or hopeful" about efficacy. Only 36 campaigns mentioned risks. The personal narratives are effective in raising funds, but oversight of the accuracy of claims is often lacking.

Questions for Discussion

1. Do you think stem cell banks should be regulated, and if so, by whom and how?
2. What information should companies offering to store stem cells post on their websites?
3. Do you think that advertisements for cord blood storage services that have quotes and anecdotal reports, but do not mention that most people who receive stem cell transplants do not receive their own cells, are deceptive? Or do you think it is the responsibility of the consumer to discover this information?
4. Several companies store stem cells extracted from baby teeth, although a use for such stem cells has not yet been found. (The use of DNA from a tooth in the chapter opener was to diagnose a child posthumously.) Suggest a different way to obtain stem cells that have the genome of a living child.
5. Should crowdfunding sites seeking support for stem cell treatments be required to provide evidence of efficacy, or would this counter freedom of speech?

2.5 The Human Microbiome

Many of the cells within a human body are not actually human—our bodies are vast ecosystems for microscopic life. The average person consists of about 30 trillion cells, but a healthy human body also includes about 39 trillion cells that are bacterial, fungal, or protozoan, as well as many viruses. The cells within and on us that are not actually *of* us constitute the human microbiome.

The term *microbiome* borrows from the ecological term *biome* to indicate all of the species in an area. Trillions of bacteria can fit into a human body because their cells, which are prokaryotic, are much smaller than ours. The Human Microbiome Project has revealed our bacterial residents, although the first to notice our bacteria was inventor of the light microscope Antonie van Leeuwenhoek, who in 1676 described the "animalcules" in his mouth. Today, researchers look at DNA sequences to identify the bacterial species that live in and on our parts. Knowledge of the human microbiome will be increasingly applied to improving health, once we learn that a particular collection of microbes causes or contributes to an illness or is a result of it. For example, some bloodstream infections once thought to come from inserting contaminated intravenous lines actually come from bacteria that have leaked from the intestines.

Each of us has a shared "core microbiome" of bacterial species, but also many other species that reflect our differing genomes, environments, habits, ages, diets, and health. Different body parts house distinctive communities of microbes, altered by experience. A circumcised penis has a different microbiome than an uncircumcised one, for example, and the vaginal microbiome of a mother differs from that of a woman who hasn't had a child. Different communities of bacteria inhabit a person's armpits, groin, navel, bottoms of the feet, spaces between fingers, and the space between the buttocks. The microbiome changes with experience and environmental exposures, as **figure 2.20** shows.

The 10 trillion bacteria that make up the human "gut microbiome" have long been known to help us digest certain foods. The mouth alone houses more than 600 species of bacteria. Analysis of their genomes yields practical information. For example, the genome sequence of the bacterium *Treponema denticola* reveals how it survives amid the films other bacteria form in the mouth and how it causes gum disease.

The more distant end of our digestive tract is easy to study in feces. The large intestine alone houses 6,800 bacterial species. One of the first microbiome studies examined soiled diapers from a few babies during their first year—one the child of the chief investigator—chronicling the establishment of the gut bacterial community.

Newborns start out with clean intestines. After various bacteria come and go, very similar species remain from baby to baby by the first birthday.

Investigators recognize three stages of the establishment of the gut microbiome in childhood. During the "developmental" phase from 3 to 14 months, the number of bacterial species increases. The types of bacteria become established during the "transitional" stage from 15 to 30 months. Finally, the gut microbiome is in a stable phase, when species stay constant, from 31 to 46 months. The most important factor influencing the gut microbiome is the mode of feeding. Infants who are breastfed have large populations of two bacterial species (*Bifidobacterium breve* and *B. bifidum*) that digest human milk and are components of probiotic products that many people take. When breastfeeding ceases, the maturation rate of the microbiome accelerates.

Illness can alter the bacterial populations within us, and we use knowledge of our gut microbiome to improve health. Probiotics are live microorganisms, such as bacteria and yeasts, that, when ingested, confer a health benefit by altering the microbiome. For example, certain *Lactobacillus* strains added to yogurt help protect against *Salmonella* foodborne infection.

Probiotics typically consist of one or several microbial species. In contrast, "fecal transplantation" is a treatment based on altering the microbiome that replaces hundreds of bacterial species at once. In the procedure, people with recurrent infection from *Clostridium difficile*, which causes severe diarrhea, receive feces from a healthy donor. The bacteria transferred in the donated feces reconstitute a healthy gut microbiome, treating the infection. Fecal transplantation has been performed by enema in cattle for a century, and since 1958 in humans. The

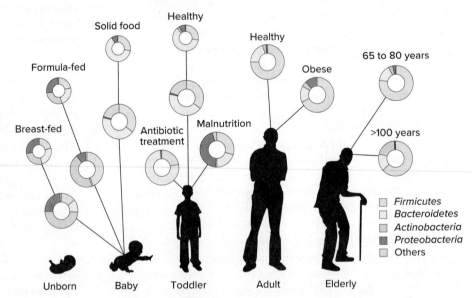

Figure 2.20 The human microbiome. Our habits, experiences, and environmental exposures influence the makeup of our microbiomes. The colors indicate different species of bacteria.

Source: From "The function of our microbiota: who is out there and what do they do?" by N. Ottman, H. Smidt, W. M. de Vos, and C. Belzer. Front. Cell. Inf. Microbio. 2:104 (2012). doi:10.3389/fcimb.2012.00104. Courtesy U.S. National Library of Medicine.

material, fresh or frozen, is typically introduced by enema. Clinical trials are testing delivery through a tube placed in the nose and snaked to the small intestine, or in a capsule.

Researchers have been studying the human microbiome for a short time, but already have learned or confirmed the following facts.

1. Certain skin bacteria cause acne, but others keep skin clear.
2. Circumcision protects against certain viral infections, including HIV.
3. Lowered blood sugar following weight-loss surgery is partly due to a changed gut microbiome.
4. An altered microbiome hastens starvation in malnourished children.
5. Antibiotics temporarily alter the gut microbiome.
6. The birth of agriculture 10,000 years ago introduced the bacteria that cause dental caries.
7. Microbiome imbalances may contribute to or cause asthma, cancers, obesity, psoriasis, Crohn's disease, gum disease, and other conditions.
8. Babies born by Cesarean section (surgically) have different microbiomes than babies born vaginally.
9. People living in isolated areas have different microbiomes than city residents. Researchers are preserving microbiomes from a diversity of environments, especially populations that have been isolated for long times.

Our genomes have evolved along with our microbiomes. Certain modern practices, such as use of antibiotic drugs, sanitation, and use of antimicrobial soap, may affect the composition of our personal microbiomes.

Key Concepts Questions 2.5

*a. Define *microbiome*.
b. Give an example of a body part that houses a unique microbiome.
c. Cite one way the microbiome affects health.
*d. What type of infection can a fecal transplant treat?

Study Guide

Summary

2.1 Introducing Cells

1. Cells are the basic units of life and comprise the human body. Inherited traits and illnesses can be understood at the cellular and molecular levels.
2. All cells share certain features, but they are also specialized because they express different subsets of genes.
3. Somatic (body) cells are **diploid**, and **germ cells** (sperm and egg cells) are **haploid**. Stem cells produce new cells.

2.2 Cell Components

4. A **prokaryotic cell** does not have a nucleus or other **organelles**. A **eukaryotic cell** has organelles, including a nucleus.
5. Cells consist primarily of water and several types of macromolecules: **carbohydrates**, **lipids**, **proteins**, and **nucleic acids**.
6. Organelles sequester related biochemical reactions, improving the efficiency of life functions and protecting the cell. The cell also consists of **cytoplasm** and chemicals.
7. The nucleus contains DNA and a **nucleolus**, a site of ribosome synthesis. **Ribosomes** provide scaffolds for protein synthesis; they exist free in the cytoplasm or complexed with the rough **endoplasmic reticulum (ER)**. In protein synthesis, **messenger RNA** (mRNA) carries the information held in a gene's DNA sequence and **transfer RNA** (tRNA) connects the mRNA to protein building blocks (amino acids).
8. In secretion, the rough ER is where protein synthesis begins. Smooth ER is the site of lipid synthesis, transport, and packaging; the **Golgi apparatus** packages secretions into **vesicles,** which exit through the **plasma membrane**. **Exosomes** ferry molecules between cells.
9. **Lysosomes** contain **enzymes** that dismantle debris through **autophagy**. **Endosomes** are vesicles derived from the plasma membrane that ferry materials to lysosomes. **Peroxisomes** house enzymes that catalyze a variety of chemical reactions.
10. Enzymes in **mitochondria** extract energy from nutrients.
11. The plasma membrane is a protein-studded phospholipid bilayer. It controls which substances exit and enter the cell (**signal transduction**) and how the cell interacts with other cells (**cellular adhesion**).
12. The **cytoskeleton** is a protein framework of hollow **microtubules**, made of tubulin, and solid **microfilaments**, which consist of actin. **Intermediate filaments** are made of more than one protein type.

2.3 Cell Division and Death

13. Coordination of cell division (**mitosis**) and cell death (**apoptosis**) maintains cell numbers, enabling structures to enlarge during growth and development but preventing abnormal growth.
14. The **cell cycle** indicates whether a cell is dividing (mitosis) or not (**interphase**). Interphase consists of G_1 and G_2 **phases**, when proteins and lipids are produced, and a synthesis **(S) phase**, when DNA is replicated. In G_0, a cell may remain specialized, divide, or die.

15. A **spindle** forms from microtubules organized into **centrioles**, which join other proteins to form paired **centrosomes** that hug the nucleus and pull chromosome pairs apart.

16. Mitosis proceeds in four stages. In **prophase**, replicated chromosomes consisting of two sister **chromatids** condense, the spindle assembles, the nuclear membrane breaks down, and the nucleolus is no longer visible. In **metaphase**, replicated chromosomes align along the center of the cell. In **anaphase**, the **centromeres** part, equally dividing the now unreplicated chromosomes into two daughter cells. In **telophase**, the new cells separate. **Cytokinesis** apportions organelles and other components into the daughter cells.

17. Internal and external factors control the cell cycle. Checkpoints are times when proteins regulate the cell cycle. **Telomere** (chromosome tip) length determines how many more times mitosis will occur. Crowding, **hormones**, and growth factors signal cells from the outside; the interactions of cyclins and kinases trigger mitosis from inside the cell.

18. In apoptosis, a receptor on the plasma membrane receives a death signal, which activates caspases that tear apart the cell in an orderly fashion. Membrane surrounds the pieces, preventing inflammation.

2.4 Stem Cells

19. Stem cells **self-renew**, producing daughter cells that are also stem cells, and yield daughter cells that specialize. **Progenitor cells** give rise to slightly more specialized progenitor cells and to more differentiated cells.

20. A fertilized ovum is totipotent. Some stem cells are pluripotent and some are multipotent. Cells are connected through lineages.

21. The three sources of human stem cells are **embryonic stem (ES) cells**, **induced pluripotent stem (iPS) cells**, and adult stem cells.

22. Stem cell technology enables researchers to observe the origins of diseases and to devise new types of treatments.

2.5 The Human Microbiome

23. About 60 percent of the cells in a human body are microorganisms. They are our microbiome.

24. Different body parts house different communities of microbes.

25. Genes interact with the microbiome and with environmental factors to mold most traits.

Key Terms

anaphase (ănə-fāz′) Stage of mitosis when the centromeres of replicated chromosomes part.

apoptosis (āpō-tō′sis) A form of cell death that is a normal part of growth and development.

autophagy (au-toph′-a-gy) A process in which a cell dismantles its own debris.

carbohydrate (kar″bo-hī′-drāt) An organic compound that consists of carbon, hydrogen, and oxygen in a 1:2:1 ratio. A sugar or a starch.

cell cycle A cycle of events describing a cell's preparation for division and division itself.

cellular adhesion A series of interactions among the proteins that connect cells.

centriole (sĕn′-trē-ohl) A structure that organizes microtubules into the mitotic spindle.

centromere (sĕn′-trō-mîr) The largest constriction in a chromosome, at a specific site in each chromosome type.

centrosome (sĕn′-trō-sōm) A structure built of centrioles and proteins that organizes microtubules into a spindle during cell division.

chromatid (krō′ mə-tĭd) A single, very long DNA molecule and its associated proteins, forming a longitudinal half of a replicated chromosome.

cytokinesis (sī-tō-kin-ē′-sis) Division of the cytoplasm and its contents.

cytoplasm (sī′-tō-plăzm) Cellular contents other than organelles.

cytoskeleton (sī-tō-skĕl′ĭ-tn) A framework of protein tubules and rods that supports the cell and gives it a distinctive form.

diploid (dip′ loid) A cell containing two sets of chromosomes.

embryonic stem (ES) cell (ĕmbrē-ŏn′ĭk) A cell derived in laboratory culture from inner cell mass cells of very early embryos that can self-renew and produce some daughter cells that can differentiate as any cell type.

endoplasmic reticulum (ĕn′-dō-plaz-mik rĕtik′-u-lum) An organelle consisting of a labyrinth of membranous tubules on which proteins, lipids, and sugars are synthesized.

endosome (en-dō-sōm) A vesicle that buds inward from the plasma membrane.

enzyme (ĕnzīm) A protein that speeds the rate of a specific biochemical reaction.

eukaryotic cell (yōō-kar′ē-ŏt′ĭk sel) A complex cell containing organelles, including a nucleus.

exosome (x-ō-sōm) A vesicle that carries molecules from cell to cell.

G_0 phase An offshoot of the cell cycle in which the cell remains specialized but does not replicate its DNA or divide.

G_1 phase The stage of the cell cycle following mitosis in which the cell resumes synthesis of proteins, lipids, and carbohydrates.

G_2 phase The stage of the cell cycle following S phase but before mitosis, when certain proteins are synthesized.

germ cells Cells that give rise to sperm or eggs.

Golgi apparatus (gōl′jē) An organelle, consisting of flattened, membranous sacs, that packages secretion components.

haploid (hap′ loid). A cell containing one set of chromosomes.

hormone (hor′ mōn) A biochemical produced in a gland and carried in the blood to a target organ, where it exerts an effect.

induced pluripotent stem (iPS) cells Somatic cells that are reprogrammed toward an alternative developmental fate by altering their gene expression.

intermediate filament A type of cytoskeletal component made of different proteins in different cell types.

interphase (ĭn′tər-fāz′) Stage of the cell cycle when a cell is not dividing.

lipid (lĭp′ĭd) A type of organic molecule that has more carbon and hydrogen atoms than oxygen atoms. Can be a fat or an oil.

lysosome (lī′sō-sōm) A saclike organelle containing enzymes that degrade debris.

meiosis (mī-ō′sĭs) A form of cell division that halves the number of chromosomes, forming haploid gametes.

messenger RNA (mRNA) A molecule of RNA complementary in sequence to the template strand of a gene that specifies a protein product.

metaphase (mĕtə-fāz) The stage of mitosis when chromosomes align along the center of the cell.

microfilament A solid rod of actin protein that forms part of the cytoskeleton.

microtubule (mīkrō-tōōbyōōl) A hollow structure built of tubulin protein that forms part of the cytoskeleton.

mitochondrion (mītō-kŏn′drē-ən) An organelle consisting of a double membrane that houses enzymes that catalyze reactions that extract energy from nutrients.

mitosis (mī-tōsĭs) Division of somatic (non-sex) cells.

nucleic acid (nōō-klē′ĭk) DNA or RNA.

nucleolus (nōō-klē′ə-ləs) A structure in the nucleus where ribosomes are assembled from ribosomal RNAs and proteins.

organelle (ŏr′gə-nĕl′) A specialized structure in a eukaryotic cell that carries out a specific function.

peroxisome (pə-rŏk′sī-sōm) An organelle with a single outer membrane that is studded with proteins and that houses enzymes that catalyze diverse reactions.

plasma membrane (plăz′mə mĕm′brān) The selective barrier around a cell, consisting of proteins, glycolipids, glycoproteins, and lipids on or in a phospholipid bilayer.

progenitor cell A cell whose descendant daughter cells can follow any of several developmental pathways, but not all.

prokaryotic cell (prō-kāre - ŏṭ′ik sĕl) A cell that does not have a nucleus or other organelles. Members of the domains Bacteria or Archaea.

prophase (prō′fāz) The first stage of mitosis or meiosis, when chromatin condenses.

protein A type of macromolecule that is the direct product of genetic information; a chain of amino acids.

ribosome (rī′bō sōm) An organelle consisting of RNA and protein that is a scaffold and catalyst for protein synthesis.

S phase The stage of interphase when DNA replicates.

self-renewal Defining property of a stem cell; the ability to yield a daughter cell like itself.

signal transduction A series of biochemical reactions and interactions that pass information from outside a cell to inside, triggering a response.

spindle A structure composed of microtubules that pulls sets of chromosomes apart in a dividing cell.

telomere (tĕl′ə-mîr) A chromosome tip.

telophase (tĕlə-fāz) The stage of mitosis or meiosis when daughter cells separate.

transfer RNA (tRNA) A type of RNA that connects mRNA to an amino acid during protein synthesis.

vesicle (ves-ə-kul) Bubble-like membrane-bounded organelle that participates in secretion.

Review Questions

1. Match each organelle to its function.

   Organelle
   a. lysosome
   b. rough ER
   c. nucleus
   d. smooth ER
   e. Golgi apparatus
   f. mitochondrion
   g. peroxisome

   Function
   1. lipid synthesis
   2. houses DNA
   3. energy extraction
   4. dismantles debris
   5. detoxification
   6. protein synthesis
   7. processes secretions

2. Describe a disease or symptoms that result when each of the following organelles or structures malfunctions.
   a. lysosome
   b. peroxisome
   c. mitochondrion
   d. cytoskeleton
   e. ion channel

3. What advantage does compartmentalization provide to a large and complex cell?

4. Distinguish between an exosome and an endosome.

5. Give two examples of how the plasma membrane functions as more than just a covering of the cell's insides.

6. List four types of controls on cell cycle rate.

7. How can all of a person's cells contain exactly the same genetic material, yet specialize as bone cells, nerve cells, muscle cells, and connective tissue cells?

8. Distinguish between
   a. a bacterial cell and a eukaryotic cell.
   b. interphase and mitosis.
   c. mitosis and apoptosis.
   d. rough ER and smooth ER.
   e. microtubules and microfilaments.
   f. a stem cell and a progenitor cell.
   g. totipotent and pluripotent.

9. How are intermediate filaments similar to microtubules and microfilaments, and how are they different?

10. List the events of each stage of mitosis.

11. How is the definition of a stem cell as a cell that "turns into any cell type" incorrect?

12. Distinguish among ES cells, iPS cells, and adult stem cells.

13. Define *microbiome*.

14. "I heard about an organic, gluten-free, nonfat, low-calorie, high-energy way to remove all of the bacteria in our intestines!" enthused the young woman in the health food store to her friend. Is this a good idea or not? Cite a reason for your answer.

Applied Questions

1. The chapter opener describes diagnosing Rett syndrome, which affects neurons in the brain, from DNA extracted from a saved baby tooth. Explain why a mutation detected in one cell type can be used to help diagnose an inherited condition that affects a different cell type.

2. If you wanted to create a synthetic organelle to test new drugs for toxicity, which natural organelle's function would you try to replicate?

3. An inherited form of migraine is caused by a mutation in a gene *(SCN1A)* that encodes a sodium channel in neurons. What is a sodium channel, and in which cell structure is it located?

4. In a form of Parkinson disease, a protein called alpha synuclein accumulates because of impaired autophagy. Which organelle is implicated in this form of the disease?

5. Why wouldn't a cell in an embryo likely be in phase G_0?

6. Describe two ways to derive stem cells without using human embryos.

7. Two roommates regularly eat 100-calorie packages of cookies. After a semester of each roommate eating one package a night, and following similar diet and exercise plans otherwise, one roommate has gained 12 pounds, but the other's weight has stayed the same. If genetics does not account for the difference in weight gain, what other factor discussed in the chapter might?

8. A "telomere health DNA test" sold online promises to "track your cellular age based on your telomere length" so that you can "improve it." The company sells supplement packages to help in the quest for biological youth. What further information might be helpful in deciding whether to purchase this type of genetic test and the suggested supplements?

Forensics Focus

1. A man who owned a company that provides human tissues was sentenced to serve many years in prison for trafficking in body parts taken, without consent, from dismembered corpses from funeral homes. Thousands of parts from hundreds of bodies were used in surgical procedures, including hip replacements and dental implants. The most commonly used product was a bone paste. Many family members of the tissue sources testified at the trials. Explain why cells from bone tissue can be matched to blood or cheek lining cells from blood relatives.

Case Studies and Research Results

1. A boy with "congenital skin crease Kunze type" has rings of creases on his arms and legs, unusual facial features, intellectual disability, slow growth, and a cleft palate. He has a mutation in the gene *TUBB,* and his condition is termed a "tubulinopathy." What part of his cells is affected?

2. Members of a large, three-generation family in Italy do not notice burns, broken bones, or even red hot chili pepper powder injected under the skin. What might their abnormality be on a cellular level?

3. Researchers isolated stem cells from fat removed from people undergoing liposuction, a procedure to remove fat. The stem cells can give rise to muscle, fat, bone, and cartilage cells.
    a. Are the stem cells totipotent or pluripotent?
    b. Are these stem cells ES cells, iPS cells, or adult stem cells?

CHAPTER 3

Meiosis, Development, and Aging

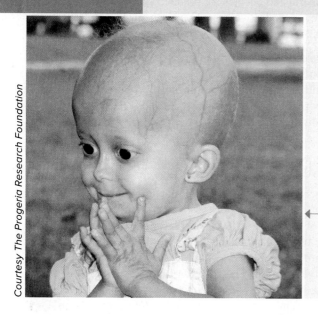

This child has progeria, a rare genetic syndrome in which some signs of aging are greatly accelerated. A worldwide effort to find all affected children has enabled testing of a drug that, so far, improves some symptoms, extends life, and may lead to treatments of common diseases in the general aging population.

Learning Outcomes

3.1 The Reproductive System
1. Describe the structures of the male and female reproductive systems.

3.2 Meiosis
2. Explain why meiosis is necessary to reproduce.
3. Summarize the events of meiosis.

3.3 Gametes Mature
4. List the steps in sperm and oocyte formation.

3.4 Prenatal Development
5. Describe early prenatal development.
6. Explain how the embryo differs from the fetus.

3.5 Birth Defects
7. Define *critical period*.
8. List some teratogens.

3.6 Maturation and Aging
9. Describe common diseases that begin in adulthood.
10. Explain how rapid aging syndromes occur.

The BIG Picture

Our reproductive systems enable us to start a new generation. First our genetic material must be halved, so it can combine with that of a partner to reconstitute a full diploid genome. Then genetic programs operate as the initial cell divides and its daughter cells specialize. The forming tissues fold into organs and the organs interact, slowly building a new human body. Mutations can cause disease at any stage of development or age.

Progress for Progeria

A child with Hutchinson-Gilford progeria syndrome (HGPS) appears old. Shortly after birth, weight gain begins to slow and hair to thin. The gums remain smooth, as teeth do not erupt. Joints stiffen and bones weaken. Skin wrinkles as the child's chubbiness melts away too quickly, and a cherubic toddler becomes increasingly birdlike in appearance.

Beneath the child's toughening skin, blood vessels stiffen with premature atherosclerosis, fat pockets shrink, and connective tissue hardens. Inside cells, chromosome tips whittle down at an accelerated pace, marking time too quickly. However, some organs remain healthy, and intellect is spared. The child usually dies during adolescence, from atherosclerosis that may cause a heart attack or stroke.

About 350 to 400 children worldwide are known to have this strange illness, first described in 1886. When researchers in 2003 identified the mutation in a gene called *LMNA* that causes HGPS, they realized, from the nature of the molecular defect, that several existing drugs might alleviate certain symptoms.

Researchers tested a failed cancer drug, tracking children's weight to assess whether the drug had an effect. These children usually stop growing at age 3, but 9 of 25 children in the initial clinical trial had a greater than 50 percent increase in the rate of weight gain. The artery stiffening and blockages of atherosclerosis not only stopped progressing, but reversed! For some children, hearing improved and bones became stronger. Sixty children have participated in clinical trials, and so far the drug has increased life expectancy (average survival) by at least 1.6 years. Because of the extended survival, progeria is now said to affect children and young adults. A program is enabling patients from all over the world to obtain the drug from their physicians.

The Progeria Research Foundation is trying to locate all people with progeria and involve them in testing several drugs, both old and new. One promising combination is the first drug, which blocks formation of the toxic, disease-causing protein, and a second drug that helps cells to expel the protein. The drugs discovered to help in progeria might also be useful in the many people who develop the symptoms as part of advancing age.

3.1 The Reproductive System

Genes orchestrate our physiology from a few days after conception through adulthood. Expression of specific sets of genes sculpts the differentiated cells that interact, aggregate, and fold, forming the organs of the body. Abnormal gene functioning can affect health at all stages of development. Certain single-gene mutations act before birth, causing broken bones, dwarfism, cancer, and other conditions. Some mutant genes exert their effects during childhood, appearing as early developmental delays or loss of skills. Inherited forms of heart disease and breast cancer can appear in early or middle adulthood, which is earlier than complex forms of these conditions. Pattern baldness is a common single-gene trait that may not become obvious until well into adulthood. This chapter explores the stages of the human life cycle that form the backdrop against which genes function.

The first cell that leads to development of a new individual is a fertilized ovum, called a **zygote**. It forms when a **sperm** cell from a male and an **oocyte** (also called an egg cell) from a female join. Sperm and oocytes are **gametes**, or sex cells. They mix genetic material from past generations. Because we have thousands of genes, some with many variants, each person (except for identical twins) has a unique combination of inherited traits.

Sperm and oocytes are produced in the reproductive system. The reproductive organs are organized similarly in the male and female. Each system has

- paired structures, called **gonads**, where the sperm and oocytes are manufactured;
- tubular structures that transport these cells; and
- hormones and secretions that control reproduction (discussed in chapter 6).

The Male

Sperm cells develop within a 125-meter-long network of seminiferous tubules, which one researcher describes as "a massive tangle of spaghetti." The seminiferous tubules are packed into paired, oval organs called **testes** (testicles) (**figure 3.1**). The testes are the male gonads. They lie outside the abdomen within a sac, the scrotum. This location keeps the testes cooler than the rest of the body, which is necessary for sperm to develop. Leading from each testis is a tightly coiled tube, the epididymis, in which sperm cells mature and are stored. Each epididymis continues into another tube, the ductus deferens (formerly called the vas deferens). Each ductus deferens bends behind the bladder and joins the urethra, which is the tube that carries sperm and urine out of the body through the penis.

Along the sperm's path, three glands add secretions. The ductus deferentia pass through the prostate gland, which produces a thin, milky, acidic fluid that activates the sperm to swim. Opening into the ductus deferens is a duct from the seminal vesicles, which secrete fructose (an energy-rich sugar) and hormonelike prostaglandins, which may stimulate contractions

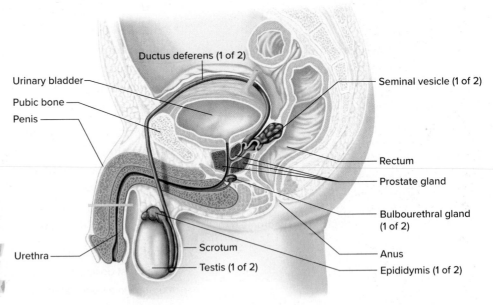

Figure 3.1 **The human male reproductive system.** Sperm cells are manufactured in the seminiferous tubules, which wind tightly within the testes and descend into the scrotum. The prostate gland, seminal vesicles, and bulbourethral glands add secretions to the sperm cells to form seminal fluid. Sperm mature and are stored in the epididymis and exit through the ductus deferens. The paired ductus deferentia join in the urethra, which transports seminal fluid from the body.

in the female that help sperm and oocyte meet. The bulbourethral glands, each about the size of a pea, join the urethra where it passes through the body wall. They secrete an alkaline mucus that coats the urethra before sperm are released. All of these secretions combine to form the seminal fluid that carries sperm. The fluid is alkaline.

During sexual arousal, the penis becomes erect, which enables it to penetrate and deposit sperm in the female reproductive tract. At the peak of sexual stimulation, a pleasurable sensation called orgasm occurs, accompanied by rhythmic muscular contractions that eject the sperm from each ductus deferens through the urethra and out the penis. The discharge of sperm from the penis, called ejaculation, delivers 200 million to 600 million sperm cells.

The Female

The female sex cells develop in paired organs in the abdomen called **ovaries** (**figure 3.2**), which are the female gonads. Within each ovary of a newborn girl are about a million immature oocytes. Each individual oocyte nestles within nourishing follicle cells, and each ovary houses oocytes in different stages of development. After puberty, about once a month, one ovary releases the most mature oocyte. Beating cilia sweep the mature oocyte into the fingerlike projections of one of two uterine (also called fallopian) tubes. The uterine tube carries the oocyte into a muscular, saclike organ called the uterus, or womb.

The oocyte released from the ovary may encounter a sperm, usually in a uterine tube. If the sperm enters the oocyte and the DNA of the two gametes merges into a new nucleus, the result is a fertilized ovum, the zygote. After a day, this first cell divides while moving through the uterine tube. It settles into the lining of the uterus, where it may continue to divide as it develops into an embryo. If fertilization does not occur, the oocyte, along with much of the uterine lining, is shed as the menstrual flow. Hormones coordinate the monthly menstrual cycle.

The lower end of the uterus narrows and leads to the cervix, which opens into the tubelike vagina. The vaginal opening is protected on the outside by two pairs of fleshy folds. At the upper juncture of both pairs is the 2-centimeter-long clitoris, which is anatomically similar to the penis. Rubbing the clitoris triggers female orgasm. Hormones control the cycle of oocyte maturation and the preparation of the uterus to nurture a fertilized ovum.

Key Concepts Questions 3.1

*a. In what body part do sperm develop?

*b. In what body part do oocytes develop?

3.2 Meiosis

Gametes form from special cells, called germline cells, in a type of cell division called meiosis that halves the chromosome number. A further process, maturation, sculpts the distinctive characteristics of sperm and oocyte.

The organelle-packed oocyte has 90,000 times the volume of the streamlined sperm.

Gametes contribute 23 different chromosomes, constituting one copy of the genome, to a fertilized ovum. In contrast, somatic cells contain 23 *pairs* of chromosomes, or 46 in total. One member of each pair comes from the person's mother and one comes from the father. The chromosome pairs are called **homologous pairs**, or *homologs* for short. Homologs have the same genes in the same order but may carry different alleles, or variants, of the same gene. Recall from chapter 2 that gametes are haploid ($1n$), which means that they have only one of each type of chromosome, and somatic cells are diploid ($2n$), with two copies of each chromosome type.

It is the meeting of sperm and oocyte that restores two full genomes. Without meiosis, the sperm and oocyte would each contain 46 chromosomes, and the fertilized ovum would have twice the normal number of chromosomes, or

Figure 3.2 The human female reproductive system. Oocytes mature in the paired ovaries. Once a month after puberty, an ovary releases one oocyte, which is drawn into a nearby uterine tube. If a sperm fertilizes the oocyte in the uterine tube, the fertilized ovum continues into the uterus, where for 9 months it divides and develops. If the oocyte is not fertilized, the body expels it, along with the built-up uterine lining, as the menstrual flow.

92 chromosomes. Such a genetically overloaded cell, with more than two sets of chromosomes, is called a **polyploid**. It usually undergoes only a few cell divisions before stopping.

In addition to producing gametes, meiosis mixes up trait combinations. For example, a person may produce one gamete containing alleles encoding blue eyes and freckles, yet another gamete with alleles encoding brown eyes and no freckles. Meiosis explains why siblings differ genetically from each other and from their parents.

In a much broader sense, meiosis, as the mechanism of sexual reproduction, provides genetic diversity. A population of sexually reproducing organisms is made up of individuals with different genotypes and phenotypes. Genetic diversity may enable a population to survive an environmental challenge. A population of asexually reproducing organisms, such as bacteria or genetically identical crops, consists of individuals with the same genome sequence. Should a new threat arise, such as an infectious disease that kills only organisms with a certain genotype, then the entire asexual population could perish. In a sexually reproducing population, individuals that inherited a certain combination of genes may survive. This differential survival of certain genotypes is the basis of evolution by natural selection, discussed in chapter 16.

Meiosis is actually two divisions of the genetic material. The first division is called a **reduction division** (or meiosis I) because it reduces the number of replicated chromosomes from 46 to 23. The second division, called an **equational division** (or meiosis II), produces four cells from the two cells formed in the first division by splitting the replicated chromosomes. **Figure 3.3** shows an overview of meiosis. The colors represent the contributions of the two parents, whereas size distinguishes different chromosomes. Only two types of chromosomes are depicted due to space constraints. In the body there would be 23 pairs of replicated chromosomes as meiosis begins.

As in mitosis, meiosis occurs after an interphase period when DNA is replicated (doubled) (**table 3.1**). After interphase, prophase I (so called because it is the prophase of meiosis I) begins as the replicated chromosomes condense and become

Table 3.1	Comparison of Mitosis and Meiosis
Mitosis	**Meiosis**
One division	Two divisions
Two daughter cells per cycle	Four daughter cells per cycle
Daughter cells genetically identical	Daughter cells genetically different
Chromosome number of daughter cells same as that of parent cell ($2n$)	Chromosome number of daughter cells half that of parent cell ($1n$)
Occurs in somatic cells	Occurs in germline cells
Occurs throughout life cycle	In humans, completes after sexual maturity
Used for growth, repair, and asexual reproduction	Used for sexual reproduction, producing new gene combinations

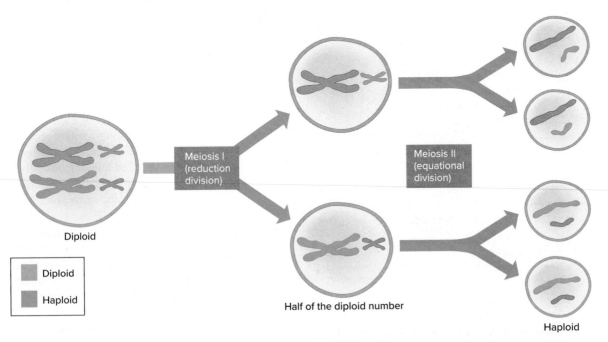

Figure 3.3 Overview of meiosis. Meiosis gives rise to haploid gametes. This simplified illustration follows the fate of two chromosome pairs rather than the true 23 pairs. In actuality, the first meiotic division reduces the number of chromosomes to 23, all in the replicated form. In the second meiotic division, the cells essentially undergo mitosis. The result of the two meiotic divisions (in this illustration and in reality) is four haploid cells. In this figure, homologous pairs of chromosomes are indicated by size, and parental origin of chromosomes by color.

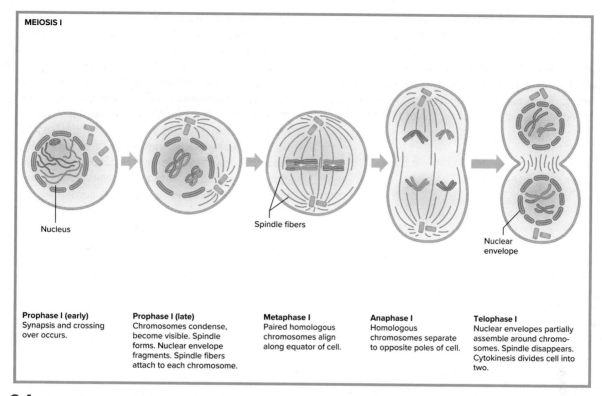

Figure 3.4 **Meiosis I.** The reduction division of meiosis I halves the number of replicated chromosomes from 46 to 23 in a human cell. (The figure depicts a representative two homologous chromosome pairs, with crossing over.)

visible when stained (**figure 3.4**). A spindle forms. Toward the middle of prophase I, the homologs line up next to one another, gene by gene, in an event called synapsis. A mixture of RNA and protein holds the chromosome pairs together.

During synapsis, the homologs exchange parts, or **cross over** (**figure 3.5**). All four chromatids that comprise each homologous chromosome pair align as exchanges occur. After crossing over, each homolog bears some genes from each parent. Prior to this, all of the genes on a homolog were derived from one parent. New gene combinations arise from crossing over when the parents carry different alleles. Toward the end of prophase I, the synapsed chromosomes separate but remain attached at a few points along their lengths.

To understand how crossing over mixes trait combinations, consider a simplified example. Suppose that homologs carry genes for hair color, eye color, and finger length. One of the chromosomes carries alleles for blond hair, blue eyes, and short fingers. Its homolog carries alleles for black hair, brown eyes, and long fingers. After crossing over, one of the chromosomes might bear alleles for blond hair, brown eyes, and long fingers, and the other might bear alleles for black hair, blue eyes, and short fingers.

Meiosis continues in metaphase I, when the homologs align down the center of the cell. Each member of a homologous pair attaches to a spindle fiber at an opposite pole. The pattern in which the chromosomes align during metaphase I generates genetic diversity. For each homologous pair, the pole the maternally or paternally derived member goes to is random. It is a little like the number of different ways that 23 boys and 23 girls can line up in opposite-sex pairs.

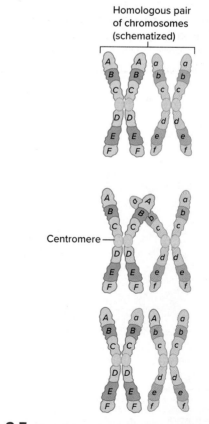

Figure 3.5 **Crossing over recombines genes.** Crossing over generates genetic diversity by recombining genes, mixing parental traits. The capital and lowercase forms of the same letter represent different variants (alleles) of the same gene. A chromosome has hundreds to thousands of genes.

Chapter 3 *Meiosis, Development, and Aging*

The greater the number of chromosomes, the greater the genetic diversity generated in metaphase I. For two pairs of homologs, four (2^2) different metaphase alignments are possible. For three pairs of homologs, eight (2^3) different alignments can occur. Our 23 chromosome pairs can line up in 8,388,608 (2^{23}) different ways. This random alignment of chromosomes causes **independent assortment** of the genes that they carry.

Figure 3.6 Independent assortment. The pattern in which homologs randomly align during metaphase I determines the combination of maternally and paternally derived chromosomes in the daughter cells. Two pairs of chromosomes can align in two ways to produce four possibilities in the daughter cells. The potential variability that meiosis generates for all 23 chromosome pairs, especially considering crossing over, is great.

Independent assortment means that the fate of a gene on one chromosome is not influenced by a gene on a different chromosome (**figure 3.6**). Independent assortment accounts for a basic law of inheritance discussed in chapter 4.

Homologs separate in anaphase I and move to opposite poles by telophase I. These movements establish one half of the original diploid number of chromosomes with a set of still-replicated chromosomes at each end of the stretched-out cell. Unlike in mitosis, the centromeres of each replicated chromosome in meiosis I remain together. During a second interphase, chromosomes unfold into thin threads. Proteins are manufactured, but DNA is not replicated a second time. The single DNA replication, followed by the double division of meiosis, halves the chromosome number. The cell is now haploid.

Prophase II marks the start of the second meiotic division (**figure 3.7**). The chromosomes are again condensed and visible. In metaphase II, the replicated chromosomes align down the center of the cell. In anaphase II, the centromeres part, and the newly formed chromosomes, each now in the unreplicated form, move to opposite poles. In telophase II, nuclear envelopes form around the four nuclei, which then separate into individual cells. The net result of meiosis is four haploid cells, each carrying a new assortment of genes and chromosomes that hold a single copy of the genome.

Meiosis generates astounding genetic variety. Any of a person's more than 8 million possible combinations of chromosomes can meet with any of the more than 8 million combinations of a partner, raising potential variability to more than 70 trillion ($8,388,608^2$) genetically unique individuals! Crossing over contributes almost limitless genetic diversity.

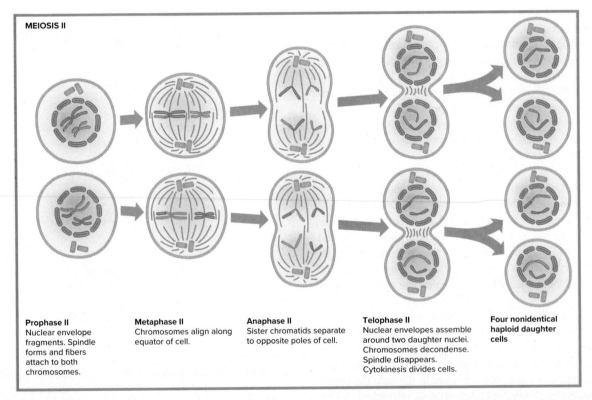

Figure 3.7 Meiosis II. The equational division of meiosis II pulls apart the replicated chromosomes, producing 46 chromosomes in the unreplicated form, with 23 chromosomes in each resulting cell. (This figure depicts two representative chromosome types.)

> **Key Concepts Questions 3.2**
>
> a. Distinguish between haploid and diploid.
> b. Explain how meiosis maintains the chromosome number over generations and mixes gene combinations.
> c. Discuss the two mechanisms that generate genotypic diversity.
> *d. Define *independent assortment*.

3.3 Gametes Mature

Meiosis happens in both sexes, but different distributions of cell components create the distinctions between sperm and oocytes. The gametes of the maturing male and female proceed through similar stages as they form, but with vastly different timetables. A male begins manufacturing sperm at puberty and continues throughout life, whereas a female begins meiosis when she is a fetus. Meiosis in the female completes only if a sperm fertilizes an oocyte.

Sperm Form

Spermatogenesis, the formation of sperm cells, begins in a diploid stem cell called a **spermatogonium (figure 3.8)**. This cell divides mitotically, yielding two daughter cells. One cell continues to specialize into a mature sperm. The other daughter cell self-renews as a stem cell, maintaining the ability to produce more sperm.

Bridges of cytoplasm attach several spermatogonia, and their daughter cells enter meiosis together. As these spermatogonia mature, they accumulate cytoplasm and replicate their DNA, becoming primary spermatocytes.

During reduction division (meiosis I), each primary spermatocyte divides, forming two equal-sized haploid cells called secondary spermatocytes, which are in the replicated form (see figure 2.12). In meiosis II, each secondary spermatocyte divides to yield two equal-sized spermatids in the replicated form. Each spermatid then develops the characteristic sperm tail, or flagellum. After spermatid differentiation, some of the cytoplasm connecting the cells falls away, leaving mature, tadpole-shaped **spermatozoa** (singular, *spermatozoon*), or sperm. **Figure 3.9** presents an anatomical view showing the stages of spermatogenesis within the seminiferous tubules.

A sperm, which is a mere 0.006 centimeter (0.0023 inch) long, must travel about 18 centimeters (7 inches) to reach an oocyte. Each sperm cell consists of a tail, body or midpiece, and a head region (**figure 3.10**).

The midpiece has many mitochondria, which will split ATP molecules, releasing energy that enables the sperm cell to swim inside the female reproductive tract. A membrane-covered area on the front end of a sperm, the acrosome, contains enzymes that help the sperm cell penetrate the protective layers around the oocyte. Within the large sperm head, DNA

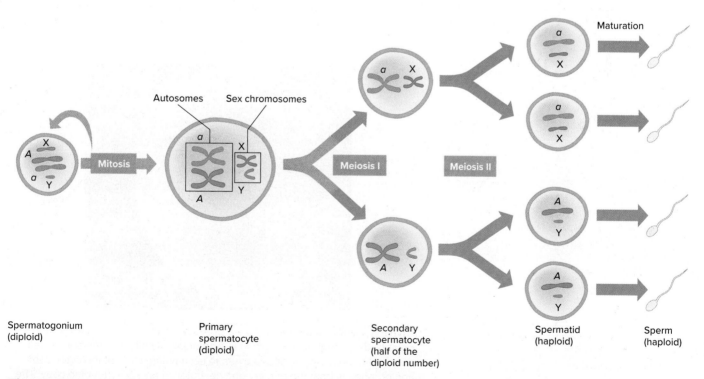

Figure 3.8 Sperm formation (spermatogenesis). Primary spermatocytes have the normal diploid number of 23 chromosome pairs. The large pair of chromosomes represents autosomes (non-sex chromosomes). The X and Y chromosomes are sex chromosomes. The spermatogonium self-renews.

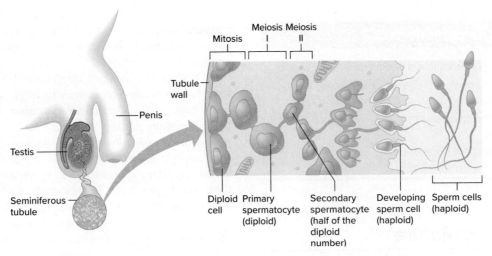

Figure 3.9 Meiosis and maturation produce sperm cells. Diploid cells divide through mitosis in the linings of the seminiferous tubules. Some daughter cells then undergo meiosis, producing haploid secondary spermatocytes, which differentiate into spermatids and then into mature sperm cells.

wraps around proteins. The sperm's DNA at this time is genetically inactive. A male manufactures trillions of sperm in his lifetime. Although many of these will come close to an oocyte, few will actually touch one.

Meiosis in the male has built-in protections that help prevent sperm from causing some birth defects. Spermatogonia that are exposed to toxins tend to be so damaged that they never mature into sperm. More mature sperm cells exposed to toxins are often so damaged that they cannot swim.

Oocytes Form

Meiosis in the female, called **oogenesis** (egg making), begins with a diploid cell, an **oogonium**, that self-renews and also yields a lineage whose cells differentiate. Unlike male cells, oogonia are not attached to each other. Instead, follicle cells surround each oogonium. As each oogonium grows, cytoplasm accumulates, DNA replicates, and the cell becomes a primary oocyte. The ensuing meiotic division in oogenesis, unlike the male pathway, produces cells of different sizes.

In meiosis I, the primary oocyte divides into two cells: a small cell with very little cytoplasm, called a first **polar body**, and a much larger cell called a secondary oocyte (**figure 3.11**). Each cell is haploid, with one chromosome

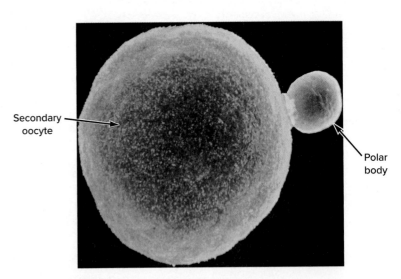

Figure 3.10 Sperm. A sperm has distinct regions that assist in delivering DNA to an oocyte.

Figure 3.11 Meiosis in a female produces a secondary oocyte and a polar body. Unequal division and apportioning of cell parts enable the cell destined to become a fertilized ovum to accumulate most of the cytoplasm and organelles from the primary oocyte, but with only one genome copy. The oocyte receives most of the cytoplasm that would have gone into the meiotic product that became the polar body if the division had been equal. *Prof. P.M. Motta/Univ. "La Sapienza", Rome/Science Source*

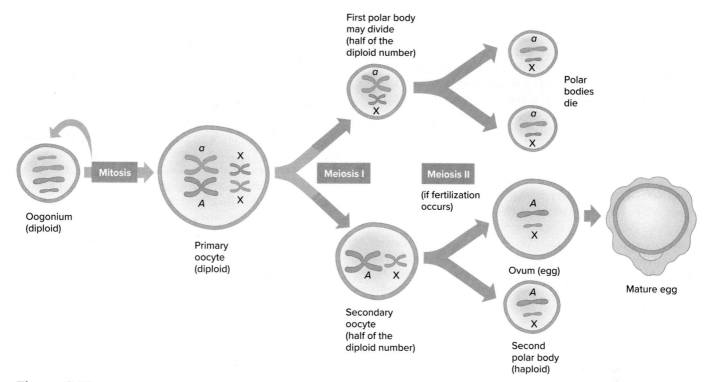

Figure 3.12 Ovum formation (oogenesis). Primary oocytes have the diploid number of 23 chromosome pairs. Meiosis in females concentrates most of the cytoplasm into one large cell called an oocyte (or egg). The other products of meiosis, called polar bodies, contain the other three sets of chromosomes and normally degenerate.

set and the chromosomes in replicated form. In meiosis II, the tiny first polar body may divide to yield two polar bodies of equal size, with unreplicated chromosomes, or the first polar body may decompose. The secondary oocyte, however, divides unequally in meiosis II to produce another small polar body, with unreplicated chromosomes, and the mature egg cell, or ovum, which contains a large volume of cytoplasm. **Figure 3.12** summarizes meiosis in the female, and **figure 3.13** provides an anatomical view of the process.

Most of the cytoplasm among the four meiotic products in the female ends up in only one cell, the ovum. The woman's

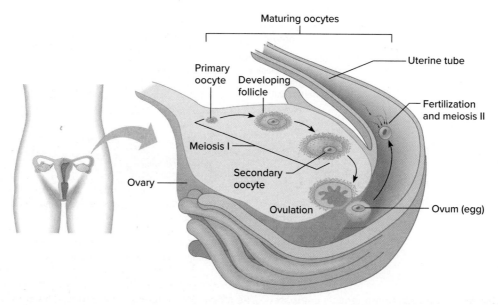

Figure 3.13 The making of oocytes. Oocytes develop within the ovary in follicles. An ovary contains many oocytes in various stages of maturation. After puberty, the most mature oocyte in one ovary bursts out each month, in an event called ovulation.

Chapter 3 *Meiosis, Development, and Aging*

body absorbs the polar bodies, which normally play no further role in development. Rarely, a sperm fertilizes a polar body. When this happens, the woman's hormones respond as if she is pregnant, but a disorganized clump of cells that is not an embryo grows for a few weeks, and then leaves the woman's body. This event is a type of miscarriage called a "blighted ovum."

Before birth, a female has a million or so oocytes arrested in prophase I. (This means that when your grandmother was pregnant with your mother, the oocyte that would be fertilized and eventually become you was already there.) By puberty, about 400,000 oocytes remain. After puberty, meiosis I continues in one or several oocytes each month, but halts again at metaphase II. In response to specific hormonal cues each month, one ovary releases a secondary oocyte; this event is ovulation. The oocyte drops into a uterine tube, where waving cilia move it toward the uterus. Along the way, if a sperm penetrates the oocyte membrane, then female meiosis completes, and a fertilized ovum forms. If the secondary oocyte is not fertilized, it degenerates and leaves the body in the menstrual flow, meiosis never completed.

A female ovulates about 400 oocytes between puberty and menopause. Most oocytes degrade, because fertilization is so rare. Furthermore, only one in three of the oocytes that do meet and merge with a sperm cell will continue to grow, divide, and specialize to eventually form a new individual.

Figure 3.14 compares meiosis and maturation of gametes in the male and female.

Meiosis and Mutations

The gametes of older people are more likely to have new mutations (that is, not *inherited* mutations) than the gametes of younger people. Older women are at higher risk of producing oocytes that have an extra or missing chromosome. If fertilized, such oocytes lead to offspring with the chromosomal conditions described in chapter 13. Older men are also more likely to produce gametes that have genetic errors, but sperm tend to have single-gene mutations rather than chromosome-level changes. The "paternal age effect" usually causes dominant single-gene diseases. That is, only one copy of the mutant gene is necessary to cause the condition.

The different timetables of meiosis explain why oocytes originate chromosome imbalances whereas sperm introduce single-gene conditions. In females, oocytes exist on the brink of meiosis I for years, and the cells complete meiosis II only if they are fertilized. Mistakes occur when gametes are active, and that is when they are distributing their chromosomes. If a homologous pair doesn't separate, the result could be an oocyte with an extra or missing chromosome.

Sperm develop in only 74 days. "Paternal age effect" conditions, in which fathers tend to be older, arise from spermatogonia stem cells in the testis that divide every 16 days, from puberty on, offering many opportunities for DNA replication (discussed in chapter 9) to make a mistake, generating a dominant mutation. **Table 3.2** describes some paternal age effect conditions. They occur in genes of the fibroblast growth

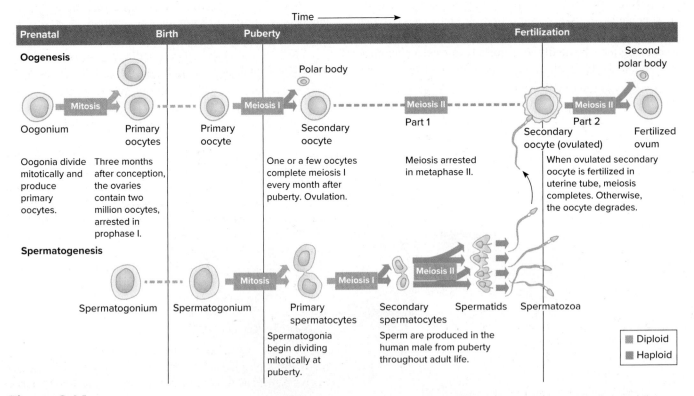

Figure 3.14 Oogenesis and spermatogenesis compared. The steps of sperm and oocyte formation are similar, and the products of each are haploid. But the male and female gametes look very different, and they form according to very different timetables.

Table 3.2	Paternal Age Effect Conditions
Disease	**Phenotype**
Achondroplasia	Short-limbed dwarfism (see figure 5.1a)
Crouzan syndrome	Premature fusion of skull bones in infancy, causing wide-spaced and bulging eyes, beaked nose, short upper lip, small upper jaw, and jutting lower jaw
Hutchinson-Gilford progeria syndrome	Thin hair, weak bones, tough and wrinkled skin, stiff joints and blood vessel linings
Multiple endocrine neoplasia 2	Cancers of thyroid, parathyroid, and adrenal glands
Pfeiffer syndrome	Premature fusion of skull bones in infancy, short and fused fingers and toes
Thanatophoric dysplasia	Severe short-limbed dwarfism

factor receptor (FGFR) family and affect skeletal growth. *FGFR* genes that mutate in the testicles as a man ages skew meiosis so that more spermatogonia give rise to cells with the mutation than to cells without it. The result is a mosaic cell population, or what has been called a "testicular time bomb."

The maternal age effect in causing chromosomal imbalances has been recognized since the nineteenth century, when physicians noticed that babies with trisomy 21 Down syndrome tended to be the youngest children in large families. The paternal age effect has only recently been recognized. Researchers examined testes from elderly men who had died and donated their bodies to science. Each testis was divided into 6 slices of 32 pieces, and probed for mutations that cause multiple endocrine neoplasia (an inherited cancer syndrome). Mutations were found in discrete sections of the testes, indicating that stem cells perpetuated the mutation as they divided. The older the man at death, the more testis cells had the mutations.

Key Concepts Questions 3.3

*a. Name the characteristic of spermatogonia and oogonia that perpetuates their ability to produce gametes.

b. List the steps of sperm and oocyte formation and specialization.

*c. Explain how the timetables of spermatogenesis and oogenesis differ.

d. Explain how paternal and maternal age effect conditions differ.

3.4 Prenatal Development

A prenatal human is considered an **embryo** for the first 8 weeks, when rudiments of all body parts form. The embryonic period begins when the fertilized ovum divides by mitosis. During the first week, the embryo is in a "preimplantation" stage because it has not yet settled into the uterine lining. Prenatal development after the eighth week is the period when structures grow and specialize. From the start of the ninth week until birth, the prenatal human organism is a **fetus**.

Sperm and Oocyte Meet at Fertilization

Hundreds of millions of sperm cells are deposited in the vagina during sexual intercourse. A sperm cell can survive in a woman's body for up to 3 days, but the oocyte can only be fertilized in the 12 to 24 hours after ovulation.

The woman's body helps sperm reach an oocyte. A process in the female called capacitation chemically activates sperm, and the oocyte secretes a chemical that attracts sperm. Contractions of the female's muscles and moving of sperm tails propel the sperm. Still, only 200 or so sperm get near the oocyte.

A sperm first contacts a covering of follicle cells, called the corona radiata, that guards a secondary oocyte. The sperm's acrosome then bursts, releasing enzymes that bore through a protective layer of glycoprotein (the zona pellucida) beneath the corona radiata. Fertilization, or conception, begins when the outer membranes of the sperm and secondary oocyte meet and a protein on the sperm head contacts a different type of protein on the oocyte (**figure 3.15**). A wave of electricity spreads physical and chemical changes across the oocyte surface, which keep other sperm out. More than one sperm can enter an oocyte, but the resulting cell has too much genetic material for development to follow. Usually only the sperm's head enters the oocyte.

Within 12 hours of the sperm's penetration, the ovum's nuclear membrane disassembles, and the two sets of chromosomes, called pronuclei, approach one another. Within each pronucleus, DNA replicates. Fertilization completes when the two genetic packages meet and merge, forming a complete set of genetic instructions. The **Bioethics** reading discusses why cloning, which is a way to start development without a fertilized egg, does not produce an identical copy of an individual.

The Embryo Cleaves and Implants

A day after fertilization, the zygote divides by mitosis, beginning a period of frequent cell division called **cleavage** (**figure 3.16**). The resulting early cells are called **blastomeres**. When the blastomeres form a solid ball of sixteen or more cells, the embryo is called a **morula** (Latin for "mulberry," which it resembles).

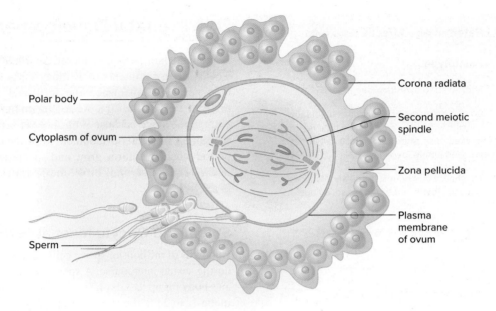

Figure 3.15 Fertilization. Entry of a sperm cell induces the oocyte (arrested in metaphase II) to complete meiosis. Before fertilization occurs, the sperm's acrosome bursts, spilling enzymes that help the sperm's nucleus enter the oocyte.

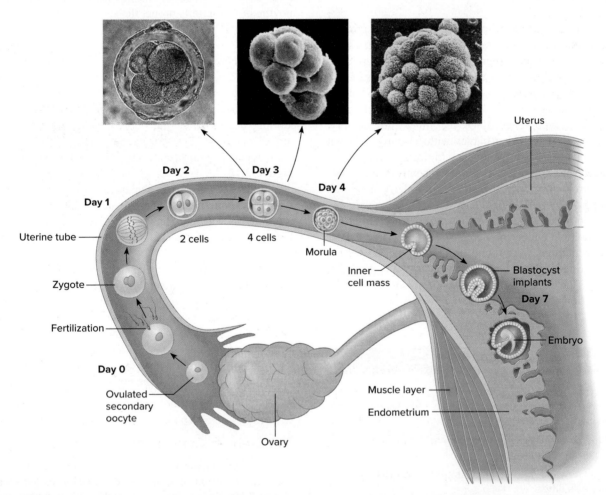

Figure 3.16 Cleavage: From ovulation to implantation. The zygote forms in the uterine tube when a sperm nucleus fuses with the nucleus of an oocyte. The first divisions proceed while the embryo moves toward the uterus. By day 7, the embryo, now called a blastocyst, begins to implant in the uterine lining. *(left): Petit Format/Nestle/Photo Researchers/Science Source; (middle): PROFESSORS P.M. MOTTA and J. VAN BLERKOM/Science Source; (right): PETIT FORMAT/Science Source/Science Source*

Bioethics

Why a Clone Is Not an Exact Duplicate

Cloning creates a genetic replica of an individual. In contrast, normal reproduction and development combine genetic material from two individuals. In fiction, scientists have cloned Nazis, politicians, dinosaurs, children, and organ donors. Real scientists have cloned sheep, mice, cats, pigs, dogs, monkeys, and amphibians.

Identical multiples, such as twins and triplets, are natural clones. Armadillos naturally always give birth to identical quadruplets (**figure 3A**). *Somatic cell nuclear transfer*, discussed in section 2.4, transfers a nucleus from a somatic cell into an oocyte whose nucleus has been removed, creating a clone of the nucleus donor.

Cloning cannot produce an exact replica of a person, for several reasons:

- **Premature cellular aging.** In some species, telomeres of chromosomes in the donor nucleus are shorter than those in the recipient cell.
- **Altered gene expression.** In normal development, for some genes, one copy is turned off, depending on which parent transmits it. This phenomenon is called genomic imprinting (see section 6.5). In cloning, genes in a donor nucleus skip passing through a sperm or oocyte, and thus are not imprinted. Lack of imprinting causes cloned animals to be unusually large.
- **More mutations.** DNA from a donor cell has had years to accumulate mutations. A mutation might not be noticeable in one of millions of somatic cells in a body but it could be devastating if that nucleus is used to program development of a new individual.
- **X inactivation.** At a certain time in early prenatal development in female mammals, one X chromosome is inactivated. Whether that X chromosome is from the mother or the father occurs at random in each cell, creating an overall mosaic pattern of expression for genes on the X chromosome. The pattern of X inactivation of a female clone would probably not match that of her nucleus donor, because X inactivation normally occurs in the embryo, not in the first cell (see section 6.4).
- **Mitochondrial DNA.** A clone's mitochondria descend from the recipient oocyte, not from the donor cell, because mitochondria are in the cytoplasm, not the nucleus (see figure 5.10).

The environment is another powerful factor in why a clone isn't an identical copy. For example, coat color patterns differ in cloned calves and cats. When the animals were embryos, cells destined to produce pigment moved in a unique way in each individual, producing different color patterns. In humans, experience, nutrition, stress, exposure to infectious disease, and many other factors join our genes in molding who we are.

Questions for Discussion

1. Which of your characteristics do you think could not be duplicated in a clone, and why?
2. What might be a reason to clone humans?
3. What are potential dangers of cloning humans? Of cloning pets?
4. Do human clones exist naturally?

Figure 3A Armadillos are clones. These armored mammals always give birth to four genetically identical offspring—that is, quadruplet clones. *Bianca Lavies/National Geographic Creative*

During cleavage, organelles and molecules from the secondary oocyte's cytoplasm still control cellular activities, but some of the embryo's genes begin to function. The ball of cells hollows out, and its center fills with fluid, creating a **blastocyst**. (The term *cyst* refers to the fluid-filled center.) Some of the cells form a clump on the inside lining called the **inner cell mass**. Its formation is the first event that distinguishes cells from each other by their relative positions. The inner cell mass continues developing, forming the embryo.

A GLIMPSE OF HISTORY

Home pregnancy tests for the hormone hCG cap a long history of ways to tell if a woman was expecting. Etchings on an Egyptian papyrus from 1350 B.C.E. show a woman whose menstrual period was late urinating on wheat and barley seeds. If wheat sprouted it would be a girl, if barley grew it was a boy, and if nothing happened, then the woman wasn't with child. The effect on seeds might have been due to estrogens in the urine. In the ensuing centuries, "piss prophets" attempted to divine the pregnant state from the shades and textures of urine.

In the early twentieth century, researchers seeking ways to confirm pregnancy measured levels of the hormones that fluctuate during the menstrual cycle and the body parts that they affect. They injected the urine of possibly pregnant women into rabbits, rats, mice, or frogs, leading to the once-famous announcement of impending parenthood, *"The rabbit died!"* Actually, all the rabbits injected with human urine died, as doctors probed the ovaries for the swelling that would indicate presence of the telltale human pregnancy hormone.

In the 1970s, increasing interest in women's health sped development of a pregnancy test for the active part of the hCG molecule using chemicals in a test tube, rather than sacrificing rabbits. At first doctors sent urine to labs for analysis. Then in 1978 came the first at-home "early pregnancy test," sold at pharmacies and revolutionary at the time. Today's home pregnancy tests provide quick, accurate results that can take much of the mystery and anxiety out of confirming what a woman may have already strongly suspected.

A week after fertilization, the blastocyst nestles into the uterine lining. This event, called implantation, takes about a week. As implantation begins, the outermost cells of the blastocyst, called the trophoblast, secrete human chorionic gonadotropin (hCG), a hormone that prevents menstruation. This hormone detected in a woman's urine or blood is one sign of pregnancy.

The Embryo Forms

During the second week of prenatal development, a space called the amniotic cavity forms between the inner cell mass and the outer cells anchored to the uterine lining. Then the inner cell mass flattens into a two-layered embryonic disc. The layer nearest the amniotic cavity is the **ectoderm**; the inner layer, closer to the blastocyst cavity, is the **endoderm**. Shortly after, a third layer, the **mesoderm**, forms in the middle. This three-layered, curved, sandwich-like structure is called the primordial embryo, or the **gastrula** (**figure 3.17**).

The cells that form the layers of the primordial embryo, called **primary germ layers**, become "determined," or fated, to develop as specific cell types. The cells specialize by the progressive switching off of the expression of genes important and active in the early embryo as other genes begin to be expressed. Gene expression shuts off when a small molecule called a methyl (CH_3; a carbon bonded to three hydrogen atoms) binds certain genes and proteins. In the early embryo, methyls bind the proteins (histones) that hold the long DNA molecules in a coil (see figure 9.13 and figure 11.5). In the later embryo, methyls bind to and silence specific sets of genes. The pattern of methyl group binding that guides the specialization of the embryo establishes the epigenome. The term **epigenetic** refers to changes to DNA expression, not to DNA nucleotide base sequence.

Each layer of the embryo retains stem cells as the organism develops. Under certain conditions, stem cells may produce daughter cells that can specialize as many cell types.

As development progresses, each primary germ layer gives rise to certain structures. Cells in the ectoderm become skin, nervous tissue, or parts of certain glands. Endoderm cells form parts of the liver and pancreas and the linings of many organs. The middle layer of the embryo, the mesoderm, forms many structures, including muscle, connective tissues, the reproductive organs, and the kidneys.

Genes called homeotics control how the embryo develops parts in the right places. Mutations in these genes cause some very interesting conditions, including forms of intellectual disability, autism, blood cancer, and blindness. **Clinical Connection 3.1** describes a human disease that results from mutations in homeotic genes. **Table 3.3** summarizes the stages of early human prenatal development.

Supportive Structures Form

As an embryo develops, structures form that support and protect it. These include chorionic villi, the placenta, the yolk sac, the allantois, the umbilical cord, and the amniotic sac.

By the third week after fertilization, fingerlike outgrowths called chorionic villi extend from the area of the embryonic disc close to the uterine wall and project into pools of the woman's blood. Her blood system and the embryo's are separate, but nutrients and oxygen diffuse across the chorionic villi from her circulation to the embryo. Wastes leave the embryo's circulation and enter the woman's circulation, to be excreted.

By 10 weeks, the placenta is fully formed from the chorionic villi. The placenta links the woman and her fetus for the rest of the pregnancy. It continues to secrete hormones that maintain pregnancy and sends nutrients to the fetus.

Other structures nurture the developing embryo. The yolk sac manufactures blood cells, as does the allantois, a membrane surrounding the embryo that gives rise to the umbilical blood vessels. The umbilical cord forms around these vessels and attaches to the center of the placenta. Toward the end of the embryonic period, the yolk sac shrinks, as the amniotic sac swells with fluid that cushions the embryo and maintains a constant temperature and pressure. The amniotic fluid contains fetal urine and a few fetal cells.

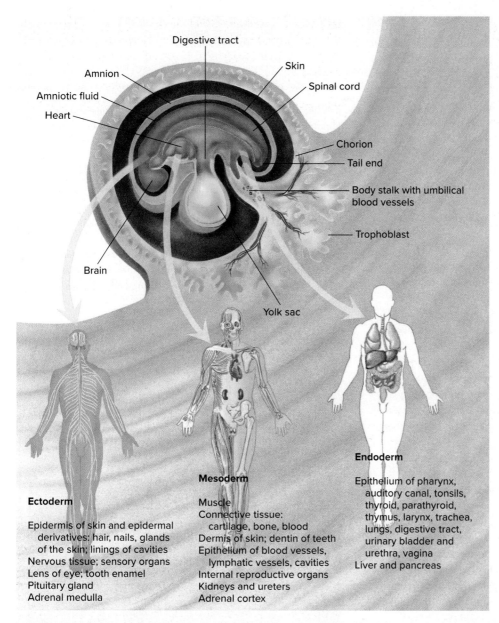

Figure 3.17 The primordial embryo. When the three primary germ layers of the embryo form at gastrulation, many cells become "fated" to follow a specific developmental pathway.

Table 3.3	Stages and Events of Early Human Prenatal Development	
Stage	**Time Period**	**Principal Events**
Fertilized ovum	12 to 24 hours following ovulation	Oocyte fertilized; zygote has 23 pairs of chromosomes and is genetically distinct
Cleavage	30 hours to third day	Mitosis increases cell number
Morula	Third to fourth day	Solid ball of cells
Blastocyst	Fifth day through second week	Hollowed fluid-filled ball forms trophoblast (outside) and inner cell mass, which implants and flattens to form embryonic disc
Gastrula	End of second week	Primary germ layers form

Two of the supportive structures that develop during pregnancy provide material for prenatal tests, discussed in chapters 13 and 22. **Chorionic villus sampling** examines chromosomes from cells snipped off the chorionic villi at 10 weeks. Because the villi cells and the embryo's cells come from the same fertilized ovum, an abnormal chromosome in villi cells should also be in cells of the embryo. In **amniocentesis**, a sample of amniotic fluid is taken and fetal cells in it are examined for biochemical, genetic, and chromosomal anomalies. However, chorionic villus sampling and amniocentesis may be replaced with collection and analysis of cell-free fetal DNA that is in the pregnant woman's bloodstream. This DNA is outside blood cells, in the liquid part of the blood. Maternal DNA pieces are in the bloodstream too, but by 8 weeks after conception, about 10 percent of the cell-free DNA in the maternal circulation comes from the fetus or the placenta.

The umbilical cord is another prenatal structure that has medical applications. Its stem cells are used to treat a respiratory disease of newborns. The cord stem cells give rise to lung cells that secrete surfactant during the seventh month of fetal development, which is the chemical that inflates the microscopic air sacs, enabling the newborn to breathe. Stem cells from umbilical cords are abundant, easy to obtain and manipulate, and can give rise to almost any cell type.

Multiples

Twins and other multiples arise early in development. Fraternal, or **dizygotic (DZ), twins** result when two sperm fertilize two oocytes. This can happen if ovulation occurs in two ovaries in the same month, or if two oocytes leave the same ovary and are both fertilized. DZ twins are no more alike than any two siblings, although they share a very early environment in the uterus. The tendency to ovulate two oocytes in a month, leading to conception of DZ twins, can run in families.

Clinical Connection 3.1

When an Arm Is Really a Leg: Homeotic Mutations

Flipping the X-ray showed Stefan Mundlos, MD, that his hunch was correct—the patient's arms were odd-looking and stiff because the elbows were actually knees! The condition, Liebenberg syndrome, had been described in 1973 among members of a five-generation white South African family (**figure 3B**). Four males and six females had stiff elbows and wrists and short fingers that looked misplaced. A trait that affects both sexes in every generation is autosomal dominant inheritance—each child of a person with the unusual limbs has a 50:50 chance of having the condition, too.

In 2000, a medical journal described a second family with Liebenberg syndrome. Several members had restricted movements because they couldn't bend their misshapen elbows. Then in 2010, a report appeared on identical twin girls with stiff elbows and long arms, with fingers that looked like toes.

In 2012, Dr. Mundlos noted that the muscles and tendons of the elbows, as well as the bones of the arms, weren't quite right in his patient. The doctor, an expert in the comparative anatomy of limb bones of different animals, realized that the stiff elbows were acting like knees. The human elbow joint hinges and rotates, but the knee extends the lower leg straight out. Then an X-ray scan of the patient's arm fell to the floor. He suddenly realized that the entire limb looked like a leg.

Genes that switch body parts are termed *homeotic*. They are well studied in organisms as evolutionarily diverse as fruit flies, flowering plants, and mice, affecting the positions of larval segments, petals, legs, and more. Assignment of body parts begins in the early embryo, when cells look alike but are already fated to become specific structures. Gradients (increasing or decreasing concentrations) of "morphogen" proteins in an embryo program a region to develop a certain way. Mix up the messages, and an antenna becomes a leg, or an elbow a knee.

Homeotic genes include a 180-base-long DNA sequence, called the homeobox, which enables the encoded protein to bind other proteins that turn on sets of other genes, crafting an embryo, section by section. Homeotic genes line up on their chromosomes in the order in which they're deployed in development, like chapters in an instruction manual to build a body.

The human genome has four clusters of homeotic genes, and mutations in them cause disease. In certain lymphomas, a homeotic mutation sends white blood cells along the wrong developmental pathway, resulting in too many of some blood cell types and too few of others. The abnormal ears, nose, mouth, and throat of DiGeorge (22q11.2 deletion) syndrome echo the abnormalities in *Antennapedia*, a fruit fly mutant that has legs on its head. Extra and fused fingers and various bony alterations also stem from homeotic mutations.

The search for the mutation behind the arm-to-leg Liebenberg phenotype began with abnormal chromosomes. Affected members of the three known families were each missing 134 DNA bases in the same part of the fifth largest chromosome. The researchers identified a gene called *PITX1* that controls other genes that oversee limb development. In the Liebenberg families, the missing DNA places an "enhancer" gene near *PITX1*, altering its expression in a way that mixes up developmental signals so that the forming arm instead becomes a leg.

Questions for Discussion

1. What is the genotype and phenotype of Liebenberg syndrome?
2. Explain the molecular basis of a homeotic mutation and the resulting phenotype.
3. Describe a human disease other than Liebenberg syndrome that results from a homeotic mutation.

Figure 3B The hands of a person with Liebenberg syndrome resemble feet; arms resemble legs. *Dr. Malte Spielmann/Institut für Medizinische Genetik*

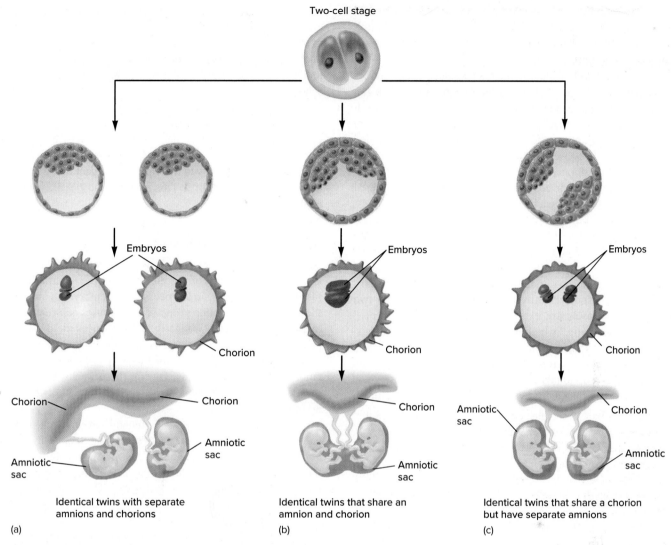

Figure 3.18 Types of identical twins. Identical (MZ) twins originate at three points in development. **(a)** In about one-third of identical twins, separation of cells into two groups occurs before the trophoblast forms on day 5. These twins have separate chorions and amnions. **(b)** About 1 percent of identical twins share a single amnion and chorion, because the tissue splits into two groups after these structures have already formed. **(c)** In about two-thirds of identical twins, the split occurs after day 5 but before day 9. These twins share a chorion but have separate amnions. Fraternal (DZ) twins result from two sperm fertilizing two secondary oocytes. These twins develop their own amniotic sacs, yolk sacs, allantois, placentas, and umbilical cords.

Identical, or **monozygotic (MZ), twins** descend from a single fertilized ovum and therefore are genetically identical. They are natural clones. Environmental influences can create telling differences between MZ twins, however.

A study compared 52-year-old identical twins who differ greatly in their physical fitness level. "TT" is a long-time triathlete, "UT" a sedentary truck driver. Because they share all their genes, differences in physical fitness reflect what the twins do and have done, not what they inherited.

The brothers underwent a three-day battery of tests to assess fitness. The athletic twin had lower resting heart rate, blood pressure, body fat, cholesterol, triglycerides and blood sugar, and greater endurance and aerobic capacity than the sedentary twin. The investigators are assessing the men at five-year intervals to track how long they live in good health.

Three types of MZ twins form, depending on when the fertilized ovum or very early embryo splits (**figure 3.18**). This difference in timing determines which supportive structures the twins share. About a third of MZ twins have completely separate chorions and amnions, and about two-thirds share a chorion but have separate amnions. Slightly fewer than 1 percent of MZ twins share both amnion and chorion. These differences may expose the different types of MZ twins to slightly different uterine environments. For example, if one chorion develops more attachment sites to the maternal circulation, one twin may receive more nutrients and gain more weight than the other.

Semi-identical twins result when two sperm fertilize an ovum. Twinning occurs at the first cleavage division, sending different paternal genomes into each twin. The twins share the maternal genome.

In 1 in 50,000 to 100,000 pregnancies, an embryo divides into twins after the point at which the two groups of cells can develop as two individuals, between days 13 and 15. The result is conjoined or "Siamese" twins. The latter name comes from Chang and Eng Bunker, who were born in Thailand, then called Siam, in 1811. They were joined by a band of tissue from the navel to the breastbone, and could easily have been separated today. Chang and Eng lived for 63 years, attached. They fathered 22 children and divided each week between their wives.

For Abigail and Brittany Hensel, shown in **figure 3.19**, the separation occurred after day 9 of development, but before day 14. Biologists know this because the shared organs have derivatives of ectoderm, mesoderm, and endoderm; that is, when the early embryo divided incompletely, the three primary germ layers had not yet fully sorted themselves out into two bodies. The Hensels are extremely rare "incomplete twins." They are "dicephalic," which means that they have two heads. Despite the fact that anatomically they are not as distinct as Chang and Eng were from each other, the two young women are very much individuals.

Each twin has her own neck, head, heart, stomach, gallbladder, lungs, and nervous system. Each has one leg and one arm, and a third arm between their heads was surgically removed. The twins share a large liver, a single bloodstream, three kidneys, a large ribcage and diaphragm muscle, and organs below the navel. Because at birth Abigail and Brittany were strong and healthy, doctors suggested surgery to separate them. But their parents, aware of other cases where only one child survived separation, declined surgery.

Figure 3.19 Conjoined twins. Abigail and Brittany Hensel are the result of incomplete twinning during the first 2 weeks of prenatal development. *Courtesy Brittany, Abby and Patty Hensel*

The Hensel twins have learned to work together, from crawling as babies to biking and driving today. As teens, Abigail and Brittany enjoyed sports and developed distinctive tastes in clothing and in food. They graduated from college and had their own reality TV series. Today they are elementary school teachers.

MZ twins occur in 3 to 4 pregnancies per 1,000 births worldwide. In North America, twins occur in about 1 in 81 pregnancies, which means that 1 in 40 of us is a twin. However, not all twins survive to be born. One study of twins detected early in pregnancy showed that up to 70 percent of the eventual births are of a single child. This is called the "vanishing twin" phenomenon.

The Embryo Develops

As prenatal development proceeds, different rates of cell division in different parts of the embryo fold the forming tissues into intricate patterns. In a process called embryonic induction, the specialization of one group of cells causes adjacent groups of cells to specialize. Gradually, these changes mold the three primary germ layers into organs and organ systems. Organogenesis is the transformation of the simple three layers of the embryo into distinct organs. During these weeks, the embryo is sensitive to environmental influences, such as exposure to toxins and viruses.

During the third week of prenatal development, a band called the primitive streak appears along the back of the embryo. Some nations designate day 14 of prenatal development and primitive streak formation as the point beyond which they ban research on the human embryo. The reasoning is that the primitive streak is the first sign of a nervous system and day 14 is when implantation completes.

The primitive streak gradually elongates to form an axis that other structures organize around as they develop. The primitive streak eventually gives rise to connective tissue progenitor cells and the notochord, which is a structure that forms the basic framework of the skeleton. The notochord induces a sheet of overlying ectoderm to fold into the hollow **neural tube**, which develops into the brain and spinal cord (central nervous system).

If the neural tube does not completely close by day 28, a neural tube defect (NTD) occurs, in which parts of the brain or spinal cord protrude from the open head or spine, and body parts below the defect cannot move. Surgery can correct some NTDs. Deficiency of the B vitamin folic acid can cause NTDs in embryos with a genetic susceptibility. For this reason, the U.S. government adds the vitamin to grains, and pregnant women take supplements. A blood test during the fifteenth week of pregnancy detects a substance from the fetus's liver called alpha fetoprotein (AFP) that moves at an abnormally rapid rate into the woman's circulation if there is an open NTD. Using hints from more than 200 genes associated with NTDs in other vertebrates, investigators are searching the genome sequences of people with NTDs to identify patterns of gene variants that could contribute to risk for or cause these conditions.

Appearance of the neural tube marks the beginning of organ development. Shortly after, a reddish bulge containing cells that will develop into a heart appear. These cells begin to beat in unison around day 18, and this is easily detectable by day 22. Soon the central nervous system starts to form.

The fourth week of embryonic existence is one of spectacularly rapid growth and differentiation (**figure 3.20**). Arms and legs begin to extend from small buds on the torso. Blood cells form and fill primitive blood vessels. Immature lungs and kidneys begin to develop.

By the fifth and sixth weeks, the embryo's head appears too large for the rest of its body. Limbs end in platelike structures with tiny ridges, and gradually apoptosis sculpts the fingers and toes. The eyes are open, but they do not yet have lids or irises. By the seventh and eighth weeks, a skeleton composed of cartilage forms. The embryo is now about the length and weight of a paper clip. At the end of 8 weeks of gestation, the prenatal human has tiny versions of all structures that will be present at birth. It is now a fetus.

Figure 3.20 Human embryos at (a) 28 days and (b) 49 days. *(a): Petit Format/Nestle/Science Source; (b): Petit Format/Science Source*

The Fetus Grows

During the fetal period, body proportions approach those of a newborn. Initially, the ears lie low and the eyes are widely spaced. Bone begins to replace the softer cartilage. As nerve and muscle functions become coordinated, the fetus moves.

Sex is determined at conception, when a sperm bearing an X or Y chromosome meets an oocyte, which always carries an X chromosome. An individual with two X chromosomes is a female, and one with an X and a Y is a male. A gene on the Y chromosome, called *SRY* (for "sex-determining region of the Y"), determines maleness.

Anatomical differences between the sexes appear at week 6, after the *SRY* gene is expressed in males. Male hormones stimulate male reproductive organs and glands to differentiate from existing, bipotential structures. In a female, the bipotential structures of the early embryo develop as female organs and glands, under the control of other genes. Differences may be noticeable on ultrasound scans by 12 to 15 weeks. Chapter 6 discusses sexual development further.

By week 12, the fetus sucks its thumb, kicks, makes fists and faces, and has the beginnings of teeth. It breathes amniotic fluid in and out, and urinates and defecates into it. The first trimester (3 months) of pregnancy ends.

By the fourth month, the fetus has hair, eyebrows, lashes, nipples, and nails (**figure 3.21**). By 18 weeks, the vocal cords have formed, but the fetus makes no sound because it doesn't breathe air. By the end of the fifth month, the fetus curls into a head-to-knees position. It weighs about 454 grams (1 pound). During the sixth month, the skin appears wrinkled because there isn't much fat beneath it, and turns pink as capillaries fill with blood. By the end of the second trimester, the fetus kicks and jabs and may even hiccup. It is now about 23 centimeters (9 inches) long.

Figure 3.21 A 16-week fetus. A fetus at this stage has hair, eyebrows, lashes, and other features that look human. *Nestle/Petit Format/Science Source*

In the final trimester, fetal brain cells rapidly link into networks as organs elaborate and grow. A layer of fat forms beneath the skin. The digestive and respiratory systems mature last, which is why infants born prematurely often have difficulty digesting milk and breathing, due to the lack of surfactant. Approximately 266 days after a single sperm burrowed its way into an oocyte, a baby is ready to be born.

The birth of a healthy baby is against the odds. Of every 100 secondary oocytes exposed to sperm, 84 are fertilized. Of these 84, only 69 implant in the uterus, 42 survive 1 week or longer, 37 survive 6 weeks or longer, and 31 are born alive. Of the fertilized ova that do not survive, about half have chromosomal abnormalities that cause problems too severe for development to proceed.

> **Key Concepts Questions 3.4**
>
> *a. At what point in prenatal development does an embryo become a fetus?
>
> b. List the events of fertilization.
>
> c. Define *zygote*, *morula*, and *blastocyst*.
>
> *d. What is the significance of the formation of the inner cell mass?
>
> e. Explain how some genes are silenced as development proceeds.
>
> *f. Name the primary germ layers.
>
> g. Describe the supportive structures that enable an embryo to develop.
>
> h. Explain how twins arise.
>
> i. What can a fetus at 12 weeks do?

3.5 Birth Defects

Certain genetic abnormalities or toxic exposures can affect development in an embryo or fetus, causing birth defects. Only a genetic birth defect can be passed to future generations. Although development can be abnormal in many ways, 97 percent of newborns appear healthy at birth.

The Critical Period

The specific nature of a birth defect reflects the structures developing when the damage occurs. The time when genetic abnormalities, toxic substances, or viruses can alter a specific structure is its **critical period** (**figure 3.22**). Some body parts, such as fingers and toes, are sensitive for short periods of time. In contrast, the brain is sensitive throughout prenatal development, and connections between nerve cells continue to change throughout life. Because of the brain's continuous critical period, many birth defect syndromes include learning or intellectual disabilities.

About two-thirds of all birth defects arise from a disruption during the embryonic period. More subtle defects, such as learning disabilities, that become noticeable only after infancy, are often caused by interventions during the fetal period. A disruption in the first trimester might cause severe intellectual disability; in the seventh month of pregnancy, it might cause difficulty in learning to read.

Some birth defects are caused by a mutation that acts at a specific point in prenatal development. In a rare inherited condition called phocomelia, for example, a mutation halts limb development from the third to the fifth week of the embryonic period, severely stunting arms and legs. The risk that a genetically caused birth defect will affect a particular family member can be calculated.

Many birth defects are caused by toxic substances the pregnant woman encounters. These environmentally caused problems will not affect other family members unless they, too, are exposed to the environmental trigger. Chemicals or other agents that cause birth defects are called **teratogens** (Greek for "monster-causing"). While it is best to avoid teratogens while pregnant, some women may need to continue to take potentially teratogenic drugs to maintain their own health.

Teratogens

Most drugs are not teratogens. Whether exposure to a particular drug causes birth defects may depend on a woman's genes. For example, certain variants of a gene that control the body's use of an amino acid called homocysteine affect whether the medication valproic acid causes birth defects. Valproic acid is used to prevent seizures and symptoms of bipolar disorder. Rarely, it can cause NTDs, heart defects, hernias, and clubfoot.

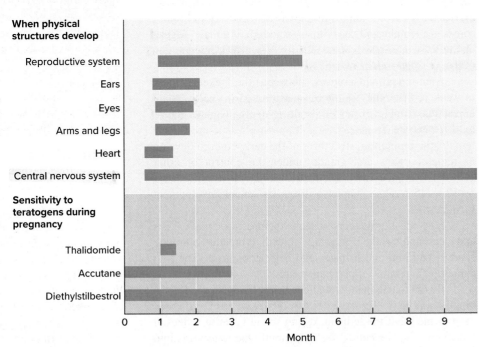

Figure 3.22 Critical periods of development. The nature of a birth defect resulting from drug exposure depends on which structures were developing at the time of exposure. Isotretinoin (Accutane) is an acne medication that causes cleft palate and eye, brain, and heart defects. Diethylstilbestrol (DES) was used in the 1950s to prevent miscarriage. It caused vaginal cancer in some "DES daughters." Thalidomide was used to prevent morning sickness.

Women who have this gene variant (*MTHFR C677T*) can use a different drug to treat seizures or bipolar disorder when they try to conceive.

Cocaine

When cocaine use soared in the late 1980s and 1990s, physicians predicted that children would be born who would have intellectual impairment and poor social skills from prenatal exposure to the drug. Although severe problems did not arise, continuing research has associated prenatal cocaine exposure with poor self-control, difficulty processing information, and attention deficit in childhood, with higher risk of language and memory problems and anxiety in adolescence. However, it is difficult to isolate effects specific to the drug, when problems associated with poverty, such as malnutrition, coexist.

Cigarettes

Chemicals in cigarette smoke stress a fetus. Carbon monoxide crosses the placenta and prevents the fetus's hemoglobin molecules from adequately binding oxygen. Nicotine is associated with sudden infant death syndrome, low birth weight, and hearing problems in childhood. Other chemicals in smoke block transfer of nutrients to the fetus. Smoke-exposed placentas are deficient in certain growth factors, causing poor growth before and after birth.

Electronic cigarettes have only been in use since 2007, so many studies that are providing information on associated birth defects are in the animal research stage. These studies have linked e-cigarettes to face and mouth defects. "Vaping" delivers nicotine (in lesser amounts), flavorings that may be dangerous, and other toxins, but because e-cigarettes do not emit carbon monoxide, they were initially described as safer than cigarettes. Now vaping is associated with a type of lung injury that can be lethal. Drugs to stop smoking are dangerous to take during pregnancy.

Thalidomide

The idea that the placenta protects the embryo and fetus from teratogens was tragically disproven between 1957 and 1961, when 10,000 children were born in Europe with what seemed, at first, to be phocomelia. This genetic disease is rare, and therefore couldn't be the cause of the sudden problem. Instead, the mothers had all taken a mild tranquilizer, thalidomide, to alleviate nausea early in pregnancy, during the critical period for limb formation. Many "thalidomide babies" were born with incomplete or missing legs and arms.

The United States was spared from the thalidomide disaster because an astute government physician noted the drug's adverse effects on laboratory monkeys. Still, several "thalidomide babies" were born in South America in 1994, where pregnant women were given the drug. In spite of its teratogenic effects, thalidomide is used to treat leprosy, AIDS, and certain blood and bone marrow cancers.

Alcohol

Greek philosopher Aristotle noticed problems in children of women with alcoholism more than 23 centuries ago. A pregnant woman who has just one or two alcoholic drinks a day, or perhaps a large amount at a single crucial time, risks a fetal alcohol spectrum disorder in her unborn child. Prenatal alcohol exposure can cause behavior problems, communication challenges, abnormal cognition, hyperactivity, and short attention span. People with fetal alcohol spectrum disorders are also more likely to have hearing and/or visual loss. According to the National Institutes of Health, 1 to 5 percent of children in the United States have some effects of prenatal alcohol exposure.

Fetal alcohol spectrum disorders include different symptoms. The most severe manifestation of alcohol exposure before birth is fetal alcohol syndrome (FAS). Most children with FAS have small heads and flat faces (**figure 3.23**). Growth is slow before and after birth. Teens and young adults are short and have small heads. Alcohol-related neurodevelopmental disorders affect intellect and learning. Alcohol-related birth defects damage the heart, kidneys, bones, and sense of hearing.

Many adults with FAS function at early grade-school level. They never develop social and communication skills and cannot understand the consequences of actions, form friendships, take initiative, and interpret social cues.

Tests for variants of genes that encode proteins that regulate alcohol metabolism may be able to identify women whose fetuses are at elevated risk for developing fetal alcohol spectrum disorders. Until these tests are validated, pregnant women are advised to avoid all alcoholic beverages.

Nutrients

Certain nutrients ingested in large amounts, particularly vitamins, act as drugs. The acne medicine isotretinoin (Accutane)

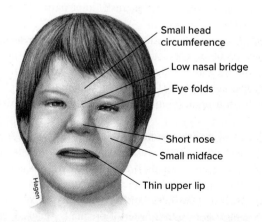

Figure 3.23 Fetal alcohol syndrome. Some children whose mothers drank alcohol during pregnancy have characteristic flat faces.

is a vitamin A derivative that causes spontaneous abortion and defects of the heart, nervous system, and face in exposed embryos. Another vitamin A–based drug, used to treat psoriasis, as well as excesses of vitamin A itself, also cause birth defects. Vitamin A and its derivatives may accumulate enough to harm embryos because they are stored in body fat for up to 3 years.

Excess vitamin C can harm a fetus that becomes accustomed to the large amounts the woman takes. After birth, when the vitamin supply suddenly plummets, the baby may develop symptoms of vitamin C deficiency, such as bruising and becoming infected easily.

Malnutrition threatens a fetus. Clinical Connection 11.1 describes the effects of starvation on embryos during the bleak "Dutch Hunger Winter" of 1944 to 1945. Poor nutrition later in pregnancy affects the development of the placenta and can cause low birth weight, short stature, tooth decay, delayed sexual development, and learning disabilities.

Viral Infection

Viruses are small enough to cross the placenta and reach a fetus. Some viruses that cause mild symptoms in an adult, such as the chickenpox virus, may devastate a fetus. Men can transmit viral infections to an embryo or fetus during sexual intercourse. Zika virus, for example, is carried in semen.

Zika virus was discovered in Uganda in 1947, but came to the world's attention in 2015, after it was associated with a dramatic increase in the incidence of the birth defect microcephaly (a very small head) in Brazil. About 11 percent of women exposed to the virus during the first trimester of pregnancy have a child with microcephaly or another brain abnormality, whether or not the women had symptoms of viral infection. Babies affected by prenatal exposure to Zika virus have a spectrum of neurological problems, including difficulty with mobility, seizures, sleep problems, and poor hearing and vision. However, school-age children do not have developmental delay. Mosquitoes spread Zika virus, which causes a mild illness of fever, rash, joint pain, and red eyes in some adults.

HIV can reach a fetus through the placenta or infect a newborn via blood contact during birth. The risk of transmission is significantly reduced if a pregnant woman takes anti-HIV drugs. Fetuses of HIV-infected women are at higher risk for low birth weight, prematurity, and stillbirth if the woman's health is failing.

German measles (rubella) is a well-known viral teratogen. In the United States, in the early 1960s, an epidemic of the usually mild illness caused 20,000 birth defects and 30,000 stillbirths. Exposure during the first trimester of pregnancy caused cataracts, deafness, and heart defects. Exposure during the second or third trimester of pregnancy caused learning disabilities, speech and hearing problems, and type 1 diabetes mellitus. Incidence of these problems, called congenital rubella syndrome, has dropped markedly since vaccination eliminated the disease in the United States.

> **Key Concepts Questions 3.5**
>
> *a. Define *critical period*.
> b. Explain why most birth defects that develop during the embryonic period are more severe than problems that arise during fetal development.
> c. Define *teratogen* and give an example of one.

3.6 Maturation and Aging

"Aging" means moving through the life cycle, despite advertisements for products that promise to reverse the process. As we age, the limited life spans of cells are reflected in the waxing and waning of biological structures and functions. Although some aspects of our anatomy and physiology peak very early—such as the number of brain cells or hearing acuity, which do so in childhood—age 30 seems to be a turning point for decline. Some researchers estimate that, after this age, the human body becomes functionally less efficient by about 0.8 percent each year.

Many diseases that begin in adulthood, or are associated with aging, have genetic components. These diseases tend to be complex, because it takes many years for environmental exposures to alter gene expression in ways that noticeably affect health. Following is a closer look at how genes may impact health throughout life.

Adult-Onset Inherited Diseases

Prenatal development is a highly regulated program of genetic switches that are turned on in specific body parts at specific times. Environmental factors can affect how certain genes are expressed before birth in ways that create risks that appear much later. A powerful environmental influence is nutrition. Adaptations that enable a fetus to grow despite near-starvation become risk factors for certain common conditions of adulthood, when conserving calories is no longer needed. Such diseases include coronary artery disease, obesity, stroke, hypertension, schizophrenia, and type 2 diabetes mellitus. A malnourished fetus has intrauterine growth retardation (IUGR), and though born on time, is quite small. Premature infants, in contrast, are small but are born early, and are not predisposed to conditions resulting from IUGR.

Low birth weight due to IUGR correlates to increased incidence of cardiovascular disease and diabetes later in life. Much of the data come from war records that enable the study of the effects of prenatal malnutrition as people age.

How can poor nutrition before birth cause disease decades later? Perhaps to survive, the starving fetus redirects its circulation to protect vital organs, as muscle mass and hormone production change to conserve energy. Growth-retarded babies have too little muscle tissue, and because muscle is the primary site of insulin action, glucose metabolism changes. Thinness at birth, and the accelerated weight gain in childhood that often compensates, increases risk for coronary heart disease and type 2 diabetes much later.

In contrast to the delayed effects of fetal malnutrition, symptoms of single-gene diseases can begin anytime. Most inherited conditions that affect children are recessive. Even a fetus can have symptoms of inherited disease, such as the broken bones of osteogenesis imperfecta ("brittle bone disease").

Most dominantly inherited conditions start to affect health in early to middle adulthood. In polycystic kidney disease, for example, cysts that may have been present but undetected in the kidneys during one's twenties begin causing bloody urine, high blood pressure, and abdominal pain in the thirties. The joint destruction of osteoarthritis may begin in one's thirties, but not be painful for 20 years.

Mutations in single genes cause from 5 to 10 percent of Alzheimer disease cases, producing initial symptoms in the forties and fifties. Noninherited Alzheimer disease typically begins later in life.

Whatever the age of onset, Alzheimer disease starts gradually. Mental function declines steadily for 3 to 10 years following the first symptoms of depression and short-term memory loss. Cognitive skills ebb. A person may forget how to put one foot in front of the other, or, confused and forgetful, wander away. Finally, the patient cannot recognize loved ones and can no longer perform basic functions such as speaking or eating.

The brains of Alzheimer disease patients contain deposits of a protein, called amyloid beta, in learning and memory centers. The brains of people with Alzheimer disease also contain structures called neurofibrillary tangles, which consist of a protein called tau. Tau binds to and disrupts microtubules in nerve cell branches, destroying the shape of the cell, impairing its ability to communicate. Clinical Connection 5.1 discusses Alzheimer disease further.

Syndromes That Resemble Accelerated Aging

Genes control aging both passively (as structures break down) and actively (by initiating new activities). A group of "premature aging" inherited disorders is rare, but may hold clues to how genes control aging in all of us, as discussed in the chapter opener.

Syndromes that accelerate aging and shorten life, or that make a person appear older than they are but do not affect life expectancy, are termed progeroid. (Life expectancy is the average age at death for a species. Life span refers to the oldest known individual of a species.) The most severe progeroid syndromes are the progerias, which shorten life expectancy. Most accelerated aging conditions are caused by the inability of cells to adequately repair DNA, which is discussed in section 12.7. With poor DNA repair, mutations that would ordinarily be corrected persist. Over time, the accumulation of mutations destabilizes the genomes of somatic cells, and more mutations occur. Changes associated with aging ensue.

Table 3.4 lists the more common progeroid syndromes. Life expectancy and symptom severity vary. People with Rothmund-Thomson syndrome, for example, may have a normal life expectancy, but develop gray hair or baldness, cataracts, cancers, and osteoporosis at a young age. Trichothiodystrophy causes sparse, brittle hair; scaly skin; short stature; intellectual disability; and eye and tooth abnormalities.

Table 3.4 Premature Aging Syndromes

Disease	Incidence	Average Life Expectancy
Ataxia telangiectasia	1/60,000	19–25
Cockayne syndrome	1/100,000	20
Hutchinson-Gilford progeria syndrome	<1/1,000,000	13
Rothmund-Thomson syndrome	<1/100,000	Normal
Trichothiodystrophy	1/1,000,000	?
Werner syndrome	<1/100,000	50

Death often comes before the age of 10 from infection. Werner syndrome becomes apparent before age 20, causing death before age 50 from diseases associated with aging. Young adults with Werner syndrome develop atherosclerosis, type 2 diabetes, hair graying and loss, osteoporosis, cataracts, and wrinkled skin. They are short because they skip the growth spurt of adolescence.

Hutchinson-Gilford progeria syndrome is the most severe rapid aging condition. In addition to the symptoms the chapter opener describes are abnormalities at the cellular level. Normal cells growing in culture divide about 50 times before dying. Cells from Hutchinson-Gilford progeria syndrome patients die in culture after only 10 to 30 divisions.

A mutation in a gene, *LMNA*, that encodes a protein, lamin A, lies behind Hutchinson-Gilford progeria syndrome. The route to premature activation of genes associated with aging is complex. A sequence of molecular interactions at the inner face of the nuclear membrane accelerates the formation of a sticky form of lamin A, called progerin, that triggers rapid programmed cell death (apoptosis, see figure 2.16). As the infoldings of the nuclear membrane touch the telomeres, the chromosome tips shrink faster. As a result, the cell ages so quickly that the nuclear membrane collapses in a characteristic way (**figure 3.24**). Although we all produce some progerin as we age, in Hutchinson-Gilford progeria, the sticky protein accumulates rapidly enough to cause early cell death. The new drug that is helping some of the children with this condition stabilizes the nuclear membranes, delaying progression of the disease.

Genes and Longevity

Aging reflects gene activity plus a lifetime of environmental influences. Families and genetically isolated populations with many aged members have a lucky collection of gene variants plus shared environmental influences such as good nutrition, excellent health care, and devoted relatives. Genome comparisons

Figure 3.24 **The nuclear membrane of a cell in the throes of apoptosis due to progeria is distinctive.** It collapses in a characteristic way. *Source: Gordon P Scaffidi P, L. T. Misteli/PLOS Biology Journal*

between people who've passed their hundredth birthdays (centenarians) to those who have died of the common illnesses of older age are revealing genes that influence longevity.

Most centenarians enjoy excellent health and are socially active, then succumb rapidly to diseases that typically claim people decades earlier. They fall into three broad groups—about 20 percent never get the diseases that kill most people; 40 percent get these diseases, but at much older ages; and the other 40 percent live with and survive the more common diseases of aging: heart disease, stroke, cancers, type 2 diabetes, and dementias. People past 110 are called supercentenarians; about 80 are known worldwide.

The environment seems to play an important role in the deaths of people ages 60 to 90, but past that age, genetic effects predominate. That is, someone who dies at age 68 of lung cancer can probably blame a lifetime of cigarette smoking. But a smoker who dies at age 101 of the same disease probably had gene variants that protected against lung cancer. Centenarians have higher levels of large lipoproteins that carry cholesterol (high-density lipoprotein [HDL]) than other people, which researchers estimate adds 20 years of life.

Children and siblings of centenarians tend to be long-lived as well, supporting the idea that longevity is inherited. Brothers of centenarians are 17 times as likely to live past age 100 as the average man, and sisters are 8.5 times as likely. Unhealthy habits among some centenarians hint that genes are protecting them. One researcher suggests that the saying, "The older you get, the sicker you get" be replaced with "The older you get, the healthier you've been."

Centenarians have inherited two types of gene variants—those that directly protect them and wild type alleles of genes that, when mutant, cause disease.

Genes related to these functions can affect longevity control:

- immune system functioning;
- insulin secretion and glucose metabolism;
- response to stress;
- the cell cycle;
- DNA repair;
- lipid (including cholesterol) metabolism;
- nutrient metabolism; and
- production of antioxidant enzymes.

Researchers estimate that at least 130 genes have variants that influence how long a person is likely to live.

One well-studied aging gene encodes the growth hormone receptor, which is a protein that enables a cell to respond to signals to enlarge, a key part of early development. People who inherit two recessive mutations in the gene have Laron syndrome. They are very short—adults stand under 4 feet tall—but they do not develop cancer or diabetes. Because of the near absence of these common diseases of aging, people with Laron syndrome live long lives. In El Oro Province in southern Ecuador, more than 100 people have Laron syndrome. They descend from Eastern European Jewish people who fled to the area to escape the Spanish Inquisition in the fifteenth century, and avoiding churches led them to the unoccupied, rural region.

An isolated population such as the one in El Oro, where people mate with close relatives and produce offspring, creates a situation in which mutations are overrepresented. However, the genomes are relatively uniform, making gene variants easier to identify than in more outbred populations. Laron syndrome somehow protects these people from two major diseases of older age. **Table 3.5** lists other single genes that are associated with longevity in isolated populations. Environmental factors are also important in longevity. The very-long-lived people of Calabria, an area of southern Italy, for example, eat mostly fruits and vegetables, which may contribute to their excellent health.

Several research projects are collecting and curating genetic information from volunteers in the general population. The New England Centenarian Study, which began in 1995 to amass information on the families of the oldest citizens then known in the United States, compiles information from people aged 103 or older or aged 100 with living siblings. These people tend not to get sick until near the ends of their lives. Longevity runs in their families, and the genetic influence increases with age, which seems counter to the prominent role of the environment in the middle years in the general population. The oldest old have the same disease-causing gene variants as those who

Table 3.5 Longevity Genes

Gene	Protects Against	Population Studied
Apolipoprotein C3 (APOC-3)	Hypertension, diabetes	Ashkenazi Jews
	Cardiovascular disease	Amish
Bitter taste receptor (TAW2R16)	Poisoning, digestive problems	Calabria, Italy
Cholesteryl ester transfer protein (CETP)	Cardiovascular disease	Ashkenazi Jews
Forkhead box O3 (FOXO3)	Cancer, cardiovascular disease	Japanese-Americans
Growth hormone receptor (GHR)	Diabetes, cancer	Ecuador, Israel
Uncoupling proteins (UCP 2, 3, 4)	Oxidative damage, poor energy use	Calabria, Italy

die younger, suggesting that other genes protect them. Another study is investigating genomes of nursing home residents.

A program in California is probing the genomes of the "wellderly". So far, these people share never having had heart disease and never having smoked. Several very-long-lived people have had cancer, indicating that cancers are survivable. Researchers at the University of Pittsburgh have identified places in the genome that harbor "successful aging genes" with variants that preserve cognition. Yet another study, of gene expression in centenarians, found that their cells were more adept at autophagy - disposing of debris. Considered together, these studies, plus data on health and longevity pouring in from people taking consumer genetics tests, may provide information that will help the majority of us who have not been fortunate enough to have inherited longevity gene variants.

> **Key Concepts Questions 3.6**
>
> a. Explain how starvation before birth sets the stage for later disease.
> *b. What percentage of people with Alzheimer disease inherit the condition?
> c. Describe a single-gene disorder that speeds aspects of aging.
> *d. What is the role of DNA repair in rapid-aging syndromes?
> *e. List three characteristics of centenarians.

Study Guide

Summary

3.1 The Reproductive System

1. The male and female reproductive systems include paired **gonads** and networks of tubes in which **sperm** and **oocytes** are made.
2. Male **gametes** originate in seminiferous tubules within the **testes**, then pass through the epididymis and ductus deferentia, where they mature before exiting the body through the urethra during sexual intercourse. The prostate gland, the seminal vesicles, and the bulbourethral glands add secretions.
3. Female gametes originate in the **ovaries**. Each month after puberty, one ovary releases an oocyte into a uterine tube. The oocyte then moves to the uterus for implantation (if fertilized) or expulsion.

3.2 Meiosis

4. Meiosis reduces the chromosome number in gametes from diploid to haploid, maintaining the chromosome number between generations. Cells that have more than two sets of chromosomes are **polyploid.** Meiosis ensures genetic variability by **independently assorting** combinations of genes into gametes as **homologous pairs** of chromosomes randomly align and **cross over**.
5. Meiosis I, a **reduction division**, halves the number of chromosomes. Meiosis II, an **equational division**, produces four cells from the two that result from meiosis I, without another DNA replication.
6. Crossing over occurs during prophase I. It mixes up paternally and maternally derived genes.
7. Chromosomes segregate and independently assort in metaphase I, which determines the distribution of genes from each parent.

3.3 Gametes Mature

8. **Spermatogenesis** begins with **spermatogonia**, which accumulate cytoplasm and replicate their DNA, becoming primary spermatocytes. After meiosis I, the cells become haploid secondary spermatocytes. In meiosis II, the secondary spermatocytes divide, each yielding two spermatids, which then differentiate into **spermatozoa.**

9. In **oogenesis**, some **oogonia** grow and replicate their DNA, becoming primary oocytes. In meiosis I, the primary oocyte divides to yield one large secondary oocyte and a small **polar body**. In meiosis II, the secondary oocyte divides to yield the large ovum and another polar body. Female meiosis is completed at fertilization.

10. Genetic errors in sperm from older men are usually dominant single-gene mutations. Genetic errors in oocytes from older women are typically extra or absent chromosomes.

3.4 Prenatal Development

11. In the female, sperm are capacitated and drawn toward a secondary oocyte. One sperm burrows through the oocyte's protective layers with acrosomal enzymes. Fertilization occurs when the sperm and oocyte fuse and their DNA combines in one nucleus, forming the **zygote**. Electrochemical changes in the egg surface block additional sperm from entering. **Cleavage** begins and continues until a 16-celled **morula** forms. Between days 3 and 6, the morula arrives at the uterus and hollows, forming a **blastocyst** made up of **blastomeres**. The trophoblast and **inner cell mass** form. Around day 6 or 7, the blastocyst implants, and trophoblast cells secrete hCG, which prevents menstruation.

12. During the second week, the amniotic cavity forms as the inner cell mass flattens. **Ectoderm** and **endoderm** form, and then **mesoderm** appears, establishing the **primary germ layers** of the **gastrula**. **Epigenetic** effects oversee differentiation as cells in each germ layer develop into organs.

13. During the third week, the placenta, yolk sac, allantois, and umbilical cord begin to form as the amniotic cavity swells with fluid. **Chorionic villus sampling** examines the chromosomes in cells collected from chorionic villi. **Amniocentesis** examines chromosomes in fetal cells in the amniotic fluid.

14. **Monozygotic (MZ) twins** result when one fertilized ovum splits. **Dizygotic (DZ) twins** result from fertilization of two ova at the same time.

15. Organs form throughout the embryonic period. The primitive streak, notochord and **neural tube**, arm and leg buds, heart, facial features, and skeleton develop.

16. At the eighth week, the **embryo** becomes a **fetus**, with all structures present but not fully grown. Organ rudiments grow and specialize. The developing organism moves and reacts. In the last trimester, the brain develops rapidly, and fat is deposited beneath the skin. The digestive and respiratory systems mature last.

3.5 Birth Defects

17. Birth defects can result from a mutation or an environmental influence.

18. A substance that causes birth defects is a **teratogen**. Environmentally caused birth defects are not transmitted to future generations.

19. The time when a structure is sensitive to damage from an abnormal gene or environmental intervention is its **critical period**.

3.6 Maturation and Aging

20. Genes cause or predispose us to illness throughout life. Single-gene diseases that strike early tend to be recessive; most adult-onset single-gene conditions are dominant.

21. Malnutrition before birth can alter gene expression in ways that cause illness later.

22. Progeroid syndromes are single-gene disorders that speed aging-associated changes. The progeria subtype shortens life.

23. Long life is due to genetics and environmental influences. Researchers identify longevity genes by studying genetically isolated populations with many long-lived members.

Key Terms

amniocentesis (ăm′nē-ō-sĕn-tē′sĭs) A prenatal diagnostic technique that examines fetal chromosomes and biochemicals in amniotic fluid.

blastocyst (blăs′tō-sĭst′) A fluid-filled ball of cells descended from a fertilized ovum.

blastomere (blăstō′-mēr′) A cell of a blastocyst.

chorionic villus sampling (CVS) (kôrē-ŏn′ĭk vĭl-us) A prenatal diagnostic technique that analyzes chromosomes in chorionic villus cells, which, like the fetus, descend from the fertilized ovum.

cleavage (klēvĭj) A series of rapid mitotic cell divisions after fertilization.

critical period The time during prenatal development when a structure is sensitive to damage from a mutation or an environmental intervention.

crossing over An event during prophase I when homologs exchange parts.

dizygotic twins (DZ) (dīzī-gŏt′ĭk) Twins that originate as two fertilized ova; fraternal twins.

ectoderm (ĕktō-dûrm) The outermost primary germ layer of the primordial embryo.

embryo (ĕm′brē-ō′) In humans, the prenatal organism from fertilization until the end of the eighth week, when all basic structures are present.

endoderm (ĕn′-dō-dûrm) The innermost primary germ layer of the primordial embryo.

epigenetic (ĕpē-jə-nĕt′ĭk) Any effect that modifies gene expression without changing the DNA sequence, such as methylation.

equational division The second meiotic division, yielding four cells from two.

fetus (fē′təs) The prenatal human after the eighth week of development, when structures grow and specialize.

gamete (găm′ēt) A sex cell.

gastrula (găstrə-lə) A three-layered embryo.

gonads (gō′-nadz) Paired structures in the reproductive system where sperm or oocytes are manufactured.

homologous pair (hō-mŏl′ə-gəs) Two chromosomes with the same gene sequence.

independent assortment Inheritance of a gene on one chromosome does not influence inheritance of a gene on

a different chromosome (Mendel's second law) because of the random arrangement of homologous chromosome pairs, in terms of maternal and paternal origin, down the center of a cell in metaphase 1.

inner cell mass A clump of cells on the inside of the blastocyst that will continue developing into an embryo.

mesoderm (mĕz-ō-dûrm) The middle primary germ layer.

monozygotic twins (MZ) (mŏnō-zī-gŏt'ĭk) Twins that originate as a single fertilized ovum; identical twins.

morula (môr' yə-lə) The very early prenatal stage that resembles a mulberry.

neural tube (nōōr'əl) A structure in the embryo that develops into the brain and spinal cord.

oocyte (ō'ə-sīt) The female gamete (sex cell). An egg.

oogenesis (ōə-jĕn'ĭ-sĭs) Oocyte (egg) formation.

oogonium (o-o-gō'-ni-um) A cell in the ovary that gives rise to an oocyte, in meiosis.

ovaries (o'və-rēz) The female gonads.

polar body A product of female meiosis that contains little cytoplasm and does not continue to develop into an oocyte.

polyploid (pŏl'ē-ploid) A cell with one or more extra sets of chromosomes.

primary germ layers The three layers of an embryo.

reduction division The first meiotic division, which halves the chromosome number.

sperm The male sex cell (gamete).

spermatogenesis (spər-măt'ə-jĕn'ĭ-sis) Sperm cell formation and differentiation.

spermatogonium (sper"mah-to-gō' ne-um) An undifferentiated cell in a seminiferous tubule that can give rise to a sperm cell following meiosis and maturation.

spermatozoon (spər-măt'ə-zō'ŏn) (sperm) A mature male reproductive cell (meiotic product).

teratogen (tə-răt'ə-jən) A substance that causes a birth defect.

testes (tes'tēz) The male gonads.

zygote (zī'gōt) A prenatal human from the fertilized ovum stage until formation of the primordial embryo, at about 2 weeks.

Review Questions

1. How many sets of human chromosomes are in each of the following cell types?
   a. an oogonium
   b. a primary spermatocyte
   c. a spermatid
   d. a cell from either sex during anaphase of meiosis I
   e. a cell from either sex during anaphase of meiosis II
   f. a secondary oocyte
   g. a polar body derived from a primary oocyte

2. List the structures and functions of the male and female reproductive systems.

3. A dog has 39 pairs of chromosomes. Considering only the random alignment of chromosomes, how many genetically different puppies are possible when two dogs mate? Is this number an underestimate or overestimate of the actual total? Why?

4. How does meiosis differ from mitosis?

5. What do oogenesis and spermatogenesis have in common, and how do they differ?

6. How does gamete maturation differ in the male and female?

7. Why is it necessary for spermatogenesis and oogenesis to generate stem cells?

8. Describe the events of fertilization.

9. Define *epigenome*.

10. Write the time sequence in which the following structures begin to develop: notochord, gastrula, inner cell mass, fetus, zygote, morula.

11. Why does exposure to teratogens produce more severe health effects in an embryo than in a fetus?

12. The same birth defect syndrome can be caused by a mutant gene or exposure to a teratogen. How do the consequences of each cause differ for future generations?

13. List three teratogens, and explain how they disrupt prenatal development.

14. Cite two pieces of evidence that genes control aging.

Applied Questions

1. Progeria is among the rarest of rare diseases. Some people question the utility of developing treatments for extremely uncommon conditions. Suggest one way that finding a treatment for one specific aspect of progeria might help many others.

2. Up to what stage, if any, do you think it is ethical to experiment on a prenatal human? Cite reasons for your answer.

3. Under a microscope, a first and second polar body look alike. What structure would distinguish them?

4. In about 1 in 200 pregnancies, a sperm fertilizes a polar body instead of an oocyte. A mass of tissue that is not an embryo develops. Why can't a polar body support the development of an entire embryo?

5. "Congratulations! You have a 97 percent chance of having a healthy baby!" says Jill to Janna, who has just

read the results of her early pregnancy test. How is Jill's encouragement statistically incorrect?

6. Should a woman be held legally responsible if she drinks alcohol, smokes, or abuses drugs during pregnancy and it harms her child? Should liability apply to all substances that can harm a fetus, or only to those that are illegal?

7. Explain how a man can transmit Zika virus infection to offspring.

8. What are possible benefits and dangers of predicting how long a person will live from analyzing his or her genome sequence?

Case Studies and Research Results

1. A mutation in the "testis-expressed 11" gene causes male infertility. The protein that the gene encodes is normally made in spermatids. Explain how the mutation may cause infertility.

2. Human embryonic stem cells can be derived and cultured from an eight-celled cleavage embryo and from a cell of an inner cell mass. Explain the difference between these stages of human prenatal development.

3. A pediatrician with training in anatomy and embryology was examining the jaw of a little girl with auriculocondylar syndrome. The child had tiny ears, a small head and mouth, and an unusual jaw. The doctor realized from studying X-rays that the lower jaw looked exactly like an upper jaw—and the girl's mother looked just like the child. What type of mutation does this family have? Describe in general terms how it causes the phenotype.

4. A few children have been reported who have thin sparse hair, bony facial features, little body fat, and appear aged. What further information would help to confirm a diagnosis of a progeroid syndrome?

5. A study of 202 individuals over the age of 90 living in Calabria in southern Italy is comparing the siblings of these people to their spouses. How could this strategy reveal the genetic underpinnings of longevity?

PART 2 Transmission Genetics

CHAPTER 4

Single-Gene Inheritance

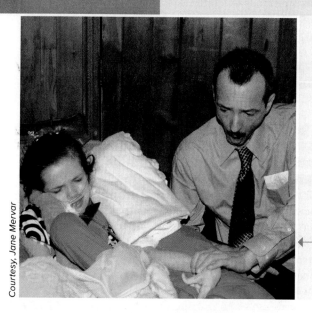

Courtesy, Jane Mervar

Karli Mervar and her father Karl suffered from Huntington disease at the same time. Karl's gaunt look results from the constant movements that are part of the disease—his cells used energy rapidly.

Learning Outcomes

4.1 Following the Inheritance of One Gene
1. List the characteristics that distinguish single-gene diseases from other types of diseases.
2. Describe how Mendel deduced that recessive traits seem to disappear in hybrids.
3. Define and distinguish *heterozygote* and *homozygote*; *dominant* and *recessive*; *phenotype* and *genotype*.
4. Explain how the law of segregation reflects the events of meiosis.
5. Indicate how a Punnett square is used to track inheritance patterns.

4.2 Single-Gene Inheritance Is Rare
6. Explain how a gene alone may not solely determine a trait.
7. Distinguish between autosomal recessive and autosomal dominant inheritance.

4.3 Following the Inheritance of More Than One Gene
8. Explain how Mendel's experiments followed the inheritance of more than one gene.
9. Explain how the law of independent assortment reflects the events of meiosis.

4.4 Pedigree Analysis
10. Explain how pedigrees are used to show transmission of single genes.
11. Explain how exome and genome sequencing can reveal whether a sick child with healthy parents inherited two autosomal recessive mutations or has a new, dominant mutation.

The BIG Picture

The laws underlying the inheritance of single-gene traits and illnesses have their origins in experiments using pea plants that Gregor Mendel conducted more than a century ago. Mendel's laws reverberate in the interpretation of DNA tests done today, from investigating single genes to sequencing exomes and genomes, which allows geneticists to distinguish inherited mutations from new mutations.

Juvenile Huntington Disease: The Cruel Mutation

Huntington disease (HD) typically begins in early adulthood, causing uncontrollable movements and changes in behavior and thinking (cognition), with death 10 to 20 years later. HD is dominant, which means that each child of an affected individual need inherit only one copy of the mutant gene to develop the disease. However, 10 percent of people who have HD are under age 20. They have juvenile Huntington disease.

The Mervar family is unusual in that an adult (Karl) and his young children were or are affected. Karl's first symptoms started when he and his wife Jane's youngest daughter, Karli, began to have trouble paying attention in school. Karl had become abusive, paranoid, and unstable on his feet, due to the disease. By age 5 Karli's left side stiffened and her movements slowed. Soon she could no longer skip, hop, or jump, and she developed a racing heartbeat, itching, pain, and unintelligible speech. Karl was diagnosed 6 weeks after Karli in 2002. He was 35, she just 6. Karli's disease progressed rapidly. By age 9 she required a feeding tube, suffered seizures, and barely slept. Karl became so violent that he spent his last years in a nursing home. Father and daughter died within weeks of each other, in early 2010.

An inherited disease affects each child independently, and so the fact that Karli was sick did not spare her sisters. One was diagnosed in 2004 at age 13 and another in 2007 at age 17. This family illustrates classic autosomal dominant inheritance—affecting every generation and both sexes. Jane has cared for them all.

4.1 Following the Inheritance of One Gene

Online Mendelian Inheritance in Man is a compendium of all known human genes and what they do. Before the internet, *OMIM* was an annual huge book. Now it is a continually updated database, with researchers reporting new gene-disease assignments as more human genomes are sequenced and mutations matched with phenotypes. OMIM describes about 8,000 single-gene diseases and traits, corresponding to about 23,000 genes. Not all of the information is in. The World Health Organization estimates that about 10,000 diseases are caused by mutations in single genes, which are called **monogenic**. About 30 million (1 in 10) people in the United States has a rare disease, as do 350 million people globally, and about 80 percent of rare diseases are monogenic. Many single-gene diseases affect fewer than 1 in 10,000 individuals. Most familiar illnesses arise from variants in one or more genes plus environmental influences.

Characteristics of Single-Gene Diseases

Inherited illnesses caused by mutations in single genes differ from other types of illnesses in several ways. In families, we can deduce the probability that a particular person has inherited a single-gene disease by considering how he or she is related to an affected relative. The sisters in the family described in the chapter opener, for example, each had a 1 in 2 chance of inheriting their father's Huntington disease. In contrast, an infectious disease requires that a pathogen pass from one person to another, which is much less predictable.

A second distinction of single-gene diseases is that tests can sometimes predict the risk of developing symptoms. This is possible because all cells harbor the mutation, if the person has inherited it. A person with a parent who has Huntington disease can have a blood test that detects the mutation at any age, even though the disease affects the brain, not the blood. Inheriting an HD mutation predicts illness with near certainty, but this is not true for many conditions. This predictive power is a characteristic of a particular gene. For a gene with less predictive power, a genetic test may indicate increased risk of developing symptoms but not the near certainty that occurs with the HD mutation. This predictive ability, called penetrance, is discussed further in chapter 5.

A third feature of single-gene diseases is that they may be much more common in some populations than others. Genes do not like or dislike certain types of people; rather, mutations stay in certain populations when people have children with people of the same ancestral group. A fourth characteristic of a genetic disease is that it may be "fixable" by compensating for the abnormal instructions, such as by providing a missing enzyme or clotting factor, or even by adding a working copy of the gene to affected cells.

Single-gene diseases such as HD and cystic fibrosis (CF) affect families in patterns, termed **modes of inheritance**. Knowing these patterns makes it possible to predict the risks that people related in particular ways have inherited the family's mutation. HD is **autosomal dominant**, which means that it affects both sexes and appears every generation. In contrast, CF is **autosomal recessive**, which means that the disease affects both sexes and can "skip" generations through carriers, who do not have symptoms.

Today, genetic tests, including exome and genome sequencing, reveal which single-gene health conditions we have, carry, or may develop. Tests of "trios" consisting of sick children and their parents can reveal whether the child inherited two disease-causing mutations from carrier parents, or whether a dominant mutation arose anew, termed *de novo*, in the child. **Section 4.4** revisits this clinically important distinction.

Single-gene traits and diseases are also called "Mendelian" in honor of Gregor Mendel, who derived the two laws of inheritance that determine how these traits are transmitted from one generation to the next. Even though Mendel worked in the late nineteenth century, his contributions are more important than ever as the sequencing of millions of human genomes is leading to discovery and description of many single-gene conditions. Mendel's tale begins with pea plants growing in a long-ago garden in what is now the Czech Republic.

Mendel's Experiments

Gregor Mendel was the first investigator to probe the underlying rules of logic that make it possible to predict inheritance of specific traits. Fascinated by the patterns in the natural variation in plants, Mendel bred pea plants to describe units of inheritance, "elementen," that pass traits from generation to generation. He could not see the elementen, but he inferred their existence from the appearances of his plants. He knew nothing of DNA, genes, chromosomes, meiosis, or even cells. Mendel's laws apply to any diploid species.

From 1857 to 1863, Mendel crossed and cataloged traits in 24,034 plants, through several plant generations. He deduced that consistent ratios of traits in the offspring indicated that the plants transmitted distinct units. His two laws—hypotheses until they had been demonstrated in several species—proposed how this happens. Mendel published his findings in 1866, but few people read the paper. In 1901, when three botanists independently read Mendel's paper, they credited him with deducing the two basic laws of inheritance. Soon Mendel became known as the "father of genetics."

Peas are ideal for probing heredity because they are easy to grow, develop quickly, and have many traits that take one of two easily distinguishable forms. **Figure 4.1** illustrates the seven traits that Mendel followed. When analyzing genetic crosses, the first generation is the parental generation, or P_1; the second generation is the first filial generation, or F_1; the next generation is the second filial generation, or F_2, and so on.

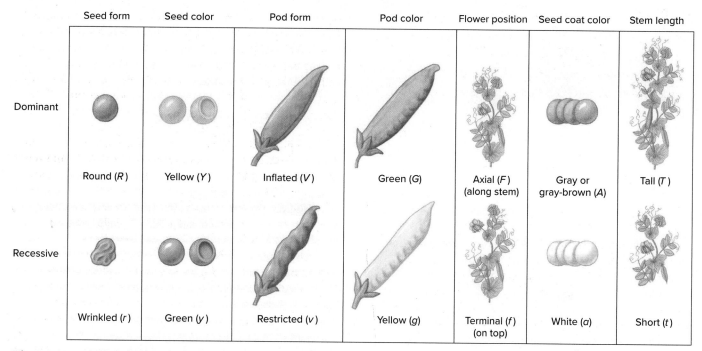

Figure 4.1 Traits Mendel studied. Gregor Mendel studied the transmission of seven traits in the pea plant. Each trait has two easily distinguished expressions, or phenotypes.

Mendel's first experiments followed single traits with two expressions, such as "short" and "tall." He began with plants that had been cultivated over many generations to ensure that they always "bred true," producing the same expression of the trait. He set up all possible combinations of artificial pollinations, crossing tall with tall, short with short, and tall with short plants. This last combination, plants with one trait variant crossed to plants with the alternate, produces hybrids, which are offspring that inherit a different gene variant (allele) from each parent. He also self-crossed plants.

Mendel noted that short plants crossed to other short plants were "true-breeding," always producing the same phenotype, in this case short plants. The crosses of tall plants to each other were more confusing. Some tall plants were true-breeding, but others crossed with each other yielded short plants in about one-quarter of the next generation. In some plants, tallness appeared to mask shortness. One trait that masks another is dominant; the masked trait is recessive.

Mendel conducted up to 70 hybrid self-crosses for each of the seven traits. This experiment is called a monohybrid cross because it follows one trait and the self-crossed plants are hybrids. The progeny were in the ratio of one-quarter short to three-quarters tall plants (**figure 4.2**). In further crosses, two-thirds of the tall plants in the F_2 generation were non-true-breeding, and the remaining third were true-breeding. (Consider only the tall plants at the bottom of the figure.)

These experiments confirmed that hybrids hide one expression of a trait—short, in this case—which reappears when hybrids are self-crossed. Mendel tried to explain how this happened: Gametes distribute "elementen" because these cells physically link generations. Paired sets of elementen separate as gametes form. When gametes join at fertilization, the elementen combine anew. Mendel reasoned that each elementen was packaged in a separate gamete. If opposite-sex gametes combine at random, he could mathematically explain the different ratios of traits produced from his pea plant crosses. Mendel's idea that elementen separate in the gametes would later be called the law of **segregation**.

A GLIMPSE OF HISTORY

The son of a farmer and grandson of a gardener, Mendel learned early how to tend fruit trees. At age 10 he left home to attend a special school for bright students, supporting himself by tutoring. He eventually became a friar at a monastery where the friars were teachers and did research in natural science. Here, Mendel learned how to artificially pollinate crop plants to control their breeding. As a university student, Mendel became interested in applying statistics to plant breeding. He thought about a question that had confounded other plant breeders: *Why did certain traits disappear in one generation, yet reappear in the next?*

Gregor Mendel
Pixtal/AGE Fotostock

Figure 4.2 A monohybrid cross. When Mendel crossed true-breeding tall plants with short plants, the next generation plants were all tall. When he self-crossed the F_1 plants, one-quarter of the plants in the next generation, the F_2, were short, and three-quarters were tall. Of the tall plants in the F_2, one-third were true-breeding, and the other two-thirds were not true-breeding. He could tell this by crossing the tall plants to short plants, to see which bred true. The monohybrid cross is the F_1 generation.

When Mendel's ratios were seen in several species in the early 1900s, just when chromosomes were being discovered, it became apparent that elementen and chromosomes had much in common. Both paired elementen and pairs of chromosomes separate at each generation and are transmitted—one from each parent—to offspring. Both are inherited in random combinations. Chromosomes provided a physical mechanism for Mendel's hypotheses. In 1909, English embryologist William Bateson renamed Mendel's elementen *genes* (Greek for "give birth to"). In the 1940s, scientists began investigating the gene's chemical basis, discussed in chapter 9.

In the twentieth century, researchers discovered the molecular basis of certain traits that Mendel studied. "Short" and "tall" plants result from the expression of a gene that enables a plant to produce the hormone gibberellin, which elongates the stem. One tiny change to the DNA, and a short plant results. Likewise, "round" and "wrinkled" peas arise from the *R* gene, whose encoded protein connects sugars into branching polysaccharide molecules. Seeds with a mutant *R* gene cannot attach the sugars. As a result, water exits the cells and the peas wrinkle.

Terms and Tools to Follow Segregating Genes

The law of segregation explains the actions of chromosomes and the genes that they carry during meiosis. Because a gene is a long sequence of DNA, it can vary in many ways. An individual with two identical alleles for a gene is **homozygous** for that gene. An individual with two different alleles is **heterozygous**—what Mendel called "non-true-breeding" or "hybrid."

For a gene with two alleles, the dominant one is shown as a capital letter and the recessive with the corresponding small letter. If both alleles are recessive, the individual is homozygous recessive, shown with two small letters. An individual with two dominant alleles is homozygous dominant, and is indicated by two capital letters. One dominant and one recessive allele, such as *Tt* for non-true-breeding tall pea plants, indicates a heterozygote. Most genes have more than two alleles and more than two variations of the associated trait. An individual with two different recessive alleles for the same gene is termed a **compound heterozygote**. Some people who have inherited diseases are compound heterozygotes.

An organism's appearance does not always reveal its alleles. Both a *TT* and a *Tt* pea plant are tall, but *TT* is a homozygote and *Tt* a heterozygote. The genotype describes the organism's alleles, and the phenotype describes the outward expression of an allele combination. A **wild type** phenotype is the most common expression of a particular allele combination in a population. The wild type allele may be recessive or dominant. A **mutant** phenotype is a variant of a gene's expression that arises when the gene undergoes a change, or mutation.

Mendel was observing the events of meiosis. When a gamete is produced, the two copies of a gene separate with the homologs that carry them. In a plant of genotype *Tt*, for example, gametes carrying either *T* or *t* form in equal numbers during anaphase I and the ratio persists when sister chromatids separate in anaphase II. Gametes combine at random. A *t*-bearing oocyte is neither more nor less attractive to a sperm than is a *T*-bearing oocyte. These two factors—equal allele distribution into gametes and random combinations of gametes—underlie Mendel's law of segregation (**figure 4.3**).

When Mendel crossed short plants (*tt*) with true-breeding tall plants (*TT*), the seeds grew into F_1 plants that were all tall (genotype *Tt*). Next, he self-crossed the F_1 plants. The progeny were *TT, tt,* and *Tt*. A *TT* individual resulted when a *T* sperm fertilized a *T* oocyte; a *tt* plant resulted when a *t* oocyte met a *t* sperm; and a *Tt* individual resulted when either a *t* sperm fertilized a *T* oocyte, or a *T* sperm fertilized a *t* oocyte.

Because two of the four possible gamete combinations produce a heterozygote, and each of the others produces a homozygote, the genotypic ratio expected of a monohybrid

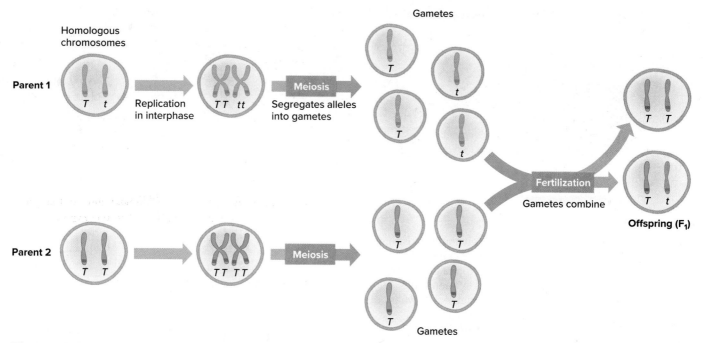

Figure 4.3 **Mendel's first law—gene segregation.** During meiosis, homologous pairs of chromosomes and their genes separate and are packaged into separate gametes. At fertilization, gametes combine at random. Pink and blue denote different parental origins of the chromosomes. This cross yields offspring of genotypes *TT* and *Tt*.

Table 4.1	Mendel's Data Demonstrating Segregation			
Experiment	**Total**	**Dominant**	**Recessive**	**F$_2$ Phenotypic Ratios**
1. Seed form	7,324	5,474	1,850	2.96:1
2. Seed color	8,023	6,022	2,001	3.01:1
3. Seed coat color	929	705	224	3.15:1
4. Pod form	1,181	882	299	2.95:1
5. Pod color	580	428	152	2.82:1
6. Flower position	858	651	207	3.14:1
7. Stem length	1,064	787	277	2.84:1
				Average = 2.98:1

cross is 1*TT*:2*Tt*:1*tt*. The corresponding phenotypic ratio is three tall plants to one short plant, a 3:1 ratio. Mendel saw these results for all seven traits that he studied (**table 4.1**). A diagram called a **Punnett square** shows these ratios (**figure 4.4**). A Punnett square represents how genes in gametes join if they are on different chromosomes. Experimental crosses yielded numbers of offspring that approximate these ratios.

Mendel set up additional crosses to distinguish the two genotypes resulting in tall progeny—*TT* from *Tt* (**figure 4.5**). He bred tall plants of unknown genotype with short (*tt*) plants.

If a tall plant crossed with a *tt* plant produced both tall and short progeny, it was genotype *Tt*; if it produced only tall plants, it must be *TT*.

Crossing an individual of unknown genotype with a homozygous recessive individual is called a **test cross**. The logic is that the homozygous recessive is the only genotype that can be identified by its phenotype—that is, a short plant is always *tt*. The homozygous recessive is a "known" that can reveal the unknown genotype of another individual to which it is crossed.

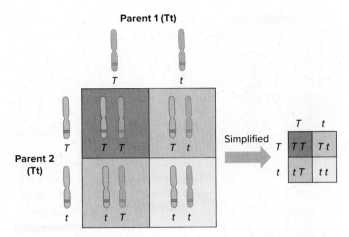

Figure 4.4 **A Punnett square.** A Punnett square illustrates how alleles combine in offspring. The different types of gametes of one parent are listed along the top of the square, with those of the other parent listed on the left-hand side. Each compartment displays the genotype that results when gametes that correspond to that compartment join.

Figure 4.5 **Test cross.** Breeding a tall pea plant with homozygous recessive short plants reveals whether the tall plant is true-breeding (*TT*) or non-true-breeding (*Tt*). Punnett squares usually indicate only the alleles.

Key Concepts Questions 4.1

a. List four ways that monogenic diseases differ from other types of diseases.

b. How did Mendel deduce that units of inheritance for height segregate, then combine at random with those from the opposite gamete at fertilization?

c. Distinguish between a homozygote and a heterozygote; dominant and recessive.

*d. What is the genotypic ratio of a monohybrid cross?

*e. What is the phenotypic ratio of a monohybrid cross?

f. How do Punnett squares display expected genotypic and phenotypic ratios among progeny?

*g. Define *test cross*.

4.2 Single-Gene Inheritance Is Rare

Single-gene inheritance may be more difficult to interpret than a pea plant having green or yellow peas because many phenotypes associated with single genes are influenced by other genes as well as by environmental factors. That is, the single gene controls trait transmission, but other genes and the environment affect the degree of the trait or severity of the illness. Eye color provides a good example of how our view of single-gene traits has evolved with increasing knowledge of our genomes.

Eye Color

Most people have brown eyes; blue and green eyes are almost exclusively in people of European ancestry. The color of the iris is due to melanin pigments, which come in two forms—the dark brown/black eumelanin, and the red-yellow pheomelanin. In the eye, cells called melanocytes produce melanin, which is stored in structures called melanosomes in the outermost layer of the iris. People differ in the amount of melanin and number of melanosomes, but have about the same number of melanocytes in their eyes.

Nuances of eye color—light versus dark brown, clear blue versus greenish or hazel—arise from the distinctive peaks and valleys at the back of the iris. Thicker regions darken appearance of the pigments, rendering brown eyes nearly black in some parts and blue eyes closer to purple. The bluest eyes have thin irises with very little pigment. The effect of the iris surface on color is a little like the visual effect of a rough-textured canvas on paint.

A single gene on chromosome 15, *OCA2*, confers eye color by controlling melanin synthesis. If this gene is missing, albinism results, causing pale skin and red eyes (**figure 4.6**; see also figure 4.15). A recessive allele of this gene confers blue color and a dominant allele confers brown. But inheritance of eye color is more complicated than this. Near the *OCA2* gene on chromosome 15 is a second gene, *HERC2*, that controls expression of the *OCA2* gene. A recessive allele of *HERC2* abolishes the control over *OCA2*, and blue eyes result. A person must inherit two copies of the recessive allele in *HERC2* to have blue eyes.

Figure 4.6 Eye color. At least two genes determine eye color in humans. Inheriting two particular recessive alleles in *HERC2* abolishes the effect of *OCA2*, causing blue eyes. The two genes are near each other on chromosome 15.

Modes of Inheritance

Knowing the mode of inheritance makes it possible to calculate the probability that a couple will have a child who inherits a particular condition. The way that a trait is passed depends on whether the gene that determines it is on an autosome or on a sex chromosome, and whether the allele is recessive or dominant. **Table 4.2** compares two of the four modes of inheritance, which apply to transmission of genes on the autosomes (chromosomes 1 through 22). Table 6.2 covers traits transmitted through the sex chromosomes (X and Y).

In autosomal dominant inheritance, a trait can appear in either sex because an autosome carries the gene. If a child has the trait, at least one parent also has it. Autosomal dominant traits do not skip generations because if no offspring inherit the mutation in one generation, transmission stops. Huntington disease is an autosomal dominant condition. The Punnett square in **figure 4.7** depicts inheritance of an autosomal dominant trait or condition. Many autosomal dominant diseases do not cause symptoms until adulthood.

In autosomal recessive inheritance, a trait can appear in either sex. Affected individuals have a homozygous recessive genotype, whereas in heterozygotes (carriers) the wild type allele masks expression of the mutant allele. A person with cystic fibrosis, for example, inherits a mutant allele from each carrier parent.

Mendel's first law can be used to calculate the probability that an individual will have either of two phenotypes. The probabilities of each possible genotype are added. For example, the chance that a child whose parents are both carriers of cystic fibrosis will *not* have the condition is the sum of the probability that she has inherited two normal alleles (1/4) plus the chance that she herself is a heterozygote (1/2), or 3/4 in total. Note that this also equals 1 minus the probability that she is homozygous recessive and has the condition (1/4).

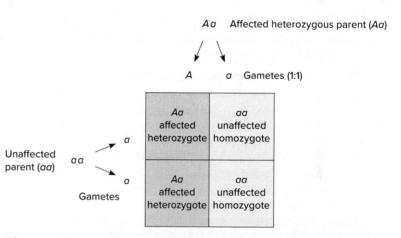

Figure 4.7 Autosomal dominant inheritance. When one parent has an autosomal dominant condition and the other does not, each offspring has a 50 percent probability of inheriting the mutant allele and the condition. In the family from the chapter opener, Karl was the "*Aa*" parent and Jane the "*aa*" parent. Each of their daughters faced the 1 in 2 probability of inheriting Huntington disease.

Table 4.2	Comparison of Autosomal Dominant and Autosomal Recessive Inheritance
Autosomal Dominant	**Autosomal Recessive**
Males and females affected, with equal frequency	Males and females affected, with equal frequency
Successive generations affected until no one inherits the mutation	Can skip generations
Affected individual has an affected parent, unless he or she has a *de novo* mutation	Affected individual has parents who are affected or are carriers (heterozygotes)

The ratios that Mendel's first law predicts for autosomal recessive inheritance apply to each offspring anew. If a couple has a child with an autosomal recessive illness, each of their future children faces the same 25 percent risk of inheriting the condition.

Most autosomal recessive conditions appear unexpectedly in families, because they are transmitted silently, through heterozygotes (carriers). **Clinical Connection 4.1** discusses one of them, cystic fibrosis. However, a situation in which an autosomal recessive condition is more likely to recur is when blood relatives have children together. The higher risk of having a child with an autosomal recessive condition is because the related parents may carry the same alleles inherited from an ancestor that they have in common, such as a great-grandparent (see figure 1.9). Marriage between relatives introduces **consanguinity**, which means "shared blood"—a figurative description, because genes are not passed in blood. Alleles inherited from shared ancestors are said to be "identical by descent."

Consanguinity is part of many cultures. For example, marriage between first cousins occurs in about a third of the Pakistani community in England, a population that researchers have been following. In this group, consanguinity doubles the birth defect rate, but it is still low—less than 4 percent of births. Consanguinity may happen unknowingly in close-knit communities where people do not realize they are related, and among families using the same sperm donor. People taking consumer genetics ancestry tests may discover half-siblings that they never knew they had.

Logic explains why consanguinity raises risk of inheriting autosomal recessive diseases. An unrelated man and woman have eight different grandparents, but first cousins have only six, because they share one pair through their parents, who are siblings (see figure 4.14c). That is, the probability of two relatives inheriting the same disease-causing recessive allele is greater than that of two unrelated people having the same allele by chance. However, studies show that cousins tend to inherit different parts of the shared ancestor's genome, which explains why populations like the Pakistani community in England are healthier than might be expected given the high frequency of consanguinity.

The nature of the phenotype is important when evaluating the transmission of single-gene traits. Some diseases are too severe for people to live long enough or feel well enough to have children. For example, each adult sibling of a person who is a known carrier of Tay-Sachs disease has a two-thirds chance of also being a carrier because only two genotypes are possible for an adult—homozygous for the wild type allele or a carrier who inherits the mutant allele from either parent. A homozygous recessive individual for this brain disease would not have survived childhood.

Geneticists who study human traits and illnesses can hardly set up crosses as Mendel did to demonstrate modes of inheritance, but they can pool information from families whose members have the same trait or illness to deduce the mode of inheritance. Consider a simplified example of 50 couples in whom both partners are carriers of sickle cell disease. If 100 children are born, about 25 of them would be expected to have sickle cell disease. Of the remaining 75, theoretically 50 would be carriers like their parents, and the remaining 25 would have two wild type alleles.

Solving a Problem in Following a Single Gene

Using Mendel's laws to predict phenotypes and genotypes requires a careful reading of the problem to organize relevant information. Common sense is useful, too. The following general steps can help to solve a problem based on the inheritance of a single-gene trait:

1. List all possible genotypes and phenotypes for the trait.
2. Determine the genotypes of the individuals in the first (P_1) generation. Use information about those individuals' parents.
3. After deducing genotypes, derive the possible alleles in gametes each individual produces.
4. Unite these gametes in all combinations to reveal all possible genotypes. Calculate ratios for the F_1 generation.
5. To extend predictions to the F_2 generation, use the genotypes of the specified F_1 individuals and repeat steps 3 and 4.

As an example, consider curly hair. If C is the dominant allele, conferring curliness, and c is the recessive allele, then CC and Cc genotypes confer curly hair. A person with a cc genotype has straight hair.

Wendy has beautiful curls, and her husband Rick has straight hair. Wendy's father is bald, but once had curly hair, and her mother has stick-straight hair. What is the probability that Wendy and Rick's child will have straight hair? Steps 1 through 5 solve the problem:

1. State possible genotypes: CC, Cc = curly; cc = straight.
2. Determine genotypes: Rick must be cc, because his hair is straight. Wendy must be Cc, because her mother has straight hair and therefore gave her a c allele.
3. Determine gametes: Rick's sperm carry only c. Half of Wendy's oocytes carry C, and half carry c.
4. Unite the gametes:

5. Conclusion: Each child of Wendy and Rick's has a 50 percent chance of having curly hair (Cc) and a 50 percent chance of having straight hair (cc).

On the Meaning of Dominance and Recessiveness

Knowing whether an allele is dominant or recessive is critical in determining the risk of inheriting a particular condition (phenotype). Dominance and recessiveness arise from the genotype, and reflect the characteristics or abundance of a protein.

Clinical Connection 4.1
"65 Roses": Progress in Treating Cystic Fibrosis

Young children who cannot pronounce the name of their disease call it "65 roses." Cystic fibrosis (CF), although still a serious illness, is a genetic disease success story thanks to a new class of drugs that correct the underlying problem in protein folding that causes many cases.

Recall from figures 1.5 and 1.6 that CF affects ion channels that control chloride movement out of cells in certain organs. In the lungs, thick, sticky mucus accumulates and creates an environment hospitable to certain bacteria that are uncommon in healthy lungs. A clogged pancreas prevents digestive secretions from reaching the intestines, impairing nutrient absorption. A child with CF has trouble breathing and maintaining weight.

Life with severe CF is challenging. In summertime, a child must avoid water from hoses, which harbor lung-loving *Pseudomonas* bacteria. A bacterium called *Burkholderia cepacia* easily spreads in summer camps. Cookouts spew lung-irritating particulates. Too much chlorine in pools irritates lungs whereas too little invites bacterial infection. New infections arise, too. In the past few years, multidrug-resistant *Mycobacterium abscessus*, related to the pathogen that causes tuberculosis, has affected up to 10 percent of CF patients in the United States and Europe. Other bacteria that can't grow in a laboratory infect CF patients. Sequenced bacterial DNA in lung fluid from children with CF reveals more than 60 species of bacteria.

CF is inherited from two carrier parents, and affects more than 30,000 people in the United States and more than 70,000 worldwide, according to the Cystic Fibrosis Foundation. Many other people may have cases so mild that they are not diagnosed with CF but instead with frequent respiratory infections. More than 1,700 variants are known in the cystic fibrosis transmembrane regulator (*CFTR*) gene, which encodes the chloride channel protein. About half of the variants are known to affect health. Today pregnant women are tested for the more common mutations to see if they are carriers, and if they are, their partners are tested. If both parents-to-be are carriers, then the fetus has a 25 percent chance of having inherited CF and may be tested as well. In most states all newborns receive a genetic test for CF. Before genetic testing became available, diagnosis typically took months and was based on observing "failure to thrive," salty sweat, and foul-smelling stools.

When CF was recognized in 1938, life expectancy was only 5 years; today median survival is 47 years, according to the UK CF Registry, with many patients living longer, thanks to several types of treatments. Inhaled antibiotics control respiratory infections and daily "bronchial drainage" exercises shake mucus from the lungs (**figure 4A**). Other forms of exercise help too, such as dance. A vibrating vest worn for half-hour periods two to

Figure 4A An inhaler delivers drugs that open the airways, prevent infections, and break up mucus. Newer treatments are pills that are taken daily. *Getty Images*

four times a day loosens mucus. Digestive enzymes mixed into soft foods enhance nutrient absorption, although some patients require feeding tubes.

New drugs to treat CF work in various ways: correcting misfolded CFTR protein, restoring liquid on airway surfaces, breaking up mucus, improving nutrition, and fighting inflammation and infection. A drug called ivacaftor (Kalydeco) that became available in 2012 at first helped only the 5 percent of patients with a mutation called G551D, in whom the ion channel proteins reach the plasma membrane but once there, remain closed, like locked gates. The drug binds the proteins in a way that opens the channels. Kalydeco given with a second drug helps other people with CF whose mutations cause dismantling of CFTR in the endoplasmic reticulum and people in whom the protein reaches the plasma membrane and folds into an ion channel but doesn't stay there long enough to work. A third new drug enables protein synthesis to ignore mutations that shorten the protein. With three drugs now available, 90 percent of patients can be treated. All of these drugs are pills that are taken daily.

Questions for Discussion

1. Which parts of the cell are affected in CF?
2. If a child with CF has two parents who do not have the disease, what is the risk that a future sibling will inherit CF?
3. How does the drug Kalydeco work?

Mendel based his definitions of dominance and recessiveness on what he could see—one allele masking the effect of the second one. Today we can often add a cellular or molecular explanation.

The basis of an inborn error of metabolism is easy to picture. These diseases are typically recessive because the half normal amount of the enzyme that a carrier produces is usually sufficient to maintain health. The one normal allele, therefore, compensates for the mutant one, to which it is dominant.

A recessive trait is said to arise from a "loss-of-function" because the recessive allele usually prevents the production or activity of the normal protein. In contrast, some dominantly inherited diseases are said to be due to a "gain-of-function," because they result from the action of an abnormal protein that interferes with the function of the normal protein. Huntington disease results from a gain-of-function in which the dominant mutant allele encodes an abnormally long protein that prevents the normal protein from functioning in certain brain cells. Huntington disease is a gain-of-function because individuals who are missing one copy of the gene do not have the illness. That is, the protein encoded by the mutant HD allele must be abnormal, not absent, to cause the disease. The gain-of-function nature of HD is why people with one mutant allele have the same phenotype as the rare individuals with two mutant alleles (**figure 4.8**).

Recessive diseases tend to be more severe, and produce symptoms earlier, than dominant diseases. Disease-causing recessive alleles remain in populations because healthy heterozygotes pass them to future generations. In contrast, if a dominant mutation arises that harms health early in life, people who have the allele are either too ill or do not survive long enough to reproduce. The allele eventually becomes rare in the population unless it arises anew by mutation in the gametes of a person who does not have the disease. Dominant diseases whose symptoms do not appear until adulthood, or that do not severely disrupt health, remain in a population because they do not prevent a person from having children and passing on the mutation.

Under certain circumstances, for some genes, a heterozygous individual (a carrier) can develop symptoms. This is the case for sickle cell disease. Carriers can develop a life-threatening breakdown of muscle if exposed to the combination of environmental heat, intense physical activity, and dehydration. Several college athletes died from these symptoms, prompting sports authorities to begin testing athletes for sickle cell disease carrier status. Testing is now routine.

> **Key Concepts Questions 4.2**
>
> *a. Explain how humans inherit eye color.
>
> *b. Define *mode of inheritance*.
>
> *c. Name the two factors that determine the mode of inheritance.
>
> d. Distinguish the inheritance patterns of autosomal dominant traits versus autosomal recessive traits.
>
> e. What is the effect of people having children with their relatives on inherited trait transmission?
>
> f. How are Mendel's first law and logic used to solve genetics problems?
>
> g. Explain how recessive traits or illnesses can result from a loss-of-function, whereas dominant traits can result from a gain-of-function.

4.3 Following the Inheritance of More Than One Gene

In a second set of experiments, Mendel examined the inheritance of two traits at a time. Today, his classic experiments on the inheritance of more than one trait are more relevant than ever, thanks to technologies that consider many genes simultaneously.

Mendel's Second Law

The law of independent assortment states that for two genes on different chromosomes, the inheritance of one gene does not influence the chance of inheriting the other gene. The two genes are said to "independently assort" because they are packaged into gametes at random (**figure 4.9**). Two genes that are far apart on the same chromosome also appear to independently assort, because so many crossovers take place between them that it is as if they are part of separate chromosomes (see figure 3.5).

Mendel looked at seed shape, which was either round or wrinkled (determined by the *R* gene), and seed color, which was either yellow or green (determined by the *Y* gene). When he crossed true-breeding plants that had round, yellow seeds

Figure 4.8 **A mutation can confer a loss-of-function or a gain-of-function.** A loss-of-function mutation is typically recessive and results in a deficit or absence of the gene's protein product. A gain-of-function mutation is typically dominant and alters the encoded protein, introducing a new activity.

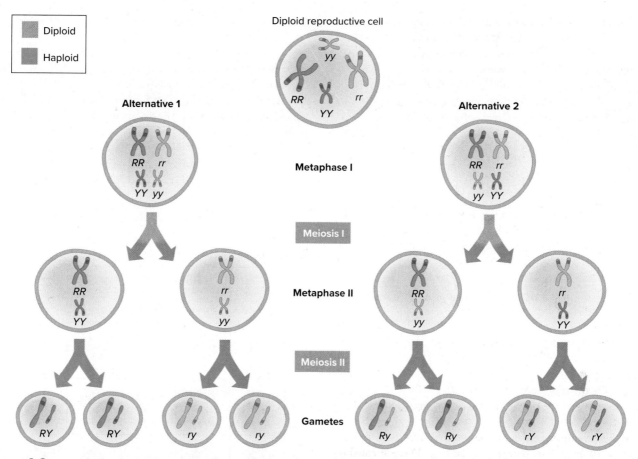

Figure 4.9 Mendel's second law—independent assortment. The independent assortment of genes carried on different chromosomes results from the random alignment of chromosome pairs during metaphase of meiosis I. An individual of genotype *RrYy*, for example, manufactures four types of gametes, containing the dominant alleles of both genes (*RY*), the recessive alleles of both genes (*ry*), and a dominant allele of one with a recessive allele of the other (*Ry* or *rY*). The allele combination depends upon which chromosomes are packaged together in a gamete—and this happens at random.

to true-breeding plants that had wrinkled, green seeds, all the progeny had round, yellow seeds (**figure 4.10**). These offspring were double heterozygotes, or dihybrids, of genotype *RrYy*. From their appearance, Mendel deduced that round is dominant to wrinkled, and yellow to green.

Next, he self-crossed the dihybrid plants in a **dihybrid cross**, so named because two genes and traits are followed. Mendel found four types of seeds in the next, third generation: 315 plants with round, yellow seeds; 108 plants with round, green seeds; 101 plants with wrinkled, yellow seeds; and 32 plants with wrinkled, green seeds. These classes appeared in a ratio of 9:3:3:1.

Mendel then crossed each plant from the third generation to plants with wrinkled, green seeds (genotype *rryy*). These test crosses established whether each plant in the third generation was true-breeding for both genes (genotypes *RRYY* or *rryy*), true-breeding for one gene but heterozygous for the other (genotypes *RRYy*, *RrYY*, *rrYy*, or *Rryy*), or heterozygous for both genes (genotype *RrYy*). Mendel could explain the 9:3:3:1 proportion of progeny classes only if one gene does not influence transmission of the other. Each parent would produce equal numbers of four different types of gametes: *RY*, *Ry*, *rY*, and *ry*. Each of these combinations has one gene for each trait. A Punnett square for this cross shows that the four types of seeds:

1. round, yellow (*RRYY*, *RrYY*, *RRYy*, and *RrYy*),
2. round, green (*RRyy* and *Rryy*),
3. wrinkled, yellow (*rrYY* and *rrYy*), and
4. wrinkled, green (*rryy*)

are present in the ratio 9:3:3:1, just as Mendel found.

Solving a Problem in Following Multiple Genes

A Punnett square for three genes has 64 boxes; for four genes, 256 boxes. An easier way to predict genotypes and phenotypes in multigene crosses is to use the mathematical laws of probability that are the basis of Punnett squares. Probability predicts the likelihood of an event.

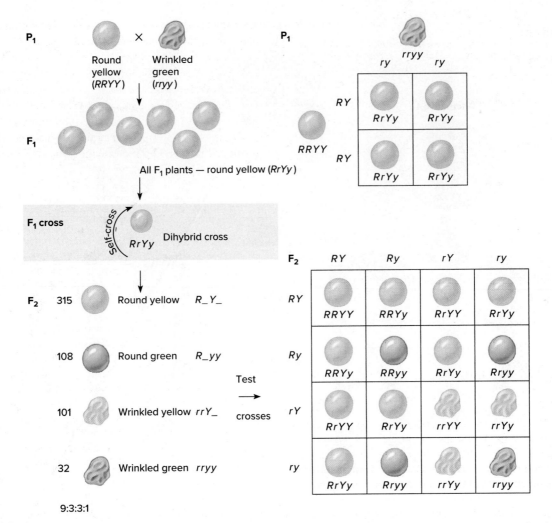

Figure 4.10 Plotting a dihybrid cross. A Punnett square can represent the random combinations of gametes produced by dihybrid individuals. An underline in a genotype (in the F_2 generation) indicates that either a dominant or a recessive allele is possible. The numbers in the F_2 generation are Mendel's experimental data. Test crosses with *rryy* plants revealed the genotypes of the F_2 generation, depicted in the 16-box Punnett square.

An application of probability theory called the product rule can predict the chance that parents with known genotypes can produce offspring of a particular genotype. The product rule states that the chance that two independent events will both occur equals the product of the chances that either event will occur alone. Consider the probability of obtaining a plant with wrinkled, green peas (genotype *rryy*) from dihybrid (*RrYy*) parents. Do the reasoning for one gene at a time, then multiply the results (**figure 4.11**).

A Punnett square depicting a cross of two *Rr* plants indicates that the probability of producing *rr* progeny is 25 percent, or 1/4. Similarly, the chance of two *Yy* plants producing a *yy* plant is 1/4. Therefore, the chance of dihybrid parents (*RrYy*) producing homozygous recessive (*rryy*) offspring is 1/4 multiplied by 1/4, or 1/16. Now consult the 16-box Punnett square for Mendel's dihybrid cross again (see figure 4.10). Only one of the 16 boxes is *rryy*, just as the product rule predicts. **Figure 4.12** shows how these tools can be used to predict offspring genotypes and phenotypes for three human traits simultaneously.

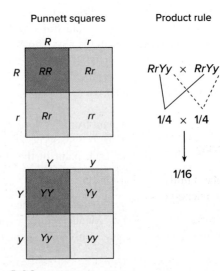

Figure 4.11 The product rule.

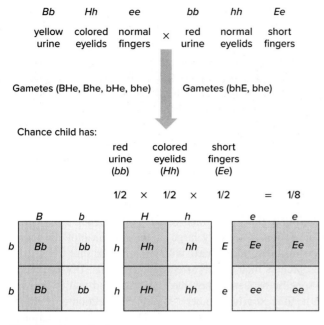

Figure 4.12 Using probability to track three traits in people. A man has normal urine, colored eyelids, and normal fingers. His partner has red urine after she eats beets, and has normal eyelids and short fingers. The chance that their child will have red urine after eating beets, colored eyelids, and short fingers is 1/8.

Key Concepts Questions 4.3

*a. State Mendel's second law.

b. Explain how meiotic events underlie independent assortment for two or more genes.

*c. Explain how Mendel demonstrated that genes on different chromosomes independently assort.

d. How are Punnett squares and probability used to follow inheritance of more than one trait?

4.4 Pedigree Analysis

For genetics researchers and genetic counselors, the bigger the family the better—the more children in a generation, the easier it is to deduce a mode of inheritance. Charts called pedigrees are used to display family relationships and depict which relatives have specific phenotypes and, sometimes, genotypes. A human pedigree serves the same purpose as one for purebred dogs or cats or thoroughbred horses—it represents relationships and sometimes traits. A pedigree in genetics differs from a family tree in genealogy, and from a genogram in social work, in that it indicates inherited diseases or traits as well as relationships and ancestry. Pedigrees may also include molecular data, test results, and information on variants of multiple genes.

In a pedigree, lines link shapes. Vertical lines represent generations, and horizontal lines that connect two shapes at their centers depict partners. Shapes connected by vertical lines that are joined horizontally represent siblings. Squares indicate males; circles, females; and diamonds, individuals of unspecified sex. Roman numerals designate generations. Arabic numerals or names indicate individuals. **Figure 4.13** shows

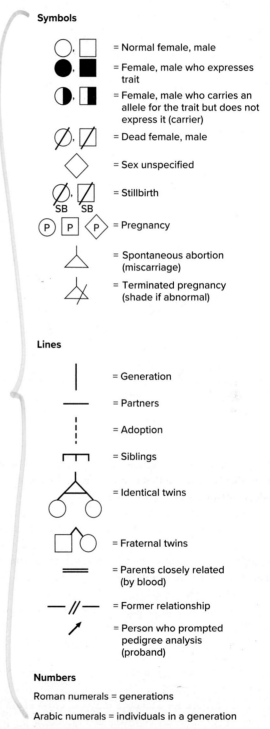

Figure 4.13 Pedigree components. Symbols representing individuals are connected to form pedigree charts, which display the inheritance patterns of traits.

these and other commonly used pedigree symbols. Colored or shaded shapes indicate individuals who express a trait, and half-filled shapes are known carriers.

Pedigrees Then and Now

The earliest pedigrees were strictly genealogical, not indicating traits. **Figure 4.14**a shows such a pedigree for a highly inbred branch of the ancient Egyptian royal family tree. The term *pedigree* arose in the fifteenth century, from the French *pie de grue*, which means "crane's foot." Pedigrees at that time, typically depicting large families, showed parents linked by curved lines to their many offspring. The overall diagram often resembled a bird's foot.

One of the first pedigrees to trace an inherited illness was an extensive family tree of several European royal families, indicating which members had the clotting disease hemophilia B, also called factor IX deficiency (see figure 6.7). The mutant gene probably originated in Queen Victoria of England in the nineteenth century, or in a female ancestor of the queen. In 1845, a genealogist named Pliny Earle constructed a pedigree of a family with colorblindness using musical notation—half notes for unaffected females, quarter notes for colorblind females, and filled-in and squared-off notes to represent the many colorblind males. In the early twentieth century, eugenicists tried to use pedigrees to show that traits such as criminality, feeblemindedness, and promiscuity were the consequence of faulty genes (see Section 16.7).

Today, pedigrees are important both for helping families identify the risk of transmitting an inherited illness and as starting points for identifying and describing, or annotating, a gene from the human genome sequence. People who have kept meticulous family records, such as the Mormons and the Amish, are invaluable in helping researchers follow the inheritance of particular genes. (See Clinical Connection 16.1.) Very large pedigrees can provide information on many individuals with a particular rare disease. The researchers then search affected individuals' DNA to identify a specific sequence they have all inherited that is not found in healthy family members. Within this section of a chromosome lies the causative mutation. Discovery of a mutation that causes an early-onset form of Alzheimer disease, for example,

Figure 4.14 Unusual pedigrees. (a) A partial pedigree of Egypt's Ptolemy dynasty shows only genealogy, not traits. It appears almost ladderlike because of the extensive consanguinity. From 323 B.C. to Cleopatra's death in 30 B.C., the family experienced one pairing between cousins related through half-brothers (generation III), four brother-sister pairings (generations IV, VI, VIII, and X), and an uncle-niece relationship (generations VI and VII). Cleopatra married her brother, Ptolemy XIII, when he was 10 years old! These marriage patterns were an attempt to preserve the royal blood. **(b)** A family with polydactyly (extra fingers and toes) extends laterally, with many children, because the phenotype doesn't affect reproduction. **(c)** This pedigree shows marriage of first cousins. They share one set of grandparents and, therefore, risk passing on the same recessive alleles to offspring.

took researchers to a remote village in Colombia, where the original mutation present today in a 1,000-plus-member family came from a Spanish settler who had arrived in the seventeenth century. This origin story is an example of the founder effect, discussed in section 16.4.

Pedigrees Display Mendel's Laws

Visual learners can easily "see" a mode of inheritance in a pedigree. Consider a pedigree for an autosomal recessive trait, albinism. Homozygous recessive individuals in the third (F_2) generation lack an enzyme necessary to manufacture the pigment melanin and, as a result, display pale hair and skin color (**figure 4.15**). Their parents are inferred to be heterozygotes (carriers). One partner from each pair of grandparents must also be a carrier. For some diseases, carrier status is detectable with a blood or urine test that shows half the normal amount of the gene product.

Recall that an autosomal dominant trait does not skip generations and can affect both sexes. A typical pedigree for an autosomal dominant trait has some squares and circles filled in to indicate affected individuals in each generation. Figure 4.14b is a pedigree for an autosomal dominant trait, extra fingers and toes (polydactyly), which is shown in figure 1.8a.

A pedigree may be inconclusive, which means that either autosomal recessive or autosomal dominant inheritance can explain the pattern of filled-in symbols. **Figure 4.16** shows one such pedigree, for a type of hair loss called alopecia areata. According to the pedigree, this trait can be passed in an autosomal dominant mode because it affects both males and females and is present in every generation. However, the pedigree can also depict autosomal recessive inheritance if the individuals represented by unfilled symbols are carriers. Inconclusive pedigrees tend to arise when families are small and the trait is not severe enough to impair fertility.

Pedigrees may be difficult to construct or interpret for reasons that have nothing to do with DNA. People may hesitate to supply information because the symptoms embarrass them. Families with adoption, children born out of wedlock, serial

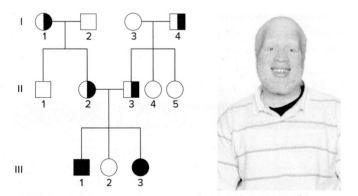

Figure 4.15 **Albinism is autosomal recessive.** Albinism affects males and females and can skip generations, as it does in generations I and II in this pedigree. Homozygous recessive individuals lack an enzyme needed to produce melanin, which colors the eyes, skin, and hair. The carrier could be either grandparent. *Photo: Paul Burns/Digital Vision/Getty Images*

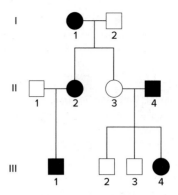

Figure 4.16 **An inconclusive pedigree.** This pedigree could account for an autosomal dominant trait or an autosomal recessive trait that does not prevent affected individuals from having children. (Unfilled symbols could represent carriers.)

relationships, blended families, and use of assisted reproductive technologies (see chapter 22) may not fit easily into the rules of pedigree construction. Many people cannot trace their families back far enough to reveal a mode of inheritance.

Solving a Problem in Conditional Probability

Pedigrees and Punnett squares can be used to trace a conditional probability, which is when an offspring's genotype depends on the parents' genotypes, which may not be obvious from their phenotypes. For example, if a person has an autosomal recessive trait or condition, his or her parents are inferred to be carriers (unless a new mutation has arisen in the affected individual).

The family represented in **figure 4.17** illustrates conditional probability. Deshawn has sickle cell disease. His unaffected parents, Kizzy and Ike, must each be heterozygotes (carriers). Deshawn's sister, Taneesha, is also healthy, and she is expecting her first child. Taneesha's husband, Antoine, has no family history of sickle cell disease. What is the probability that Taneesha's child will inherit her mutant allele and be a carrier?

Taneesha's concern raises two questions. First, what is the probability that she is a carrier? Because she is the product of a monohybrid cross, and she is not homozygous recessive (sick), she has a 2 in 3 chance of being a carrier. If so, the chance that she will transmit the mutant allele is 1 in 2, because she has two copies of the gene, and only one allele goes into each gamete. To calculate the overall risk to her child, multiply the probability that she is a carrier (2/3) by the chance that if she is, she will transmit the mutant allele (1/2). The result is 1/3.

Exome and Genome Sequencing Clarify Pedigrees

An increasingly common clinical scenario occurs when a child has a syndrome (a collection of symptoms) not seen before, and both parents are unaffected. Two explanations are possible. Either both parents are carriers (heterozygotes) for recessive mutations in the same gene for a condition that has

Figure 4.17 Making predictions. Taneesha's brother Deshawn has sickle cell disease **(a)**. Taneesha wonders if her fetus has inherited the sickle cell allele. First, she must calculate the chance that she is a carrier. The Punnett square in **(b)** shows that this risk is 2 in 3. (She must be genotype *SS* or *Ss* but cannot be *ss* because she does not have the disease.) The risk that the fetus is a carrier, assuming that the father is not a carrier, is half Taneesha's risk of being a carrier, or 1 in 3 **(c)**.

not yet been identified, or a *de novo* (new) dominant mutation arose spontaneously in the sick child, in a sperm or an oocyte. The distinction is important to a family because a mutation inherited from carrier parents could be transmitted to siblings of the affected child, according to Mendel's first law. In contrast, a mutation originating in a single gamete would *not* affect a sibling, unless the parent was a mosaic and has other gametes with the mutation. However, a *de novo* dominant mutation *can* be passed on to the *next* generation from the affected child, according to Mendel's first law, with a probability of 1 in 2. Chapter 12 discusses *de novo* mutations further.

Exome and genome sequencing of parents-child "trios" can identify a mutant gene in the affected child and distinguish whether the condition is autosomal recessive and inherited, or autosomal dominant and new. Algorithms compare the DNA sequences of the affected child and unaffected parents, searching for variants of "candidate genes" whose function, or malfunction, might explain the specific symptoms. The challenge is a little like finding a needle in a haystack. Exome sequencing typically detects 70 to 175 *de novo* gene variants, but fewer than 3 and often none of them change the encoded protein in a way that can affect the phenotype. The numbers of identified gene variants are higher for sequencing the entire genome.

When Millie McWilliams, described in Clinical Connection 1.1, and her parents had their genomes sequenced, the analysis revealed her *de novo* mutation in the *ASXL3* gene on chromosome 18. She is missing two bases in the gene, which causes the rare Bainbridge-Ropers syndrome. Her parents are wild type for this gene (**figure 4.18**). The case of Millie and thousands of others with genetic syndromes, many of them children, illustrate the marriage of genome sequencing and the more traditional genetic tools of observing families and charting their traits—all based on Mendel's laws of inheritance. The **Bioethics** reading revisits Huntington disease, the subject of the chapter opener, in its more common form of an adult onset.

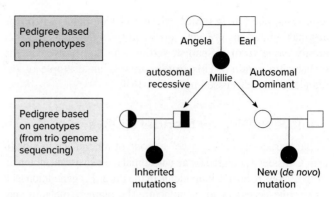

Figure 4.18 Parent and child trios that have their exomes or genomes sequenced can enable geneticists to distinguish Mendelian inheritance of a recessive mutation from carrier parents from a new, dominant mutation in the child. A pedigree based on phenotypes only reveals that two healthy parents have a sick child. The pedigree based on genotypes can distinguish inheritance of a mutation from carrier parents from a *de novo* mutation in the child.

Key Concepts Questions 4.4

*a. What information does a pedigree provide?

b. How do pedigrees and Punnett squares apply Mendel's laws to predict the recurrence risks of inherited diseases or traits?

c. Explain how exome and genome sequencing can be used to distinguish autosomal recessive inheritance in a "parents-child trio" from a new dominant mutation affecting the child.

Bioethics

The Right Not to Know Genetic Information: Huntington Disease

Peter and Martha were 28 and expecting their first child when Peter's mother, Bristol, began to show the first symptoms of Huntington disease (HD): confusion choosing items in a familiar grocery store, flashes of extreme irritability, and persistent small movements of her hands and lower legs.

Bristol had been adopted, with no known family medical history, but her primary care physician noted the unusual movements and referred her to a neurologist. By the time of that appointment, the swaying circles of arm motions and fidgeting feet were more frequent, Bristol told the neurologist, who ordered a battery of tests, to which Bristol agreed. One test detected the extended huntingtin gene that lies behind HD. Bristol's movements would increase in frequency and intensity as her cognition declined. The disease is ultimately fatal.

Each child of a person with HD faces a 50:50 chance of inheriting it. If so, she or he has a near 100 percent probability of eventually developing symptoms.

Peter refused to take the test to identify the DNA repeat, which might have found him to be "pre-manifest." He did not want to know, stating that knowing he would some day have the disease that he was watching in his mother would ruin his life. But then Martha insisted that they test the fetus. Peter objected. If the fetus had the mutation, then so did he.

A genetic counselor explained that it was typical to only test people for this particular disease if they were over age 18, when they were considered old enough to understand the long-term implications of having such knowledge. The genetic counselor also told Peter about several clinical trials that are testing drugs that block the RNA molecules that take the message to manufacture mutant huntingtin from the nucleus into the cytoplasm. Early results were showing a correction of the disease at the molecular level. He might sign up for one of those trials, or join another in which 20,000 people from families with HD are being followed to identify the very earliest changes indicating the disease.

Peter, encouraged by the news of the studies, felt even more strongly that he did not wish to know his status. If he was diagnosed in the future, treatment might be possible. Martha agreed not to test the fetus.

Questions for Discussion

1. What types of information and which professionals should advise a person on whether to take a predictive test for HD?
2. What issues arise in genetic counseling for an autosomal dominant condition that differ from the issues that arise for an autosomal recessive condition?
3. What are arguments for and against participating in a clinical trial to evaluate a treatment for an inherited disease that has no treatment?
4. What is the conditional probability that the child of Peter and Martha has inherited HD?

Study Guide

Summary

4.1 Following the Inheritance of One Gene

1. **Monogenic** (single-gene) diseases differ from other types of diseases in the ability to predict risk of occurrence in a relative, possibility of presymptomatic testing, increased prevalence in specific populations, and the possibility of correcting or compensating for the underlying mutation.
2. **Modes of inheritance** are patterns of the probability that people related in a certain way will inherit a trait or illness.
3. An **autosomal dominant** trait affects males and females and does not skip generations.
4. An **autosomal recessive** trait affects males or females and may skip generations.
5. Mendel's laws, based on pea plant crosses, derive from how chromosomes act during meiosis. The laws apply to all diploid organisms.
6. Mendel used statistics to investigate why some traits vanish in hybrids. The law of **segregation** states that alleles of a gene go into separate gametes during meiosis. Mendel demonstrated segregation of seven traits in pea plants using **monohybrid crosses**.
7. A diploid individual with two identical alleles of a gene is a **homozygote**. A **heterozygote** has two different alleles of a gene. A **compound heterozygote** has two different recessive alleles in the same gene. A gene may have many alleles.
8. A dominant allele masks the expression of a recessive allele.
9. An individual may be homozygous dominant, homozygous recessive, or heterozygous for a given pair of alleles.
10. The alleles for a gene in an individual constitute the genotype, and their expression the phenotype.
11. The most common allele in a population is the **wild type**, and variants are **mutant**. Mutation is a change in a gene.
12. When Mendel crossed two true-breeding plants, then bred the resulting hybrids to each other, the two variants of the trait appeared in a 3:1 phenotypic ratio. Crossing these progeny revealed a genotypic ratio of 1:2:1.
13. A **Punnett square** follows the transmission of alleles and is based on probability. A **test cross** breeds an individual of unknown genotype with a homozygous recessive individual to deduce the unknown genotype from observing the traits of the offspring.

4.2 Single-Gene Inheritance Is Rare

14. Eye color illustrates how a monogenic trait can be affected by other genes.
15. Autosomal recessive conditions are more likely in families where there is **consanguinity** (marriage between relatives).
16. Genetic problems can be solved by tracing alleles as gametes form and then combine in a new individual.
17. How alleles affect the abundance or activity of the gene's protein product may determine whether the effect of an allele is dominant or recessive. Loss-of-function mutations lead to a missing protein. Gain-of-function mutations alter a protein's function.

4.3 Following the Inheritance of More Than One Gene

18. Mendel's second law, the law of independent assortment, follows transmission of two or more genes on different chromosomes.
19. **Dihybrid crosses** showed that a random assortment of maternally and paternally derived chromosomes during meiosis yields gametes with different gene combinations.
20. The product rule states that the chance that two independent genetic events occur equals the product of the probabilities that each event will occur. The product rule is useful in following the inheritance of genes on different chromosomes.

4.4 Pedigree Analysis

21. A **pedigree** is a chart that depicts family relationships and patterns of inheritance for traits. A pedigree can be inconclusive if a dominant trait or illness does not impair fertility. An individual with a *de novo* dominant mutation or two autosomal recessive alleles inherited from carrier parents can appear the same in a pedigree.
22. Pedigrees and Punnett squares can predict the phenotypic and genotypic classes of offspring in cases of conditional probability in which knowing a parent's genotype restricts the possible genotypic classes of offspring.
23. Comparing exome or genome sequences in parents-child trios when the child has an unrecognized syndrome and the parents are unaffected can distinguish autosomal recessive inheritance from the parents from a *de novo* dominant mutation in the child.

Key Terms

autosomal dominant (ôtə-sōməl) The mode of inheritance in which one autosomal allele causes a phenotype. Such a trait can affect males and females and does not skip generations.

autosomal recessive (ôtə-sōməl) The mode of inheritance in which two autosomal alleles are required to cause a phenotype. Such a trait can affect males and females and can skip generations.

compound heterozygote An individual with two different recessive alleles in the same gene.

consanguinity (kŏnsăn-gwĭn'ĭ-tē) Blood relatives having children together.

dihybrid cross Breeding individuals heterozygous for two traits.

heterozygous (hĕtə-rō-zī'gəs) Having two different alleles of a gene.

homozygous (hōmō-zī'gəs) Having two identical alleles of a gene.

mode of inheritance The pattern in which a gene variant passes from generation to generation. It may be dominant or recessive, autosomal, or X- or Y-linked.

monogenic (mono-jēn'ĭk) A trait caused by mutation in a single gene.

monohybrid cross (mŏn'ō-hībrĭd) A cross of two individuals who are heterozygous for a single gene.

mutant (myōōt'nt) An allele that differs from the normal or most common allele in a population and alters the phenotype.

Punnett square A diagram that follows and combines parental gene contributions to offspring.

segregation Mendel's first law, which states that the alleles of a gene are distributed into separate gametes during meiosis.

test cross Breeding an individual of unknown genotype with a homozygous recessive individual to deduce the unknown genotype from observing the traits of the offspring.

wild type The most common phenotype in a population for a particular gene.

Review Questions

1. List four ways that single-gene diseases such as Huntington disease and cystic fibrosis differ from an infectious disease, such as malaria or influenza.
2. How does meiosis explain Mendel's laws?
3. Describe two ways to inherit blue eyes.
4. Compare how an autosomal recessive condition affects a family to how an autosomal dominant condition does so.
5. Discuss how Mendel derived the two laws of inheritance without knowing about chromosomes.
6. Distinguish between
    a. autosomal recessive and autosomal dominant inheritance.
    b. Mendel's first and second laws.
    c. a homozygote and a heterozygote.
    d. a monohybrid and a dihybrid cross.
    e. a loss-of-function mutation and a gain-of-function mutation.
    f. a Punnett square and a pedigree.
7. Explain how a dominant disease can occur if its symptoms prevent reproduction.
8. Why would Mendel's results for the dihybrid cross have been different if the genes for the traits he followed were near each other on the same chromosome?
9. Why are extremely rare autosomal recessive diseases more likely to appear in families in which blood relatives have children together?

10. How does the pedigree of the ancient Egyptian royal family in figure 4.14a differ from a pedigree a genetic counselor might use today?

11. What is the probability that two individuals with an autosomal recessive trait, such as albinism, will have a child with the same genotype and phenotype as they have?

12. Explain why discovering a *de novo* mutation in a child who has a syndrome does not pose an elevated risk for her siblings, but does to her children.

Applied Questions

1. Calculate the probability that all three children in the Mervar family would inherit Huntington disease.

2. Predict the phenotypic and genotypic ratios for crossing the following pea plants:
   a. short × short
   b. short × true-breeding tall
   c. true-breeding tall × true-breeding tall

3. What are the genotypes of the pea plants that would have to be bred to yield one plant with restricted pods for every three plants with inflated pods?

4. If pea plants with all white seed coats are crossed, what are the possible phenotypes of their progeny?

5. Pea plants with restricted yellow pods are crossed to plants that are true-breeding for inflated green pods and the F_1 plants are crossed. Derive the phenotypic and genotypic ratios for the F_2 generation.

6. More than 100 genes cause deafness when mutant. What is the most likely mode of inheritance in families in which all children and the parents were born deaf?

7. The MacDonalds raise Labrador retrievers. In one litter, two of eight puppies, a male and a female, have a condition called exercise-induced collapse. After about 15 minutes of intense exercise, the dogs wobble about, develop a fever, and their hind legs collapse. The parents are healthy. What is the mode of inheritance?

8. Chands syndrome is autosomal recessive and causes curly hair, underdeveloped nails, and abnormally shaped eyelids. In the following pedigree, which individuals must be carriers?

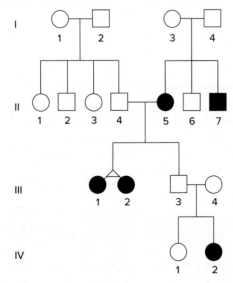

Chands syndrome

9. Lorenzo has a double row of eyelashes, which he inherited from his mother as a dominant trait. His maternal grandfather is the only other relative to have it. Fatima, who has normal eyelashes, marries Lorenzo. Their first child, Nicola, has normal eyelashes. Fatima is pregnant again and hopes for a child with double eyelashes. What chance does the child have of inheriting double eyelashes?

10. In peeling skin syndrome, the outer skin layers fall off on the hands and feet. The pedigrees depict three affected families. What do the families share that might explain the appearance of this otherwise rare condition?

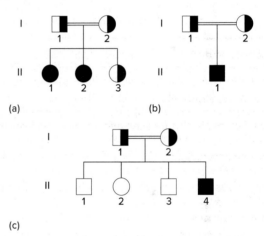

Peeling skin syndrome

11. The child referenced in figure 4.12 who has red urine after eating beets, colored eyelids, and short fingers is genotype *bbHhEe*. The genes for these traits are on different chromosomes. If he has children with a woman who is a trihybrid, what are the expected genotypic and phenotypic ratios for their offspring?

Forensics Focus

1. A woman, desperate to complete her family tree for an upcoming family reunion, cornered a stranger in a fast-food restaurant. Her genealogical research had identified him as a distant cousin, and she needed his DNA. He refused to cooperate, looked scared, and ran off. The woman took his discarded coffee cup and collected DNA from traces of saliva, which she sent on a swab to a DNA ancestry testing company. (This is a true story.)
   a. Do you think the woman was justified in her action? Why or why not?
   b. What are the strengths and limitations of using genealogical information (family records, word-of-mouth) versus DNA testing to construct a pedigree?

2. A young woman walking to her car in a parking lot late at night was attacked and raped. She recalled that the rapist wore a green silk shirt, a belt with a metal buckle that left a bruise on her abdomen, jeans, and running shoes. She also remembered that he had white skin, a shaved head, and startling blue eyes. She recalled footsteps approaching and then someone yanking the man off her, but her head was let go and hit the pavement, knocking her unconscious.

   A few weeks after the rape, a naked male body washed up in a nearby river, and police found a belt buckle farther downstream that matched the pattern of the woman's bruise. The body, however, was headless. How can police determine the eye color of the corpse to help in identifying or ruling him out as the rapist?

Case Studies and Research Results

1. Several class-action lawsuits are being filed against physicians who used their own sperm to inseminate dozens of patients. The practice began in the 1950s, when medical students or young physicians would provide sperm for a fee. Today it continues with infertile couples ordering sperm from the same donor from a sperm bank, although the number of donations per man is limited. As a result of past practices, thousands of people today, ranging greatly in age, are discovering that they have half-siblings by doing DNA ancestry testing. What is the possible consequence of the use of sperm from the same donor on many women, in terms of future prevalence of single-gene diseases? Explain your answer.

2. Hannah Sames (see Clinical Connection 2.2) inherited two identical mutations that delete part of the gigaxonin gene. A boy who received gene therapy to treat his hereditary blindness (see section 20.3) has two different recessive mutations in the *RPE65* gene. Which child is a compound heterozygote?

3. On the daytime drama *The Young and the Restless*, several individuals, now adults, suffered as children from SORAS, which stands for "soap opera rapid aging syndrome." In SORAS, a young child is sent off to boarding school and returns three months later as an angry teenager. In the Newman family, siblings Nicholas and Victoria SORASed from childhood to adolescence in mere months. Half-sister Abby, whose mother is Ashley, matured on a normal timetable. Nicholas and Victoria's parents, Nikki and Victor, were not affected. What is the most likely mode of inheritance of SORAS, and how do you know this?

4. The Cleaver family awaited the birth of Claudette's puppies with great anticipation. But shortly after the standard poodle gave birth to six pups, two of them, a male and a female, developed seizures and died. A surviving pup, Sylvester, was bred when he was a year old to Minuette, who was an offspring of Sylvester's father, Otis. Alas, one of Minuette's four pups also died of the canine seizure disorder. What is the mode of inheritance of this condition?

CHAPTER 5

Beyond Mendel's Laws

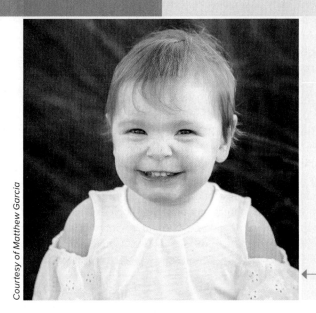

At least nine genes can interact to specify red hair color.

Learning Outcomes

5.1 When Gene Expression Appears to Alter Mendelian Ratios
1. Explain the effect of lethal alleles on Mendelian ratios.
2. State how DNA structure underlies multiple gene variants.
3. Distinguish among complete dominance, incomplete dominance, and codominance.
4. Distinguish epistasis from interactions of alleles of the same genes.
5. Describe how penetrance, expressivity, and pleiotropy affect gene expression.
6. Explain how a phenocopy can appear to be inherited.

5.2 Mitochondrial Genes
7. Describe the mode of inheritance of a mitochondrial trait.
8. Explain how mitochondrial DNA differs from DNA in the nucleus.

5.3 Linkage
9. Explain how inheritance of linked traits differs from inheritance of genes on different chromosomes.
10. Discuss the basis of linkage in meiosis.
11. Explain how linkage is used to derive genetic maps.

The BIG Picture

The modes of inheritance that Gregor Mendel's elegant experiments on peas revealed can be obscured when genes have many variants, interact with each other or the environment, are in mitochondria, or are linked on the same chromosome.

The Complex Inheritance of Red Hair

A pea is either green or yellow, determined by a single gene with two variants. Hair color in humans was also thought to be straightforward: blond, red, light brown, dark brown, or black. Until recently, having red hair was attributed to inheriting mutations in both copies of a gene called *MC1R*. An analysis of nearly 350,000 people who have submitted their DNA to the UK Biobank revealed that the inheritance of red hair is more complicated.

In the scalp, cells called melanocytes produce melanin pigments, which are transferred inside vesicles (melanosomes) to nearby skin-like cells in the "bulge" region of a hair follicle. The pigment spreads up the hair shaft. Eumelanin confers dark color, and a form called pheomelanin provides a reddish-yellow color.

Determining the alleles (gene variants) of *MC1R* among the people who contributed DNA to the UK Biobank revealed that 93 percent of redheads have two *MC1R* mutations that encode pheomelanin. However, these people comprise only 15 percent of the number of people who have two mutations in the gene. That is, mutations in *MC1R* do not always result in red hair color. Other genes are implicated.

The researchers identified eight additional genes that either control *MC1R*'s production of eumelanin or pheomelanin, or produce reddish pigment another way. In addition, the genes influence each other's activity. Yet other genes affect the texture and curliness of hair.

Red hair is considered a monogenic trait, but one that is modified by the actions of other genes. In contrast, blonde hair is a polygenic trait, the consequence of small actions from at least 200 genes. Red hair color is a perfect example of a single-gene trait with nuances dictated by the actions of other genes—the topic of this chapter.

5.1 When Gene Expression Appears to Alter Mendelian Ratios

Single genes seldom completely control a phenotype in the straightforward way that Mendel's experiments with peas suggested. Genes interact with each other and with environmental influences. When transmission patterns of a visible trait do not exactly fit autosomal recessive or autosomal dominant inheritance, Mendel's laws are still operating. The underlying genotypic ratios persist, but other factors affect the phenotypes.

This chapter considers three general phenomena that seem to be exceptions to Mendel's laws, but are actually not: gene expression, mitochondrial inheritance, and linkage. In several circumstances, phenotypic ratios appear to contradict Mendel's laws, but they do not. Chapter 11 revisits gene expression from a molecular point of view.

Lethal Allele Combinations

A genotype (allele combination) that causes death is, by strict definition, lethal. Death from genetic disease can occur at any stage of development or life. Tay-Sachs disease is lethal by age 3 or 4, whereas Huntington disease may not be lethal until late middle age. In a population genetics and evolutionary sense, a lethal genotype has a more specific meaning—it causes death before the individual can reproduce, which prevents passage of mutations to the next generation.

In organisms used in experiments, such as fruit flies, pea plants, or mice, lethal allele combinations remove an expected progeny class following a specific cross. For example, in a cross of heterozygous flies carrying lethal alleles in the same gene, homozygous recessive progeny die as embryos, leaving only heterozygous and homozygous dominant adult fly offspring. In humans, early-acting lethal alleles cause spontaneous abortion, sometimes before a woman even knows she is pregnant. When both parents carry a recessive lethal allele for the same gene, each pregnancy has a 25 percent chance of spontaneously aborting—that is, 25 percent of embryos are homozygous recessive. They do not develop further, and therefore this genotype is not seen in any person.

An example of a lethal genotype in humans is achondroplastic dwarfism, which has the distinct phenotype of a long trunk, short limbs, and a large head bearing a flat face (**figure 5.1**). It is an autosomal dominant trait, but is most often the result of a spontaneous (new) mutation. Each child of two people with achondroplasia has a one in four chance of inheriting both mutant alleles; however, because such homozygotes are not seen, this genotype is presumed to be lethal. Each child in such a family therefore faces a 2/3 probability of having achondroplasia and a 1/3 probability of being of normal height, illustrating conditional probability. Homozygotes for achondroplasia mutations in other species cannot breathe because the lungs do not have room to inflate. The cause of achondroplasia is a gain-of-function mutation in a gene (*FGFR3*) that normally inhibits growth of the long bones of the arms and legs.

Multiple Alleles

A person has two alleles for any autosomal gene—one part of each of a pair of homologous chromosomes. However, a gene can exist in more than two allelic forms in a population because it can mutate in many ways, just as a typo can appear anywhere in a document. Different allele combinations can produce variations in the phenotype. The more alleles, the more variations possible. Recall from chapter 4 that an individual with two different mutant recessive alleles for the same gene is a *compound heterozygote*.

For some inherited diseases, knowing the genotype enables a physician to predict the general course of an illness, or which of several symptoms are likely to develop. This is the case for cystic fibrosis (CF) (see figures 1.4 and 1.6 and Clinical Connection 4.1). Some CF genotypes cause frequent, severe respiratory infections; highly congested lungs; and poor weight gain. A different CF genotype increases susceptibility to bronchitis, sinusitis, and pneumonia. Yet another genotype causes only male infertility.

Different Dominance Relationships

In complete dominance, only one allele is expressed; the other is not. In **incomplete dominance**, the heterozygous phenotype is intermediate between that of either homozygote.

Enzyme deficiencies in which a threshold level is necessary for health illustrate both complete and incomplete dominance. For example, Tay-Sachs disease displays complete dominance because the heterozygote (carrier) is as healthy as a homozygous dominant individual. However, the heterozygote has an intermediate level of enzyme between the homozygous dominant (full enzyme level) and homozygous recessive (no enzyme). Half the normal amount of enzyme is sufficient for health, which is why at the whole-person level, the wild type allele is completely dominant.

Familial hypercholesterolemia (FH) is an example of incomplete dominance that can be observed in carriers on

Figure 5.1 Lethal alleles. (a) A person with achondroplasia has inherited a mutant dominant allele. Inheriting two such alleles is lethal to embryos. **(b)** If two people with achondroplasia have children, each child has a 2/3 chance of inheriting achondroplasia and 1/3 of being normal height. This is a conditional probability because one in four conceptions does not survive. *(a): Kevin Winter/Getty Images*

both the molecular and whole-body levels. A person with two disease-causing alleles does not have receptors on liver cells that take up the low-density lipoprotein (LDL) form of cholesterol from the bloodstream, so it builds up. A person with one disease-causing allele has half the normal number of receptors. Someone with two wild type alleles has the normal number of receptors. **Figure 5.2** shows the concentration range of plasma cholesterol for the three genotypes. The phenotypes parallel the number of receptors—individuals with two mutant alleles die in childhood of heart attacks, those with one mutant allele may suffer heart attacks in young adulthood, and those with two wild type alleles do not develop this inherited form of heart disease.

Different alleles that are both expressed in a heterozygote are **codominant**. The ABO blood group system is based on the expression of codominant alleles.

Blood types are determined by the patterns of molecules on the surfaces of red blood cells. Most of these molecules are proteins embedded in the plasma membrane with attached sugars that extend from the cell surface. The sugar is an **antigen**, a molecule that the immune system recognizes and responds to. People in blood group A have an allele that encodes an enzyme that adds a piece to a certain sugar attached to the plasma membrane, producing antigen A. In people with blood type B, the allele

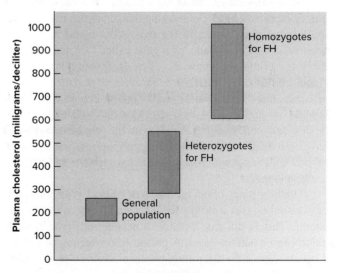

Figure 5.2 Incomplete dominance. A heterozygote for familial hypercholesterolemia (FH) has approximately half the normal number of cell surface receptors in the liver for LDL cholesterol. An individual with two mutant alleles has the severe form of FH, with liver cells that totally lack the receptors. As a result, serum cholesterol level is very high. Cholesterol is also deposited in joints and many other body parts.

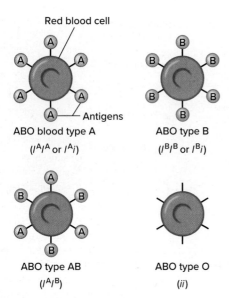

Figure 5.3 ABO blood types illustrate codominance. ABO blood types are based on antigens on red blood cell surfaces. This depiction greatly exaggerates the size of the A and B antigens. Genotypes are in parentheses.

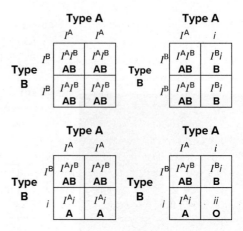

Figure 5.4 Codominance. The I^A and I^B alleles of the *I* gene are codominant, but they follow Mendel's law of segregation. These Punnett squares show the genotypes that could result when a person with type A blood has children with a person with type B blood.

and its encoded enzyme are slightly different, which places a different piece on the sugar, producing antigen B. People in blood group AB have both antigen types. Blood group O results from yet a third allele of this gene. It is missing just one DNA nucleotide, but this changes the encoded enzyme in a way that removes the sugar chain from its final piece (**figure 5.3**). As a result, type O red blood cells do not have either A or B antigens.

The A and B alleles are codominant, and both are completely dominant to O. Considering the genotypes reveals how these interactions occur. In the past, ABO blood types have been described as variants of a gene called "*I*," which stands for isoagglutinin. The three alleles are I^A, I^B, and *i*. People with blood type A have antigen A on the surfaces of their red blood cells, and may be of genotype $I^A I^A$ or $I^A i$. People with blood type B have antigen B on their red blood cell surfaces, and may be of genotype $I^B I^B$ or $I^B i$. People with the rare blood type AB have both antigens A and B on their cell surfaces, and are genotype $I^A I^B$. People with blood type O have neither antigen, and are genotype *ii*.

Fiction plots often misuse ABO blood type terminology, assuming that a child's ABO type must match that of one parent. This is not true because a person with type A or B blood can be heterozygous. A person who is genotype $I^A i$ and a person who is $I^B i$ can jointly produce offspring of any ABO genotype or phenotype, as **figure 5.4** illustrates.

Until recently, human blood was typed for transfusions by detecting the binding of antibody proteins in the blood serum of the recipient to antigens on the red blood cells of the donor. If the preparation clumped, the match was incompatible. Today blood typing for transfusions detects the genes that encode the blood group antigens. The DNA testing approach is more accurate and less costly.

Epistasis

Mendel's laws can appear not to operate when one gene masks or otherwise affects the phenotype of another. This phenomenon is called **epistasis**. It refers to interaction between *different* genes, not between the alleles of the same gene. A gene that affects expression of another is called a modifier gene. The modifier gene is "epistatic to" the gene whose expression it affects.

In epistasis, the blocked gene is expressed (transcribed into RNA) normally, but the protein product of the modifier gene inactivates it, removes a structure needed for it to contribute to the phenotype, or otherwise counteracts its effects.

A blood type called the Bombay phenotype also illustrates epistasis. It results from an interaction between a gene called *H* and the *I* gene that confers ABO blood type. The *H* gene controls the placement of a molecule to which antigens A and B attach on red blood cell surfaces. A person of genotype *hh* can't make that molecule, so the A and B antigens cannot attach to red blood cell surfaces. The A and B antigens fall off and the person tests as type O, but may be any ABO genotype. The *H* gene is epistatic to the *I* gene.

Epistasis can explain why siblings who inherit the same disease can suffer to differing degrees. One study examined siblings who both inherited spinal muscular atrophy (SMA) type 1, in which nerves cannot signal muscles. The muscles weaken and atrophy, usually proving fatal in early childhood. The mutation encodes an abnormal protein that shortens axons, which are the extensions on nerve cells that send messages. Some siblings who inherited the SMA genotype, however, never developed symptoms. They can thank a variant of another gene, *plastin 3*, which is epistatic to the mutant gene that causes SMA by increasing production of the cytoskeletal protein actin that extends axons. Because the healthy siblings inherited the ability to make extra long axons, the axon-shortening effects of SMA were not harmful.

Penetrance and Expressivity

The same genotype can produce different degrees of a phenotype in different individuals because of influences of other genes, as well as environmental influences such as nutrition, exposure to toxins, and stress. For example, for two people who have the same *CF* genotype, one may be much sicker because she also inherited gene variants predisposing her to develop asthma and respiratory allergies.

Two terms describe the degrees of expression of a single gene. **Penetrance** refers to the percentage of individuals who have a particular *genotype* who have the associated phenotype. **Expressivity** refers to variability in severity of a *phenotype*, or the extent to which the gene is expressed. That is, penetrance describes the gene, whereas expressivity describes the person.

An allele combination (genotype) that produces a phenotype in everyone who inherits it is completely penetrant. Huntington disease (see the opener to chapter 4) is nearly completely penetrant. Almost all people who inherit a mutant allele will develop symptoms if they live long enough. Complete penetrance is rare.

A genotype is incompletely penetrant if some individuals do not express the phenotype (have no symptoms). Polydactyly (see figure 1.8a) is incompletely penetrant. Some people who inherit the dominant mutant allele have more than five digits on a hand or foot. Yet others who must have inherited the allele because they have an affected parent and child have ten fingers and ten toes. Penetrance is described numerically. If 80 of 100 people who inherit the dominant mutant polydactyly allele have extra digits, the genotype is 80 percent penetrant.

A phenotype is variably expressive if symptoms vary in intensity among different people. One person with polydactyly might have an extra digit on both hands and a foot, but another might have just one extra fingertip. Polydactyly is both incompletely penetrant and variably expressive.

Pleiotropy

A single-gene disease with many symptoms, or a gene that controls several functions or has more than one effect, is termed **pleiotropic**. Such conditions can be difficult to trace through families because people with different subsets of symptoms may appear to have different diseases.

On a molecular level, pleiotropy occurs when a single protein affects different body parts, participates in more than one biochemical reaction, or has different effects in different amounts. Consider Marfan syndrome. The most common form of this autosomal dominant condition is a defect in an elastic connective tissue protein called fibrillin. The protein is abundant in the lens of the eye, in the aorta (the largest artery in the body, leading from the heart), and in bones of the limbs, fingers, and ribs. The symptoms are lens dislocation, long limbs, spindly fingers, and a caved-in chest. The most serious symptom is a weakening in the aorta, which can suddenly burst. If the weakening is detected early, a synthetic graft can replace the section of artery wall and save the person's life.

Figure 5.5 Genetic heterogeneity. Mutations in several genes whose protein products maintain the retina cause blindness. The retina, a multilayered structure at the back of the eye, includes the rods and cones, which are cells that signal incoming light to the brain (photoreceptors). Here, in a form of blindness called a retinal dystrophy, specific mutations (in italics) disrupt the functioning of the rods, which provide black and white vision, and the retinal pigment epithelium that cleans up debris.

Genetic Heterogeneity

Mutations in different genes that produce the same phenotype lie behind **genetic heterogeneity**. It can occur when genes encode enzymes or other proteins that are part of the same biochemical pathway, or when proteins affect the same body part. Genetic heterogeneity may make it appear that Mendel's laws are not operating, even though they are. For example, mutations in more than 20 genes cause retinal dystrophies, because a mutation can disrupt the functioning of the rods and cones, the cells that provide vision, or their supportive cells, in many ways (**figure 5.5**). If a man who is homozygous recessive for a mutation in one of the genes that causes the condition has a child with a woman who is homozygous recessive for a *different* gene, then the child would not inherit either form of blindness because he or she would be heterozygous for both genes.

Discovering additional genes that can cause a known disease is happening more often as human genomes are analyzed, and such discoveries can have practical repercussions. Consider osteogenesis imperfecta, in which abnormal collagen causes fragile bones. Before a second causative gene was discovered, some parents of children who were brought to the hospital with frequent fractures and who did not have a mutation in the one known gene were accused of child abuse. Today at least eight genetically distinct forms of the disease are recognized.

Clinical Connection 5.1 describes a common disease whose familial forms are genetically heterogeneic—Alzheimer disease.

Phenocopies

An environmentally caused trait that appears to be inherited is a **phenocopy**. Such a trait can either produce symptoms that

Clinical Connection 5.1
The Roots of Familial Alzheimer Disease

"What's that on that tiny screen?" asked 82-year-old Ginny for the fifth time in half an hour as her son glanced down at his phone.

"Mom, don't you know you've asked me that several times since I got here? It's my wallpaper, you know, my cat Trouser," he said, showing her the photo of the orange cat on his mobile phone and trying not to become impatient.

"No, I've never asked you that before." She paused, looking puzzled. "What did you say his name is?"

In the months following that conversation, Ginny's short-term memory declined further. She could rarely concentrate long enough to finish reading an article or follow a conversation. In the grocery store, she couldn't find items she'd been buying for decades. Aware of her growing deficits, she became depressed. Finally, her son suggested she have a complete neurological exam. By the time she could see a physician, other signs had emerged. Ginny couldn't recall her zip code or the name of the small town where she grew up. Sometimes she couldn't remember where things belonged—she put a cantaloupe in the bathtub.

Ginny had "mild cognitive impairment" (MCI). It could be the start of Alzheimer disease, which affects 26 million people worldwide. After ruling out a vitamin deficiency and lack of oxygen to the brain as causes of Ginny's forgetfulness, a neurologist started her on a drug to slow breakdown of the neurotransmitter acetylcholine in the brain. The doctor also prescribed an antidepressant, which revived Ginny enough so that she was more willing to leave her apartment. If the MCI progressed to Alzheimer disease, Ginny would continue to lose thinking, reasoning, learning, and communicating abilities, and one day would not recognize her loved ones. Eventually, she would no longer speak and smile, and would cease walking, not because her legs wouldn't function, but because she would forget how to walk. Yet the haze would sometimes seem to lift from her eyes and the old Ginny would return for a precious moment or two.

In Alzheimer disease, certain brain parts—the amygdala (seat of emotion) and hippocampus (the memory center)—become buried in two types of protein. Amyloid precursor protein normally converts iron into a safe form. In Alzheimer disease, the protein is cut into unusually sized amyloid beta peptides, which aggregate to form "plaques" outside brain cells because cells cannot remove the material fast enough. At the same time, unsafe iron accumulates inside brain neurons, and zinc builds up in the plaques outside them. Ginny's brain scan showed plaques accumulating. The second type of protein, called tau, was building up, too, clumping into "tangles" inside brain neurons (**figure 5A**). The telltale plaques and tangles, present to a lesser extent in everyone, could cause the symptoms of Alzheimer disease or result from them. Either

Figure 5A In Alzheimer disease, amyloid plaques form outside neurons and neurofibrillary tangles form inside the cells. *Marc Phares/Science Source*

way, their abundance in a cerebrospinal fluid test, combined with Ginny's cognitive problems, strongly suggested Alzheimer disease.

Fewer than 1 percent of Alzheimer cases are familial (inherited), caused by mutations in any of several genes that affect amyloid accumulation or clearance (**table 5A**). However, some variants of genes associated with Alzheimer disease are actually protective. One variant of the gene *APOE* increases the risk of developing a late-onset form slightly in a heterozygote and more so, up to 15-fold, in a homozygote. People with two high-risk *APOE4* variants who live to very old age and do not development Alzheimer disease may have a variant of a different gene, *RAB10*, that is protective.

A 31-gene test provides a *polygenic risk score* that identifies people at higher risk of developing Alzheimer disease. Each gene contributes a small degree to the risk. Polygenic traits are discussed in chapter 7.

Questions for Discussion

1. Why is Alzheimer disease difficult to diagnose, especially in the early stage?
2. How is the genetics of Alzheimer disease complex?
3. Describe the protein abnormalities that occur in Alzheimer disease.
4. How common are inherited forms of Alzheimer disease?

(Continued)

Table 5A — Genes Associated with Alzheimer Disease

Causative Gene	Chromosome	Mechanism
Amyloid precursor protein (APP)	21	Unusually sized pieces aggregate outside brain cells.
Calcium homeostasis modulator 1 (CALHM1)	10	Controls amyloid cutting.
Presenilin 1 (PSEN1)	14	Forms part of secretase (enzyme) that cuts APP.
Presenilin 2 (PSEN2)	1	Forms part of secretase (enzyme) that cuts APP.
Risk Gene		
Apolipoprotein E4 (APOE4)	19	Unusually sized pieces add phosphates to tau protein, making it accumulate and impairing microtubule binding.
Triggering receptor expressed on myeloid cells (TREM)	6	Promotes brain inflammation.

resemble those of a known single-gene disease or mimic inheritance patterns by affecting certain relatives. For example, the limb birth defect caused by the drug thalidomide, discussed in chapter 3, is a phenocopy of the rare inherited illness phocomelia. Physicians recognized the environmental disaster when they began seeing many children born with what looked like phocomelia. A birth defect caused by exposure to a widely used teratogen was more likely than a sudden increase in incidence of a rare inherited disease.

An infection can be a phenocopy if it affects more than one family member. Children who have AIDS may have parents who also have the disease, but these children acquired AIDS by viral infection, not by inheriting a mutation that causes immune deficiency.

Common symptoms may resemble those of an inherited condition, but be due to an environmental situation. For example, an underweight child who has frequent colds may show some signs of cystic fibrosis, but may instead suffer from malnutrition.

The Human Genome Sequence Adds Perspective

As researchers describe more genes from human genomes, some of the phenomena just discussed are turning out not to be rare, as had once been thought. Terms such as *epistasis* and *genetic heterogeneity* are beginning to overlap and blur. For example, most people who have Marfan syndrome have a mutation in the fibrillin gene. However, some people with the syndrome instead have a mutation in the gene that encodes the transforming growth factor beta receptor (TGFβR). Fibrillin and TGFβR are part of the same biochemical pathway. So, Marfan syndrome fits the definition of genetic heterogeneity because mutations in different genes cause identical symptoms. Yet the forms of the syndrome are also epistatic because a variant TGFβR blocks the activity of fibrillin.

Gene interactions also underlie penetrance and expressivity because even genes that do not directly interact, in space or time, can affect each other's expression. This is the case for Huntington disease, in which cells in a certain part of the brain die. Siblings who inherit the same HD mutation may differ in the number of cells that they have in the affected brain area because of variants of other genes that affected the division rate of neural stem cells in the brain during embryonic development. As a result, an individual who inherits HD, but also extra brain cells, might develop symptoms much later in life than a brother or sister who does not have such a built-in reserve supply. If the delay is long enough that death comes from another cause, the HD mutation would then be considered nonpenetrant.

Technologies that reveal gene expression patterns in different tissues are painting detailed portraits of pleiotropy, showing that inherited diseases may affect more tissues or organs than are obvious from symptoms. Finally, more cases of genetic heterogeneity are being discovered as researchers identify genes with redundant or overlapping functions.

Table 5.1 summarizes phenomena that appear to alter single-gene inheritance. Our definitions and designations are changing as improving technology enables us to describe and differentiate diseases in greater detail. Phenomena such as variable expressivity, incomplete penetrance, epistasis, pleiotropy, and genetic heterogeneity, once considered unusual characteristics of single genes, may turn out to be the norm.

Gregor Mendel derived the two laws of inheritance working with traits conferred by genes located on different chromosomes in the nucleus. When genes are on the same chromosomes, however, the associated traits may not appear in Mendelian ratios. The rest of this chapter considers two types

Table 5.1 — Factors That Alter Single-Gene Phenotypic Ratios

Phenomenon	Effect on Phenotype	Example
Lethal alleles	A phenotypic class does not survive to reproduce.	Achondroplasia
Multiple alleles	Many variants or degrees of a phenotype are possible.	Cystic fibrosis
Incomplete dominance	A heterozygote's phenotype is intermediate between those of the two homozygotes.	Familial hypercholesterolemia
Codominance	A heterozygote's phenotype is distinct from and not intermediate between those of the two homozygotes.	ABO blood types
Epistasis	One gene masks or otherwise affects another's phenotype.	Bombay phenotype
Penetrance	Some individuals with a particular genotype do not have the associated phenotype.	Polydactyly
Expressivity	A genotype is associated with a phenotype of varying intensity.	Polydactyly
Pleiotropy	The phenotype includes many symptoms, with different subsets in different individuals.	Marfan syndrome
Phenocopy	An environmentally caused condition has symptoms and a recurrence pattern similar to those of a known inherited trait.	Infection
Genetic heterogeneity	Genotypes of different genes cause the same phenotype.	Osteogenesis imperfecta

of gene transmission that are not Mendelian—mitochondrial inheritance and linkage.

Key Concepts Questions 5.1

*a. Define *lethal genotype*.
b. Distinguish among complete dominance, incomplete dominance, and codominance.
*c. What is epistasis?
d. Distinguish between penetrance and expressivity.
*e. What is pleiotropy?
f. How can genetic heterogeneity affect inheritance patterns?
*g. What is a phenocopy?

5.2 Mitochondrial Genes

Mitochondria are the cellular organelles that house the reactions that derive energy from nutrients (see figure 2.6). Each of the hundreds to thousands of mitochondria in each human cell contains several copies of a "mini-chromosome" that carries just 37 genes. Genes encoded in mitochondrial DNA (mtDNA) act in the mitochondrion, but the organelle also requires the activities of certain genes from the nucleus.

The inheritance patterns and mutation rates for mitochondrial genes differ from those for genes in the nucleus. Rather than being transmitted equally from both parents, mitochondrial genes are maternally inherited. They are passed only from an individual's mother because the sperm head, which enters an oocyte at fertilization, does not include mitochondria, which are found in the sperm midsection, where they provide energy for moving the tail (see figure 3.10). In the rare instances when mitochondria from sperm enter an oocyte, they are usually selectively destroyed early in development. Pedigrees that follow mitochondrial genes therefore show a woman passing the trait to all her children, whereas a male cannot pass the trait to any of his children (**figure 5.6**).

DNA in the mitochondria differs functionally from DNA in the nucleus in several ways (**figure 5.7**):

- MtDNA does not cross over.
- It mutates faster than DNA in the nucleus because fewer types of DNA repair are available and DNA-damaging oxygen free radicals are produced in the energy reactions.
- Mitochondrial genes are not wrapped in proteins.
- Mitochondrial genes are not "interrupted" by DNA sequences that do not encode protein.
- A cell has many mitochondria, and each mitochondrion contains several copies of the mitochondrial genome.
- Mitochondria with different alleles for the same gene can reside in the same cell.

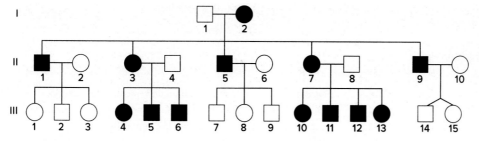

Figure 5.6 **Inheritance of mitochondrial genes.** Mothers pass mitochondrial genes to all offspring. Fathers do not transmit mitochondrial genes because sperm only rarely contribute mitochondria to fertilized ova. If mitochondria from a male do enter, they are almost always destroyed.

Mitochondrial Diseases

Mitochondrial genes encode proteins that participate in protein synthesis and energy production. Twenty-four of the 37 genes encode RNA molecules: 22 transfer RNAs (tRNAs) and 2 ribosomal RNAs (rRNAs), that help assemble proteins. (Chapter 10 explains the actions of these RNAs.) The other 13 mitochondrial genes encode proteins that function in cellular respiration, which is the process that uses energy from digested nutrients to synthesize ATP, the biological energy molecule.

A class of diseases results from mutations in mitochondrial genes. They are called mitochondrial myopathies and have specific names, but news reports often lump them together as "mitochondrial disease." Symptoms arise from tissues whose cells normally have many mitochondria, such as skeletal muscle. They include great fatigue, weak and flaccid muscles, and intolerance to exercise. Skeletal muscle fibers appear "red and ragged" when stained and viewed under a light microscope because their abundant abnormal mitochondria are visible beneath the plasma membrane. Diseases considered to be mitochondrial may also result from mutations in genes in the nucleus that encode proteins that are essential for mitochondria to function.

A mutation in a mitochondrial gene that encodes a tRNA or rRNA can be devastating because it impairs the organelle's general ability to manufacture proteins. Consider the case of a once active and articulate dental hygienist. In her forties, Lindzy began to slow down at work. She heard a buzzing and developed difficulty talking and walking. Then her memory began to fade in and out, she became lost easily in familiar places, and her conversation made no sense. Her condition worsened, and she developed diabetes, seizures, and pneumonia and became deaf and demented. She was finally diagnosed with MELAS, which stands for "mitochondrial myopathy encephalopathy lactic acidosis syndrome." Her muscle cells had the telltale red-ragged fibers. Lindzy died. Her son and daughter may develop the condition because they inherited her mitochondria.

About 1 in 200 people has a mutation in a mitochondrial gene that could cause disease. However, mitochondrial diseases are rare, affecting about 1 in 6,500 people, apparently because of a weeding-out process during egg formation. A mitochondrial mutation may disrupt energy acquisition so greatly in an oocyte that it cannot survive.

Theoretically, a woman with a mitochondrial disease can avoid transmitting it to her children if her mitochondria can be replaced with healthy mitochondria from a donor. **Bioethics** describes two ways to do this, and notes the controversy over creating "three-parent" children.

Figure 5.7 **Mitochondrial DNA.** A mitochondrion contains several rings of DNA. Different alleles can reside on different copies of the mitochondrial "mini-chromosome." A typical cell has hundreds to thousands of mitochondria, each of which has many copies of its mini-chromosome.

Heteroplasmy

The fact that a cell contains many mitochondria makes possible a condition called **heteroplasmy**, in which a mutation is in some copies of mitochondrial chromosomes but not others. At each cell division, the mitochondria are distributed at random into daughter cells. Over time, the chromosomes within a mitochondrion tend to be all wild type or all mutant for any particular gene, but different mitochondria can have different alleles predominating. As an oocyte matures, the number of mitochondria drops from about 100,000 to 100 or fewer. If the woman is heteroplasmic for a mutation, by chance, she can produce an oocyte that has mostly mitochondria that are wild type, mostly mitochondria that have the mutation, or anything

Chapter 5 Beyond Mendel's Laws

Bioethics

Replacing Mitochondria

The tiny mitochondrial genome of 16,569 DNA base pairs can cause severe illness when any of its 37 genes, which encode a few proteins, rRNAs, and tRNAs, is mutant. A way to prevent a mitochondrial disease, theoretically, is to intervene at conception and replace the mitochondria in an oocyte from a woman who has a mitochondrial mutation with mitochondria that do not have mutations.

Researchers replace mitochondria in several ways: They (1) transfer the male and female pronuclei from an *in vitro* fertilized ovum into an oocyte donated from a healthy woman that has had its nucleus removed, so that the pronuclei share a cytoplasm with donor mitochondria; (2) transfer the female pronucleus into a donated oocyte with its nucleus removed and then fertilize that cell *in vitro* or (3) transfer the spindle apparatus, to which mitochondria cling, from a donor cell to a fertilized ovum. However it happens, the resulting first cell has two haploid genomes plus the small amount of mitochondrial DNA from the donor, a third individual. In 2016 a boy was born using the second method, and so far is healthy. Two siblings had died from Leigh syndrome.

Mitochondrial replacement has worked in monkeys but in other animals has led to developmental problems, including accelerated aging, poor growth, lowered fertility, and early death. Some researchers caution that the rarity of mitochondrial diseases, compared to the high mutation rate of mitochondrial DNA, suggests that embryos that inherit the diseases are naturally weeded out. Another complicating factor may be epigenetic effects from a cytoplasm on a nucleus from a different individual.

When the U.S. Food and Drug Administration held a meeting to discuss the science behind mitochondrial manipulation before the first child was born from such a procedure, a public comment session dealt with the bioethics of creating a "three-parent" embryo. Another objection came from a young woman who has suffered with a mitochondrial disease for more than a decade. She suggested that research regarding mitochondrial disease try to help those who are severely affected by it.

Questions for Discussion

1. Do you think that a person resulting from mitochondrial replacement would actually have three parents, as many popular accounts of the technology claim? If so, is creation of such an individual unethical, and why?
2. What is an alternative to mitochondrial replacement?
3. What are the potential risks and benefits of mitochondrial replacement?

in between (**figure 5.8**). In this way, a woman who does not have a mitochondrial disease, because the mitochondria bearing the mutation are either rare or not abundant in affected cell types, can nevertheless pass the associated condition to a child. Therefore, mitochondrial inheritance is both complex and unpredictable.

Heteroplasmy has several consequences on phenotypes. Expressivity may vary widely among siblings, depending upon how many mutation-bearing mitochondria were in the oocyte that became each individual when fertilized. Severity of symptoms reflects which tissues have cells whose mitochondria bear the mutation. This is the case for a

Figure 5.8 Mitochondrial inheritance. Mitochondria and their genes are passed only from the mother. Cells have many mitochondria. If an oocyte is heteroplasmic, differing numbers of copies of a mitochondrial mutation may be transmitted. The phenotype arises from the proportion of mitochondria bearing the mutation.

family with Leigh syndrome, which affects the enzyme that directly produces ATP. Two boys died of the severe form of the disease, because the brain regions that control movement rapidly degenerated. Another sibling was blind. Several relatives, however, had only mild impairment of their peripheral vision. The more severely affected family members had more brain cells that received the mutation-bearing mitochondria.

The most severe mitochondrial illnesses are heteroplasmic. This is presumably because *homo*plasmy—when all mitochondria bear the mutant allele—too severely impairs protein synthesis or energy production for embryonic development to complete. Often, severe heteroplasmic mitochondrial diseases do not produce symptoms until adulthood, because it takes many cell divisions, and therefore years, for a cell to receive enough mitochondria bearing mutant alleles to cause symptoms. **A Glimpse of History** describes the effect of heteroplasmy on solving a royal mystery.

About 1 in 10 of the DNA bases in the mitochondrial genome is heteroplasmic—that is, the bases at the same site can differ within an individual. The mutations that generate these single-base variations in mtDNA probably occur all the time, but only those that originate early in development have a chance to accumulate enough to be detectable.

Mitochondrial DNA Reveals the Past

Interest in mtDNA extends beyond the medical. In forensics it is used to link suspects to crimes. The forensic usefulness of mtDNA is based on the fact that it is more likely than nuclear DNA to remain after extensive damage to a body, because cells have many copies of it.

A GLIMPSE OF HISTORY

On a July night in 1918, Tsar Nicholas II of Russia and his family, the royal Romanovs, were shot, their bodies damaged with acid, and buried. In another July, in 1991, two amateur historians found the grave. DNA testing revealed the number of buried people, Y chromosome sequences distinguished males from females, and mitochondrial DNA sequences identified a mother (the tsarina) and three daughters.

When researchers consulted the DNA of modern-day relatives to link the remains to the royal family, they encountered a problem: The remains of the suspected tsar and his living great-grandniece Xenia differed at nucleotide position 16169 in the mitochondrial DNA. More puzzling, retesting the remains showed that at this site in the mitochondrial genome, the purported tsar had in some samples thymine (T), yet in others cytosine (C).

In yet another July, in 1994, researchers dug up the body of Nicholas's brother, the Grand Duke Georgij Romanov, and solved the mystery. Mitochondrial DNA nucleotide position 16169 in bone cells from the grand duke also had either T or C. The family had heteroplasmy.

Analysis of mtDNA can also identify war dead and support or challenge historical records. For example, sequencing mtDNA identified the son of Marie Antoinette and Louis XVI, who supposedly died in prison at age 10. In 1845, the boy was given a royal burial, but some people thought the buried child was an imposter. His heart had been stolen at the autopsy, and through a series of bizarre events, wound up, dried out, in the possession of the royal family. A few years ago, researchers compared mtDNA sequences from cells in the boy's heart to corresponding sequences in heart and hair cells from Marie Antoinette (her decapitated body identified by her fancy underwear), two of her sisters, and living relatives Queen Anne of Romania and her brother. The mtDNA evidence showed that the buried boy was indeed the prince, Louis XVII. Chapter 17 discusses how researchers consult mtDNA sequences to reconstruct ancient migration paths.

Key Concepts Questions 5.2

*a. Describe the inheritance pattern of mitochondrial DNA.

b. Why is mitochondrial DNA more prone to mutation than DNA in the nucleus?

*c. Define *heteroplasmy*.

d. Explain why mitochondrial DNA is useful in forensic investigations.

5.3 Linkage

The genes responsible for the traits that Mendel studied in pea plants were on different chromosomes. When genes are close to each other on the same chromosome, they usually do not segregate at random during meiosis and therefore their expression does not support Mendel's predictions. Instead, genes close on a chromosome are packaged into the same gametes and are said to be "linked" (**figure 5.9**).

Linkage in genetics refers to the transmission of genes on the same chromosome. Linked genes do *not* assort independently and do *not* produce Mendelian ratios for crosses tracking two or more genes. Understanding and using linkage as a mapping tool helped to identify many disease-causing genes before genome sequencing became possible. Today, the principle behind genetic linkage lies behind ancestry testing. The closer relatives are, the more segments of the same linked gene variants they have in common (see section 18.3).

Discovery in Pea Plants

William Bateson and R. C. Punnett first observed the unexpected ratios indicating linkage in the early 1900s, again in pea plants. They crossed true-breeding plants with purple flowers

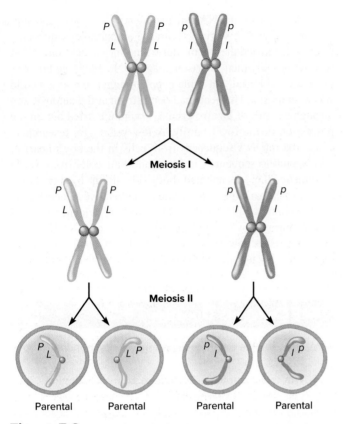

Figure 5.9 Inheritance of linked genes. Genes linked closely to one another are usually inherited together when the chromosome is packaged into a gamete.

Figure 5.10 Expected results of a dihybrid cross. (a) Unlinked genes assort independently. The gametes represent all possible allele combinations. The expected phenotypic ratio of a dihybrid cross is 9:3:3:1. **(b)** If genes are linked, only two allele combinations are expected in the gametes. The phenotypic ratio is 3:1, the same as for a monohybrid cross.

and long pollen grains (genotype *PPLL*) to true-breeding plants with red flowers and round pollen grains (genotype *ppll*). The plants in the next generation, of genotype *PpLl*, were then self-crossed. But this dihybrid cross did not yield the expected 9:3:3:1 phenotypic ratio that Mendel's second law predicts **(figure 5.10)**.

Bateson and Punnett noticed that two types of third-generation peas, those with the parental phenotypes *P_L_* and *ppll*, were more abundant than predicted, while the other two progeny classes, *ppL_* and *P_ll*, were less common (the blank indicates that the allele can be dominant or recessive). The more prevalent parental allele combinations, Bateson and Punnett hypothesized, could be due to genes that are transmitted on the same chromosome and that therefore do not separate during meiosis. The two less common offspring classes could also be explained by a meiotic event, crossing over. Recall that crossing over is an exchange between homologs that mixes up maternal and paternal gene combinations without disturbing the sequence of genes on the chromosome **(figure 5.11)**.

Progeny that exhibit this mixing of maternal and paternal alleles on a single chromosome are called **recombinant**. *Parental* and *recombinant* are context-dependent terms. Had the parents in Bateson and Punnett's crosses been of genotypes *ppL_* and *P_ll*, then *P_L_* and *ppll* would be recombinant rather than parental classes.

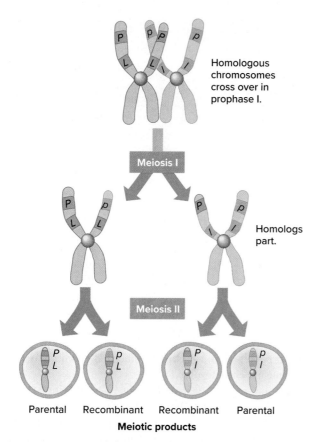

Figure 5.11 Crossing over disrupts linkage. The linkage between two genes may be interrupted if the chromosome they are on crosses over with its homolog between the two genes. Crossing over packages recombinant groupings of the genes into gametes.

P = Purple flowers
p = Red flowers
L = Long pollen grains
l = Round pollen grains

Figure 5.12 Allele configuration is important. Parental chromosomes can be distinguished from recombinant chromosomes only if the allele configuration of the two genes is known—they are either in *cis* **(a)** or in *trans* **(b)**.

Two other terms describe the configurations of linked genes in dihybrids. Consider a pea plant with genotype *PpLl*. These alleles can be part of the chromosomes in either of two ways. If the two dominant alleles are on one chromosome and the two recessive alleles on the other, the genes are in *"cis."* In the opposite configuration, with one dominant and one recessive allele on each chromosome, the genes are in *"trans"* (**figure 5.12**). Whether alleles in a dihybrid are in *cis* or *trans* is important in distinguishing recombinant from parental progeny classes in specific crosses in species used in experiments like peas and mice, and in clinical cases for people.

Linkage Maps

As Bateson and Punnett were discovering linkage in peas, geneticist Thomas Hunt Morgan and his coworkers in Columbia University's "fly room" were doing the same using the fruit fly, *Drosophila melanogaster*. They assigned genes to relative positions on chromosomes and compared progeny class sizes to assess whether traits were linked. The pairs of traits fell into four groups. Within each group, crossed dihybrids did not produce offspring classes according to Mendel's second law. Also, the number of linkage groups—four—matched the number of chromosome pairs in the fly. Coincidence? No. The traits fell into four groups because their genes are inherited together on the same chromosome.

The genius of the work on linkage in fruit flies was twofold. First, the researchers used test crosses (see figure 4.5) to follow parental versus recombinant progeny. Second, the fly room investigators translated their data into actual maps depicting positions of genes on chromosomes.

The idea for a genetic map grew out of Morgan's observation that the sizes of the recombinant classes varied for different genes. In 1911 he hypothesized that the farther apart two genes are on a chromosome, the more likely they are to cross over simply because more physical distance separates them (**figure 5.13**). He proposed that the size of a recombinant class is directly proportional to the distance between the genes on the chromosome they share. Then undergraduate student Alfred Sturtevant devised a way to represent the correlation between crossover frequency and the distance between genes as a **linkage map**. These diagrams showed the order of genes on chromosomes and the relative distances between them. The distance was represented using "map units" called **centimorgans** (abbreviated cM), where 1 centimorgan equals 1 percent recombination. Linkage maps of human genes were

Figure 5.13 Breaking linkage. Crossing over is more likely between widely spaced linked genes *A* and *B*, or between *A* and *C*, than between more closely spaced linked genes *B* and *C* because there is more room for an exchange to occur.

Table 5.2 Shared Centimorgans Indicate Relationships

Range of Shared DNA (centimorgans)	Possible Relationship	% Genome Shared
3,400–3,475	Parent to child; full siblings	50
2,400–2,800	Full sibling (including DZ twins)	50
1,450–2,050	Half sibling, aunt, uncle, grandparent	25
680–1,150	First cousin, great-grandparent	12.5
200–620	Second cousin	3.12
90–180	Third cousin	0.781
20–85	Fourth cousin	0.195
6–20	Distant cousins	<0.059

Source: Data from AncestryDNA

important in the initial sequencing of the human genome and are still used to estimate genetic distance along a chromosome. **Table 5.2** shows how the extent of shared centimorgans is used to estimate relationships between individuals in ancestry testing.

The frequency of a crossover between any two linked genes is inferred from the proportion of offspring from a cross that are recombinant. Frequency of recombination is based on the percentage of meiotic divisions that break the linkage between—that is, separate—two parental alleles. Genes at opposite ends of the same chromosome cross over often, generating a large recombinant class. Genes lying very close on the chromosome would only rarely be separated by a crossover. The probability that genes on opposite ends of a chromosome cross over approaches the probability that, if on different chromosomes, they would independently assort—about 50 percent. **Figure 5.14** illustrates this distinction.

The situation with linked genes is like a street lined with stores on both sides. There are more places to cross the street between stores at opposite ends on opposite sides than between two stores in the middle of the block on opposite sides of the street. Similarly, more crossovers, or progeny with recombinant genotypes, are seen when two genes are farther apart on the same chromosome.

Geneticists at Columbia University mapped several genes on all four fruit fly chromosomes, while researchers in other labs assigned genes to the human X chromosome. Localizing genes on the X chromosome was easier than doing so on the autosomes, because X-linked traits follow an inheritance pattern that is distinct from the one all autosomal genes follow. In human males, with their single X chromosome, recessive alleles on the X are expressed and observable. Chapter 6 returns to this point.

By 1950, geneticists began to think about mapping genes on the 22 human autosomes. To start, a gene must be matched to its chromosome. This became possible when geneticists identified people with a particular inherited condition or trait and an unusual chromosome.

In 1968, researcher R. P. Donohue was looking at chromosomes in his own white blood cells when he noticed a dark area consistently located near the centromere of one member of his largest chromosome pair (chromosome 1). He examined chromosomes from several family members for the dark area, noting also whether each family member had a blood type called Duffy. (Recall that blood types refer to the patterns of sugars on red blood cell surfaces.) Donohue found that the Duffy blood type was linked to the chromosome variant. That is, he could predict a relative's Duffy blood type by whether or not the chromosome had the dark area. This was the first assignment of a trait in humans to an autosome.

Finding a chromosomal variation linked to a family trait like Donohue did was unusual. More often, researchers mapped genes in experimental organisms, such as fruit flies, by calculating percent recombination (crossovers) between two genes with known locations on a chromosome. However, because human parents do not have hundreds of offspring, nor do they produce a new generation every 10 days, getting enough information to establish linkage relationships for us requires observing the same traits in many families and pooling the results. Today, even though we can sequence human genomes, linkage remains a powerful tool to track disease-associated genes.

Solving Linkage Problems Uses Logic

To illustrate determining the degree of linkage by percent recombination using families, consider the traits of Rh blood type and a form of anemia called elliptocytosis. An Rh-positive phenotype corresponds to genotypes *RR* or *Rr*. (This is simplified.) The anemia corresponds to genotypes *EE* or *Ee*.

In 100 one-child families, one parent is Rh negative with no anemia (*rree*), and the other parent is Rh positive with anemia (*RrEe*), and the *R* and *E* (or *r* and *e*) alleles are in *cis*. Of

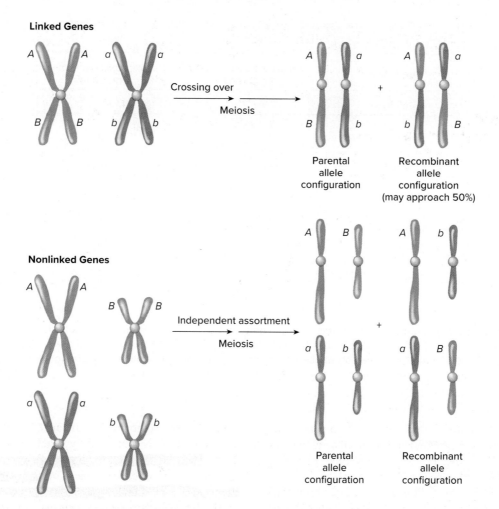

Figure 5.14 Linkage versus nonlinkage (independent assortment). When two genes are widely separated on a chromosome, the likelihood of a crossover is so great that the recombinant class may approach 50 percent—which may appear to be the result of independent assortment.

the 100 offspring, 96 have parental genotypes (*re/re* or *RE/re*) and four are recombinants for these two genes (*Re/re* or *rE/re*). Percent recombination is therefore 4 percent, and the two linked genes are 4 centimorgans apart.

Another pair of linked genes in humans is the gene that underlies nail-patella syndrome, a rare autosomal dominant disease that causes absent or underdeveloped fingernails and toenails and painful arthritis in the knee and elbow joints, and the *I* gene that determines the ABO blood type, on chromosome 9. These two genes are 10 map units apart, which geneticists determined by pooling information from many families. The information is used to predict genotypes and phenotypes in offspring, as in the following example.

Greg and Susan each have nail-patella syndrome. Greg has type A blood. Susan has type B blood. What is the chance that their child inherits normal nails and knees and type O blood? A genetic counselor deduces their allele configurations using information about the couple's parents (**figure 5.15**).

Greg's mother has nail-patella syndrome and type A blood. His father has normal nails and type O blood. Therefore, Greg must have inherited the dominant nail-patella syndrome allele (*N*) and the I^A allele from his mother, on the same chromosome. We know this because Greg has type A blood and his father has type O blood—therefore, he couldn't have gotten the I^A allele from his father. Greg's other chromosome 9 must carry the alleles *n* and *i*. His alleles are therefore in *cis*.

Susan's mother has nail-patella syndrome and type O blood, and so Susan inherited *N* and *i* on the same chromosome. Because her father has normal nails and type B blood, her homolog from him bears alleles *n* and I^B. Her alleles are in *trans*.

Determining the probability that Susan and Greg's child could have normal nails and knees and type O blood is the easiest question the couple could ask. The only way this genotype can arise from theirs is if an *ni* sperm (which occurs with a frequency of 45 percent, based on pooled data) fertilizes an *ni* oocyte (which occurs 5 percent of the time). The result—according to the product rule—is a 2.25 percent chance of producing a child with the *nnii* genotype.

Calculating other genotypes for their offspring is more complicated, because more combinations of sperm and oocytes

	Greg	Susan
Phenotype	nail-patella syndrome, type A blood	nail-patella syndrome, type B blood
Genotype	NnI^Ai	$NnI^B\_$
Allele configuration	$\dfrac{N\ I^A}{n\ i}$	$\dfrac{N\ i}{n\ I^B}$
Gametes:	sperm — frequency — oocytes	
Parental	$N\ I^A$ — 45% — $N\ i$	
	$n\ i$ — 45% — $n\ I^B$	
Recombinants	$N\ i$ — 5% — $N\ I^B$	
	$n\ I^A$ — 5% — $n\ i$	

N = nail-patella syndrome
n = normal

Figure 5.15 Inheritance of nail-patella syndrome. Greg inherited the N and I^A alleles from his mother; that is why the alleles are on the same chromosome. His n and i alleles must therefore be on the homolog. Susan inherited alleles N and i from her mother, and n and I^B from her father. Population-based probabilities are used to calculate the likelihood of phenotypes in the offspring of this couple. Note that in this case, map distances are known and are used to predict outcomes.

could account for them. For example, a child with nail-patella syndrome and type AB blood could arise from all combinations that include I^A and I^B as well as at least one N allele (assuming that NN has the same phenotype as Nn).

The Rh blood type and elliptocytosis, and nail-patella syndrome and ABO blood type, are examples of linked gene pairs. A linkage map begins to emerge when percent recombination is known between all possible pairs of three or more linked genes, just as a road map with more landmarks provides more information on distance and direction. Consider genes x, y, and z (**figure 5.16**). If the percent recombination between x and y is 10, between x and z is 4, and between z and y is 6, then the order of the genes on the chromosome is x-z-y, the only order that accounts for the percent recombination data. It is a little like deriving a geographical map from knowing the distances between cities.

Genetic maps derived from percent recombination between linked genes accurately reflect the order of the genes on the chromosome, but the distances are estimates because crossing over is not equally likely across the genome. Some DNA sequences are nearly always inherited together, like two inseparable friends. This nonrandom association between DNA sequences is called **linkage disequilibrium (LD)**. The human genome consists of many LD blocks, where stretches of alleles stick together, interspersed with areas where crossing over is prevalent. Chapters 7, 16, and 17 discuss the use of LD blocks, called **haplotypes**, to track genes in populations.

From Linkage to Genome-Wide Associations

The first human genes mapped to their chromosomes encoded blood proteins, because these were easy to study. In 1980, researchers began using DNA sequences near genes of interest as landmarks called **genetic markers**. These markers need not encode proteins that cause a phenotype. They might be DNA sequence differences that alter where a DNA cutting enzyme cuts, differing numbers of short repeated sequences of DNA with no obvious function, or single sites where the base varies among individuals. The term "genetic marker" is used more generally to mean any DNA sequence that is associated with a phenotype, usually one affecting health.

Algorithms tally how often genes and their markers are inherited together. The "tightness" of linkage between a marker and a gene of interest is represented as a LOD score, where *LOD* stands for "logarithm of the odds." A LOD score indicates the likelihood that particular crossover frequency data indicate linkage, rather than the inheritance of two alleles by chance. The higher the LOD score, the closer the two genes on a chromosome.

Figure 5.16 Recombination mapping. If we know the percent recombination between all possible pairs of three genes, we can determine their relative positions on the chromosome.

A LOD score of 3 or greater signifies linkage. It means that the observed data are 1,000 (10^3) times more likely to have occurred if the two DNA sequences (a disease-causing allele and its marker) are linked than if they reside on different chromosomes and just happen to often be inherited together by chance. It is like deciding whether two coins tossed together 1,000 times always come up both heads or both tails by chance, or because they are taped together side by side in that position, as are linked genes. If the coins land with the same sides up in all 1,000 trials, it indicates they are likely taped.

Before sequencing of the first human genomes, genetic markers were used to predict which individuals in some families were most likely to have inherited a particular disease, before symptoms began. To predict Huntington disease, detecting a linked marker was done from 1983 until 1993, when the *huntingtin* gene was discovered (see the Opener and Bioethics reading in chapter 4). Such indirect analyses are no longer necessary, because tests directly detect disease-causing genes, allowing more accurate predictions.

Genetic markers are still used to distinguish parts of chromosomes. In pedigrees, designations for markers linked into haplotypes are sometimes placed beneath the symbols to further describe chromosomes. In the family with cystic fibrosis depicted in **figure 5.17**, each set of numbers beneath a symbol represents a haplotype. Knowing the haplotype of individual II–2 reveals which chromosome in parent I–1 contributed the mutant allele. Because Mr. II–2 received haplotype 3233 from his affected mother, his other haplotype, 2222, came from his father. Because Mr. II–2 is affected and his father is not, the father must be a heterozygote, and 2222 must be the haplotype linked to the mutant *CFTR* allele.

Today, entire human genomes can be sequenced in under a day. However, researchers are still filling in the orders of genes on chromosomes and determining gene functions and interactions.

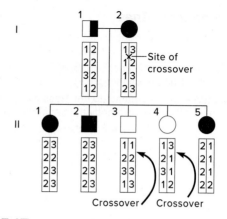

Figure 5.17 Haplotypes. The numbers in bars beneath pedigree symbols enable researchers to track specific chromosome segments with markers. Disruptions of a marker sequence indicate crossover sites.

Key Concepts Questions 5.3

*a. Define *genetic linkage*.
b. Why are linked genes inherited in different patterns than unlinked genes?
c. What is the relationship between crossover frequency and relative positions of genes on chromosomes?
*d. What is a linkage map?
*e. What is linkage disequilibrium?
*f. What is a genetic marker?
*g. What is a LOD score?
*h. Define *haplotype*.

Study Guide

Summary

5.1 When Gene Expression Appears to Alter Mendelian Ratios

1. Homozygosity for lethal recessive alleles stops development before birth, eliminating an offspring class.
2. A gene can have multiple alleles because its DNA sequence can vary in many ways. Different allele combinations may produce different expressions of the phenotype.
3. Heterozygotes of **incompletely dominant** alleles have phenotypes intermediate between those associated with the two homozygotes. **Codominant** alleles are both expressed in the phenotype.
4. An **antigen** is a molecule on a cell surface to which the immune system responds.
5. In **epistasis**, one gene affects the phenotype of another.
6. An incompletely **penetrant** genotype is not expressed in all individuals who inherit it. A phenotype has variable **expressivity** if it varies in intensity among individuals.
7. **Pleiotropic** genes have several expressions due to functioning in shared pathways, processes, or structures.
8. In **genetic heterogeneity**, two or more genes specify the same phenotype.
9. A **phenocopy** is a characteristic that appears to be inherited but is environmentally caused.

Chapter 5 *Beyond Mendel's Laws* 107

5.2 Mitochondrial Genes

10. Cells have many mitochondria, and each mitochondrion has many copies of the mitochondrial genome.
11. Only females transmit mitochondrial genes; males can inherit such a trait but cannot pass it on.
12. Mitochondrial chromosomes do not cross over, and mtDNA mutates more frequently than nuclear DNA.
13. The 37 mitochondrial genes encode tRNA, rRNA, or proteins involved in protein synthesis or energy reactions.
14. Many mitochondrial diseases are due to **heteroplasmy**, in which mitochondria in a single cell have mutations in different alleles.

5.3 Linkage

15. Genes on the same chromosome are **linked** and, unlike genes that independently assort, produce many individuals with parental genotypes and a few with **recombinant** genotypes.
16. **Linkage maps** depict the order of genes on a chromosome and the distances between them. **Centimorgans** measure distance along a chromosome in terms of the percent crossing over.
17. **Linkage disequilibrium** refers to stretches of alleles along a chromosome that are transmitted together.
18. We can predict the probabilities that certain genotypes will appear in progeny if we know crossover frequencies from pooled data and whether linked alleles are in *cis* or in *trans*.
19. **Genetic markers** are DNA sequences near a gene of interest that are inherited with it due to linkage.
20. A LOD score indicates how tightly a gene and a genetic marker are linked.
21. Groups of linked genes that are always inherited together and not separated by crossing over are termed **haplotypes**.

Key Terms

antigen (ăn′tē-jən) A molecule that elicits an immune response.

centimorgan A unit that indicates percent recombination between two loci linked on a chromosome, serving as a measure of relative distance.

codominant A form of gene expression in which both alleles are fully expressed in a heterozygote.

epistasis (ĕpē-stā-sis) A gene masking or affecting the expression of another gene.

expressivity Severity or degree of a phenotype.

genetic heterogeneity A phenotype that can be caused by variants of any of several genes.

genetic marker A DNA sequence near a gene of interest that is co-inherited unless separated by a crossover.

haplotype (hăp′ lō tīp) A series of genes linked on a chromosome that do not separate by crossing over.

heteroplasmy (hĕt′ə-rō-plăz-mē) Mitochondria in the same cell that have different alleles of a gene.

incomplete dominance A form of gene expression that results in a heterozygote intermediate in phenotype between either homozygote.

linkage Genes on the same chromosome.

linkage disequilibrium (LD) A consequence of extremely tight linkage between DNA sequences in which two genes or DNA sequences are nearly always inherited together.

linkage maps Diagrams that show gene order on chromosomes, determined from crossover frequencies between pairs of genes.

penetrance Percentage of individuals with a genotype who have an associated phenotype.

phenocopy (fē′ nō-kŏp′ē) An environmentally caused trait that occurs in a familial pattern, mimicking inheritance.

pleiotropic (plēə-trōpĭk) A single-gene disease with several symptoms or a gene that controls several functions or has more than one effect, causing different symptom subsets in different individuals.

recombinant (rē-kŏm′bə-nənt) A series of alleles on a chromosome that differs from the series of either parent.

Review Questions

1. Explain how each of the following phenomena can disrupt Mendelian phenotypic ratios.
    a. lethal alleles
    b. multiple alleles
    c. incomplete dominance
    d. codominance
    e. epistasis
    f. incomplete penetrance
    g. variable expressivity
    h. pleiotropy
    i. a phenocopy
    j. genetic heterogeneity
2. How does the relationship between dominant and recessive alleles of a gene differ from epistasis?
3. Why can transmission of an autosomal dominant trait with incomplete penetrance look like autosomal recessive inheritance?
4. How does inheritance of ABO blood types exhibit both complete dominance and codominance?
5. How could two people with albinism have a child who has normal skin pigment?
6. Name two diseases discussed in the chapter that demonstrate genetic heterogeneity.
7. Explain ways that genome sequencing is showing that phenomena that we once thought were exceptions to Mendel's laws and quite rare are actually more common.

8. How can epistasis explain incomplete penetrance?

9. The lung condition emphysema may be caused by lack of an enzyme or by smoking. In which situation is the condition a phenocopy?

10. List three ways that mtDNA differs from DNA in a cell's nucleus.

11. Describe why inheritance of mitochondrial DNA and linkage do not follow Mendel's laws.

12. Explain how a pedigree for a maternally inherited trait differs from one for an autosomal dominant trait.

Applied Questions

1. For each of the diseases described in situations (a) through (i), indicate which of the following phenomena (A–H) is at work. More than one may apply.

   A. lethal alleles
   B. multiple alleles
   C. epistasis
   D. incomplete penetrance
   E. variable expressivity
   F. pleiotropy
   G. a phenocopy
   H. genetic heterogeneity

   a. A woman with severe neurofibromatosis type 1 has brown spots on her skin and several large tumors beneath her skin. A genetic test shows that her son has the disease-causing autosomal dominant allele, but he has no symptoms.
   b. A man would have a widow's peak, if he wasn't bald.
   c. A man and woman have six children. They also had two stillbirths—fetuses that stopped developing shortly before birth.
   d. Mutations in a gene that encodes a muscle protein called titin cause 22 percent of cases of inherited dilated cardiomyopathy, a form of heart disease. Other single genes cause the other cases.
   e. A woman with dark brown skin uses a bleaching cream that darkens her fingertips and ears, making her look like she has the inherited disease alkaptonuria.
   f. In Labrador retrievers, the B allele confers black coat color and the b allele brown coat color. The E gene controls the expression of the B gene. If a dog inherits the E allele, the coat is golden no matter what the B genotype is. A dog of genotype ee expresses the B (black) phenotype.
   g. Two parents are heterozygous for genes that cause albinism, but each gene specifies a different enzyme in the biochemical pathway for skin pigment synthesis. Their children thus do not face a 25 percent risk of having albinism.
   h. All children with Aicardi syndrome have defects in their retinas that look like white craters (chorioretinal lacunae), but only some of the 4,000 recognized cases also have absence of the corpus callosum (the band of nerve fibers that joins the halves of the brain).
   i. Two young children in a family have very decayed teeth. Their parents think it is genetic, but the true cause is a babysitter who puts them to sleep with juice bottles in their mouths.

2. A man who has type O blood has a child with a woman who has type A blood. The woman's mother has AB blood, and her father, type O. What is the probability that the child has type O blood?

3. Blood banks use enzymes to remove the A and B antigens from blood types A and B, making the blood type O. Does this process alter the phenotype or the genotype?

4. What is the chance that Greg and Susan, the couple with nail-patella syndrome, could have a child with normal nails and type AB blood?

5. A Martian creature called a gazook has 17 chromosome pairs. On the largest chromosome are genes for three traits—round or square eyeballs (R or r); a hairy or smooth tail (H or h); and 9 or 11 toes (T or t). Round eyeballs, hairy tail, and 9 toes are dominant to square eyeballs, smooth tail, and 11 toes. A trihybrid male has offspring with a female who has square eyeballs, a smooth tail, and 11 toes on each of her three feet. She gives birth to 100 little gazooks, who have the following phenotypes:

   - 40 have round eyeballs, a hairy tail, and 9 toes
   - 40 have square eyeballs, a smooth tail, and 11 toes
   - 6 have round eyeballs, a hairy tail, and 11 toes
   - 6 have square eyeballs, a smooth tail, and 9 toes
   - 4 have round eyeballs, a smooth tail, and 11 toes
   - 4 have square eyeballs, a hairy tail, and 9 toes

   a. Draw the allele configurations of the parents.
   b. Identify the parental and recombinant progeny classes.
   c. What is the crossover frequency between the R and T genes?

Forensics Focus

1. "Earthquake McGoon" was 32 years old when the plane he was piloting over North Vietnam was hit by groundfire on May 6, 1954. Of the five others aboard, only two survived. McGoon, actually named James B. McGovern, was well known for his flying in World War II, and for his jolliness. Remains of a man about his height and age at death were discovered in late 2002 but could not be identified by dental records. However, DNA sampled from a leg bone enabled forensic scientists to identify him. Name the type of DNA likely analyzed, and describe the further, nontechnical information that was needed to make the identification.

Case Studies and Research Results

1. Uncombable hair syndrome causes silvery or straw-colored dry, frizzy, wiry hair that sticks out in all directions and is impossible to comb. Mutations in any of three genes cause the syndrome: *TCHH* (which encodes a protein that forms the hair shaft) and *PAD13* and *TGM3* (which encode enzymes that trim the TCHH protein into an active form). Albert Einstein may have had the condition. A study of five affected children had these results:

   Patient 1 has identical alleles of *PAD13*.
   Patient 2 has identical alleles of *PAD13* different from the one that patient 1 has.
   Patient 3 has identical alleles of *TCHH*.
   Patient 4 has different alleles of *TGM3*.
   Patient 5 has the same *TCHH* alleles as patient 3 but has much more abnormal hair.

   a. Which patient(s) is a compound heterozygote?
   b. What is the evidence for genetic heterogeneity for uncombable hair syndrome?
   c. What is the evidence for variable expressivity?

2. Suzanne is deaf. In early childhood, she began having fainting spells, especially when she became excited. When she fainted while opening Christmas gifts, her parents took her to the hospital, where doctors couldn't identify a problem. As the spells continued, Suzanne became able to predict the attacks, telling her parents that her head hurt. Finally a neurologist checked Suzanne's heart and diagnosed long QT syndrome with deafness, which is a severe form of inherited heartbeat irregularity. Ten genes can cause long QT syndrome. The doctor told them of a case from 1856: A young girl, called at school to face the headmaster for an infraction, became so agitated that she suddenly died. The parents had lost two other children to great excitement.

   The family visited a medical geneticist, who discovered that each parent had a mild heartbeat irregularity that did not produce symptoms. Suzanne's parents had normal hearing. Her younger brother Frank was also hearing impaired and suffered night terrors but had so far not fainted. Like Suzanne, he had the full syndrome. Vanessa, still a baby, did not have either form of the family's illness; her heartbeat was normal.

   Today, Suzanne and Frank are treated with beta blocker drugs, and each has an implantable defibrillator to correct a potentially fatal heartbeat. Suzanne's diagnosis may have saved her brother's life.

   a. Which of the following applies to the condition in this family?
      i. genetic heterogeneity
      ii. pleiotropy
      iii. variable expressivity
      iv. incomplete dominance
      v. a phenocopy
   b. How is the inheritance pattern of this form of long QT syndrome consistent with that of familial hypercholesterolemia?
   c. Explain how it is possible that Vanessa did not inherit either the serious or the asymptomatic form of the disease.
   d. Do the treatments for the disease affect the genotype or the phenotype?

3. A 5-year-old girl who tired very easily and was losing strength was diagnosed with a mitochondrial disease. Genetic testing revealed two mutations in a gene that is on chromosome 1, but no mutations in mitochondrial genes. Explain how the risk to other children in this family differs from the risk of inheriting a mitochondrial disease caused by mutation in a mitochondrial gene.

CHAPTER 6

Matters of Sex

In 2016, the government of China officially ended the 35-year one-child policy that has profoundly skewed the sex ratio, with effects on population composition that will persist for years.

Learning Outcomes

6.1 Our Sexual Selves
1. Describe the factors that contribute to whether we are and feel male or female.
2. Distinguish between the X and Y chromosomes.
3. Discuss how manipulating the sex ratio can affect societies.

6.2 Traits Inherited on Sex Chromosomes
4. Distinguish between Y linkage and X linkage.
5. Compare and contrast X-linked recessive inheritance and X-linked dominant inheritance.

6.3 Sex-Limited and Sex-Influenced Traits
6. Discuss the inheritance pattern of a trait that appears in only one sex.
7. Define *sex-influenced trait*.

6.4 X Inactivation
8. Explain why X inactivation is necessary.
9. Explain how X inactivation is an epigenetic change.
10. Discuss how X inactivation affects the phenotype in female mammals.

6.5 Parent-of-Origin Effects
11. Explain the chemical basis of silencing the genetic contribution from one parent.
12. Explain how differences in the timetables of sperm and oocyte formation can lead to parent-of-origin effects.

The BIG Picture

Sex affects our lives in many ways. Which sex chromosomes we are dealt at conception sets the developmental program for the organs of a male or female body, but gene expression before and after birth greatly influences how that program unfolds.

China's One-Child Policy Created an Aged Population

China's one-child policy began with a letter from the Communist Party in September 1980 announcing a "voluntary" program to limit family size, to counter overpopulation. Couples having only one child enjoyed financial incentives from the government that were revoked if they had a second child.

Prenatal diagnosis was widely used to identify XX fetuses. Many families wanting their only child to be a boy did not continue or report pregnancies carrying girls. Forced pregnancy terminations and sterilizations and, rarely, infanticide of girl babies, all contributed to an unnaturally skewed sex ratio favoring males. The reasoning was societal: A son would care for his aging parents, but a daughter would care for her in-laws.

China's one-child policy prevented hundreds of millions of births. By 2005, 121 boys were being born for every 100 girls. In 2007, a committee of experts asked the government to end the policy because it created social problems, but the end did not come until 2016. Then began a two-child policy, but the effort may have been too late: the birth rate is still low, because there are too few women of childbearing age.

Today, many children in China have few siblings, cousins, aunts, or uncles. In rural "bachelor villages," young men cannot find brides.

Only-children dominate the younger generation with about 33 million more men than women, according to the Council on Foreign Relations.

By 2050, 25 percent of the population of China will be older than 65, most of them male, with few young people to care for them. China's one-child policy has been called the world's most radical social experiment.

6.1 Our Sexual Selves

Biological maleness or femaleness is determined at conception, when she inherits two X chromosomes or he inherits an X and a Y (figure 6.1). Another level of sexual identity comes from the control that hormones exert over the development of reproductive structures. Finally, both biological factors and social cues influence sexual feelings, including the strong sense of whether we are female or male, or something nonbinary—our gender identity.

Sexual Development

Physical gender differences become apparent around the ninth week of prenatal development. During the fifth week, all embryos develop two unspecialized gonads, which are organs that will develop as either testes or ovaries. Each such "bipotential" gonad forms near two sets of ducts that offer two developmental options. If one set of tubes, called the Müllerian ducts, continues to develop near potential ovaries, female sexual structures form. If the other set, the Wolffian ducts, persists near developing testes, male sexual structures form.

The option to follow either developmental pathway occurs during the sixth week, depending upon the sex chromosomes present and the actions of certain genes. If a gene on the Y chromosome called *SRY* (for "sex-determining region of the Y") is activated, hormones steer development along the male route. In the absence of *SRY* activation, a female develops (figure 6.2).

Femaleness was once considered a "default" option in human development, defined as the absence of maleness. Sex determination is more accurately described as a fate imposed

Figure 6.1 **The sex chromosomes.** A human female has two X chromosomes and a human male has one X chromosome and one Y chromosome. *Biophoto Associates/Photo Researchers/Science Source*

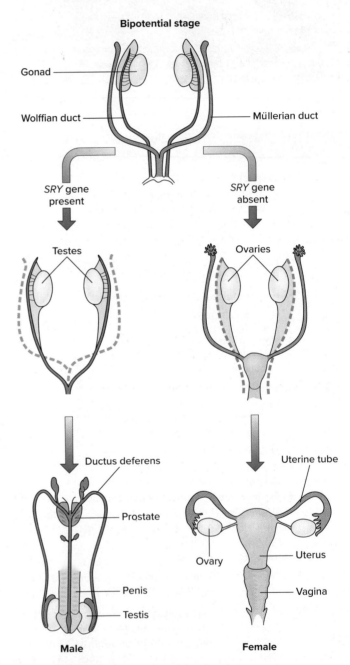

Figure 6.2 **Male or female?** The paired duct systems in the early human embryo may develop into male *or* female reproductive organs. The red tubes represent female structures and the blue tubes, male structures.

on ambiguous precursor structures. Several genes in addition to *SRY* guide early development of reproductive structures. A mutation in the gene *Wnt4*, for example, causes an XX female to have high levels of male sex hormones and lack a vaginal canal and uterus. Ovaries do not develop properly and secondary sex characteristics do not appear. Hence, *Wnt4* is essential for development and maturation as a female.

Sex Chromosomes

The sex with two different sex chromosomes is called the **heterogametic** sex, and the other, with two of the same sex chromosome, is the **homogametic** sex. In humans, males are heterogametic (XY) and females are homogametic (XX). Some other species are different. In birds and snakes, for example, males are ZZ (homogametic) and females are ZW (heterogametic).

The sex chromosomes differ in size and capacity. The human X chromosome has more than 1,500 genes and is much larger than the Y chromosome, which has 231 protein-encoding genes. In meiosis in a male, the X and Y chromosomes act as if they are a pair of homologs.

Identifying genes on the human Y chromosome has been difficult. Before human genomes were easily sequenced, researchers inferred the functions and locations of Y-linked genes by observing how men missing parts of the chromosome differed from normal. Creating linkage maps for the Y chromosome was not possible, because it does not cross over along all of its length and because of its highly unusual organization. About 95 percent of the chromosome harbors 22 male-specific genes, and many DNA segments here are palindromes. In written languages, palindromes are sequences of letters that read the same in both directions—"Madam, I'm Adam," for example. This symmetry in a DNA sequence destabilizes DNA replication so that during meiosis, sections of a Y chromosome attract each other, which can loop out parts in between. The resulting deletions in the Y account for some cases of male infertility.

The Y chromosome has a very short arm and a long arm (**figure 6.3**). At both tips are regions, called PAR1 and PAR2, which are termed "pseudoautosomal" because they can cross over with counterparts on the X chromosome. The pseudoautosomal regions comprise only 5 percent of the Y chromosome and include 63 pseudoautosomal genes that encode proteins that function in both sexes. These genes control bone growth, cell division, immunity, signal transduction, the synthesis of hormones and receptors, fertility, and energy metabolism.

Most of the Y chromosome is the male-specific region, or MSY, which lies between the two pseudoautosomal regions. Many of the genes in the MSY are essential to fertility, including *SRY*. The MSY consists of three classes of DNA sequences. About 10 to 15 percent of the MSY sequence is almost identical to parts of the X chromosome. Protein-encoding genes are scarce here. Another 20 percent of the MSY is somewhat similar to X chromosome sequences, and may be remnants of an ancient autosome that gave rise to the X chromosome. The rest of the MSY is riddled with palindromes. The MSY has many repeats and the specified proteins combine in different ways,

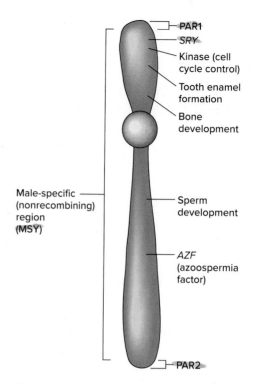

Figure 6.3 **Anatomy of the Y chromosome.** The Y chromosome has two pseudoautosomal regions (PAR1 and PAR2) and a large central area (MSY) that comprises about 95 percent of the chromosome and has only 22 genes. *SRY* determines male sex. *AZF* encodes a protein essential to producing sperm; mutations in it cause infertility.

making it possible for males to have 20, 21, or 22 genes in this region of the Y chromosome.

The Phenotype Forms

The *SRY* gene encodes an important type of protein called a **transcription factor**, which controls the expression of other genes. The SRY transcription factor stimulates male development by sending signals to yet another transcription factor called Sox9 that, in turn, stimulates development of male structures from the bipotential gonads while inhibiting further development of female structures.

Prenatal sexual development is a multistep process, and mutations can intervene at several points, causing "disorders of sex development," which affect the gonads or genitalia (**figure 6.4**). In androgen insensitivity syndrome, for example, a mutation in a gene on the X chromosome blocks formation of receptors for androgens (the male sex hormones testosterone and dihydrotestosterone [DHT]). As a result, cells in early reproductive structures do not receive signals to develop as male. The person looks female, but is XY.

Several terms are used to describe individuals whose genetic/chromosomal sex and physical structures, both internal and external, are not consistent with one gender. *Hermaphroditism* is an older and more general term for an individual with male and female sexual structures. *Intersex* refers to individuals whose

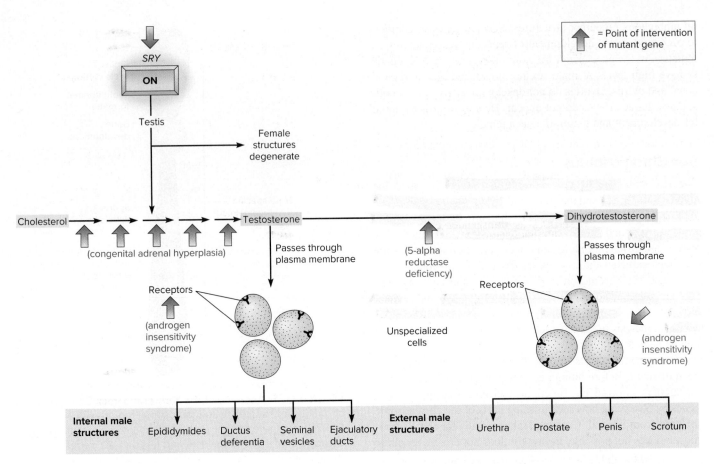

Figure 6.4 Mutations that affect male sexual development. In normal male prenatal development, activation of a set of genes beginning with *SRY* sends signals that destroy female rudiments, while activating the biochemical pathway that produces testosterone and dihydrotestosterone, which promote the development of male structures. The green arrows indicate where mutations disrupt normal sexual development. Diseases are identified in red. Note that the enzyme disruption in congenital adrenal hyperplasia results in excess testosterone and dihydrotestosterone, but the other two conditions lower the levels of the male hormones.

internal structures are inconsistent with external structures, or whose genitalia are ambiguous. It is the preferred term outside of medical circles. *Pseudohermaphroditism* refers to the presence of both female and male structures but at different life stages.

Living with pseudohermaphroditism can be confusing. Consider people who have 5-alpha reductase deficiency, which is autosomal recessive. Affected individuals have a normal Y chromosome, a wild type *SRY* gene, and testes. The internal male reproductive tract develops and internal female structures do not, so the male anatomy is present on the inside. But the child looks like a girl on the outside. Without 5-alpha reductase, which normally catalyzes the reaction of testosterone to form DHT, a penis cannot form. At puberty, when the adrenal glands, which sit atop the kidneys, start to produce testosterone, this XY individual, who thought she was female, starts to exhibit the signs of maleness—deepening voice, growth of facial hair, and the sculpting of muscles into a masculine physique. Instead of the girl developing breasts and menstruating, the clitoris enlarges into a penis. Usually sperm production is normal. XX individuals with 5-alpha reductase deficiency look female.

The degree to which pseudohermaphroditism disturbs the individual depends as much on societal acceptance as it does on genetics, illustrated by what happened in the Dominican Republic. In the 1970s, 22 young girls reached the age of puberty and began to change into boys. They had a form of 5-alpha reductase deficiency that was fairly prevalent in the population due to consanguinity (blood relatives having children). The parents were happy that they had sons after all, and so these special adolescents were given their own gender name—*guevedoces,* for "penis at age 12." They were fully accepted as whatever they wanted to be. This isn't always the case. A realistic novel, *Middlesex,* tells the story of a young man with this condition who grew up as a female.

In a more common form of pseudohermaphroditism, congenital adrenal hyperplasia (CAH) due to 21-hydroxylase deficiency, an enzyme block causes testosterone and DHT to accumulate. It is autosomal recessive, and both males and females are affected. The higher levels of androgens cause precocious puberty in males or male secondary sex characteristics to develop in females. A boy may enter puberty as early as 3 years of age, with well-developed musculature, small testes, and an enlarged penis. At birth, a girl may have a swollen clitoris that appears to be a small penis. The person is female internally, but as puberty begins the voice deepens, facial hair grows, and menstruation never happens.

Prenatal tests that detect chromosomal sex have changed the way that pseudohermaphroditism is diagnosed, from phenotype to genotype. Before, the condition was detected only after puberty, when a girl began to look like a boy. Today pseudohermaphroditism is suspected when a prenatal chromosome check reveals an X and a Y chromosome, but the newborn looks like a girl.

A person who identifies as transgender has a gender identity that does not match the gender assigned at birth based on genitalia. Chromosomal sex remains binary (female or male), but gender identity is a more "fluid" phenomenon. **Bioethics** describes a study that identified gene variants associated with transgender identity.

Table 6.1 Sexual Identity

Level	Events	Timing
Chromosomal/genetic sex	XY = male XX = female	Fertilization
Gonadal sex	Undifferentiated structure begins to develop as testis or ovary	6 weeks after fertilization
Phenotypic sex	Development of external and internal reproductive structures continues as male or female in response to hormones	8 weeks after fertilization, at puberty
Gender identity	Strong feelings of being male or female develop	From childhood, possibly earlier
Sexual orientation	Attraction to same or opposite sex	From childhood

Source: Courtesy Professor Jennifer A. Marshall-Graves, Australian National University.

Is Homosexuality Inherited?

In homosexuality, a person's phenotype and genotype are consistent, and physical attraction is toward members of the same sex. Homosexuality is seen in all human cultures and has been observed for thousands of years. It has been documented in more than 1500 animal species.

Homosexuality reflects complex input from genes and the environment. The genetic influence may be seen in the strong feelings that homosexual individuals have as young children, long before they know of the existence or meaning of the term. Other evidence comes from identical twins, who are more likely to both be homosexual than are both members of fraternal twin pairs. Twin studies compare a trait between identical and fraternal twin pairs, to estimate the rough proportion of the variability of a trait that can be attributed to genes. Chapter 7 discusses this approach further.

Experiments in the 1990s identified genetic markers on the X chromosome that were more often identical among pairs of homosexual brothers than among other pairs of brothers. This finding led to the idea that a single gene, or a few genes, dictates sexual preference, but these results could not be confirmed. Further studies on twins have indicated what many people have long suspected—that the roots of homosexuality are complex. Homosexuality likely reflects the input of several genes and environmental factors, and may arise in a variety of ways.

Another approach to searching for genes that contribute to homosexuality asked people who had DNA testing for the UK Biobank and a consumer genetics firm one question: *"Have you ever had sex with someone of the same sex?"* The researchers searched for single-base variant markers throughout the genome that were more common among the 26,890 individuals who reported at least one homosexual experience than among the 450,939 respondents who were exclusively heterosexual. They identified four gene variants, on different autosomes—not the X chromosome implicated in the twin studies in 1993. The researchers concluded that "many tiny genetic effects" may contribute to homosexuality. In other words, there is no one specific gene that determines attraction to the same or opposite sex.

Table 6.1 summarizes the components of sexual identity.

Sex Ratio

Mendel's law of segregation predicts that populations should have approximately equal numbers of male and female newborns. That is, male meiosis should yield equal numbers of X-bearing and Y-bearing sperm. After birth, societal and environmental factors may favor survival of one sex over the other.

The proportion of males to females in a human population is called the **sex ratio**. It is calculated as the number of males divided by the number of females multiplied by 1,000, for people of a particular age. (Some organizations describe sex ratio based on 1.0.) A sex ratio of equal numbers of males and females would be designated 1,000. The sex ratio at conception is called the primary sex ratio. In the United States for the past six decades, newborn boys have slightly outnumbered newborn girls, with the primary sex ratio averaging 1,050. Boys are slightly in excess because Y-bearing sperm weigh slightly less than X-bearing sperm, and so they may reach the oocyte faster and be more likely to enter.

The sex ratio at birth is termed secondary, and at maturity it is called tertiary. Sex ratio can change markedly with age due to illnesses and environmental factors that affect the sexes differently, such as participating in wars and giving birth. When a society intentionally alters the sex ratio, results can be drastic, as the chapter opener discusses.

At the other end of the human life cycle, the sex ratio favors females in most populations. For people over the age of 65 in the United States, for example, the sex ratio is 720, meaning that there are 720 men for every 1,000 women. The ratio

Bioethics

Probing a Genetic Basis of Transgender Identity

For people who identify as transgender, the sex assigned at birth based on the appearance of the genitalia does not match the very deep feelings of maleness, femaleness, or something in between.

The *Diagnostic and Statistical Manual of Mental Disorders (DSM)*, which is the standard compendium of psychiatric conditions, defines "gender dysphoria" as "a conflict between a person's physical or assigned gender and the gender with which he/she/they identify." The designation *gender dysphoria*, begun in 2013, denotes individuals who are distressed by the mismatch between their gender identity and anatomy. It was formerly called *gender identity disorder*. Although transgender identity may now be viewed more as a variant than a disorder, being listed in the *DSM* may facilitate having medical procedures to alter appearance to transition, such as hormone therapy and gender affirmation surgery.

The genetic or physical basis of transgender identity is not fully understood. One of the first studies showed that variants in 19 genes may provide clues to how the disconnect between phenotypic sex and gender identity happens.

Graham Theisen, MD, from the Medical College of Georgia and his team sequenced the exomes of 30 volunteers who identify as transgender. Thirteen were female-to-male and 17 male-to-female, the first gender assigned based on the genitalia at birth. The investigators looked for variants of genes that seemed logical choices for affecting a sexual phenotype, such as regulating the metabolism of steroid sex hormones in the developing brain. They also screened for rare gene variants, found in fewer than 1 in 10,000 individuals in the general population, because a rare variant might be associated with an uncommon trait. If gene variants that are extremely rare in the general population are considerably more prevalent among people who identify as transgender, then perhaps specific combinations of those variants contribute to the unshakable feeling that sex at birth doesn't align with gender identity.

The team selected 30 variants representing the 19 genes, for further investigation. Nine variants changed the encoded amino acid in genes involved in development of the nervous system and/or sex steroid metabolism, and 11 others altered an amino acid but might affect gender identity in another way.

The discovery of 19 genes that may play a role in gender identity is a beginning. It is an association that may or may not turn out to be a correlation or even a cause. Further data may indicate that identifying as transgender is neither a mental illness nor a lifestyle choice, but a variation of being human.

Questions for Discussion

1. In what ways is listing transgender identity in the DSM helpful or harmful?
2. If a genetic profile of transgender identity could ever be developed, what might be the consequences of using such a test on newborns?
3. What are different ways that people might react to the discovery of genotypes that increase the likelihood that a person identifies as transgender?
4. Should a consumer genetics test for transgender gene variants be offered? What are the potential benefits and risks?

---

among older people is the result of diseases that are more likely to be fatal in men, as well as behaviors that may shorten their life expectancies compared to those of women.

Key Concepts Questions 6.1

a. Distinguish the sex chromosome constitutions of human females and males.
b. Describe how male and female structures develop from bipotential forms.
*c. What characteristic of Y chromosome structure made analysis of its genes challenging?
*d. What is the function of the *SRY* gene?
e. Describe conditions that disrupt sexual development.
f. Explain how genes and the environment contribute to homosexuality.
*g. Define *sex ratio*.

6.2 Traits Inherited on Sex Chromosomes

Genes on the Y chromosome are **Y-linked** and genes on the X chromosome are **X-linked**. Y-linked traits are rare because the chromosome has few genes, and many of its genes have counterparts on the X chromosome. Y-linked traits are passed from male to male, because a female does not have a Y chromosome. No other Y-linked traits besides infertility (which obviously can't be passed on) are yet clearly defined. Some traits at first attributed to the Y chromosome are actually due to genes that have been inserted into that chromosome from other chromosomes, such as a deafness gene.

A disproportionate number of X-linked genes cause illness when mutant. The chromosome includes 4 percent of all the genes in the human genome, but accounts for about 10 percent of Mendelian (single-gene) diseases.

Genes on the X chromosome have different patterns of expression in females and males because a female has two X chromosomes and a male just one. In females, X-linked traits are passed just like autosomal traits—that is, two copies are required for expression of a recessive allele and one copy for a dominant allele. In males, however, a single copy of an X-linked allele causes expression of the trait or illness because there is no copy of the gene on a second X chromosome to mask the effect. A man inherits an X-linked trait only from his mother, and can pass it only to daughters. The human male is considered **hemizygous** for X-linked traits, because he has only one set of X-linked genes.

Understanding how sex chromosomes are inherited is important in predicting phenotypes and genotypes in offspring. A male inherits his Y chromosome from his father and his X chromosome from his mother (**figure 6.5**). A female inherits one X chromosome from each parent. If a mother is heterozygous for a particular X-linked gene, her son or daughter has a 50 percent chance of inheriting either allele from her. X-linked traits are always passed on the X chromosome from a mother to a son or from either parent to a daughter. X-linked traits are not passed male-to-male.

X-Linked Recessive Inheritance

An X-linked recessive trait is expressed in females if the causative allele is present in two copies. Many times, an X-linked trait passes from an unaffected heterozygous mother to an affected son. **Table 6.2** summarizes the transmission of an X-linked recessive trait compared to transmission of an X-linked dominant trait, discussed in the next section.

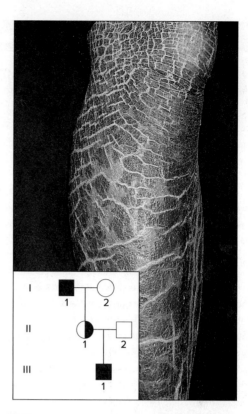

Figure 6.6 **An X-linked recessive trait.** One form of ichthyosis is transmitted as an X-linked recessive trait. A grandfather and grandson were affected in this family. *Courtesy Dr. Mark A. Crowe*

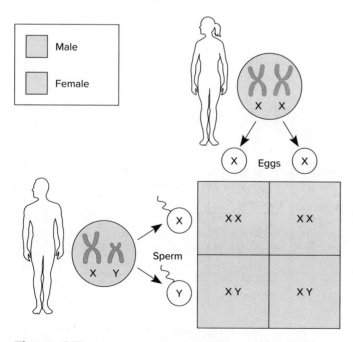

Figure 6.5 **Sex determination in humans.** An oocyte has a single X chromosome. A sperm cell has either an X or a Y chromosome. If a Y-bearing sperm cell with a functional *SRY* gene fertilizes an oocyte, then the zygote is a male (XY). If an X-bearing sperm cell fertilizes an oocyte, then the zygote is a female (XX).

If an X-linked condition is not lethal, a man may be healthy enough to transmit it to offspring. Consider the small family depicted in the pedigree in **figure 6.6**. A middle-aged man who had rough, brown, scaly skin did not realize his condition was inherited until his daughter had a son. By a year of age, the boy's skin resembled his grandfather's. In the condition, a form of ichthyosis, an enzyme deficiency blocks removal of cholesterol from skin cells. The upper skin layer cannot peel off as it normally does, and appears brown and scaly. A test of the daughter's skin cells revealed that she produced half the normal amount of the enzyme, indicating that she was a carrier.

Colorblindness is another X-linked recessive trait that does not hamper the ability of a man to have children. About 8 percent of males of European ancestry are colorblind, as are 4 percent of males of African descent. Only 0.4 percent of females in both groups are colorblind. **Clinical Connection 6.1** looks at this interesting trait.

Figure 6.7 shows part of an extensive pedigree for another X-linked recessive trait, the blood-clotting disease hemophilia B. It is also known as Christmas disease and factor IX ("FIX") deficiency. Part (a) depicts the combination of pedigree symbols and a Punnett square to trace transmission of the trait. Dominant and recessive alleles are indicated by superscripts to the X and Y chromosomes. Part (b) shows how the mutant allele was first deduced to have been in one of Queen Victoria's X chromosomes and then passed to other members of the royal families of England, Germany, Spain, and Russia.

Clinical Connection 6.1

Colorblindness and Tetrachromacy

English chemist John Dalton and his brother saw objects differently from most people. Sealing wax that appeared red to other people was as green as a leaf to the Dalton brothers. Pink wildflowers were blue. The brothers had X-linked recessive colorblindness.

Curious about the cause of his colorblindness, John Dalton asked his physician, Joseph Ransome, to dissect his eyes after he died. When that happened, Ransome snipped off the back of one eye, removing the retina, where the cone cells that provide color vision are nestled among the more abundant rod cells that impart black-and-white vision. Because Ransome could see red and green normally when he peered through the back of his friend's disembodied eyeball, he concluded that it was not an abnormal filter in front of the eye that altered color vision. He stored the eyes in dry air, enabling researchers at the London Institute of Ophthalmology to analyze DNA in Dalton's eyeballs in 1994. Dalton's remaining retina did not have one of the three types of photopigments that enable cone cells to capture certain ranges of wavelengths of light.

Color Vision Basics

Our three types of cone cells correspond to three types of photopigments, making us "trichromats." Primates other than apes, such as monkeys, are dichromats. An object appears colored because it reflects certain wavelengths of light, and each photopigment captures a specific range of wavelengths. The brain interprets the information as a visual perception, much as an artist mixes the three primary colors to create many hues.

Each photopigment has a vitamin A–derived portion called retinal and a protein portion called opsin. The three types of opsins correspond to short, middle, and long wavelengths of light. Mutations in opsin genes cause three types of colorblindness. A gene on chromosome 7 encodes shortwave opsins, and mutations in it produce rare "blue" colorblindness. Dalton had deuteranopia (green colorblindness); his eyes did not have the middle-wavelength opsin. The third type, protanopia (red colorblindness), does not have long-wavelength opsin. Deuteranopia and protanopia are X-linked.

A Molecular View

Jeremy Nathans of Johns Hopkins University also personally contributed to understanding color vision. He used a cow version of a protein called rhodopsin that provides black-and-white vision to identify the human version of the gene. He then used the human rhodopsin gene to search his own DNA. He found three genes with similar sequences: one on chromosome 7 and two on the X chromosome.

Nathans can see colors, but his opsin genes provided a clue to how colorblindness arises. His X chromosome has one red opsin gene and two green genes, instead of the normal one of each. Because the red and green genes have similar sequences, they can misalign during meiosis in the female (**figure 6A**). The resulting oocytes have either two or none of one opsin gene type, instead of one. An oocyte missing an opsin gene would, when fertilized by a Y-bearing sperm, give rise to a colorblind male.

A mutation in an opsin gene can also be beneficial. In some women mispairing during meiosis yields an X chromosome that

Figure 6A **How colorblindness arises.** (a) The sequence similarities among the opsin genes responsible for color vision may cause chromosomes to misalign during meiosis in the female. Offspring may inherit too many, or too few, opsin genes. A son inheriting an X chromosome missing an opsin gene will be colorblind. A daughter would have to inherit two mutations to be colorblind. (b) A missing gene causes X-linked colorblindness.

(Continued)

has three opsins: the normal red and green, as well as another that has part red and part green DNA sequences. Combined with her chromosome 7 blue opsin, she has four opsin genes, and is termed a tetrachromat. If she is also artistic and has a good vocabulary (both environmental influences), she may perceive colors that others cannot and be able to describe them. One tetrachromat, for example, can see undertones that ruin the color of clothing, becomes frustrated with automated paint mixers because they are not precise enough, and even detected changes in the sky just before tornado-strength winds began. A cosmetic company hired her to help design products.

Questions for Discussion

1. Discuss how the three types of colorblindness differ in genotype and phenotype.
2. Explain how Dr. Nathans discovered the mutations that cause colorblindness.
3. How does a tetrachromat differ from a trichromat?

Figure 6.7 Hemophilia B (factor IX deficiency). (a) This X-linked recessive disease usually passes from a heterozygous woman (designated $X^H X^h$, where h is the hemophilia-causing allele) to heterozygous daughters or hemizygous sons. The father has normal blood clotting. (b) The disease has appeared in the royal families of England, Germany, Spain, and Russia. The modern royal family in England does not carry hemophilia.

Chapter 6 *Matters of Sex*

The queen either had a new mutation or she inherited it. Either way, she passed the mutation to carrier daughters and one mildly affected son.

For many years, historians assumed that the royal families of Europe suffered from the more common form of hemophilia, type A. However, researchers in 2009 tested bits of bone from Tsarina Alexandra and Crown Prince Alexei for the wild type and mutant alleles for the genes that cause hemophilia A as well as B. Both genes are on the X chromosome. The result: Mother Alexandra was a carrier of the rarer hemophilia B, and her son Crown Prince Alexei had the disease, which had been known from documents describing his severe bleeding episodes.

The transmission pattern of hemophilia B is consistent with the criteria for an X-linked recessive trait listed in table 6.2. A daughter can inherit an X-linked recessive disease or trait if her father is affected and her mother is a carrier, because the daughter inherits one mutation-bearing X chromosome from each parent. Without a biochemical test, though, an unaffected woman will not know she is a carrier for an X-linked recessive trait unless she has an affected son.

A woman whose brother has hemophilia B has a 1 in 2 risk of being a carrier. Both her parents are healthy, but her mother must be a carrier because her brother is affected. Her risk is the chance that she has inherited the X chromosome bearing the hemophilia allele from her mother. The chance of the woman conceiving a son is 1 in 2, and of that son inheriting hemophilia is 1 in 2. Using the product rule, the risk that she is a carrier and will have a son with hemophilia, out of all the possible children she can conceive, is 1/2 × 1/2 × 1/2, or 1/8.

X-Linked Dominant Inheritance

Dominant X-linked conditions and traits are rare. Again, gene expression differs between the sexes (see table 6.2). A female who inherits a dominant X-linked allele or in whom the mutation originates has the associated trait or illness, but a male who inherits the allele is usually more severely affected because he has no other allele to mask its effect.

The children of a man who has a normal X chromosome and a woman who has a dominant mutation on one X chromosome face the risks summarized in **figure 6.8**. However, for a severe condition such as Rett syndrome (see the chapter 2 opener), females may be too disabled to have children. Rett syndrome and similar X-linked dominant conditions affect only girls, because sons would inherit the X chromosome bearing the mutation from their mothers and not survive. Because an X-linked dominant condition is lethal in males and severely affects females, most cases are the result of new mutations, rather than inheriting a mutation.

An X-linked dominant condition that could be mild enough to be passed to children is incontinentia pigmenti (IP; see figure 6.8b). In affected females, swirls of skin pigment arise where melanin penetrates the deeper skin layers. A newborn girl with IP has yellow, pus-filled vesicles on her limbs that come and go over the first few weeks. Then the lesions become warty and eventually become brown splotches that may remain for life, although they fade with time. Males who inherit the

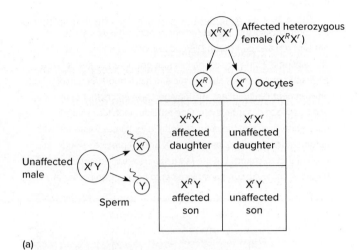

Figure 6.8 X-linked dominant inheritance. (a) A female who has an X-linked dominant trait has a 1 in 2 probability of passing it to her offspring, male or female. Males are generally more severely affected than females. **(b)** Note the characteristic patchy pigmentation on the leg of a girl who has incontinentia pigmenti. *Courtesy Richard Alan Lewis M.D., M.S., Baylor College of Medicine*

mutant allele from their heterozygous mothers do not survive to be born, which is why women with the disease who become pregnant have a miscarriage rate of about 25 percent.

Solving a Problem of X-Linked Inheritance

Mendel's first law (segregation) applies to genes on the X chromosome. The same logic is used to solve problems of X-linked inheritance as to trace traits transmitted on autosomes, but considering the X and Y chromosomes in Punnett squares. Follow these steps:

1. Look at the pattern of inheritance. Different frequencies of affected males and females in each generation may suggest X linkage. For an X-linked recessive trait:
   - An affected male has a carrier mother.
   - An unaffected female with an affected brother has a 50 percent (1 in 2) chance of being a carrier.
   - An affected female has a carrier or affected mother *and* an affected father.

- A carrier (female) has a carrier or affected mother or an affected father.

For an X-linked dominant trait:
- There may be no affected males, because they die early.
- An affected female has an affected mother.

2. Draw the pedigree.
3. List all genotypes and phenotypes and their probabilities.
4. Assign genotypes and phenotypes to the parents. Consider clues in the phenotypes of relatives.
5. Determine how alleles separate into gametes for the genes of interest on the X and Y chromosomes.
6. Unite the gametes in a Punnett square.
7. Determine the phenotypic and genotypic ratios for the F_1 generation.
8. To predict further generations, use the F_1 genotypes and repeat steps 4 through 6.

Consider as an example Kallmann syndrome, which causes a poor or absent sense of smell and small testes or ovaries. It is X-linked recessive. Brenda does not have Kallmann syndrome, but her brother Jamal and her maternal cousin Malcolm (her mother's sister's child) have it. Brenda's and Malcolm's parents are unaffected, as is Brenda's husband Sam. Brenda and Sam wish to know the risk that a son will inherit the condition. Sam has no affected relatives.

Solution

Mode of inheritance: The trait is X-linked recessive because males are affected through carrier mothers.

K = wild type; k = Kallmann syndrome

Genotypes	Phenotypes
$X^K X^K$, $X^K X^k$, $X^K Y$	normal
$X^k X^k$, $X^k Y$	affected

Individual	Genotype	Phenotype	Probability
Brenda	$X^K X^k$ or $X^K X^K$	normal (carrier)	50% each
Jamal	$X^k Y$	affected	100%
Malcolm	$X^k Y$	affected	100%
Sam	$X^K Y$	normal	100%

Brenda's gametes

if she is a carrier: X^K X^k

Sam's gametes: X^K Y

Punnett square

	X^K	X^k
X^K	$X^K X^K$	$X^K X^k$
Y	$X^K Y$	$X^k Y$

Interpretation: If Brenda is a carrier, the probability that their son will have Kallmann syndrome is 50 percent, or 1 in 2. This is a conditional probability. The chance that any son will have the condition is actually 1 in 4, because Brenda also has a 50 percent chance of being genotype $X^K X^K$ and therefore not a carrier.

Table 6.2 Comparison of X-Linked Recessive and X-Linked Dominant Inheritance

X-Linked Recessive Trait	X-Linked Dominant Trait
Always expressed in the male	Expressed in females in one copy
Expressed in a female homozygote and very rarely in a female heterozygote	Much more severe effects in males
Affected male inherits trait from heterozygote or homozygote mother	High rates of miscarriage due to early lethality in males
Affected female inherits trait from affected father and affected or heterozygote mother	Passed from male to all daughters but to no sons

Key Concepts Questions 6.2

a. Explain how males and females differ in their expression of X-linked traits.

*b. Explain why X-linked dominant conditions are rare in males.

c. Discuss how Mendel's first law can be applied to solve problems involving transmission of genes on the X chromosome.

6.3 Sex-Limited and Sex-Influenced Traits

An X-linked recessive trait generally is more prevalent in males than females. Other situations, however, can affect gene expression in the sexes differently.

Sex-Limited Traits

A **sex-limited trait** affects a structure or function of the body that is present in only males or only females. The gene for such a trait may be X-linked or autosomal.

Understanding sex-limited inheritance is important in animal breeding. For example, a New Zealand cow named Marge has a mutation that makes her milk very low in saturated fat; she is founding a commercial herd. Males play their part by transmitting the mutation, even though they do not make milk. In humans, beard growth is sex-limited. A woman does not grow a beard, because she does not manufacture the hormones required for facial hair growth. She can, however, pass to her sons the genes specifying heavy beard growth.

Curiously, a father-to-be can affect the health of the mother-to-be. Although an inherited disease that causes symptoms associated with pregnancy is obviously sex-limited, the male genome may play a role by contributing to the development of supportive structures, such as the placenta. This is the case for preeclampsia, a sudden rise in blood pressure late in pregnancy. It kills 50,000 women worldwide each year. A study of 1.7 million pregnancies in Norway found that if a man's first wife had preeclampsia, his second wife had double the risk of developing it. Another study found that women whose mothers-in-law developed preeclampsia when pregnant with the women's husbands had twice the rate of developing the condition themselves. Perhaps a gene from the father affects the placenta in a way that elevates the pregnant woman's blood pressure.

Sex-Influenced Traits

With a **sex-influenced trait**, an allele is dominant in one sex but recessive in the other. Such a gene may be X-linked or autosomal. The difference in expression can be caused by hormonal differences between the sexes. For example, an autosomal gene for hair growth pattern has two alleles, one that produces hair all over the head and another that causes pattern baldness. The baldness allele is dominant in males but recessive in females, which is why more men than women are bald. A heterozygous male is bald, but a heterozygous female is not. A bald woman is homozygous recessive. Even a bald woman tends to have some wisps of hair, whereas an affected male may be completely hairless on the top of his head.

Pattern baldness is a classic example of a sex-influenced trait controlled by one gene. A more recently identified example is response to treatment for glioblastoma, a type of brain cancer. Clinicians had long noted that female patients live significantly longer than male patients following the standard treatment of surgery, radiation, and use of a drug, temozolomide. When researchers looked at which genes are expressed in the cancer cells of treated patients, considering sex, they discovered a difference. Cancer cells from women inhibited the expression of 9 genes that participate in cell-cell signaling, whereas cancer cells from men dampened expression of 15 genes that regulate the cell cycle. These differences affect how people respond to the drug. If the findings are confirmed, perhaps men can take a different drug that is more likely to extend life.

Key Concepts Questions 6.3

*a. Define *sex-limited trait*.

*b. Define *sex-influenced trait*.

6.4 X Inactivation

Females have two alleles for every gene on the X chromosomes, whereas males have only one. In mammals, a mechanism called **X inactivation** balances this apparent inequality in the expression of genes on the X chromosome.

Equaling Out the Sexes

By the time the embryo of a female mammal consists of eight cells, about 75 percent of the genes on one X chromosome in each cell are inactivated, and the remaining 25 percent are expressed to different degrees in different females. Which X chromosome is mostly turned off in each cell—the one inherited from the mother or the one from the father—is usually random. As a result, a female mammal expresses the X chromosome genes inherited from her father in some cells and those from her mother in others. She is, therefore, a mosaic for expression of most genes on the X chromosome (**figure 6.9**).

The gene *XIST* in part of the X chromosome called the X inactivation center controls X inactivation. *XIST* encodes an RNA that binds to a specific site on the same (inactivated) X chromosome. From this site to the chromosome tip, most of the X chromosome is then silenced, like a row of falling dominoes.

Once an X chromosome is inactivated in one cell, all its daughter cells have the same X chromosome shut off. Because the inactivation occurs early in development, the adult female has patches of tissue that differ in their expression of X-linked genes. With each cell in her body having only one active X chromosome, she is roughly equivalent to the male in gene expression.

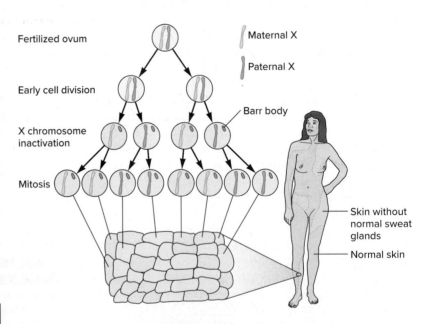

Figure 6.9 **X inactivation.** A female is a mosaic for expression of genes on the X chromosome because of the random inactivation of either the maternal or paternal X in each cell early in prenatal development. In anhidrotic ectodermal dysplasia, a woman has patches of skin that do not have sweat glands and hair. (Colors distinguish cells with the inactivated X; they do not depict skin color.)

X inactivation can alter the phenotype (gene expression), but not the genotype. The change is reversed in germline cells destined to become oocytes, and this is why a fertilized ovum does not have an inactivated X chromosome. X inactivation is an example of an epigenetic change—a change in gene expression that is passed from one cell generation to the next, but that does not alter the DNA base sequence.

We can observe X inactivation at the cellular level because the turned-off X chromosome absorbs a stain much faster than the active one. This differential staining occurs because inactivated DNA has chemical methyl (CH_3) groups bound to it that prevent it from being transcribed into RNA and also enable it to absorb stain.

X inactivation can be used to check the sex of an individual. The nucleus of a cell of a female, during interphase, has one dark-staining X chromosome called a Barr body. A cell from a male has no Barr body because his one X chromosome remains active.

Effect of X Inactivation on the Phenotype

The consequences of X inactivation on the phenotype can be interesting. For homozygous X-linked genotypes, X inactivation has no effect. Whichever X chromosome is turned off, a copy of the same allele is left to be expressed. For heterozygotes, however, X inactivation leads to expression of one allele or the other. This doesn't affect health if enough cells express the functional gene product (protein). However, traits that affect skin or fur reveal X inactivation. The swirls of skin color in incontinentia pigmenti (IP) patients are a consequence of patterns of X inactivation in skin cells (see figure 6.8b). Where the normal allele for melanin pigment is shut off, pale swirls develop. Where pigment is produced, brown swirls result.

A familiar example of X inactivation is the coat colors of tortoiseshell and calico cats. An X-linked gene confers brownish-black (dominant) or yellow-orange (recessive) color. A female cat heterozygous for this gene has patches of each color, forming a tortoiseshell pattern that arises from different cells expressing either of the two alleles (**figure 6.10**). The earlier the X inactivation, the larger the patches, because more cell divisions can occur after the event, producing more daughter cells. White patches may form due to epistasis by an autosomal gene that shuts off pigment synthesis. A cat with colored patches against such a white background is a calico. Tortoiseshell (small patches) and calico (large patches) cats are nearly always female. A male can have one of these coat patterns only if he inherits an extra X chromosome.

X inactivation has clinical implications. A female who is heterozygous for an X-linked recessive gene can express the associated condition if the normal allele is inactivated in the tissues that the illness affects. Consider a carrier of hemophilia A (factor VIII deficiency). If the X chromosome carrying the normal allele for the clotting factor is turned off in her liver, where clotting factors are made, then the blood will clot slowly enough to cause mild hemophilia. Luckily for her, slowed clotting time also greatly reduces her risk of cardiovascular disease caused by blood clots blocking circulation.

A carrier of an X-linked trait who expresses the phenotype, such as a heterozygote for hemophilia A or B, is called a

Figure 6.10 Visualizing X inactivation. X inactivation is obvious in a calico cat. X inactivation is rarely observable in humans because most cells do not remain together during development, as a cat's skin cells do. *G.K. & Vikki Hart/Getty Images*

manifesting heterozygote. Whether or not X inactivation causes a phenotype depends on how adept cells are at sharing. Two lysosomal storage diseases illustrate how cell–cell interactions mediate effects of X inactivation in manifesting heterozygotes. A lysosomal storage disease is a deficiency of a specific enzyme that normally dismantles cellular debris in lysosomes.

In carriers of Hunter syndrome (also called mucopolysaccharidosis II), cells that make the enzyme readily send it to neighboring cells that do not, essentially correcting the defect in cells that can't make the enzyme. Carriers of Hunter syndrome do not have symptoms because cells get enough enzyme. Boys with Hunter syndrome are deaf, intellectually disabled, have dwarfism and abnormal facial features, heart damage, and an enlarged liver and spleen. In contrast, in Fabry disease, cells do not readily release the enzyme alpha-galactosidase A, so a female who is a heterozygote may have cells in the affected organs deficient in the enzyme. She may develop mild symptoms of this disease that causes skin lesions, abdominal pain, and kidney failure in boys.

In humans, X inactivation can be used to identify carriers of some X-linked diseases. This is the case for Lesch-Nyhan syndrome, in which an affected boy has cerebral palsy; bites his fingers, toes, and lips to the point of mutilation; is intellectually disabled; and passes painful urinary stones.

Mutation results in defective or absent HGPRT, an enzyme. A woman who carries Lesch-Nyhan syndrome can be detected when hairs from widely separated parts of her head are tested for HGPRT. (Hair is used for the test because it is accessible and produces the enzyme.) If some hairs contain HGPRT but others do not, she is a carrier. The hair cells that do not have the enzyme have turned off the X chromosome that carries the normal allele; the hair cells that manufacture the normal enzyme have turned off the X chromosome that carries the disease-causing allele. The woman is healthy because her brain has enough HGPRT, but each son has a 50 percent chance of inheriting the disease.

Theoretically, X inactivation evens out the sexes for expression of X-linked genes. In actuality, however, a female may *not* be equivalent, in gene expression, to a male because she has two cell populations, whereas a male has only one. One of a female's two cell populations has the X she inherited from her father active, and the other has the X chromosome she inherited from her mother active. For heterozygous X-linked genes, she has some cells that manufacture the protein encoded by one allele, and some cells that produce the protein specified by the other allele. Although the alleles of most heterozygous genes are about equally represented, sometimes X inactivation can be skewed. In such a case, most cells express the X inherited from the same parent. This unequal X inactivation pattern can happen if the two X chromosomes have different alleles for a gene that controls cell division rate, giving certain cells a survival advantage.

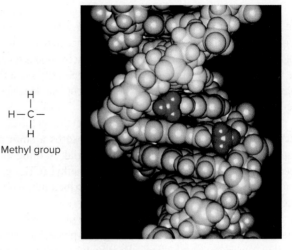

Figure 6.11 Methyl (CH_3) groups (red) "silence" certain genes. *Designed by Mark Sherman. Provided by Arthur Riggs and Craig Cooney.*

> **Key Concepts Questions 6.4**
>
> *a. What is X inactivation?
> *b. When in prenatal development does X inactivation begin?
> c. Explain how X inactivation is an epigenetic and not a genetic change.
> *d. What is a manifesting heterozygote?

6.5 Parent-of-Origin Effects

In Mendel's pea experiments, it didn't matter whether a trait came from the male or female parent. For certain genes in mammals, however, parental origin influences age of onset of a disease or symptom severity. Two mechanisms of parent-of-origin effects are genomic imprinting and differences between the developmental timetables of sperm and oocytes.

Genomic Imprinting

In **genomic imprinting,** methyl (CH_3) groups cover a gene or several linked genes and prevent them from being accessed (transcribed and translated) to synthesize protein (**figure 6.11**).

For an imprinted gene, the copy inherited from either the father or the mother is always covered with methyls, even in different individuals. That is, a particular gene might function if it came from the father, but not if it came from the mother, or vice versa. This parental effect on gene expression is seen as diseases that are always inherited from the mother or father. For example, central precocious puberty is always inherited from the father. A gene called *MKRN3* from the father is covered in methyls, and therefore imprinted. In this condition, girls reach puberty before age 8 and boys before age 9.

Imprinting is an epigenetic alteration, in which a layer of meaning is stamped on a gene without changing its DNA sequence. The imprinting pattern is passed from cell to cell in mitosis, but not from individual to individual through meiosis. When silenced DNA is replicated during mitosis, the pattern of blocked genes is exactly placed, or imprinted, on the new DNA, covering the same genes as in the parental DNA (**figure 6.12**). In this way, the "imprint" of inactivation is perpetuated, as if each such gene "remembers" which parent it came from.

In meiosis, imprints are removed and reset. As oocyte and sperm form, the CH_3 groups shielding their imprinted genes are stripped away and new patterns are set down, depending on whether the fertilized ovum is chromosomally male (XY) or female (XX). In this way, women can have sons and men can have daughters without passing on sex-specific parental imprints.

The function of genomic imprinting isn't well understood; however, because many imprinted genes take part in early development, particularly of the brain, imprinting may be a way to finely regulate the abundance of key proteins in the embryo. The fact that some genes lose their imprints after birth supports this idea of early importance. Also, imprinted genes are in clusters along a chromosome, and are controlled by other regions of DNA called imprinting centers. Perhaps

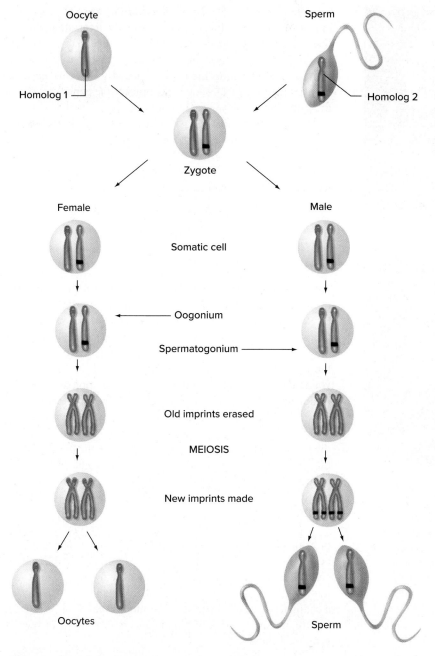

Figure 6.12 Genomic imprinting. Imprints are erased during meiosis, then reset according to the sex of the new individual. Chromosomes are depicted in the same color to highlight the imprinted areas of the representative pair of autosomes.

Researchers created fertilized mouse ova that contained two male pronuclei or two female pronuclei, instead of one from each. Results were strange. When the fertilized ovum had two male genomes, a normal placenta developed, but the embryo was tiny and quickly stopped developing. A zygote with two female pronuclei developed into an embryo, but the placenta was highly abnormal. Therefore, the male genome controls placenta development, and the female genome, embryo development.

The mouse results echoed abnormalities of human development. When two sperm fertilize an oocyte and the female pronucleus degenerates, an abnormal growth of placenta-like tissue called a hydatidiform mole forms. Conversely, if a fertilized ovum contains two female genomes but no male genome, a mass of random differentiated tissue, called a teratoma, grows. A teratoma consists of a strange mix of tissues, such as skin and teeth. When a hydatidiform mole or a teratoma develops, there is no embryo.

Genomic imprinting can explain incomplete penetrance, in which an individual is known to have inherited a genotype associated with a particular phenotype, but has no signs of the trait. This is the case for a person with normal fingers whose parent and child have polydactyly. An imprinted gene silences the dominant mutant allele.

Imprinting may be an important concern in assisted reproductive technologies that manipulate gametes to treat infertility. For example, the otherwise rare Beckwith-Wiedemann syndrome is more prevalent among the offspring of people who used *in vitro* fertilization and intracytoplasmic sperm injection (discussed in chapter 23) to become pregnant.

Imprinting Diseases in Humans

The number of imprinted genes in the human genome exceeds 156, and at least 60 of them affect health when abnormally expressed. The effects of genomic imprinting are noticeable only when an individual has one copy of a normally imprinted (silenced) allele and the other, active allele is inactivated or deleted.

Imprinting diseases can be dramatic, such as two syndromes that arise from small deletions in the same region of chromosome 15 (**figure 6.13**). A child with Prader-Willi syndrome is small at birth and in infancy has difficulty gaining weight. Between ages 1 and 3, the child develops an obsession with eating and a very slow metabolism. Parents

one gene in a cluster is essential for early development, and the others become imprinted simply because they are nearby—a bystander effect.

Genomic imprinting has implications for understanding early human development. It suggests that for mammals, two opposite-sex parents are necessary to produce a healthy embryo and placenta. This apparent requirement for opposite-sex parents was discovered in the 1980s, through experiments on early mouse embryos and examination of certain rare pregnancy problems in humans.

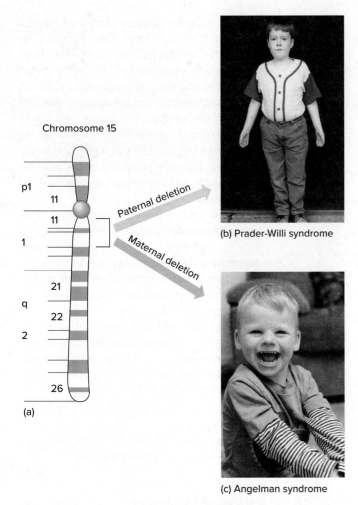

Figure 6.13 Prader-Willi and Angelman syndromes result when the nonimprinted copy of a gene is deleted. (a) Two distinct syndromes result from missing genetic material in the same region of chromosome 15. (b) This child has Prader-Willi syndrome, due to a deletion in the copy of the chromosome he inherited from his father. Note his small hands. (c) This boy has Angelman syndrome, caused by a deletion in the chromosome 15 that he inherited from his mother. *(b): Carla D. Kipper; (c): Dan James/Caters News Agency*

Symptoms of Prader-Willi syndrome arise because several paternal genes that are not normally imprinted (that is, that are normally active) are missing. In Angelman syndrome, it is a normally active single maternal gene that is deleted. This part of chromosome 15 is especially unstable because highly repetitive DNA sequences bracket the genes that cause the symptoms.

Abnormal imprinting is associated with forms of diabetes mellitus, autism, Alzheimer disease, schizophrenia, and male homosexuality. A clue that indicates a condition is associated with genomic imprinting is differing severity depending on whether it is inherited from the father or mother.

Different Timetables in Sperm and Oocyte Formation

Altered imprinting is not the only mechanism of a parent-of-origin effect on the phenotype of a genetic disease. Huntington disease (HD) has a younger age of onset, with faster progression and more severe symptoms, if a person inherits the mutation from his or her father. This is the case for the family discussed in the opener to chapter 4, in which three young daughters inherited their father's mutation and became ill during childhood or adolescence. In HD, the effect may be due to the different timetables of sperm and egg production. A female at puberty has about 400,000 eggs, each halted on the brink of completing meiosis, when the replication slippage that expands the gene could happen. But a man has many more chances for the gene to be miscopied and grow over a reproductive lifetime because each ejaculation contains a quarter of a billion sperm!

Another example of a paternal age effect is increased risk of Noonan syndrome in a child of an older father. The mutation occurs in stem cells that give rise to sperm, giving those stem cells a faster division rate than other stem cells. Over time, mutation-bearing sperm accumulate. Noonan syndrome causes a characteristic face, heart defects, intellectual disability, short stature, and higher risk of blood cancers. Mutations in several genes can cause it.

lock kitchen cabinets and refrigerators to keep these children from eating themselves to death by bursting digestive organs. The other condition resulting from the small deletion in chromosome 15, Angelman syndrome, causes autism and intellectual disability, an extended tongue, large jaw, poor muscle coordination, and convulsions that make the arms flap.

In many cases of Prader-Willi syndrome, only the mother's chromosome 15 region is expressed; the father's chromosome is deleted in that region. In Angelman syndrome, the reverse occurs: The father's gene (or genes) is expressed, and the mother's chromosome has the deletion.

Key Concepts Questions 6.5

*a. What is genomic imprinting?

*b. What is the role of the binding of methyl groups to DNA in genomic imprinting?

c. How can abnormal imprinting cause disease?

d. How can differences in the timetables of sperm and oocyte formation cause parent-of-origin effects?

Study Guide

Summary

6.1 Our Sexual Selves

1. A human male is **heterogametic** with an X and a Y chromosome, and a human female is **homogametic** with two X chromosomes.
2. Two sets of bipotential gonads and other structures are present until the sixth week of prenatal development. After that one set persists, depending on gene expression.
3. The human Y chromosome has two pseudoautosomal regions and a large, male-specific region (MSY) that does not recombine. Some Y-linked genes correspond to X-linked genes. Some are similar to them and some are unique. The Y chromosome is highly symmetrical, with palindromic DNA sequences and inverted repeats.
4. If the SRY **transcription factor** activates the Sox9 transcription factor, bipotential gonads develop as testes. If *SRY* is not expressed, the gonads develop as ovaries, under the direction of *Wnt4* and other genes.
5. Starting 8 weeks after fertilization, the testes secrete a hormone that prevents development of female structures and testosterone, which, once converted to dihydrotestosterone (DHT), stimulates development of male structures.
6. Mutations in genes that control sexual development cause conditions in which the sex chromosome constitution may not match the phenotype.
7. Evidence suggests that small influences from several genes may contribute to homosexuality.
8. Sexual identity consists of chromosomal sex, gonadal sex, phenotypic sex, gender identity, and sexual orientation.
9. The **sex ratio** is the number of males divided by the number of females multiplied by 1,000, for people of a particular age. Interfering with pregnancy outcomes can skew sex ratios.

6.2 Traits Inherited on Sex Chromosomes

10. **Y-linked** traits are passed from fathers to sons only.
11. Males are **hemizygous** for genes on the X chromosome and express phenotypes associated with these genes because they do not have another allele of the gene on a homolog. An **X-linked** trait passes from mother to son because he inherits his X from his mother and his Y from his father.
12. An X-linked allele may be dominant or recessive. X-linked dominant traits are more devastating to males.

6.3 Sex-Limited and Sex-Influenced Traits

13. **Sex-limited traits** may be autosomal or sex-linked, but they affect only one sex because of anatomical or hormonal gender differences.
14. A **sex-influenced trait** is dominant in one sex but recessive in the other.

6.4 X Inactivation

15. **X inactivation** shuts off one X chromosome in each cell in female mammals, making them mosaics for heterozygous X-linked genes. It somewhat equalizes the dosages of genes on the sex chromosomes between the sexes. X inactivation is an epigenetic change because it does not alter the DNA sequence.
16. A female who expresses the phenotype corresponding to an X-linked gene she carries is a **manifesting heterozygote**.

6.5 Parent-of-Origin Effects

17. In parent-of-origin effects, the phenotype corresponding to a genotype differs depending upon whether the gene was inherited from the mother or father.
18. **Genomic imprinting** temporarily suppresses expression of some genes, leading to parent-of-origin effects.
19. Imprints are erased during meiosis and reassigned based upon whether the fertilized ovum is XY or XX.
20. Methyl groups that bind to DNA and temporarily suppress gene expression are the physical basis of genomic imprinting.
21. Differences in the timetables of sperm and oocyte formation explain some parent-of-origin effects.

Key Terms

genomic imprinting (jē nō'm ĭk) The covering of a gene or several linked genes by methyl groups, which prevents expression and affects the phenotype, depending upon which parent transmits an allele.

hemizygous (hĕm' ē-zī' gəs) The sex that has half as many X-linked genes as the other; a human male is hemizygous.

heterogametic (hĕt'ə-rō-gə-mē'tĭk) The sex with two different sex chromosomes; a human male.

homogametic (hō'mō-gə-mē'tĭk) The sex with two of the same sex chromosomes; a human female.

manifesting heterozygote (hĕt'ə-rō-z ī'gōt) A carrier of an X-linked recessive trait who expresses the phenotype because the normal allele is inactivated in some tissues.

sex ratio Number of males divided by number of females multiplied by 1,000, for people of a certain age in a population.

sex-influenced trait Phenotype caused when an allele is recessive in one sex but dominant in the other.

sex-limited trait A trait that affects a structure or function present in only one sex.

transcription factor A protein that activates the transcription of certain genes.

X inactivation A mechanism that turns off most of the genes on one X chromosome in each cell of a female mammal, occurring early in embryonic development.

X-linked Genes on an X chromosome.

Y-linked Genes on a Y chromosome.

Review Questions

1. Distinguish the five levels of sexual identity listed in Table 6.1.
2. Discuss how genes in the pseudoautosomal regions of the Y chromosome differ from genes in the male-specific region (MSY).
3. Describe the phenotypes of
   a. a person with a deletion of the *SRY* gene.
   b. an XX individual.
   c. an XY individual with a block in testosterone synthesis.
4. List the events required for a fetus to develop as a female.
5. State two reasons why males with X-linked dominant conditions are rare.
6. Explain the basis of the sex ratio at birth.
7. Traits that appear more frequently in one sex than the other may be caused by genes that are inherited in an X-linked, sex-limited, or sex-influenced fashion. How might you distinguish among these possibilities in a given individual?
8. Why are male calico cats rare?
9. Explain how X inactivation equalizes the "doses" of X-linked genes between the sexes.
10. Cite evidence that genetic contributions from both parents are necessary for normal prenatal development.

Applied Questions

1. To answer the following questions, consider these population data on sex ratios:

Selected sex ratios at birth		Selected sex ratios after age 65	
Nation	Sex ratio	Nation	Sex ratio
Costa Rica	970	Rwanda	620
Tanzania	1,000	South Africa	630
Liechtenstein	1,010	France	700
South Africa	1,020	United States	720
United States	1,050	Qatar	990
Sweden	1,060	Montserrat	1,060
Italy	1,070	Bangladesh	1,160
China	1,150	Nigeria	990

   a. In Rwanda, South Africa, France, and the United States, males die, on average, significantly younger than females do. What types of information might explain the difference?
   b. In Costa Rica, how many males at birth are there for every 100 females?
   c. Name a country in which more males than females live past 65.

2. In severe Hunter syndrome, deficiency of the enzyme iduronate sulfate sulfatase leads to buildup of certain carbohydrates, swelling the liver, spleen, and heart. In mild cases, deafness may be the only symptom. Intellect is usually normal, and life span can be normal. Hunter syndrome is X-linked recessive. A man with a mild case has a child with a carrier.
   a. What is the probability that a son inherits Hunter syndrome?
   b. What is the chance that a daughter inherits Hunter syndrome?
   c. What is the chance that a daughter is a carrier?

3. Amelogenesis imperfecta is X-linked dominant. Affected males have extremely thin enamel on the teeth. Female carriers have grooved teeth from uneven deposition of enamel. Why might the phenotype differ between the sexes?

4. Explain how X inactivation might enable a woman to pass Rett syndrome to a son.

5. A drug used to treat cancer also shows promise in treating Angelman syndrome. The drug removes methyl groups from the paternal copy of the gene *UBE3A* that is part of the imprinted region of chromosome 15. Explain how the drug might work.

Case Studies and Research Results

1. For each case description, identify the principle at work from the list that follows. More than one answer may apply to a given case.

   A. Y-linked inheritance
   B. X-linked recessive
   C. X-linked dominant inheritance
   D. sex-limited inheritance
   E. sex-influenced inheritance
   F. X inactivation or manifesting heterozygote
   G. imprinting abnormality

a. A child with Russell-Silver syndrome is small with a large head, a triangular face, and digestive problems. In some cases, a gene on chromosome 11 that is normally methylated in the father is not methylated.

b. Six-year-old LeQuan inherited Fabry disease from his mother. She is a heterozygote for the mutant allele of the gene on the X chromosome that encodes a lysosomal enzyme. LeQuan would have died before age 50 of heart failure, kidney failure, or a stroke, but he receives twice-monthly infusions of the enzyme. His mother recently began experiencing recurrent fevers, a burning pain in her hands and feet, a rash, and sensitivity to cold. She is experiencing mild Fabry disease.

c. The Chandler family has many male members who have a form of retinitis pigmentosa (RP), which causes gradual visual loss. Several women in the family are presumed to be carriers because they have affected sons. The women are tested for RP genes on chromosomes 1, 3, 6, and the X, but do not carry mutations in these genes. Years ago, Rachel married her cousin Ross, who has the family's form of RP. They had six children. The three sons are all affected, but their daughters have normal vision.

d. Simon's mother and his mother's sister are breast cancer survivors, and their mother died of the disease. Simon's sister learns that she has inherited the *BRCA1* mutation that increases risk of developing certain cancers. Simon has two daughters but doesn't want to be tested because he thinks a man cannot transmit a trait that affects a body part that is more developed in females.

e. Tribbles are extraterrestrial mammals that long ago invaded a starship on the television program *Star Trek*. A gene called *frizzled* causes kinky hair in female tribbles who inherit just one allele. However, two mutant alleles must be inherited for a male tribble to have kinky hair. The gene is on an autosome.

f. Devon died at age 16 of Lowe syndrome. He had slight intellectual disability, visual problems (cataracts and glaucoma), seizures, poor muscle tone, and progressive kidney failure, which was ultimately fatal. His sister Lily is pregnant, and wonders whether she is a carrier of the disease that killed her brother. She remembers a doctor saying that her mother Drucilla was a carrier. Lily's physician determines that she is a carrier because she has cataracts, which is a clouding of the lenses. It has not yet affected her vision. When a prenatal test reveals that Lily's fetus is a female, her doctor tells her not to worry about Lowe syndrome.

g. Mating among Texas field crickets depends upon females responding to a male mating call. The sounds must arrive at a particular frequency to excite the females, who do not sing back in response. However, females can pass on an allele that confers the "correct" frequency of singing.

h. When Winthrop was a baby, he was diagnosed with "failure to thrive." At 14 months of age, he suddenly took an interest in food, and his parents couldn't feed him fast enough. By age 4, Winthrop was obese, with disturbing behavior. He was so hungry that after he'd eaten his meal and everyone else's leftovers, he'd hunt through the garbage for more. Finally a psychiatrist who had a background in genetics diagnosed Prader-Willi syndrome. Testing showed that the allele for the Prader-Willi gene that Winthrop had inherited from his father was abnormally methylated.

i. Certain breeds of dogs have cryptorchidism, in which the testicles do not descend into the scrotum. The trait is passed through females.

2. Reginald has mild hemophilia A that he can control by taking a clotting factor. He marries Lydia, whom he met at the hospital where he and Lydia's brother, Marvin, receive the same treatment. Lydia and Marvin's mother and father, Emma and Clyde, do not have hemophilia. What is the probability that Reginald and Lydia's son will inherit hemophilia A?

3. Harold works in a fish market, but the odor does not bother him because he has anosmia, an X-linked recessive lack of the sense of smell. Harold's wife, Shirley, has a normal sense of smell. Harold's sister, Maude, also has a normal sense of smell, as does her husband, Phil, and daughter, Marsha, but their identical twin boys, Alvin and Simon, cannot detect odors. Harold and Maude's parents, Edgar and Florence, can smell normally. Draw a pedigree for this family, indicating people who must be carriers of the anosmia mutation.

4. When a South African sprinter won the 800-meter race at the World Championships in 2009, with a time significantly improved over past events, the International Association of Athletics Federations asked her to take a chromosome test. The implication of the much-publicized request was that she was really a male. For reasons of privacy, the results of the test were never released. Which gene would have been tested to determine the sprinter's chromosomal sex? Is this the same as gender?

CHAPTER 7

Complex Traits

Should parents submit their children's DNA for genetic testing to determine which sports the youngsters should play?

Learning Outcomes

7.1 Genes and the Environment Mold Traits
1. Distinguish between single-gene and polygenic traits.
2. Define *complex traits*.

7.2 Polygenic Traits Are Continuously Varying
3. Explain how continuously varying traits reflect genetic and environmental influences.

7.3 Methods to Investigate Complex Traits
4. Explain how determining empiric risk differs from calculating a Mendelian frequency.
5. Define *heritability*.
6. Discuss how studies on adopted individuals and twins can indicate the extent to which a trait is inherited.
7. Explain the type of information that a genome-wide association study can reveal.

7.4 A Closer Look: Body Weight
8. Discuss how the methods discussed in the chapter can be applied to investigate variation in body weight.

The BIG Picture

Who we are and how we feel arise from an intricate interplay between our genes and environmental influences. Understanding genetic contributions to traits and illnesses can suggest how we can alter our environments to improve our lives.

The Complex Genetics of Athletics

A website offers a single-gene test that purportedly provides parents and coaches early information on a child's future success in team or individual sports requiring speed/power or endurance. The test is for variants of a gene that encodes a protein called actinin 3 that is expressed in skeletal muscle. One genotype is overrepresented among elite sprinters, another among world-class endurance runners. The test takes a simplistic view of a complex trait.

Many genes as well as environmental influences determine at which activities or sports a person may excel. Any single gene is unlikely to have a great influence. Environmental factors include exposure to pollution and toxins, as well as opportunities to participate in sports. The ability to work well on a team cannot be explained by a string of DNA letters that encodes a protein.

The idea to market tests for athletic genes may have come from knowledge of rare mutations in other genes that bestow great physical prowess. Members of a German family with a mutation in a "double muscle" gene are amazing weight lifters; a Scandinavian family of Olympic skiers has a mutation that increases the number of red blood cells, bringing more oxygen to their muscles. Genes also influence metabolic rate, bone mineral density, fat storage, glucose use, and lung function.

Genetic testing to predict athletic success might stress a child with no interest in competing, or discourage a child who loves a particular sport not to do it. However, tests of fitness-related genes might identify health risks that a child might face in the future, enabling parents to intervene with a healthy diet and regular exercise plan.

7.1 Genes and the Environment Mold Traits

A woman who is a prolific writer has a daughter who becomes a successful novelist. An overweight man and woman have obese children. A man whose father suffers from alcoholism has the same problem. Are these characteristics—writing talent, obesity, and alcoholism—inherited or learned? Such traits, and nearly all others, are not the result of an "either/or" mechanism, but reflect the input of many genes as well as environmental influences. Even single-gene diseases are modified by environmental factors and/or other genes.

A trait can be described as either single-gene (Mendelian or monogenic) or **polygenic,** caused by more than one gene. A monogenic or polygenic characteristic is called a **complex trait** if it is influenced by an environmental factor. (Complex traits are also called multifactorial traits.).

Lung cancer is a complex trait (**figure 7.1**). Variants of genes that increase the risk of becoming addicted to nicotine and of developing cancer come into play—but may not ever be expressed if a person never smokes and breathes only fresh air. Purely polygenic traits—those not influenced by the environment at all—are rare. Eye color, discussed in chapter 4, is close to being purely polygenic.

Polygenic complex traits include common ones, such as height, skin color, body weight, many illnesses, and behavioral conditions and tendencies. Behavioral traits are not inherently different from other types of traits; they involve the functioning of the brain, rather than another organ. Chapter 8 discusses them. Despite the name "complex," the genes underlying such traits are not more complicated than others. They follow Mendel's laws, but contribute only partly to a phenotype and are therefore more difficult to track.

A polygenic complex trait or condition is the consequence of contributions of several genes. Each gene confers a degree of susceptibility, but not necessarily equally. A rare allele may exert a large influence, but common alleles may each contribute only slightly to a trait. For example, three genes contribute significantly to the risk of developing type 2 diabetes mellitus, but other genes exert smaller effects.

Different genes may contribute different parts of a phenotype that was once thought to be due to the actions of a single gene. This is the case for migraine. A gene on chromosome 1 contributes sensitivity to sound; a gene on chromosome 5 produces the pulsating headache and sensitivity to light; and a gene on chromosome 8 is associated with nausea and vomiting. In addition, environmental influences trigger migraine in some people, such as eating certain foods, stress, or certain weather conditions. **Clinical Connection 7.1** takes a closer look at risk factors for a common type of illness, heart disease.

Figure 7.1 Genetic and environmental factors contribute to lung cancer risk. Mutations raise lung cancer risk in several ways: impairing DNA repair, promoting inflammation, blocking detoxification of carcinogens, keeping telomeres long, and promoting addiction. These genetic risk factors interact with each other and with environmental influences, such as smoking, breathing polluted air, and diet. *Akbudak Rimma/Shutterstock*

Key Concepts Questions 7.1

a. Explain how polygenic traits differ from single-gene (Mendelian) traits.
*b. Define *complex trait.*
c. Are the contributions of individual genes to a polygenic complex trait necessarily equal?

Clinical Connection 7.1
Many Genes Control Heart Health

Many types of cells and processes must interact for the heart and blood vessels (the cardiovascular system) to circulate blood, and many genes maintain the system. The outside environment has strong effects, too. Cardiovascular disease affects one in three individuals.

Genes control the transport of lipids, blood clotting, blood pressure, and how well white blood cells stick to blood vessel walls. Lipids can only circulate when bound to proteins, forming lipoproteins.

Maintaining a healthy cardiovascular system requires a balance between having enough lipids inside cells and not having an excess outside cells. Genes specify enzymes that process lipids, the proteins that transport them, and receptor proteins that admit lipids into cells. Atherosclerosis is the buildup of fatty material, called plaque, on the interior linings of arteries.

Several genes encode the protein parts of lipoproteins, which are called apolipoproteins. Some types of lipoproteins carry lipids in the blood to tissues, where they are used, and other types of lipoproteins take lipids to the liver, where they are broken down into biochemicals that the body can excrete. An allele of the apolipoprotein E gene (*APO E4*) increases the risk of a heart attack threefold in people who smoke.

The fluidity of the blood is also critical to health. Overly active clotting factors or extra sticky white blood cells (leukocytes) can block blood flow, usually in blood vessels in the heart or in the legs. Poor clotting causes dangerous bleeding. Because clotting factors are proteins, clotting is genetically controlled.

An enzyme, lipoprotein lipase, lines the walls of the smallest blood vessels, where it breaks down fat packets released from the small intestine and liver. Lipoprotein lipase is activated by high-density lipoproteins (HDLs), and it breaks down low-density lipoproteins (LDLs). Because low LDL levels are associated with a healthy cardiovascular system in many families, low LDL is used as a biomarker of heart health. High HDL is generally a biomarker for good health, but in some families with no heart problems, HDL is low. People with elevated LDL levels that diet and exercise cannot control may take statin drugs, which block the enzyme the liver requires to synthesize cholesterol. Food provides cholesterol but the liver also synthesizes it.

Elevated cholesterol in the blood illustrates the complex nature of cardiovascular disease. People who inherit one mutant allele of the LDL receptor gene typically suffer heart attacks in early adulthood. The rare individuals who inherit two mutant alleles have fatal heart attacks much earlier (see figure 5.2). However, high cholesterol can arise from the actions of at least 12 other genes.

Genetic test panels detect alleles of genes that contribute to risk of developing cardiovascular disease. More than 500 genes regulate blood pressure, and more than 95 contribute to inherited variation in blood cholesterol and triglyceride levels. Tests of gene expression can indicate which cholesterol-lowering drugs are most likely to be effective and tolerable for a particular individual. Computer analysis of multigene tests accounts for controllable environmental factors, such as exercising, not smoking, and maintaining a healthy weight (**table 7A**). **Figure 7A** shows an artery blocked by fatty plaque. Diet, regular exercise, and medication can help to counter an inherited tendency to deposit cholesterol-rich material on the interior linings of arteries.

Table 7A Risk Factors for Cardiovascular Disease

Uncontrollable	Controllable
Age	Fatty diet
Male sex	Hypertension
Genes	Smoking
Lipid metabolism	High serum cholesterol
Apolipoproteins	Stress
Lipoprotein lipase	Diabetes
Blood clotting	
Fibrinogen	
Clotting factors	
Inflammation	
C-reactive protein	
Homocysteine metabolism	
Leukocyte adhesion	

Figure 7A Cardiovascular disease. Accumulation of fatty plaque blocks circulation of the blood through arteries.

Questions for Discussion

1. What are three ways a person can counteract having inherited gene variants that increase the likelihood of developing cardiovascular disease?
2. List four ways that genes control cardiovascular health.
3. Do you think a person who refuses to follow dietary and exercise advice to lower cholesterol should be prescribed a statin?

7.2 Polygenic Traits Are Continuously Varying

For a polygenic trait, the combined action of many genes often produces a "shades of gray" or "continuously varying" phenotype, also called a quantitative trait. DNA sequences that contribute to polygenic traits are called **quantitative trait loci (QTLs)**. A complex trait is continuously varying if it is also polygenic. That is, it is the multigene component of the trait that contributes the continuing variation of the phenotype. The individual genes that confer a polygenic trait follow Mendel's laws, but together they do not produce single-gene phenotypic ratios. They all contribute to the phenotype, but without being dominant or recessive to each other. Single-gene traits are instead discrete or qualitative, often providing an "all-or-none" phenotype such as "normal" versus "affected."

A polygenic trait varies in populations, as our many nuances of hair color, body weight, and cholesterol levels demonstrate. Some genes contribute more to a polygenic trait than others. Within genes, alleles can have differing impacts depending upon exactly how they alter an encoded protein and how common they are in a population. For example, a mutation in the LDL receptor gene greatly raises blood serum cholesterol level. But because fewer than 1 percent of individuals in most populations have this mutation, it contributes very little to variation in cholesterol level in a population; however, the mutation has a large impact on the person who has it.

Although the expression of a polygenic trait is continuous, we can categorize individuals into classes and calculate the frequencies of the classes. When we do this and plot the frequency for each phenotype class, a bell-shaped curve results. Even when different numbers of genes affect the trait, the curve takes the same shape, as figures 7.2, 7.3, and 7.4 show.

Fingerprint Patterns

Skin on the fingertips is folded into raised patterns called dermal ridges that align to form loops, whorls, and arches. This pattern is a fingerprint. A technique called dermatoglyphics ("skin writing") compares the number of ridges that comprise these patterns to identify and distinguish individuals (**figure 7.2**). Dermatoglyphics is part of genetics because certain diseases (such as Down syndrome) produce unusual ridge patterns. Forensic fingerprint analysis is also an application of dermatoglyphics.

Genes largely determine the number of ridges in a fingerprint, but the environment can affect them, too. During weeks 6 through 13 of prenatal development, the ridge pattern can be altered as the fetus touches the finger and toe pads to the wall of the amniotic sac. This early environmental effect explains why the fingerprints of identical twins, who share all genes, are in some cases not exactly alike.

A fingerprint is quantified with a measurement called a total ridge count, which tallies the numbers of ridges in whorls, loops, or arches. The average total ridge count in a male is 145, and in a female, 126. Plotting total ridge count yields the bell curve of a continuously varying trait.

Height

The effect of the environment on height is obvious—people who do not eat enough do not reach their genetic potential for height. Students lined up according to height, but raised in two different decades and under different circumstances, illustrate the effects of genes and the environment on this continuously varying trait. Students from 1920 were on average considerably shorter than students from recent years. The tallest individuals from 1920 were 5'9", whereas the tallest now are 6'5".

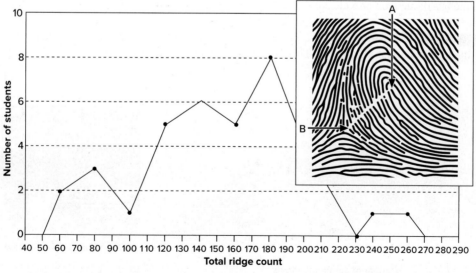

Figure 7.2 Anatomy of a fingerprint. Total ridge counts for a number of individuals, plotted on a bar graph, form an approximately bell-shaped curve. The number of ridges between landmark points A and B on this loop pattern is 12. Total ridge count includes the number of ridges on all fingers.
(*Source:* Data from Gordon Mendenhall, Thomas Mertens, and Jon Hendrix, "Fingerprint Ridge Count" in The American Biology Teacher, vol. 51, no. 4, April 1989, pp. 204–206).

Figure 7.3 **The inheritance of height.** Genetics students at the University of Notre Dame lined up by height in inches, revealing the continuously varying nature of height. *David Hyde/Wayne Falda/McGraw-Hill Education*

The difference in heights between then and now is attributed to improved diet, better overall health, and the practical effect that taller soldiers were more likely to have been killed in the trench warfare of the first world war, leaving behind a slightly shorter population. A purely genetic effect so dramatic would not show up in the course of just one or two generations. At least 50 genes affect variation in height, some with large effects and some with lesser effects. **Figure 7.3** shows a bell curve for height.

Skin Color and Race

More than 100 genes affect pigmentation in skin, hair, and the irises (see the chapter opener to chapter 5). Melanin pigments color the skin to different degrees in different individuals. In the skin, as in the iris (see figure 4.6), melanocytes are cells that contain melanin in packets called melanosomes. Melanocytes extend between the tilelike skin cells, distributing pigment granules through the skin layers. Some melanin exits the melanocytes and enters the hardened cells in the skin's upper layers. Here the melanin breaks into pieces, and as the skin cells are pushed up toward the skin's surface as stem cells beneath them divide, the bits of melanin provide color. The pigment protects against DNA damage from ultraviolet radiation. Exposure to sunlight increases melanin synthesis. **Figure 7.4**a shows a three-gene model for human skin color. This is an oversimplification, but it illustrates how several genes can contribute to a variable trait.

We all have about the same number of melanocytes per unit area of skin. People have different skin colors because they vary in melanosome number, size, and density of pigment distribution. Different skin colors arise from the number and distribution of melanin pieces in cells in the uppermost skin layers.

Skin color is one physical trait that has been used to distinguish race. However, defining race based on skin color or another physical trait is a social construct rather than a biological one, because skin color is only one of thousands of traits whose frequencies vary in different human populations and is influenced by environmental factors, such as exposure to the sun. From a genetic perspective, when referring to nonhumans, races are groups within species that are distinguished by different allele frequencies. Humans are actually a lot less variable in appearance than other mammals. We may classify people by skin color because it is a visual way to distinguish individuals, but this trait is *not* a reliable indicator of ancestry.

The concept of race based on skin color falls apart when considering many genes. That is, two people with very dark skin may be less alike than either is to another person with very light skin. For example, sub-Saharan Africans and Australian aborigines have dark skin, but are dissimilar in other inherited characteristics. Their dark skins may reflect adaptation to life in a sunny, tropical climate rather than recent shared ancestry. Overall, 93 percent of varying inherited traits are no more common in people of one skin color than any other.

Some of our knowledge about human skin color has been biased by the focus of studies on Europeans. A recent study traced pale skin tones among more than 6,000 people living in Latin America (Mexico, Peru, Colombia, Brazil, and Chile) to a variant of a gene, *MFSD12,* that was brought in with the ancestors of Native Americans across the Bering Strait long before a European set foot on American soil. The variant is today seen among East Asians and Native Americans, who contributed to the Latin American gene pool studied. Underexpression of the gene produces dark skin and overexpression produces lighter skin tones.

Figure 7.4 **Variations in skin color.** **(a)** A model of three genes, with two alleles each, can explain broad hues of human skin. In actuality, this trait likely involves many more than three genes. **(b)** Humans come in a great variety of skin colors. Skin color genes can assort in interesting ways. These beautiful young ladies, Marcia and Millie, are twins! Their father is Jamaican with dark skin and tight dark curls and their mother is European with fair skin and golden-brown hair. *(b): SWNS/South West News Service Ltd.*

Testing DNA indicates that it makes more biological sense to classify people by ancestry than by skin color. Consumer genetic testing companies that provide reports on ancestry compare DNA data between pairs of individuals to deduce relatedness, not physical traits. Section 18.3 describes the reasoning behind ethnicity estimates from consumer genetic ancestry tests.

In a genetic sense the concept of race based on skin color has little meaning, but in a practical sense, racial groups *do* have different incidences of certain diseases. Part of this is due to the tendency to choose partners within a group, which retains certain alleles. Racial differences in disease prevalence may also result from social inequities, such as access to good nutrition or health care. Still, genetic information can be considered along with socially defined race to make treatment decisions. For example, a new class of cholesterol-lowering drugs was developed by testing European populations for a specific variant of the gene *PCSK9*. A different variant of the gene has the opposite effect on blood cholesterol level in African Americans, who would benefit from different drugs.

It is more accurate to prescribe drugs based on genotypes that determine whether a drug will work or have side effects, than by skin color. When researchers cataloged 23 markers for genes that control drug metabolism in 354 people representing blacks (Bantu, Ethiopian, and Afro-Caribbean), whites (Norwegian, Armenian, and Ashkenazi Jews), and Asians (Chinese and New Guinean), the genetic markers fell into four distinct groups that predict which of several blood thinners, chemotherapies, and painkillers would be effective—and these response groups did *not* match the groups based on skin color. Chapter 22 continues discussion of selecting drugs based on genotype.

Key Concepts Questions 7.2

*a. Define *quantitative trait loci*.

*b. What does "continuously varying" mean?

c. Explain how a bell curve describes the distribution of phenotype classes of any polygenic trait.

d. Explain how fingerprint patterns, height, and skin color are complex traits.

7.3 Methods to Investigate Complex Traits

Predicting recurrence risks for polygenic traits is much more challenging than doing so for single-gene traits. Researchers use several strategies to investigate polygenic traits.

Empiric Risk

Using Mendel's laws, it is possible to predict the risk that a single-gene trait will recur in a family from knowing the mode of inheritance—such as autosomal dominant or recessive. To predict the chance that a polygenic complex trait will occur in a particular individual, geneticists use **empiric risk,** which is based on incidence in a specific population. **Incidence** is the rate at which a certain event occurs, such as the number of new cases of a disease diagnosed per year in a population of known size. **Prevalence** is the proportion or number of individuals in a population who have a particular disease at a specific time, such as during one year.

Empiric risk is not a calculation, but a population statistic based on observation. The population might be broad, such as an ethnic group or community, or genetically more well defined, such as families that have cystic fibrosis.

Empiric risk increases with severity of the disease, number of affected family members, and how closely related a person is to affected individuals. Consider empiric risk used to predict the likelihood of a child being born with a neural tube defect (NTD). In the United States, the overall population risk of carrying a fetus with an NTD is about 1 in 1,000 (0.1 percent). For people of English, Irish, or Scottish ancestry, the risk is about 3 in 1,000 (0.3 percent). However, if a sibling has an NTD, for any ethnic group, the risk of recurrence increases to 3 percent, and if two siblings are affected, the risk for a third child is even greater.

If a trait has an inherited component, then it makes sense that the closer the relationship between two individuals, one of whom has the trait, the greater the probability that the second individual has the trait, too, because they share more genes. Studies of empiric risk support this logic. **Table 7.1** summarizes empiric risks for relatives of individuals with cleft lip (**figure 7.5**).

Because empiric risk is based solely on observation, it is useful to derive risks for diseases with poorly understood transmission patterns. For example, certain complex diseases affect one sex more often than the other. Pyloric stenosis, an overgrowth of muscle at the juncture between the stomach and the small intestine, is five times more common among males than females. The condition must be corrected surgically shortly after birth, or the newborn will be unable to digest foods. Empiric data show that the risk of recurrence for the brother of an affected brother is 3.8 percent, but the risk for the brother of

Table 7.1	Empiric Risk of Recurrence for Cleft Lip
Relationship to Affected Person	**Empiric Risk of Recurrence**
Identical twin	40.0%
Sibling	4.1%
Child	3.5%
Niece/nephew	0.8%
First cousin	0.3%
General population risk (no affected relatives)	0.1%

Figure 7.5 Cleft lip. Cleft lip is more likely in a person who has a relative with the condition. Corrective surgery is highly effective.

an affected sister is 9.2 percent. Empiric risk is based on real-world observations, even if the cause of the illness or why it is more common in one gender than another is not known.

Heritability

A measurement called **heritability,** designated H, estimates the proportion of the phenotypic variation for a trait that is due to genetic differences in a certain population at a certain time. The distinction between empiric risk and heritability is that empiric risk could be affected by nongenetic influences, whereas heritability focuses on the genetic component of the variation in a trait. Heritability refers to the degree of *variation* in a trait due to genetics, and not to the proportion of the trait itself attributed to genes.

Figure 7.6 outlines the factors that contribute to observed variation in a trait. Heritability equals 1.0 for a trait whose variability is completely the result of gene action, such as in a population of laboratory mice who share the same environment. Without environmental variability, genetic differences alone determine expression of the trait in the population. Variability of most traits, however, is due to differences among genes

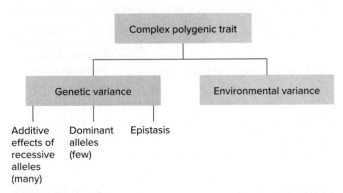

Figure 7.6 Heritability estimates the genetic contribution to the variability of a trait. Observed variance in a complex polygenic trait or illness reflects genetic and environmental contributions.

Table 7.2	Heritabilities for Some Human Traits
Trait	Heritability
Clubfoot	0.8
Height	0.8
Blood pressure	0.6
Body mass index	0.4–0.7
Verbal aptitude	0.7
Mathematical aptitude	0.3
Spelling aptitude	0.5
Total fingerprint ridge count	0.9
Intelligence	0.5–0.8
Total serum cholesterol	0.6

and environmental influences. Genes contribute to differing degrees to trait variability. For height, for example, gene variants that are prevalent in a mostly European population were found to contribute 25.7 percent to heritability, and rare variants contribute 1.7 percent. That means, overall, that genes are responsible for about a quarter of the variability in height in this population. **Table 7.2** lists some traits and their heritabilities.

Heritability changes as the environment changes. For example, the heritability of skin color is higher in the winter months, when sun exposure is less likely to increase melanin synthesis. The same trait may be highly heritable in two populations, but certain variants much more common in one group due to long-term environmental differences. Populations in equatorial Africa, for example, have darker skin than sun-deprived Scandinavians.

Missing Heritability

For many traits and illnesses, genes or other DNA sequences cannot account for all of the variability of the phenotype. This situation is termed *missing heritability*. For example, genes account for only 7 percent of the variation in human life expectancy, but heritability in past studies ranges from 20 to 30 percent. Several of the disorders discussed in the next chapter (major depressive disorder, bipolar disorder, schizophrenia, and autism) also show missing heritability, which means that environmental influences are significant. The inability of identified gene variants to explain variation in a trait could be partly due to interactions of genes that are difficult to measure or assess.

Estimating Heritability

Researchers use several statistical methods to estimate heritability. One way is to compare the actual proportion of pairs of people related in a certain manner who share a particular trait

Table 7.3	Coefficient of Relatedness for Pairs of Relatives		
Relationship		**Degree of Relationship**	**Percent Shared Genes (Coefficient of Relatedness)**
Sibling to sibling		1°	50% (1/2)
Parent to child		1°	50% (1/2)
Uncl/aunt to niece/nephew		2°	25% (1/4)
Half-siblings		2°	25% (1/4)
Grandparent to grandchild		2°	25% (1/4)
First cousin to first cousin		3°	12.5% (1/8)

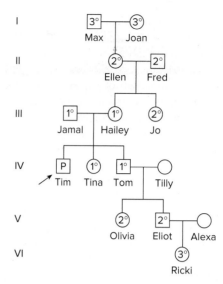

Figure 7.7 Tracing relatives. Tim has an inherited illness. A genetic counselor drew this pedigree to explain the approximate percentage of genes Tim shares with relatives. This information can be used to alert certain relatives to their risks. ("P" is the proband, or affected individual who initiated the study. See table 7.3 for definitions of 1°, 2°, and 3° relationships.)

to the expected proportion of pairs that would share it if it were inherited in a Mendelian fashion. The expected proportion is derived by knowing the blood relationships of the individuals and using a measurement called the **coefficient of relatedness**, which is the proportion of genes that two people related in a certain way share (**table 7.3**).

A parent and child share 50 percent of their genes because of the mechanism of meiosis. Siblings share on average 50 percent of their genes because they have a 50 percent chance of inheriting each allele for a gene from each parent. Genetic counselors use the designations primary (1°), secondary (2°), and tertiary (3°) relatives when calculating risks (see table 7.3; **figure 7.7**). For extended or complicated pedigrees, the value of 1 in 2 (or 50 percent) between siblings and between parent-child pairs can be used to trace and calculate the percentage of genes shared between people related in other ways.

If the heritability of a trait is very high, then of a group of 100 sibling pairs, nearly 50 would be expected to have the same phenotype, because siblings share on average 50 percent of their genes. Returning to the example of height, for which heritability reflects the environmental influence of nutrition, of 100 sibling pairs in a population, for example, 40 are expected to be the same number of inches tall. This is because heritability for height is 0.40/0.50, or 80 percent. That value is the observed phenotypic variation divided by the expected phenotypic variation if environment had no influence.

Genetic variance for a polygenic trait is mostly due to the additive effects of recessive alleles of different genes. For some traits, a few dominant alleles can greatly influence the phenotype, but because they are rare, they do not contribute greatly to heritability. This is the case for heart disease caused by a faulty LDL receptor. Heritabilities for some traits or diseases may be underestimated if only mutations that change DNA base sequences are compared, without also considering **copy number variants (CNVs)**, which are differences in the numbers of copies of a DNA sequence.

Epistasis (interaction between alleles of different genes) can also influence heritability. To account for the fact that different genes affect a phenotype to differing degrees, geneticists calculate a "narrow" heritability that considers only additive recessive effects and a "broad" heritability that also considers the effects of rare dominant alleles and epistasis. For LDL cholesterol level, for example, the narrow heritability is 0.36 but the broad heritability is 0.96, indicating that a rare dominant allele has a large impact.

Understanding inheritance of complex traits is important in agriculture. A breeder needs to know whether genetic or environmental influences contribute to variability in such traits as birth weight, milk yield, and egg hatchability. It is also valuable to know whether the genetic influences are additive or epistatic. The breeder can control the environment by adjusting the conditions under which animals are raised and crops grown and control genetic effects by setting up crosses between particular individuals.

Studying complex traits in humans is difficult because information must be obtained from many families. Two special types of people, however, can help geneticists to tease apart the genetic and environmental components of the variability of complex traits—adopted individuals and twins.

Adopted Individuals

An adopted person typically shares environmental influences, but not many gene variants, with the adoptive family. Conversely, adopted individuals share genes, but not the exact environment, with their biological parents. Therefore, similarities between adopted people and adoptive parents are mostly due to environmental influences, whereas similarities between adoptees and their biological parents are mostly genetic influences. Information on both sets of parents can sometimes indicate how heredity and the environment contribute to a trait.

Adoption studies go back many years. An early investigation used a database of all adopted children in Denmark and their families from 1924 to 1947. One study examined correlations between causes of death among biological and adoptive parents and adopted children. If a biological parent died of infection before age 50, the child he or she gave up for adoption was five times more likely to die of infection at a young age than a similar person in the general population. This may be because inherited variants in immune system genes increase susceptibility to certain infections. In support of this hypothesis, the risk that an adopted individual would die young from infection did *not* correlate with adoptive parents' deaths from infection before age 50. Researchers concluded that genetics mostly determines length of life, with environmental influences. For example, if adoptive parents died before age 50 of cardiovascular disease, their adopted children were three times as likely to die of heart and blood vessel disease as a person in the general population. What environmental factors might explain this finding?

Twins

Studies that use twins to separate the genetic from the environmental contribution to a phenotype provide more meaningful information than studies of adopted individuals.

A trait that occurs more frequently in both members of identical (monozygotic, or MZ) twin pairs than in both members of fraternal (dizygotic, or DZ) twin pairs is at least partly controlled by heredity. Geneticists calculate the **concordance** of a trait as the percentage of pairs in which both twins express the trait among pairs of twins in whom at least one has the trait. Twins who differ in a trait are said to be discordant for it.

In one study, 142 MZ twin pairs and 142 DZ twin pairs took a "distorted tunes test," in which 26 familiar songs were played, each with at least one note altered. A person was considered "tune deaf" if he or she could not detect the mistakes in three or more tunes. Concordance for "tune deafness" was 67 percent for MZ twins but only 44 percent for DZ twins, indicating a considerable inherited component in the ability to accurately perceive musical pitch. **Table 7.4** compares twin types for a variety of hard-to-measure traits. (Figure 3.18 shows how DZ and MZ twins arise.)

Diseases caused by single genes that approach 100 percent penetrance, whether dominant or recessive, also approach 100 percent concordance in MZ twins. That is, if one identical twin has the disease, so does the other. However, among DZ twins, concordance generally is 50 percent for a dominant trait and 25 percent for a recessive trait. These are the Mendelian values that apply to any two nontwin siblings. For a polygenic trait with little environmental input, concordance values for MZ twins are significantly greater than for DZ twins. A trait

Table 7.4 Concordance Values for Some Traits in Twins

Trait	MZ (Identical) Twins	DZ (Fraternal) Twins
Acne	14%	14%
Alzheimer disease	78%	39%
Anorexia nervosa	55%	7%
Autism	90%	4.5%
Bipolar disorder	33%–80%	0%–8%
Cleft lip with or without cleft palate	40%	3%–6%
Hypertension	62%	48%
Schizophrenia	40%–50%	10%

molded mostly by the environment exhibits similar concordance values for both types of twins.

Comparing twin types assumes that both types of twins share similar experiences. In fact, MZ twins are often closer emotionally than DZ twins. This discrepancy between the closeness of the two types of twins can lead to misleading results. A study from the 1940s, for example, concluded that tuberculosis is inherited because concordance among MZ twins was higher than among DZ twins. Actually, the infectious disease more readily passed between MZ twins because their parents kept them closer. However, we do inherit susceptibilities to some infectious diseases. MZ twins would share such genes, whereas DZ twins would only be as likely as any sibling pairs to do so.

A more informative way to assess the genetic component of a complex trait is to study MZ twins who were separated at birth, then raised in different environments. A documentary film, *Three Identical Strangers,* tells the remarkable tale of triplets separated in infancy who met for the first time at age 19, in 1980. Their matching faces, mannerisms, behaviors, and quirks reverberate throughout the film and are astonishing because the triplets were raised in economically diverse families. However, the film pays much less attention to what makes them different. More scientific is the Minnesota Twin Family Study at the University of Minnesota, which has used this "twins reared apart" approach.

Since 1989, thousands of sets of twins and triplets who were separated at birth have visited the laboratories of Thomas Bouchard at the university. For a week or more, the twins and triplets are tested for physical and behavioral traits, including 24 blood types, handedness, direction of hair growth, fingerprint pattern, height, weight, organ system function, intelligence, allergies, and dental patterns. The participants provide DNA samples. Researchers videotape facial expressions and body movements in different circumstances and probe participants' fears, interests, and superstitions.

Figure 7.8 Identical twins have much in common. In addition to physical traits, MZ twins may share tastes, preferences, and behaviors. Studying MZ twins and triplets separated at birth and reunited is a way to assess which traits are inherited. *Bob Kreisel/Alamy Stock Photo*

Twins and triplets separated at birth provide natural experiments for distinguishing nature from nurture. Many of their common traits can be attributed to genetics, especially if their environments have been very different. Their differences tend to come from differences in upbringing, because their genes are identical (MZ twins and triplets) or similar (DZ twins and triplets).

Some MZ twins separated at birth and reunited later are remarkably similar, even when they grow up in very different adoptive families (**figure 7.8**). Idiosyncrasies are particularly striking. One pair of twins who met for the first time when they were in their thirties responded identically to questions; each paused for 30 seconds, rotated a gold necklace she was wearing three times, and then answered the question. Coincidence, or genetics?

The "twins reared apart" approach is not an ideal way to separate nature from nurture. MZ twins and other multiples share an environment in the uterus and possibly in early infancy that may affect later development. Siblings, whether adoptive or biological, do not always share identical home environments. Differences in sex, general health, school and peer experiences, temperament, and personality affect each individual's perception of such environmental influences as parental affection and discipline.

Genome-Wide Association Studies

A **genome-wide association study (GWAS)** compares many landmarks (genetic markers) across the genome between two large groups of people—one with a particular trait or disease and one without it. Identifying parts of the genome that are much more common among the people with the trait or illness can enable researchers to identify genes that contribute to the phenotype.

Genetic Markers in GWAS

Genome-wide association studies use several types of genetic markers (**table 7.5**). **Single nucleotide polymorphisms (SNPs)** and **copy number variants (CNVs)** describe the DNA base sequence. A SNP is a site in the genome that has a different DNA base in at least 1 percent of a population (**figure 7.9**). A SNP may be in a gene or in the vast genomic regions between genes. A CNV is a DNA sequence that repeats a different number of times in different individuals (**figure 7.10**). A CNV does not provide information in the same way as a gene that encodes protein does, but comparing CNVs is another way to distinguish individuals. CNVs are useful in forensic applications, discussed in section 18.2.

Gene expression patterns are also used in genome-wide association studies. These patterns indicate which proteins are overproduced or underproduced in people with the trait or illness, compared to unaffected controls. Yet another way to compare genomes is by the sites to which methyl (CH_3) groups bind, shutting off gene expression. This is an epigenetic change because it doesn't affect the DNA base sequence. DNA methylation is important in the variation of blood pressure in a population, for example.

To achieve statistical significance, a GWAS must include at least 100,000 markers. The association of markers to a trait or disease provides the information (**figure 7.11**). Typically, a GWAS uses a million or more SNPs, grouped into half a million or so haplotypes. A specific "tag SNP" is used as a stand-in for many linked SNPs to identify a haplotype.

Table 7.5	Types of Information Used in Genome-Wide Association Studies
Marker Type	**Definition**
SNP	A single nucleotide polymorphism is a site in the genome that has a different DNA base in >1% of a population.
CNV	A copy number variant is a tandemly repeated DNA sequence, such as CGTA CGTA CGTA.
Gene expression	The pattern of genes that are overexpressed and/or underexpressed in people with a particular trait or disease.
	Epigenetic signature of methyl groups binding DNA.

Figure 7.9 **SNPs are sites of variability.** The DNA base pair in red is a SNP—a site that differs in more than 1 percent of a population. (The percentage may change as SNPs are identified in more individuals.) In this case, 98% of individuals might have GC at the indicated position **(a)**, and 2% have TA **(b)**.

Polygenic Risk Scores

A GWAS is a stepwise focusing-in on parts of a genome that are associated to some degree with a particular trait or illness (**figure 7.12**). The technique has rapidly evolved from a research tool helpful in identifying genes to having practical applications.

Machine-learning algorithms identify SNP patterns, such as the 500,000 tag SNPs, that are present in a large group of people who have the trait or illness of interest but are not seen in an equally large group of people who do *not* have the

GATTACA	Allele 1
GATTACAGATTACA	Allele 2
GATTACAGATTACAGATTACA	Allele 3
GATTACAGATTACAGATTACAGATTACA	Allele 4

Figure 7.10 **Copy number variants provide a different type of genetic information.** For copy number variants, different numbers of repeats of a short DNA base sequence are considered to be different alleles.

trait or illness. Then, a new sample can be tested for the telltale pattern to predict appearance of the characteristic. Data sets are "trained" and then "validated."

SNP patterns can be converted into **polygenic risk scores** that can be used to predict illness. For Alzheimer disease, researchers tracked SNPs in 31 genes for 17,008 people with the disease and for 37,154 people who did not have it. Then they tested the SNP panel on 1,081 people from the National Alzheimer's Coordinating Center who are cognitively normal but have affected relatives, as well as 571 people with mild cognitive impairment (MCI), which can progress to Alzheimer disease. The polygenic risk score predicted progression to Alzheimer disease, even for people who do not have the well-studied gene variants discussed in Clinical Connection 5.1. A low score predicts onset of symptoms by age 85, whereas a high score predicts symptom onset by age 78. However, polygenic risk scores need to be validated in populations that are not represented in the GWAS data on which the scores are based.

SNP patterns are compared in ancestry testing to identify relatives. The approximately 770,000 SNPs that one company uses were first trained on a large number of sets of known first cousins. If a person sends a DNA sample to a consumer testing company (see figure 1.1), she or he may "match" with first

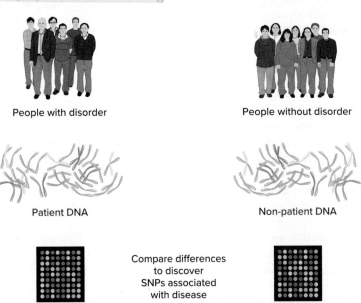

Figure 7.11 **Tracking genes in groups.** Genome-wide association studies seek DNA sequence variants that are shared with much greater frequency among individuals with the same illness or trait than among others. The squares are DNA microarrays, which display short, labeled DNA pieces (see figure 20.7). Different patterns indicate different alleles.

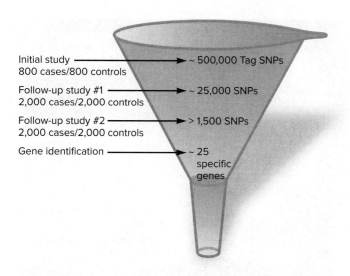

Figure 7.12 A stepwise approach to gene discovery. Genome-wide association study results must be validated in several different populations. Further research is necessary to go beyond association to demonstrate correlation and cause.

cousins based on the number of tag SNPs that first cousins typically share. Comparing SNP patterns has also been used to predict response to antidepressant drugs and the risk of preterm birth. It is a technique with many applications.

Study Designs in GWAS

Several study designs are used in genome-wide association studies (**table 7.6**). In a cohort study, researchers follow a large group of individuals over time and measure many aspects of their health. The most famous is the Framingham Heart Study, which began tracking thousands of people and their descendants in Massachusetts in 1968. Nine thousand of them are participating in a GWAS.

In a case-control study, each individual in one group is matched to an individual in another group who shares as many characteristics as possible, such as age, sex, activity level, and environmental exposures. SNP differences are then associated with the presence or absence of the disease or trait. For example, if 5,000 individuals with hypertension (high blood pressure) have particular DNA bases at six sites in the genome, and 5,000 matched individuals who do not have hypertension have different bases at only these six sites, then these genome regions may include genes whose protein products control blood pressure.

An **affected sibling pair study** follows the logic that because siblings share 50 percent of their genes, a trait or condition that many siblings share is likely to be inherited. Researchers scan genomes for SNPs that most siblings who have the same condition share, but that siblings who do not both have the condition do not often share. Such genome regions may have genes that contribute to the condition.

A variation on the affected sibling pair strategy is **homozygosity mapping,** which is carried out with families that are consanguineous—that is, the parents are related. The genomes of children whose parents share recent ancestors have more homozygous regions than do other children, and therefore greater likelihood that they have inherited two copies of a susceptibility or disease-causing mutation.

Common characteristics, such as height and body mass index, are also investigated with genome-wide association studies. One study examined facial features. Researchers measured eye and nose positions and dimensions, 20 lip descriptors, and length of space between the nose and upper lip. The study identified variants of several genes already known to be mutant in specific syndromes that disrupt development of facial features.

Limitations of GWAS

Genome-wide association studies provide less information than sequencing genomes, but they help researchers to locate genes. A conceptual limitation of genome-wide association studies is that they reveal associations between sets of information, and not causes. An *association* only means that one event or characteristic occurs at the same time as another. A *correlation* is a directional association: If one measurement increases, so does the other, such as stress and blood pressure. In contrast, establishing a cause requires that a specific mechanism explains how one event makes another happen: *How* does stress elevate blood pressure? An association study does not provide direct information about a gene's function—it is more a discovery tool.

A practical limitation of genome-wide association studies is that they often identify parts of the genome that contribute

Table 7.6	Study Designs for Complex Traits
Type of Study	**Definition**
Cohort	Researchers follow many people over time and measure several traits.
Case-control	People in two groups are individually matched for several characteristics, and differences in SNP patterns are identified.
Affected sibling pair	Tests identify SNPs that siblings with the same condition share but that siblings who do not both have the condition do not share.
Homozygosity mapping	Disease-causing mutations are identified in homozygous genome regions that children inherit from parents who are related to each other.

Chapter 7 *Complex Traits* **141**

only slightly to the risk of developing a disease. A genetic test that indicates a 1 percent increase in risk of developing lung cancer, for example, would not matter much to a smoker whose environmental risk is much higher.

The way in which researchers select a patient population can bias a GWAS. Samples drawn from clinics, for example, would not include the very mildly affected who are not ill enough to show up, or those who have died. Another source of error is that individuals in the control population might not actually be healthy. They might have problems other than the one being investigated.

GWAS do not adequately represent all human populations, particularly small and/or underserved ones. An analysis of the individuals in the GWAS catalog, which is a database maintained by the National Human Genome Research Institute in the United States and the European Bioinformatics Institute of all such studies, found that about 80 percent of people included are European, 10 percent are Asian, and only 2 percent are African.

The complicating factors discussed in chapter 5 also affect the accuracy of genome-wide association studies. Recall that a phenocopy is a trait or illness that resembles an inherited one, but has an environmental cause. Placing a person with anemia due to a reaction to a drug in a group with people who have an inherited anemia would be misleading. Genetic heterogeneity, in which different genes cause the same trait or condition, could also be a source of error. Epistasis, when one gene masks the effect of another, also confounds these studies.

Another limitation of genome-wide association studies is that people who share symptoms and a SNP pattern may share something *else* that accounts for the association, such as an environmental exposure that can generate a false positive result. For example, mutations contribute to atherosclerosis risk, but so do infection, smoking, lack of exercise, and a fatty diet. These environmental factors are so common that if the group studied isn't large enough, a GWAS might not correctly identify a genetic influence.

The success of a GWAS may depend on the quality of the question asked. The technique was helpful, for example, in explaining why some people who live on the Solomon Islands have blond hair (**figure 7.13**). Most people living on these equatorial islands have dark hair and skin, similar to people who live in equatorial Africa. A case-control genome-wide association study on 43 blond Solomon Islanders and 42 dark-haired islanders clearly showed that the blonds were much more likely to have a particular SNP on chromosome 9. When researchers consulted the human genome sequence, they discovered in that interval a gene called *tyrosine-related protein 1* (*TYRP1*). Its protein product controls melanin pigmentation in all vertebrate animals and the mutant gene causes a form of human albinism. A single DNA base change is responsible for the unusual blond hair of some Solomon Islanders.

Often, the old and the new techniques for dissecting complex traits work well together. This is the case for the investigation of stuttering. Concordance for MZ twins ranges from 20 to 83 percent, and for DZ twins, from 4 to 9 percent, suggesting a large inherited component. The risk of a first-degree relative of a person who stutters also stuttering is 15 percent based on empiric evidence, compared

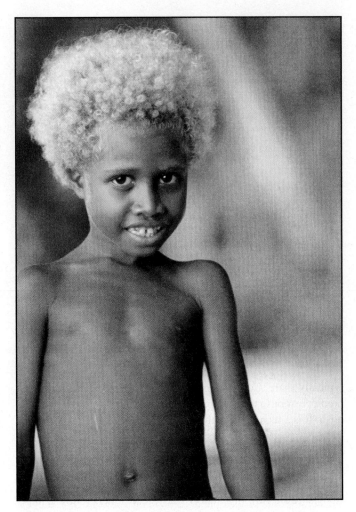

Figure 7.13 A striking phenotype. Blond hair among the residents of the Solomon Islands is due to a single base difference in a single gene. *Darryl Leniuk/age fotostock*

to the lifetime risk of stuttering in the general population of 5 percent, although part of that increase could be due to imitating an affected relative. A GWAS on 100 families who have at least two members who stutter identified candidate genes on three chromosomes that contribute to the trait.

Table 7.7 reviews terms used to study complex traits. Section 7.4 probes an example of such a trait—body weight.

Key Concepts Questions 7.3

*a. Define *empiric risk*.
*b. Define *heritability*.
c. Explain how the coefficient of relatedness is used to calculate heritability.
d. Describe how adopted individuals and twins are used to study environmental and inherited components of traits.
*e. What type of information do genome-wide association studies provide?

Table 7.7	Terms Used in Evaluating Complex Traits
Coefficient of relatedness	The proportion of genes shared by two people related in a particular way. Used to calculate heritability.
Concordance	The percentage of twin pairs in which both twins express a trait.
Empiric risk	The risk of recurrence of a trait or illness based on known incidence in a population.
Genome-wide association study (GWAS)	Detecting any association between marker patterns and increased risk of a condition.
Heritability	The percentage of phenotypic variation for a trait that is attributable to genetic differences. It equals the ratio of the observed phenotypic variation to the expected phenotypic variation for a population of individuals.

7.4 A Closer Look: Body Weight

Weight is a complex trait that can be studied using the methods discussed in this chapter. The rate at which food is taken in versus the rate at which the body uses it for fuel determines weight.

Scientific studies of body weight use a measurement called body mass index (BMI), which is weight in proportion to height (**figure 7.14**). BMI makes sense—a person who weighs 170 pounds and is 6 feet tall is slim, whereas a person of the same weight who is 5 feet tall is obese. The tall person's BMI is 23; the short person's is 33.5.

Heritability for BMI ranges in various studies from 0.4 to 0.7, which leaves room for environmental influences on our appetites and sizes. Dozens of genes affect how much we eat, how we use calories, and how fat is distributed in the body. The biochemical pathways and hormonal interactions that control weight may reveal points for drug intervention (**table 7.8**).

Genes That Affect Weight

Genetics became prominent in obesity research in 1994, when Jeffrey Friedman at Rockefeller University discovered a gene that encodes the protein hormone leptin in mice and in humans. Normally, eating stimulates fat cells (adipocytes) to secrete leptin, which travels in the bloodstream to a region of the brain's hypothalamus, where it binds to receptors on neurons. Leptin binding signals the neurons to release another type of hormone that binds yet other types of receptors, which ultimately function as an appetite "brake," while speeding digestion of food already eaten. When a person hasn't eaten in several hours, leptin levels fall, which triggers the release of an appetite "accelerator."

The discovery of genes and proteins that affect appetite sparked interest in targeting them with drugs to help people lose or gain weight. When Friedman gave mice extra leptin, they ate less and lost weight. Headlines soon proclaimed the new magic weight-loss elixir, a biotech company paid $20 million for rights to the hormone, and clinical trials ensued. The idea was to give obese people leptin, assuming that they had a deficiency, to trick them into feeling full. Only about 15 percent of the people lost weight, but the other 85 percent didn't actually lack leptin. Instead, most of them had leptin resistance, which is a diminished ability to recognize the hormone due to defective leptin receptors. Giving these people leptin had no effect on their appetites. However, the discovery helped a few severely obese children with true leptin deficiency attain normal weights after years of daily leptin injections.

The stomach is another source of weight-related proteins. Ghrelin is a peptide (small protein) hormone produced in the stomach that responds to hunger, signaling the hypothalamus to produce more of the appetite accelerator. One of the ways that weight-loss surgery may work is to decrease ghrelin secretion by making the stomach smaller.

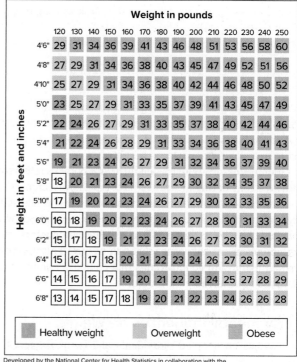

Figure 7.14 Body mass index (BMI). BMI equals weight/height^2, with weight measured in kilograms and height measured in meters. This chart provides a shortcut—the calculations have been done and converted to the English system of measurement. Squares that are not filled in indicate the person is underweight.
(Source: CDC.gov)

Table 7.8 Some Sites of Genetic Control of Body Weight

Protein	Function	Effect on Appetite
Leptin	Stimulates cells in hypothalamus to decrease appetite and metabolize nutrients	↓
Leptin transporter	Enables leptin to cross from bloodstream into brain	↓
Leptin receptor	Binds leptin on hypothalamus cell surfaces, triggering hormone's effects	↓
Neuropeptide Y	Produced in hypothalamus when leptin levels are low and the individual loses weight	↑
Melanocortin-4 receptor	Activated when leptin levels are high and the individual gains weight	↓
Ghrelin	Signals hunger from stomach to brain in the short term, stimulating neuropeptide Y	↑
PYY	Signals satiety from stomach to brain	↓
Stearoyl-CoA desaturase-1	Controls whether body stores or uses fat	↑

While leptin acts in the long term to maintain weight, the stomach's appetite control hormones function in the short term. All of these hormonal signals are integrated to finely control appetite in a way that maintains weight.

Identifying single genes that influence body weight paved the way for considering the trait to be complex. Researchers are investigating how combinations of genes control weight. One study looked at 21 genes in which mutations cause syndromes that include obesity, as well as 37 genes whose products participate in biochemical pathways related to weight. This approach identified many rare gene variants that could, in combination, explain the tendency to gain weight.

Environmental Influences on Weight

Many studies on adopted individuals and twins suggest that obesity has a heritability of 75 percent. Because the heritability for BMI is lower than this, the discrepancy suggests that genes play a larger role in those who tend to gain weight easily. The role of genes in obesity is seen when populations that have an inherited tendency to easily gain weight experience a large and sudden plunge in the quality of the diet.

On the tiny island of Naura, in Western Samoa, the residents' lifestyles changed greatly when they found a market for the tons of bird droppings on their island as commercial fertilizer. The money led to inactivity and adopting a high-calorie, high-fat diet, replacing an agricultural lifestyle and diet of fish and vegetables. Within a generation, two-thirds of the population had become obese, and a third had type 2 diabetes mellitus.

The Pima people provide another example of environmental effects on body weight. They separated into two populations during the Middle Ages, one group settling in the Sierra Madre mountains of Mexico, the other in southern Arizona. By the 1970s, the Arizona Indians no longer farmed nor ate a low-calorie, low-fat diet, but instead consumed 40 percent of their calories from fat. With this extreme change in lifestyle, they developed the highest prevalence of obesity of any population on earth. Half of the Arizona group had diabetes by age 35, weighing, on average, 57 pounds (26 kilograms) more than their southern relatives, who still eat a low-fat diet and are very active.

The Pima Indians demonstrate that future obesity is not sealed in the genes at conception, but instead is much more likely to occur if the environment provides too many calories and too much fat. Geneticist James Neel expressed this idea as the "thrifty gene hypothesis" in 1962. He suggested that long ago, the hunter-gatherers who survived famine had genes that enabled them to store fat. Today, with food plentiful, the genetic tendency to retain fat is no longer healthful, but harmful.

Another environmental influence on weight is the "gut microbiome," the types of bacteria that normally live in our digestive tracts (see section 2.5). The actions of certain types of bacteria affect the number of calories that we extract from particular foods. An obese person has a different gut microbiome than a person who easily stays thin. The gut microbiome changes dramatically after weight-loss surgery, and finding a way to recreate this changed microbiome might one day provide an alternative to the surgery.

Perhaps nowhere are the complexities and challenges of gene-environment interactions more profound than in behavioral characteristics, nuances, quirks, and illnesses. Chapter 8 looks at a few of these.

Key Concepts Questions 7.4

a. Explain how leptin, ghrelin, and other proteins affect weight.

*b. What is the significance of the difference in heritability for BMI and obesity?

c. What can populations that suddenly become sedentary and switch to a high-calorie diet reveal about environmental influences on body weight?

Study Guide

Summary

7.1 Genes and the Environment Mold Traits

1. A **polygenic** trait is determined by more than one gene. Complex traits reflect influences of the environment and genes.
2. Single-gene traits are rare. For most traits, many genes contribute to a small, but not necessarily equal, degree.

7.2 Polygenic Traits Are Continuously Varying

3. Genes that contribute to polygenic traits are called **quantitative trait loci**. The frequency distribution of phenotypes for a polygenic trait forms a bell curve.
4. Fingerprint patterns, height, and skin color are continuously varying, polygenic traits.

7.3 Methods to Investigate Complex Traits

5. **Empiric risk** measures the likelihood that a complex trait will recur based on **incidence**, which is the number of new cases of a disease during a certain time in a population. **Prevalence** is the proportion of cases.
6. Emperic risk rises with genetic closeness, severity, and number of affected relatives. It is based on observation only.
7. **Heritability** estimates the proportion of variation in a complex trait due to genetics in a particular population at a particular time.
8. Missing heritability occurs when identified genes cannot account for all of the variability in a trait or illness.
9. The **coefficient of relatedness** is the proportion of genes that two people related in a certain way share.
10. Characteristics shared by adopted people and their biological parents are mostly inherited, whereas similarities between adopted people and their adoptive parents reflect environmental influences.
11. **Concordance** measures the frequency of expression of a trait in both members of MZ or DZ twin pairs. The more influence genes exert over a trait, the higher the differences in concordance between MZ and DZ twins.
12. **Genome-wide association studies** correlate patterns of genetic markers (**single nucleotide polymorphisms** and/or **copy number variants**) to increased disease risk.
13. Machine learning is used to train and validate GWAS data, for use in predicting disease risks. A **polygenic risk score** predicts a trait or disease based on SNP data.
14. A GWAS may use a cohort study to follow a large group over time, or a case-control study on matched pairs.
15. An **affected sibling pair study** identifies homozygous regions that may include genes of interest. **Homozygosity mapping** identifies mutations in genome regions that are homozygous because the parents shared recent ancestors.

7.4 A Closer Look: Body Weight

16. Leptin and associated proteins affect appetite. Fat cells secrete leptin in response to eating, which decreases appetite.
17. Populations that switch to a high-fat, high-calorie diet and a less-active lifestyle reveal effects of the environment on body weight.

Key Terms

affected sibling pair study A gene identification approach that looks for single nucleotide polymorphisms (SNPs) that siblings with a particular trait share, but that siblings who do not have the trait do not share.

coefficient of relatedness The proportion of genes that two people related in a certain way share.

complex trait A phenotype that is either monogenic or polygenic, and influenced by the environment.

concordance (kən-kôr′dəns) A measure indicating the degree to which a trait is inherited; percentage of twin pairs in which both members express a trait.

copy number variant (CNV) A DNA sequence present in different numbers of copies (repeats) in different individuals.

empiric risk Probability that a trait will recur based upon its incidence in a population, based on observation.

genome-wide association study (GWAS) Comparison of thousands or millions of variants (single nucleotide polymorphisms or copy number variants) that form haplotypes between people with a condition and unaffected individuals to identify parts of the genome that might contribute to a phenotype.

heritability An estimate of the proportion of phenotypic variation in a population that is due to genetic differences.

homozygosity mapping (hōmō-zī-gəs′-ĭ-tē) An approach to gene discovery that correlates stretches of homozygous DNA base sequences in the genomes of related individuals to certain traits or disorders.

incidence The number of new cases of a disease during a certain time in a population.

polygenic (pŏlē-jēn′ ĭk) A trait determined by more than one gene.

polygenic risk score A test of single base differences in several genes derived from people with a specific condition that is then used to identify people in another group who are at high risk of developing the condition.

prevalence The number of cases of a disease in a population at a particular time.

quantitative trait loci (QTLs) DNA sequences that contribute to polygenic traits.

single nucleotide polymorphism (SNP) (nōōklēō-tīd pŏlē-môr′ fĭz′əm) A site in the DNA that has a different base in at least 1 percent of a population.

Review Questions

1. Explain how Mendel's laws apply to complex traits.
2. Explain the difference between a Mendelian complex trait and a polygenic complex trait.
3. Do all genes that contribute to a polygenic trait do so to the same degree?
4. Explain why the curves shown in figures 7.2, 7.3, and 7.4 have the same bell shape, even though they represent different traits.
5. How can skin color have a different heritability at different times of the year?
6. Blood pressure is a trait that has "missing heritability." What does this mean?
7. Explain how the twins in figure 7.4 can have such different skin colors.
8. In a large, diverse population, why are medium brown skin colors more common than very white or very black skin?
9. Which has a greater heritability—eye color or height? State a reason for your answer.
10. What is empiric risk?
11. How are adopted individuals and twins used to study complex traits?
12. What does a polygenic risk score evaluate?
13. What are limitations of a genome-wide association study?

Applied Questions

1. The incidence of obesity in the United States has doubled over the past two decades. Is this due more to genetic or environmental factors? Cite a reason for your answer.
2. One way to calculate heritability is to double the difference between the concordance values for MZ versus DZ twins. For multiple sclerosis, concordance for MZ twins is 30 percent, and for DZ twins, 3 percent. What is the heritability? What does the heritability suggest about the relative contributions of genes and the environment in causing MS?
3. In chickens, high body weight is a complex trait. Several genes contribute small effects additively, and a few genes exert a great effect. Do the several genes provide broad heritability and the few genes narrow heritability, or vice versa?
4. The environmental epigenetics hypothesis states that early negative experiences, such as neglect, abuse, and extreme stress, increase the risk of developing depression, anxiety disease, addictions, and obesity later in life, through effects on gene expression that persist. Suggest an experiment to test this hypothesis.
5. Pharmaceutical companies are developing drugs based on a few cases of people who have extremely low LDL levels, due to a homozygous recessive mutation in a gene called *proprotein convertase subtilisin/kexin type 9* (*PCSK9*). The idea is that mimicking the effects of the mutation in people who don't have it will lower LDL level enough to combat elevated risk of heart disease. What types of information are important in evaluating whether one of the new drugs can prevent heart disease?
6. Many genes contribute to lung cancer risk, especially among people who smoke. These genes include *p53, IL1A* and *IL1B, CYP1A1, EPHX1, TERT,* and *CRR9*. Search for one of these genes on the internet and describe how variants of it may contribute to causing lung cancer.

Case Studies and Research Results

1. Researchers asked 49 MZ and 50 DZ twin pairs in which only one had Parkinson disease about lifetime exposure to six chemical solvents that had previously been associated with the disease. Average age at diagnosis was 66, with symptoms starting 10 to 40 years after exposure began. The researchers found that long-term exposure to trichloroethylene increases Parkinson disease risk sixfold and long-term exposure to perchloroethylene ninefold, in both types of twins. Explain how the design and results of this study support either of two hypotheses: that Parkinson disease is inherited, or that it is caused by environmental exposure. Do you think that the results are good news or bad news, and why?
2. Researchers developed a test to predict which women are at high risk for a prevalent type of breast cancer (estrogen receptor positive) by considering combinations of 313 SNPs for 94,075 women with the cancer and the same SNPs for 75,017 women who did not have the cancer. They validated the predictive power in an additional 11,428 cases and 18,323 controls and for 190,040 women from the UK Biobank. Women in the top 1 percent of risk face 4.37-fold increased probability of developing the cancer compared to an average woman in the population, and women in the lowest 1 percent of risk face a 0.16-fold probability.

   a. What is this type of analysis called?
   b. How might the findings be used in the diagnosis and treatment of breast cancer?
   c. How does a test derived from the SNP panel analysis differ from a genetic test for mutations in the *BRCA1* gene?

3. Concordance for the eating disorder anorexia nervosa for MZ twins is 55 percent and for DZ twins, 7 percent. Ashley and Maggie are DZ twins. Maggie has anorexia nervosa. Should Ashley worry about an inherited tendency to develop the condition? Explain your answer.

4. A study looked at 200,000 SNPs throughout the genome for 1,820 people with premature graying of the hair and 1,820 without this trait. Those with the trait shared several SNPs on chromosome 9. What type of study is this?

5. An affected sibling pair study identified areas of chromosomes 1, 14, and 20 that are likely to harbor genes that predispose individuals to developing schizophrenia. Explain how such an investigation is conducted. (Section 8.6 discusses schizophrenia.)

6. A study in England tested 20,430 people for alleles of 12 genes known to increase the risk of becoming obese. Each person had 24 alleles assessed, 12 from each parent. Although the number of risk alleles correlated to BMI, the more a person exercised, the lower the BMI. Do these findings support a genetic control of the trait of body weight, or not? Cite a reason for your answer.

7. A researcher applied for a grant to carry out a GWAS that included several small indigenous populations, as well as large, more diverse populations. One reviewer criticized the application because a population group with only a few hundred or thousand individuals could not generate robust statistics to compare to larger groups, such as Europeans and Asians. Cite a practical (clinical) problem that could arise from eliminating small population groups from GWAS.

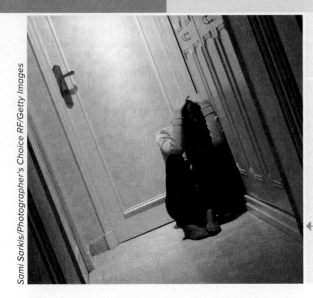

CHAPTER 8

Genetics of Behavior

Child abuse and PTSD. *Inheriting a variant of a particular gene may influence how a person handles memories of child abuse throughout her or his lifetime.*

Learning Outcomes

8.1 Genes and Behavior
1. Identify the physical basis of behavioral traits in the brain.
2. Explain how genes affect behavior.

8.2 Sleep
3. Describe how studies on dogs helped to find a narcolepsy gene.
4. Explain how mutations in the *period 2* gene disrupt sleep-wake cycles.

8.3 Intelligence and Intellectual Disability
5. Describe intelligence tests.
6. List causes of intellectual disability.

8.4 Drug Addiction
7. State the two identifying characteristics of drug addiction.
8. Discuss evidence that drug addiction is a biological phenomenon.
9. Describe two types of receptors that are implicated in drug addiction.

8.5 Mood Disorders
10. Explain how genes may affect risk of developing major depressive disorder.
11. Explain how the genetics of bipolar disorder is complex.

8.6 Schizophrenia
12. Explain how schizophrenia differs from the mood disorders.
13. Discuss how synaptic pruning may explain the origin of schizophrenia.

8.7 Autism
14. List risk factors for autism.
15. Explain three ways to use DNA sequence data to discover genes that contribute to autism.
16. Discuss the role of neurexins and neuroligins in autism.

The BIG Picture

Genes that affect brain structure and function give rise to such behavioral traits and tendencies as sleep patterns, addictions, and intelligence. Variants of such genes, with environmental factors, contribute to or cause mood disorders, schizophrenia, and autism. Conditions that affect how we think and communicate, once considered part of psychology and psychiatry, are becoming the province of neurology and genetics.

Posttraumatic Stress Disorder: Who Is Predisposed?

A young woman who was physically and emotionally abused as a child becomes a warm, caring mother with a successful career. Another woman with a similar upbringing grows into an anxious adult who suffers from posttraumatic stress disorder (PTSD).

PTSD is a reaction to traumatic physical harm, or the perceived threat of harm, which persists long after the triggering event has passed. The person suffers from anxiety, flashbacks, and avoidance of memories of the event, resulting from inappropriate activation of the stress hormones that flooded the bloodstream during the original trauma. PTSD is a response to many types of disturbing experiences, including war, immigration, a serious illness, natural disasters, violent crimes, mistreatment, and other disturbing events.

Many factors contribute to the development of PTSD. One study of 2,000 people who had been severely traumatized from abuse, either as children or adults, found that those abused as children who developed PTSD were significantly more likely to have a specific variant of the gene *FKBP5* than other individuals. The effect is both genetic and epigenetic. In people with the gene variant who were abused as children and developed PTSD, methyl groups on the gene are absent—an effect not seen in the people who did not develop PTSD. As a result of the decreased methylation, receptors for stress hormones (glucocorticoids) malfunction, and the body cannot react and adapt to stress. Chapter 11 discusses epigenetics.

The study found this epigenetic change in white blood cells, but other experiments on cells growing in the laboratory showed that it also occurs and persists in neural progenitor cells that differentiate into brain nerve cells that will last a lifetime. In a person, the epigenetic change happened at a critical period in development, setting the stage in a child for an inability to appropriately use stress hormones later in life. The result: PTSD.

8.1 Genes and Behavior

The methods used to analyze the complex traits that are the subject of chapter 7 are also useful to understand behavior and diagnose and treat behavioral disorders. Behavior is a complex continuum of emotions, moods, intelligence, and personality that drives how we function on a daily basis. We are, to an extent, defined and judged by our many behaviors. They control how we communicate, cope with negative feelings, and react to stress. Behavioral disorders are common, with wide-ranging and sometimes overlapping symptoms. Past chapters have referred to inherited *diseases,* which are conditions for which the precise cause is known. Behavioral conditions are termed *disorders* because many causes are possible, and they may not be known.

Many of our behaviors occur in response to environmental factors, but *how* we respond may have genetic underpinnings, as the chapter opener describes for flashing back to abuse during childhood. Understanding the biology behind behavior can help scientists develop treatments for behavioral disorders.

Behavioral genetics considers nervous system function and variation, including the hard-to-define qualities of mood and mind. The human brain weighs about 3 pounds and resembles a giant gray walnut, but with the appearance and consistency of pudding. It consists of 100 billion nerve cells, or **neurons,** and at least a trillion other cells called **neuroglia,** which support and nurture the neurons.

Brain neurons connect and interact in complex networks that control all of the body. Branches from each of the 100 billion neurons in the brain form close associations, called **synapses,** with 1,000 to 10,000 other neurons. Neurons communicate across these tiny spaces using chemical signals called **neurotransmitters**. Networked neurons oversee broad functions such as sensation and perception, memory, reasoning, and muscular movements.

Genes affect behavior by encoding diverse proteins that control the production and distribution of neurotransmitters. Genes also affect the number and pattern of connections that neurons make. The term *neural plasticity* refers to the constant changes in nervous system structure and function. **Figure 8.1** indicates the points where genes act in sending and receiving nervous system information.

Enzymes oversee the synthesis of neurotransmitters and their transport from the sending, or presynaptic, neuron across the synapse to receptors on the receiving, or postsynaptic, neuron. Proteins called transporters ferry neurotransmitters from sending to receiving neurons, and proteins also form the subunits of the receptors that bind neurotransmitters. Genes

Figure 8.1 Neurotransmission. Many genes that affect behavior encode proteins that affect neurotransmission, which is sending a chemical signal from one neuron to another across a synapse.

control the synthesis of myelin, a substance consisting of fats and proteins that coats neuron extensions called axons. Myelin, which accumulates in and causes the plasma membranes of neuroglia to balloon out, forms a sheath that insulates the neuron, which speeds neurotransmission. Myelin-rich neuroglia wrap around an axon like a bandage wound around a finger.

Many behavioral traits and disorders, like other characteristics, are due to a major influence from a single gene, perhaps one whose protein product takes part directly in neurotransmission, but also small inputs from variants of hundreds of other genes. Inheriting certain subsets of variants of these genes makes an individual susceptible to developing a certain disorder in the presence of a particular environmental stimulus. People who have different behavioral disorders, such as depression, addiction, and schizophrenia, may share underlying gene variants. Human behavior is genetically quite complex!

Deciphering the genetic components of behavioral traits and diseases uses traditional empiric risk determination, adoptee and twin study data, and exome and genome sequencing to identify responsible gene variants. Considering genes may make it possible to subtype behavioral conditions in a biologically meaningful way so that diagnoses will be personalized and effective treatments begun sooner.

Table 8.1 lists the 12-month prevalence of some behavioral disorders, which refers to people who have had symptoms in the year prior to being asked. In contrast, lifetime prevalence indicates the proportion of individuals who report symptoms at any time over a lifetime.

Key Concepts Questions 8.1

*a. What are the two types of brain cells?
*b. Define *synapse*.
*c. How do genes affect behavior?

8.2 Sleep

Sleep has been called "a vital behavior of unknown function," and, indeed, without sleep, animals die. We spend a third of our lives in this mysterious state.

Genes influence sleep characteristics. When asked about sleep duration, schedule, quality, nap habits, and whether they are "night owls" or "morning people," MZ (identical) twins report significantly more in common than do DZ twins. This is true even for MZ twins separated at birth. Twin studies of brain wave patterns through four of the five stages of sleep confirm a hereditary influence. The fifth stage, rapid eye movement or REM sleep, is associated with dreaming and therefore may indicate the input of experience more than genes.

Narcolepsy

Researchers discovered the first gene related to sleep in 1999, for a condition called "narcolepsy with cataplexy" in dogs. Mutations in several genes cause narcolepsy in humans.

A person (or dog) with narcolepsy falls asleep suddenly several times a day. Extreme daytime sleepiness greatly disrupts daily activities. People with narcolepsy have a 10-fold higher rate of car accidents. Another symptom is sleep paralysis, which is the inability to move for a few minutes after awakening. The most dramatic manifestation of narcolepsy is cataplexy. During these short and sudden episodes of muscle weakness, the jaw sags, the head drops, knees buckle, and the person falls to the ground. This often occurs during a bout of laughter or excitement. Narcolepsy with cataplexy affects only 0.02 to 0.06 percent of the general populations of North America and Europe, but the fact that it is much more common in certain families suggests a genetic component.

Studies on dogs led the way to discovery of one human narcolepsy gene. In 1999, researchers discovered mutations in a gene that encodes a receptor for a neuropeptide (a short protein that functions as a neurotransmitter) called hypocretin. In Doberman pinschers and Labrador retrievers, the receptor does not reach the surfaces of certain brain cells, blocking signals to

Table 8.1 12-Month Prevalence of Some Behavioral Disorders in the US Adult Population

Condition	Prevalence (%)
Anxiety	18.0
Social phobia	6.8
Posttraumatic stress disorder	3.5
Generalized anxiety disorder	3.1
Obsessive compulsive disorder	1.0
Panic disorder	2.7
Autism spectrum disorders (children)	1.5
Eating disorders	3.0
Anorexia nervosa	female 0.9, male 0.3
Binge eating	1.2
Bulimia nervosa	0.3
Mood disorders	9.5
Major depressive disorder	6.7
Bipolar disorder	2.6
Schizophrenia	1.1

Source: Data from the National Institute of Mental Health

Figure 8.2 Letting sleeping dogs lie. These Doberman pinschers have inherited narcolepsy. They suddenly fall into a short but deep sleep while playing. Research on dogs with narcolepsy led to the discovery of a gene that affects sleep in humans and development of a drug to treat insomnia.
Stanford University Center for Narcolepsy

promote wakefulness. **Figure 8.2** shows a still frame of a film of narcoleptic dogs playing. Suddenly, they all collapse. A minute later, they get up and resume their antics. To induce a narcoleptic episode in puppies, researchers let them play with each other. Feeding older dogs meat excites them so much that they can take a while to finish a meal because they fall down in delight so often. Getting narcoleptic dogs to breed is difficult because mating is apparently even more exciting than playing or eating.

The gene for the neuropeptide hypocretin that binds to the hypocretin receptor turned out to be the same as a gene discovered a year earlier that encodes a neuropeptide that controls eating, called orexin. Neurons that produce the neuropeptide in different parts of the brain account for the two effects, sleep and appetite. Researchers study hypocretin/orexin and the molecules with which they interact to develop drugs to treat insomnia. At least three other genes are associated with insomnia. The most severe condition is "fatal familial insomnia," which is caused by infectious protein misfolding (see section 10.4).

Familial Advanced Sleep Phase Syndrome

Daily rhythms such as the sleep-wake cycle are set by cells that form a "circadian pacemaker" in two clusters of neurons in the brain called the suprachiasmatic nuclei. *Circadian* means "about a day." In these cells, certain "clock" genes are expressed in response to light or dark in the environment.

The function of clock genes is startling in families that have a mutation. For example, five generations of a family in Utah have familial advanced sleep phase syndrome. Affected people fall asleep at 7:30 P.M. and awaken at 4:30 A.M. They have a single DNA base mutation in a gene on chromosome 2, *period 2*, that interferes with synchronization of the sleep-wake cycle with daily sunrise and sunset, which is an example of a circadian rhythm. The condition is autosomal dominant, and each generation has several affected individuals.

Mutation at a different part of the *period 2* gene affects how early we tend to awaken. People with two adenines at this site (genotype *AA*) awaken an hour earlier than people with two guanines (genotype *GG*). Heterozygotes with genotype *AG*, awaken at times between those of the two homozygotes.

> **Key Concepts Questions 8.2**
>
> *a. What is the evidence that sleep habits are inherited?
> b. Describe two mutations that alter sleep.

8.3 Intelligence and Intellectual Disability

Intelligence is a highly variable complex trait that is subject to many genetic and environmental influences. In general, intelligence refers to the ability to reason, learn, remember, connect ideas, deduce, and create.

The first intelligence tests, developed in the late nineteenth century, assessed sensory perception and reaction times to various stimuli. In 1904, Alfred Binet in France developed a test with questions based on language, numbers, and pictures to predict the success of developmentally disabled youngsters in school. The test was later modified at Stanford University to assess white, middle-class Americans. An average score on this "intelligence quotient," or IQ test, is 100, with two-thirds of all people scoring between 85 and 115 in a bell-shaped curve, or normal distribution (**figure 8.3**). An IQ between 50 and 70 is considered mild intellectual disability and below 50, severe intellectual disability. IQ has been a fairly accurate predictor of success in school and work, but it does not solely represent biological influences on intelligence. Low IQ correlates with societal situations, such as poverty, not graduating from high school, incarceration (males), and having a child out of wedlock (females).

Psychologists use a general intelligence ability, called "*g*," to represent the four basic skills that IQ encompasses. These are verbal fluency, mathematical reasoning, memory, and spatial visualization skills. The general intelligence measure was developed to reduce the effect of unequal opportunities on assessment of intelligence.

Traditional methods of investigating complex traits, such as adoption and twin studies, show a heritability of 0.5 to 0.8 for intelligence, suggesting a significant contribution from genetics. New types of studies confirm the heritability established with older techniques, and also identify specific genes that contribute to or affect intelligence.

Researchers used a variation of a genome-wide association study (see section 7.3) to evaluate change in intelligence over a lifetime. They considered general intelligence measured with a standard test given to nearly 2,000 people in Scotland when they were 11 years old and again when they were older (age 65, 70, or 79). The researchers found that smart 11-year-olds became smart senior citizens.

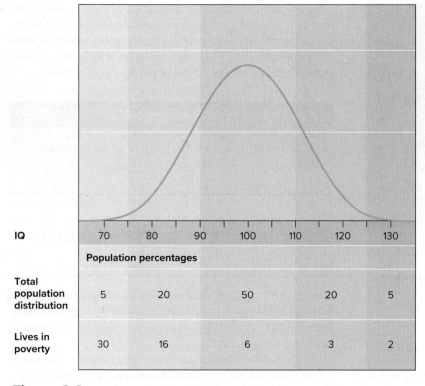

Figure 8.3 Success and IQ. IQ scores predict success in school and the workplace in US society. The bell-shaped curve for IQ indicates that most people fall between 85 and 115. However, those living in poverty tend to have lower IQs.

Genetics can explain some cases of intellectual disability, which affects 3 in 100 individuals in the United States. (Intellectual disability is a broad term that includes mental retardation.) Causes are many: noninherited birth defects, abnormal genes and chromosomes, but also environmental exposure to toxins, infections and malnutrition. Probably at least 1,000 of our 20,325 or so genes affect the brain, with variants that impact intelligence.

Among the best-studied genetic causes of intellectual disability are Down syndrome (see Bioethics in chapter 13) and fragile X syndrome (see Clinical Connection 12.1). One of the most familiar environmental causes is fetal alcohol syndrome (see figure 3.23). Exome and genome sequencing have been helpful in assigning causative genes in children with intellectual disability who have not been diagnosed with these or other more common conditions.

Key Concepts Questions 8.3

*a. What is intelligence?

*b. How does using measurement of general intelligence improve upon using IQ score to assess the inheritance of intelligence?

c. How are newer genetic technologies being used to discover gene variants that cause or contribute to intellectual disability?

8.4 Drug Addiction

One person sees a loved one battling lung cancer and never smokes again. Another person actually has lung cancer, yet takes frequent breaks from using her oxygen tank to smoke. Evidence is mounting that genes play a large role in making some individuals prone to dependency on certain substances, and others not.

Substance Use Disorders

Our ancient ancestors must have discovered that ingesting certain natural substances, particularly from plants, provided a feeling of well-being. That tendency persists today. Substance use disorders can develop in individuals who repeatedly ingest certain psychoactive chemical compounds, even if the dangers are clear. The most severe type of substance use disorder is drug addiction. It is defined as compulsively seeking and taking a drug despite knowing its dangers.

The two identifying characteristics of drug addiction are tolerance and dependence. Tolerance is the need to take more of the drug to achieve the same effects as time goes on. Dependence is the onset of withdrawal symptoms upon stopping use of the drug. Both tolerance and dependence contribute to the biological and psychological components of craving the drug. The behavior associated with drug addiction can be extremely difficult to break.

Whether or not a particular chemical compound is addictive depends on which receptors on brain cells it binds to and where in the brain those cells are located. Drug addiction produces long-lasting changes in brain structure and function. Craving and high risk of relapse remain even after a person has abstained for years. Heritability of drug addiction is 0.4 to 0.6, with a two- to threefold increase in risk among adopted individuals who have one affected biological parent. Twin studies also indicate an inherited component to drug addiction.

Brain imaging techniques localize the "seat" of drug addiction in the brain by highlighting the cell surface receptors that bind neurotransmitters when a person craves the drug. The brain changes that contribute to addiction occur in parts called the nucleus accumbens, the prefrontal cortex, and the ventral tegmental area, which are part of a larger set of brain structures called the limbic system (**figure 8.4**). The effects of cocaine seem to be largely confined to the nucleus accumbens, whereas alcohol affects the prefrontal cortex.

The specific genes and proteins that are implicated in addiction to different substances may vary, but several general routes of interference in brain function are at play. Proteins involved in drug addiction are those that

- produce neurotransmitters, such as enzymes required for neurotransmitter synthesis;

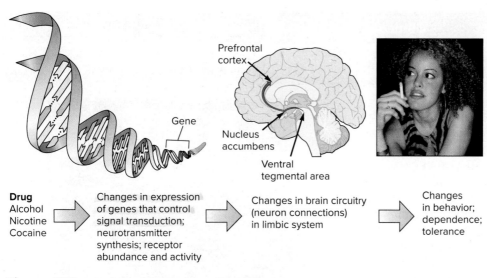

Figure 8.4 The events of addiction. Addiction is manifest at several levels: at the molecular level, in neuron-neuron interactions in the brain, and in behavioral responses.
Ryan McVay/Photodisc/Getty Images

- remove excess neurotransmitters from the synapse (called reuptake transporters);
- form receptors on the postsynaptic (receiving) neuron that are activated or inactivated when specific neurotransmitters bind; and
- convey chemical signals in the postsynaptic neuron.

Addictive Drugs

Many addictive drugs are derived from plants, including cocaine, opium, and tetrahydrocannabinol (THC), the main psychoactive component of cannabis (marijuana). These substances bind to receptors on human neurons, which indicates that our bodies have versions of these plant-derived compounds.

The human equivalents of the opiates are the endorphins and enkephalins, which are short peptides that dull the perception of pain and elevate mood. Endorphins explain why some people addicted to opiate drugs, such as heroin and oxycontin, experience withdrawal symptoms when they stop taking the drug. At first the body interprets the frequent binding of the drug to its endorphin receptors as an excess of endorphins. To lower the level, the body slows its own production of endorphins. Then, when the person stops taking the drug, the body becomes short of opiates. The result is pain. Opiate drugs are powerfully addicting when taken repeatedly in the absence of a medical condition causing pain. These same drugs are very useful in dulling the pain from surgery or a terminal illness.

THC is the equivalent of anandamide in a human body. Anandamide modulates how brain cells respond to stimulation by binding to neurotransmitter receptors on presynaptic (sending) neurons, and are involved in response to stress and in social behavior. In contrast, neurotransmitters bind to receptors on postsynaptic neurons. **Clinical Connection 8.1** discusses a second component of cannabis that is used to treat seizures and other conditions.

LSD (lysergic acid diethylamide) binds to receptors on neurons that normally bind neurotransmitters called trace amines, which are found throughout the brain at low levels. LSD causes effects similar to the symptoms of schizophrenia (see section 8.6), suggesting that the trace amine receptors, which are proteins, may be implicated in the illness. LSD produces hallucinations, distorted perception, and heightened states of awareness, but is not addictive.

Amphetamines are highly addictive drugs when misused. Known since 1887, amphetamines have been used to treat nasal congestion, attention deficit hyperactivity disorder, narcolepsy, and other conditions. They are used recreationally for performance enhancement and abused because they promote euphoria, wakefulness, cognitive control, increased strength, and increased sex drive. Amphetamines promote the release of the neurotransmitter dopamine from vesicles in presynaptic neurons and enable dopamine to accumulate in synapses by controlling the transporter proteins that carry the neurotransmitter. Addiction to amphetamines can lead to psychosis, which is a loss of contact with reality, causing delusions and hallucinations.

People addicted to various drugs share certain gene variants that must be paired with environmental stimuli for addiction to occur. For example, people who are homozygous for the *A1* allele of the gene that encodes the dopamine D(2) receptor are overrepresented among those with alcoholism and other addictions. Genome-wide association studies have found more than 50 chromosomal regions that may include genes that contribute to drug craving. Studies of gene expression flesh out this picture by providing a real-time view of biochemical changes that happen when a person seeks a drug, and then takes it.

Discovering the genetic underpinnings of nicotine addiction is increasing our knowledge of addiction in general and may have practical consequences. Genetics may explain how the nicotine in tobacco products causes lung cancer and why nicotine is so highly addictive.

Each year, 35 million people try to quit smoking, yet only 7 percent of them succeed. It is easy to see on a whole-body level how this occurs: Nicotine levels peak 10 seconds after an inhalation and the resulting pleasurable release of the neurotransmitter dopamine in brain cells fades away in minutes. To keep the good feeling, smoking must continue.

On a molecular level, nicotine does damage through a five-part assembly of proteins called a nicotinic receptor. The receptor normally binds the neurotransmitter acetylcholine, but it also

Clinical Connection 8.1

Cannabis: Addictive Drug and Seizure Treatment

The plant *Cannabis sativa* includes more than 500 chemical compounds, 113 of which are classified as cannabinoids because they bind the same family of receptors (**figure 8A**). The cannabinoid that provides the psychoactive and addictive effects of marijuana (the flowers of the plant) is tetrahydrocannabinol (THC). Genome-wide association studies (see figure 7.11) identify three regions in the human genome that include genes that control dependency on THC.

Another abundant cannabinoid is cannabidiol (CBD). Plants that are very high in THC tend to be low in CBD, and vice versa. CBD binds to several types of receptors and does not produce dependency or psychoactive effects. However, CBD, given as an oil, alleviates the intractable seizures of a genetic condition called Dravet syndrome. CBD is currently in dozens of clinical trials to test treatments for other seizure disorders as well as anxiety, schizophrenia, irritable bowel syndrome, attention deficit hyperactivity disorder, and pain from multiple sclerosis.

Plant-based pharmaceuticals have long histories. In the nineteenth century, marijuana was used to treat epilepsy, and small studies have tested it over the years. CBD came to public attention in 2013, with the story of a 5-year-old who has Dravet syndrome. She was having three to four short seizures per hour, interspersed with long ones that would last 2 to 4 hours. Her heart had stopped many times, she couldn't eat or walk or talk, and she'd often been close to death. After no standard treatments for seizures were effective, the family turned to medical marijuana, which required support from two physicians.

The parents found growers who had cultivated a strain of cannabis that wouldn't sell as a recreational drug because it had very low THC, but it had high CBD. When the child drank the plant oil, her seizures ceased—for a week! Over time, with continuing treatment, her seizure frequency plummeted from 300 episodes a week to 2 to 3 per month.

Today the child takes the oil twice a day mixed with food. The Food and Drug Administration approved the use of cannabidiol in 2018, to treat Dravet syndrome and a nongenetic form of epilepsy.

Questions for Discussion

1. Possession of *Cannabis sativa* has been illegal in the United States until recently, when a few states have allowed use of small amounts of recreational marijuana

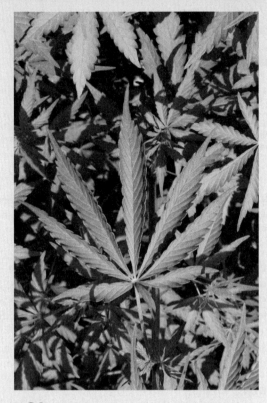

Figure 8A *Cannabis sativa* travelab india/Alamy Images

and have expanded uses for medical marijuana. Yet *Cannabis sativa* contains an addictive substance. How can products derived from the plant be allowed, yet the risk of addiction minimized?

2. Some of the gene variants linked to cannabis addiction are also implicated in elevating risk of developing major depressive disorder and schizophrenia, discussed later in the chapter. How might these connections explain cannabis dependency?

3. Discuss possible reasoning behind the fact that possession of *Cannabis sativa* still is a crime in many states, and its addictive nature only recently understood, while tobacco, known since the 1960s to cause lung cancer and to be highly addictive, has been legal.

---

binds the similarly shaped nicotine molecule (**figure 8.5**). Certain variants of the receptor bind nicotine very strongly, which triggers a nerve impulse that, in turn, stimulates the pleasurable dopamine release into the synapse. That may explain the addiction. These receptors are also located on several types of lung cells, where they bind carcinogens. So the nicotine in tobacco causes addiction and susceptibility to lung cancer, and the smoke delivers the carcinogens right to the sensitive lung cells.

A different gene encodes each of the five parts of the nicotinic receptor. If two of the five parts are certain variants, then a person experiences desire to continue smoking after the first cigarette.

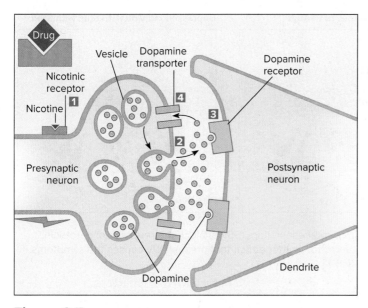

Figure 8.5 Nicotine's effects at the cellular level.
(1) Binding of nicotine to nicotinic receptors, which also bind the neurotransmitter acetylcholine, triggers release of dopamine **(2)** from vesicles into the synapse. Some dopamine binds receptors on the postsynaptic (receiving) neuron **(3)** and some dopamine reenters the presynaptic neuron through a dopamine transporter protein **(4)**. Uptake of dopamine into the postsynaptic cell triggers pleasurable feelings. The inset illustrates how a smoking cessation drug blocks nicotinic receptors.

Key Concepts Questions 8.4

*a. Distinguish tolerance from dependence in drug addiction.
*b. How do genes control addiction?
c. How can a human become addicted to a chemical from plants?
d. Explain the basis of nicotine addiction.

8.5 Mood Disorders

We all have moods, but mood disorders, which affect millions of people, impair the ability to function on a day-to-day basis. Context is important in evaluating extreme moods. Symptoms that may lead to a diagnosis of depression if they occur for no apparent reason are normal in the context of experiencing profound grief. The two most prevalent mood disorders are major depressive disorder and bipolar disorder (also called manic depression).

Major Depressive Disorder

Major depressive disorder is more than being "down in the dumps," nor can a person simply "snap out of it." It is a disabling collection of symptoms that go beyond sadness. A person with depression no longer enjoys favorite activities. Fatigue may be great, and the person has trouble making decisions, concentrating, and recalling details. Motivation, energy, appetite, and productivity lag, as low self-esteem and hopelessness rise. Crying may start unexpectedly, often several times a day. People with depression may avoid social interactions. They have difficulty falling and staying asleep, which intensifies the daytime fatigue. Routine, simple tasks may become overwhelming, and working or studying may become difficult. Weight may drop as interest in food, as in most everything else, diminishes. Fifteen percent of people hospitalized for severe, recurrent depression ultimately end their lives. A subtype is persistent depressive disorder, in which the low mood persists, nearly every day, for more than two years.

Genes and the environment contribute about equally to depression. Risk is greater for people with an affected first-degree relative (parent or sibling). Children of parents who are first cousins face a threefold increase in the risk of developing depression.

Researchers initially discovered genes that increase risk of depression by focusing on disturbed sleep. They looked for genes known to control circadian rhythms (sleep, hormone levels, and body temperature), and discovered eight that show altered expression in the brains of people who had depression compared to brains from people who hadn't been depressed. A GWAS found 44 sites in human genomes that are near or include genes that have variants that contribute to depression. About twice as many women as men suffer from depression.

A likely cause of depression is deficiency of the neurotransmitter serotonin in synapses. Serotonin affects mood, emotion, appetite, and sleep. Antidepressant drugs called selective serotonin reuptake inhibitors (SSRIs) prevent presynaptic neurons from taking up serotonin from the synapse, leaving more of it available to stimulate the postsynaptic cell (**figure 8.6**). This action apparently offsets the neurotransmitter deficit. The genetic connection is that a protein in the plasma membrane of the presynaptic neuron, called the serotonin transporter, may function too efficiently in depressed individuals, removing too much serotonin from the synapse. SSRIs lock the transporter in an "outward-only" position that prevents serotonin from binding to the presynaptic plasma membrane receptor. The drugs may begin to produce effects after 1 week, often enabling a person with moderate or severe depression to return to some activities, but full response can take up to 8 weeks. Other drugs, serotonin-norepinephrine reuptake inhibitors (SNRIs), target serotonin and the neurotransmitter norepinephrine.

In the past, to treat patients who have depression, physicians would try one antidepressant drug after another, based on experience with other patients. This trial-and-error approach would often take months. As the genes in the human genome are identified and their variants described, researchers are establishing genetic profiles that are associated with response to a particular drug, which can save time and minimize adverse effects. A hint that treatment can be individualized came from people who have a subtype of major depressive disorder, called psychotic depression. They need anti-psychotic drugs as well as an SSRI or SNRI. Psychosis adds agitation, hypochondria, constipation, and unexplained anger to the clinical picture.

Figure 8.6 How an antidepressant works. Selective serotonin reuptake inhibitors (SSRIs) block the reuptake of serotonin, leaving more of the neurotransmitter in synapses. This corrects a neurotransmitter deficit that presumably causes the symptoms. Overactive or overabundant reuptake receptors can cause the deficit.

Bipolar Disorder

Bipolar disorder is rarer than depression. It affects 2.6 percent of the population and has a general population lifetime risk of 0.5 to 1 percent. Weeks or months of depression alternate with periods of mania, when the person is hyperactive and restless and may experience a rush of ideas and excitement. A person who is normally frugal might, when manic, spend lavishly. In one subtype, the "up" times are termed hypomania, and they seem more a temporary reprieve from depression than full mania. Bipolar disorder with hypomania may appear to be depression. Distinguishing between major depressive disorder and bipolar disorder with hypomania in a clinical situation is important because different drugs are used to treat the conditions.

Many gene variants contribute to developing bipolar disorder. Early genetic studies looked at large Amish families, in whom the manic phase was quite obvious amid their restrained lifestyle. But studies in other families implicated different genes. Studies disagree because many gene variant combinations cause or contribute to the distinctive phenotype of bipolar disorder, but only a few such variants are seen in any one family. A systematic scrutiny of all studies of the genes behind bipolar disorder suggests as much as 10 percent of the genome—that is, hundreds of genes—are part of the clinical picture.

Key Concepts Questions 8.5

*a. What are the symptoms of major depressive disorder?

*b. What is the evidence that genes contribute to major depressive disorder?

c. What types of genes may affect the risk of developing depression?

*d. What are the symptoms of bipolar disorder?

e. How is the role of genes in bipolar disorder complex?

8.6 Schizophrenia

Schizophrenia is a debilitating loss of the ability to organize thoughts and perceptions, which leads to a withdrawal from reality. It is a form of psychosis, which is a disease of thought and sense of self. In contrast, the mood disorders are emotional, and the dementias are cognitive. Various forms of schizophrenia together affect about 1 percent of the world's population. Ten percent of affected individuals commit suicide.

Identifying genetic contributions to schizophrenia illustrates the difficulties in analyzing a behavioral condition:

- Some of the symptoms are also associated with other illnesses.
- Variants of many genes cause or contribute to schizophrenia.
- Several environmental factors mimic schizophrenia.

Signs and Symptoms

The first signs of schizophrenia often affect thinking. In late childhood or early adolescence, a person might develop trouble paying attention in school. Learning may become difficult as memory falters and information-processing skills diminish.

Traditionally, symptoms of psychosis have begun between ages 17 and 27 for males and 20 and 37 for females. These include delusions and hallucinations, sometimes heard, sometimes seen. A person with schizophrenia may hear a voice giving instructions. What others recognize as irrational fears, such as being followed by monsters, are real to a person with schizophrenia. Speech reflects the garbled thought process, skipping from topic to topic with no obvious thread of logic. Responses may be inappropriate, such as laughing at sad news. More recently, schizophrenia in older women, nearing and going through menopause, has been identified as another vulnerable population group.

Figure 8.7 Schizophrenia alters thinking. People with schizophrenia communicate the disarray of their thoughts in characteristically disjointed drawings. *BSIP/UIG/Getty Images*

Artwork by a person with schizophrenia can display the characteristic fragmentation of the mind (**figure 8.7**). Schizophrenia means "split mind," but it does not cause a split or multiple personality. In many patients the course of schizophrenia plateaus (evens out) or becomes episodic. It is not a continuous decline, as is the case for dementia.

A Disorder of Synaptic Pruning

A heritability of 0.9 and empiric risk values indicate a strong role for genes in causing schizophrenia (**table 8.2**). However, it is possible to develop some of the symptoms, such as disordered thinking, from living with and imitating people who have schizophrenia. Although concordance is high, a person who has an identical twin with schizophrenia has a 52 percent chance of *not* developing it. Therefore, the condition has a significant environmental component, too.

Several environmental factors may increase the risk of developing schizophrenia. These include birth complications, fetal oxygen deprivation, herpesvirus infection at birth, and traumatic brain injury or malnutrition in the mother (see Clinical Connection 11.1). Infection during pregnancy is a well-studied environmental factor.

When a pregnant woman is infected, her immune system bathes the brain of the embryo or fetus with cytokines, which are proteins that cause inflammation and can subtly alter brain development. It isn't the infection that sets the stage for later schizophrenia, but the maternal immune response to it. Investigators have long noted a tendency of people with schizophrenia to have been born during the winter, especially during years of flu pandemics. Research supports this connection between schizophrenia and response to maternal infection. Blood stored from 12,000 pregnant women during the 1950s and 1960s shows an association between high levels of the cytokine interleukin-8 and having a child who developed schizophrenia.

Evidence from many studies has strengthened the role of an overactive or inappropriate immune response operating at key times in development in causing schizophrenia. Such critical periods include the time of prenatal development and again during late adolescence, when symptoms typically become noticeable. The immune response introduces both an environmental component—infection—and a genetic component in the deployment of immune system proteins.

Genetic research has gradually focused in on a narrower set of candidate genes behind schizophrenia, with one emerging as a strong factor. At first, genome-wide screens of families with schizophrenia revealed at least 24 sites where affected siblings share alleles much more often than the 50 percent of the time that Mendel's first law (segregation) predicts. Then genome-wide association studies (see figure 7.11) implicated 108 sites in the genome that could include schizophrenia genes. One part stood out: the major histocompatibility complex, an extensive, gene-packed part of chromosome 6 (see section 19.1). Researchers traced the "signal" to a pair of nearly identical genes called *C4 complement component 4*. The genes are called *C4A* and *C4B*. "Complement" is a series of interacting proteins that are part of the immune response (see section 19.2).

The *C4* genes are highly variable—that is, people have different sizes of these genes, different numbers of copies of the genes, and single nucleotide polymorphisms (SNPs) within them. Certain variants of *C4A* are much more common among individuals with schizophrenia. Consideration of what the encoded protein does makes sense: It destroys synapses.

In late adolescence and in the early twenties, in all human brains, the dense bushes of dendrites emanating from brain neurons are trimmed, as part of normal development (**figure 8.8**). In people who have the *C4A* gene variant

Table 8.2	Empiric Risks for Schizophrenia
Relationship	**Risk**
MZ twin	48%
DZ twin	17%
Child	13%
Sibling	9%
Parent	6%
Half-sibling	6%
Grandchild	5%
Niece/nephew	4%
Aunt/uncle	2%
First cousin	2%
General population	1%

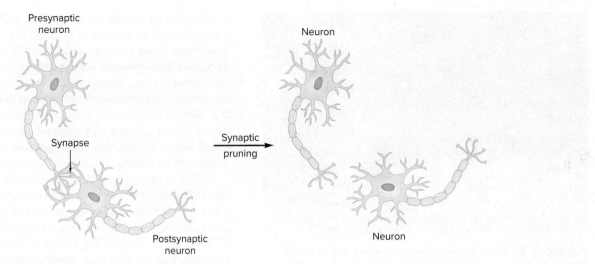

Figure 8.8 **Synaptic pruning.** Microglia bind to *C4A* protein "tags" (not shown) on dendrites, breaking the connections between neurons. The result: fewer synapses and some loss of brain function.

associated with schizophrenia, this "pruning" is too extensive, leaving too few connections in the parts of the brain that control thinking, planning, and the ability to link thoughts into accurate perceptions of the world (**figure 8.9**). Cells called microglia do the pruning, in response to *C4A* protein "tags" on the targeted synapses. Microglia are a type of neuroglia that had been thought to function mainly in cleaning up cellular debris. When they are too active, perhaps in response to infection, a person loses too many synapses, neuron-to-neuron communication in the brain suffers, and the symptoms of schizophrenia appear.

> **Key Concepts Questions 8.6**
>
> *a. What are symptoms of schizophrenia?
> b. How does schizophrenia differ from the mood disorders?
> c. Explain how a role of the immune system in raising risk of or causing schizophrenia accounts for environmental and genetic factors.
> d. Explain how excessive synaptic pruning may set the stage for development of schizophrenia.

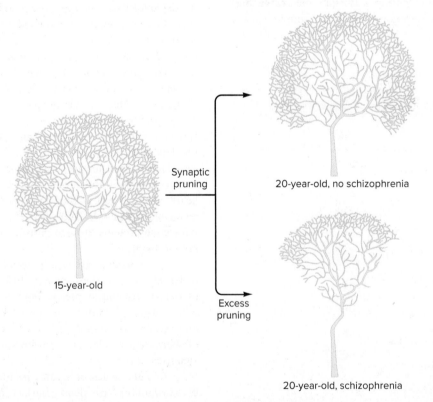

Figure 8.9 **Schizophrenia may result from overactive synaptic pruning.** Losing up to 2 billion brain synapses a day is normal as adulthood begins, but an acceleration in the process may lie behind schizophrenia.

8.7 Autism

Autism spectrum disorders impair socialization and communication skills, with symptoms appearing before age 3. One in 40 children in the United States has autism, and boys outnumber girls four to one. About 25 percent of affected children have seizures as they grow older. Although 70 percent of people with autism have intellectual disability, others may be very intelligent, and many are talented in science or the arts (**table 8.3**).

In severe autism, a toddler may lose vocabulary words and cease starting conversations. He or she stops playing with other children, preferring to be alone. Rocking back and forth or clutching a treasured object and using repetitive motions are soothing behaviors for some children. The child may refer to herself by her name, rather than "me," not make eye contact, and appear oblivious to nonverbal cues such as facial expressions or tone of voice. Special education programs that provide a rigid routine can help an individual with autism to connect with the world. People who have one type of autism (Asperger syndrome) may have the socialization effects but not the language deficits.

In most cases of autism, many genes contribute to different degrees against a backdrop of environmental influences. Environmental triggers include prenatal exposures to rubella (German measles) and to the drug valproate. Evidence is accumulating that exposure to folic acid supplements late in gestation can cause epigenetic changes that increase the risk of autism. A better way to identify environmental risk factors may be to consider different genetic subtypes, so that studies can compare individuals with the same underlying problem.

Heritability of autism is high: about 0.90 (90% of variability due to genetics). Although siblings of affected children are at a 15 percent risk of being affected, compared to the less than 1.5 percent for the general population, in some MZ (identical) twin pairs, one twin has autism and one does not. More than 30 susceptibility or causative genes have been identified.

In about 15 percent of people with autism, the condition is part of a syndrome, including single-gene disorders such as Rett syndrome (see the chapter 2 opener), mitochondrial diseases (see section 5.2), and chromosomal conditions such as fragile X syndrome (see Clinical Connection 12.1) and Down syndrome (see Bioethics in chapter 13). A diagnostic workup for autism includes tests for these conditions as well as chromosomal microarray analysis, which detects copy number variants (repetitions of DNA sequences that vary in number among individuals).

Our understanding of the genetics of autism parallels development of genetic technology. At first, the rare individuals who had autism as well as an abnormal chromosome pointed the way to a place in the genome—the abnormal chromosome—that likely harbors a predisposing or causative mutation. Then genome-wide association studies highlighted genome regions that include genes that increase risk of developing autism.

The fact that autism is part of so many genetic syndromes suggests that it might be a common response to variants of multiple genes. Autism tends to occur in syndromes that also include unusual facial features, developmental delay, and/or intellectual disability. Each genetic syndrome accounts for less than 1 percent of autism cases, but a significant percentage of individuals with each condition has autistic behaviors.

Today exome and whole-genome sequencing projects are enabling researchers to identify many more genes behind autism. Experiments focus on candidate genes within the huge datasets that these technologies generate. **Figure 8.10** illustrates three general strategies for isolating autism genes (or other genes for a shared phenotype) from exome or genome sequencing data.

The first approach, "multiple unrelated affected subjects," compares the exome or genome sequences of many unrelated individuals with autism to "reference" genomes from healthy people compiled in a population database. In which genes do affected children share variants that are not in unaffected children? In one study, researchers compared genome sequences from affected members of 55 families that each had several people with autism to reference genomes from nonaffected families. They identified 153 parts of the genome. Then the researchers made probes—copies of those identified DNA sequences that incorporate a fluorescent label—and applied them to the DNA of 2,175 children with autism and 5,801 children without autism. In those 153 genome regions, the investigators narrowed down 15 sites, with a total of 24 gene variants, seen in the affected children and associated with a twofold increase in the risk of autism.

Table 8.3	Famous People Who Had Autistic Behaviors
Person	**Characteristics**
Albert Einstein	Did not like to be touched
	Difficulty learning in school
	Awkward in social situations
Charles Darwin	Preferred letter writing to face-to-face interactions
	Fixated on certain objects and topics
Emily Dickinson	Heightened sense of smell
	Preferred white clothing
	Reclusive
	Fascinated with flowers
Michelangelo	Limited interests but obsessed with work
	Poor communication and social skills
	Relied on rigid routines

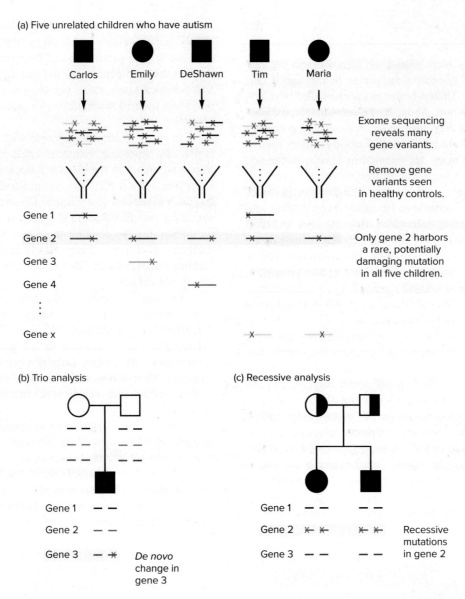

Figure 8.10 Three approaches to using exome sequencing to discover genes that contribute to autism. (a) Comparison to a reference genome identifies genes that unrelated people with autism share. (b) Trio (parents-child) analysis detects *de novo* mutations in a child. (c) Recessive analysis detects mutations in families with more than one affected child.

The second approach is *trio analysis,* which compares the DNA in parents-child trios. Figure 4.18 and Clinical Connection 1.1 discuss the trio analysis of a girl whose genetic syndrome includes autism. These studies either find that parents are carriers of autosomal recessive or X-linked conditions, or that the autism in the child arose *de novo* (new).

The third approach, "recessive analysis," compares DNA sequences in families in which more than one child is affected but the parents are not, seeking genes that are mutant in two copies in the children with autism. One study considered three large families from the Middle East that have many cousin-cousin marriages, which are at higher risk for passing along autosomal recessive conditions due to inheritance of mutations from shared ancestors. When the researchers discovered mutations in three genes already known to cause other diseases, they hypothesized that some cases of autism are actually mild or atypical versions of recognized syndromes. The researchers found seven other known autosomal recessive conditions that also cause autism. These diseases are pleiotropic—they have several manifestations, and in some children, autism is the only symptom.

After identifying mutations that might cause or contribute to autism based on their occurrence in people with the condition, the next step is to consider what these genes do. How might malfunction of the gene impair socialization and communication skills? For example, abnormalities in either of two types of cellular adhesion proteins found at synapses, called

Bioethics

Finding Fault for Autism

The behaviors that we call autistic have been around for centuries and likely longer, unnamed but perhaps not unnoticed. An early reference is to Victor, the "wild boy of Avalon," who lived alone in the woods. In 1910 Swiss psychiatrist Paul Bleuler coined the term "autismus" (in German), from the Greek *autos* for "self" and *ismos* for "state"—the definition was "morbid self-absorption."

American child psychiatrist Leo Kanner is often credited with the first modern report on autism, published in 1943 on 11 children with what he termed "infantile autism." The children shared obsession with certain objects, inability to interact with others, and a "need for sameness." In 1944 pediatrician Hans Asperger, at the University of Vienna, published a report on a similar group of young patients, but his work was unnoticed for many years.

Before autism was recognized as a medical condition, children with more severe manifestations tended to be ignored or abused, some even institutionalized or subjected to barbaric treatments. In 1959, psychologist Bruno Bettelheim published a controversial article, "Joey, a Mechanical Boy," in *Scientific American,* in which he blamed a deficiency of maternal affection for the child's behavior. Dr. Kanner at first agreed, using the term "refrigerator mother," but eventually recognized the accumulating evidence of a biological explanation of autistic behaviors. The maternal connection could not explain why families with multiple children had only one child with autism.

Once the term *autism* became well recognized, some desperate parents were drawn in by promises of miracle cures. Yet in some individuals with autism, astonishing talents tended to overshadow the unusual behaviors.

In the 1980s British psychiatrist Lorna Wing rediscovered Dr. Asperger's contribution and described the part of the spectrum that takes his name to denote individuals who may be quite verbal but unable to recognize subtleties of language and who may have remarkable skills or gifts. Autism was added to the *Diagnostic and Statistical Manual* in 1980. Since the Individuals with Disability Education Act of 1990, public schools in the United States have provided services to children with autism spectrum disorders.

Questions for Discussion

1. Some people argue that autism is an extreme of normal behavior that has been "medicalized," or is a manifestation of "neurodiversity." What type of evidence supports the view that autism is a medical condition, versus an extreme of normal behavior?
2. What are the risks and benefits of autism being considered a disease?
3. Research famous people who were or are "on the spectrum." What are possible benefits of studying these individuals?
4. Why might Dr. Bettelheim have blamed mothers, and what type of evidence from everyday life disproved his hypothesis?

---

neurexins and **neuroligins**, are responsible for a small percentage of cases of autism.

Three types of neurexins emanate from the "sending" presynaptic neuron, and a neuroligin extends toward the neurexin from the "receiving" postsynaptic neuron. When the two protein types meet, they draw the neurons closer together so that one can signal the other, using the neurotransmitter glutamate. Mutations misfold the two proteins so that they cannot interact as they normally do, which disrupts the sending of signals from one nerve cell to another (**figure 8.11**). If this interference happens in early childhood, when synapses naturally form as a child experiences the environment, the behaviors of autism could emerge from fewer or impaired synapses in brain parts that oversee learning and memory. **Bioethics** looks at the evolution of treatments for and attitudes toward people with autism spectrum disorders.

Identifying the genes behind autism and the other behavioral (also known as neuropsychiatric) disorders discussed in this chapter can have clinical benefits. Genetic information can add precision to a traditional diagnosis based on symptoms, define clinical subtypes of a condition detectable with existing diagnostic tests, and provide a way to predict risk that relatives will develop the condition, which is greater than the general population risk. Finally, identifying genes behind behavioral disorders is important in drug discovery research. One day we may have genetically based personalized treatments for subtypes of these complex conditions that we have traditionally considered together based on their phenotypes.

Key Concepts Questions 8.7

a. What is autism?

*b. What is the evidence that genes contribute to autism risk?

*c. What are some environmental factors that contribute to the development of autism?

d. Explain how exome and genome sequencing have revealed genes implicated in autism.

e. How do interactions between neurexins and neuroligins explain some cases of autism?

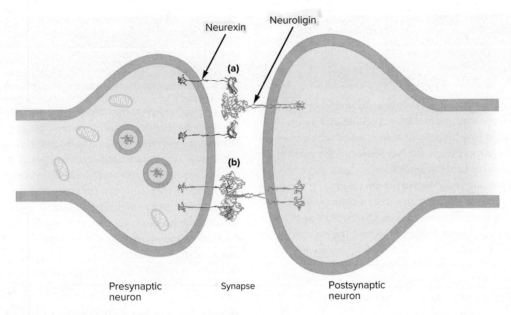

Figure 8.11 Understanding autism. In some cases of autism, neurexins and neuroligins cannot interact strongly enough at synapses (a). Normally the two kinds of proteins can interact as shown in (b).

Study Guide

Summary

8.1 Genes and Behavior

1. Genes affect how the brain responds to environmental stimuli. **Neurons** and **neuroglia** are the two major types of brain cells.
2. Candidate genes for behavioral traits and disorders affect **neurotransmitters** crossing **synapses** (neurotransmission).
3. Old and new genetic tools are being used to describe the biological causes of various behavioral disorders.
4. Variants of hundreds of genes contribute in different degrees to behavioral disorders.

8.2 Sleep

5. Twin studies and single-gene disorders that affect the sleep-wake cycle reveal a large inherited component to sleep-wake behavior.
6. Clock genes such as *period 2* enable a person to respond to day and night environmental cues.

8.3 Intelligence and Intellectual Disability

7. The general intelligence ability (*g*) measures the inherited portion of IQ that may underlie population variance in IQ test performance.
8. Studies indicate that general intelligence is maintained throughout life. The fact that many syndromes that result from abnormal chromosomes affect intelligence suggests high heritability.
9. Exome and genome sequencing are identifying mutations that cause or contribute to intellectual disability, which has many causes.

8.4 Drug Addiction

10. Drug addiction is a severe substance use disorder that arises from developing tolerance to and dependence on a drug. Addiction produces stable changes in certain brain parts.
11. Proteins involved in drug addiction affect neurotransmitter synthesis and neurotransmitter removal from synapses, and receptor structure and function.
12. Candidate genes for drug addiction encode the dopamine D(2) receptor and nicotinic receptor parts.

8.5 Mood Disorders

13. Major depressive disorder is common and associated with mutations in genes that regulate circadian rhythms, and deficits of serotonin.
14. Bipolar disorder includes depressive periods and periods of mania or hypomania. Combinations of predisposing variants in any of hundreds of genes in the presence of certain environmental influences can lead to bipolar disorder.

8.6 Schizophrenia

15. Schizophrenia greatly disrupts the ability to think and perceive the world. Onset is typically in early adulthood. The course is episodic or it plateaus.
16. Empiric risk values and a heritability of 0.9 indicate a large genetic component to schizophrenia.
17. An overactive or inappropriate immune response during prenatal development or in late adolescence may increase the risk of developing schizophrenia.

18. Inheriting a variant of a gene (*C4A*) that encodes an immune system protein increases schizophrenia risk by accelerating synaptic pruning during late adolescence or young adulthood.

8.7 Autism

19. Autism is a loss of communication and social skills beginning in early childhood.
20. Exposure to certain pathogens and chemicals can increase the risk of autism.
21. Autism is part of many genetic syndromes, but these account for a small percentage of autism cases.
22. Three approaches are used to identify gene variants that contribute to autism. The "multiple unrelated affected subjects" approach seeks gene variants in individuals with autism that are not in population databases of unaffected children. "Trio analysis" distinguishes carrier parents from *de novo* mutations in children. "Recessive analysis" searches for gene variants that two or more affected siblings share.
23. **Neurexins** and **neuroligins** are proteins embedded in the plasma membranes of presynaptic and postsynaptic neurons, respectively. In some cases of autism they cannot interact normally at synapses.

Key Terms

neurexin A protein in the presynaptic membrane of a neuron that uses the neurotransmitter glutamate; when misfolded, it is involved in autism.

neuroglia Several types of cells in the nervous system that support and nurture neurons.

neuroligin A protein in the postsynaptic membrane of a neuron that uses the neurotransmitter glutamate; when misfolded, it is involved in autism.

neuron (nōrin′) A nerve cell.

neurotransmitter A molecule that transmits messages in the nervous system across synapses.

synapse The space between two neurons that a neurotransmitter must cross to transmit a message.

Review Questions

1. What are the two major types of cells in the brain, and what do they do?
2. Why are mood disorders, schizophrenia, and autism considered to be complex traits?
3. Name a pathway or mechanism involving a protein that, when absent or atypical, affects behavior.
4. Describe how variants of the following genes affect the risk of developing specific behavioral disorders.
   a. the gene that encodes the orexin receptor
   b. the *period 2* gene
   c. the gene for the dopamine D(2) receptor
   d. the gene for nicotinic receptor parts
   e. the gene for the serotonin transporter
   f. *C4 complement component 4*
   g. genes that encode neurexins and neuroligins
5. What evidence would indicate that a family with familial advanced sleep phase syndrome has a genetic condition rather than just all the members becoming used to keeping unusual hours?
6. What are the two defining characteristics of addiction?
7. What is the evidence that our bodies have their own uses for cocaine, THC, and opium?
8. How are SSRIs thought to work to treat depression?
9. How can genetic subtyping of major depressive disorder improve clinical outcomes?
10. Explain how both environmental and genetic factors can cause schizophrenia.
11. Distinguish among the three approaches to autism gene discovery that figure 8.10 shows.
12. Explain how abnormal interactions between neurexins and neuroligins at synapses can cause autism.

Applied Questions

1. Identify a gene variant and an environmental factor that together might significantly increase the risk of an individual developing posttraumatic stress disorder. (See Forensics Focus Question 1.)
2. The drug Belsomna inhibits orexin. What condition does it treat?
3. Many older individuals experience advanced sleep phase syndrome. Even though this condition is probably a normal part of aging, how might research on the Utah family with an inherited form of the condition help researchers develop a drug to assist the elderly in sleeping through the night and awakening later in the morning?
4. Some technical terms are being replaced so that they do not make people uncomfortable. "Mental retardation" is now considered one condition of "intellectual disability," and "drug addiction" is one manifestation of "drug

use disease." Do you think that these redefinitions are helpful, harmful, or just confusing? Cite a reason for your answer.

5. How does the subunit structure of the nicotinic receptor provide a mechanism for epistasis? (Epistasis is an interaction in which one gene masks or affects the expression of another; see chapter 5.)

6. What might be the advantages and disadvantages of a genetic test done at birth that indicates whether a person is at high risk for developing a drug addiction?

7. What aspect of bipolar disorder may account for the fact that the condition is overrepresented among very creative individuals?

8. Who is at higher risk for schizophrenia, the grandchild of a person with the condition or a niece/nephew?

9. In the United States, the incidence of autism has dramatically increased since 2012.
    a. Does this finding better support a genetic cause or an environmental cause for autism?
    b. What is a nongenetic factor that might explain the increased incidence of autism?

10. A consumer genetic testing company offers a test for risk of depression and response to particular drugs, based on results of a genome-wide association study. What further information is necessary to determine the value of such a test for an individual?

Forensics Focus

1. When 43-year-old Francis discovered his soon-to-be ex-wife in bed with her boyfriend, he shot and killed them both. The defense ordered a pretrial forensic psychiatric workup that included genotyping for the gene for the enzyme monoamine oxidase A (*MAOA*), which breaks down the neurotransmitters serotonin, dopamine, and norepinephrine, and for the gene for the serotonin transporter (*SLC6A4*). A "high-activity" allele for *MAOA* is associated with violence in people who also suffered child abuse. Inheriting one or two "short" alleles of *SLC6A4* is associated with depression and suicidal thoughts, in people who have experienced great stress. Francis had the high-activity *MAOA* genotype, two short *SLC6A4* alleles, and a lifetime of abuse and stress. However, the judge ruled that the science was not far enough along to admit the genotyping results as evidence.
    a. Under what conditions or situations do you think it is valid to include genotyping results in cases like this?
    b. In a Dutch family, a mutation disables *MAOA*, causing a syndrome of intellectual disability and abnormal behavior. Family members had committed arson, attempted rape, and shown exhibitionism. How can both high and low levels of an enzyme cause behavioral problems?
    c. What is a limitation to use of behavioral genotyping in a criminal trial?
    d. The media often run articles calling the MAOA gene the "warrior gene." What might be a negative consequence of using this nickname?

Case Studies and Research Results

1. The drug MDMA, known on the street as ecstasy, X, E, and disco biscuit, has been evaluated in clinical trials to treat PTSD in 26 combat veterans, firefighters, and police officers. When paired with psychotherapy, the drug reduced the severity of PTSD. What further information is needed before the Food and Drug Administration approves use of the drug to treat PTSD?

2. In a short e-book by Karen Russell called *Sleep Donation*, overuse of lit screens leads to a pandemic of an infectious, deadly insomnia due to abnormal expression of the gene encoding a protein mentioned in the chapter. Name the implicated gene, and state whether it would have to be overexpressed or underexpressed to cause insomnia. (The book is fiction.)

3. Many studies have shown that the relatives of people who have bipolar disorder or schizophrenia are overrepresented among the creative professions, including writers, actors, dancers, musicians, and visual artists. Researchers in Iceland analyzed genetic information on hundreds of thousands of people and found that gene variants associated with the two conditions were also found in family members that did not have any psychiatric diagnoses, but who were members of Iceland's artistic societies.
    a. How might the same gene variants cause bipolar disorder or schizophrenia in some family members, but creativity in others?
    b. Do you think that in light of these findings, either of these disorders, or both, should be considered extremes of "normal" behavior? How might this designation be advantageous or dangerous?

PART 3: DNA and Chromosomes

CHAPTER 9
DNA Structure and Replication

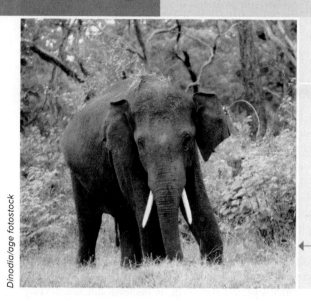

Comparing DNA sequences is a way to solve problems, including exposing crimes such as poaching wildlife.

Learning Outcomes

9.1 Experiments Identify and Describe the Genetic Material
1. Describe the experiments that demonstrated that DNA is the genetic material and that protein is not.
2. Explain how Watson and Crick deduced the three-dimensional structure of DNA.

9.2 DNA Structure
3. List the components of a DNA nucleotide.
4. Explain how two chains of nucleotides form a DNA molecule.
5. Describe how a long DNA molecule folds and loops.

9.3 DNA Replication—Maintaining Genetic Information
6. Explain the semiconservative mechanism of DNA replication.
7. List the steps of DNA replication.
8. Explain how the polymerase chain reaction is used to amplify DNA outside cells.

9.4 Sequencing DNA
9. Explain the basic strategy used to determine the base sequence of a DNA molecule.
10. Explain how next-generation sequencing improves upon Sanger sequencing.

The BIG Picture

DNA is the basis of life because of three qualities: It holds information, it copies itself, and it changes. DNA sequences can be used to identify individuals and as the basis of diagnostic tests for genetic diseases.

Elephant Forensics

Poachers take more than 50 tons of ivory from the tusks of elephants in Africa each year. The animals, which now number about 500,000, will soon become extinct.

To identify the geographical sources of poached ivory, researchers collected dung from 1,350 elephants in 71 locations in 29 African nations—1,001 animals from the savanna and 349 from the forest. They genotyped the DNA using short repeats from 16 genome regions. Then they genotyped ivory from 28 large seizures—nearly half a ton, representing 20 percent of all large seizures from 1996 through 2014.

By matching the DNA patterns from the two sources—dung and tusks—the researchers deduced where the crimes occurred. The savanna elephant tusks came from Mozambique and Tanzania, and the forest elephant tusks came from Gabon, the Republic of Congo, and the Central African Republic. DNA analysis also revealed that ivory is moved from storage sites to shipping sites.

The DNA of tusks is in the outer cementum layer. The researchers pulverized samples, then used chemicals to extract the minerals and proteins, leaving enough DNA to determine genotypes.

9.1 Experiments Identify and Describe the Genetic Material

"A genetic material must carry out two jobs: duplicate itself and control the development of the rest of the cell in a specific way," wrote Francis Crick, co-discoverer with James Watson of the three-dimensional structure of DNA in 1953 (*Nature*, April 25, 1953: 737–738). DNA is the only molecule that fulfills these requirements.

DNA was first described in the mid-nineteenth century, when Swiss physician and biochemist Friedrich Miescher isolated nuclei from white blood cells in pus on soiled bandages. He discovered in the nuclei an unusual acidic substance containing nitrogen and phosphorus. He and others then found it in cells from a variety of sources. Because the material resided in cell nuclei, Miescher named it *nuclein* in an 1871 paper; subsequently, it was called a nucleic acid. Few people appreciated the importance of Miescher's discovery at the time, when inherited disease was widely attributed to abnormal protein.

In 1902, English physician Archibald Garrod was the first to link inherited disease and protein. He noted that people who had certain inborn errors of metabolism did not have certain enzymes. Other researchers added evidence of a link between heredity and enzymes from other species, such as fruit flies with unusual eye colors and bread molds with nutritional deficiencies. Both organisms had absent or malfunctioning specific enzymes. As researchers wondered about the connection between enzymes and heredity, they returned to Miescher's discovery of nucleic acids.

DNA *Is* the Hereditary Molecule

In 1928, English microbiologist Frederick Griffith took the first step in identifying DNA as the genetic material. He was studying bacterial pneumonia in the years after the 1918 flu pandemic, which killed more people than influenza alone. Griffith noticed that mice with a certain form of pneumonia harbored one of two types of *Streptococcus pneumoniae* bacteria. Type R bacteria were rough in texture. The mouse immune system recognizes the rough bacteria and kills them. Type S bacteria were smooth because they were enclosed in a polysaccharide (a type of carbohydrate) capsule. The mouse immune system cannot recognize the bacteria beneath the polysaccharide coats, and therefore mice injected with type R bacteria did not develop pneumonia (**figure 9.1***a*), but mice injected with type S did (figure 9.1*b*). The polysaccharide-shielded bacteria caused severe (virulent) infection. Injecting mice with unaltered type R or type S bacteria served as control experiments, which represent the situation without the experimental intervention.

When type S bacteria were heated, which killed them, they no longer could cause pneumonia in mice (figure 9.1*c*). However,

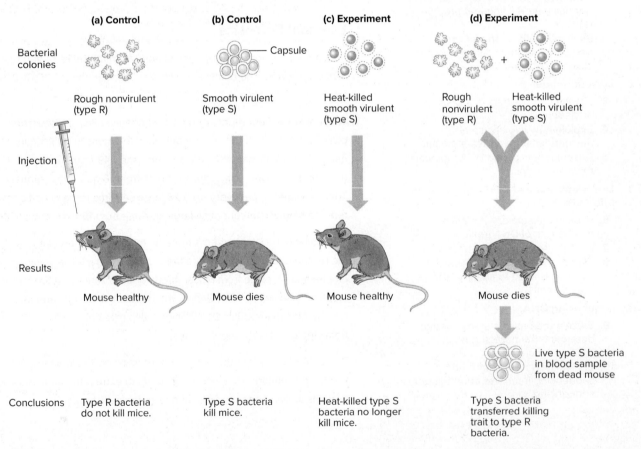

Figure 9.1 Discovery of bacterial transformation. Griffith's experiments showed that a molecule in a lethal type of bacteria can transform nonkilling (nonvirulent) bacteria into killers (virulent).

when Griffith injected mice with a mixture of type R bacteria plus heat-killed type S bacteria—neither of which, alone, was deadly to the mice—the mice died of pneumonia (figure 9.1d). Their bodies contained live type S bacteria, encased in polysaccharide. Griffith termed the apparent conversion of one bacterial type into another "transformation." How did it happen? What component of the dead, smooth bacteria transformed type R to type S? He couldn't answer the question with further experiments because he was killed by a German bomb in London in 1941 during World War II.

U.S. physicians Oswald Avery, Colin MacLeod, and Maclyn McCarty did the next logical set of experiments. They hypothesized that a nucleic acid might be Griffith's "transforming principle." The trio of researchers observed that treating broken-open type S bacteria with a protease—an enzyme that dismantles protein—did not prevent the transformation of a nonvirulent to a virulent strain, but treating such bacteria with deoxyribonuclease (or DNase), an enzyme that dismantles DNA only, did disrupt transformation. In 1944, they confirmed that DNA transformed the bacteria. They isolated DNA from heat-killed type S bacteria and injected it with type R bacteria into mice (**figure 9.2**). The mice died, and their bodies contained active type S bacteria. The conclusion: DNA passed from type S bacteria into type R, enabling the type R to manufacture the smooth coat necessary for infection. Once type R bacteria encase themselves in smooth coats, they are no longer type R.

Protein *Is Not* the Hereditary Molecule

Scientists seek to explain biological phenomena by designing and carrying out experiments that enable them to eliminate explanations. Such testable hypotheses provide evidence, but not proof, because conclusions can change when new data become available. In the series of experiments to identify the genetic material, researchers also had to show that protein does *not* transmit genetic information. To do this, in 1953, U.S. microbiologists Alfred Hershey and Martha Chase infected *Escherichia coli* bacteria with a virus that consisted largely of a protein "head" or coat surrounding DNA. Viruses infect bacterial cells by injecting their DNA (or RNA) into them. Infected bacteria then produce many more viruses. The viral protein coats remain outside the bacterial cells.

Researchers can analyze viruses by growing them on culture medium that contains a radioactive chemical that the viruses take up. If that chemical is part of DNA or RNA, then a "labeled" viral nucleic acid emits radiation, which can be detected in several ways. Hershey and Chase cleverly took advantage of the observation that protein contains sulfur but not phosphorus, and that nucleic acids contain phosphorus but not sulfur. Radioactive forms exist for both elements. Hershey and Chase hypothesized that only the viral genetic material need to enter a bacterial cell to transmit infection.

When Hershey and Chase grew viruses in the presence of radioactive sulfur, the viral *protein coats* took up and emitted radioactivity, but when they ran the experiment using radioactive phosphorus, the viral *DNA* emitted radioactivity. If protein is the genetic material, then the infected bacteria would have radioactive sulfur. But if DNA is the genetic material, then the bacteria would have radioactive phosphorus.

Hershey and Chase grew one batch of virus in a medium containing radioactive sulfur (designated ^{35}S) and another in a medium containing radioactive phosphorus (designated ^{32}P). The viruses grown on sulfur had their protein marked, but not their DNA, because protein incorporates sulfur but DNA does not. Conversely, the viruses grown on labeled phosphorus had their DNA marked, but not their protein, because this element is found in DNA but not protein.

After allowing several minutes for the virus particles to bind to the bacteria and inject their genetic material into them, Hershey and Chase agitated each mixture in a blender, shaking free the empty virus protein coats. The contents of each blender were collected in test tubes, then spun at high speed in a centrifuge. This forced the larger bacteria to the bottom of the tube, forming a solid "pellet," while the lighter virus coats remained in the fluid above the pellet.

At the end of the procedure, Hershey and Chase examined fractions containing the virus coats from the top of each test tube and the infected bacteria that had settled to the bottom (**figure 9.3**). In the tube containing viruses labeled with sulfur, the virus coats in the fluid were radioactive, but the bacterial pellets were not. In the other tube, where the virus had incorporated radioactive phosphorus, the bacterial pellets were radioactive, but the virus coats in the fluid were not. Therefore, the part of the virus that could enter bacteria and direct them to mass produce more virus was the part that had incorporated phosphorus—the DNA. The genetic material is DNA, and not protein.

Discovering the Structure of DNA

The experiments that revealed the structure and function of DNA toggle back and forth in time because they came from different disciplines, and the field of molecular genetics did not yet exist. In 1909, Russian-American biochemist Phoebus Levene identified the 5-carbon sugar **ribose** as part of some nucleic acids. In 1929 he discovered a similar sugar—**deoxyribose**—in

Figure 9.2 DNA is the "transforming principle." Avery, MacLeod, and McCarty identified DNA as Griffith's transforming principle. By adding enzymes that either destroy proteins (protease) or DNA (deoxyribonuclease or DNase) to bacteria that were broken apart to release their contents, the researchers demonstrated that DNA transforms bacteria—and that protein does not.

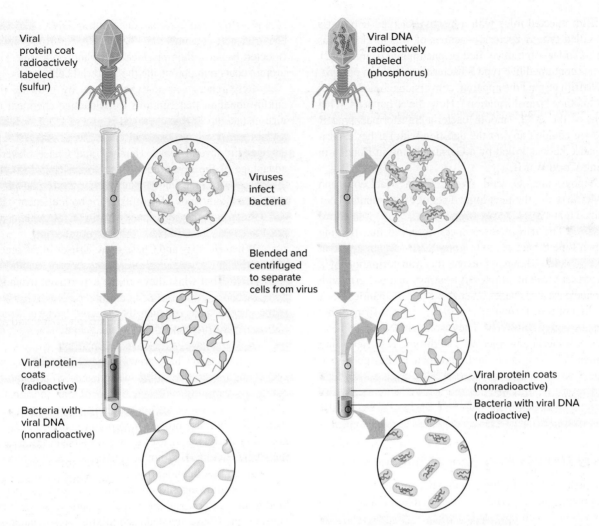

Figure 9.3 DNA is the hereditary material; protein is not. Hershey and Chase used different radioactive molecules to distinguish the viral protein coat from the genetic material (DNA). These "blender experiments" showed that the virus transfers DNA, and not protein, to the bacterium. Therefore, DNA is the genetic material. The blender experiments used particular types (isotopes) of sulfur and phosphorus atoms that emit detectable radiation.

other nucleic acids. He had revealed a major chemical distinction between RNA and DNA: RNA has ribose, and DNA has deoxyribose. (Recall that RNA serves as a carrier of some of the information in a DNA molecule that instructs the cell to manufacture a particular protein.)

Levene then discovered that the three parts of a nucleic acid—a sugar, a nitrogen-containing (nitrogenous) base, and a phosphorus-containing component—are present in approximately equal proportions. He deduced that a nucleic acid building block must contain one of each component. Furthermore, although the sugar and phosphate portions of nucleic acids were always the same, the nitrogen-containing bases were of four types. Scientists at first thought that the bases were present in equal amounts. If this were so, however, DNA could not encode as much information as it could if the number of each base type varied. Imagine how much less useful a written language would be if it had to use all the letters with equal frequency.

In the early 1950s, two lines of experimental evidence converged to provide the direct clues that finally revealed DNA's structure. Austrian-American biochemist Erwin Chargaff showed that DNA in several species contains equal amounts of the bases adenine (A) and thymine (T) and equal amounts of the bases guanine (G) and cytosine (C). Next, English physicist Maurice Wilkins and English chemist Rosalind Franklin bombarded DNA with X rays using a technique called X-ray diffraction. The patterns in which the X-rays were deflected from the molecule revealed some aspects of its structure, including that DNA is a twisted ladder, a helix. Wilkins' and Franklin's data showed the distances between the atoms making up the twisted ladder.

Rosalind Franklin provided the pivotal clue that there were two forms of DNA. She distinguished a dry, crystalline "A" form from the wetter type seen in cells, the "B" form. It took her 100 hours to obtain the now famous "photo 51" of the B form in May 1952 (**figure 9.4***a*). When a graduate student showed photo 51 to Wilkins, who showed it to Watson at the end of January in 1953, the men realized that the remarkable symmetry of the molecule, indicated by the positions of the phosphates in the photograph, best fits the shape of a highly regular helix. "It was just a beautiful helix!" Watson wrote.

Figure 9.4 Deciphering DNA structure. **(a)** Rosalind Franklin's "photo 51" of B DNA was critical to Watson and Crick's deduction of the three-dimensional structure of the molecule. The "X" in the center indicates a helix, and the darkened regions reveal symmetrically organized subunits. **(b)** Franklin died very young, of cancer. *(a): Science Source/Science Source; (b): National Library of Medicine/Science Source*

(a)

(b) Rosalind Franklin 1920–1958

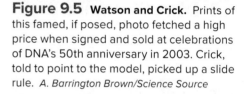

Figure 9.5 Watson and Crick. Prints of this famed, if posed, photo fetched a high price when signed and sold at celebrations of DNA's 50th anniversary in 2003. Crick, told to point to the model, picked up a slide rule. *A. Barrington Brown/Science Source*

The race was on. During February, famed biochemist Linus Pauling suggested a triple helix structure for DNA. This was incorrect. Meanwhile, Watson and Crick, certain of the sugar-phosphate backbone largely from photo 51, turned their attention to the bases. They found the answer not through sophisticated chemistry or crystallography, but using cardboard cutouts of the DNA components.

On Saturday morning, February 28, Watson arrived early for a meeting with Crick. While he was waiting, he played with cardboard cutouts of the four DNA bases, pairing A with A, then A with G. When he assembled A next to T, and G next to C, he noted the similar shapes, and suddenly all the pieces fit. He had been modeling the chemical attractions (hydrogen bonds) between the DNA bases that create the "steps" of the double helix. When Crick arrived 40 minutes later, the two quickly realized they had solved the puzzle of DNA's structure (**figure 9.5**). They built a more detailed and traditional "ball-and-stick" metal model in front of which they posed, and published their findings in the April 25, 1953 issue of *Nature* magazine, without ever having done an experiment.

Watson, Crick, and Wilkins eventually received the Nobel Prize. In 1958, Franklin died at the age of 37 from ovarian cancer, and the Nobel can only be awarded to a living person. However, she has become a heroine for her long-underappreciated role in deciphering the structure of DNA. In 2010, researchers found six boxes of lost correspondence from Crick, mostly to Wilkins. The letters, photos, and postcards reveal how upset the elder Wilkins was to be in a race with the young upstarts Watson and Crick.

Table 9.1 summarizes some of the experiments that led to the discovery of the structure of the genetic material.

Table 9.1 The Road to the Double Helix

Investigator	Contribution	Timeline
Friedrich Miescher	Isolated nuclein in white blood cell nuclei	1869
Frederick Griffith	Transferred killing ability between types of bacteria	1928
Oswald Avery, Colin MacLeod, and Maclyn McCarty	Discovered that DNA transmits killing ability in bacteria	1940s
Alfred Hershey and Martha Chase	Determined that the part of a virus that infects and replicates is its nucleic acid and not its protein	1950
Phoebus Levene, Erwin Chargaff, Maurice Wilkins, and Rosalind Franklin	Discovered DNA components, proportions, and positions	1909–early 1950s
James Watson and Francis Crick	Elucidated DNA's three-dimensional structure	1953
James Watson	Had his genome sequenced	2008

However, researchers have yet to visualize an actual single DNA molecule. What we know about DNA structure and function has been deduced and inferred from larger-scale experiments and observations.

Key Concepts Questions 9.1

*a. List the two requirements for a genetic material.

b. Describe the experiments that revealed the nature of the genetic material.

c. Describe the experiments that revealed the three-dimensional structure of the genetic material.

9.2 DNA Structure

The structure of the genetic material can be considered at several levels. These include individual bits of information (the bases), genes and chromosomes, and the parts of the genome that loop, fold, touch, and interact as genes are expressed.

A gene is a section of a DNA molecule whose order of bases specifies the sequence of amino acids in a protein. The activity of the protein imparts the phenotype. The fact that different building blocks combine in sequences to form nucleic acids enables them to carry information, as the letters of an alphabet combine to form words.

Inherited traits are diverse because proteins have diverse functions. Malfunctioning or inactive proteins can devastate health. Most of the amino acids that are assembled into proteins come from the diet or from breaking down proteins in the cell. The body synthesizes the others.

Nucleotide Building Blocks

A single building block of DNA is a **nucleotide** (**figure 9.6**). It consists of one deoxyribose sugar, one phosphate group (a phosphorus atom bonded to four oxygen atoms), and one

Figure 9.6 **Nucleotides.** A nucleotide of a nucleic acid consists of a 5-carbon sugar, a phosphate group, and an organic, nitrogenous base (G, A, C, or T).

Figure 9.7 **DNA bases are the informational parts of nucleotides.** Adenine and guanine are purines, each composed of a six-membered and a five-membered ring. Cytosine and thymine are pyrimidines, each having a single six-membered ring. (Within the molecules C, H, N, and O are atoms of carbon, hydrogen, nitrogen, and oxygen, respectively.)

nitrogenous base. The bases adenine (A) and guanine (G) are **purines,** which have a two-ring structure. Cytosine (C) and thymine (T) are **pyrimidines,** which have a single-ring structure (**figure 9.7**).

The nitrogenous bases are the information-containing parts of DNA because they form sequences. DNA sequences are measured in numbers of base pairs. The terms kilobase (kb) and megabase (mb) stand for a thousand and a million DNA bases, respectively. Genes range from hundreds to millions of nucleotides long.

Polynucleotide Chains Are Antiparallel

Nucleotides join into long "polynucleotide" chains when strong attachments, called phosphodiester bonds, form between the deoxyribose sugars and the phosphates. These links create a continuous **sugar-phosphate backbone** (**figure 9.8**). Two such polynucleotide chains align head-to-toe, as **figure 9.9**a depicts.

The opposing orientation of the two nucleotide chains in a DNA molecule is called **antiparallelism.** The head-to-toe configuration derives from the structure of the sugar-phosphate backbone, which we can follow by assigning numbers to the five carbons of the sugars based on their relative positions in the molecule (**figure 9.10**).

The carbons of deoxyribose are numbered from 1 to 5, starting with the carbon found by moving clockwise from

Figure 9.8 A chain of nucleotides. A single DNA strand consists of a polynucleotide chain that forms when the deoxyribose sugars (green) and phosphates (yellow) bond to form a sugar-phosphate backbone. The bases A, C, G, and T are blue.

the oxygen, as shown in figure 9.10. In **figure 9.11,** one chain runs from the number 5 carbon (top of the figure) to the number 3 carbon, but the chain aligned with it runs from the number 3 to the number 5 carbon. These ends are labeled 5′ and 3′, pronounced "5 prime" and "3 prime."

Complementary Base Pairing

The symmetrical DNA double helix forms when nucleotides containing A pair with those containing T, and nucleotides containing G pair with those carrying C. Because purines have two rings and pyrimidines one ring, the consistent pairing of a purine with a pyrimidine ensures that the double helix has the same width throughout. This is what Watson discovered using cardboard cutouts. The specific purine-pyrimidine couples are called **complementary base pairs**.

Chemical attractions called hydrogen bonds hold the DNA base pairs together. In a hydrogen bond, a hydrogen atom on one molecule is attracted to an oxygen atom or nitrogen atom on another molecule, as **figure 9.12** depicts. The locations of the hydrogen bonds that form between A and T and between G and C create the complementary base pairs and generate the highly symmetrical DNA double helix.

Hydrogen bonds are weak individually, but over the many bases of a DNA molecule, they impart great strength. Two hydrogen bonds join each A and T, and three hydrogen bonds join each G and C. Finally, DNA forms a double helix when the antiparallel, base-paired strands twist about one another in a regular fashion. A single strand of DNA will not form a helix.

Figure 9.9 DNA consists of two chains of two polynucleotide chains in an antiparallel configuration. **(a)** Hydrogen bonds (dotted lines) hold the nitrogenous bases of one strand to the nitrogenous bases of the second strand. The sugars point in opposite directions—that is, the strands are antiparallel. **(b)** These zebras are assuming an antiparallel stance. *(b): Mark Boulton/Science Source*

Figure 9.10 Numbering the five carbons of deoxyribose. The antiparallel nature of the DNA double helix becomes apparent when the carbons in the sugar are numbered. The 5′ (prime) and 3′ carbons establish the directionality of each DNA strand.

Chapter 9 *DNA Structure and Replication* **171**

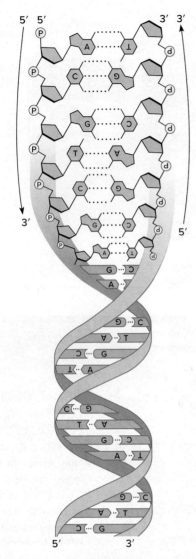

Figure 9.11 DNA is directional. Antiparallelism in a DNA molecule arises from the orientation of the deoxyribose sugars. One strand of the double helix runs in a 5′ to 3′ direction, and the other strand runs in a 3′ to 5′ direction.

Figure 9.12 DNA base pairs. The key to the constant width of the DNA double helix is the pairing of purines with pyrimidines. Two hydrogen bonds join adenine and thymine; three hydrogen bonds link cytosine and guanine. The dotted lines indicate the hydrogen bonds.

DNA Configuration in the Nucleus

DNA molecules are extremely long. The DNA of the smallest human chromosome, if stretched out, would be 14 millimeters (thousandths of a meter) long, but it is packed into a chromosome that, during cell division, is only 2 micrometers (millionths of a meter) long. To fit inside the nucleus, the DNA molecule must fold so tightly that its compacted length shrinks by a factor of 7,000.

Several types of proteins compress DNA without damaging or tangling it. Scaffold proteins form frameworks that guide DNA strands. On a smaller scale, the DNA coils around proteins called **histones,** forming structures that resemble beads on a string. A DNA "bead" is called a **nucleosome**. The compaction of a molecule of DNA is a little like wrapping a very long, thin piece of thread around your fingers, to keep it from unraveling and tangling.

DNA wraps at several levels, until it is compacted into a chromatid (a chromosome consisting of one double helix, in the unreplicated form) (**figure 9.13**). Specifically, a nucleosome is composed of eight histone proteins (a pair of each of four types) plus the 147 nucleotides of DNA entwined around them. A fifth type of histone protein anchors nucleosomes to short "linker" regions of DNA. The linkers tighten the nucleosomes into fibers 30 nanometers (nm) in diameter. As a result, at any given time, only small sections of the DNA double helix are exposed. Chemical modification of the histones is one way of controlling when particular DNA sequences unwind and become accessible for the cell to use to guide protein synthesis. (This is discussed in chapter 11.) DNA unwinds further in localized regions when it is replicated.

Altogether, the chromosome substance is called **chromatin,** which means "colored material." Chromatin is not

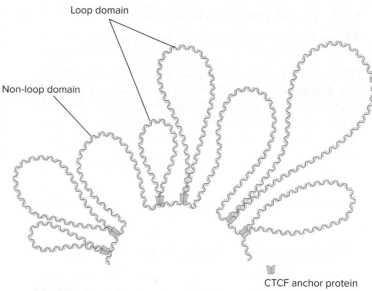

Figure 9.14 DNA looping. Proteins called CTCF bring together parts of a long DNA molecule that include genes whose expression is coordinated.

Figure 9.13 DNA is highly condensed. DNA spools around octets of four types of histone proteins, forming nucleosomes. A fifth type of protein anchors nucleosomes to connecting "linker" regions, which tighten the nucleosomes into fibers 30 nanometers in diameter. Further compaction squeezes the DNA into the nucleus. The chromosomes are in the replicated form, consisting of two chromatids. The two DNA molecules are joined at the centromere.

just DNA; it is about 30 percent histone proteins, 30 percent scaffold proteins and other proteins that bind DNA, 30 percent DNA, and 10 percent RNA.

When chromatin is loose (not condensed into chromosomes that are visible upon staining), it forms loops at about 10,000 places in the genome. An "anchor" protein called CTCF brings together parts of the DNA sequence within the same long DNA molecule to form the overall "loop-ome" structure (**figure 9.14**). Loop formation is not random—the genome parts drawn together include genes that function together. Chromatin loops rarely overlap, and affect swaths of the DNA sequence that are smaller than 2 million base pairs. Researchers are investigating chromatin loops and folds that function as "hidden switches" that might trigger diseases, such as cancers, when altered. Another level of looping and gene expression arises from the attachment of loops to the inner face of the nuclear membrane, which places specific chromosome parts in particular locations in the nucleus. Sites of chromatin attachment to the nuclear membrane are not random, and the placement may control which genes a cell is using to make proteins. Progeria, the genetic disorder that resembles rapid aging, described in the chapter 3 opener, disrupts chromatin binding to the nuclear envelope.

Key Concepts Questions 9.2

*a. What are the components of the DNA double helix?

*b. Explain the basis of antiparallelism.

*c. Explain how complementary base pairing results in a symmetrical double helical structure.

d. How are very long DNA molecules wound so that they fit inside cell nuclei?

e. What is the significance of DNA loops?

Chapter 9 *DNA Structure and Replication*

9.3 DNA Replication—Maintaining Genetic Information

DNA must be copied, or replicated, so that the information it holds can be maintained and passed to future cell generations, even as that information is accessed to guide the manufacture of proteins. As soon as Watson and Crick deciphered the structure of DNA, its mechanism for replication became obvious. They ended their report on the structure of DNA with the statement, *"It has not escaped our notice that the specific pairing we have postulated immediately suggests a possible copying mechanism for the genetic material."*

DNA Replication Is Semiconservative

Watson and Crick envisioned the two strands of the DNA double helix unwinding and separating. Then the exposed unpaired bases would attract their complements from free, unattached nucleotides available in the cell from nutrients. In this way, two identical double helices would form from one original, parental double helix. This route to replication is called **semiconservative** because each new DNA double helix conserves half of the original. However, separating the long strands poses a great physical challenge.

In 1957, two young researchers, Matthew Meselson and Franklin Stahl, demonstrated the semiconservative mechanism of DNA replication with a series of "density shift" experiments. They labeled replicating DNA from bacteria with a dense, heavy form of nitrogen, for a number of cell generations, then shifted the bacteria into a medium with less dense, "light" nitrogen for one cell generation. After replication, each double helix consisted of one "heavy" and one "light" strand (**figure 9.15**). Alternative mechanisms that the experiments ruled out were replicating a daughter DNA double helix built of entirely "heavy" labeled nucleotides (a conservative mechanism) or a daughter double helix in which both strands were composed of joined pieces of "light" and "heavy" nucleotides (a dispersive mechanism).

Steps of DNA Replication

DNA replication occurs during S phase of the cell cycle (see figure 2.11). When DNA is replicated, it first unwinds and then locally separates. The hydrogen bonds holding the base pairs together break, and two identical nucleotide chains are built from one, as the bases form new complementary pairs (**figure 9.16**). A site where DNA is locally opened, resembling a fork, is called a **replication fork**.

DNA replication begins when an unwinding protein called a helicase breaks the hydrogen bonds that connect a base pair (**figure 9.17**). The helicase opens up a localized area, enabling other enzymes to guide assembly of a new DNA strand. Binding proteins hold the two single strands apart. Another enzyme, primase, then attracts complementary RNA nucleotides to build a short piece of RNA, called an RNA primer, at the start of each segment of DNA to be replicated. The RNA primer is required because the major replication enzyme, **DNA polymerase (DNAP)**, can only add bases to an existing nucleic acid strand. (A polymerase is an enzyme

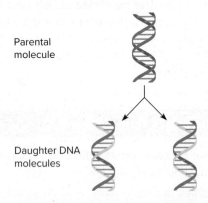

Figure 9.15 **DNA replication is semiconservative.** The two daughter double helices are identical to the original parental double helix in DNA sequence. However, the light blue helix halves of the daughter DNA molecules indicate that each consists of one strand of parental (heavy) DNA and one strand of newly replicated (light) DNA.

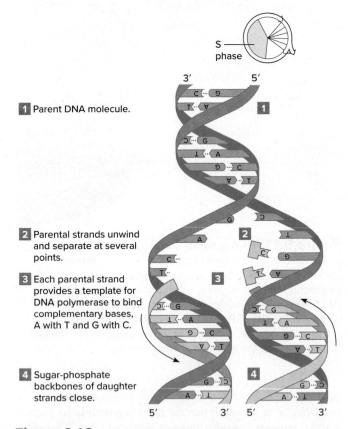

Figure 9.16 **Overview of DNA replication.** After experiments demonstrated the semiconservative nature of DNA replication, the next challenge was to decipher the steps of the process.

Figure 9.17 Activities at the replication fork. DNA replication takes many steps. (The continuous strand is also called the leading strand, and the discontinuous strand is the lagging strand.)

that builds a polymer, which is a chain of chemical building blocks.) Primase has no such limitation and can thus "prime" DNA synthesis in the cell.

Next, the RNA primer binds to DNAP, which brings in DNA nucleotides complementary to the exposed bases on the parental strand; the parental strand serves as a mold, or template. New bases are added one at a time, starting at the RNA primer. The new DNA strand grows as hydrogen bonds form between the complementary bases and the sugar-phosphate backbone links the newly incorporated nucleotides into a strong chain. DNA polymerase carries out both of these activities, and also removes the RNA primer once replication is underway and replaces it with the correct DNA bases. Nucleotides are abundant in cells, and are synthesized from dietary nutrients. The base-pairing rules ensure that each base on the parental DNA strand pulls in its complement.

Because of the antiparallel configuration of the DNA molecule, DNAP works directionally. It adds nucleotides to the exposed 3′ end of the sugar in the growing strand. Replication adds nucleotides in a 5′ to 3′ direction, because this is the only chemical configuration in which DNAP can function. How can the growing fork proceed in one direction, when both parental strands must be replicated and run in opposite directions? The answer is that on one strand, replication is discontinuous. It is accomplished in small pieces from the inner part of the fork outward, in a pattern similar to backstitching. Next, an enzyme called a **ligase** seals the sugar-phosphate backbones of the pieces, building the new strand. These pieces, up to 150 nucleotides long, are called Okazaki fragments, after their discoverer (see figure 9.17).

DNA polymerase also "proofreads" as it goes, excising mismatched bases and inserting correct ones. Yet another enzyme, called an annealing helicase, rewinds any sections of the DNA molecule that remain unwound. Finally, ligases seal the entire sugar-phosphate backbone. *Ligase* comes from a Latin word meaning "to tie."

Human DNA replicates at a rate of about 50 bases per second. To accomplish the job in about an hour rather than the month it would take if starting from only one point, a human chromosome replicates simultaneously at hundreds of points, called replication bubbles. Then the pieces are joined (**figure 9.18**). As a human body grows to trillions of cells, DNA replication occurs about 100 quadrillion times. However, the tips of the telomeres (chromosome ends) do not replicate, and so chromosomes shrink with each cell division (see figure 2.15).

Polymerase Chain Reaction

DNA replication is necessary in a cell to perpetuate genetic information. Researchers use DNA replication conducted outside cells in a biotechnology called DNA amplification. It has diverse applications.

Chapter 9 *DNA Structure and Replication*

Figure 9.18 DNA replication bubbles. DNA replication occurs simultaneously along the 150 million base pairs of the average human chromosome. The sites of replication resemble bubbles that coalesce as the daughter double helices form.
© James Cavallini/Science Source

The first and best-known DNA amplification technique is the **polymerase chain reaction (PCR)**. PCR uses DNA polymerase to rapidly replicate a specific DNA sequence in a test tube.

PCR has been used to amplify DNA from:

- Extinct animals.
- Microorganisms that cannot grow in a laboratory.
- The brain of a 7,000-year-old human mummy.
- The digestive tracts of carnivores, to reveal food web interactions.
- Roadkills and carcasses washed ashore, to identify locally threatened species.
- Genetically altered bacteria that are released in field tests, to follow their dispersion.
- Poached moose meat in hamburger.
- Remains in Jesse James's grave, to make a positive identification.
- The intestines of genital crab lice on a rape victim, which matched the DNA of the suspect.

Figure 9.19 presents the steps in amplifying DNA using PCR to identify a specific DNA sequence in a virus causing respiratory illness. The parenthetical numbers correspond to the numbers in the figure.

PCR requires the following:

- *The target DNA sequence* (1). The target sequence from the suspected pathogen must be known.
- *Two different sequences of lab-made single-stranded short pieces of DNA called primers* (2). These are complementary in sequence to opposite ends of the target sequence and serve the same function as RNA primers in natural DNA replication.
- *Many nucleotides* (3). Multiple copies of each of the four types of DNA nucleotides are required.

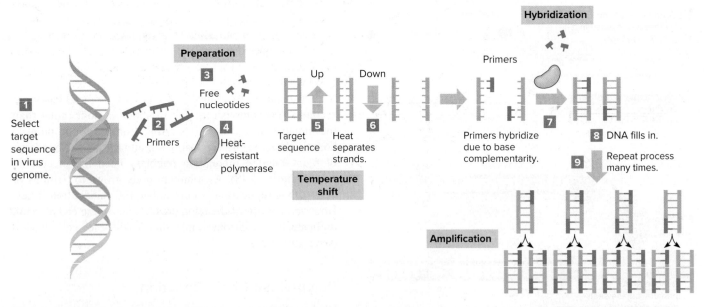

Figure 9.19 PCR amplifies a specific nucleic acid sequence. Many types of viruses cause upper respiratory infections. This figure illustrates how PCR is used on secretions to identify an adenovirus, which has double-stranded DNA as its genetic material. Influenza viruses have RNA as the genetic material, and a different form of PCR, called reverse transcription PCR, is used to analyze them.

■ *Taq1, a DNA polymerase from a bacterium that lives in hot springs* (4). The enzyme eases PCR because it withstands heat, which is necessary to separate the DNA strands.

In the first step of PCR, the temperature is raised (5) and then lowered (6), which separates the two strands of the target DNA. Next (7), many copies of the two short DNA primers and Taq1 DNA polymerase are added. Primers bind by complementary base pairing to the separated target strands. The polymerase then fills in the bases opposite their complements, creating two daughter strands from each separated target double helix (8). The two double helices resulting from the first round of amplification then serve as the target sequences for the next round (9), generating four double helices, and the process continues by again raising the temperature.

Pieces of identical DNA accumulate exponentially. The number of amplified pieces of DNA equals 2^n, where n is the number of temperature cycles. After 30 cycles, PCR yields more than one billion copies of the target DNA sequence. PCR is useful in forensic investigations to amplify small DNA samples (see **Bioethics**).

> ### Key Concepts Questions 9.3
>
> *a. What does DNA replication accomplish?
> *b. Define *semiconservative replication*.
> c. Explain how the steps of semiconservative DNA replication differ from the steps of the two other possible mechanisms.
> d. Explain how DNA can be replicated fast enough to sustain a cell.
> e. Discuss how the polymerase chain reaction is based on DNA replication.

9.4 Sequencing DNA

In 1977, Frederick Sanger, a British biochemist and two-time Nobel Prize winner, invented a way to determine the base sequence of a small piece of DNA. *Sanger sequencing* remains the conceptual basis for techniques that today can sequence an entire human genome in a day. Sanger's method is still used to sequence individual genes or to check the accuracy of a selected sequenced part of a genome. The ability to sequence millions of small DNA pieces at once is the basis of newer methods, called *next-generation sequencing,* that can handle much larger DNA molecules much faster.

Sanger sequencing generates a series of DNA fragments of identical sequence that are complementary to the DNA sequence of interest, which serves as a template strand. The fragments differ in length from each other by one end base, as follows:

```
Sequence of interest:       T A C G C A G T A C
Complementary sequence:     A T G C G T C A T G
Series of fragments:        T G C G T C A T G
                              G C G T C A T G
                                C G T C A T G
                                  G T C A T G
                                    T C A T G
                                      C A T G
                                        A T G
                                          T G
                                            G
```

DNA sequencing technologies are based on clever applications of chemistry. Sanger sequencing uses an approach called *chain termination*. In today's version, PCR replicates a sequence of DNA many times, incorporating chemically altered bases that bear fluorescent tags, a different color for A, T, C, and G. When a tagged base is incorporated into a growing DNA chain, the DNA replication halts because another base cannot be added. The result is a set of partial DNA molecules.

When chain termination happens many times, at different points in the replication of many copies of the same sequence, a collection of partial molecules results. The end bases reveal the sequence. The pieces are separated by size. A flash of laser light excites the four fluorescent tags and produces signals captured in a readout (**figures 9.20** and **9.21**). An algorithm deduces and displays the DNA sequence as a series of wavelengths and as the letters A, T, C, and G.

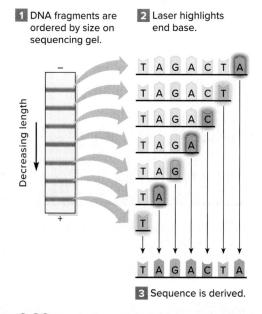

Figure 9.20 Reading a DNA sequence. A computer algorithm detects and records the end base from a series of size-ordered DNA fragments.

Bioethics

Infidelity Testing

Bridgette came home a day early from a business trip to find her husband Roy drinking coffee in their kitchen with Tiffany, his business associate. They were laughing so hard that it took a few moments for them to notice Bridgette standing there. When they did, Tiffany blushed and Roy quickly withdrew his hand from her arm, knocking over her coffee mug.

Bridgette went upstairs to unpack. Flinging her purse on the bed, she noticed several strands of red hair on her pillow. Bridgette's hair was dark brown. She also noticed a crumpled tissue on the floor, partway under the bed.

Bridgette had read an article on the plane about companies that test "abandoned DNA." She went back downstairs for some plastic bags and picked up Tiffany's coffee mug, carrying it all back upstairs. In the bedroom, she quickly collected evidence—the telltale hairs, the discarded used tissue, and a cotton swab rubbed along the inside rim of the lipstick-stained mug. Bridgette emailed gotchaDNA.com and received a cheek swab collection kit a few days later, which she used to send in her own DNA along with the hair, tissue, and coffee cup samples she had collected, for comparison, plus the $600 fee. Then she waited.

The technicians at gotchaDNA.com extracted the DNA from the samples and amplified it. First they checked for Y chromosome markers, which they found on the crumpled tissue. Then they identified patterns of thousands of SNPs (see figure 7.9) that indicated that the DNA on the mug that Tiffany had used and in the hair cells matched each other and not Bridgette's DNA. Tiffany, or at least her red hair, had somehow found its way onto Bridgette's pillow.

Cells use DNA to manufacture protein. People use DNA to identify people. Dozens of companies offer "infidelity DNA testing." Although a few websites provide documents for attesting that the samples are given willingly, many do not—and even list suggested sources of DNA for "adultery tracing." These sources include underwear, toothbrushes, dental floss, nail clippings, gum, cigarette butts, and razor clippings.

Questions for Discussion

1. In the United Kingdom, a law was enacted to prohibit sampling of a celebrity's DNA after someone tried to steal hair from Prince Harry to determine whether or not Prince Charles is his biological father. The United States has no such law. Do you think that one is warranted?
2. Do you think that DNA data obtained without consent should be admissible in a court of law? State a reason for your answer.
3. Discuss one reason in support of infidelity testing with DNA and one reason against it.
4. Identify the individuals in the scenario whom you believe behaved unethically.

Figure 9.21 DNA sequence data. In automated first-generation DNA sequencing, a readout of sequenced DNA is a series of wavelengths that represent the terminal DNA base labeled with a fluorescent molecule.

In 2007, new DNA sequencing "platforms" began to become available to researchers, and these are improved regularly, with storage of megadata in clouds. The sequencing technologies follow the basic approach of Sanger sequencing, but use different chemistries and materials as well as different ways to display the DNA pieces. The new techniques still fragment the human genome into millions of DNA strands, but then capture the pieces onto a solid surface such as glass or beads, and then make identical copies of each fragment on its unique bead or position on the glass. A reiterative process called "sequencing by synthesis" then reads one base at a time on the tens of millions of clusters or beads, yielding 40 billion to 600 billion bases of sequence information per experiment.

Several companies provide variations of next-generation sequencing. One product attaches short DNA pieces in a flow cell, which resembles a microscope slide. Another device uses tiny DNA-fringed beads in a water-oil mixture. A laser reads off the bases corresponding to the fluorescently-tagged bases that are added and stream past the strands, binding their complements.

Another approach uses nanomaterials to distinguish how each of the four nucleotide bases disrupts an electrical field as a DNA molecule passes through nanopores at about 1,000 bases per second. One such material is a one-atom-thick sheet of carbon called graphene that is strong, very thin, and conducts electricity. DNA is threaded through nanopores in it. Each of the four DNA base types disrupts an electrical field in a slightly different way. An algorithm converts the voltage changes into the DNA sequence that they represent.

The new ways to sequence DNA, some of which are company secrets, build on Sanger's invention, but differ in scale. Next-generation sequencing reads millions of sequences at once, and so the general approach is called "massively parallel DNA sequencing." The sequenced short pieces are called "reads." A sequencing experiment will yield many copies of each and every subset of contiguous bases that makes up a target sequence or genome. The more times any particular site in the target is represented among the pieces, the more accurate the deduced sequence—like learning more from reading this book over and over, from different starting points. For example, 30-fold coverage of a genome sequence is more accurate than 20-fold coverage. Thanks to next-generation sequencing, a human genome can be sequenced in hours—the first sequencing took 10 years!

Key Concepts Questions 9.4

a. Explain how Sanger sequencing works.

b. How are many copies of DNA pieces generated to obtain the sequence?

*c. Why must several copies of a genome be cut up to sequence it?

*d. What are next-generation sequencing technologies?

Study Guide

Summary

9.1 Experiments Identify and Describe the Genetic Material

1. DNA encodes information that the cell uses to synthesize protein. DNA can also be replicated, passing on its information during cell division.

2. Many scientists contributed to the discovery of DNA as the hereditary material. Miescher identified DNA in white blood cell nuclei. Garrod connected heredity to enzyme abnormalities. Griffith identified a "transforming principle" that transmitted infectiousness in pneumonia-causing bacteria; Avery, MacLeod, and McCarty discovered that the transforming principle is DNA; and Hershey and Chase confirmed that the genetic material is DNA and not protein.

3. Levene described the three components of a DNA building block (**nucleotide**) and found that they appear in DNA in equal amounts, and that nucleic acids include the sugar **ribose** or **deoxyribose**. Chargaff discovered that the amount of adenine (A) equals the amount of thymine (T), and the amount of guanine (G) equals that of cytosine (C). Franklin used X-ray diffraction to provide crucial clues to the molecular structure. Watson and Crick deduced that DNA is a double helix.

9.2 DNA Structure

4. A gene encodes a protein. A nucleotide building block of a gene consists of a deoxyribose, a phosphate, and a nitrogenous base. A and G are **purines;** C and T are **pyrimidines**.

5. The rungs of the DNA double helix consist of hydrogen-bonded **complementary base pairs** (A with T, and C with G). The rails are chains of alternating sugars and phosphates (the **sugar-phosphate backbone**) that run **antiparallel** to each other. DNA is highly coiled, and complexed with protein to form **chromatin**. DNA winds around **histone** proteins and forms beadlike structures called **nucleosomes**. Loops enable different regions of the same DNA molecule to interact.

9.3 DNA Replication—Maintaining Genetic Information

6. Meselson and Stahl demonstrated the **semiconservative** nature of DNA replication with density shift experiments.

7. During replication, the DNA unwinds locally at several sites. **Replication forks** form as hydrogen bonds break between base pairs. Primase builds short RNA primers, which DNA sequences eventually replace. Next, **DNA polymerase** fills in DNA bases, and **ligase** seals remaining gaps, filling in the sugar-phosphate backbone.

8. Replication proceeds in a 5′ to 3′ direction, so the process must be discontinuous over short stretches on one strand.

9. Nucleic acid amplification (often by the **polymerase chain reaction**, or **PCR**) uses the power and precision of DNA replication enzymes to selectively mass produce selected DNA sequences. In PCR, primers corresponding to a DNA sequence of interest direct polymerization of supplied nucleotides to make many copies of that sequence.

9.4 Sequencing DNA

10. Sanger sequencing deduces a DNA sequence by aligning pieces of different sizes that differ from each other at the end base. Variations on this theme label, cut, and/or immobilize the DNA pieces in different ways, greatly speeding the process.

11. Next-generation sequencing uses a massively parallel approach, and different types of materials on which to immobilize DNA pieces. These techniques read and overlap millions of pieces at once, greatly speeding DNA sequencing.

Key Terms

antiparallelism The head-to-toe orientation of the two nucleotide chains of the DNA double helix.

chromatin (krō′ mə-tĭn) DNA and its associated proteins.

complementary base pairs The pairs of DNA bases that form hydrogen bonds; adenine bonds to thymine and guanine bonds to cytosine.

deoxyribose (dē-ŏksē-rī′bōs) A 5-carbon sugar that is part of a DNA nucleotide.

DNA polymerase (DNAP) (pə-lim′ər-ās) An enzyme that adds new bases to replicating DNA and corrects mismatched base pairs.

histone (hĭs′tōn) A type of protein around which DNA coils in a regular pattern.

ligase (lī ′gās) An enzyme that catalyzes the formation of covalent bonds in the sugar-phosphate backbone of a nucleic acid.

nucleosome (nōō- klē-ō-sōm) A unit of chromatin structure consisting of DNA coiled around an octet of histone proteins.

nucleotide (nōō-klēō-tīde) The building block of a nucleic acid, consisting of a phosphate group, a nitrogenous base, and a 5-carbon sugar.

polymerase chain reaction (PCR) (pŏlə′-mə-r′ās) A nucleic acid amplification technique in which a DNA sequence is replicated in a test tube. It is used to rapidly produce many copies of a specific DNA sequence.

purine (pyōō r′ēn) A nucleic acid base with a two-ring structure; adenine and guanine are purines.

pyrimidine (pĭ-rĭm′ĭ-dēn) A nucleic acid base with a single-ring structure; cytosine, thymine, and uracil are pyrimidines.

replication fork A locally opened portion of a replicating DNA double helix.

ribose (rī′bōs) A 5-carbon sugar that is part of RNA.

semiconservative replication DNA synthesis along each separated strand of the double helix.

sugar-phosphate backbone The "rails" of a DNA double helix, consisting of alternating deoxyribose and phosphate groups, oriented opposite one another.

Review Questions

1. List the components of a nucleotide.
2. How does a purine differ from a pyrimidine?
3. Distinguish between the locations of the phosphodiester bonds and the hydrogen bonds in a DNA molecule.
4. Why must DNA be replicated?
5. Why would a DNA structure in which each base type could form hydrogen bonds with any of the other three base types not produce a molecule that is easily replicated?
6. What part of the DNA molecule encodes information?
7. Explain how DNA is a directional molecule in a chemical sense.
8. What characteristic of the DNA structure does the following statement refer to? "New DNA forms in the 5′ to 3′ direction, but the template strand is read in the 3′ to 5′ direction."
9. Match each experiment in 1–5 to the appropriate description or conclusion from a–e.

   1. Density shift experiments
   2. Discovery on dirty bandages of an acidic substance that includes nitrogen and phosphorus
   3. "Blender experiments" showing that the part of a virus that infects bacteria contains phosphorus, but not sulfur
   4. Determination that DNA contains equal amounts of guanine and cytosine, and of adenine and thymine
   5. Discovery that bacteria can transfer a "factor" that transforms a harmless strain into a lethal one

   a. First experiments in identifying hereditary material.
   b. Complementary base pairing is part of DNA structure and maintains a symmetrical double helix.
   c. Identification of nuclein.
   d. DNA, not protein, is the hereditary material.
   e. DNA replication is semiconservative.

10. Place the following enzymes in the order in which they function in DNA replication: ligase, primase, helicase, and DNA polymerase.
11. How can very long DNA molecules fit into a cell's nucleus?
12. Place in increasing size order: nucleosome, histone protein, and chromatin.
13. Explain how loop formation enables gene-gene interactions.
14. How are very long strands of DNA replicated without becoming twisted into a huge tangle?
15. List the steps in DNA replication.
16. Why must DNA be replicated continuously as well as discontinuously?
17. Explain how RNA participates in DNA replication.
18. Is downloading a document from the Internet analogous to replicating DNA? Cite a reason for your answer.

Applied Questions

1. The Rhino DNA Indexing System (RhODIS) was established in South Africa to track rhinos, which are hunted for their horns. In some Asian countries, rhino horn is believed to cure many illnesses and is collected as a status symbol. The black rhino became extinct in 2011, and three of the remaining five species are critically endangered. Explain how law enforcement officials can use RhODIS to capture poachers.

2. Write the sequence of a strand of DNA replicated using each of the following base sequences as a template:
   a. T C G A G A A T C T C G A T T
   b. C C G T A T A G C C G G T A C
   c. A T C G G A T C G C T A C T G

3. People often use the phrase "the gene for" to describe traits that do not necessarily or directly arise from a protein's actions, such as "the gene for jealousy" or "the gene for acting." How would you explain to them what a gene actually is?

4. The first several DNA bases in the gene that encodes serum albumin, which is a protein that is abundant in blood as well as in egg white, are as follows:

   A G C T T T T C T C T T C T G T C A A C

   This is the strand that holds the information the cell uses to construct the protein. Write the sequence of the complementary DNA strand.

5. To diagnose a rare form of encephalitis (brain inflammation), a researcher needs a million copies of a viral gene. She decides to use PCR on a sample of the patient's cerebrospinal fluid. If one cycle takes 2 minutes, how long will it take to perform a millionfold amplification?

Case Studies and Research Results

1. Which do you think was the more far-reaching accomplishment, determining the structure of DNA or sequencing human genomes? State a reason for your answer.

2. Find further information on one of the inventors or researchers mentioned in the chapter, and describe his or her inspiration or thought process.

3. Read a book or article about Rosalind Franklin and discuss the challenges she faced as a female scientist. Compare her situation to that of a female physicist or geneticist working today.

4. The heath hen is an extinct bird that resembled a prairie chicken. It lived along the east coast of the United States, but by the mid-nineteenth century, was found only on Martha's Vineyard, an island off Cape Cod, Massachusetts. The last heath hen succumbed in 1932. A nonprofit organization is proposing to "bring back" the heath hen by identifying genes unique to the bird samples from the Martha's Vineyard museum, synthesize and introduce those genes into the genomes of modern prairie chickens, and then breed heath hens.

   a. Explain how the plan to bring back the heath hen differs from bringing back the Przewalskii's horse. This "wild ass" was thought to have become extinct in the 1960s but conventional breeding of a few of them discovered roaming the plains of Mongolia brought back a small population.
   b. Do you think "de-extinction" by "genetic rescue" (manipulating DNA) is a good or a bad idea?
   c. How would you recreate a woolly mammoth?

CHAPTER 10

Gene Action: From DNA to Protein

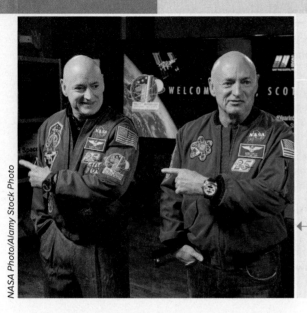

What does being in space do to a human genome? The experience likely causes transient changes in gene expression, as any experience does. Longer-term studies that compare the genomes of the twin who went into space and his earthbound brother may reveal more.

Learning Outcomes

10.1 The Importance of Proteins
1. Explain the relationship between proteins and amino acids.
2. Explain the relationship between genes and proteins.

10.2 Transcription Copies the Information in DNA
3. List the major types of RNA molecules and their functions.
4. Explain the importance of transcription factors.
5. List the steps of transcription.

10.3 Translation of a Protein
6. Discuss how researchers deduced the genetic code.
7. List the steps of protein synthesis.

10.4 Processing a Protein
8. Define the four components of a protein's shape.
9. Explain the importance of protein folding.

The BIG Picture

DNA sequences are the blueprints of life. Cells must maintain this information, yet also access it to manufacture proteins. RNA acts as the go-between, linking DNA sequences to the amino acid sequences of proteins.

Altering Gene Expression in Space

Scott and Mark Kelly are identical twins. When Scott Kelly returned from spending a year on the International Space Station, he and his brother underwent a battery of medical and biochemical tests. What had changed in the astronaut's body during his time above the Earth, compared to his brother?

The media was quick to report that his DNA had changed, perhaps based on the news release from the National Aeronautics and Space Administration (NASA). The agency claimed that the comparison of Scott and Mark Kelly is "the perfect nature versus nurture study."

The twins' DNA hadn't changed. What had been altered from a year in space was Scott's gene *expression,* which refers to the transcription of genes into molecules of messenger RNA. Other types of RNA molecules then assist the cell in translating that message into a protein—a sequence of amino acids. Altering the DNA sequence of a gene is a type of genetic change, or mutation, the subject of chapter 12.

Scott's highly unusual environment had temporarily altered his DNA *function,* not its *structure*—as most any environmental change will do. The expression of thousands of genes continuously rises and falls as we go about being alive. Smell pizza, and salivary amylase floods the mouth. Blood clots. Hair and nails grow. Endorphins soothe. Muscles contract. Hormones release an egg. Cells take up glucose from the bloodstream.

In the months following Scott's return from space, researchers identified transient changes in the expression of genes controlling proteins in urine and in inflammation. His chromosomes' telomeres extended slightly, then shrunk back. Scott's intestinal microbiome changed when he returned to Earth and stopped eating packaged space food.

A longer-term look at DNA shed into the twins' bloodstreams may reveal actual genetic changes. Researchers will determine which organs released the DNA, identify mutations, compare the genetic changes to Earthbound genome sequencing, and compare the rate and nature of the mutations that differ between the brothers.

Figure 10.1 Amino acid structure. An amino acid consists of a central carbon atom bonded to a hydrogen atom, an amino group, an acid group, and a distinctive R group. This amino acid is arginine.

10.1 The Importance of Proteins

Genes provide cells with instructions for synthesizing specific protein molecules. Proteins comprise one of the three major types of nutrients and are abundant in meats, eggs, and legumes. When we eat protein molecules, our digestive systems break them down into **amino acids**, which are molecules small enough to enter the bloodstream through the interior lining of the small intestine. From there, amino acids enter cells, where RNA molecules transcribed from genes guide their assembly into new protein molecules. The muscle protein myosin from a hamburger does not directly become the myosin in a muscle of the person who eats it. Instead, the cow myosin is broken down into its constituent amino acids. These are absorbed in the small intestine and then linked, following the instructions of a gene, to form a protein that is part of the human body, perhaps even the human version of myosin, that is part of the contractile fibers of our muscles.

More than 500 types of amino acids are chemically possible, but only 20 are required for the body to produce its own protein molecules. Nine of the 20 are considered essential because the human body cannot synthesize them; the essential amino acids must come from the diet. Another 6 amino acids are conditionally essential, which means that a human body can manufacture them except for certain circumstances, such as being born prematurely or having a metabolic disease. The other 5 amino acids are nonessential; the body can make them.

The 20 biological amino acids have similar frameworks. Each has a central carbon atom that bonds to an amino group (NH_2), an acid group (COOH), a hydrogen atom (H), and an "R" group (**figure 10.1**). It is the R group that distinguishes the 20 types of amino acids the human body uses. The amino acids link into chains when the amino group of one amino acid attaches to the acid group of another, via a peptide bond. Figure 10.19 shows how amino acids bond. As a protein molecule forms, the many linked amino acids fold into a three-dimensional structure that is essential to the protein's function. A protein consists of one or more long chains of amino acids called **polypeptides**. Shorter chains of amino acids are called peptides. A protein's three-dimensional shape is called its **conformation**, discussed further in section 10.4.

It is impossible to imagine a living human body that cannot manufacture protein. Without proteins, blood couldn't clot, muscles couldn't contract, antibodies couldn't fight infection, and we'd have no hair, skin, or connective tissue, nor would any biochemical reactions that require enzymes proceed fast enough to keep us alive.

The inability to make just 1 of the 20,325 or so types of proteins that the human exome encodes is devastating. A child with giant axonal neuropathy (see Clinical Connection 2.2) can no longer walk because her motor neurons cannot manufacture gigaxonin protein. People who have cystic fibrosis (see figures 1.5 and 1.6 and Clinical Connection 4.1) have respiratory and digestive problems because ion channel proteins cannot fold properly.

The production of proteins and then their actions and interactions translate genotype into phenotype. The process begins with copying a gene's information into a more accessible form.

Key Concepts Questions 10.1

*a. Name the parts of an amino acid.

b. Explain how a protein's shape forms.

*c. What is the connection between a gene and a protein?

d. List some functions of proteins in the human body.

10.2 Transcription Copies the Information in DNA

A cell uses two processes to manufacture proteins using genetic instructions. Transcription first synthesizes an RNA molecule that is complementary to one strand of the DNA double helix for a particular gene. The RNA copy is taken out of the nucleus and into the cytoplasm. There, the process of translation uses the information in the RNA to manufacture a protein by aligning and joining the specified amino acids. Finally, the protein folds into a specific three-dimensional form necessary for its function.

Accessing the genome to manufacture protein is a huge, constant task. Cells replicate their DNA only during S (synthesis) phase of the cell cycle, but transcription and translation occur continuously, except during M phase. The two processes supply the proteins essential to be alive, as well as those that give a cell its specialized characteristics.

Shortly after Watson and Crick published their structure of DNA in 1953, they described the relationship between nucleic acids and proteins as a directional flow of information called the "central dogma" (**figure 10.2**). As Francis Crick explained in 1957, *"The specificity of a piece of nucleic acid is expressed solely by the sequence of its bases, and this sequence is a code for the amino acid sequence of a particular protein."* This statement inspired more than a decade of intense research to discover exactly how cells make proteins, and to decipher the code that takes gene to protein. The next two subsections introduce the participants in transcription and the following two subsections describe the steps in the complex process.

RNA Structure and Types

RNA bridges gene and protein, as **figure 10.3** depicts. The bases of an RNA sequence are complementary to those of one strand of the double helix, which is called the **template strand**. An enzyme, **RNA polymerase,** builds an RNA molecule. The other, nontemplate strand of the DNA double helix is called the **coding strand**.

RNA and DNA have similarities and differences (**figure 10.4** and **table 10.1**). Both are nucleic acids, consisting of sequences of nitrogen-containing bases joined by sugar-phosphate backbones. However, their structures are different.

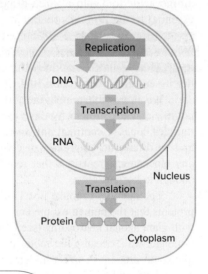

Figure 10.2 DNA to RNA to protein. Some of the information stored in DNA is copied to RNA (transcription), some of which is used to assemble amino acids into proteins (translation). DNA replication perpetuates genetic information. This figure repeats throughout the chapter, with the part under discussion highlighted.

Table 10.1	How DNA and RNA Differ
DNA	**RNA**
1. Usually double-stranded	1. Usually single-stranded
2. Thymine as a base	2. Uracil as a base
3. Deoxyribose as the sugar	3. Ribose as the sugar
4. Maintains protein-encoding information	4. Carries protein-encoding information and controls how information is used
5. Cannot function as an enzyme	5. Can function as an enzyme
6. Persists	6. Short-lived

Figure 10.3 The relationship among RNA, the DNA template strand, and the DNA coding strand. The RNA sequence is complementary to the DNA template strand and is the same sequence as the DNA coding strand, with uracil (U) in place of thymine (T).

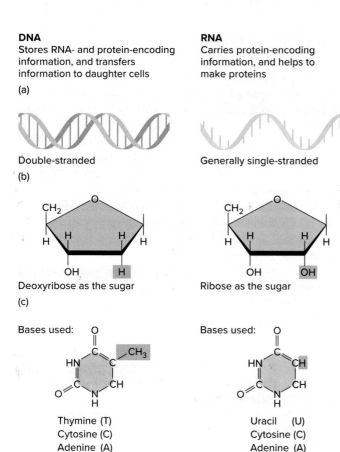

Figure 10.4 DNA and RNA differences. (a) DNA and RNA have different functions. (b) DNA is double-stranded; RNA is usually single-stranded. (c) DNA nucleotides include deoxyribose; RNA nucleotides have ribose. (d) Finally, DNA nucleotides include the pyrimidine thymine, whereas RNA nucleotides have uracil.

RNA is usually single-stranded, whereas DNA is double-stranded. Also, RNA has the pyrimidine base uracil where DNA has thymine. RNA (*ribo*nucleic acid) nucleotides include the sugar ribose. DNA (*deoxyribo*nucleic acid) nucleotides include the sugar deoxyribose.

DNA and RNA differ in function, too. DNA stores genetic information, whereas RNA controls how that information is used. The presence of the OH group at the 2′ position of ribose makes RNA much less stable than DNA. This distinction is critical in its function as a short-lived carrier of genetic information.

As RNA is synthesized along DNA, it folds into its conformation, which arises from complementary base pairing within the same RNA molecule. For example, a sequence of AAUUCC might hydrogen bond to a sequence of UUAAGG—its complement—elsewhere in the same molecule, a little like touching elbows to knees. Conformation is very important for RNA's functioning. The three major types of RNA, which have distinctive conformations, are messenger RNA, ribosomal RNA, and transfer RNA (**table 10.2**). Other classes of RNA control which genes are expressed (transcribed and translated) under specific circumstances (see table 11.2 for their descriptions).

Messenger RNA (mRNA) carries the information for a particular protein. Each set of three consecutive mRNA bases forms a genetic code word, or **codon,** that specifies a certain amino acid. Because genes vary in length, so do mature mRNA molecules. Most mRNAs are 500 to 4,500 bases long.

Differentiated cells can carry out specialized functions because they express certain subsets of genes—that is, they produce certain mRNA molecules, which are also called transcripts. The information in the transcripts is then used to manufacture the encoded proteins. A muscle cell, for example, has many mRNAs that correspond to the contractile proteins actin and myosin, whereas a skin cell contains many mRNAs that specify scaly keratin proteins.

To use the information in an mRNA sequence, a cell requires the two other major classes of RNA. **Ribosomal RNA (rRNA)** molecules range from 100 to nearly 3,000 nucleotides long. Ribosomal RNAs associate with certain proteins to form a ribosome. Recall from chapter 2 that a ribosome is an organelle made up of many different protein and RNA subunits. Overall, a ribosome functions like a machine to assemble and link amino acids to form proteins (**figure 10.5**).

Figure 10.5 The ribosome. A ribosome from a eukaryotic cell has two subunits; together, they consist of 79 proteins and 4 rRNA molecules.

Table 10.2	Major Types of RNA	
Type of RNA	**Size (number of nucleotides)**	**Function**
Messenger RNA (mRNA)	500 to 4,500+	Encodes amino acid sequence
Ribosomal RNA (rRNA)	100 to 3,000	Associates with proteins to form ribosomes, which structurally support and catalyze protein synthesis
Transfer RNA (tRNA)	75 to 80	Transports specific amino acids to the ribosome for protein synthesis

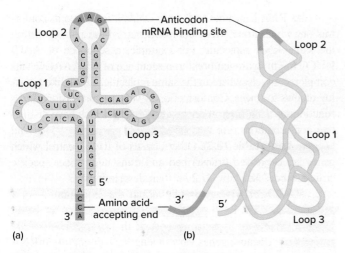

Figure 10.6 Transfer RNA. (a) Certain nucleotide bases within a tRNA hydrogen bond with each other, giving the molecule a "cloverleaf" conformation if viewed in two dimensions. The darker bases at the top form the anticodon, the sequence that binds a complementary mRNA codon. Each tRNA ends with CCA, where a particular amino acid covalently bonds. (b) A three-dimensional view of a tRNA shows the loops that interact with the ribosome.

A ribosome has two subunits that are separate in the cytoplasm but join at the site of initiation of protein synthesis. The larger ribosomal subunit has three types of rRNA molecules, and the small subunit has one. Ribosomal RNA, however, is more than a structural support. Certain rRNAs catalyze the formation of the peptide bonds between amino acids. Such an RNA with enzymatic function is called a ribozyme. Other rRNAs help to align the ribosome and mRNA.

The third major type of RNA molecule, transfer RNA (tRNA), binds an mRNA codon at one end and a specific amino acid at the other. A tRNA molecule is only 75 to 80 nucleotides long. Some of its bases form weak chemical bonds with each other, folding the tRNA into loops in a characteristic cloverleaf shape (**figure 10.6**). One loop of the tRNA has three bases in a row that form the **anticodon,** which is complementary to an mRNA codon. The end of the tRNA opposite the anticodon strongly bonds to a specific type of amino acid, and a tRNA with a particular anticodon sequence always carries the same type of amino acid. For example, a tRNA with the anticodon sequence GAA always picks up the amino acid phenylalanine. Enzymes attach amino acids to tRNAs that bear the appropriate anticodons (**figure 10.7**).

Figure 10.7 A tRNA with a specific anticodon sequence always binds the same type of amino acid.

Transcription Factors

If all of the genes in the human genome were being transcribed and translated at the same time, chaos would result. It would be like trying to open all of the files and programs on a computer at once. Instead, accessing genetic information is selective and efficient.

Different cell types express different subsets of genes. To manage this, proteins called transcription factors come together and interact, forming an apparatus that binds DNA at certain sequences and initiates transcription at specific sites on chromosomes. The transcription factors respond to signals from outside the cell, such as hormones and growth factors, and form a pocket for RNA polymerase to bind and begin building RNA molecules at specific genes.

Transcription factors include regions called binding domains that guide them to the genes they control. The DNA binding domains have colorful names, such as *helix-turn-helix, zinc fingers,* and *leucine zippers,* that describe their distinctive shapes.

Types of transcription factors are few, but they work in combinations, exerting exquisite control over gene expression. Yet overall, the effect of transcription factors is to link the genome to the environment. For example, lack of oxygen, such as from choking or smoking, sends signals that activate transcription factors to turn on dozens of genes that enable cells to handle the stress of low-oxygen conditions.

Mutations in transcription factor genes can be devastating. Rett syndrome (see the chapter 2 opener) and the homeotic mutations described in Clinical Connection 3.1 result from mutations in transcription factor genes. Overexpressed transcription factors cause many types of cancer. Transcription factors are themselves controlled by each other and by other classes of molecules.

Steps of Transcription

Transcription is described in three steps: initiation, elongation, and termination. The process is called *transcription* because it copies the DNA information into RNA, but keeps the information in the genetic language of DNA nucleotide bases.

Transcription factors and RNA polymerase use clues in the DNA sequence to "know" where to bind to DNA to begin transcribing a specific gene. In transcription initiation, transcription factors and RNA polymerase bind to a **promoter,** which is a DNA sequence that signals the start of the gene, like a capital letter at the start of a sentence.

Figure 10.8 illustrates transcription factor binding, which sets up a site to receive RNA polymerase. The first transcription factor to bind, called a TATA binding protein, is chemically attracted to a DNA sequence called a TATA box—the base sequence TATA surrounded by long stretches of G and C. Once the first transcription factor binds, others bind in groups. Finally, RNA polymerase joins the complex, binding just in front of the start of the gene sequence. The assembly of these components is transcription initiation.

In the next stage, transcription elongation, enzymes unwind the DNA double helix locally, and free RNA

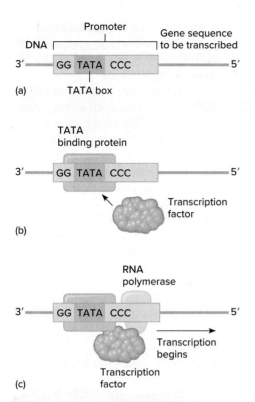

Figure 10.8 Setting the stage for transcription to begin. (a) Proteins that initiate transcription recognize specific sequences in the promoter region of a gene. (b) A binding protein recognizes the TATA region and binds to the DNA. This allows other transcription factors to bind. (c) The bound transcription factors form a pocket that allows RNA polymerase to bind and begin making RNA.

from different strands of the double helix. The coding strand of the DNA is so-called because its sequence is identical to that of the RNA, except with thymine (T) in place of uracil (U).

Several RNAs may be transcribed from the same DNA template strand simultaneously (**figure 10.10**). Because mRNA is short-lived, with about half of it degraded every 10 minutes, a cell must constantly transcribe certain genes to maintain supplies of essential proteins.

To determine the sequence of RNA bases transcribed from a gene, write the RNA bases that are complementary to the template DNA strand, using uracil opposite adenine. For example, a DNA template strand that has the sequence

$$3´\ C\ C\ T\ A\ G\ C\ T\ A\ C\ 5´$$

is transcribed into RNA with the sequence

$$5´\ G\ G\ A\ U\ C\ G\ A\ U\ G\ 3´$$

and the coding DNA sequence is

$$5´\ G\ G\ A\ T\ C\ G\ A\ T\ G\ 3´$$

nucleotides bond with exposed complementary bases on the DNA template strand (see figure 10.3). RNA polymerase adds the RNA nucleotides in the sequence the DNA specifies, moving along the DNA strand in a 3′ to 5′ direction, synthesizing the RNA molecule in a 5′ to 3′ direction. A terminator sequence in the DNA indicates where the gene's RNA-encoding region ends, like the period at the end of a sentence. When this spot is reached, the third stage, transcription termination, occurs (**figure 10.9**). A typical rate of transcription in humans is 20 bases per second.

RNA is transcribed using only the gene sequence on the template strand. However, different genes on the same chromosome may be transcribed

Figure 10.9 Transcription of RNA from DNA. Transcription occurs in three stages: initiation, elongation, and termination. Initiation is the control point that determines which genes are transcribed. RNA nucleotides are added during elongation. A terminator sequence signals the end of transcription.

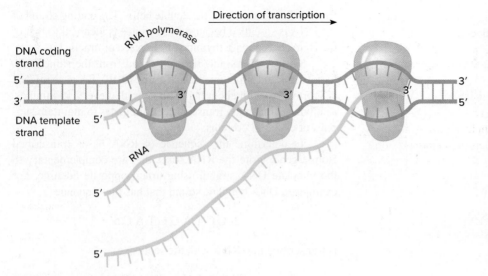

Figure 10.10 Many identical copies of RNA are transcribed simultaneously. Usually 100 or more DNA bases lie between RNA polymerases.

RNA Processing

In bacteria, RNA is translated into protein as soon as it is transcribed from DNA because a nucleus does not physically separate the two processes. In eukaryotic cells like ours, mRNA must first exit the nucleus and enter the cytoplasm, where ribosomes are located. Messenger RNA is altered in several steps before it is translated in these more complex cells.

First, after mRNA is transcribed, a short sequence of modified nucleotides, called a cap, is added to the 5′ end of the molecule. The cap consists of a backwardly inserted guanine (G), which attracts an enzyme that adds methyl groups (CH_3) to the G and one or two adjacent nucleotides. This methylated cap is a recognition site for protein synthesis. At the 3′ end, a special polymerase adds about 200 adenines, forming a "poly A tail." The poly A tail is necessary for protein synthesis to begin, and may also stabilize the mRNA so that it stays intact longer.

Further changes occur to the capped, poly A tailed mRNA before it is translated into protein. Parts of the mRNA called **introns** (short for "intervening sequences") are removed and the ends of the remaining molecule spliced together before translation into protein. The parts of mRNA that remain and are translated into amino acid sequences (protein) are called **exons** (figure 10.11).

Once introns are spliced out, enzymes check, or proofread, the remaining mRNA. Messenger RNAs that are too short or too long may be held in the nucleus. Proofreading also monitors tRNAs, ensuring that they assume the correct cloverleaf shape.

Prior to intron removal, the mRNA is called pre-mRNA. Introns control their own removal. They associate with certain proteins to form small nuclear ribonucleoproteins (snRNPs), or "snurps." Four snurps form a structure called a spliceosome that cuts introns out and attaches exons to each other to form the mature mRNA that exits the nucleus.

Introns range in size from 65 to 10,000 or more bases; the average intron is 3,365 bases. The average exon, in contrast, is only 145 bases long. The number, size, and organization of introns vary from gene to gene. The coding portion of the average human gene is 1,340 bases, whereas the average total size of a gene is 27,000 bases. The dystrophin gene is 2,500,000 bases, but its corresponding mRNA sequence is only 14,000 bases! The gene contains 80 introns.

The discovery of introns in the 1970s surprised geneticists, who had thought genes were like sentences in which all of the information has meaning. At first, some geneticists called introns *junk DNA*. Then researchers discovered functions of many sequences thought to be "junk." Some introns encode RNAs that control gene expression, whereas others are actually exons on the complementary strand of DNA. Introns may also be vestiges

Figure 10.11 Messenger RNA processing—the maturing of the message. Several steps process pre-mRNA into mature mRNA. First, a large region of DNA containing the gene is transcribed. Then a modified nucleotide cap and poly A tail are added and introns are spliced out. Finally, the intact, mature mRNA is sent out of the nucleus.

of ancient genes that have lost their original function, or are remnants of the DNA of viruses that once integrated into chromosomes.

The intron/exon organization of genes maximizes genetic information. Different combinations of exons of about 95 percent of human genes encode different versions of a protein, termed **isoforms**. The mechanism of combining exons of a gene in different ways is called **alternative splicing**. In this way, cell types can use versions of the same protein in slightly different ways in different tissues. We return to alternative splicing in chapter 11.

Key Concepts Questions 10.2

*a. How does RNA differ from DNA in structure?
*b. How does RNA differ from DNA in function?
c. Explain how messenger RNA transmits instructions to build proteins.
*d. What is the function of transfer RNAs in synthesizing proteins?
e. List the steps of transcription.
*f. Define *alternative splicing*.

10.3 Translation of a Protein

Translation assembles a protein using the information in the mRNA sequence. Particular mRNA codons correspond to particular amino acids (**figure 10.12**). This correspondence between the chemical languages of mRNA and protein is the **genetic code** (**table 10.3**).

In the 1960s, researchers described and deciphered the genetic code. They used logic and clever experiments on simple

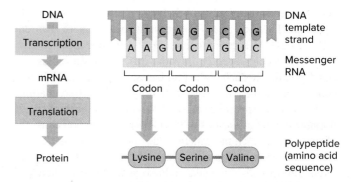

Figure 10.12 From DNA to RNA to protein. Messenger RNA is transcribed from a locally unwound portion of DNA. In translation, transfer RNA matches mRNA codons with amino acids.

Table 10.3 The Genetic Code*

First Letter		Second Letter				Third Letter
		U	**C**	**A**	**G**	
U	UUU	Phenylalanine (Phe)	UCU — Serine (Ser)	UAU — Tyrosine (Tyr)	UGU — Cysteine (Cys)	U
	UUC	Phenylalanine (Phe)	UCC — Serine (Ser)	UAC — Tyrosine (Tyr)	UGC — Cysteine (Cys)	C
	UUA	Leucine (Leu)	UCA — Serine (Ser)	UAA "stop"	UGA "stop"	A
	UUG	Leucine (Leu)	UCG — Serine (Ser)	UAG "stop"	UGG Tryptophan (Trp)	G
C	CUU	Leucine (Leu)	CCU — Proline (Pro)	CAU — Histidine (His)	CGU — Arginine (Arg)	U
	CUC	Leucine (Leu)	CCC — Proline (Pro)	CAC — Histidine (His)	CGC — Arginine (Arg)	C
	CUA	Leucine (Leu)	CCA — Proline (Pro)	CAA — Glutamine (Gln)	CGA — Arginine (Arg)	A
	CUG	Leucine (Leu)	CCG — Proline (Pro)	CAG — Glutamine (Gln)	CGG — Arginine (Arg)	G
A	AUU	Isoleucine (Ile)	ACU — Threonine (Thr)	AAU — Asparagine (Asn)	AGU — Serine (Ser)	U
	AUC	Isoleucine (Ile)	ACC — Threonine (Thr)	AAC — Asparagine (Asn)	AGC — Serine (Ser)	C
	AUA	Isoleucine (Ile)	ACA — Threonine (Thr)	AAA — Lysine (Lys)	AGA — Arginine (Arg)	A
	AUG	Methionine (Met) and "start"	ACG — Threonine (Thr)	AAG — Lysine (Lys)	AGG — Arginine (Arg)	G
G	GUU	Valine (Val)	GCU — Alanine (Ala)	GAU — Aspartic acid (Asp)	GGU — Glycine (Gly)	U
	GUC	Valine (Val)	GCC — Alanine (Ala)	GAC — Aspartic acid (Asp)	GGC — Glycine (Gly)	C
	GUA	Valine (Val)	GCA — Alanine (Ala)	GAA — Glutamic acid (Glu)	GGA — Glycine (Gly)	A
	GUG	Valine (Val)	GCG — Alanine (Ala)	GAG — Glutamic acid (Glu)	GGG — Glycine (Gly)	G

*The genetic code consists of mRNA triplets and the amino acids that they specify.

genetic systems such as viruses, and synthetic DNA molecules, to discover which mRNA codons correspond to which amino acids.

The Genetic Code

Before researchers could match mRNA codons to the amino acids they encode, they had to establish certain requirements for such a code.

1. The Code Is Triplet.

The number of different protein building blocks (20 amino acids) exceeds the number of different mRNA building blocks (4 bases). Therefore, each codon must include more than one mRNA base. If a codon consisted of only one mRNA base, then codons could specify only four different amino acids, one corresponding to each of the four bases: A, C, G, and U. If each codon consisted of two bases, then only 16 (4^2) different amino acids could be specified, one corresponding to each of the 16 possible combinations of two RNA bases. If a codon consisted of three bases, then the genetic code could specify as many as 64 (4^3) different amino acids, sufficient to encode the 20 different amino acids that make up biological proteins. Therefore, the minimum number of bases in a codon is three.

Francis Crick and his coworkers showed that the code is triplet by adding or removing one, two, or three bases to or from a viral gene with a well-known sequence and protein product. Altering the DNA sequence by one or two bases produced a different amino acid sequence. This happened because the change disrupted the **reading frame,** which is the sequence of amino acids encoded from a certain starting point in a DNA sequence. However, adding or deleting three contiguous bases added or deleted only one amino acid in the protein without disrupting the reading frame. The rest of the amino acid sequence was retained. The code, the researchers deduced, is triplet (**figure 10.13**). A mutation that alters the reading frame is called a **frameshift mutation.**

Further experiments confirmed the triplet nature of the genetic code. Adding a base at one point in the gene and deleting a base at another point disrupted the reading frame only between these sites. The result was a protein with a stretch of the wrong amino acids, like a sentence with a few misspelled words in the middle after one letter is added or removed.

2. The Code Does Not Overlap.

Consider a hypothetical mRNA sequence:

AUGCCCAAG

If the genetic code is triplet and a DNA sequence is "read" in a nonoverlapping manner, then this sequence has only three codons and specifies three amino acids:

AUGCCCAAG

AUG (methionine)

CCC (proline)

AAG (lysine)

Figure 10.13 Three at a time. Adding or deleting one or two nucleotides in a DNA sequence results in a frameshift that disrupts the encoded amino acid sequence. Adding or deleting three bases does not disrupt the reading frame because the code is triplet. This is a simplified view of the Crick experiment.

If the DNA sequence is overlapping, however, the sequence specifies seven codons:

AUGCCCAAG

AUG (methionine)

UGC (cysteine)

GCC (alanine)

CCC (proline)

CCA (proline)

CAA (glutamine)

AAG (lysine)

In an overlapping DNA sequence, certain amino acids would always follow certain others, constraining protein structure. For example, AUG would always be followed by an amino acid whose codon begins with UG. This does not happen in nature. Therefore, the protein-encoding DNA sequence is not overlapping.

Even though the genetic code is nonoverlapping, any DNA or RNA sequence can be read in three different reading frames, depending upon the "start" base. **Figure 10.14** depicts the three reading frames for the sequence just discussed, slightly extended. It encodes three different trios of amino acids.

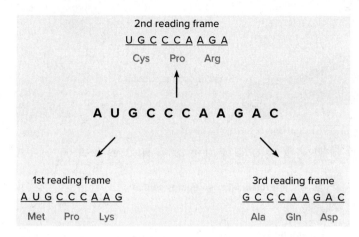

Figure 10.14 Reading frames—where the sequence begins. A sequence of DNA has three reading frames.

3. The Code Includes Controls.

The genetic code includes directions for starting and stopping translation. The codon AUG signals "start," and the codons UGA, UAA, and UAG signify "stop." A sequence of DNA that does not include a stop codon is called an **open reading frame,** and is a sign of a protein-encoding gene. If a DNA sequence was random, assuming equal numbers of each base, a stop codon would arise about every 21 codons.

Another form of "punctuation" in the genetic code is a short sequence of bases at the start of each mRNA that enables the mRNA to form hydrogen bonds with rRNA in a ribosome. It is called a leader sequence.

4. The Code Is the Same in All Species.

All species use the same mRNA codons to specify the same amino acids, and therefore the same genetic code. The code is the correspondence and not the sequence. References to the "human genetic code" usually mean a human genome sequence.

The simplest explanation for the "universality" of the genetic code, the fact that it is the same in all species, is that all life evolved from a common ancestor. No other mechanism as efficient at directing cellular activities has emerged and persisted. The only known exceptions to the universality of the genetic code are a few codons in mitochondria and in certain single-celled eukaryotes (ciliated protozoa).

Once researchers knew that a codon consists of three bases and the code is nonoverlapping, the next step was to match codons to amino acids. **A Glimpse of History** explains the clever experiments that revealed the genetic code. Sixty of the possible 64 codons specify amino acids, three codons indicate "stop," and one encodes both the amino acid methionine and "start." Some amino acids are specified by more than one codon. For example, UUU and UUC encode phenylalanine.

Different codons that specify the same amino acid are termed **synonymous codons,** just as synonyms are words with the same meaning. The genetic code is termed *degenerate* because most amino acids are not uniquely specified. Several synonymous codons differ by the base in the third position. The corresponding base of a tRNA's anticodon is called the *wobble* position because it can bind to more than one type of base in synonymous codons. The degeneracy of the genetic code protects against mutation, because changes in the DNA that substitute a synonymous codon do not alter the protein's amino acid sequence. Codons that encode different amino acids are called **nonsynonymous codons**.

Sequencing human genomes added to the genetic code experiments of the 1960s by identifying the DNA sequences that are transcribed into tRNAs. Sixty-one different tRNAs could exist, one for each codon that specifies an amino acid (the 64 triplets minus 3 stop codons). However, only 49 different genes encode tRNAs. This is because the same type of tRNA can detect synonymous codons that differ only in whether the wobble (third) position has U or C, such as the codons UUU and UUC that specify phenylalanine.

Building a Protein

Protein synthesis requires several participants: mRNA, tRNA molecules carrying amino acids, ribosomes, energy-storing molecules such as adenosine triphosphate (ATP) and guanosine triphosphate (GTP), and protein factors. These pieces meet during translation initiation (**figure 10.15**). Chemical bonds hold the components together.

A GLIMPSE OF HISTORY

In 1961, Marshall Nirenberg and Heinrich Matthaei at the National Institutes of Health deciphered the genetic code. They synthesized simple mRNA molecules and added them to test tubes containing all of the molecules needed for translating a genetic message into a string of amino acids. The materials came from bacterial cells.

Methodically, Nirenberg and Matthaei deduced which amino acid each synthetic RNA specified. For example, the mRNA sequence UUUUUUUUU . . . was translated into a peptide consisting entirely of the amino acid phenylalanine. Therefore, UUU specifies phenylalanine. To reveal more complex codon–amino acid pairs, the researchers made chains of alternating bases. AUAUAU . . . introduced codons AUA and UAU, and specified the amino acid sequence of alternating isoleucines and tyrosines. But did AUA encode isoleucine and UAU tyrosine, or the reverse? Another experiment provided the answer. The mRNA UUUAUAUUUAUA encoded alternating phenylalanines and isoleucines. Because the first experiment had shown that UUU encodes phenylalanine, then AUA must encode isoleucine and therefore UAU must encode tyrosine. In this way, the researchers built up the information in table 10.3.

By the end of the 1960s, the mix of experiments and logic had revealed the entire genetic code. The code-breakers, who were mostly physical scientists, formed the "RNA tie club." They inducted a member whenever he added a piece to the puzzle of the genetic code, anointing him (there were no female members) with a tie and pin emblazoned with the structure of the specified amino acid.

Figure 10.15 Translation begins as the initiation complex forms. Initiation of translation brings together a small ribosomal subunit, mRNA, and an initiator tRNA, and aligns them in the proper orientation to begin translation.

First, the mRNA leader sequence forms hydrogen bonds with a short sequence of rRNA in a small ribosomal subunit. The first mRNA codon to specify an amino acid is always AUG, which attracts an initiator tRNA that carries the amino acid methionine (abbreviated Met). This methionine signifies the start of a polypeptide. The small ribosomal subunit, the mRNA bonded to it, and the initiator tRNA with its attached methionine, form the initiation complex at the appropriate AUG codon of the mRNA.

To begin elongation, a large ribosomal subunit bonds to the initiation complex. The codon adjacent to the initiation codon (AUG), which is GGA in **figure 10.16,** then bonds to its complementary anticodon, which is part of a free tRNA that carries the amino acid glycine. The two amino acids (Met and Gly in the example), still attached to their tRNAs, align.

The part of the ribosome that holds the mRNA and tRNAs together can be described as having two sites. The positions of the sites on the ribosome remain the same with respect to each other as translation proceeds, but they cover different parts of the mRNA as the ribosome moves. The P ("peptide") site holds the growing amino acid chain, and the A ("acceptor") site next to it holds the next amino acid to be added to the chain. In figure 10.16a, when the forming protein consists of only the first two amino acids, Met occupies the P site and Gly the A site.

The amino acids link by peptide bonds, with the help of rRNA that functions as a ribozyme (an RNA with enzymatic activity). Then the first tRNA is released to pick up another amino acid of the same type and be used again, a little like a shopping cart ferrying bags of the same type of food over and over. Special enzymes ensure that tRNAs always bind the correct amino acids. The ribosome moves down the mRNA by one codon, so that the former A site is now the P site (figure 10.16b). The attached mRNA is now bound to a single tRNA, with two amino acids extending from it at the P site. This is the start of a polypeptide.

Next, the ribosome moves down the mRNA by another codon, and a third tRNA brings in its amino acid (Cys in figure 10.16b). This third amino acid aligns with the other two and forms a peptide bond to the second amino acid in the growing chain. The tRNA attached to the second amino acid is released and recycled. The polypeptide builds one amino acid at a time, each piece brought in by a tRNA whose anticodon corresponds to a consecutive mRNA codon as the ribosome moves down the mRNA (figure 10.16c).

Elongation halts when the A site of the ribosome has a "stop" codon (UGA, UAG, or UAA), because no tRNA molecules correspond to it. A protein release factor starts to free the polypeptide. The last tRNA leaves the ribosome, the ribosomal subunits separate and are recycled, and the new polypeptide is released (**figure 10.17**).

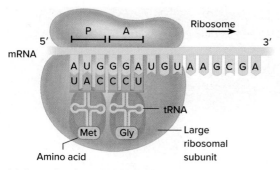

(a) Second amino acid joins initiation complex.

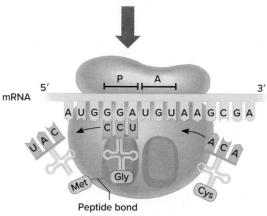

(b) First peptide bond forms as new amino acid arrives.

(c) Amino acid chain extends.

Figure 10.16 **Building a polypeptide.** **(a)** A large ribosomal subunit binds to the initiation complex, and a tRNA bearing a second amino acid (glycine, in this example) forms hydrogen bonds between its anticodon and the mRNA's second codon at the A site. The first amino acid, methionine, occupies the P site. **(b)** The methionine brought in by the first tRNA forms a peptide bond with the amino acid brought in by the second tRNA, and a third tRNA arrives, in this example carrying the amino acid cysteine, at the temporarily vacant A site. **(c)** A fourth and then fifth amino acid link to the growing polypeptide chain. The process continues until reaching a stop codon.

Protein synthesis is efficient. A cell can produce large numbers of a particular protein molecule from just one or two copies of a gene. A plasma cell in the immune system, for example, manufactures 2,000 identical antibody molecules per second. To mass produce proteins at this rate, RNA, ribosomes, enzymes,

(a) Ribosome reaches stop codon.

(b) Components disassemble.

Figure 10.17 **Terminating translation.** **(a)** A protein release factor binds to the stop codon, releasing the completed polypeptide from the tRNA and **(b)** freeing all of the components of the translation complex.

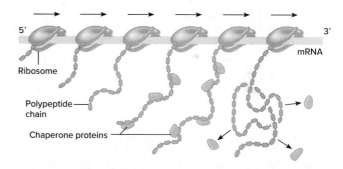

Figure 10.18 **Making many copies of a protein.** Several ribosomes can simultaneously translate a protein from a single mRNA. These ribosomes hold different-sized polypeptides—the closer to the end of a gene, the longer the polypeptide. Proteins called chaperones help fold the polypeptide.

and other proteins are continually recycled. In addition, transcription produces many copies of a particular mRNA, and each mRNA may bind dozens of ribosomes, as **figure 10.18** shows. Many protein molecules are made from the same mRNA.

Some proteins undergo further alterations, called post-translational modifications, before they can function. For example, insulin is initially translated as the 80-amino-acid-long polypeptide proinsulin, but enzymes cut it to 51 amino acids. Some proteins must have sugars attached or sometimes polypeptides must aggregate to become active. **Clinical Connection 10.1** describes a genetic disease that results from a cascade of enzyme deficiencies.

Clinical Connection 10.1

A Missing Enzyme Knocks Out Nine Others: Multiple Sulfatase Deficiency

Enzymes are an important class of proteins because they enable specific biochemical reactions to proceed quickly enough to keep a body functioning. Clinical Connection 2.1 introduced several "inborn errors of metabolism" that result from a missing or malfunctioning enzyme. Multiple sulfatase deficiency is another inborn error that is particularly severe because the impaired enzyme controls nine other enzymes. Here is one child's story.

Willow Cannan was born on August 21, 2013. At first she seemed healthy, except for difficulty nursing. "Nothing in itself was significant, but a combination of things over time got worse. She crawled a little late. She walked late, at 18 months. She had frequent ear infections and dry skin. She never talked, no words, not even *dada*. She just made a few sounds," recalled her mother, Amber Olsen. By her second birthday, Willow still couldn't speak, despite speech therapy and adenoid surgery.

"She was walking and running, but she was not progressing like a normal toddler. The pediatrician finally agreed to an endocrinology consult because Willow hadn't grown," Amber said. When hormone tests came back normal, a pediatric neurologist was consulted. "He said there was nothing significant, but Willow had started having balance issues and trouble walking. When I said that I had started to see skills undoing, he said OK, we'd better do a genetic test panel," Amber recalled.

Genetic testing identified small deletions (missing DNA) in both copies of a gene, *SUMF1*, that normally encodes a protein called sulfatase modifying factor 1. The little girl is 1 of about 50 people in the world with multiple sulfatase deficiency (MSD). The absence of the enzyme results in her missing nine other enzymes that break down mucopolysaccharides—components of mucus that are mostly sugars with some proteins (**figure 10A**). In an effect like falling dominoes, each missing enzyme triggered deficiencies in yet other enzymes. The sticky mucopolysaccharides were building up in her cells' lysosomes—MSD is a severe lysosomal storage disease (see section 2.2). (The exact number of affected enzyme types isn't known, because some of them are in tiny bubbles near lysosomes, rather than in the lysosomes.)

Doctors advised Willow's parents to take her home and continue general supportive care, such as physical and speech therapy, but that she would not survive. Willow was nearing her third birthday.

The buildup of sticky material in Willow's lysosomes would cloud her corneas, impairing her vision; clog her ears and cause hearing loss; narrow her windpipe, making swallowing difficult; and enlarge her liver and spleen. She would continue to lose developmental skills and would need a feeding tube. Willow's coarse facial features would resemble those of other children with inborn errors that cause mucopolysaccharide buildup.

Like many other parents of children with rare genetic diseases, Amber found and networked with other families, attended scientific meetings, started a not-for-profit organization to raise funds, and began collaborating with researchers who are now working on a gene therapy (see Clinical Connection 2.1 and chapter 21) for MSD.

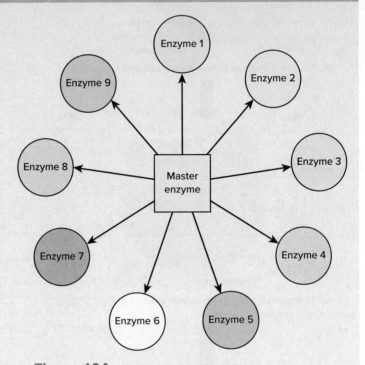

Figure 10A Multiple sulfatase deficiency.
In multiple sulfatase deficiency, mutations in both copies of the gene that encodes a master enzyme halts synthesis of nine others, causing several symptoms.

Questions for Discussion

1. Why is MSD more severe than some other inborn errors of metabolism?
2. Explain why Willow's cells cannot make sulfatase modifying factor 1.
3. A family friend learns that MSD affects the levels of nine proteins and suggests that Amber mix protein powder into the child's drinks. Explain why this intervention will not correct Willow's enzyme deficiencies.
4. Name the organelle that is the cellular site of the disease.

> **Key Concepts Questions 10.3**
>
> *a. List the general characteristics of the genetic code.
> *b. List the steps of translation.
> c. Explain how translation is efficient.

10.4 Processing a Protein

Translation takes place on free ribosomes in the cytoplasm and on ribosomes associated with the endoplasmic reticulum. Proteins fold into one or more conformations. This folding arises from chemistry: attraction and repulsion between atoms of the proteins as well as interactions of proteins with chemicals in the immediate environment. For example, thousands of water molecules surround a growing chain of amino acids. Because some amino acids are attracted to water and some are repelled by it, the protein contorts in water. Sulfur atoms also affect protein conformation by bridging the two types of amino acids that contain them (cysteine and methionine).

The conformation of a protein is described at several levels (**figure 10.19**). The amino acid sequence of a polypeptide chain is its **primary (1°) structure**. Chemical attractions between amino acids that are close together in the 1° structure fold the polypeptide chain into its **secondary (2°) structure,** which may form loops, coils, barrels, helices, or sheets. Two common secondary structures are an alpha helix and a beta-pleated sheet. Secondary structures wind into larger **tertiary (3°) structures**

H_2N—Ala—Thr—Cys—Tyr—Glu—Gly—COOH

(a) **Primary structure**—the sequence of amino acids in a polypeptide chain

(b) **Secondary structure**—loops, coils, sheets, or other shapes formed by hydrogen bonds between neighboring carboxyl and amino groups

(c) **Tertiary structure**—three-dimensional forms shaped by bonds between R groups, interactions between R groups and water

(d) **Quaternary structure**—protein complexes formed by bonds between separate polypeptides

Figure 10.19 Four levels of protein structure. (a) The amino acid sequence of a polypeptide forms the primary structure. Each amino acid has an amino end (H_2N) and a carboxyl end (COOH), and an "R" group distinguishes each of the 20 types of amino acids. (b) Hydrogen bonds between non–R groups create secondary structures such as helices and sheets. (c) The tertiary structure arises when R groups interact with other R groups or with water, folding the polypeptide into a unique shape. (d) A protein consisting of more than one polypeptide has a quaternary structure.

as more widely separated amino acids attract or repel in response to water molecules. Finally, proteins consisting of more than one polypeptide form a **quaternary (4°) structure**. Hemoglobin, the blood protein that carries oxygen, has four polypeptide chains (see figure 11.1). The liver protein ferritin has 20 identical polypeptides of 200 amino acids each. In contrast, the muscle protein myoglobin is a single polypeptide chain.

Mutations may alter the primary structure of a protein if the genetic change is nonsynonymous, which means that it changes the amino acid. In contrast, more than one tertiary or quaternary structure may be possible for a protein if an amino acid chain can fold in different ways.

Protein Folding and Misfolding

Proteins begin to fold within a minute after the amino acid chain winds away from the ribosome. A small protein might contort into its final, functional form in one step, taking microseconds. Larger proteins may initially fold into a series of short-lived intermediates.

Proteins start to move toward their destinations as they are being synthesized. In some proteins, part of the start of the amino acid chain forms a tag that helps direct the protein in the cell. The first few amino acids in a protein that will be secreted or lodge in a membrane form a "signal sequence" that leads it and the ribosome to which it binds into a pore in the ER membrane. Once in the ER, the protein enters the secretory network (see figure 2.4). Proteins destined for the mitochondria bear a different signal sequence.

Signal sequences are not found on proteins synthesized on free ribosomes in the cytoplasm. These proteins may function right where they are made; examples include the protein tubules and filaments of the cytoskeleton (see figure 2.8) and enzymes that take part in metabolism. Proteins destined for the nucleus, such as transcription factors, are also synthesized on free ribosomes.

Various proteins assist in precise folding, whatever the destination. **Chaperone proteins** stabilize partially folded regions in their correct form, and prevent a protein from getting "stuck" in a useless intermediate form. Chaperone proteins are being developed as drugs to treat diseases that result from misfolded proteins. Other proteins help new chemical bonds to form as the final shape arises, and yet others monitor the accuracy of folding. If a protein misfolds, an "unfolded protein response" slows or even stops protein synthesis while accelerating transcription of genes that encode chaperone proteins and the other folding proteins. Proper protein folding is quickly restored.

If a protein misfolds despite these protections, cells have ways to either refold the protein correctly, or get rid of it. Misfolded proteins are sent out of the ER back into the cytoplasm, where they are "tagged" with yet another protein, called ubiquitin. A misfolded protein bearing just one ubiquitin tag may straighten and refold correctly, but a protein with more than one tag is taken to another cellular machine called a **proteasome** (figure 10.20). A proteasome is a tunnel-like multiprotein structure. As a misfolded protein moves through the proteasome opening, it is stretched out, chopped up, and its peptide pieces degraded into amino acids, a little like a wood chipper. The amino acids may be recycled to build new proteins.

Proteasomes also destroy properly folded proteins that are in excess or no longer needed. For example, a cell must dismantle excess transcription factors, or the genes that they control may remain activated or repressed for too long. Proteasomes also destroy proteins from pathogens, such as viruses.

Diseases of Protein Misfolding

Proteins misfold in two ways: either from a mutation or by having more than one conformation. A mutation may change the amino acid sequence in a way that alters the attractions and repulsions

Figure 10.20 Proteasomes provide quality control. Ubiquitin binds to a misfolded protein and escorts it to a proteasome. The proteasome, which is composed of several proteins, encases the misfolded protein, straightening and dismantling it.

between parts of the protein, contorting it. In many cases of cystic fibrosis, CFTR protein misfolds in the endoplasmic reticulum, attracts too many chaperones, and is sent to the cytoplasm, never reaching the plasma membrane where normally it would form a chloride channel (see figures 1.5 and 1.6 and Clinical Connection 4.1).

Some proteins can have the same amino acid sequence but fold into more than one conformation. In these proteins, rarely, one conformation becomes "infectious," converting molecules with other conformations into more copies of itself. An infectious protein is called a **prion** (pronounced "pree-on"). The term is short for *proteinaceous infectious agent*. Passage of an infectious form of a protein to another individual causes an illness that is transmissible, rather than inherited.

In several diseases that affect the brain, misfolded proteins aggregate, forming masses that clog the proteasomes and block them from processing any malformed proteins. Different proteins are affected in different diseases. In Huntington disease, for example, extra glutamines in the protein huntingtin cause it to obstruct proteasomes. Misfolded proteins that clog proteasomes also form in the diseases listed in **table 10.4**, but it isn't always clear whether the accumulated proteins cause the disease or are a response to it.

Prion Diseases

A subgroup of protein misfolding conditions are the prion diseases. They were first described in sheep, which develop a disease called scrapie when they eat prion-infected brains of other sheep. The name denotes the way the sick sheep rub against things to scratch themselves. Their brains became riddled with holes. More than 85 animal species develop similar diseases. **Table 10.5** describes three prion diseases in humans that are caused by either a mutation in the prion-related protein gene *(PRPN)* or transmissible prion protein (sporadic) in the rare conformation. Abnormal prions may be present for many years before symptoms begin, but once they do, death can come in months or a few years. The three diseases described in the table have mutations in different parts of the same gene.

The first prion disease recognized in humans was kuru, which affected the native Fore people who lived in the remote mountains of Papua New Guinea (**figure 10.21**). *Kuru* means "to shake." The disease began with wobbly legs, then trembling

Table 10.5	Prion Diseases of Humans
Disease	**Symptoms**
Creutzfeldt-Jakob disease	Memory loss, nervousness, trembling hands, blank facial expression, jerky movements, unsteady gait, personality changes, anxiety, depression, impaired thinking, insomnia, visual loss, difficulty speaking and swallowing, coma, with death from heart failure, pneumonia, or infections within a year.
Fatal familial insomnia	Progressive inability to sleep and vivid dreams during scant sleep. Physical and mental deterioration, coma, death within a few months.
Gerstmann-Straussler disease	Memory loss, dementia, ataxia (uncoordinated movement), amyloid plaques in the brain. Death within 7 years.

Table 10.4	Diseases Associated with Protein Misfolding
Disease	**Misfolded Protein(s)**
Alzheimer disease	Amyloid beta precursor protein, tau proteins
Familial amyotrophic lateral sclerosis	Superoxide dismutase, TDP-43
Frontotemporal dementia	Tau proteins, TDP-43
Hereditary ATTR amyloidosis	Transthyretin
Huntington disease	Huntingtin
Parkinson disease	Alpha synuclein
Lewy body dementia	Alpha synuclein
PKU	Phenylalanine hydroxylase
Prion diseases	Prion protein

(All but Huntington disease are genetically heterogeneic; that is, abnormalities in different proteins cause similar syndromes.)

Figure 10.21 Kuru. Kuru affected the Fore people of New Guinea until they gave up a cannibalism ritual that spread infectious prion protein. *The Nobel Foundation, 1976.*

Figure 10.22 Prions change shape. Scrapie is a disease of sheep that was the first recognized prion disease. A single scrapie prion protein (PrP) contacts a normal PrP and changes it into the scrapie conformation. As the change spreads, disease results. Accumulated scrapie prion proteins clog brain tissue, eventually causing symptoms.

and whole-body shaking. Uncontrollable laughter led to the name "laughing disease." Speech slurred, thinking slowed, and the person became unable to walk or eat. Death came within a year. The disease, which affected mostly women and children, was traced to a ritual in which the people ate their war heroes. When the women and children prepared the brains, prions entered skin cuts and abrasions and they became infected. At first, the disease was thought to be inherited because it affected several members of families.

Not many people knew about the plight of the Fore in the 1950s. Then in the mid-1990s prion diseases dominated the headlines in the United Kingdom when more than 120 people ate infectious prions in beef and died. Popularly called *mad cow disease*, the condition is considered a variant of the brain disease Creutzfeldt-Jakob disease (CJD). The disease was also transmitted to a few hundred children with short stature who received human growth hormone pooled from the pituitary glands of thousands of donors, a few of whom were unknowingly incubating CJD.

Prions cause disease both by spreading the alternate form (infectious or mutant) and by aggregation of the protein. The normal function of prion protein is not well understood. It is on the surfaces of neurons and is present in many animal species, suggesting that it may be important in protecting the nervous system. These aggregates are also seen in more familiar diseases, such as beta amyloid plaques and tau protein neurofibrillary tangles in Alzheimer disease, and alpha synuclein deposits in Parkinson disease (see table 10.4).

The rare prion diseases like kuru, scrapie, and mad cow disease, as well as the more common ones, are all diseases of protein folding. The infectious forms have beta-pleated sheets in places where their normal counterparts have alpha helices. **Figure 10.22** depicts schematically how normal proteins become prions, which can directly destroy brain areas and aggregate. In rare cases of these conditions, a mutation misfolds the protein. Drug discovery efforts are being directed toward refolding misfolded proteins.

Key Concepts Questions 10.4

a. List and describe the four levels of protein structure.
b. How does a cell handle misfolded proteins?
*c. Define *proteasome*.
d. How can a protein that has more than one conformation cause a disease?

Study Guide

Summary

10.1 The Importance of Proteins

1. Dietary proteins are digested into **amino acids**. Genes control the linking of amino acids into proteins.

2. Biological proteins consist of 20 types of amino acids. Each amino acid has an amino group, an acid group, a hydrogen atom, and an R group, bonded to a central carbon atom. Amino acids are linked with peptide bonds

into long chains called **polypeptides**. A protein molecule consists of one or more polypeptides. A protein's three-dimensional shape is its **conformation**.

10.2 Transcription Copies the Information in DNA

3. Transcription copies DNA information into RNA. Translation uses the information in RNA to connect amino acids into proteins.
4. RNA is transcribed from the **template strand** of DNA. The other DNA strand is called the **coding strand**.
5. RNA is a single-stranded nucleic acid similar to DNA but with uracil and ribose rather than thymine and deoxyribose.
6. Messenger RNA (mRNA) carries a protein-encoding gene's information in the form of three-base units called **codons**. **Ribosomal RNA (rRNA)** associates with certain proteins to form ribosomes, on which proteins are synthesized. Transfer RNA (tRNA) is cloverleaf-shaped, with a three-base **anticodon** that is complementary to an mRNA on one end and bonds to a specific amino acid on the other end.
7. Transcription factors regulate which genes are transcribed in a particular cell type under particular conditions.
8. Transcription begins when transcription factors help **RNA polymerase** bind to a gene's **promoter** and then adds RNA nucleotides to a growing chain, in a sequence complementary to the DNA template strand.
9. After a gene is transcribed, the mRNA receives a "cap" of modified nucleotides at the 5′ end and a poly A tail at the 3′ end.
10. For many mRNAs, **introns** are removed and **exons** are then translated into protein. Introns may outnumber and outsize exons. **Alternative splicing** combines exons in different ways, increasing the number and diversity of proteins transcribed from a single gene. **Isoforms** are different forms (lengths) of the same protein resulting from alternative splicing.

10.3 Translation of a Protein

11. Each three consecutive mRNA bases form a codon that specifies a certain amino acid. The **genetic code** is the correspondence between each codon and the amino acid it specifies. Of the 64 possible codon types, 60 specify amino acids, one specifies the amino acid methionine and "start," and three signal "stop." More than one type of codon may encode an amino acid. The genetic code is triplet and nonoverlapping, includes controls, and is universal. **Synonymous codons** encode the same amino acid, and **nonsynonymous codons** encode different amino acids.
12. The **reading frame** is the starting point of a DNA sequence that encodes a protein. A **frameshift mutation** adds or deletes one or two bases, which disrupts the reading frame, leading to a shortened or incorrect protein. An **open reading frame** is a stretch of DNA without a stop codon that might indicate a protein-encoding gene.
13. In translation, an initiation complex forms when mRNA, a small ribosomal subunit, and a tRNA carrying methionine join. The amino acid chain elongates when a large ribosomal subunit joins the small one. Next, a second tRNA binds by its anticodon to the next mRNA codon, and its amino acid bonds with the first amino acid. Transfer RNAs add more amino acids. The ribosome moves down the mRNA as the chain grows. The P site bears the growing amino acid chain, and the A site holds the next amino acid to be added. When the ribosome reaches a "stop" codon, it falls apart into its two subunits and is released. The new polypeptide breaks free.
14. After translation, some polypeptides are cleaved, have sugars added, or aggregate. The cell uses or secretes the protein.

10.4 Processing a Protein

15. A protein must fold into a specific **conformation** to function. A protein's **primary (1°) structure** is its amino acid sequence. Its **secondary (2°) structure** forms as amino acids close in the primary structure attract. **Tertiary (3°) structure** appears as more widely separated amino acids attract or repel in response to water molecules. **Quaternary (4°) structure** forms when a protein consists of more than one polypeptide.
16. **Chaperone proteins** help the correct conformation arise. Other proteins help new bonds form and oversee folding accuracy. Ubiquitin attaches to misfolded proteins and enables them to refold or escorts them to **proteasomes** for dismantling. Protein misfolding causes diseases. This can happen by mutation or if a protein can fold into several conformations, some of which are "infectious" and can cause disease.
17. At least one conformation of a **prion** protein is infectious.

Key Terms

alternative splicing Assembling subsets of the exons of a gene, which increases the number and diversity of proteins it encodes.

amino acid (ə-mē′nō) A small organic molecule that is a protein building block.

anticodon (ăntē-kō′dŏn) A three-base sequence on one loop of a transfer RNA molecule that is complementary to an mRNA codon and connects the appropriate amino acid and its mRNA.

chaperone protein A protein that binds a polypeptide and guides folding.

coding strand The strand of the DNA double helix for a particular gene from which RNA is not transcribed.

codon (kō′dŏn) A continuous triplet of mRNA that specifies a particular amino acid.

conformation The three-dimensional shape of a molecule.

exon (x-on) A part of a gene and its corresponding mRNA that encodes amino acids.

frameshift mutation (myōō-tā′shən) A mutation that alters a gene's reading frame.

genetic code (jə-nĕt′ĭk) The correspondence between specific mRNA triplets and the amino acids they encode.

intron (ĭn trŏn) Part of a gene that is transcribed but is excised from the mRNA before translation into protein.

isoform An alternate version of a protein that arises from a certain combination of translated exons.

nonsynonymous codon (kō'don) A codon that encodes a different amino acid from another codon.

open reading frame A sequence of DNA that does not include a stop codon.

polypeptide A long chain of amino acids. A protein consists of one or more polypeptides.

primary (1°) structure The amino acid sequence of a protein.

prion An infectious protein.

promoter A control DNA sequence that signals the start of a gene.

proteasome (prō-tēə-sōm) A multiprotein cellular structure with a tunnel-like shape through which misfolded or excess proteins pass and are dismantled.

quaternary (4°) structure A protein that has more than one polypeptide subunit.

reading frame The point in a DNA sequence from which contiguous triplets encode amino acids of a protein. A DNA sequence has three reading frames.

ribosomal RNA (rRNA) (rī'bōs-ō'məl) RNA that, with proteins, comprises ribosomes.

RNA polymerase (RNAP) (pŏl'ə-mə-rās) An enzyme that adds nucleotides to a growing RNA chain.

secondary (2°) structure Folds in a polypeptide caused by attractions between amino acids close together in the primary structure.

synonymous codons (kō d ŏnz) DNA triplets that specify the same amino acid.

template strand The DNA strand carrying the information to be transcribed.

tertiary (3°) structure Folds in a polypeptide caused by interactions between amino acids and water. This draws together amino acids that are far apart in the primary structure.

Review Questions

1. Explain how complementary base pairing is responsible for
   a. the structure of the DNA double helix.
   b. DNA replication.
   c. transcription of RNA from DNA.
   d. the attachment of mRNA to a ribosome.
   e. codon-anticodon pairing.
   f. tRNA conformation.

2. State the functions of these proteins.
   a. RNA polymerase
   b. ubiquitin
   c. a chaperone protein
   d. a transcription factor

3. Explain where a hydrogen bond forms and where a peptide bond forms in the transmission of genetic information.

4. List the differences between RNA and DNA.

5. Identify where in a cell DNA replication, transcription, and translation occur.

6. Explain how transcription controls cell specialization.

7. How can the same mRNA codon be at an A site on a ribosome at one time but at a P site at another time?

8. Describe the events of transcription initiation.

9. List the three major types of RNA and their functions.

10. Describe three ways RNA is altered after it is transcribed.

11. What are the components of a ribosome?

12. Why would an overlapping genetic code be restrictive?

13. Why would two-nucleotide codons be insufficient to encode the number of amino acids in biological proteins?

14. How are the processes of transcription and translation efficient?

15. What factors determine how a protein folds into its characteristic conformation?

16. How do a protein's primary, secondary, and tertiary structures affect conformation? Which is the most important determinant of conformation?

17. Explain how a protein can be infectious.

Applied Questions

1. Can starting a new exercise program "change your genes," as one ad promises?

2. List the RNA sequence transcribed from the DNA template sequence TTACACTTGCTTGAGAGTC.

3. Reconstruct the corresponding DNA template sequence from the partial mRNA sequence GCUAUCUGUCAUAAAGAGGA.

4. Using table 10.3, list three mRNA sequences that could encode the amino acid sequence histidine-alanine-arginine-serine-leucine-valine-cysteine.

5. Write a DNA sequence that would encode the amino acid sequence valine-tryptophan-lysine-proline-phenylalanine-threonine.

6. In the film *Jurassic Park,* which is about cloned dinosaurs, a cartoon character named Mr. DNA talks about the billions of genetic codes in DNA. Why is this statement incorrect?

7. Titin is a muscle protein named for its size—its gene has the largest known coding sequence of 80,781 DNA bases. How many amino acids long is it?

8. An extraterrestrial life form has a triplet genetic code with five different bases. How many different amino acids can this code specify, assuming no degeneracy?

9. In malignant hyperthermia, a person develops a life-threateningly high fever after taking certain types of anesthetic drugs. In one family, mutation deletes three contiguous DNA bases in exon 44. How many amino acids are missing from the protein?

10. The protein that serves as a receptor that allows insulin to enter cells has a different number of amino acids in a fetus and in an adult. Explain how this may happen.

11. Cataract is a clouding of vision that is more common among older people. Proteins called lens crystallins form the lens, but with age, some of the proteins misfold and, over time, more of them do so. Name a disease mentioned in the chapter that has a similar pathology.

Case Studies and Research Results

1. Five patients meet at a clinic for families with early-onset Parkinson disease. This condition causes rigidity, tremors, and other motor symptoms. Only 2 percent of cases of Parkinson disease are inherited. The five patients all have mutations in a gene that encodes the protein parkin, which has 12 exons. For each patient, indicate whether the mutation shortens, lengthens, or does not change the size of the protein, or if it isn't possible to tell what the effect might be.
    1. Manny's *parkin* gene is missing exon 3.
    2. Frank's *parkin* gene has a duplication in intron 4.
    3. Theresa's *parkin* gene lacks six contiguous nucleotides in exon 1.
    4. Elyse's *parkin* gene has an altered splice site between exon 8 and intron 8.
    5. Scott's *parkin* gene is deleted.

2. The human genome sequence encodes many more mRNA transcripts than there are genes. Why isn't the number the same?

3. Francis Crick envisioned "20 different kinds of adaptor molecule, one for each amino acid, and 20 different enzymes to join the amino acid to their adaptors." What type of molecule was Dr. Crick describing?

CHAPTER 11

Gene Expression and Epigenetics

Forensic entomologists once relied on cataloging the developmental stages of blow flies on a corpse to estimate the time of death. Today they can also look at patterns of gene expression in the victim in the hours after death.

Learning Outcomes

11.1 Gene Expression Through Time and Tissue
1. Define *epigenetic*.
2. Explain how development of the placenta, globin chain switching, development of organs, and the types of proteins cells make over time illustrate gene expression.

11.2 Control of Gene Expression
3. Explain how small molecules binding to histone proteins control gene expression by remodeling chromatin.
4. Explain how microRNAs and microproteins control transcription.

11.3 Maximizing Genetic Information
5. Explain how division of genes into exons and introns maximizes the number of encoded proteins.

11.4 Most of the Human Genome Does *Not* Encode Protein
6. Discuss how viral DNA, noncoding RNAs, and repeated DNA sequences account for large proportions of the human genome.

The BIG Picture

Discovering the nature of the genetic material, determining the structure of DNA, deciphering the genetic code, and sequencing the human genome led to today's challenge: learning how genes are expressed and interact through tissue and time.

Tracking RNA to Pinpoint Time of Death

Crime scene investigators have traditionally extrapolated from the developmental stages of insect larvae feeding on human corpses to estimate the time since the person's death—the postmortem interval. The larger and more developed the insect parts, the longer the body has been dead. More recently, researchers have turned to the dynamics of the human mRNA in parts of the victim.

Analyzing the mRNAs in individual cells under defined circumstances is an application of a biotechnology called transcriptomics. It can reveal the gene expression changes that are part of development and life—and death. Researchers at the Broad Institute analyzed thousands of samples from body donors, including several organs, salivary glands, and skin with and without sun exposure, to identify messenger RNAs whose abundance changes as a body cools after death.

Transcription of DNA into RNA continues, for awhile, after cardiac and brain death. For the first 7 hours, many genes are expressed more than they were before death, which may be a reaction to trauma. Gene expression fluctuates in the 7-to-14 hour interval, with thousands of genes transcribed much more or much less than they were when the person was alive, as if the system of DNA-to-RNA-to protein is going haywire as it shuts down. From 14 through 24 hours after death, some genes are still being transcribed at a faster or slower rate than they were when the person was alive.

The waves of changing gene expression paint a panorama of death at the microscopic level. Mitosis halts. White blood cell count plummets, infection no longer a threat. Necrosis (cell death from injury) ensues, while blood clotting, DNA repair, and the stress response increase. Ion transport, carbohydrate metabolism, and cholesterol synthesis halt. Cells shift from aerobic respiration to the anaerobic route, as the lungs no longer bring in oxygen nor does the circulation deliver it.

The researchers developed a machine learning tool that can tell the time elapsed since death from the RNAs in just four tissues: lung, subcutaneous fat, thyroid, and sun-exposed skin on the lower leg. When they tested it on 129 bodies with time of death known but blinded, the results were accurate to within 10 minutes of the actual time of demise. Precision in determining time of death is important for forensic investigations. The technology may also be useful in identifying tissues and organs viable enough for transplant.

11.1 Gene Expression Through Time and Tissue

A genome is like an orchestra. Just as not all of the musical instruments play with the same intensity at every moment, not all genes are expressed continually at the same levels. Some genes are always transcribed and translated, in all cells. Because they keep cells running, these genes are called "housekeeping genes." Other genes have more specialized roles, becoming active as a cell differentiates or challenges arise.

Before the field of genomics began in the 1990s, the study of genetics proceeded one gene at a time, like hearing the separate contributions of a violin, a viola, and a flute. Many genetic investigations today, in contrast, track the crescendos of gene activity that parallel events in an organism's life, down to the single cell level. This new view introduced the element of time to genetic analysis. Unlike the gene maps of old, which ordered genes linearly on chromosomes, new types of maps are more like networks that depict the timing of gene expression in unfolding programs of development and response to the environment.

The discoveries of the 1950s and 1960s about DNA structure and function answered some questions about the control of gene expression, while raising many more: How does a bone cell "know" to transcribe the genes that control the synthesis of collagen and not to transcribe genes that specify muscle proteins? What shifts the proportions of blood cell types into leukemia? How do chemical groups "know" to shield a specific DNA sequence from transcription in one circumstance, yet expose it in others? The collection of mRNAs in a cell constitutes its **transcriptome**. A research technique called single-cell transcriptomics catalogs the mRNA molecules made in a specific cell type at a specific time.

Changes to the chemical groups that associate with DNA greatly affect which parts of the genome are accessible to transcription factors and under which conditions. Such epigenetic changes to the molecules that bind to DNA are transmitted to daughter cells when the cell divides. Epigenetic changes do not alter the DNA base sequence and are reversible. However, at a few sites in the genome, an epigenetic change may persist through meiosis to a third generation, but this appears to be rare. Figure 6.11 shows how methyl groups bind to DNA, causing epigenetic changes.

This chapter presents a few of the nearly limitless examples of how gene expression changes in response to the environment both inside and outside the body.

Gene Expression in the Placenta

Analyzing the RNAs made in a particular type of cell in a tissue or organ provides a molecular view of a biological process at one point in time, such as development or response to an injury. Researchers use the polymerase chain reaction (see figure 9.19) to obtain enough mRNAs to study. The collection is the transcriptome.

Single-cell transcriptomics has provided new details on the functioning of the placenta, the organ that connects pregnant woman to fetus. A placenta looks like a pink pancake that whooshes out of the woman's body just after the baby is born, but it is an intricate collection of cell populations. Understanding the physiology of the placenta at various times using mRNA patterns is revealing functions that might be assessed to help in preventing some pregnancy losses or stillbirths.

The genes expressed in human placentas fall into five groups. Only one represents the woman, and the mRNAs indicate that the genes in that group dampen her immune response, which might otherwise reject the fetus. The other four groups of genes encode proteins that invade the uterine lining, build the blood supply, and release hormones necessary to maintain the pregnancy. The most important finding is that cells that ultimately specialize as the same type, by turning on and off transcription of the same suites of genes, do so in different orders—a little like several cars taking different routes to the same destination. This might be a protective mechanism to ensure that at least some cells differentiate appropriately.

Globin Chain Switching

Single-cell transcriptomics is a relatively new technology. A classic illustration of control of gene expression is the switching of components of the blood protein hemoglobin during embryonic and fetal development and after birth.

A hemoglobin molecule in the blood of an adult consists of four polypeptide chains, each wound into a globular conformation (**figure 11.1**). Two of the chains are 146 amino acids long and are called "beta" (β). The other two chains are 141 amino acids long and are termed "alpha" (α). Genes on different chromosomes encode the alpha and beta globin polypeptide chains.

As a human develops, different globin polypeptide chains are used to make molecules of hemoglobin in a process called

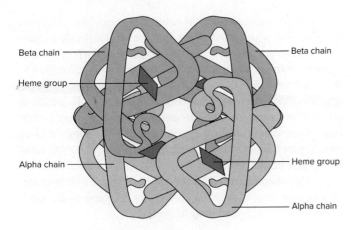

Figure 11.1 **The structure of hemoglobin.** A hemoglobin molecule is made up of two globular protein chains from the beta (β) globin group and two from the alpha (α) globin group. Each globin surrounds an iron-containing chemical group called a heme.

group. About 4 percent of the hemoglobin in the embryo includes "adult" beta chains. This percentage gradually increases. Globin chains are manufactured first in the yolk sac in the embryo, then in the liver and spleen in the fetus, and finally primarily in the bone marrow after birth.

As the embryo develops into a fetus, the epsilon and zeta globin polypeptide chains decrease in number as gamma (γ) and alpha chains accumulate. Hemoglobin consisting of two gamma and two alpha chains is called fetal hemoglobin. Because the gamma globin subunits bind very strongly to oxygen released from maternal red blood cells into the placenta, fetal blood carries 20 to 30 percent more oxygen than an adult's blood. As the fetus matures, beta chains gradually replace the gamma chains. At birth, however, the hemoglobin is not fully of the adult type—fetal hemoglobin (two gamma and two alpha chains) comprises from 50 to 85 percent of the blood. By 4 months of age, the proportion drops to 10 to 15 percent, and by 4 years, less than 1 percent of the child's hemoglobin is the fetal form.

globin chain switching. The different forms of hemoglobin are necessary because of changes in blood oxygen levels that happen when a newborn begins breathing and no longer receives oxygen through the placenta. The promoters of the globin genes include binding sites for transcription factors, which orchestrate the changes in hemoglobin molecules through prenatal development. Other DNA sequences in the globin gene clusters turn off expression of genes no longer needed. **Figure 11.2** tracks the changes to the globin proteins before and after birth.

The chemical basis for globin chain switching is that different globin polypeptide chains attract oxygen molecules to different degrees. In the embryo, as the placenta forms, hemoglobin consists first of two epsilon (ε) chains, which are in the beta globin group, and two zeta (ζ) chains, which are in the alpha globin

Building Tissues and Organs

Blood is a structurally simple tissue that is easy to obtain and study because its components are easily separated. A solid gland or organ is more complex. It has a distinctive three-dimensional form and is constructed from specialized types of cells and commonly more than one type of tissue. The specific, solid organization of an organ must persist throughout a lifetime of growth, repair, and changing external conditions, while its cells must maintain their specializations.

In all tissues and organs, genes are turned on and off during development, as stem cells self-renew and yield more specialized daughter cells. The organization of the pancreas is simple enough to reveal how its cells specialize. It is a

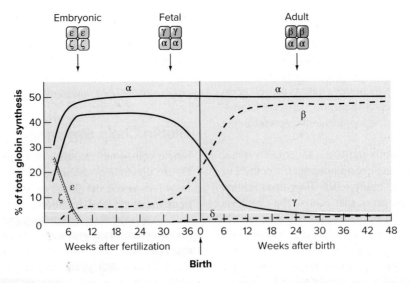

Figure 11.2 **Globin chain switching.** The subunit composition of human hemoglobin changes as the concentration of oxygen in the environment changes. Each globin quartet has two polypeptide chains encoded by genes in the alpha (α) globin cluster (chromosome 16) and two polypeptide chains from the beta (β) globin cluster (chromosome 11). When the oxygen source switches from the placenta to the newborn's lungs, beta globin begins to replace gamma (γ) globin.

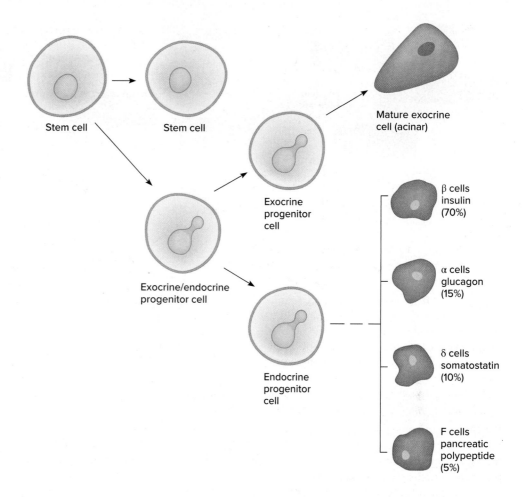

Figure 11.3 Building a pancreas. A single type of stem cell theoretically divides to give rise to an exocrine/endocrine progenitor cell that in turn divides to yield more restricted progenitor cells that give rise to both mature exocrine and endocrine cells. The endocrine progenitor cell in turn divides, yielding specialized daughter cells that produce specific hormones.

dual gland, with two types of cell clusters. The exocrine part releases digestive enzymes into ducts. The endocrine part secretes directly into the bloodstream polypeptide hormones that control nutrient use. The endocrine cell clusters are called pancreatic islets.

The dual nature of the pancreas unfolds in the embryo, when ducts form. Within duct walls reside rare stem cells and progenitor cells (see figure 2.17). A transcription factor is activated and controls expression of other genes in a way that stimulates some progenitor cells to divide. Certain daughter cells follow an exocrine pathway and will produce digestive enzymes. Other progenitor cells respond to different signals and divide to yield daughter cells that follow the endocrine pathway. **Figure 11.3** shows the differentiated cell types that form from the two cell lineages. The most familiar pancreatic hormone is insulin, which the beta cells of the pancreas secrete. The absence of insulin, or the inability of cells to recognize it, causes diabetes mellitus.

Shared Gene Expression Connects Diseases

A more complete portrait of gene expression emerges through transcriptomics, as well as through **proteomics,** which identifies all the proteins a cell manufactures under specific conditions. However, identifying and grouping the mRNAs and proteins in a cell is only a first step in understanding how cells specialize. The next challenge is to determine how proteins with related functions interact, which can reveal links between medical conditions. **Figure 11.4** shows part of a huge disease map called a "diseasome." It connects diseases that share genes that have altered expression. Some of the links and clusters are well known, such as obesity, hypertension, and diabetes. Others are surprises, such as Duchenne muscular dystrophy and heart attacks. Following the discovery of this link using gene expression information, researchers are now testing heart drugs on boys with DMD. In other cases, the association of a disease with genes whose expression goes up or down can suggest drugs to try or develop.

Transcriptomics and proteomics are also useful in better understanding basic biology. One proteomic experiment, for example, compared the relative contributions of major categories of proteins from conception through birth and from conception through old age. The differences in the levels of proteins made at these times make sense. Transcription factors are more abundant before birth because of the extensive cell differentiation of this period, as organs form. During the prenatal period, enzymes are less abundant, perhaps because the fetus receives some enzymes through the placenta. Antibodies appear after birth, when the immune system begins to function by responding to environmental stimuli.

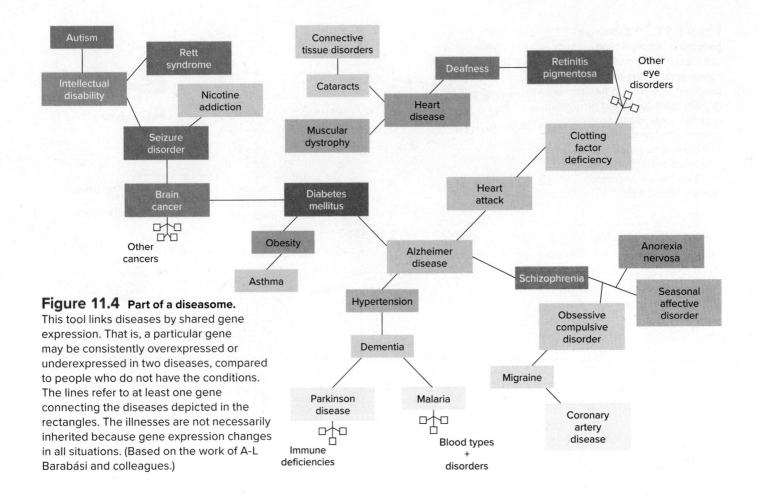

Figure 11.4 Part of a diseasome. This tool links diseases by shared gene expression. That is, a particular gene may be consistently overexpressed or underexpressed in two diseases, compared to people who do not have the conditions. The lines refer to at least one gene connecting the diseases depicted in the rectangles. The illnesses are not necessarily inherited because gene expression changes in all situations. (Based on the work of A-L Barabási and colleagues.)

Gene expression profiles for different cell types under various conditions are the basis for many new medical tests that assess risk, diagnose disease, or monitor response to treatment. For example, 55 genes are overexpressed and 480 are underexpressed in cells of a prostate cancer that has a high likelihood of spreading, but not in a prostate cancer that will probably not spread. A test based on such findings can assist physicians in distinguishing patients who can safely delay or avoid invasive and risky treatment from those whose cancers are more likely to spread.

Key Concepts Questions 11.1

*a. What does single-cell transcriptomics assess?

b. Explain how single-cell transcriptomics reveals events in a placenta.

c. How does gene expression change over time for the polypeptide components of hemoglobin?

d. How does gene expression change as a pancreas develops?

e. How do transcriptomics and proteomics differ?

11.2 Control of Gene Expression

In a human genome, the blueprints for building proteins take up significantly less DNA than the instructions for when and where the proteins are synthesized and function. A protein-encoding gene contains controls of its own expression, and is also controlled by the configuration of the chromatin of which it is a part.

An internal control of a gene's expression is the promoter sequence. Recall from chapter 10 that the promoter is the part of the gene where RNA polymerase and transcription factors bind, marking the start point of transcription. Variations in the promoter sequence of a gene can affect how quickly the encoded protein is synthesized. In a form of early-onset Alzheimer disease, for example, a mutation in the promoter for the gene that encodes amyloid precursor protein (see Clinical Connection 5.1) causes the sticky protein to accumulate in the brain twice as fast as normal because it isn't cleared quickly enough. A second way that expression of a gene can exceed the normal pace is if a person has more than one copy of it.

Stepping back from a single-gene perspective, cells specialize as different combinations of transcription factors bind to the many genes that provide the cell's characteristics. Transcription factors can bind at a gene's promoter or to a DNA sequence away from the gene, providing long-distance regulation, too.

Control of gene expression happens in steps. In **chromatin remodeling,** the histone proteins associated with DNA interact with other chemical groups in ways that expose some sections of DNA to transcription factors and shield other sections, blocking their expression. Later in the protein production process, small RNAs called, appropriately, **microRNAs,** bind to certain mRNAs, preventing their translation into protein. Overall, these two processes—chromatin remodeling and blocking of translation by microRNAs—determine the ebb and flow of different proteins, enabling cells to adapt to changing conditions. Proteins also control each other beyond the major role of transcription factors. Even tiny proteins, called **microproteins,** can control the functioning of larger proteins.

Chromatin Remodeling

Recall from figure 9.13 that DNA associates with proteins and RNA to form chromatin. For many years, biologists thought that the histone proteins that wind long DNA molecules into nucleosomes were little more than tiny spools. However, histones do much more. Enzymes add or delete small organic chemical groups to histones, affecting expression of the protein-encoding genes that these proteins bind.

The three major types of small molecules that bind to histones are acetyl groups, methyl groups, and phosphate groups (**figure 11.5**). The key to the role histones play in controlling gene expression lies in acetyl groups (CH_3CO_2). They bind to specific sites on certain histones, particularly to the amino acid lysine.

Figure 11.6 shows how acetyl binding shifts histone interactions in a way that allows transcription to begin. A series of proteins moves the histone complex away from a DNA sequence called a TATA box, exposing it enough for RNA polymerase to bind and transcription to begin (see figure 10.8). First, a group of proteins called an enhanceosome attracts the enzyme (acetylase) that adds acetyl groups to specific lysines on specific histones, which neutralizes the histones' positive charge. Because DNA carries a negative charge, this change moves the histones away from the DNA, making room for transcription factors to bind and begin transcription. Enzymes

Figure 11.6 Acetylated histones allow transcription to begin. Once acetyl groups are added to particular amino acids in the tails of certain histones, the TATA box becomes accessible to transcription factors. H3 and H4 are histone types. The small and large protein complexes ease movement of the TATA box away from the histones, enabling transcription to proceed.

called deacetylases remove acetyl groups, which shuts off gene expression.

In most cell types, only about 2 percent of the chromatin is "open" and "easy to read" (**figure 11.7**). Researchers can separate these regions from the rest of the DNA and identify the genes being transcribed in a particular cell type under particular conditions.

Methyl groups (CH_3) are also added to or taken away from histones. When CH_3 binds to a specific amino acid in a specific histone type, a protein is attracted that prevents transcription. As CH_3 groups are added, methylation spreads from the tail of one histone to the adjacent histone, propagating the gene silencing. Methyl groups also turn off transcription by binding to DNA directly, to cytosines at about 16,000 places in the genome. Therefore, methyl groups bind both histones *and* DNA. The *methylome* is the collection of all the methylated DNA sites in the genome. Different cell types have different genes that are methylated and turned off.

Figure 11.5 Chromatin remodeling. Chromatin remodeling adds or removes certain small chemical groups to or from histones. The pattern of binding controls whether the DNA wrapped around the histones is transcribed or not. A gene is expressed when acetyl groups and phosphate groups are added and when methyl groups are removed.

Chapter 11 *Gene Expression and Epigenetics*

Figure 11.7 **Chromatin may be closed and not accessible or open and accessible.** The red section of DNA represents DNA that is either hidden and hard to read or exposed and easy to read. Only 146 base pairs of DNA wrap around a nucleosome.

Figure 11.8 **Control of gene expression.** Chromatin opens to allow transcription factors to bind, whereas microRNAs bind to specific mRNAs, blocking their translation into protein.

The modified state of chromatin can be passed on when DNA replicates. These changes in gene expression are heritable from cell generation to cell generation, but they do not alter the DNA base sequence—that is, they are epigenetic. Addition and removal of acetyl groups, methyl groups, or phosphate groups are all epigenetic changes. Effects of DNA methylation can sometimes be seen when MZ (identical) twins inherit the same disease-causing genotype, but only one twin is sick. The reason for the discordance may be different patterns of methylation of one or more genes.

Enzymes that add or delete acetyl, methyl, and phosphate groups must be in a balance that controls which genes are expressed and which are silenced. Upset this balance, and disease can result. At least 44 Mendelian conditions arise from abnormal functioning of the "epigenetic machinery." Nearly all of them include intellectual disability, which makes sense considering that many genes control brain function. One limitation to altering chromatin remodeling to treat inherited disease is that this action could affect the expression of many genes—not just the gene (or genes) implicated in the disease. **Clinical Connection 11.1** discusses an epigenetic explanation for schizophrenia—starvation before birth.

MicroRNAs

Chromatin remodeling determines which genes are transcribed. MicroRNAs act later in gene expression, preventing the translation of mRNA transcripts into protein. If chromatin remodeling is considered as an on/off switch to transcription, then the control of microRNAs is more like that of a dimmer switch—fine-tuning gene expression at a later stage. **Figure 11.8** schematically compares chromatin remodeling and the actions of microRNAs, and **figure 11.9** places these events in the overall flow of gene expression.

MicroRNAs are so-named because they are small—just 21 or 22 bases long. They are cut from precursor RNA molecules. MicroRNAs are a type of noncoding RNA, which means that they do not encode an amino acid sequence. The human genome encodes at least 2,555 distinct sequences of microRNAs that regulate at least one-third of the protein-encoding genes. A typical human cell has from 1,000 to 200,000 individual microRNA molecules.

Figure 11.9 **A summary of the events of gene expression.** Chromatin remodeling determines which genes are transcribed. Alternative splicing creates different forms of a protein. MicroRNAs bind to mRNAs by complementary base pairing, blocking translation. After translation, a protein must fold a certain way. Certain polypeptides are shortened, attached to sugars, and/or aggregated.

Clinical Connection 11.1

The Dutch Hunger Winter: One Route to Schizophrenia

When a fetus is starving, control of gene expression changes in ways that enable immediate survival, but may harm health decades later. The Dutch Famine Study is tracing the effects of prenatal starvation during World War II on health decades later, publishing findings since 1976.

During the "Dutch Hunger Winter" from November 1944 through May 1945, the Nazis blocked food supplies from entering six large cities in western Holland. As severe malnutrition weakened and killed people, a cruel experiment took place, in retrospect. By consulting food ration records that showed exact daily caloric intakes, time during pregnancy for specific women, and diagnoses from psychiatric registries and military induction records for the people exposed before birth, researchers at Columbia University and Leiden University in the Netherlands discovered clear links between prenatal malnutrition and health problems years later. They compared the health of people born during the Dutch Hunger Winter to their same-sex siblings born during better times.

Prenatal malnutrition affects an adult phenotype because starvation alters the pattern in which methyl groups (CH_3) bind to DNA, selectively silencing genes. The DNA of the "Dutch Hunger Winter children," when studied 60 years later, had different methylation patterns in genes that control metabolism than their siblings. The exposed children, when grown, had poor glucose control, high body mass index, and elevated serum cholesterol—clearly adaptations to prenatal starvation. They also have a strikingly increased risk of developing schizophrenia that perhaps is related to altered expression of the gene that encodes insulin-like growth factor 2 (IGF-2), which controls genes that affect thinking. Researchers hypothesize that schizophrenia may develop in some people born into famine when *IGF-2* has too few methyl groups and is overexpressed in the brain.

Questions for Discussion

1. Explain why changes in gene expression are not the same as genetic mutations.
2. How did the Dutch Famine Study conclude that prenatal malnutrition was the cause of increased schizophrenia risk?
3. Schizophrenia incidence has more than doubled in the United States and Europe in the past 20 years. At the same time, the concentration of THC in cannabis products (see Clinical Connection 8.1) has increased, more states have legalized use of cannabis products, and the products are available in more forms. What can researchers learn from the Dutch Famine Study to design investigations to identify or clarify a link between increasing cannabis use and increasing incidence of schizophrenia?

Each type of microRNA binds to complementary sequences within a particular set of mRNAs. When a microRNA binds a "target" mRNA, it prevents translation. Because a single type of microRNA has many targets, it controls the expression of sets of genes. In turn, a single type of mRNA can bind several different microRNAs.

Within the patterns of microRNA function may lie clues to new ways to fight disease or injury, because discovering the mRNAs that a certain microRNA targets may reveal genes that interact. Discovering these relationships can suggest new uses for and ways to combine existing drugs. The first practical applications might be in cancer treatment because certain microRNAs are more or less abundant in cancer cells than in healthy cells of the same type from which the cancer cells formed. Restoring the levels of microRNAs that normally suppress the too-rapid cell cycling of cancer, or blocking production of microRNAs that are too abundant in cancer, could help to restore a normal cell division rate.

Another application of microRNAs arises from the fact that they are transported between cells in bubble-like exosomes. MicroRNAs are being tested as markers to identify cases of concussion that are likely to persist as more serious "neurocognitive disorder due to traumatic brain injury." Researchers identified 15 microRNAs that differ in levels in the cerebrospinal fluid between people with concussion who go on to develop the complication and others with concussion who do not. The microRNAs can be tested in blood or urine, and their small size enables them to escape degradation by nucleases in the body fluids. This set of 15 microRNAs targets nearly 2,500 genes, many of which control repair of damaged neurons and how the nerve cells send and receive signals. Different microRNA patterns tracked with specific concussion symptoms, such as fatigue and memory loss. If validated in many injured people, a microRNA "spit test" could help identify athletes at elevated risk of suffering prolonged concussion symptoms.

Microproteins

Proteins can control each other. The transcription factors, discussed in Chapter 10, control suites of genes. Transcription factors are small, but not nearly as tiny as the microproteins, which are fewer than 100 amino acids long. Microproteins can fit inside the nooks and crannies of large proteins, such as cell surface receptors and ion channels, affecting their functioning.

The genes that encode microproteins may lie within other genes, but the microproteins are not just parts of larger proteins, nor are they encoded by microRNAs. They act later in the utilization of genetic information than the processes depicted in figure 11.9.

Researchers are just beginning to discover and describe the microproteins encoded in human genomes. One type of microprotein is like a garbage disposal, mopping up mRNA molecules that are no longer needed. In mice, microproteins control muscles. Microproteins in the venoms of scorpions, spiders, and centipedes block ion channels in the victim, causing paralysis.

Microproteins were only recently discovered because when researchers began assigning proteins to genes in the 1990s, they arbitrarily set 100 amino acids as a lower size limit for proteins. That approach missed these important, tiny molecules.

Key Concepts Questions 11.2

*a. What is the role of histone proteins in controlling gene expression?

b. How do acetyl, phosphate, and methyl groups control transcription?

c. Explain how microRNAs contribute to controlling gene expression.

d. How do microproteins control the activities of other proteins?

11.3 Maximizing Genetic Information

A human genome maximizes information in the 20,325 genes that encode about 100,000 mRNAs, which in turn specify more than a million proteins. Figure 11.9 depicts this increase in information from gene to RNA to protein on the left and lists the mechanisms that maximize the information on the right.

Several events account for the fact that proteins outnumber genes. Alternative splicing of the exons and introns of a gene may encode several isoforms of a protein (**figure 11.10**). Recall from figure 10.11 that alternative splicing of a single gene can generate multiple transcripts. In a broad sense this mechanism expands the information content of the mRNA. For example, a part of chromosome 22 that harbors 245 genes yields 642 mRNA transcripts. Drilling down to the single-gene level also reveals the expansion of information that alternative splicing affords.

Consider the Hippo pathway, which is a network of biochemical pathways that regulate the sizes of organs. The enzymes and other proteins that are part of the Hippo pathway do this by communicating with many components of cell function: adhesion, shape, stiffness, and division rate. A mutation in a gene of the Hippo pathway may cause cancer by disrupting these processes. Alternative splicing affects many of the genes. One (*MAP4K*) appears as 15 transcripts that encode 5 proteins. Another (*LATS1*) is transcribed into 7 mRNAs and 4 proteins. Yet another (*MOB1*) has 6 splice variants but a single mRNA, which encodes a 216 amino acid protein.

Another way that the genome holds more information than it may appear to is that a DNA sequence that is an intron in one context may encode protein (function as an exon) in another. Consider prostate specific antigen (PSA). It is a protein on certain cell surfaces in the prostate gland that liquifies semen and is overproduced when the gland is enlarged, which could indicate a benign growth or cancer. PSA level in the blood is regularly measured in men over age 50, as a biomarker of possible cancer.

The gene encoding PSA has five exons and four introns. It is alternatively spliced to encode seven isoforms. One of them, called PSA-linked molecule (PSA-LM), is encoded by the first exon and the fourth intron (**figure 11.11**). The two proteins (PSA and PSA-LM) work against one another. When the level of one is high, the other is low. Blood tests that measure levels of both proteins may more accurately assess the risk of developing prostate cancer than testing PSA alone.

Yet another way that introns may increase the number of proteins over the number of genes is that a DNA sequence that

Figure 11.10 Exons provide flexibility. Alternative splicing enables a cell to manufacture different versions (isoforms) of a protein by adding or deleting parts. Introns are removed and exons retained.

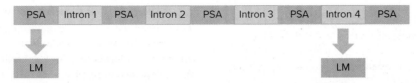

Figure 11.11 Two genes from one. Embedded in the PSA gene are two protein-encoding sequences. The PSA portion consists of five exons, and the PSA-LM part consists of two exons, one of which lies within an intron of the gene. (Not drawn to scale; introns are much larger than exons.)

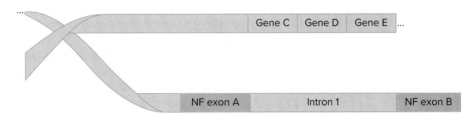

Figure 11.12 Genes in introns. An intron of the neurofibromin gene harbors three genes on the opposite DNA strand.

is an intron in a gene on the template strand may encode protein on the coding strand. This is so for the gene that encodes neurofibromin (NF). Mutations in the gene cause neurofibromatosis type 1, an autosomal dominant condition that causes benign tumors beneath the skin and spots on the skin surface (**figure 11.12**). Within an intron of the gene, but on the coding strand, are instructions for three other genes. Finding such dual meaning in a gene is a little like reading a novel backward and discovering a second story!

Genome information is maximized after translation when a protein is modified into different forms by adding sugars or lipids to create glycoproteins and lipoproteins, respectively. Many proteins on cell surfaces have these additions, marking them as part of a particular tissue, organ, and individual.

Another way that one gene can encode more than one protein is if the protein is cut in two. This happens in dentinogenesis imperfecta, which is an autosomal dominant condition that causes discolored, misshapen teeth with peeling enamel. The dentin, which is the bonelike substance beneath the enamel that forms the bulk of the tooth, is abnormal in this condition.

Dentin is a mixture of extracellular matrix proteins, 90 percent of which are forms of collagen. However, two proteins are unique to dentin: the abundant dentin phosphoprotein (DPP) and the rare dentin sialoprotein (DSP). DSP regulates mineral deposition into dentin, and DPP controls maturation of the mineralized dentin. A single gene encodes both proteins. DPP and DSP are translated from a single mRNA molecule as the precursor protein dentin sialophosphoprotein (DSPP) (**figure 11.13**). Then they are

Figure 11.13 Sculpting two genes from one, in a different way than in figure 11.11. In dentinogenesis imperfecta, a large precursor protein is cut in two, but one of the products is degraded much faster than the other.

separated. The two proteins are present in differing amounts because one (DSP) is degraded much faster than the other (DPP). The disease dentinogenesis imperfecta results from abnormal comparative levels of the two final proteins, DSP and DPP.

Table 11.1 summarizes mechanisms that maximize genetic information.

Table 11.1 Mechanisms That Maximize Genetic Information

Mechanism	Example
Alternative splicing	Cell surface and secreted forms of antibodies
An intron encoding one isoform is an exon in another	Prostate specific antigen (PSA) and PSA-linked molecule (PSA-LM)
An intron on one DNA strand is an exon on the other	Neurofibromin and three other genes
Adding sugars to form glycoproteins or lipids to form lipoproteins	Cell surface molecules important in cell-cell recognition
Precursor protein is cut to yield two proteins	Dentinogenesis imperfecta

Key Concepts Questions 11.3

*a. List ways that the number of different proteins encoded by the genome differs from the number of genes.

b. Explain how alternative splicing, introns that encode protein, protein modifications, and cutting a precursor protein maximize the number of proteins that DNA encodes.

11.4 Most of the Human Genome Does *Not* Encode Protein

Only about 1.5 percent of human DNA encodes protein. The rest of the genome includes viral sequences, sequences that encode RNAs other than mRNA, introns, promoters and other control sequences, and repeated sequences. Most of the genome is transcribed—a DNA sequence is not "junk" if it does not encode protein or we have not discovered a function.

Viral DNA

Our genomes include DNA sequences that represent viruses. Viruses are nonliving infectious particles that consist of a nucleic acid (DNA or RNA) encased in a protein coat (see Clinical Connection 19.1). A virus mass produces itself using the infected cell's transcriptional and translational machinery. New viruses may exit the host cell, or the viral nucleic acid may remain in the cell. A DNA virus may insert its DNA into a host chromosome or remain within the nucleus in a circle of DNA called an episome. A class of RNA viruses called **retroviruses,** which includes HIV, first uses an enzyme called **reverse transcriptase** to copy its genetic material into DNA, which then inserts into a host chromosome.

About 100,000 sequences in our DNA, of varying lengths and comprising about 8 percent of the genome, were once retroviruses. The name refers to a retrovirus's direction of genetic information transfer, which is opposite DNA to RNA to protein. Retroviral sequences in our chromosomes are termed *endogenous* because they are carried from generation to generation of the host, rather than acquired as an acute infection. The retroviruses whose genetic material is in our chromosomes are called human endogenous retroviruses, or HERVs.

HERVs likely infected our ancestors' genomes about 5 million years ago and have been retained in our own. Researchers deduced this by comparing HERV sequences to similar viruses in other primates. Most HERV sequences have exchanged parts (recombined) and mutated to the extent that they no longer make us sick. Harmless HERVs silently pass from human generation to generation as parts of our chromosomes, increasing in number with time. However, several small studies show that people with certain diseases (forms of amyotrophic lateral sclerosis, multiple sclerosis, and melanoma) have overexpressed HERVs.

Noncoding RNAs

Much more of a human genome is transcribed than would be predicted based on the number and diversity of proteins in a human body. The two general classes of RNAs are coding (the mRNAs) and noncoding (ncRNAs), which include everything else. The best-studied noncoding RNAs are the tRNAs and rRNAs. Table 11.2 lists noncoding RNAs.

The rate of transcription of a cell's tRNA genes is attuned to cell specialization. The proteins of a skeletal muscle cell, for example, require different numbers of certain amino acids than the proteins of a white blood cell, and therefore different numbers of the corresponding tRNAs. Human tRNA genes are dispersed among the chromosomes in clusters. Altogether, our 500 or so types of tRNA genes account for 0.1 percent of the genome.

The 243 types of rRNA genes are grouped on six chromosomes, each cluster harboring 150 to 200 copies of a 44,000-base repeat sequence. Once transcribed from these clustered genes, the rRNAs go to the nucleolus, where another type of ncRNA called small nucleolar RNA (snoRNA) cuts them into their final forms.

Hundreds of thousands of noncoding RNAs are not tRNA, rRNA, snoRNAs, microRNAs, nor the other less abundant types described in table 11.2. Some noncoding RNAs correspond to DNA sequences called **pseudogenes**. A pseudogene is similar in sequence to a protein-encoding gene that may be transcribed, but it is not translated into protein, because it is altered in sequence from an ancestral gene in a way that may impair its translation or folding. Pseudogenes may be remnants of genes that functioned in the past, variants that diverged from the normal sequence too greatly to encode a working protein. The alpha and beta globin gene complexes (see figure 11.1) include pseudogenes.

Another class of RNA molecules are the 12,000 **long noncoding RNAs,** which are each 200 or more nucleotides long. They are transcribed from exons, introns, and regions between genes, and they associate with chromatin and control gene expression. A third of the long noncoding RNAs are found only

Table 11.2 Some Non-Protein-Encoding Parts of the Human Genome

Type of Sequence	Function or Characteristic
Viral DNA	Evidence of past infection
Noncoding RNA genes	
tRNA genes	Connect mRNA codons to amino acids
rRNA genes	Parts of ribosomes
Long noncoding RNAs	Control of gene expression
Pseudogenes	DNA sequences very similar to known genes that are not translated
Piwi-interacting RNA (piRNA)	Keeps transposons out of germline cells
Large intergenic noncoding RNAs	Between genes
Small nucleolar RNAs (snoRNAs)	Process rRNAs in nucleolus
Small nuclear RNAs (snRNAs)	Parts of spliceosomes
Telomerase RNA	Adds bases to chromosome tips
Xist RNA	Inactivates one X chromosome in cells of females
Introns	Parts of genes that are cut out of mRNA
Promoters and other control sequences	Guide enzymes that carry out DNA replication, transcription, or translation
Small interfering RNAs (siRNAs)	Control translation
MicroRNAs (miRNAs)	Control translation of many genes
Repeats	
Transposons	Repeats that move around the genome
Telomeres	Protect chromosome tips
Centromeres	Largest constrictions in chromosomes, providing attachment points for spindle fibers
Segmental duplications	Repeated DNA sequences of >1,000 bases long
Simple sequences (Alu and others)	Functions unknown

in the genomes of primates, and many are transcribed only in the brain. These findings suggest that the long noncoding RNAs may hold clues to what makes us human—a topic discussed in chapter 17. More practically, low levels of a noncoding RNA called lnc13 have been correlated to the overactive immune response that lies behind gluten intolerance (celiac disease).

Repeats

Human genomes are riddled with highly repetitive sequences that hold a different type of information than a protein's amino acid sequence. The most abundant type of repeat is a sequence of DNA that can move about the genome. It is called a transposable element, or **transposon** for short. Geneticist Barbara McClintock originally identified transposons in corn in the 1940s, and they were rediscovered in bacteria in the 1960s.

Transposons comprise about 45 percent of a human genome sequence, and typically are present in many copies. Some transposons include parts that encode enzymes that cut them out of one chromosomal site and integrate them into another. Unstable transposons may lie behind inherited diseases that have several symptoms, because they insert into different genes. This is the case for Rett syndrome (see the chapter opener for chapter 2).

An example of a specific type of repeat is an Alu sequence. Each Alu repeat is about 300 bases long, and a human genome may contain 300,000 to 500,000 of them. Alu repeats comprise 2 to 3 percent of the genome, and they have been increasing in number over time because they can copy themselves. We don't know exactly what these common repeats do, if anything. They may serve as attachment points for proteins that bind newly replicated DNA to parental strands before anaphase of mitosis, when replicated chromosomes pull apart.

Rarer classes of repeats are the telomeres, centromeres, duplications of 10,000 to 300,000 bases, copies of pseudogenes, and simple repeats of one, two, or three bases. In fact, the entire human genome may have duplicated once or even twice.

Repeats may make sense in light of evolution, past and future. Pseudogenes are likely vestiges of genes that functioned in our nonhuman ancestors. Perhaps the repeats that seem to have no obvious function today will serve as raw material from which future genes may arise by mutation. **Figure 11.14** summarizes parts of a human genome.

Key Concepts Questions 11.4

a. How are viruses part of our genomes?
*b. What does reverse transcriptase do?
c. List five types of non-protein-encoding RNAs.
*d. Define *transposon*.

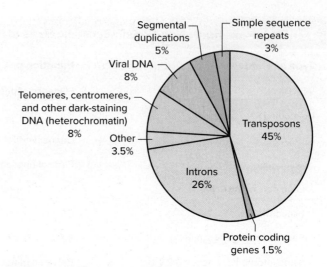

Figure 11.14 **Parts of a human genome.** A pie chart shows the major parts, as labeled.

Study Guide

Summary

11.1 Gene Expression Through Time and Tissue

1. Changes in gene expression occur over time at the molecular and organ levels.
2. The **transcriptome** is the collection of mRNAs in a cell. Single-cell transcriptomics catalogs the mRNAs in a cell and **proteomics** catalogs the proteins in a cell.

11.2 Control of Gene Expression

3. In **chromatin remodeling**, acetylation and phosphorylation of certain histone proteins enables the transcription of associated genes, whereas methylation selectively prevents transcription. These changes are epigenetic.
4. **MicroRNAs** are short RNAs that bind to certain sets of mRNAs, blocking their translation. **Microproteins** are very small proteins that can fit into the nooks and crannies of larger proteins, affecting their function.

11.3 Maximizing Genetic Information

5. A small part of the genome encodes protein, but the number of proteins is much greater than the number of genes.
6. Alternative splicing, use of introns, protein modification, and cutting of a precursor protein into two proteins contribute to protein diversity.

11.4 Most of the Human Genome Does *Not* Encode Protein

7. The non-protein-encoding part of the genome includes viral sequences, sequences that encode noncoding RNAs, **pseudogenes**, introns, promoters and other controls, and repeats such as **transposons**.
8. A type of RNA virus called a **retrovirus** uses **reverse transcriptase** to copy DNA from RNA. This action enables viruses to insert their genetic material into human chromosomes.
9. **Long noncoding RNAs** control gene expression.

Key Terms

chromatin remodeling (krō′ mə-tĭn) Adding or removing chemical groups to or from histones, which can alter gene expression.

long noncoding RNA An RNA molecule that associates with chromatin in the nucleus and likely helps control gene expression, especially in the brain.

microRNA A 21- or 22-base-long RNA that binds to certain mRNAs, blocking their translation into protein.

proteomics (prō′tē-ōm-ix) Cataloging all of the proteins a specific cell type makes under specific conditions.

pseudogene (sōō′ dō jēn) A DNA sequence that does not encode a protein but whose sequence is highly similar to that of a known coding gene.

retrovirus A subset of viruses that have RNA for the genetic material.

reverse transcriptase An enzyme originally from viruses that copies an RNA sequence into a DNA sequence and is used in several biotechnologies.

transcriptome The collection of mRNA molecules in a cell.

transposon (trăns-pōzŏn) A transposable element, or sequence of DNA that can move within the genome.

Review Questions

1. Define *housekeeping genes*.
2. What does single-cell transcriptomics evaluate?
3. What is the environmental signal that the composition of the globin molecule is an adaptation to?
4. How does development of the pancreas illustrate differential gene expression?
5. Explain how a mutation in a promoter can affect gene expression.
6. How do histones control gene expression, yet genes also control histones?
7. What controls whether histones enable the DNA wrapped around them to be transcribed?
8. State two ways that methyl groups control gene expression.
9. Name a mechanism that silences transcription of a gene and a mechanism that blocks translation of an mRNA.
10. Explain why evaluating microRNA functions in human genomes is computationally complex.
11. How does a microprotein affect another protein?
12. List three reasons why the number of proteins exceeds the number of protein-encoding genes in human genomes.
13. Explain how alternative splicing can generate more than one type of protein from the information in a gene.
14. In the 1960s, a gene was defined as a continuous sequence of DNA that is located permanently at one place on a chromosome and specifies a sequence of amino acids from one strand. List three ways this definition has changed.
15. What is the evidence that some long noncoding RNAs may hold clues to human evolution?

Applied Questions

1. The World Anti-Doping Agency warns against gene doping, which it defines as "the non-therapeutic use of cells, genes, genetic elements, or of the modulation of gene expression, having the capacity to improve athletic performance." The organization lists the following genes as candidates for gene doping when overexpressed:

    Insulin-like growth factor (*IGF-1*)
    Growth hormone (*GH*)
    Erythropoietin (*EPO*)
    Vascular endothelial growth factor (*VEGF*)
    Fibroblast growth factor (*FGF*)

    Select one of these genes, look up information about it, and explain how its overexpression might improve athletic performance.

2. What might be the effect of a mutation in the part of the gamma globin gene that normally binds a transcription factor in an adult?

3. Researchers compared the pattern of microRNAs in breast cancer tumors from women whose disease did not spread to the brain to the pattern in tumors that did spread to the brain. Suggest a clinical application of this finding.

4. Several new drugs inhibit the enzymes that either put acetyl groups on histones or take them off. Would a drug that combats a cancer caused by underexpression of a gene that normally suppresses cell division add or remove acetyl groups?

5. Chromosome 7 has 863 protein-encoding genes, but encodes many more than this number of proteins. The average gene is 69,877 bases long, but the average mRNA is 2,639 bases long. Explain both of these observations.

6. How many different proteins encoded by two exons can be produced from a gene that has three exons?

Forensics Focus

1. Explain how gene expression profiling of humans might be used to more precisely define the postmortem interval than was possible by identifying the ages of larvae collected from a decomposing corpse.

Case Studies and Research Results

1. Kabuki syndrome is named for the resemblance of an affected individual to a performer wearing the dramatic makeup used in traditional Japanese theater called Kabuki. The face has long lashes, arched eyebrows, flared eyelids, a flat nose tip, and large earlobes. The syndrome is associated with many symptoms, including developmental delay and intellectual disability, seizures, a small head (microcephaly), weak muscle tone, fleshy fingertips, cleft palate, short stature, hearing loss, and heart problems. Both genes associated with the condition result in too many regions of closed chromatin. Drugs that inhibit histone deacetylases (enzymes that remove acetyl groups from histone proteins) are effective. Explain how the drugs work.

2. To make a "reprogrammed" induced pluripotent stem (iPS) cell (see figure 2.19), researchers expose fibroblasts taken from skin to "cocktails" that include transcription factors. The fibroblasts divide and give rise to iPS cells, which, when exposed to other transcription factors, divide and yield daughter cells that specialize in distinctive ways that make them different from the original fibroblasts. How do transcription factors orchestrate these changes in cell type?

3. Using an enzyme called DNAse 1, researchers can determine which parts of the genome are in the "open chromatin" configuration in a particular cell. How could this technique be used to develop a new cancer treatment?

CHAPTER 12

Gene Mutation

"The Blue People," copyright ©1982–2020 by Walt Spitzmiller. All Rights Reserved

The Fugate family of Troublesome Creek, Kentucky, has a mutation that affects hemoglobin. Artist Walt Spitzmiller painted this haunting image based on photographs.

Learning Outcomes

12.1 The Nature of Gene Variants
1. Distinguish between mutation and mutant.
2. Define *polymorphism*.
3. Distinguish between germline and somatic mutations.

12.2 A Closer Look at Two Mutations
4. Describe mutations in the genes that encode beta globin and collagen.

12.3 Allelic Diseases
5. Provide examples of how mutations in a single gene can cause more than one illness.

12.4 Causes of Mutation
6. Explain how mutations arise spontaneously and how they may be induced.

12.5 Types of Mutations
7. Describe the two types of single-base mutations.
8. Explain the consequences of a splice-site mutation.
9. Discuss mutations that add, remove, or move DNA nucleotides.
10. Describe how pseudogenes and transposons can cause mutations.

12.6 The Importance of Position
11. Give examples of how the location of a mutation in a gene affects the phenotype.
12. Describe a conditional mutation.

12.7 DNA Repair
13. List the types of damage that DNA repair mechanisms fix.
14. Describe the types of DNA repair.

The BIG Picture

The informational nature of DNA means that many variants of each gene are possible. Many variant combinations (genotypes) cause disease or increase susceptibility to a disease, yet some may be protective. The nature of the genetic code as well as natural DNA repair mechanisms minimize the effects of mutation and DNA damage.

The Blue People of Troublesome Creek

The globin genes that encode the four protein subunits of the hemoglobin molecule are mutant in many ways. A rare but very noticeable condition of abnormal hemoglobin affects the "blue people of Troublesome Creek." Seven generations ago, in 1820, a French orphan named Martin Fugate who settled in this area of Kentucky brought in a mutation that causes methemoglobinemia type I, also known as "blue person disease." Martin's mutation was in the *CYP5R3* gene, which encodes an enzyme (cytochrome b_5 methemoglobin reductase) that normally catalyzes a reaction that converts a type of hemoglobin with poor oxygen affinity, called methemoglobin, back into normal hemoglobin by adding an electron.

Martin was a heterozygote, and slightly bluish. His wife, Elizabeth Smith, also had one mutation, and four of their seven children were blue, those inheriting two mutations, one from each parent, bluer than those inheriting only one. After extensive inbreeding in the isolated community—their son married his aunt, for example—a large pedigree of "blue people" of both sexes arose. (Exposure to certain drugs causes a noninherited form of the condition.)

In "blue person disease," excess oxygen-poor hemoglobin causes a dark blue complexion. Heterozygotes may have bluish lips and fingernails at birth, which usually lighten. Treatment is simple: A tablet of methylene blue, a commonly used dye, or of ascorbic acid (vitamin C) for mild cases, adds the electron back to methemoglobin, converting it to normal hemoglobin.

In most members of the Fugate family, blueness was the only symptom. Normally, less than 1 percent of hemoglobin molecules are the methemoglobin form, which binds less oxygen. The Fugates had 10 to 20 percent in this form. People with the inherited condition who have more than 20 percent methemoglobin may suffer seizures, heart failure, and even death. The disease remains rare. It is still seen in the Kentucky family, among certain families in Alaska and Algeria, and among Navajo Indians.

12.1 The Nature of Gene Variants

A mutation is a change in a DNA sequence. *Mutate* refers to the process of altering a DNA sequence. Mutations range in magnitude from substitution of a single DNA base; to deletion or duplication of tens, hundreds, thousands, or even millions of bases; to deletion or duplication of entire chromosomes. This chapter discusses smaller-scale mutations, and chapter 13 considers mutations at the chromosomal level. However, the extent of mutation is a continuum.

Mutation can affect any part of the genome: sequences that encode proteins or control transcription, introns, repeats, and sites critical to intron removal and exon splicing. Not all DNA sequences are equally likely to mutate. A mutation is not always a change from wild type. A mutation associated with a phenotype can actually mutate again and revert to wild type.

The effects of mutation vary. Mutations may impair a function, have no effect, or even be beneficial. A deleterious (harmful) mutation can stop or slow production of a protein, overproduce it, or impair the protein's function—such as altering its secretion, location, or interaction with another protein. The effect of a mutation is called a "loss-of-function" when the gene's product is reduced or absent, or a "gain-of-function" when the gene's activity changes. Most mutations are recessive and cause a loss-of-function (see figure 4.8). Gain-of-function mutations tend to be dominant and are also called *toxic*.

The ability to sequence exomes and genomes has revealed people who have loss-of-function mutations but are healthy. These mutations are also called *knockouts*. Probing further into these peoples' genomes can reveal variants in other genes that counteract the loss of function, which is information that can inspire research into new treatment strategies.

Because a gene is an informational molecule and mutations alter a DNA sequence, a gene can exist in different forms, or variants. A variant present in more than one percent of a population is termed a **polymorphism**. Most of the single nucleotide polymorphisms (SNPs) used in genome-wide association studies are common and typically each contributes only a small degree to an illness or other phenotype. Therefore, they are transmitted to future generations because people who have them are healthy enough to reproduce. In contrast, mutations that cause severe diseases are rare if they impair reproduction in homozygotes or are lethal before sexual maturity.

The term *variant* refers to both polymorphisms and mutations. The **Bioethics** box at the end of this chapter discusses the clinical challenges of identifying and reporting to patients "variants of uncertain significance" that are discovered through genetic testing.

A mutation may be helpful. For example, about 1 percent of the general population is homozygous for a recessive allele that encodes a cell surface protein called CCR5 (see figure 19.11). To infect an immune system cell, HIV must bind CCR5 and another protein. Because the mutation prevents CCR5 from moving to the cell surface from inside the cell, HIV cannot bind. Homozygotes for a *CCR5* mutation are highly resistant to HIV infection and heterozygotes are partially protected. The opener to chapter 19 describes how mimicking the *CCR5* mutation treats HIV infection.

The term *mutation* refers to genotype—that is, a change at the DNA or chromosome level. The familiar term *mutant* refers to phenotype and also describes an allele. The nature of a mutant phenotype depends upon how the mutation affects the gene's product or activity, and usually connotes an abnormal or unusual characteristic. However, a mutant phenotype may also be a rare variant that is nevertheless "normal," such as red hair. The ability to digest lactose arose as a mutation (see the opener to chapter 16).

Germline and Somatic Mutation

A mutation may be present in all the cells of an individual or only in some cells. In a **germline mutation,** the change occurs

Figure 12.1 Somatic mutation causes pigment patches on golden retrievers and labrador retrievers. *All rights reserved by Sally McBurney*

during the DNA replication that precedes *meiosis*. The resulting gamete and all the cells that descend from it after fertilization have the mutation—that is, every cell in the body. Germline mutations are transmitted to the next generation of individuals.

A **somatic mutation** happens during DNA replication before *mitosis*, and is passed to the next generation of cells, but not to all the cells in the individual's body. All the cells that descend from the original changed cell are altered, but they might only comprise a small part of the body. A person with a somatic mutation has **somatic mosaicism**. Somatic mutations are more likely to occur in cells that divide often, such as skin and blood cells, because there are more opportunities for DNA replication errors.

Figure 12.1 shows how a somatic mutation creates dark patches in the fur of some golden retrievers. A dog with genotype e/e of the *MC1R* color gene (see the chapter 5 opener) is golden or cream-colored, whereas a dog who is genotype E/e is black or chocolate. A somatic mutation in a cell before birth that changes e/e to E/e can result, after many cell divisions and migration of pigmented cells, in a dog with a patch of a different pigment.

Some genetic diseases in humans are so severe that they are only seen in mosaics. That is, having the mutation in every cell is incompatible with life. A mutation occurs in one cell of an early embryo, and leads to an individual with patches of tissue that descend from the cell in which the mutation originated, like the spotted golden retriever but in a different gene. Patterns arise as affected cells migrate during development.

This is what happens in Proteus syndrome, which occurs *de novo* (not inherited) and is known to affect only a few hundred people worldwide. They have overgrowth of skin and bone, but only in the body parts that bear a mutation in the gene *AKT1* (**figure 12.2a**). The different growth rates in different parts of the body over time lead to disfigurement, apparent by 18 months of age (**figure 12.2b**). The disfigurement worsens over time.

Figure 12.2
Somatic mosaicism.
(a) In an early embryo, this mutation leads to patches of the body in which cells have the mutation in *AKT1*.
(b) Note the overgrown fingers and hands of this young man who has Proteus syndrome. His neck and spine are also enlarged. Only some of his cells have the mutation. *(b): Courtesy The Proteus Syndrome Foundation, UK*

The *AKT1* gene mutates in one cell of an early embryo.

The cells of the embryo divide, including the cell with the *AKT1* mutation. Cells migrate.

The fetus has two cell populations: A larger one with cells that have non-mutant *AKT1* genes, and a smaller population of cells bearing the mutation.

After birth, cells with the mutation grow faster, leading to body asymmetry.

Chapter 12 *Gene Mutation*

Somatic mutations occur as errors during DNA replication at about every 300 mitotic cell divisions. Researchers confirmed the fact that somatic mutations are more likely to happen in cell types that divide frequently when they autopsied a woman named Henne Holstege, who died in 2005 at the age of 115. Whole genome sequencing of Henne's white blood cells revealed 450 mutations, but her brain neurons had none. White blood cells descend from many divisions of stem cells, providing many opportunities for replication error, but neurons never divide. Her white blood cells had many opportunities for a somatic mutation to arise. Chapter 20 considers how series of somatic mutations cause and "drive" the progression of most cancers.

Sequencing the genomes of individual cells from different organs from the same individual has shown that somatic mosaicism is not as rare as had been thought. In addition to somatic mutation, mosaicism can result from having a fraternal twin that died before birth and incorporating some of its cells, and, for women, from fetal cells retained from a pregnancy. Most of us may be, in some way, genomic mosaics.

Figure 12.3 **Sickle cell disease results from a single DNA base change that substitutes one amino acid in the protein (valine replaces glutamic acid).** This changes the surfaces of the hemoglobin molecules, and they aggregate into long, curved rods that deform the red blood cell. The illustration shows the appearance of sickle cells.

> **Key Concepts Questions 12.1**
>
> *a. What is a mutation?
> *b. Define *polymorphism*.
> c. Distinguish between the consequences of a germline and a somatic mutation.
> d. Explain why some conditions are only seen as somatic mosaics.

12.2 A Closer Look at Two Mutations

Identifying how a mutation causes symptoms has clinical applications, and also reveals the workings of biology. Following are two examples of well-studied mutations that cause disease.

The Beta Globin Gene Revisited

The first genetic illness understood at the molecular level was sickle cell disease. The tiny mutation that causes it substitutes the amino acid valine for the glutamic acid that is normally the sixth amino acid in the beta globin polypeptide chain (**figure 12.3**). At the DNA level, the change is even smaller—a CTC is changed to a CAC, corresponding to RNA codons GAG and GUG. Valine at this position changes the surfaces of hemoglobin molecules so that in low-oxygen conditions they attach at many more points than they would if the wild type glutamic acid were at the site. The aggregated hemoglobin molecules form ropelike cables that make red blood cells sticky and able to deform, and then bend the cells into rigid, fragile sickle shapes. The cells lodge in narrow blood vessels, cutting off local blood supplies. Once a blockage occurs, sickling speeds up and spreads, as the oxygen level falls. The result is great pain in the blocked body parts, particularly the hands, feet, and intestines. The bones ache, and depletion of normal red blood cells causes the great fatigue of anemia. At the same time, the liver and spleen rapidly destroy the sickled cells, causing the anemia that gave the condition its original name, sickle cell anemia.

Other conditions result from mutations in the beta globin gene. In 1925, Thomas Cooley and Pearl Lee described severe anemia in Italian children, and in the decade following, others described a milder version of "Cooley's anemia," also in Italian children. The disease was named thalassemia, from the Greek for "sea," in light of its high prevalence in the Mediterranean area. The two anemias turned out to be the same. The severe form, sometimes called thalassemia major, results from a homozygous mutation in the beta globin gene at a site other than the one that causes sickle cell disease. The milder form, called thalassemia minor, affects some individuals who are heterozygous for the mutation.

Once researchers had worked out the structure of hemoglobin, and learned that different globins function in the embryo and fetus (see figure 11.2), the molecular basis of thalassemia became clear. The disease that is common in the Mediterranean is more accurately called beta thalassemia, because the symptoms result from too few beta globin chains. Without them, not enough hemoglobin molecules are assembled to effectively deliver oxygen to tissues. Fatigue and bone pain arise during the first year of life as the child depletes fetal hemoglobin, and the "adult" beta globin genes are not transcribed and translated on schedule.

As severe beta thalassemia progresses, red blood cells die because the excess of alpha globin chains prevents formation of hemoglobin molecules. Liberated iron slowly destroys the heart, liver, and endocrine glands. Periodic blood transfusions can control the anemia, but they hasten iron buildup and organ damage. Drugs called chelators that entrap the iron can extend life past early adulthood, but they are costly and not available in some nations.

Disorders of Orderly Collagen

The protein collagen is a major component of connective tissue. Collagen accounts for more than 60 percent of the protein in bone and cartilage and provides 50 to 90 percent of the dry weight of skin, ligaments, tendons, and the dentin of teeth.

Genetic control of collagen synthesis and distribution is complex; more than 35 collagen genes encode more than 20 types of collagen molecules. Other genes affect collagen, too. Mutations in the genes that encode collagen, not surprisingly, lead to a variety of medical conditions (**table 12.1**). These disorders are devastating because collagen is nearly everywhere, and the molecules have a precise conformation that is easily disrupted, even by slight alterations that might have little effect in proteins with other shapes (**figure 12.4**).

Collagen is trimmed from a longer precursor molecule called procollagen, which consists of many repeats of the amino acid sequence *glycine-proline-modified proline*. Three procollagen chains entwine. Two of the chains are identical and are encoded by one gene, and the other is encoded by a second gene and has a different amino acid sequence. The

A GLIMPSE OF HISTORY

In 1904, a young medical intern Ernest Irons noted "many pear-shaped and elongated forms" in a blood sample from a dental student in Chicago who had anemia. Irons sketched this first view of sickle cell disease at the cellular level, and reported his findings to his supervisor, physician James Herrick. Alas, Herrick published the work without including Irons and has been credited with the discovery ever since.

In 1949, Linus Pauling found that hemoglobin from healthy people and from people with the anemia, when placed in a solution in an electrically charged field, moved to different positions. Hemoglobin molecules from the parents of people with the anemia, who were carriers, moved to both positions.

The difference between the two types of hemoglobin lay in beta globin. Recall from figure 11.1 that adult hemoglobin consists of two alpha polypeptide subunits and two beta subunits. Protein chemist V. M. Ingram took a shortcut to localize the sickle cell mutation in the 146-amino-acid-long beta subunit. He cut normal and sickle hemoglobin with a protein-digesting enzyme, separated the pieces, stained them, and displayed them on filter paper. The patterns of fragments—known as peptide fingerprints—were different for the two types of beta globin. This meant, Ingram deduced, that the two molecules differ in amino acid sequence. One piece of the molecule in the fingerprint, fragment four, occupied a different position for each type of beta globin. Because this peptide was only eight amino acids long, Ingram needed to decipher only that short sequence to find the site of the mutation. It was a little like finding which sentence on a page contains a miskeyed word.

Table 12.1 Some Collagen Disorders

Disorder	Mutations (Genotype)	Signs and Symptoms (Phenotype)
Alport syndrome	Mutations in any of three genes (*COL4A3, COL4A4, COL4A5*) affect type IV collagen, which disrupts tissue boundaries.	Deafness and inflamed kidneys
Chondrodysplasia	Deletion, insertion, or missense mutation replaces Gly with bulky amino acids in *COL2A1* type II collagen gene.	Stunted growth, deformed joints
Dystrophic epidermolysis bullosa	Mutation in *COL7A1* gene that encodes type VII collagen breaks down fibrils that attach epidermis to dermis.	Skin blisters upon any touch
Ehlers-Danlos syndrome	Diverse mutations in at least a dozen genes affect collagens or the molecules to which they bind.	Stretchy, easily scarred skin, lax joints
Osteoarthritis	Missense mutation in α1 collagen gene (*COL1A1*) substitutes Cys for Arg.	Painful joints
Osteogenesis imperfecta type I	Inactivation of α1 collagen gene (*COL1A1* or *COL1A2*) reduces number of collagen triple helices by 50%.	Easily broken bones; blue eye whites; deafness
Stickler syndrome	Nonsense mutations in type II procollagen gene (*COL2A1* or *COL11A1*) reduce number of collagen molecules.	Joint pain, degeneration of vitreous gel and retina

Figure 12.4 Collagen has a precise conformation. The α1 collagen gene encodes the two blue polypeptide chains, and the α2 procollagen gene encodes the third (red) chain. The procollagen triple helix is shortened before it becomes functional, forming the fibrils and networks that comprise much of the human body. The inset shows aligned collagen fibrils. *(photo): Science Photo Library RF/Getty Images*

electrical charges and interactions of the amino acids with water coil the procollagen chains into a very regular triple helix, with space in the middle only for tiny glycine. Enzymes snip off the ragged ends of the polypeptides, forming mature collagen. The collagen fibrils continue to associate with each other outside the cell, building the fibrils and networks that hold the body together. **Figure 12.5** shows the characteristic stretchy skin that results from unassembled collagen molecules.

Figure 12.5 Collagen mutations can affect the skin. Ehlers-Danlos syndrome type I causes highly extensible joints and stretchy skin, due to inability of collagen molecules to assemble properly. *Biophoto Associates/Science Source*

Table 12.2 — How Mutations Cause Disease

Disease	Protein	Mutations (Genotype)	Signs and Symptoms (Phenotype)
Cystic fibrosis	Cystic fibrosis transmembrane regulator (CFTR)	Missing amino acid or other variant alters conformation of chloride channels in certain epithelial cell plasma membranes. Water enters cells, drying out secretions.	Frequent lung infection, pancreatic insufficiency
Duchenne muscular dystrophy	Dystrophin	Deletion eliminates dystrophin, which normally binds inner face of muscle cell to plasma membrane. Muscles weaken.	Gradual loss of muscle function
Familial hypercholesterolemia	LDL receptor	Deficient LDL receptors cause cholesterol to accumulate in blood.	High blood cholesterol, early heart disease
Hemophilia B	Factor IX	Absent or deficient clotting factor causes hard-to-control bleeding.	Slow or absent blood clotting
Huntington disease	Huntingtin	Extra bases add amino acids to the protein, which impairs certain transcription factors and proteasomes.	Uncontrollable movements, personality changes
Marfan syndrome	Fibrillin or transforming growth factor β receptor	Deficient proteins in lenses cause cataracts and in the wall of the aorta cause aneurysm (bursting).	Long limbs, weakened aorta, spindly fingers, sunken chest, lens dislocation
Neurofibromatosis type 1	Neurofibromin	Defect in protein that normally suppresses activity of a gene that causes cell division, leading to abnormal growths.	Pigmented skin patches and benign tumors of nervous tissue beneath skin

So important is the precision of collagen formation that a mutation that controls placement of a single hydroxyl chemical group (—OH⁻) on collagen causes a form of osteogenesis imperfecta ("brittle bone disease"). Other collagen mutations remove procollagen chains, kink the triple helix, and disrupt aggregation outside the cell.

Aortic aneurysm is a serious connective tissue abnormality that can occur by itself or as part of Marfan syndrome. Detection of mutations that cause Marfan syndrome before symptoms arise can be lifesaving, because frequent ultrasound exams can detect aortic weakening early enough to patch the vessel before it bursts. **Table 12.2** describes how other mutations mentioned in this and other chapters impair health.

> **Key Concepts Questions 12.2**
>
> a. Describe how mutations affect the beta globin gene.
> b. Explain why collagen genes are unusually prone to mutation.

12.3 Allelic Diseases

Analysis of human genomes is changing the way we describe single-gene diseases. In the past, geneticists were inconsistent when assigning disease names to mutations. For certain genes, mutations cause the same disease, yet for other genes, different mutations are said to cause different diseases. Compare the genes that cause cystic fibrosis and sickle cell disease when mutant. Mutations in the *CFTR* gene cause cystic fibrosis, which may include the full spectrum of impaired breathing and digestion, or just male infertility or frequent bronchitis. Yet different mutations in the beta globin gene cause sickle cell disease and beta thalassemia.

Mutations in some genes correspond to several diseases. This is the case for the gene *lamin A*, in which different mutations cause different diseases in different tissues. The diseases include the rapid aging condition Hutchinson-Gilford progeria syndrome (see the chapter 3 opener), muscular dystrophies, and a heart disease. Lamin A proteins form a network beneath the inner nuclear membrane that interacts with chromatin. Different mutations affect lamin A's interactions with chromatin in ways that cause the diverse associated diseases.

As researchers discovered more cases of different diseases arising from mutations in the same gene, it became clear that the phenomenon is not unusual, and is not merely a matter of how we name diseases. Different clinical phenotypes caused by mutations in the same gene are termed **allelic diseases**.

The same gene can underlie different diseases in different ways (**table 12.3**). A pair of allelic diseases may result from mutations in different parts of the gene, mutations that are localized (a single-base change) or catastrophic (a missing gene), or mutations that alter the protein in ways that affect its interactions with other proteins.

Allelic diseases may arise from a mutation that affects a protein that is used in different tissues. Some researchers are reclassifying cystic fibrosis as two allelic diseases, based on whether the lungs are affected. At least nine CF genotypes cause male infertility, pancreatitis, sinusitis, or a combination of these, but not respiratory impairment. The clinical view of this classic single-gene disease has certainly evolved as more mutations in *CFTR* have been identified.

> **Key Concepts Questions 12.3**
>
> a. How are the ways that cystic fibrosis and sickle cell disease are named inconsistent?
> *b. Explain how mutations in *lamin A* cause wide-ranging effects on the phenotype.
> *c. Define *allelic diseases*.
> d. Name a pair of allelic diseases and the gene from which they arise.

Table 12.3	Allelic Diseases	
Gene	**Function**	**Associated Diseases**
ATP7A	Copper transport	Menkes ("kinky hair") disease; peripheral neuropathy
DMD	Dystrophin muscle protein	Duchenne and Becker muscular dystrophy
FBN1	Encodes fibrillin-1, which forms tiny fibrils outside cells; a connective tissue protein	Marfan syndrome; systemic sclerosis (scleroderma; "stiff skin syndrome")
FGFR3	Fibroblast growth factor	Two types of dwarfism
GBA	Glucocerebrosidase	Gaucher disease; Parkinson disease
PSEN1	Presenilin 1 (enzyme part that trims membrane proteins)	Acne inversa; Alzheimer disease
RET	Oncogene (causes cancer)	Multiple endocrine neoplasia; Hirschsprung disease
TRPV4	Calcium channel	Peripheral neuropathy; spinal muscular atrophy

12.4 Causes of Mutation

A mutation can occur spontaneously or be induced by exposure to a chemical or radiation. An agent that causes mutation is called a **mutagen**.

Spontaneous Mutation

A spontaneous mutation can be a surprise. For example, two healthy people of normal height have a child with achondroplasia dwarfism due to a *de novo* (new) autosomal dominant mutation (see figure 5.1a). Siblings of the affected child have no higher risk of inheriting the dwarfism than anyone in the general population (unless the parents are mosaics), but each of the affected person's children faces a 50 percent chance of inheriting it.

A spontaneous mutation usually originates as an error in DNA replication. One trigger results from the tendency of free DNA bases to exist in two slightly different chemical structures, called tautomers. For extremely brief times, each base is in an unstable tautomeric form. If, by chance, such an unstable base is inserted into newly forming DNA, an error will be generated and perpetuated when that strand replicates. **Figure 12.6** shows how incorporation of the wrong base can happen.

Figure 12.6 Spontaneous mutation. DNA bases exist fleetingly in alternate forms. A replication fork encountering a base in its unstable form can insert a noncomplementary base. After another round of replication, one of the daughter cells has a different base pair than the one in the corresponding position in the original DNA. (This figure depicts two rounds of DNA replication.)

Spontaneous Mutation Rate

Different genes have different spontaneous mutation rates. The gene that, when mutant, causes neurofibromatosis type 1 (NF1; see figure 11.12), for example, has a very high mutation rate, arising in 40 to 100 of every million gametes. NF1 affects 1 in 3,000 births, about half in families with no prior cases. The gene's large size may contribute to its high mutability—there are more ways for its sequence to change, just as there are more opportunities for a misspelling to occur in a long sentence than in a short one. In contrast, the mutation that causes the clotting disorder hemophilia B happens in only 1 to 10 of every million gametes formed.

Spontaneous mutation lies behind many instances of somatic mosaicism, such as Proteus syndrome (see figure 12.2). Spontaneous mutation also manifests as **gonadal mosaicism**. In this situation, a parent has a mutation in some sperm or oocytes, because a spontaneous mutation occurred in the developing testis or ovary and was transmitted only to the cells descended from the original cell bearing the mutation. Gonadal mosaicism is suspected when more than one child in a family has a genetic condition but both parents do not have the mutation in the cells sampled for genetic testing. That is, two siblings may have cystic fibrosis, yet their parents are not carriers based on blood tests.

Comparing the genome sequences of parents and children has revealed that spontaneous mutations are fairly common. At least 10 percent of heart defects in newborns, for example, are due to mutations not present in the parents. Each of us has about 175 spontaneously mutated alleles. Mitochondrial genes mutate at a higher rate than genes in the nucleus because mitochondria cannot repair their DNA (see section 12.7).

Estimates of the spontaneous mutation rate for a particular gene are usually derived from observations of new, dominant conditions, such as achondroplasia. This is possible because a new dominant mutation is detectable simply by observing the phenotype. In contrast, a new recessive mutation is not obvious until two heterozygotes have a homozygous recessive child with a noticeable phenotype.

The spontaneous mutation rate for autosomal genes can be estimated using this formula: number of *de novo* cases/$2X$, where X is the number of individuals examined. The denominator has a factor of 2 to account for the nonmutated homologous chromosome.

Spontaneous mutation rates in human genes are difficult to assess because our generation time is long—usually 20 to 30 years. In bacteria, a new generation arises every half hour or so, and mutation is therefore much more frequent. The genetic material of some viruses also spontaneously mutates rapidly, because they reproduce quickly and do not repair DNA errors.

Mutational Hot Spots

In some genes, mutations are more likely to occur in regions called hot spots, where sequences are repetitive. It is as if the molecules that guide and carry out replication become "confused" by short repeated sequences, much as an editor scanning a manuscript might miss the spelling errors in the words "hipppopotamus" and "bananana" (**figure 12.7**).

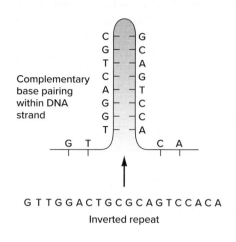

Figure 12.7 **DNA symmetry may increase the likelihood of mutation.** These examples show repetitive and symmetrical DNA sequences that may "confuse" replication enzymes, causing errors.

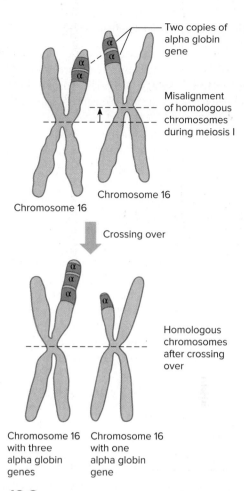

Figure 12.8 **Gene duplication and deletion.** The repeated alpha globin genes are prone to mutation by mispairing during meiosis.

The increased incidence of mutations in repeats has a physical basis. Within a gene, when DNA strands locally unwind to replicate in symmetrical or repeated sequences, bases located on the same strand may pair. A stretch of ATATAT might pair with TATATA elsewhere on the same strand, creating a loop that interferes with replication and repair enzymes. Errors may result. This is the case for mutations in the gene for clotting factor IX, which causes hemophilia B. Mutations occur 10 to 100 times as often at any of 11 sites in the gene that have extensive direct repeats of CG than they do elsewhere in the gene.

Small additions and deletions of DNA bases are more likely to occur near sequences called palindromes (figure 12.7). These sequences read the same, in a 5′ to 3′ direction, on complementary strands. Put another way, the sequence on one strand is the reverse of the sequence on the complementary strand. Palindromes probably increase the spontaneous mutation rate by disturbing replication.

The blood disease alpha thalassemia illustrates the confusing effect of direct (as opposed to inverted) repeats of an entire gene. A person who does not have the disease has four genes that specify alpha globin chains, two next to each other on each chromosome 16. Homologs with repeated genes can misalign during meiosis when the first sequence on one chromosome lies opposite the second sequence on the homolog. If such a misalignment happens, crossing over can result in a sperm or oocyte that has one or three alpha globin genes instead of the normal two (**figure 12.8**). Fertilization with a normal gamete then results in a zygote with one extra or one missing alpha globin gene. At least three dozen conditions result from this unequal crossing over, including colorblindness (see Clinical Connection 6.1).

The number of alpha globin genes affects health. A person with only three alpha globin genes produces enough hemoglobin, and is a healthy carrier. Individuals with only two copies of the gene are mildly anemic and tire easily, and a person with a single alpha globin gene is severely anemic. A fetus lacking all alpha globin genes does not survive. Spontaneous mutations are also more likely to occur near regions where crossovers occur, and are more common in genomes of males. Section 14.3 describes how this was recently discovered.

Induced Mutation

Researchers can infer a gene's normal function by observing what happens when a mutation alters it. Because the spontaneous mutation rate is much too low to be a practical source of genetic variants for experiments, researchers make mutants. They use mutagens on "model" organisms such as mice and fruit flies. These types of experiments can yield insights into human health because many genes are alike in DNA sequence among species.

Intentional Use of Mutagens

Chemicals or radiation are used to induce mutation. Alkylating agents, for example, are chemicals that remove a DNA base, which is replaced with any of the four bases—three of which are a mismatch against the complementary strand. Dyes called acridines add or remove a single DNA base. Because the DNA sequence is read three bases in a row, adding or deleting a single base can destroy a gene's information, altering the amino acid sequence of the encoded protein. Several other mutagenic chemicals alter base pairs, so that an A-T replaces a G-C, or vice versa. X-rays and other forms of radiation delete a few bases or break chromosomes.

Researchers have several ways to test the mutagenicity of a substance. The best known, the Ames test, assesses how likely a substance is to harm the DNA of rapidly reproducing bacteria. One version of the test uses a strain of *Salmonella* that cannot grow when the amino acid histidine is absent from its medium. If exposure to a substance enables bacteria to grow on the deficient medium, then a gene has mutated that allows it to do so.

In a variation of the Ames test, researchers exposed human connective tissue cells growing in culture to liquefied cigarette smoke. The chemicals from the smoke cut chromosomes through both DNA strands. Broken chromosomes can join with each other in different ways that can activate cancer-causing genes. Hence, the experiment may have modeled one way that cigarettes cause cancer. Because many mutagens are also carcinogens (cancer-causing agents), the substances that the Ames test identifies as mutagens may also cause cancer. Common products that contain mutagens are hair dye, smoked meats, certain flame retardants used in children's sleepwear, and food additives.

A limitation of using a mutagen is that it cannot cause a specific mutation. In contrast, a technique called site-directed mutagenesis changes a gene in a desired way. A gene is mass produced (see section 9.3), but the copies include an intentionally substituted base, just as an error in a manuscript is printed in every copy of a book. Site-directed mutagenesis is faster and more precise than waiting for nature or a mutagen to produce a useful gene variant. The gene editing techniques described in sections 21.4 and 22.4 can also rapidly introduce changes to a DNA sequence.

Accidental Exposures to Mutagens

Some mutagen exposure is unintentional. Such exposure occurs from workplace contact before the danger is known, from industrial accidents, from medical treatments such as chemotherapy and radiation, from exposure to weapons that emit radiation, and from natural disasters that damage radiation-emitting equipment. One of the most famous examples of an accident that triggered mutations was the explosion on April 25, 1986, between 1:23 and 1:24 A.M., at Reactor 4 at the Chernobyl Nuclear Power Station in Ukraine. A great plume of radioactive isotopes burst into the air and spread for thousands of miles. The reactor had been undergoing a test, its safety systems temporarily disabled, when it overloaded and rapidly flared out of control. Two people died at the scene and another 29 died from acute radiation sickness over the ensuing three months. Today the level of radioactivity has diminished by 75 percent. Tourists visit limited areas of the site, and vegetation has returned.

Evidence of a mutagenic effect of acute radiation poisoning is the 10-fold increased rate of thyroid cancer among children who were living near the Chernobyl plant in Belarus when the disaster happened. The thyroid glands of young people soaked up iodine that, in a radioactive form, bathed Belarus in the days after the explosion. The World Health Organization predicts that 5,000 cases of thyroid cancer in Russia, Belarus, and Ukraine due to the Chernobyl accident will be diagnosed in the next 20 years.

One way that researchers tracked mutation rates after the Chernobyl explosion was to compare the lengths of short DNA repeats, called minisatellite sequences, in children born in 1994 and in their parents, who lived in the exposed district at the time of the accident and have remained there. Minisatellites are the same length in all cells of an individual. A minisatellite size in a child that does not match the size of either parent indicates that a mutation occurred in a parent's gamete. Such a mutation was twice as likely to occur in exposed families as in families living elsewhere. Because mutation rates of non-repeated DNA sequences are too low to provide useful information on the effects of radiation exposure, investigators track minisatellites as a sensitive test of genetic change.

Natural Exposure to Mutagens

Simply being alive exposes us to radiation that can cause mutation. Natural environmental sources of radiation account for 81 percent of our exposure, including cosmic rays, sunlight, and radioactive substances in the Earth's crust, such as radium, which produces radon. Medical X-rays and occupational radiation hazards add risk. Job sites with increased radiation exposure include weapons facilities, research laboratories, health care facilities, nuclear power plants, and certain manufacturing plants.

Most of the potentially mutagenic radiation we are exposed to is ionizing, which means that it has sufficient energy to remove electrons from atoms. Unstable atoms that emit ionizing radiation exist naturally, and chemists synthesize them. Ionizing radiation breaks the DNA sugar-phosphate backbone. A type of ionizing radiation, gamma rays, penetrate the body, damaging tissues. Plutonium and cesium isotopes used in weapons emit gamma rays, and this form of radiation is used to kill cancer cells.

X-rays are the major source of exposure to human-made radiation. They have less energy and do not penetrate the body to the extent that gamma rays do.

The effects of radiation damage to DNA depend upon the functions of the mutated genes. Mutations in oncogenes or tumor suppressor genes, discussed in chapter 20, can cause cancer. Radiation damage can be widespread, too. Exposing cells to radiation and then culturing them causes a genome-wide destabilization, so that mutations may occur even after the cell has divided a few times. Cell culture studies have also identified a "bystander effect," when radiation harms cells not directly exposed.

Chemical mutagens are in the environment, too. Evaluating the risk that a specific chemical exposure will cause a mutation is difficult, largely because people vary greatly in inherited susceptibilities, and are exposed to many chemicals. The risk that exposure to a certain chemical will cause a mutation is often less than the natural variability in susceptibility within a population, making it nearly impossible to track the true source and mechanism of any mutational event.

> **Key Concepts Questions 12.4**
>
> *a. Describe the chemical mechanism behind spontaneous mutation.
>
> *b. What characteristic of a DNA sequence increases the likelihood of mutation?
>
> c. What are some mutagens?
>
> d. Why are the effects of mutagens encountered in the environment difficult to assess?

12.5 Types of Mutations

Mutations are classified by whether they remove, alter, or add a function, or by how they structurally alter DNA. The same single-gene disease can result from different types of mutations. **Table 12.4** summarizes types of mutations using an analogy to an English sentence.

Point Mutations

A **point mutation** is a change in a single DNA base. A **transition** is a point mutation that replaces a purine with a purine (A to G or G to A) or a pyrimidine with a pyrimidine (C to T or T to C).

A **transversion** replaces a purine with a pyrimidine, or vice versa (A or G to T or C) and is a greater change than a transition because purines and pyrimidines differ in the number of rings.

Addition or deletion of a single DNA base is also considered to be a point mutation. A point mutation can have any of several consequences—or it may have no obvious effect at all on the phenotype, acting as a silent mutation.

Missense and Nonsense Mutations

A point mutation that changes a codon that normally specifies a particular amino acid into one that codes for a different amino acid is called a **missense mutation**. If the substituted amino acid alters the protein's conformation significantly or occurs at a site critical to its function, signs or symptoms of disease or an observable variant of a trait may result. About a third of missense mutations harm health. **Figure 12.9a** shows a missense mutation in the gene encoding the low-density lipoprotein (LDL) receptor that is abnormal in familial hypercholesterolemia (see figure 5.2). Figure 12.3 depicts the missense mutation behind sickle cell disease and figure 1.6 shows one that causes cystic fibrosis.

A point mutation that changes a codon specifying an amino acid into a "stop" codon—UAA, UAG, or UGA in mRNA—is a **nonsense mutation,** shown in **figure 12.9b**. A premature stop codon is one that occurs before the natural end of the gene. It shortens the protein product, which can greatly influence the phenotype. For example, in factor XI deficiency, which is a blood clotting disorder, a GAA codon specifying glutamic acid is changed to UAA, signifying "stop." The shortened clotting factor cannot halt the profuse bleeding that occurs during surgery or from an injury. Cells have a response to shortened proteins called **nonsense-mediated decay** that recognizes and destroys mRNAs with premature stop codons. This response is protective because some shortened proteins can cause a "gain-of-function" and damage the cell beyond the absence of the normal function.

In the opposite of a nonsense mutation, a normal stop codon mutates into a codon that specifies an amino acid. The resulting protein is too long, because translation continues through what is typically a stop codon.

Point mutations can control transcription, affecting the quantity rather than the quality of a protein. For example, in 15 percent of people who have Becker muscular dystrophy, the muscle protein dystrophin is normal, but its levels are reduced. The mutation is in the promoter for the dystrophin gene. This slows transcription, and dystrophin protein is scarce. The other 85 percent of individuals who have Becker muscular dystrophy have shortened proteins, rather than too few normal-length proteins.

Splice-Site Mutations

A point mutation can greatly affect a gene's protein product if it alters a site where introns are

Table 12.4 Types of Mutations

A sentence comprised of three-letter words provides analogies to the effects of mutations on a gene's DNA sequence:

Normal	THE ONE BIG FLY HAD ONE RED EYE
Missense	THQ ONE BIG FLY HAD ONE RED EYE
Nonsense	THE ONE BIG
Frameshift	THE ONE QBI GFL YHA DON ERE DEY
Deletion	THE ONE BIG HAD ONE RED EYE
Insertion	THE ONE BIG FLY WET HAD ONE RED EYE
Duplication	THE ONE BIG FLY FLY HAD ONE RED EYE
Expanding mutation	
Generation 1	THE ONE BIG FLY HAD ONE RED EYE
Generation 2	THE ONE BIG FLY FLY FLY HAD ONE RED EYE
Generation 3	THE ONE BIG FLY FLY FLY FLY HAD ONE RED EYE

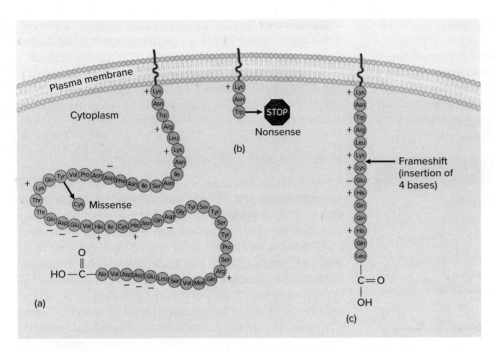

Figure 12.9 **Different mutations in a gene can cause the same disease.** In familial hypercholesterolemia, several types of mutations alter the LDL receptor normally anchored in the plasma membrane. LDL receptor **(a)** bears a missense mutation—a cysteine substitutes for a tyrosine, bending the receptor enough to impair its function. The short LDL receptor in **(b)** results from a nonsense mutation, in which a stop codon replaces a tryptophan codon. In **(c)**, a 4-base insertion adds an amino acid and then alters the reading frame; + and − indicate electrical charges on the amino acids.

normally removed from the mRNA. This type of alteration is called a **splice-site mutation**. It can affect the phenotype if an intron is translated into amino acids, lengthening the protein, or if an exon is skipped instead of being translated, shortening the protein (**figure 12.10**).

Retaining an intron is unusual because most introns have stop codons in all reading frames. However, if a stop codon is not encountered, a retained intron adds bases to the protein-coding part of an mRNA. For example, in a family with severe cystic fibrosis, a mutation alters an intron site so that it is not removed from the mRNA. The extra amino acids make the encoded ion channel protein too bulky to move to its normal position in the plasma membrane (see figure 1.5).

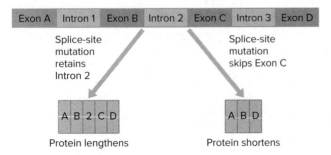

Figure 12.10 **A point mutation can alter a splice site.** The effect is retaining an intron, adding amino acids to the encoded protein, or skipping an exon, shortening the encoded protein.

The exome includes only exons. Therefore, whole exome sequencing would not detect a mutation that disrupts an intron. This was the case for a child who had a single-base mutation in an intron of a gene called *RTTN,* which encodes a protein called rotatin. The child had a very small head, large face, dwarfism, and reduced brain folds. Tests for 11 known conditions with these symptoms were negative, and exome sequencing did not find a mutation that could explain the condition. The child had a point mutation in a splice site. This led to inclusion of the amino acids encoded in part of an intron until a stop codon ended translation of the extended protein.

A missense mutation that creates an intron splicing site where there should not be one can cause **exon skipping,** which removes a few contiguous amino acids from the protein product. An entire exon may be "skipped," as if it were an intron, when the mRNA is translated into protein. It is a little like leaving out a word when cutting and pasting a sentence in a document. An exon-skipping mutation is a deletion at the mRNA level, but it is a point (single-base) mutation at the DNA level. For example, a disease called familial dysautonomia (FD) can result from exon skipping in the gene encoding an enzyme necessary for the survival of certain neurons that control sensation and involuntary responses. Symptoms include pneumonia, vomiting and retching, extremely high fevers, chills, rapid heartbeat, rashes, and seizures. FD also reduces sensations of pain, heat, and cold, and impairs motor skills, affecting eating, swallowing, and breathing.

A peculiarity of some diseases caused by exon skipping, including FD, is that some cells ignore the mutation and manufacture a normal protein from the affected gene—possible because the amino acid sequence information is still there in the DNA. Depending upon which cells actually make the full encoded protein, the phenotype may be less severe than in individuals with the same disease but with a different type of mutation in an exon. In FD, cells in which the exon is skipped are the cells that contribute to symptoms. That is, many cells from the brain and spinal cord skip the exon, but cells from muscle, lung, liver, white blood cells, and glands produce normal-length proteins. For FD, clinical trials are testing compounds that restore retention of the skipped exon in mRNAs in neurons of affected children.

The natural exon skipping seen in FD inspired development of drugs to treat Duchenne muscular dystrophy (DMD). One drug consists of synthetic DNA-like molecules that are complementary, or "antisense," to part of the huge dystrophin gene near a mutation that causes the disease by altering the reading frame (a frameshift mutation, see figure 10.13). A frameshift mutation results in a shortened protein, as shown in **figure 12.9c**. The drug causes transcription to skip the exon containing the mutation in a way that restores the reading frame, so that skeletal muscle cells produce some of the protein. It is like reading past an error in a typed sentence but understanding the meaning. The drug was approved because it leads to production of dystrophin, the missing protein. However, research continues to investigate effects on mobility in patients and how to deliver the drug to enough muscle cells to make a difference in quality of life and survival.

Deletions and Insertions

In genes, the number 3 is critical, because triplets of DNA bases specify amino acids. Adding or deleting a number of bases that is not a multiple of 3 devastates a gene's function because it disrupts the gene's reading frame. This is a frameshift mutation. Most exons are "readable" (have no stop codons) in only one of the three possible reading frames.

A **deletion** mutation removes DNA. A deletion that removes three bases or a multiple of three bases will not cause a frameshift, but can still alter the phenotype. Deletions range from a single DNA nucleotide to thousands of bases to large parts of chromosomes. An **insertion** mutation adds DNA, which can offset the reading frame. In one form of Gaucher disease, for example, an inserted single DNA base prevents production of an enzyme that normally breaks down glycolipids in lysosomes (see figure 2.5). The resulting buildup of glycolipid enlarges the liver and spleen and causes easily fractured bones and neurological impairment.

One type of insertion mutation repeats part of a gene's sequence. The insertion is usually adjacent to or close to the original sequence, like a keying error repeating a word. Two complete copies of a gene next to each other is a type of mutation called a **tandem duplication**. A form of Charcot-Marie-Tooth disease, which causes numb hands and feet, results from a 1.5-million-base-long tandem duplication.

Pseudogenes and Transposons Revisited

Pseudogenes and transposons can cause mutations. Recall from chapter 11 that a pseudogene is a DNA sequence that is very similar to the sequence of a protein-encoding gene. A pseudogene is not translated into protein, although it may be transcribed. The pseudogene may have descended from a gene sequence that was duplicated when DNA strands misaligned during meiosis (similar to the situation depicted in figure 12.8 for the alpha globin gene). When this happens, a gene and its copy end up next to each other on the chromosome. The original gene or the copy then mutates to such an extent that it is no longer functional and becomes a pseudogene. Its partner lives on as the functional gene.

Even though a pseudogene is not translated, it can interfere with the expression of the functional gene and cause a mutation. For example, some cases of Gaucher disease result from a crossover between the working gene that encodes the enzyme glucocerebrosidase and its pseudogene, which has 96 percent of the same sequence and is located 16,000 bases away (**figure 12.11**). The gene and its pseudogene each has 11 exons. The result of a crossover between the gene and the pseudogene is two rearranged genes: a gene fusion allele that is part functional gene and part pseudogene, and a partially duplicated allele. The fusion allele does not retain enough of the normal gene sequence to enable the cell to synthesize the encoded enzyme. Gaucher disease affects an offspring with a probability of 25 percent if the other parent carries a mutation in the gene. The disease is autosomal recessive.

Chapter 11 also discussed transposons. These "jumping genes" can alter gene function in several ways. They can disrupt the site they jump from, shut off transcription of the gene they jump into, or alter the reading frame if they are not a multiple of three bases. For example, a boy with X-linked

Figure 12.11 *A pseudogene causes mutation.* A pseudogene located 16,000 bases (16 kb) from the glucocerebrosidase gene is a copy, with a 55-base deletion. During meiosis, the gene and its pseudogene pair but misalign, generating a gene fusion allele with too little DNA and a partially duplicated allele with too much. Inheriting the gene fusion allele with another mutant allele causes Gaucher disease.

hemophilia A had a transposon in his factor VIII gene—a sequence that was also in his carrier mother's genome, but on her chromosome 22. Apparently, in the oocyte, the transposon jumped from chromosome 22 into the factor VIII gene on the X chromosome, eventually causing the boy's hemophilia.

Expanding Repeats

In a type of mutation called an **expanding repeat,** a gene actually grows as a small part of the DNA sequence is copied and added. The family described in the opener to chapter 4 offers an extreme example of the devastation that an expanding repeat mutation can cause.

One of the first of the 30 or so recognized expanding repeat disorders studied was myotonic dystrophy, which begins earlier and causes more severe symptoms from one generation to the next. A grandfather might experience mild weakness in his forearms, but his daughter may have more noticeable arm and leg weakness and a flat facial expression. Her children might have severe muscle weakness.

For many years, clinicians thought that the "anticipation"—the worsening of symptoms over generations—was psychological. Then, with the ability to sequence DNA, researchers found that the gene expands! The gene that causes the first recognized type of myotonic dystrophy, on chromosome 19, has an area rich in CTG repeats (GAC mRNA repeats). The wild type repeat number is 5 to 37; a person with the disease has from 50 to thousands of copies (**figure 12.12**).

The genes behind most expanding repeat disorders are not associated with symptoms if they are present in fewer than 40 copies. Larger repeats are unstable, growing with each generation and causing symptoms that are more severe and begin sooner the longer the repeats. **Clinical Connection 12.1** describes the triplet repeat condition fragile X syndrome, one of the first studied.

The mechanism behind expanding repeat diseases lies in the DNA sequence. The bases of the repeated triplets implicated in the expansion diseases, unlike others, bond to each other in ways that bend the DNA strand into shapes, such as hairpins. These shapes then interfere with DNA replication, which extends the expansion. In some expanded repeat disorders, the mRNAs are so symmetrical that they bind and disable certain other proteins, inducing additional symptoms. If repeats are translated, the extra-long proteins harm cells in several ways:

- They bind to parts of transcription factors that have stretches of amino acid repeats similar to or matching the expanded repeat.
- They block proteasomes, enabling misfolded proteins to persist.
- They trigger apoptosis.

Triplet repeat proteins may also enter the nucleus, even though their wild type versions function only in the cytoplasm, or vice versa.

The triplet repeat diseases cause a "dominant toxic gain-of-function," illustrated in figure 4.8. They cause something novel to happen, rather than removing a function, such as occurs with a recessive enzyme deficiency. The idea of a gain-of-function arose from the observation that people who have deletions of these genes do not have associated symptoms. Several triplet repeat conditions are "polyglutamine diseases" that are due to repeats of the mRNA codon CAG, which encodes the amino acid glutamine.

For some triplet repeat diseases, the mutation blocks gene expression before a protein is even made. In myotonic dystrophy type 1, the expansion is in the initial untranslated region of the gene on chromosome 19, resulting in a huge mRNA. When genetic testing became available for the disease, researchers discovered a second form of the illness in patients who had wild type alleles of the gene on chromosome 19. They have myotonic dystrophy type 2, caused by an expanding *quadruple* repeat of CCTG in a gene on chromosome 3. Affected individuals have more than a hundred copies of the repeat, compared to the normal maximum of 10 copies.

Copy Number Variants

Recall from chapter 7 that a copy number variant (CNV) is a different form of information than a DNA sequence variant. A CNV is a specific DNA sequence that varies in number of copies from person to person. Our genomes have hundreds to thousands of them. Sequencing the first human genomes missed CNVs because the technology could then detect a DNA sequence only once. It was a little like searching this book for the word "variant," and not the number of times it is used. CNVs called **short tandem repeats (STRs)** are used in forensics to distinguish individuals. STRs are 2 to 10 bases long. The number of repeats of a specific STR at a certain site in the genome is considered an allele. STRs are discussed in chapter 18.

A language analogy is useful to distinguish point mutations and single nucleotide polymorphisms (SNPs) from

Myotonic Dystrophy

Pedigree	Age of onset	Phenotype	Number of copies of GAC mRNA repeat
I	Older adulthood	Mild forearm weakness, cataracts	50–80
II	Mid-adulthood	Moderate limb weakness	80–700
III	Childhood	Severe muscle impairment, respiratory distress, early death	700+

Figure 12.12 Expanding genes explain anticipation. In some diseases, symptoms that worsen from one generation to the next—termed *anticipation*—have a physical basis: The gene is expanding as the number of repeats increases.

Clinical Connection 12.1

Fragile X Mutations Affect Boys and Their Grandfathers

Fragile X syndrome is the most common inherited form of intellectual disability and accounts for 3 percent of autism cases. In the 1940s, geneticists thought that a gene on the X chromosome caused intellectual disability, which was then termed mental retardation, because more affected individuals were male. In 1969, a clue emerged to the genetic basis of X-linked intellectual disability. Two brothers with the condition and their mother had an unusual X chromosome. The tips at one chromosome end dangled by a thin thread (**figure 12A**). This part of the X chromosome, when grown in culture medium lacking folic acid, was prone to breaking—hence, the name fragile X syndrome. Worldwide, it affects 1 in 4,000 males, accounting for 4 to 8 percent of all males with intellectual disability. One in 8,000 females is affected. Females usually have milder cases because their cells have normal X chromosomes, too.

In fragile X syndrome, brain differences appear as early as age 2 on MRI scans, and parents may first notice speech and language delay during this time. By young adulthood, the face becomes long and narrow, with a pronounced jaw and protruding ears. The testicles are large. Learning disabilities, repetitive speech, hyperactivity, shyness, social anxiety, a short attention span, language delays, and temper outbursts are also part of the syndrome. About 30 percent of affected individuals have autism and 15 percent of males and 5 percent of females have seizures.

Fragile X syndrome is inherited in an unusual pattern. The syndrome should be transmitted like any X-linked trait, from carrier mother to affected son. However, penetrance is incomplete. One-fifth of males who inherit the chromosomal abnormality have no symptoms. Because they transmit the affected chromosome to all their daughters—half of whom have some mental impairment—they are called *transmitting males*. A transmitting male's grandchildren may inherit fragile X syndrome.

A triplet repeat mutation causes fragile X syndrome. In unaffected individuals, the fragile X area contains 5 to 40 repeats of the sequence CGG, in a gene called the fragile X mental retardation gene (*FMR1*). In people who have the fragile chromosome and show its effects, this region is expanded to 200 to 2,000 CGG repeats. Transmitting males, as well as females who have mild symptoms or have affected sons, may have a "premutation" of 55 to 200 repeats. People with the premutation may develop some symptoms, such as large ears, anxiety, and depression. About a fifth of women with the premutation have infertility due to fragile X–associated primary ovarian insufficiency.

The *FMR1* gene encodes fragile X mental retardation protein (FMRP). The protein normally binds to several types of mRNA molecules whose encoded proteins are crucial for forming synapses in the brain. In fragile X syndrome, FMRI protein is not produced, although mRNA may be abundant.

Mysteries remain about fragile X syndrome. A distinct set of symptoms is seen in the maternal grandfathers of boys who have fragile X syndrome. Mothers of boys with fragile X syndrome have reported the same symptoms in their fathers—tremors, balance problems, and then cognitive or psychiatric difficulties. The grandfathers were sometimes misdiagnosed with Parkinson disease due to the tremors and their age. However, Parkinson patients can walk a straight line, but the grandfathers could not. The grandfathers' symptoms worsened with time and sometimes led to premature death. Further investigation of the grandfathers led to the description of fragile X–associated tremor/ataxia syndrome (FXTAS). (Ataxia is poor balance and coordination.)

Newborn screening for fragile X syndrome is possible, but is not being pursued in the general population until researchers can determine the percentage of individuals with certain numbers of repeats who develop symptoms. Prenatal testing is available. If a pregnant woman has a premutation or a more extensive mutation, then the fetus can be tested.

Questions for Discussion

1. Why do females have milder symptoms of fragile X syndrome than males?
2. What is a "transmitting male"?
3. Describe the mutation that causes fragile X syndrome.

— Fragile site

(a) (b)

Figure 12A Fragile X syndrome. A fragile site on the tip of the long arm of the X chromosome **(a)** is associated with intellectual disability and a characteristic long face that becomes pronounced with age **(b)**. *(b): Courtesy of Dr. Richard Simensen/Greenwood Genetic Center*

CNVs. If a wild type short sequence and a variant with two SNPs are written as:

> *The sad rat sat on a red cat* (wild type)
>
> *The sad rat sat in a red hat* (two SNPs)

then the sequence with two CNVs might be:

> *The sad sad rat sat on a red red red cat*

CNVs may contribute significantly to the differences among us. A CNV can range in size from a few DNA bases to millions, and copies may lie next to each other on a chromosome ("tandem") or might be far away—even parts of other chromosomes. CNVs can duplicate or triplicate entire genes, or delete them.

CNVs may have no effect on the phenotype, or they can disrupt a gene's function and harm health. A CNV may have a direct effect by inserting into a protein-encoding gene and offsetting its reading frame or have an indirect effect by destabilizing surrounding sequences. CNVs are particularly common among people who have behavioral disorders, such as attention deficit hyperactivity disorder (ADHD), autism, and schizophrenia.

Key Concepts Questions 12.5

*a. What is a point mutation?

b. Distinguish between a transversion and a transition.

c. Distinguish between a missense mutation and a nonsense mutation.

d. Explain how mutations outside the coding region of a gene can affect its function.

*e. How do insertions and deletions affect gene function?

f. How do pseudogenes and transposons affect gene function?

g. Describe expanding repeat mutations and their effects.

*h. What are copy number variants?

12.6 The Importance of Position

The degree to which a mutation alters a phenotype depends upon where in the gene the change occurs. The location can affect the folding, conformation, activity, or abundance of an encoded protein. A mutation that replaces an amino acid with a similar one is less likely to affect the phenotype greatly, because it wouldn't substantially change the conformation of the protein. Even substituting a very different amino acid would still be less likely to have an effect if the change is in part of the protein not crucial to its function. The effects of specific mutations are well studied in hemoglobin.

Globin Gene Variants

Hundreds of globin gene mutations are known. Mutations in these genes can cause anemia with or without sickling, or cause cyanosis (a blue pallor due to poor oxygen binding). Rarely, a mutation boosts the molecule's affinity for oxygen. Some globin gene variants exert no effect at all and are considered "clinically silent" (**table 12.5**).

Different mutations at the same site in a gene can have different effects. For example, hemoglobin S and hemoglobin C result from mutations that change the sixth amino acid in the beta globin polypeptide, but in different ways. Homozygotes for hemoglobin S have sickle cell disease. Homozygotes for hemoglobin C have hemoglobin C disease, with mild or borderline anemia, compared to the severe anemia of sickle cell disease. The difference between the two conditions is that hemoglobin S changes glutamic acid to valine, whereas hemoglobin C changes the glutamic acid at position 6 to lysine. Because valine is not charged and lysine has a positive charge, the two substitution mutations have different interactions with other molecules in the bloodstream. As a result, hemoglobin S forms long, insoluble threads, whereas hemoglobin C forms small crystals that are less likely to occlude circulation and cause severe anemia. Both types of homozygotes are resistant

Table 12.5 Globin Mutations

Associated Phenotype	Name	Mutation
Clinically silent	Hemoglobin (Hb) Wayne	Single-base deletion in alpha globin gene causes frameshift, changing amino acids 139–141 and adding amino acids
	Hb Grady	Nine extra bases add three amino acids between amino acids 118 and 119 of alpha chain
Oxygen binding	Hb Chesapeake	Change from arginine to leucine at amino acid 92 of beta chain
	Hb McKees Rocks	Change from tyrosine to STOP codon at amino acid 145 in beta chain
Anemia	Hb Constant Spring	Change from STOP codon to glutamine elongates alpha chain
	Hb S	Change from glutamic acid to valine at amino acid 6 in beta chain causes sickling
	Hb Leiden	Amino acid 6 deleted from beta chain
Protection against malaria	Hb C	Change from glutamic acid to lysine at amino acid 6 in beta chain causes mild sickling

to malaria because the unusual hemoglobin alters the shapes and surfaces of red blood cells in ways that keep out the parasite that causes the infectious illness (see figure 16.13).

Factors That Lessen the Effects of Mutation

Mutation is a natural consequence of DNA's ability to change. This flexibility is essential for evolution because it generates new gene variants, some of which may resist environmental change and enable a population or even a species to survive. However, many factors minimize the negative effects of mutations on phenotypes.

The genetic code protects against mutation to an extent. Recall from chapter 10 that synonymous codons specify the same amino acid. Mutation in the third position in a codon is called "silent" when the mutated and original codons are synonymous. For example, a change from RNA codon CAA to CAG does not change the encoded amino acid, glutamine, so a protein whose gene contains the change would not change. However, synonymous codons can affect splicing out of introns and mRNA stability. As a result, expression of genes that have mutations that result in synonymous codons can differ.

Other genetic code nuances prevent synthesis of highly altered proteins. For example, mutations in the second position in a codon sometimes replace one amino acid with another that has a similar conformation, minimizing disruption of the protein's shape. GCC mutated to GGC, for instance, replaces alanine with another small amino acid, glycine.

A **conditional mutation** affects the phenotype only under certain circumstances. This can be protective if an individual avoids the exposures that trigger symptoms. Consider a common variant of the X-linked gene *G6PD* that encodes glucose-6-phosphate dehydrogenase, an enzyme that immature red blood cells use to extract energy from glucose. One hundred million people worldwide have *G6PD* deficiency, which causes life-threatening hemolytic anemia, due to any of more than 200 known mutations in this X-linked gene. People with *G6PD* deficiency cannot make an anti-oxidant molecule that protects against destabilization of certain molecules. The various triggers produce damaging oxidants and include eating fava beans, taking certain antibiotic or antimalarial drugs, or having certain infections. (See **A Glimpse of History**.)

A GLIMPSE OF HISTORY

In the fifth century B.C.E., the Greek mathematician Pythagoras wouldn't allow his followers to eat fava beans, because he had discovered that it would sicken some of them. They had G6PD deficiency. During World War II, some soldiers taking the antimalarial drug primaquine developed hemolytic anemia. They too had G6PD deficiency. What these, and other triggering substances, have in common is that they "stress" red blood cells by exposing them to oxidants, chemicals that strip electrons from other compounds. Without the enzyme, the red blood cells burst.

Symptoms of *G6PD* deficiency are fatigue, weakness, muscle aches, and jaundice. It is the most common enzyme deficiency in the world. *G6PD*, like sickle cell disease, remains prevalent because carriers are resistant to malaria, discussed further in section 16.6.

Another protection against mutation is in stem cells. When a stem cell divides to yield another stem cell and a progenitor or differentiated cell, the oldest DNA strands segregate with the stem cell, and the most recently replicated DNA strands go to the more specialized daughter cells (see figure 2.17). This makes sense in organs where stem cells actively yield specialized daughter cells, such as the skin and small intestine. Because mutations occur when DNA replicates, this skewed distribution of chromosomes sends the DNA most likely to harbor mutations into cells that will soon be shed (from a towel rubbed on skin or in a bowel movement) while keeping mutations away from the stem cells that must continually regenerate the tissues.

Key Concepts Questions 12.6

a. What characteristics of a mutation determine its effects?

b. Describe a mutation in a globin gene.

*c. Name three phenomena that lessen the effects of mutations.

12.7 DNA Repair

Any manufacturing facility tests a product in several ways to see whether it has been assembled correctly. Mistakes in production are rectified before the item goes on the market—most of the time. The same is true for a cell's manufacture of DNA.

Damage to DNA becomes important when the genetic material is replicated, because the error is passed on to daughter cells. In response to damage, the cell may repair the error or die by apoptosis. If the error is not repaired and the cell doesn't die, cancer may result. Fortunately, DNA replication is very accurate—only 1 in 100 million or so bases is incorrectly incorporated. This is quite an accomplishment, because DNA replicates approximately 10^{16} times during an average human lifetime.

DNA polymerase as well as "DNA damage response" genes oversee the accuracy of replication. In DNA repair, a cell detects damage and then signaling systems in the cell respond by repairing the damage or signaling apoptosis to kill the cell. More than 50 DNA damage response genes have been identified. Mitochondrial DNA cannot repair itself, which contributes to its higher mutation rate.

Many types of organisms repair their DNA, some more efficiently than others. The master at DNA repair is a large, reddish microbe, *Deinococcus radiodurans*. It tolerates 1,000 times the radiation level that a person can, and it can even live amidst the intense radiation of a nuclear reactor. The bacterium realigns its radiation-shattered pieces of DNA. Then enzymes bring in new nucleotides and assemble the pieces.

Types of DNA Repair

Exposure to radiation is a fact of life. The Earth, since its beginning, has been bathed in UV radiation, the intensity varying. Volcanoes, comets, meteorites, and supernovas all depleted ozone in the atmosphere, which allowed more ultraviolet wavelengths of light to reach organisms. The longer UV wavelengths—UVA—are not dangerous, but the shorter UVB wavelengths damage DNA by forming an extra covalent bond between adjacent (same-strand) pyrimidines, particularly thymines. The linked thymines are called thymine dimers. Their extra bonds kink the double helix enough to disrupt replication and permit insertion of a noncomplementary base. For example, an A might be inserted opposite a G or C, instead of opposite a T. Thymine dimers also disrupt transcription.

UV light only penetrates one millimeter, so it typically only causes skin cancer. However, UV radiation was lethal to simple microorganisms that were the planet's first inhabitants. Early in the evolution of life, organisms that could survive UV damage had an advantage. Enzymes enabled them to do this, and because enzymes are gene-encoded, DNA repair came to persist.

In many modern species, three types of DNA repair mechanisms check the genetic material for mismatched base pairs. In the first type, enzymes called photolyases absorb energy from visible light and use it to detect and bind to pyrimidine dimers, then break the extra bonds. This type of repair, called photoreactivation, enables UV-damaged fungi to recover with exposure to sunlight. Humans do not have this type of DNA repair.

In the early 1960s, researchers discovered a second type of DNA self-mending, called **excision repair,** in mutant *E. coli* that were unable to repair UV-induced DNA damage. Enzymes cut the bond between the DNA sugar and base and excise the pyrimidine dimer and surrounding bases. Then, a DNA polymerase fills in the correct nucleotides, using the exposed template as a guide. DNA polymerase also detects and corrects mismatched bases in newly replicated DNA.

Humans have two types of excision repair. **Nucleotide excision repair** replaces up to 30 nucleotides and removes errors that result from several types of causes, including exposure to chemical carcinogens, UVB in sunlight, and oxidative damage (**figure 12.13**). Thirty different proteins carry out nucleotide excision repair.

The second type of excision repair, **base excision repair,** replaces one to five nucleotides at a time, but specifically corrects errors that result from oxidative damage.

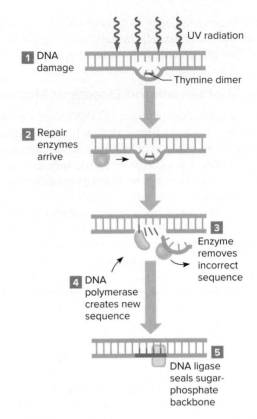

Figure 12.13 **Nucleotide excision repair.** Human DNA damaged by UV light is corrected by nucleotide excision repair, which removes and replaces the pyrimidine dimer and a few surrounding bases.

Oxygen free radicals are highly reactive forms of oxygen that arise during chemical reactions such as those of metabolism and transcription. Free radicals damage DNA. Genes that are actively transcribed face greater oxidative damage from free radicals; base excision repair targets this type of damage.

A third mechanism of DNA sequence correction is **mismatch repair**. Enzymes check newly replicated DNA for small loops that emerge from the double helix. The enzymes excise the mismatched nucleotide and replace it with the correct one (**figure 12.14**). The small loops form where the two strands do not precisely align, but instead slip and misalign. This happens where very short DNA sequences repeat. These sequences, called microsatellites, are scattered throughout the genome. Like the minisatellites mentioned in section 12.4, microsatellite lengths can vary from person to person, but within an individual, they are usually the same length at the same site in the genome. Excision and mismatch repair differ in the cause of the error—UV-induced pyrimidine dimers versus replication errors—and in the types of enzymes involved.

Excision repair and mismatch repair in human cells relieve the strain on thymine dimers or replace incorrectly inserted nucleotides. Another form of repair can heal a broken

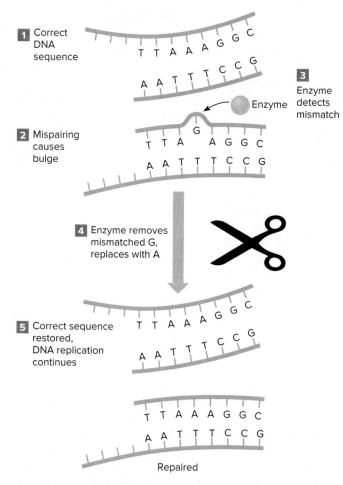

Figure 12.14 Mismatch repair. In this form of DNA repair, enzymes detect loops and bulges in newly replicated DNA that indicate mispairing. The enzymes correct the error. Highly repeated sequences are more prone to this type of error.

sugar-phosphate backbone in both DNA strands, which can result from exposure to ionizing radiation or oxidative damage. Such a double-stranded break is especially damaging because it breaks a chromosome, which can cause cancer. At least two types of multiprotein complexes reseal the sugar-phosphate backbone, either by rejoining the broken ends or recombining with DNA on the unaffected homolog. Double-stranded breaks and their natural repair form the basis of some genome editing technologies, discussed in sections 21.4 and 22.4.

In yet another type of DNA repair called damage tolerance, a "wrong" DNA nucleotide is left in place, but replication and transcription proceed. "Sloppy" DNA polymerases, with looser adherence to the base-pairing rules, read past the error, randomly inserting any other nucleotide. It is a little like retaining a misspelled word in a sentence—usually the meaning remains clear.

Figure 12.15 summarizes DNA repair mechanisms.

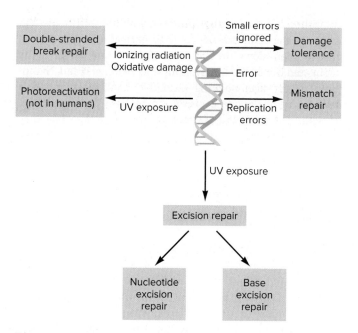

Figure 12.15 DNA repair mechanisms.

DNA Repair Disorders

The ability to repair DNA is crucial to health. If both copies of a repair gene are mutant, a disorder can result. Heterozygotes who have one mutant repair gene may be more sensitive to damage from environmental factors, such as toxins and radiation.

A well-studied gene that encodes a protein called p53 is known as "the guardian of the genome." It controls whether DNA is repaired and the cell salvaged, or the cell dies by apoptosis (see figure 20.3). Signals from outside the cell activate p53 proteins to aggregate into complexes of four molecules. These quartets bind certain genes that slow the cell cycle, enabling DNA repair to take place. If the damage is too severe, the p53 protein quartets instead increase the rate of transcription of genes that promote apoptosis, and the cell dies.

In DNA repair disorders, mutations and chromosome breaks persist. Mutations in repair genes greatly increase susceptibility to certain types of cancer following exposure to mutagens. Cancers develop as errors in the DNA sequence accumulate and are perpetuated to a much greater extent than they are in people with functioning repair systems. We conclude this chapter with a closer look at repair disorders (they are generally termed disorders and not diseases because the symptoms do not arise directly from the mutations). Chapter 20 discusses other cancers that are a consequence of disrupted DNA repair.

Trichothiodystrophy

At least five genes can cause trichothiodystrophy. At its worst, this condition causes dwarfism, intellectual disability, and brittle hair and scaly skin, both with low sulfur content. Although the child may appear to be normal for a year or two, growth soon slows dramatically, signs of premature aging begin, and life ends early. Hearing and vision may fail. Symptoms reflect accumulating oxidative damage. Individuals have faulty nucleotide excision repair, base excision repair, or both. The trichothiodystrophies are unusual in that they do not increase cancer risk. Only about 100 cases are known worldwide.

Inherited Colon Cancer

Hereditary nonpolyposis colon cancer (HNPCC), a group of seven disorders also known as Lynch syndrome, was linked to a DNA repair defect when researchers discovered different-length short repeated sequences of DNA (microsatellites) within an individual. People with this type of colon cancer have a breakdown of mismatch repair, which normally keeps a person's microsatellites all the same length. HNPCC affects 1 in 200 people and accounts for 3 percent of newly diagnosed colorectal cancers. Genetic testing in all people newly diagnosed with colon cancer is advised because if they have a germline mutation (that is, in all of their cells), their relatives can be tested to see if they inherited the mutation, too. If healthy relatives test positive, frequent colonoscopies can detect disease early, at a more treatable stage. Penetrance of HNPCC is about 45 percent by age 70—considered a high cancer risk.

Xeroderma Pigmentosum (XP)

A child with XP must stay indoors in artificial light, because even the briefest exposure to sunlight causes painful blisters. Failing to cover up and use sunblock can cause skin cancer (**figure 12.16**), which more than half of all children with XP develop before adolescence. People with XP have a 1,000-fold increased risk of developing skin cancer compared to others, and a 10-fold increased risk of developing tumors inside the body.

XP is autosomal recessive, and results from mutations in any of seven genes. It can reflect malfunction of nucleotide excision repair or deficient "sloppy" DNA polymerase, both of which allow thymine dimers to stay and block replication. One of the genes that causes XP, when mutant, also causes trichothiodystrophy and another disease, Cockayne syndrome. The different symptoms arise from the different ways that mutations disrupt the encoded protein, which is a helicase that helps unwind replicating DNA.

Only about 250 people in the world are known to have XP. A family living in upstate New York runs a special summer camp for children with XP, where night is turned into day. Activities take place at night, or in special areas where the windows are covered and light comes from low-ultraviolet incandescent lightbulbs.

Ataxia Telangiectasia (AT)

This multisymptom condition is the result of a defect in a kinase that functions as a cell cycle checkpoint (see figure 2.14). In AT, cells proceed through the cell cycle without pausing just after replication to inspect the new DNA and to repair any mispaired bases. Some cells die through apoptosis if the damage is too great to repair. Because of the malfunctioning cell cycle, individuals who have this autosomal recessive disorder have 50 times the risk of developing cancer, particularly of the blood. About 40 percent of individuals with AT have cancer by age 30. Additional symptoms include poor balance and coordination (ataxia), red marks on the face (telangiectasia), delayed sexual maturation, and high risk of infection and diabetes mellitus.

AT is rare, but heterozygotes are not. They make up from 0.5 to 1.4 percent of various populations. Carriers may have mild radiation sensitivity, which causes a two- to sixfold increase in cancer risk over that of the general population. For people who are AT carriers, dental or medical X-rays may cause cancer.

DNA's changeability, so vital for evolution of a species, comes at the cost of occasional harm to individuals. Each of us harbors about 175 new mutations, many old ones, and many polymorphisms. The **Bioethics** box considers a situation in which learning about gene variants can be confusing.

Figure 12.16 A DNA repair disorder. The marks on this person's face result from sun exposure. Xeroderma pigmentosum, an impairment of excision repair, makes the skin highly sensitive to the sun. *Wellcome Image Library/Custom Medical Stock Photo*

Key Concepts Questions 12.7

*a. What happens in nucleotide excision repair?

*b. What happens in base excision repair?

*c. Explain how mismatch DNA repair works.

d. Describe disorders that result from faulty DNA repair.

Bioethics

A Diagnostic Dilemma: What to Do with a VUS?

"I have a *what*?" Ellie asked her genetic counselor, tearing up.

"You have a variant of uncertain significance. We call it a VUS. It's a bit confusing, but please let me explain."

Ellie had taken a genetic test to find out if she had a mutation in either the *BRCA1* or *BRCA2* cancer susceptibility genes, because she has an aunt with ovarian cancer and two cousins who had breast cancer. Although the family is in a high-risk group (Ashkenazi Jewish), none of Ellie's relatives has had genetic testing. She had prepared for either a "*yes you have a mutation*" or "*no you don't*" answer—but not this.

"One of your two *BRCA1* genes has an unusual DNA sequence. It doesn't match any of the mutations that we know confers cancer susceptibility, but it is not the 'normal,' or wild type sequence. We just don't know its clinical significance, if any," said the counselor.

"What am I supposed to do with that information?" Ellie asked. She had been considering surgery to remove her breasts and ovaries if she had a mutation. Now she didn't know what to do.

All genes vary. Some variants affect the encoded protein and the phenotype, and others do not. As a result, some gene variants are more dangerous than others. The *BRCA1* and *BRCA2* mutations that are more common among the Ashkenazim delete one or two DNA bases, which offsets the reading frame and introduces inappropriate "stop" codons. The encoded proteins are too short to function as they normally would in the mismatch repair reactions that prevent cancer (see figure 20.15). It was good news that Ellie didn't have those mutations—but what *did* she have?

The genetic counselor explained that as researchers identify specific gene variants and associate them with diseases because they are found exclusively in people with the disease, databases are built. Clinical variant scientists, also called curators, categorize gene variants according to the statistical methods used in research reports and the degree to which the genetic change is predicted to alter the structure and function of the encoded protein, sometimes using machine learning trained on known mutations. The American College of Medical Genetics and Genomics, ClinVar, and other classifying organizations denote five categories of risk **(figure 12B)**. A gene variant is one of the following:

- pathogenic
- likely pathogenic
- a variant of uncertain significance
- likely benign
- benign

More than 2,000 variants are known for the *BRCA1* and *BRCA2* genes. About 15 percent of people having the genes sequenced receive the same news as Ellie did—the middle category, the mysterious VUS. Clinicians are beginning to regard a finding of "likely pathogenic" equivalent to "pathogenic," in terms of guiding whether or not to treat the condition.

Classifying gene variants is a bioinformatics challenge that requires identifying all gene variants that nearly always track with a suspected clinical condition, and almost never with people who don't have the condition. As time goes on, and perhaps gene-gene interactions become considered too, the likelihood of getting a "VUS" result should decline. But until then, the situation can be upsetting.

Questions for Discussion

1. At which of the five levels of variant interpretation should a patient be informed about the test results? Cite a reason for your answer.
2. Explain why the number of variants of uncertain significance will decrease.
3. Should a nurse, physician, or genetic counselor deliver news of a VUS? Cite a reason for your answer.
4. What degree of certainty in a genetic test result would you want to know before considering surgery or another type of procedure to prevent a genetic condition that has not yet produced symptoms?
5. Some researchers have called the situation with uncertain findings a *variant-interpretation crisis*. What are some clinical consequences of receiving a VUS test result?

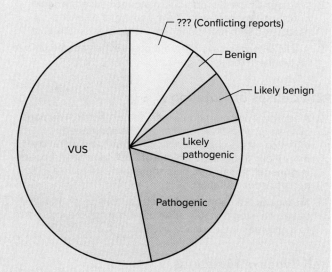

Figure 12B **Gene variants are classified as pathogenic, likely pathogenic, of uncertain significance, likely benign, or benign** The percentages of gene variant classifications change as associations with phenotypes are discovered. Findings may disagree when researchers use different database criteria to classify variants.

Study Guide

Summary

12.1 The Nature of Gene Variants

1. A mutation is a change in a gene's base sequence. It can cause a mutant phenotype. A mutation is caused by a **mutagen**.
2. A mutation can disrupt the function or abundance of a protein or introduce a new function. Most loss-of-function mutations are recessive, and most gain-of-function mutations are dominant.
3. A **polymorphism** is a gene variant present in more than 1 percent of a population. Polymorphisms are common and may not be associated with a phenotype.
4. A **germline mutation** occurs during DNA replication before meiosis, affects all cells of an individual, and can be transmitted to the next generation in gametes. A **somatic mutation** occurs before mitosis and affects a subset of cells.
5. A severe genetic disease may be seen only in individuals with **somatic mosaicism**.

12.2 A Closer Look at Two Mutations

6. Sickle cell disease results from a single-base mutation in the beta globin gene, and beta thalassemia results from too few beta globin chains.
7. Many medical conditions result from mutations in collagen genes because of the precise conformation of the protein and its pervasiveness in the body.

12.3 Allelic Diseases

8. Whether different mutations in a gene cause the same or distinct illnesses varies; the naming of diseases associated with mutations was inconsistent in the past.
9. **Allelic diseases** have different phenotypes but result from mutations in the same gene.

12.4 Causes of Mutation

10. A spontaneous mutation arises due to chemical damage or to an error in DNA replication. The spontaneous mutation rate is characteristic of a gene, and mutation is more likely to occur in regions of DNA sequence repeats. In **gonadal mosaicism**, only some gametes have a spontaneous mutation.
11. **Mutagens** are agents that delete, substitute, or add DNA bases, causing mutations. An organism may be exposed to a mutagen intentionally, accidentally, or naturally.

12.5 Types of Mutations

12. A **point mutation** alters a single DNA base. It may be a **transition** (purine to purine or pyrimidine to pyrimidine) or a **transversion** (purine to pyrimidine, or vice versa).
13. A **missense mutation** substitutes one amino acid for another. A **nonsense mutation** substitutes a "stop" codon for a codon that specifies an amino acid, shortening the encoded protein.
14. **Nonsense-mediated decay** destroys mRNAs that have premature stop codons. **Splice-site mutations** add or delete amino acids and can shift the reading frame. A missense mutation that creates an intron splicing site can cause **exon skipping**.
15. A **deletion** mutation removes genetic material and an **insertion** mutation adds it. A frameshift mutation alters the sequence of amino acids (disrupting the reading frame). A **tandem duplication** is a copy of a gene next to the original.
16. A pseudogene results when a duplicate of a gene mutates. It may disrupt chromosome pairing, causing mutation. Transposons may disrupt the functions of genes they jump into.
17. **Expanding repeats** are mutations that add stretches of the same amino acid to a protein. The DNA repeats expand because they attract each other, which causes replication errors.
18. Copy number variants are DNA sequences that are repeated a different number of times among individuals and include deletions, duplications, and repeats.

12.6 The Importance of Position

19. Mutations in the globin genes may affect the ability of the blood to transport oxygen, or they may have no effect.
20. Synonymous codons limit the effects of mutation. Changes in the second position in a codon may substitute a similarly shaped amino acid.
21. **Conditional mutations** are expressed only in response to certain environmental triggers.
22. Sending the most recently replicated DNA into cells headed for differentiation, while sending older strands into stem cells, protects against the effects of a new mutation.

12.7 DNA Repair

23. DNA polymerase checks the DNA sequence. Repair enzymes correct errors in other ways.
24. Photoreactivation repair uses light energy to split pyrimidine dimers, but humans do not have it. Humans have two types of **excision repair,** which uses enzymes to remove and replace errors in the DNA sequence.
25. **Nucleotide excision repair** replaces up to 30 nucleotides to fix errors from various sources of mutation, including UV damage. **Base excision repair** replaces up to five nucleotides to correct errors due to oxidative damage.
26. **Mismatch repair** checks newly replicated DNA for loops that indicate noncomplementary base pairing.
27. Another type of DNA repair fixes the sugar-phosphate backbone. Damage tolerance enables replication to continue beyond a mismatch.
28. Mutations in DNA repair genes break chromosomes and increase cancer risk.
29. DNA repair disorders include trichothiodystrophy, a form of inherited colon cancer, xeroderma pigmentosum, and ataxia telangiectasia.

Key Terms

allelic diseases (ə-lēl ik) Different diseases caused by mutations in the same gene.

base excision repair Replacement of one to five contiguous DNA nucleotides to correct oxidative damage.

conditional mutation A genotype that is expressed only under certain environmental conditions.

deletion A mutation that removes part of a DNA sequence or part of a chromosome.

excision repair Enzyme-catalyzed removal of pyrimidine dimers in DNA.

exon skipping A protein that is missing contiguous amino acids because a missense mutation creates an intron splice site.

expanding repeat A short DNA sequence of usually three or four nucleotides that is present in a certain range of copy numbers in wild type individuals but, when expanded, causes a disease.

germline mutation A mutation that is in every cell in an individual because it was present in the fertilized ovum.

gonadal mosaicism Having two or more genetically distinct cell populations in an ovary or testis.

insertion A mutation that adds DNA bases.

mismatch repair Checking DNA for misalignment of short, repeated segments.

missense mutation A single-base change (point mutation) that alters a codon so that it specifies a different amino acid.

mutagen An agent that causes mutation.

nonsense mutation A point mutation that changes an amino-acid-coding codon into a stop codon, prematurely terminating synthesis of the encoded protein.

nonsense-mediated decay A response that destroys mRNAs in which nonsense mutations encode shortened proteins that could have toxic effects on the cell.

nucleotide excision repair (nōō-klēō-tīd) Replacement of up to thirty nucleotides to correct DNA damage of several types.

point mutation A single-base change in DNA.

polymorphism (pŏlē-môr′ fĭz əm) A DNA base or sequence at a certain chromosomal locus that varies in a small percentage of individuals in a population.

short tandem repeat (STR) A DNA sequence of 2 to 10 bases that repeats at a specific site in a genome. Used in forensics to distinguish individuals.

somatic mosaicism A condition in which a somatic mutation causes a phenotype to affect parts of the body descended from the original cell in which the mutation occurred.

somatic mutation (sō-măt'ĭk) A genetic change in a non-sex cell.

splice-site mutation A point mutation at a site in a gene that controls intron removal, resulting in extra or absent amino acids in the protein product.

tandem duplication A copy of a DNA sequence next to the original sequence.

transition A point mutation that replaces a purine with a purine or a pyrimidine with a pyrimidine.

transversion A point mutation that replaces a purine with a pyrimidine or vice versa.

Review Questions

1. Define *mutation*.
2. Distinguish between a loss-of-function and a gain-of-function mutation.
3. Define *polymorphism*.
4. Give an example of a gene variant that is not harmful.
5. Distinguish between a germline and a somatic mutation.
6. How is a human body a genomic mosaic?
7. Explain why Henne Holstege's white blood cells had hundreds of mutations, but her brain neurons had none.
8. Explain how a mutation causes sickle cell disease.
9. Why are mutations in collagen genes especially likely to affect health?
10. Describe a pair of allelic diseases.
11. How can DNA spontaneously mutate?
12. Distinguish between
    a. a transition and a transversion.
    b. a missense mutation and a nonsense mutation.
13. List two types of mutations that can alter the reading frame.
14. List four ways that DNA can mutate without affecting the phenotype.
15. Cite two ways a transposon can disrupt gene function.
16. What is a molecular explanation for the worsening of an inherited illness over generations?
17. Explain how a copy number variant differs from a missense mutation.
18. Explain how a single-base mutation can encode a protein that is missing many amino acids.
19. Cite three ways in which the genetic code protects against the effects of mutation.
20. What is a conditional mutation?
21. How do excision repair and mismatch repair differ?

Applied Questions

1. Some members of the Fugate family described in the chapter opener have type I methemoglobinemia. They have mild symptoms of a bluish skin tone because the enzyme works, but slightly less efficiently than normal. In contrast, people with type II methemoglobinemia have severe intellectual disability and die young. Both types are due to mutations in the gene that encodes the enzyme CYP5R3. Suggest a type of mutation that might cause type I and a type of mutation that might cause the much more severe type II.

2. In McCune-Albright syndrome, fibrous connective tissue replaces bone, tan patches (café-au-lait spots) dot the skin, and hormone abnormalities cause early puberty and malfunction of the thyroid, pituitary, and adrenal glands. The phenotype is highly variable, and all patients are somatic mosaics for the mutation, which is in the gene GNAS1. Why is the condition seen only in mosaics?

3. Which of the following DNA sequences is most likely to mutate? Cite a reason for your answer.
   a. CTGAGTCGTGACTCGTACGTCAT
   b. ATGCTCAGCTACTCAGCTACGGA
   c. ATCCTAGTCGTCGTCGTCATC

4. A mutation that changes a C to a T causes a type of Ehlers-Danlos syndrome, forming a "stop" codon and resulting in shortened procollagen. Consult the genetic code (see table 10.3) and suggest one way that this can happen.

5. Part of the mRNA sequence of an exon of a gene that encodes a blood protein is: AUGACUCAUCGCUGUAGUUUACGA. Consult the genetic code to answer the following questions.
   a. What is the sequence of amino acids that this mRNA encodes?
   b. What is the sequence if a point mutation changes the 10th base from a C to an A?
   c. What is the effect of a point mutation that changes the 15th base from a U to an A?
   d. How does the encoded amino acid sequence change if a C is inserted between the fourth and fifth bases?
   e. Which would be more devastating to the encoded amino acid sequence, insertion of three bases in a row or insertion of two bases in a row?

6. Susceptibility to developing prion diseases (see section 10.4) arises from a mutation that changes aspartic acid (Asp) to asparagine (Asn). Which nucleotide base changes could make this happen?

7. A man develops Huntington disease at age 48. He is tested and his mutant gene has 54 repeats. His 20-year-old daughter is tested, and she has a gene with 68 repeats. Explain how the gene likely expanded.

8. In one family, Tay-Sachs disease stems from a four-base insertion, which changes an amino-acid-encoding codon into a "stop" codon. What type of mutation is this?

9. A biotechnology company has encapsulated DNA repair enzymes in fatty bubbles called liposomes. Why would these be a valuable addition to a suntanning lotion?

Forensics Focus

1. Late one night, a man broke into Colleen's apartment and raped her. He wore a mask and it was dark, so she couldn't see his face, but she yanked out some of his long, greasy hair. The forensic investigator, after examining the hair, asked Colleen if her boyfriend had been in the bed or with her earlier in the evening, because the hairs were of two genotypes for one of the repeated sequences that was analyzed. What is another explanation for finding two genotypes in the hair?

Case Studies and Research Results

1. Robinow syndrome causes a face that resembles that of a fetus, with a high prominent forehead, wide eyes, a flat nose, and large head and face. Other symptoms include heart defects, hearing loss, and a dense skull. The gene that is mutant, DVL1, consists of 2,941 nucleotides. It has 15 exons and encodes a 670-amino-acid-long protein. A study of eight affected individuals found that they have different mutations, but all the mutations deleted a number of nucleotides from exon 14 that is not a multiple of 3, which results in a shortened protein.
   a. What is the significance of the fact that the small deletions are not a multiple of three nucleotides?
   b. Researchers were surprised to identify mRNAs encoding 14 exons of the DVL1 gene from the eight people with Robinow syndrome. Why were they surprised?
   c. How many nucleotides are in the introns of this gene?

2. Latika and Keyshauna meet at a clinic for college students who have cystic fibrosis. Latika's mutation results in exon skipping. Keyshauna's mutation is a nonsense mutation. Which young woman probably has more severe symptoms? Cite a reason for your answer.

3. Marshall and Angela have skin cancer resulting from xeroderma pigmentosum. They meet at an event for teenagers with cancer. However, their mutations affect different genes. They decide to marry but not to have children because they believe that each child would have a 25 percent chance of inheriting XP because it is autosomal recessive. Are they correct? Why or why not?

CHAPTER 13

Chromosomes

Chromothripsis is a rare event that spontaneously shatters a chromosome—sometimes several. Discovered in 2011 in cancer cells, chromothripsis may also account for some cases of severe birth defects.

Learning Outcomes

13.1 Portrait of a Chromosome
1. List the major parts of a chromosome.
2. List the types of chromosomes based on centromere position.

13.2 Detecting Chromosomes
3. Describe ways that chromosomes are obtained, prepared, detected, and depicted in detail.

13.3 Atypical Chromosome Number
4. Explain how atypical chromosome numbers arise.
5. Distinguish polyploidy from aneuploidy.
6. Describe specific aneuploid conditions.

13.4 Atypical Chromosome Structure
7. Distinguish between the consequences of having balanced versus unbalanced chromosomes.
8. Describe deletions, duplications, and the two major types of translocation.
9. Distinguish the two types of inversions.
10. Describe the consequence of having an isochromosome.
11. Describe the consequence of having a ring chromosome.

13.5 Uniparental Disomy—A Double Dose from One Parent
12. Explain how a person could inherit both copies of a DNA sequence from one parent.
13. Describe how inheriting both copies of DNA from one parent can affect a person's health.

The BIG Picture

A human genome has 20,000-plus protein-encoding genes dispersed among 24 chromosome types. Chromosome-level illnesses reflect disruption of individual genes or their regulation. Because chromosomes determine which sets of genes are transmitted together to the next generation, detecting and analyzing chromosomes can provide clinically useful information.

Shattered!

In 1965, a case report in a medical journal described a 9-year-old girl who was the only person in the world known to have an immune condition that causes warts from human papillomavirus infection and susceptibility to bacterial infections. When a few other cases were identified, the condition was attributed to a mutation in an immune system gene, *CXCR4*. The mutation, in one copy of the gene, prevented the mobilization of a class of white blood cells in the bone marrow from entering the bloodstream to fight the pathogens.

The disease is now called WHIM syndrome. A few years ago, the first patient, in her late fifties, brought her two grown daughters to the National Institutes of Health Clinical Center. They had inherited her WHIM, but the mother had fully recovered. The disease had vanished: no infections, no warts. What had happened?

At some unknown point in time, in the woman's bone marrow, a hematopoietic (blood) stem cell had undergone an extremely rare event, previously known only to occur in cancer cells, called **chromothripsis**. Double-stranded DNA breaks had shattered one of her chromosomes, as if the chromosome had exploded. Chromothripsis is Greek for "shattered colored bodies."

When the researchers sequenced the genome of some of the woman's white blood cells, they found that one copy of her chromosome 2 was shorter than the other copy—it was missing 164 genes, including the copy of the *CXCR4* gene that had caused her disease. The chromosome had broken and sent pieces into other chromosomes, but DNA repair had healed them. The white blood cells that had descended from the altered stem cell had begun to leave the bone marrow and function, and her infections ceased. However, half of her oocytes still had the mutation, which explains how her daughters inherited WHIM.

A few cases are now known of children with severe birth defects whose mothers show evidence of past chromothripsis in all of their cells. DNA repair knit together the chromosome pieces in the mothers, who are healthy because all of their DNA is present, just rearranged. However, during meiosis, the patchwork chromosomes can mispair as pieces from homologous chromosomes attract from odd places, resulting in offspring that have extra or missing genetic material, which can cause severe birth defects. It is as if the mothers' genomes are like large jigsaw puzzles, cut up and pasted back together, but the offspring inherit too few or too many puzzle pieces.

13.1 Portrait of a Chromosome

Mutations range from single-base changes to entire extra sets of chromosomes. A mutation is considered a chromosomal aberration if it is large enough to be seen with a light microscope using stains and/or fluorescent probes to highlight missing, extra, or moved DNA sequences. A chromosome includes hundreds to thousands of genes.

In general, too little genetic material has more severe effects on health than too much. Extensive chromosome abnormalities that are present in all cells of an embryo or fetus disrupt or halt prenatal development. As a result, only 0.65 percent of all newborns have chromosomal abnormalities that produce symptoms. An additional 0.20 percent of newborns have chromosomal rearrangements in which chromosome parts have flipped or swapped, but the rearrangements do not produce symptoms unless they disrupt the structures or functions of genes that affect health.

Cytogenetics is the classical area of genetics that links chromosome variations to specific traits, including illnesses. This chapter explores several ways that chromosomes can be atypical (used synonymously with abnormal) and affect health. Actual cases introduce some of them.

Required Parts: Telomeres and Centromeres

A chromosome consists primarily of DNA and proteins with a small amount of RNA. Chromosomes are duplicated and

Figure 13.1 **Portrait of a chromosome.** Tightly wound, highly repetitive heterochromatin forms the centromere (the largest constriction) and the telomeres (the tips) of chromosomes. Elsewhere, lighter-staining euchromatin includes many protein-encoding genes. The centromere divides this chromosome into a short arm (p) and a long arm (q). This chromosome is in the replicated form.

transmitted, via cell division (mitosis or meiosis), to the next cell generation. The different chromosome types have long been described and distinguished by size and shape, using stains and dyes to contrast dark **heterochromatin** with the lighter **euchromatin** (**figure 13.1**). Heterochromatin consists mostly of highly repetitive DNA sequences, whereas euchromatin has many protein-encoding sequences.

A chromosome must include structures that enable it to replicate and remain intact. Everything else is informational: protein-encoding genes and their controls. The essential parts of a chromosome are

- telomeres;
- origin of replication sites, where replication forks begin to form; and
- the centromere.

Recall from figure 2.15 that telomeres are chromosome tips. In humans, each telomere repeats the sequence TTAGGG. In most cell types, telomeres shorten with each mitotic cell division.

The centromere is the largest constriction of a chromosome. It is where spindle fibers attach when the cell divides. A chromosome without a centromere is no longer a chromosome. It vanishes from the cell as soon as division begins because there is no way to attach to the spindle.

Centromeres, like chromosomes, consist mostly of DNA and protein. Many of the hundreds of thousands of DNA bases that form the centromere are copies of a specific 171-base DNA sequence. The size and number of repeats are similar in many species, although the sequence differs. The similarity among species suggests that these repeats have a structural role in maintaining chromosomes rather than an informational one from their sequence. Certain centromere-associated proteins are synthesized only when mitosis is imminent, forming a structure called a kinetochore that contacts the spindle fibers, enabling the cell to divide.

Centromeres replicate toward the end of S phase of the cell cycle. A protein that may control the process is centromere

Figure 13.2 Subtelomeres. The repetitive sequence of a telomere gradually diversifies toward the centromere. The centromere is depicted as a buttonlike structure to more easily distinguish it, but it is composed of DNA like the rest of the chromosome.

protein A, or CENP-A. Molecules of CENP-A stay with centromeres as chromosomes replicate, covering about half a million DNA base pairs. When the replicated (sister) chromatids separate at anaphase, each member of the pair retains some CENP-A. The protein therefore passes to the next cell generation, but it is *not* DNA. This is an epigenetic change.

Centromeres lie within vast stretches of heterochromatin. The arms of the chromosome extend outward from the centromere. Gradually, with increasing distance from the centromere, the DNA includes more protein-encoding sequences. Gene density varies greatly among chromosomes. Chromosome 21 includes a gene "desert," harboring a million-base stretch with no protein-encoding genes at all. Chromosome 22, in contrast, is a gene "jungle." Yet these two tiniest chromosomes are remarkably similar in size, which is why they are out of place in karyotypes. Chromosome 22 contains 545 genes to chromosome 21's 225! That is, chromosome 21 is actually the smallest, if size is measured by gene content.

The chromosome parts that lie between protein-rich areas and the telomeres are termed subtelomeres (**figure 13.2**). These areas extend from 8,000 to 300,000 bases inward toward the centromere from the telomeres. Subtelomeres include some protein-encoding genes and therefore bridge the gene-rich regions and the telomere repeats. The transition is gradual. Areas of 50 to 250 bases, right next to the telomeres, consist of 6-base repeats, many of them very similar to the TTAGGG of the telomeres. Then, moving inward from the 6-base zone are many shorter repeats. Their function isn't known. Finally, the sequence diversifies and protein-encoding genes appear.

At least 500 protein-encoding genes lie in the subtelomere regions of all the chromosomes combined. About half are members of multigene families (groups of genes of very similar sequence next to each other) that include pseudogenes. These multigene families may be signs of recent evolution: Apes and chimps have only one or two genes for many of the large gene families in humans.

Karyotypes Chart Chromosomes

An old but still useful tool is the chromosome chart, or karyotype. By showing which genes are transmitted together as part of the same chromosome, a karyotype can reveal certain conditions that DNA sequencing can miss. **Clinical Connection 13.1** in section 13.4 returns to this point.

Figure 13.3 A karyotype displays chromosome pairs in size order. Note the extra chromosome 21 that causes trisomy 21 Down syndrome. (Chromosomes are color-enhanced; A–G denote generalized groups.) *Biophoto Associates/Science Source*

A karyotype displays chromosomes in pairs by size and by physical landmarks that appear during mitotic metaphase, when DNA coils tightly, enabling it to be visualized. **Figure 13.3** shows a karyotype with one extra chromosome, which is called a **trisomy**.

The 24 human chromosome types are numbered from largest to smallest—1 to 22. The other two chromosomes are the X and the Y. Early attempts to size-order chromosomes resulted in generalized groupings because many of the chromosomes are of similar size. Use of dyes and stains made it easier to distinguish chromosomes by creating patterns of bands.

Centromere position is one physical feature of chromosomes. A chromosome is **metacentric** if the centromere divides it into two arms of approximately equal length. It is **submetacentric** if the centromere establishes one long arm and one short arm, and **acrocentric** if it pinches off only a small amount of material toward one end (**figure 13.4**). Some

Figure 13.4 Centromere position distinguishes chromosomes. (a) A telocentric chromosome has the centromere near one end, although telomere DNA sequences are at the tip. Humans do not have telocentric chromosomes. **(b)** An acrocentric chromosome has the centromere near an end. **(c)** A submetacentric chromosome's centromere creates a long arm (q) and a short arm (p). **(d)** A metacentric chromosome's centromere establishes more equal-size arms.

species have telocentric chromosomes that have only one arm, but humans do not. The long arm of a chromosome is designated *q*, and the short arm *p* (for "petite").

Five human chromosomes (13, 14, 15, 21, and 22) have bloblike ends, called satellites, that extend from a thin, stalk-like bridge. The stalk regions do not stain, but they carry many copies of genes encoding ribosomal RNA and ribosomal proteins. These areas coalesce to form the nucleolus, a structure in the nucleus where ribosomal building blocks are produced and assembled (see figure 2.2).

Karyotypes are useful at several levels. When a baby is born with distinctive facial characteristics of a chromosomal syndrome, a karyotype can confirm the clinical diagnosis. Within families, karyotypes are used to identify relatives with a chromosome aberration that can affect health. In one family, several adults died from a rare form of kidney cancer. Karyotypes revealed that the affected individuals all had an exchange of genetic material between chromosomes 3 and 8. When karyotypes showed that two healthy young family members had the unusual chromosomes, further testing indicated the cancer. They were successfully treated.

Karyotypes of individuals from different populations can reveal the effects of environmental toxins, if abnormalities appear only in a group exposed to a contaminant. Because chemicals and radiation that can cause cancer and birth defects often break chromosomes into fragments or rings, detecting this genetic damage can alert researchers to the possibility that certain cancers may appear in the population.

> **Key Concepts Questions 13.1**
>
> a. What are the basic parts of a chromosome?
> *b. What does a karyotype show?

13.2 Detecting Chromosomes

Chromosomes can be imaged from any type of human cell that has a nucleus (red blood cells do not). The most common application of human chromosome testing is in prenatal diagnosis. Chromosomes are also checked in relatives of people known to have atypical chromosomes, to explain infertility, or to diagnose or track changes as a cancer progresses.

Older Techniques

The technologies that identify atypical chromosomes in fetuses have evolved since the first fetal karyotype was constructed in 1966. Just in the past few years, deducing chromosome counts from short pieces of placental DNA floating in the maternal bloodstream has mostly replaced the older approaches of amniocentesis and chorionic villus sampling (CVS), which construct karyotypes from cells from structures that support the fetus.

In amniocentesis, a needle passed through the abdominal wall removes a small sample of amniotic fluid from the uterus of a pregnant woman (**figure 13.5a**). Fetal cells in the fluid are cultured for 7 to 10 days, and then 20 cells are karyotyped. Additional tests on biochemicals in the fluid and single-gene tests can reveal some inborn errors of metabolism. Ultrasound is used to place the needle and to visualize the fetus (**figure 13.5c**). Amniocentesis can be safely performed after 14 weeks. For many years amniocentesis was limited to women over age 35, when the risk of the procedure equals the risk of miscarriage, which rises with maternal age (**figure 13.6**). As safety improved, amniocentesis was offered to younger women. Then other tests were developed.

The cells sampled between weeks 10 and 12 in CVS come from the chorionic villi, which are fingerlike structures that develop into the placenta (**figure 13.5b**). Because chorionic villi cells and fetal cells descend from the fertilized ovum, it is assumed that they will have the same chromosomal content. However, rarely, a mutation can occur in either type of cell, leading to a false negative or false positive. If CVS indicates an abnormal chromosome in a villus cell that is not also in the fetus, the couple may elect termination of a chromosomally normal fetus. In the opposite situation, a chorionic villus cell might have normal chromosomes when the fetus does not, providing a false sense of security that all is well. CVS cannot detect inborn errors in biochemicals because amniotic fluid is not sampled.

As the techniques for collecting chromosomes evolved, so did the methods for distinguishing them. Early karyotypes used generalized stains, which could not distinguish chromosomes of similar size, which were considered in groups. Development of more discriminating binding chemicals led to the ability to tell chromosomes apart by banding patterns. A technique called fluorescence *in situ* hybridization (FISH) adds precision by applying pieces of DNA, called probes, which are attached to molecules that produce a flash of color when they bind their complements on the chromosomes in a tissue sample. Use of an algorithm applies a unique false color to each chromosome, "painting" a karyotype from several fluorescent dyes. Many

(a) Amniocentesis

(b) Chorionic villus sampling

(c) Ultrasound image

Figure 13.5 **Checking fetal chromosomes, directly and indirectly.** **(a)** Amniocentesis shows chromosomes from fetal cells in amniotic fluid. **(b)** Chorionic villus sampling examines chromosomes from cells of structures that develop into the placenta. **(c)** Ultrasound images the fetus and may reveal an anomaly that is associated with a specific chromosome abnormality. *(C): Dr. Najeeb Layyous/Science Source*

laboratories use FISH probes for the most common chromosomal abnormalities: an extra chromosome 13, 18, or 21, and sex chromosome anomalies. **Figure 13.7** shows the extra flash of a cell from a person with trisomy 21.

Cell-Free Fetal DNA Testing

Small pieces of DNA are normally present in the bloodstream. In a pregnant woman, about 20 percent of those pieces come from the placenta, and therefore represent the fetus **(figure 13.8)**. The DNA pieces are called "cell-free" because they are not enclosed in cells. Tests of cell-free fetal DNA are done at 10 weeks or later, and are rapidly replacing the older

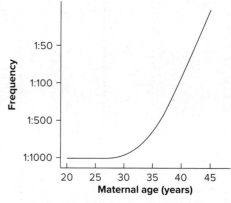

Trisomy 21 in liveborn infants

Figure 13.6 **The maternal age effect.** The risk of conceiving an offspring with trisomy 21 rises dramatically with maternal age. *Source: Color Atlas of Genetics by Eberhard Passage, p. 401. Thieme Medical Publishers, Inc.*

Figure 13.7 **FISHing for genes and chromosomes.** FISH shows three fluorescent dots that correspond to three copies of chromosome 21. Each dot represents a specific DNA sequence with which the fluorescently labeled probe forms complementary base pairs. *James King-Holmes/Science Source*

techniques. It is also called noninvasive prenatal diagnosis or testing. (Ultrasound is also noninvasive.)

Analysis of cell-free fetal DNA is based on proportions. For example, blood from a woman whose fetus has trisomy 21 Down syndrome has about 50 percent more DNA pieces from chromosome 21 than it does from the other chromosomes. That extra DNA represents the third copy of the chromosome. Amniocentesis may be offered to confirm a finding using cell-free DNA, from which a karyotype is constructed showing the extra chromosome. One advantage of the cell-free analysis is that women of all ages use it, saving older women from the more invasive older techniques, while also detecting chromosomal conditions in the fetuses of younger women that might have been missed with the older techniques. Cell-free testing is so sensitive that it has even picked up cancer cells, which also shed DNA, in women who do not yet have cancer symptoms.

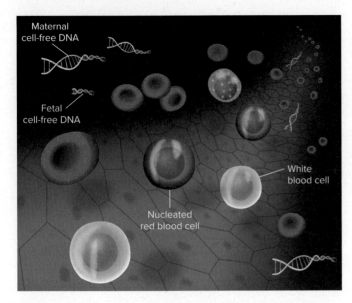

Figure 13.8 Testing cell-free fetal DNA. The fact that pieces of cell-free fetal DNA in the maternal circulation are much shorter than pieces of maternal DNA provides a way to collect the fetal material and sequence the genome by overlapping pieces.

Table 13.1	Chromosomal Shorthand
Abbreviation	**Meaning**
46,XY	Normal male
46,XX	Normal female
45,X	A female with one X
47,XXY	A male with an extra X
47,XYY	A male with an extra Y
46,XY,del (7q)	A male missing part of the long arm of chromosome 7
47,XX,+21	A female with trisomy 21 Down syndrome
46,XY,t(7;9)(p21.1; q34.1)	A male with a translocation between the short arm of chromosome 7 at band 21.1 and the long arm of chromosome 9 at band 34.1
48,XXYY	A male with an extra X and an extra Y chromosome

To detect single-gene conditions, cell-free fetal DNA must be compared to haplotypes from blood samples from the parents for the same chromosomal regions. This is done to distinguish mutations that are in the fetus from those in the pregnant woman. For example, a fetus with identical mutant *CFTR* alleles yet no wild type alleles may have indeed inherited cystic fibrosis from two carrier parents, but the sample might only have captured mutant alleles from the pregnant woman. A haplotype would show surrounding DNA sequences, enabling the distinction between the fetal and maternal genomes.

Some laboratories analyze cell-free fetal DNA only for the more common chromosomal abnormalities, but entire genomes can be reconstructed from the DNA pieces. These more complete analyses are used to distinguish *de novo* (new) mutations from inherited mutations in parent-child trios, such as the family described in Clinical Connection 1.1.

An indirect type of test indicates an elevated risk of having an abnormal chromosome number. It measures biochemicals whose levels in the blood are within a certain range in a pregnant woman carrying a fetus with the normal number of chromosomes, but lie outside that range in fetuses whose cells have an extra copy of a certain chromosome.

Chromosomal Shorthand

The information in a karyotype is abbreviated by listing the chromosome number, sex chromosome makeup, and atypical autosomes (discussed in subsequent sections of this chapter) (**table 13.1**). Symbols and shorthand describe the type of aberration, such as "del" for *deletion* and "t" for ***translocation,*** in which chromosomes exchange parts or a piece of one chromosome is moved to another. After the shorthand, numbers correspond to bands and subbands that are assigned to the markings on chromosomes that arise from staining and FISH.

A chromosomally normal male is designated 46,XY and a female 46,XX. Bands and subbands identify specific genes. For example, the gene that encodes the beta-globin subunit of hemoglobin is located at 11p15.5—the short arm of chromosome 11 at subband 15.5. Graphical representations called ideograms depict the arms, bands, and subbands of chromosomes (**figure 13.9**).

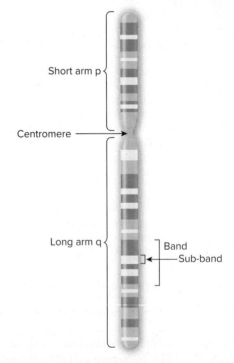

Figure 13.9 Ideogram. This ideogram depicts chromosome 7. It is approximately 160 million base pairs long. *Source: U.S. National Library of Medicine*

> **Key Concepts Questions 13.2**
>
> a. What are some cell types from which chromosomes are examined?
>
> *b. Name the two older techniques of prenatal diagnosis that sample and culture chromosomes from cells taken from the supportive structures of the fetus.
>
> c. Explain how distinguishing chromosome types has become more specific.
>
> *d. Identify the source of fetal DNA in cell-free fetal DNA testing.
>
> e. What are the symbols for the chromosomes of a normal male and female?

13.3 Atypical Chromosome Number

A human karyotype is atypical (abnormal) if the number of chromosomes in a somatic cell is not 46, or if individual chromosomes have extra, missing, or rearranged genetic material. More discriminating technologies can detect very small numbers of extra or missing nucleotides. As a result, more people are being diagnosed with chromosomal abnormalities than in the days when stains made all chromosomes look alike.

Atypical chromosomes account for at least 50 percent of spontaneous abortions, yet only 0.65 percent of newborns have them. Therefore, most embryos and fetuses with atypical chromosomes stop developing before birth. **Table 13.2** summarizes the types of chromosome variants in the order in which they are discussed.

Table 13.2	Chromosome Abnormalities
Type of Abnormality	**Definition**
Polyploidy	Extra chromosome sets
Aneuploidy	An extra or missing chromosome
Monosomy	One chromosome absent
Trisomy	One chromosome extra
Deletion	Part of a chromosome missing
Duplication	Part of a chromosome present twice
Translocation	Two chromosomes join long arms or exchange parts
Inversion	Segment of chromosome reversed
Isochromosome	A chromosome with identical arms
Ring chromosome	A chromosome that forms a ring due to deletions in telomeres, which cause ends to adhere
Chromothripsis	One or more chromosomes shatters

Polyploidy

The most extreme upset in chromosome number is an entire extra set. A cell with extra sets of chromosomes is polyploid. An individual whose cells have three copies of each chromosome is a triploid (designated 3N, for three sets of chromosomes). Two-thirds of all triploids result from fertilization of an oocyte by two sperm. The other cases arise from formation of a diploid gamete, such as when a normal haploid sperm fertilizes a diploid oocyte. Triploids account for 17 percent of spontaneous abortions (**figure 13.10**). Very rarely, an infant survives a few days, with defects in nearly all organs. However, certain human cells may be polyploid. The liver, for example, has some tetraploid (4N) and even octaploid (8N) cells.

Polyploids, although uncommon in humans, are common among flowering plants, including roses, cotton, barley, and wheat, and in some insects. Fish farmers raise triploid salmon, which cannot breed because their gametes contain different numbers of chromosomes.

Aneuploidy

Cells missing a single chromosome or having an extra chromosome are **aneuploid**, which means "not good set." Rarely, aneuploids can have more than one missing or extra chromosome, indicating abnormal meiosis in a parent. A normal chromosome number is **euploid**, which means "good set."

Figure 13.10 Polyploids in humans are nearly always lethal. Individuals with three copies of each chromosome (triploids) in every cell account for 17 percent of all spontaneous abortions and 3 percent of stillbirths and newborn deaths. *CNRI/Science Source*

Most autosomal aneuploids (with a missing or extra non-sex chromosome) are spontaneously aborted. Those that survive have specific syndromes, with symptoms depending on which chromosomes are missing or extra. Intellectual disability is common in aneuploidy because development of the brain is so complex and of such long duration that nearly any chromosome-scale disruption affects genes whose protein products affect the brain. Sex chromosome aneuploidy usually produces milder symptoms.

Most children born with a chromosome number other than 46 have an extra chromosome (a trisomy) rather than a missing one (a **monosomy**), because monosomies are typically so severe that an affected embryo ceases developing (except some cases of a female with only one X chromosome). Trisomies and monosomies are named for the chromosomes involved, and in the past the associated syndromes were named for the discoverers. Today, cytogenetic terminology is used because it is more precise. For example, Down syndrome can result from a trisomy or a translocation. The distinction is important in genetic counseling. Translocation Down syndrome, although accounting for only 4 percent of cases, has a much higher recurrence risk within a family than does trisomy 21 Down syndrome, a point we return to later in the chapter.

The meiotic error that causes aneuploidy is called **nondisjunction**. Recall that in normal meiosis, homologs separate and each of the resulting gametes receives only one member of each chromosome pair. In nondisjunction, a chromosome pair does not separate at anaphase of either the first or second meiotic division. This unequal division produces a sperm or oocyte that has two copies of a particular chromosome, or none, rather than one copy (**figure 13.11**). When such

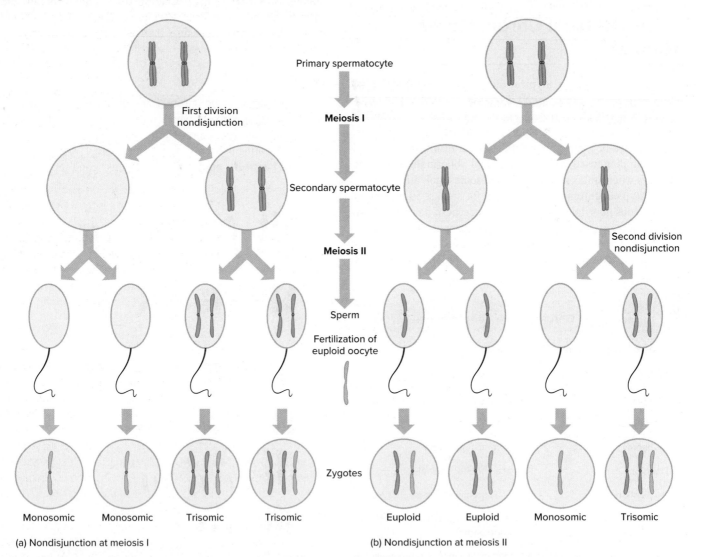

Figure 13.11 Extra and missing chromosomes—aneuploidy. Unequal division of chromosome pairs can occur at either the first or second meiotic division. **(a)** A single pair of chromosomes is unevenly partitioned into the two cells arising from meiosis I in a male. The result: Two sperm cells have two copies of the chromosome, and two sperm cells have no copies. When a sperm cell with two copies of the chromosome fertilizes a normal oocyte, the zygote is trisomic; when a sperm cell lacking the chromosome fertilizes a normal oocyte, the zygote is monosomic. **(b)** This nondisjunction occurs at meiosis II. Because the two products of the first division are unaffected, two of the mature sperm are normal and two are aneuploid. Oocytes can undergo nondisjunction as well, leading to zygotes with extra or missing chromosomes when normal sperm cells fertilize them. (Both chromosome pairs of the spermatocytes are depicted in the same color to distinguish them from the introduction of a maternal genome with the oocyte.)

a gamete meets its partner at fertilization, the resulting zygote has either 45 or 47 chromosomes, instead of the normal 46. Different trisomies tend to be caused by nondisjunction in the male or female, at meiosis I or II.

A cell can have a missing or extra chromosome in 49 ways—an extra or missing copy of each of the 22 autosomes, plus the sex chromosome combinations of Y, X, XXX, XXY, and XYY. (Some individuals have four or even five sex chromosomes.) However, only nine types of aneuploids are recognized in newborns. Others are seen in spontaneous abortions or fertilized ova intended for *in vitro* fertilization.

Most of the 50 percent of spontaneous abortions that result from extra or missing chromosomes are 45,X individuals (missing an X chromosome), triploids, or trisomy 16. About 9 percent of spontaneous abortions are trisomy 13, 18, or 21. More than 95 percent of newborns with atypical chromosome numbers have an extra chromosome 13, 18, or 21, or an extra or missing X or Y chromosome. These conditions are all rare at birth—together they affect only 0.1 percent of all children; however, nondisjunction occurs in 5 percent of recognized pregnancies.

Types of chromosome abnormalities differ between the sexes. Atypical oocytes mostly have extra or missing chromosomes, whereas atypical sperm more often have structural variants, such as inversions or translocations, discussed later in the chapter.

Aneuploidy and polyploidy also arise during mitosis, producing groups of somatic cells with the extra or missing chromosome (see figure 12.2). A mitotic abnormality that occurs early in development, so that many cells descend from the unusual one, can affect health. For example, a chromosomal mosaic for a trisomy may have a mild version of the associated condition. This is usually the case for 1 to 2 percent of people with Down syndrome; they are mosaics. The phenotype depends on which cells have the extra chromosome. Unfortunately, prenatal testing cannot reveal which cells are affected.

Autosomal Aneuploids

Most autosomal aneuploids cease developing long before birth. Following are cases and descriptions of the most common autosomal aneuploids among liveborns. The most frequently seen extra autosomes in newborns are chromosomes 21, 18, and 13, because these chromosomes carry many fewer protein-encoding genes than the other autosomes, compared to their total amount of DNA. Therefore, extra copies of these chromosomes are tolerated well enough for some fetuses with them to survive to be born (**table 13.3**).

Trisomy 21

When David was born in 1994, doctors told his 19-year-old mother, Toni, to put him into an institution, because he would never walk or talk. David had other ideas. He grew up to graduate high school.

Like other teens, David has held part-time jobs, gone to dances, and uses a computer. But he is unlike most other teens in that his cells have an extra chromosome 21, which limits his intellectual abilities. "Maybe he's not book smart, but when

Table 13.3 Comparing and Contrasting Trisomies 13, 18, and 21

Type of Trisomy	Incidence at Birth	Percentage of Conceptions That Survive 1 Year After Birth
13 (Patau)	1/12,500–1/21,700	<5%
18 (Edwards)	1/6,000–1/10,000	<5%
21 (Down)	1/800–1/826	85%

you look around at what he can do, he's smart," Toni says. His speech is difficult to understand, and he has facial features characteristic of Down syndrome, but he has a winning personality and close friends.

Sometimes David gets into unusual situations because he thinks literally. He once dialed 911 when he stubbed his toe, because he'd been told to do just that when he was hurt. Today David lives in a group home and attends community college.

The most common autosomal aneuploid among liveborns is trisomy 21, because this chromosome has the fewest genes. On an ultrasound scan, a fetus with Down syndrome has short limbs, a flattened nose, and excess fluid at the back of the neck. A person with Down syndrome is usually short and has straight, sparse hair and a tongue protruding through thick lips. The hands have an atypical pattern of creases, the joints are loose, and poor reflexes and muscle tone give a "floppy" appearance to a baby. Developmental milestones (such as sitting, standing, and walking) come slowly, and toilet training may take several years. Intelligence varies greatly. Parents of a child with Down syndrome can help their child reach maximal potential by providing a stimulating environment (**figure 13.12**).

Many people with Down syndrome have physical problems, including heart and kidney defects and hearing and visual loss. A suppressed immune system can make influenza deadly. Digestive system blockages are common and may require surgical correction. A child with Down syndrome is 15 times more likely to develop leukemia than a child who does not have the syndrome, but this is still only a 1 percent risk. However, people with Down syndrome are somewhat protected against developing solid tumors. Many of the medical problems associated with Down syndrome are treatable, so that average life expectancy is now 60 years. In 1910, life expectancy was only 9 years. Down syndrome occurs in all population groups.

Some people with Down syndrome older than 40 develop the black tau fibers and amyloid beta protein deposits in their brains characteristic of Alzheimer disease, although they usually do not develop severe dementia (see Clinical Connection 5.1). The chance of a person with trisomy 21 developing Alzheimer disease is 25 percent, compared to 6 percent for the general population. A gene on chromosome 21 causes one inherited form of Alzheimer disease. Perhaps the extra copy of the gene in trisomy 21 has a similar effect to a mutation in the gene that causes Alzheimer disease, such as causing amyloid beta buildup.

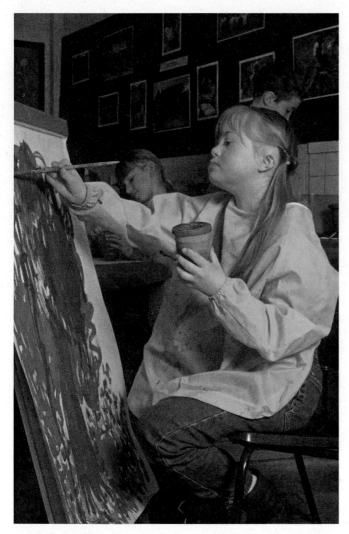

Figure 13.12 Trisomy 21. Many years ago, people with Down syndrome were institutionalized. Today, thanks to strides in both medical care and special education, people with the condition hold jobs and attend college. This young lady enjoys painting. *Angela Hampton Picture Library/Alamy Stock Photo*

Before human genome sequencing began, researchers studied people who have a third copy of only part of chromosome 21 to deduce which genes cause which symptoms. This approach led to the discovery that genes near the tip of the long arm of the chromosome contribute most of the abnormalities. Two specific genes control many aspects of Down syndrome by controlling a third gene, which encodes a transcription factor and therefore affects expression of many other genes.

A newer approach, genome editing, is being used to study Down syndrome at the cellular level. Section 21.4 discusses how genome editing cuts double-stranded DNA at a specific site on a chromosome and allows insertion of a specific DNA sequence. For trisomy 21 Down syndrome, the technique "borrows" the mechanism that shuts off one X chromosome in the cells of females. Researchers insert the DNA that encodes the long noncoding RNA sequence called XIST, which normally shuts off one X chromosome (see figure 6.9), into one chromosome 21 of induced pluripotent stem cells (see Table 2.3) made from skin cells of a boy with trisomy 21. The shut-off chromosome 21 forms a Barr body, which normally happens when XIST silences one X chromosome in a female cell. (Male cells do not normally have Barr bodies because they have only one X chromosome.) The treated male cells have Barr bodies (**figure 13.13**), meaning one of the three copies of chromosome 21 is turned off. This experiment at the cellular level may lead to new treatments.

The likelihood of giving birth to a child with trisomy 21 Down syndrome increases dramatically with the age of the mother (see figure 13.6). However, 80 percent of children with trisomy 21 are born to women under age 35. About 90 percent of trisomy 21 conceptions are due to nondisjunction during meiosis I in the female. The 10 percent of cases due to the male result from nondisjunction during meiosis I or II. The chance that trisomy 21 will recur in a family, based on empirical data (how often it actually does recur in families), is 1 percent.

The age factor in trisomy 21 Down syndrome and other trisomies may be because the older a woman is, the longer her oocytes have been arrested on the brink of completing meiosis. During this 15 to 45 years, oocytes may have been exposed to toxins, viruses, and radiation. A second explanation for the maternal age effect is that females have a pool of immature aneuploid oocytes resulting from spindle abnormalities that cause nondisjunction. As a woman ages, selectively releasing normal oocytes each month, the atypical ones remain.

The association between maternal age and Down syndrome has been recognized for a long time, because affected individuals were often the youngest children in large families. Before the chromosome connection was made in 1959, the syndrome was attributed to syphilis, tuberculosis, thyroid malfunction, alcoholism, emotional trauma, or "maternal reproductive exhaustion." The increased risk of Down syndrome correlates to maternal age, not to the number of children in the family. **Bioethics** examines the effect of prenatal diagnosis on the prevalence of people with trisomy 21 Down syndrome.

Trisomy 18

When an ultrasound scan early in pregnancy revealed a small fetus with low-set ears, a small jaw, a pocket of fluid in the brain, and a clenched fist, the parents-to-be were offered amniocentesis to confirm the extra chromosome 18 that the scan indicated might be there. Further ultrasound scans showed that only one kidney worked, the heart had holes between the chambers, and part of the intestine lay outside the stomach in a sac. The boy was delivered at 36 weeks of gestation, after his heart rate became erratic during a routine prenatal visit. He lived only 22 days.

Most individuals with trisomy 18 (**figure 13.14**) (Edwards syndrome) or trisomy 13 (Patau syndrome) are not born or die in infancy, but a few have lived into young

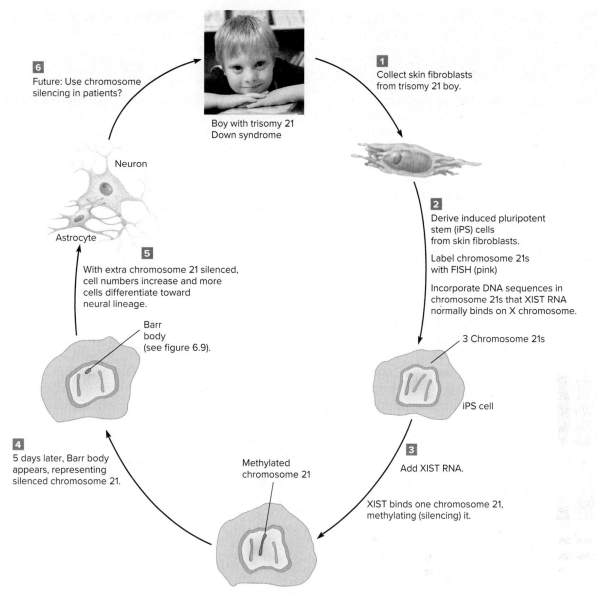

Figure 13.13 Turning off the extra chromosome 21 of Down syndrome. One application of genome editing uses the mechanism of X inactivation to "turn off" the extra chromosome of trisomy 21, silencing the extra genetic information—in induced pluripotent stem cells from an affected individual. Observing what happens as the stem cells are stimulated to differentiate as neurons and comparing this to what happens to trisomic cells may reveal new ways to treat the syndrome. *George Doyle/Stockbyte/Getty Images*

Figure 13.14 Trisomy 18. An infant with trisomy 18 clenches the fists in a characteristic way, with fingers overlapping. *Wellcome Image Library/Custom Medical Stock Photo*

adulthood. Most children who have trisomy 18 have great physical and intellectual disabilities, with developmental skills stalled at the 6-month level. Major abnormalities include heart defects, a displaced liver, growth retardation, and oddly clenched fists. Milder signs include overlapping fingers, a narrow and flat skull, low-set ears, a small mouth, unusual fingerprints, and "rocker-bottom" feet. Most cases of trisomy 18 arise from nondisjunction in meiosis II of the oocyte.

Trisomy 13

The 15-month-old is a "long-term survivor" of trisomy 13. She is small for her age, at the 5th percentile for weight, but she is happy, curious, and playful. Her physical skills, however, lag. She can finally, with great effort, sit up, but cannot yet crawl.

Bioethics

Will Trisomy 21 Down Syndrome Disappear?

In the 1960s, a newborn with Down syndrome was usually sent to an institution. In the 1970s, testing for trisomy 21 Down syndrome became available but was riskier than it is today because the only technologies available—amniocentesis and chorionic villus sampling—can cause spontaneous abortion. Only women over age 35 or with a family history of chromosome abnormalities were tested, because their risk of carrying a fetus with a detectable condition exceeded the risk of the test causing spontaneous abortion. Maternal serum marker testing in the 1980s and then cell-free fetal DNA testing starting in 2011 made it safer and easier for any pregnant woman to learn if the fetus had trisomy 21 Down syndrome.

The ability to detect trisomy 21 prenatally has led to a decrease in the population of people who have this condition. In the United Kingdom, about 90 percent of couples learning that their child will have trisomy 21 end the pregnancy. In the United States, it is about 60 percent. At the same time, some organizations representing people with Down syndrome, distressed that affected individuals will have fewer peers as time goes on and prenatal testing becomes more widespread, began to pressure pregnant women not to avoid having children with the condition. Laws have been passed in some states that ban pregnancy termination if test results indicate Down syndrome.

Questions for Discussion

1. Should state laws control the circumstances under which women can end pregnancies?
2. Look up the definition of eugenics, or read about it in chapter 16. Do you think that use of prenatal testing for trisomy 21 is a eugenic measure?
3. What impact do you think the wider availability of early, safe, and accurate testing for trisomy 21 Down syndrome will have on society? Discuss views that might come from a person with trisomy 21, a pregnant woman and her partner, a parent of an adult child with Down syndrome, and a physician who offers prenatal testing.

She has about 20 minor seizures a day, which look like startles, and has difficulty eating because of persistent acid reflux. She is also missing a rib. Early surgeries corrected a cleft lip and palate, removed an extra finger and toe, and repaired a hernia. The girl is intellectually disabled, but her parents hope she will live long enough to attend preschool. Unlike many others with trisomy 13, she can see and hear.

Trisomy 13 has different signs and symptoms than trisomy 18. Most striking is fusion of the developing eyes into one large eyelike structure in the center of the face, or a small or absent eye. Major abnormalities affect the heart, kidneys, brain, face, and limbs. The nose is often malformed, and cleft lip and/or palate is present in a small head. There may be extra fingers and toes. Ultrasound examination of an affected newborn may reveal an extra spleen, atypical liver, rotated intestines, and an abnormal pancreas.

Sex Chromosome Aneuploids: Female

People with sex chromosome aneuploidy have extra or missing sex chromosomes. **Table 13.4** indicates how these aneuploids can arise. Some conditions can result from nondisjunction in meiosis in the male *or* female. Sex chromosome aneuploids are generally associated with much less severe symptoms and characteristics than autosomal aneuploids. The

Table 13.4 How Nondisjunction Leads to Sex Chromosome Aneuploids

Situation	Oocyte	Sperm	Consequence
Normal	X	Y	46,XY normal male
	X	X	46,XX normal female
Female nondisjunction	XX	Y	47,XXY syndrome
	XX	X	47,XXX triplo-X
		Y	45,Y nonviable
		X	45,X syndrome
Male nondisjunction (meiosis I)	X		45,X syndrome
	X	XY	47,XXY syndrome
Male nondisjunction (meiosis II)	X	XX	47,XXX triplo-X
	X	YY	47,XYY syndrome
	X		45,X syndrome
Male and female nondisjunction	XX	YY	48,XXYY syndrome

syndromes are no longer commonly referred to with eponyms, but 47,XXY was called Klinefelter syndrome, 45,X was Turner syndrome (also called 45,XO), and 47,XYY was Jacobs syndrome.

45,X Syndrome

At age 17, the young woman still looked about 12. She was short, her breasts had never developed, and she had never menstruated. When she turned 16, her physician suggested that she have her chromosomes checked. The karyotype revealed only one X chromosome. The diagnosis explained her poor hearing, high blood pressure, low thyroid function, and "beauty marks" that dotted her skin.

Like the autosomal aneuploids, 45,X syndrome is more frequent among spontaneously aborted fetuses than among newborns—99 percent of 45,X fetuses are not born. The syndrome affects 1 in 2,500 female births. However, if amniocentesis or CVS was not done, a person with 45,X syndrome would likely not know she has a chromosome abnormality until she lags in sexual development and has her chromosomes checked. Two X chromosomes are necessary for normal sexual development in females.

At birth, a girl with 45,X syndrome looks normal, except for puffy hands and feet caused by impaired lymph flow. In childhood, signs of 45,X syndrome include wide-set nipples, soft nails that turn up at the tips, slight webbing at the back of the neck, short stature, coarse facial features, and a low hairline at the back of the head. About half of people with 45,X syndrome have impaired hearing and frequent ear infections due to a small defect in the shape of the coiled part of the inner ear. They cannot hear certain frequencies of sound.

At sexual maturity, sparse body hair develops, but the girls do not ovulate or menstruate, and their breasts do not develop. The uterus is very small, but the vagina and cervix are normal size. In the ovaries, oocytes mature too fast, depleting the supply during infancy. 45,X syndrome may impair the ability to solve math problems that entail envisioning objects in three-dimensional space, and may cause memory deficits, but intelligence is normal. Low doses of hormones (estrogen and progesterone) can stimulate development of secondary sexual structures for girls diagnosed before puberty, and growth hormone can maximize height.

Individuals who are mosaics (only some cells are 45,X) may have children, but their offspring are at high risk of having too many or too few chromosomes. 45,X syndrome is unrelated to the age of the mother. The effects of 45,X syndrome continue past the reproductive years. Adults with the syndrome are more likely to develop certain diseases than the general population, including osteoporosis, types 1 and 2 diabetes, and colon cancer. The many signs and symptoms of 45,X syndrome result from loss of specific genes. Life span is shortened slightly.

Triplo-X

About 1 in every 1,000 females has an extra X chromosome in each of her cells, a condition called triplo-X. The only symptoms are tall stature and menstrual irregularities. Although triplo-X females are rarely intellectually disabled, they tend to be less intelligent than their siblings who have the normal number of chromosomes. The lack of symptoms reflects the protective effect of X inactivation—all but one of the X chromosomes is inactivated.

Sex Chromosome Aneuploids: Male

Any individual with a Y chromosome is a male. A man can have one or more extra X or Y chromosomes.

47,XXY Syndrome

Looking back, the man's only indication of 47,XXY syndrome was small testes. When his extra X chromosome was detected when he was 25, suddenly his personality quirks made sense. As a child and teen, he had been very shy, had trouble making friends, and angered easily. His parents had consulted several clinicians. Psychologists, psychiatrists, and therapists diagnosed learning disabilities. One said that the man would never be able to attend college. Yet he earned two bachelor's degrees and became a successful software engineer.

About 1 in 500 males has the extra X chromosome that causes 47,XXY syndrome. Severely affected men are underdeveloped sexually, with rudimentary testes and prostate glands and sparse pubic and facial hair. They have very long arms and legs, large hands and feet, and may develop

---

A GLIMPSE OF HISTORY

In 1938, at a medical conference, a U.S. endocrinologist named Henry Turner described seven young women, aged 15 to 23, who were sexually undeveloped, short, had folds of skin on the back of their necks, and had malformed elbows. (Eight years earlier, an English physician had described the syndrome in young girls, so it is called Ullrich syndrome in the United Kingdom.) Alerted to what would become known as Turner syndrome in the United States, other physicians soon began identifying such patients. Physicians assumed that a hormonal insufficiency caused the symptoms, but a chromosomal imbalance caused the hormone deficit.

In 1954, at a London hospital, a physician discovered that cells from Turner patients do not have a Barr body, the dark spot that indicates the silenced X chromosome. By 1959, karyotyping confirmed the presence of only one X chromosome. Later, researchers learned that only 50 percent of affected individuals are 45,X. The rest are missing only part of an X chromosome or are mosaics, with only some cells missing an X.

breast tissue. Yet some men with an extra X chromosome have no associated symptoms.

Testosterone injections during adolescence can limit limb lengthening and stimulate development of secondary sexual characteristics. Boys and men with 47,XXY syndrome may be slow to learn, but they are usually not intellectually disabled unless they have more than two X chromosomes, which is rare.

47,XXY syndrome is the most common genetic or chromosomal cause of male infertility. Doctors can select sperm that contain only one sex chromosome and use the sperm to fertilize oocytes *in vitro*. However, sperm from men with 47,XXY syndrome are more likely to have extra chromosomes—usually X or Y, but also autosomes—than sperm from men who do not have 47,XXY syndrome.

47,XXYY Syndrome

The boy's problems were at first so common that it was years before a chromosome check revealed he had an extra X and an extra Y chromosome. He was late to sit, crawl, walk, and talk. In preschool, he had frequent outbursts and made inappropriate comments. He was tall and clumsy and drooled and choked easily. The boy would run about flapping his arms, then hide under a chair. Severe ulcers formed on his legs.

In the second grade, the boy's special education teacher suggested to his parents that they and their son have their chromosomes checked. The parents' chromosomes were normal. The son must have been conceived from a very unusual oocyte meeting a very unusual sperm, both arising from nondisjunction. The extra sex chromosomes explained many of the boy's problems, and even a few that hadn't been recognized, such as curved pinkie fingers, flat feet, and scoliosis. He began receiving testosterone injections so that his teen years would be more normal than his difficult childhood had been.

About 1 in 17,000 newborn boys have an extra X chromosome *and* an extra Y chromosome. These 48,XXYY males have more severe behavioral problems than males with 47,XXY syndrome and tend to develop foot and leg ulcers, resulting from poor venous circulation. Attention deficit disorder, obsessive compulsive disorder, autism, and learning disabilities typically develop by adolescence. In the teen years, testosterone level is low, development of secondary sexual characteristics is delayed, and the testes are undescended. A man with 48,XXYY syndrome is infertile.

47,XYY Syndrome

In 1965, cytogeneticist Patricia Jacobs surveyed 197 inmates at Carstairs, a high-security prison in Scotland. Of 12 men with unusual chromosomes, seven had an extra Y. Jacobs's findings were repeated for mental institutions, and soon after, Newsweek magazine ran a cover story on "congenital criminals," blaming such behavior on the extra Y chromosome. 47,XYY syndrome, also known as Jacobs syndrome, became a legal defense for committing a violent crime, but has never successfully been used as such due to the biological uncertainty.

In the early 1970s, newborn screens began in hospital nurseries in England, Canada, Denmark, and Boston. Social workers and psychologists visited 47,XYY boys to offer "anticipatory guidance" to the parents on how to deal with their toddling future criminals. By 1974, geneticists and others halted the programs, pointing out that singling out these boys on the basis of a few statistical studies was inviting self-fulfilling prophecy.

One male in 1,000 has an extra Y chromosome. Today, we know that 96 percent of males with an extra Y chromosome are *not* destined to become criminals, but may be tall and have acne. Problems with speech and understanding language are subtle, such as an inability to understand humor. Attaining skills like dressing and socializing may be delayed.

The higher prevalence of 47,XYY among mental and penal institution populations may stem more from psychology than biology. Large body size may lead teachers, employers, parents, and others to expect more of these people, and a few 47,XYY individuals may deal with this stress aggressively.

47,XYY syndrome arises from nondisjunction in the male, producing a sperm with two Y chromosomes that fertilizes a normal oocyte. Geneticists have never observed a sex chromosome constitution of one Y and no X because the X chromosome carries so many genes that a 45,X embryo stops developing very early.

> **Key Concepts Questions 13.3**
>
> *a. How many chromosomes are in a normal human karyotype?
>
> *b. What is a polyploid?
>
> *c. Explain how nondisjunction generates aneuploids.
>
> d. Compare the severities of monosomies and trisomies, and of sex chromosome and autosomal aneuploidies.
>
> e. Describe the most common aneuploid conditions.

13.4 Atypical Chromosome Structure

A chromosome can be structurally atypical in several ways. It may have too much or too little genetic material, or a stretch of DNA that is inverted or moved and inserted into a different type of chromosome (**figure 13.15**). Atypical chromosomes are balanced if they have the normal amount of genetic material and unbalanced if they have extra or missing DNA sequences.

Deletions and Duplications

The beautiful baby had a strange cry. When the child was a few weeks old, her mother was talking on the phone to her sister, a physician. The sister, overhearing the baby's cries, asked when they had gotten a cat. They hadn't. Alarmed, the

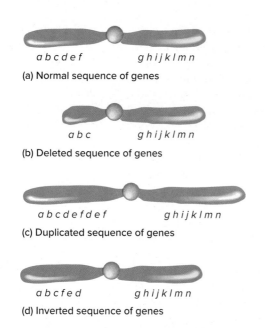

Figure 13.15 Chromosome abnormalities. If a hypothetical normal gene sequence appears as shown in (a), then (b) represents a deletion, (c) a duplication, and (d) an inversion.

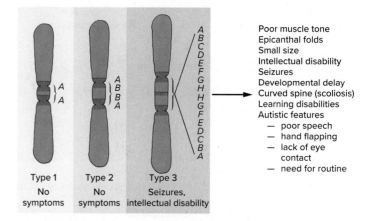

Figure 13.16 A duplication. A study of duplications of parts of chromosome 15 revealed that small duplications do not affect the phenotype, but larger ones may. The letters indicate specific DNA sequences, which serve as markers to compare chromosome regions. Note that the duplication is also inverted.

sister advised telling the child's pediatrician about the cry. The pediatrician heard the mewling and agreed. The baby's karyotype revealed what both doctors had suspected: 5p− (5p minus or deletion) syndrome, also called cri-du-chat (cat's cry) syndrome. One of the baby's copies of chromosome 5 is missing part of the short arm. The child has defied low expectations and can walk, eat, see, and hear, although she has autistic behaviors, is small, cannot use a toilet, and spikes fevers. She can't speak, but communicates quite well in other ways and goes to school.

Deletion and **duplication** mutations are missing and extra DNA sequences, respectively. They are types of copy number variants (CNVs). (Sections 7.3 and 12.5 discussed CNVs.) The more genes involved, the more severe the associated syndrome. **Figure 13.16** depicts a common duplication, of part of chromosome 15. Many deletions and duplications are *de novo* (neither parent has the abnormality).

A technique called **comparative genomic hybridization (CGH)** is used to detect very small CNVs, which are also termed **microdeletions** and **microduplications**. CGH compares the number of copies of a CNV in the same amount of DNA from two people—one with a medical condition, one healthy, to associate the duplication or deletion with the phenotype. CGH is used to help diagnose autism, intellectual disability, learning disabilities, and other behavioral conditions. For example, the technique showed that a young boy who had difficulty concentrating and sleeping and would often scream for no apparent reason had a small duplication in chromosome 7. A young girl plagued with head-banging behavior, digestive difficulties, severe constipation, and great sensitivity to sound had a microdeletion in chromosome 16. Other microdeletions cause male infertility.

Deletions and duplications can arise from chromosome rearrangements. These include translocations, inversions, and ring chromosomes.

Translocation Down Syndrome

When the child was born, while her parents marveled at her beauty, the obstetrician noted that the broad, tilted eyes and sunken nose looked like the face of a child with Down syndrome. The doctor knew that the mother had experienced two spontaneous abortions, a family history suggesting a chromosome problem. So the doctor looked for and found the telltale single crease in the palms of people with Down syndrome. She gently asked the parents for permission to take a blood sample to check their daughter's chromosomes.

A few days later, the new parents learned that their daughter has an unusual form of Down syndrome that she inherited from one of them, rather than the more common "extra chromosome" (trisomy) form. Because the father's mother and sister had also had several pregnancy losses, the exchanged chromosomes likely came from his side. Karyotypes of the parents confirmed that the father was a translocation carrier: One chromosome 14 had attached to one chromosome 21, and distribution of this unusual chromosome in meiosis had led to various imbalances, depicted in **figure 13.17**.

In a translocation, different (nonhomologous) chromosomes exchange or combine parts. Translocations can be inherited because they can be present in carriers, who have the normal amount of genetic material, but it is rearranged. A translocation can affect the phenotype if it breaks a gene or leads to duplications or deletions in the chromosomes of offspring.

There are two major types of translocations, as well as a few rarer types. In a **Robertsonian translocation**, the short arms of two different acrocentric chromosomes break, leaving sticky ends on the two long arms that join, forming a single,

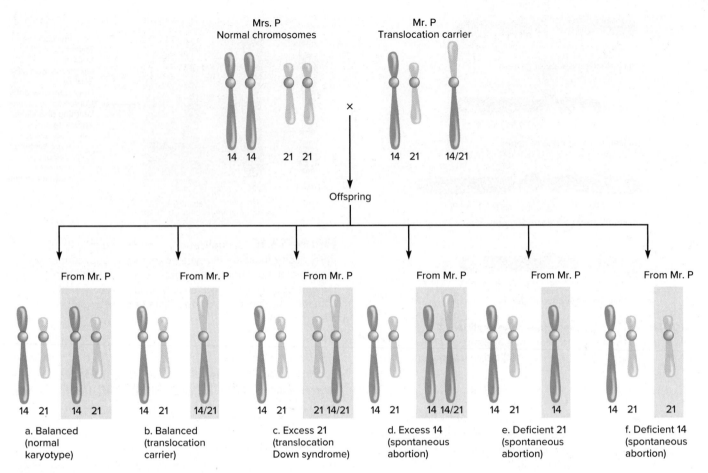

Figure 13.17 A Robertsonian translocation. Mr. P. has only 45 chromosomes because the long arm of one chromosome 14 has joined the long arm of one chromosome 21. He has no symptoms. Mr. P. makes six types of sperm. **(a)** A sperm with one normal chromosome 14 and one normal 21 yields a normal child when fertilized. **(b)** A sperm carrying the translocated chromosome produces a child who is a translocation carrier, like Mr. P. **(c)** If a sperm contains Mr. P.'s normal 21 and his translocated chromosome, the child receives too much chromosome 21 material and has Down syndrome. **(d)** A sperm containing the translocated chromosome and a normal 14 leads to excess chromosome 14 material, which is lethal in the embryo or fetus. If a sperm does not have chromosome 21 **(e)** or 14 **(f)**, a monosomy results, which is lethal prenatally. (Chromosome arm lengths are not precisely accurate.)

large chromosome with two long arms (see chromosome 14/21 in figure 13.17). The tiny short arms are lost, but their DNA sequences are repeated elsewhere in the genome, so the loss does not cause symptoms. The person with the large, translocated chromosome, called a **translocation carrier**, has 45 chromosomes instead of 46, but may not have symptoms if no crucial genes have been deleted or damaged. Even so, he or she may produce unbalanced gametes—sperm or oocytes with too many or too few genes. This can lead to spontaneous abortion or birth defects.

In 1 out of 20 cases of Down syndrome, a parent has a Robertsonian translocation between chromosome 21 and another, usually chromosome 14. That parent produces some gametes that do not have either of the involved chromosomes and some gametes that have extra material from one of the translocated chromosomes. In such a case, each fertilized ovum has a 1 in 2 chance of being spontaneously aborted, and a 1 in 6 chance of developing into an individual with Down syndrome. The risk of giving birth to a child with Down syndrome is theoretically 1 in 3, because the spontaneous abortions are not births. However, because some Down syndrome fetuses spontaneously abort, the empiric risk of a couple in this situation having a child with Down syndrome is about 15 percent. The other two outcomes—a fetus with normal chromosomes or a translocation carrier like the parent—have normal phenotypes. Either a male or a female can be a translocation carrier, and the condition is not related to age. About 1 in 1,000 individuals in the general population carries a Robertsonian translocation.

Much rarer than being a heterozygote for a Robertsonian translocation is being a homozygote, because such individuals can only arise from inheriting one copy of the unusual chromosome from each parent—which typically means the parents are related and inherited the translocation from a shared ancestor, like a common great-grandparent. Robertsonian homozygotes have 44 chromosomes rather than the normal 46. A case report describes a healthy 25-year-old man from China who has 44 chromosomes because each chromosome 14 joins a chromosome 15. His parents, both translocation carriers, were first

Clinical Connection 13.1

Distinguishing a Robertsonian Translocation from a Trisomy

A human genome is distributed over 23 pairs of chromosomes. Rearrangements of the chromosome pairs are invisible to genome sequencing if two complete genomes are present, or may lead to diagnosis of the wrong type of chromosome abnormality. However, the distribution of genes on chromosomes, and the fact that genes on the same chromosome are transmitted together unless separated during meiosis by a crossover, is important in genetic counseling of parents-to-be. Consider a case in which DNA sequencing masked a Robertsonian translocation, leading to an error in deducing the type of chromosomal abnormality.

The young couple had suffered two early pregnancy losses. The second time, the woman collected some of the tissue she had passed and gave it to her obstetrician, who sent it for genetic testing. The lab report profiled single nucleotide polymorphisms (SNPs), landmarks among the chromosomes, and detected overrepresentation of those on chromosome 21. The report did *not* provide a photograph of the actual chromosomes in the fetus, an old-fashioned karyotype. The obstetrician assumed that extra chromosome 21 material meant trisomy 21 Down syndrome, which could recur in future pregnancies at the population empiric risk of 1 in 100. The doctor reassured the patient and her partner.

Then the couple had a third pregnancy loss. Again, the SNP test revealed extra chromosome 21 material. The genetic counselor noted that the extra chromosome 21 material was more likely due to a translocation than a trisomy, because the nondisjunction that would have led to a trisomy is a random event. In contrast, translocation Down syndrome would recur with a probability of 1 in 3 (see figure 13.17). Having a translocation would mean that every conception for this couple had a 2 in 3 risk of having an imbalance of genetic material. On the genetic counselor's advice and interpretation, the physician ordered a karyotype of the products of conception. Sure enough, a Robertsonian translocation had placed some chromosome 21 material onto one chromosome 14. In addition to the two normal copies of chromosome 21, that was an overload (**figure 13A**).

Trisomy 21 Down syndrome

Translocation Down syndrome

21 21 21   14 21 21 14/21

Figure 13A Two origins of Down syndrome. Trisomy 21 is an extra chromosome 21. Translocation Down syndrome is extra genetic material from chromosome 21 resulting from an exchange of chromosome parts.

Questions for Discussion

1. Explain why the risk of recurrence of extra chromosome 21 material in this family is much higher for a Robertsonian translocation than for trisomy 21.
2. Describe the karyotype of the parent who passed on the translocated chromosome.
3. What other types of chromosomal problems might arise in future pregnancies for this couple?

---

cousins. The Chinese man's sperm carry 21 autosomes and an X or Y. He can only be fertile with a woman whose cells have the same 44 chromosome types.

Robertsonian translocation homozygotes bring up an interesting scenario. If several people with the same 44 chromosome types breed among themselves, they could, theoretically, start a new human subspecies. The probability of two carriers of the same Robertsonian translocation randomly meeting and having children together is about 1 in 4 million. **Clinical Connection 13.1** describes how genomewide SNP analysis of sequence variants can miss a Robertsonian translocation if all of the parts of the affected chromosomes are present.

In a **reciprocal translocation,** the second major type of translocation, two different chromosomes exchange parts (**figure 13.18**). About 1 in 600 people is a carrier for a reciprocal translocation. FISH can be used to highlight the involved chromosomes. If the chromosome exchange does not break any genes, then a person who has both translocated chromosomes is healthy and a translocation carrier. He or she has the normal amount of genetic material, but it is rearranged. A reciprocal translocation carrier can have symptoms if one of the two breakpoints lies in a gene, disrupting its function. For example, a translocation between chromosomes 11 and 22 causes infertility in males and recurrent pregnancy loss in females. However, a reciprocal translocation that arises *de novo* in a sperm or oocyte can affect health in the next generation if fertilization occurs and development proceeds, because the resulting individual would have extra or missing genetic material.

Reciprocal translocations usually occur in specific chromosomes that have unstable parts. Vulnerable parts of chromosomes are located where the DNA is so symmetrical in sequence that complementary base pairing occurs within the same DNA strand, folding it into loops during DNA replication. When these contortions break both DNA strands, parts of different chromosomes can switch places.

Figure 13.18 A reciprocal translocation. In a reciprocal translocation, two nonhomologous chromosomes exchange parts. In **(a)**, genes *C, D,* and *E* on the blue chromosome exchange positions with genes *M* and *N* on the red chromosome. In **(b)**, a reciprocal translocation has been highlighted using FISH. The pink chromosome with the dab of blue, and the blue chromosome with a small section of pink, are the translocated chromosomes. *(b): Courtesy Lawrence Livermore National Laboratory*

Unbalanced gametes can result when a reciprocal translocation is inherited or *de novo*, just as they can from a Robertsonian translocation. The four possibilities are (1) transmitting two normal copies of the two involved chromosomes; (2) transmitting two abnormal copies, with no effect on the phenotype of an offspring unless a vital gene is disrupted; or (3 and 4) transmitting either translocated chromosome, which introduces extra or missing genetic material and likely would affect the phenotype.

A rare type of translocation is an insertional translocation, in which part of one chromosome inserts into a nonhomologous chromosome. Symptoms may result if the inserted DNA disrupts a vital gene or if crucial DNA sequences are lost or present in excess.

A carrier of any type of translocation can produce unbalanced gametes—sperm or oocytes that have deletions or duplications of some of the genes in the translocated chromosomes. The phenotype depends on the genes that the rearrangement disrupts and whether genes are extra or missing. A translocation and a deletion can cause the same syndrome if they affect the same part of a chromosome.

A genetic counselor suspects a translocation when a family has a history of birth defects, pregnancy loss, and/or stillbirths.

Prenatal testing may reveal a translocation in a fetus, which can then be traced back to a parent who is a translocation carrier. If neither parent has the translocation in cells typically used to check chromosomes, then the translocation is *de novo*, or a parent is a somatic mosaic and only some gametes have the atypical chromosome.

Inversions

The couple was excited about getting the results of the amniocentesis because a pregnancy had not gone this far before. Then they were called to come in for the results. Expecting bad news, they were surprised to learn that the fetus had an inverted chromosome. Some of the bands that normally appear on chromosome 11 were flipped. The genetic counselor advised them to have their own chromosomes checked. Karyotyping revealed that the woman had the same inversion, so it likely wouldn't harm their daughter, although someday she, too, might experience pregnancy loss.

An inverted sequence of chromosome bands disrupts important genes and harms health in only 5 to 10 percent of cases. If neither parent has the inversion, then it arose in a gamete *de novo* or because a parental ovary or testis is mosaic. The

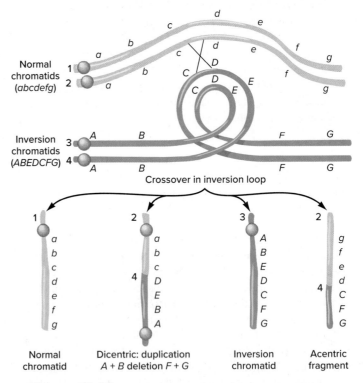

Figure 13.19 Paracentric inversion. A paracentric inversion in one chromosome leads to one normal chromatid, one inverted chromatid, one with two centromeres (dicentric), and one with no centromere (an acentric fragment) if a crossover occurs with the normal homolog. The letters *a* through *g* denote genes.

specific effects of an inverted chromosome may depend on which genes the flip disrupts. Consulting the "reference" human genome (a consensus of common sequences compiled from many genomes) can help to identify genes that might be implicated in a specific inversion.

Like a translocation carrier, an adult who is heterozygous for an inversion can be healthy, but have reproductive problems. One woman had an inversion in the long arm of chromosome 15 and had two spontaneous abortions, two stillbirths, and two children who died within days of birth. She did eventually give birth to a healthy child. How did the inversion cause these problems?

Inversions with such devastating effects can be traced to meiosis, when a crossover occurs between the inverted chromosome segment and the noninverted homolog. To allow the genes to align, the inverted chromosome forms a loop. When crossovers occur within the loop, some areas are duplicated and some deleted in the resulting recombinant chromosomes. In inversions, the atypical chromosomes result from the chromatids that crossed over.

Two types of inversions are distinguished by the position of the centromere relative to the inverted section. **Figure 13.19** shows a **paracentric inversion,** in which the inverted section does not include the centromere. A single crossover within the inverted segment gives rise to one normal, one inversion, and two highly atypical chromatids. One abnormal chromatid retains both centromeres and is termed dicentric. When the cell divides, the two centromeres are pulled to opposite sides of the cell, and the chromatid breaks, leaving pieces with extra or missing segments. The second type of atypical chromatid resulting from a crossover within an inversion loop is a small piece that lacks a centromere, called an acentric fragment. When the cell divides, the fragment is lost because a centromere is required for cell division.

A **pericentric inversion** includes the centromere within the loop. A crossover in the inversion loop produces two chromatids that have duplications and deletions, but one centromere each, plus one normal and one inversion chromatid (**figure 13.20**).

Isochromosomes and Ring Chromosomes

An **isochromosome** is the result of another meiotic error that leads to unbalanced genetic material. It is a chromosome that has identical arms. An isochromosome forms when, during division, the centromeres part in the wrong plane (**figure 13.21**). Isochromosomes are known for chromosomes 12 and 21 and for the long arms of the X and the Y. For example, a woman may have an isochromosome with the long arm of the X chromosome duplicated but the short arm absent.

Chromosomes shaped like rings form in 1 of 25,000 conceptions. Ring chromosomes may arise when telomeres are lost, leaving sticky ends that adhere (**figure 13.22**). Exposure to radiation can form rings. Any chromosome can circularize. Most ring chromosomes consist of DNA repeats and do not affect

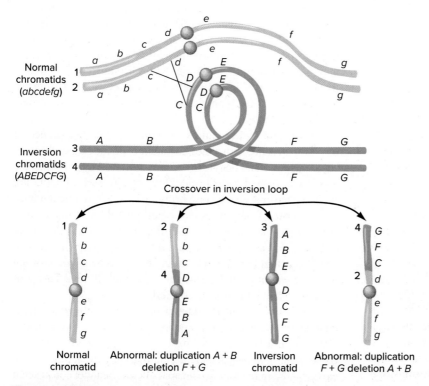

Figure 13.20 Pericentric inversion. A pericentric inversion in one chromosome leads to two chromatids with duplications and deletions, one normal chromatid, and one inverted chromatid if a crossover occurs with the normal homolog.

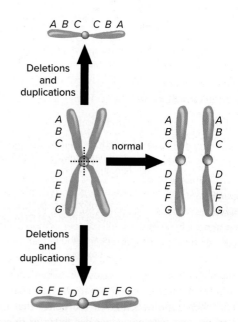

Figure 13.21 Isochromosomes have identical arms. They form when chromatids divide along the wrong plane (in this depiction, horizontally rather than vertically).

Chapter 13 Chromosomes

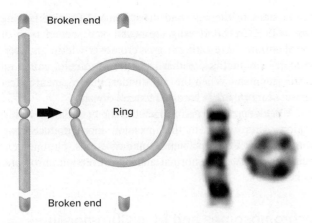

Figure 13.22 A ring chromosome. A ring chromosome may form if the chromosome's tips (telomeres) break, forming sticky ends. Genes can be lost or disrupted, possibly causing symptoms. *Courtesy Ring Chromosome 20 Foundation*

health. The few that do arise *de novo*. They all cause a small head (microcephaly), learning disabilities or intellectual disabilities, slow growth, and unusual facial features.

Table 13.5 summarizes causes of types of abnormal chromosomes. The very rare chromosomal anomaly chromothripsis, discussed in the chapter opener, shatters chromosomes resulting from double-strand breaks.

Table 13.5	Causes of Abnormal Chromosomes
Abnormalities	**Causes**
Numerical Abnormalities	
Polyploidy	Multiple fertilization or diploid gamete leads to extra sets of chromosomes
Aneuploidy	Nondisjunction (in meiosis or mitosis) leads to lost or extra chromosomes when not all chromatid pairs separate in anaphase
Structural Abnormalities	
Deletions and duplications	Translocation
	Crossover between a chromosome that has a pericentric inversion and its noninverted homolog
Translocation	Exchange between nonhomologous chromosomes at parts that are unstable due to symmetrical DNA sequences
Inversion	Breakage and reunion of fragment in same chromosome, but with wrong orientation
Dicentric and acentric chromatids	Crossover between a chromosome with a paracentric inversion and its noninverted homolog
Ring chromosome	A chromosome loses telomeres and the ends fuse, forming a circle
Chromothripsis	Up to three chromosomes spontaneously shatter

Key Concepts Questions 13.4

*a. Which chromosome rearrangements can cause deletions and duplications?

b. Distinguish between the two major types of translocations.

*c. What must occur for a translocation or inversion to cause symptoms?

d. How do isochromosomes and ring chromosomes arise?

13.5 Uniparental Disomy—A Double Dose from One Parent

If nondisjunction (unequal chromosome division) occurs in both a sperm and an oocyte that join, a pair of chromosomes (or their parts) can come solely from one parent, rather than one from each parent, as Mendel's law of segregation predicts. For example, if a sperm that does not have a chromosome 14 fertilizes an ovum that has two copies of that chromosome, then an individual with the normal 46 chromosomes results, but both copies of chromosome 14 come from the mother.

Inheriting two chromosomes or chromosome segments from one parent is called **uniparental disomy (UPD)** ("two bodies from one parent"). UPD can also arise from a trisomic embryo in which some cells lose the extra chromosome, leaving two homologs from one parent. For example, an embryo may have trisomy 21, with the extra chromosome 21 coming from the father. If in some cells the chromosome 21 from the mother is lost, then both remaining copies of the chromosome are from the father.

Because UPD requires the simultaneous occurrence of two rare events—either nondisjunction of the same chromosome in sperm and oocyte or trisomy followed by chromosome loss—it is indeed rare. In addition, many cases of UPD are probably never seen, because bringing together identical homologs inherited from one parent could give the fertilized ovum homozygous lethal alleles. Development would halt. Other cases of UPD may go undetected if they cause known recessive conditions and both parents are assumed to be carriers, when only one parent contributed to the offspring's illness. This situation was how UPD was discovered.

In 1988, Arthur Beaudet of the Baylor College of Medicine saw an unusual patient with cystic fibrosis (see Clinical Connection 4.1). In comparing *CFTR* alleles of the patient to those of her parents, Beaudet found that only the mother was a carrier—the father had two normal alleles. Beaudet constructed haplotypes for each parent's chromosome 7, which includes the *CFTR* gene, and he found that the daughter had two copies from her mother, and none from her father (**figure 13.23**). How did this happen?

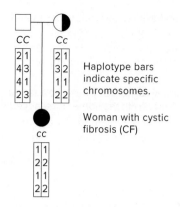

Figure 13.23 Uniparental disomy doubles part of one parent's genetic contribution. In this family, the woman with CF inherited two copies of her mother's chromosome 7 and neither of her father's. Unfortunately, it was the chromosome with the disease-causing allele that she inherited in a double dose.

Apparently, in the patient's mother, nondisjunction of chromosome 7 in meiosis II led to formation of an oocyte bearing two identical copies of the chromosome, instead of the usual one copy. A sperm that had also undergone nondisjunction and did not have a chromosome 7 then fertilized the abnormal oocyte. The mother's extra genetic material compensated for the father's deficit, but unfortunately, the child inherited a double dose of the mother's chromosome that carried the mutant *CFTR* allele. In effect, inheriting two of the same chromosome from one parent shatters the protection that combining genetic material from two individuals offers. This protection is the defining characteristic of sexual reproduction.

UPD may also cause disease if it removes the contribution of the important parent for an imprinted gene. Recall from chapter 6 that an imprinted gene is expressed if it comes from one parent, but silenced if it comes from the other (see figure 6.12). If UPD removes the parental genetic material that must be present for a critical gene to be expressed, a mutant phenotype results. The classic example of UPD disrupting imprinting is Prader-Willi syndrome and Angelman syndrome, for which UPD causes 20 to 30 percent of cases (see figure 6.13). These diseases arise from mutations in different genes that are closely linked in a region of the long arm of chromosome 15, where imprinting occurs. They both cause intellectual disability and a variety of other symptoms, but are quite distinct.

Some children with Prader-Willi syndrome have two parts of the long arm of chromosome 15 from their mothers. The disease results because the father's Prader-Willi gene must be expressed for the child to not have the associated illness. For Angelman syndrome, the situation is reversed. Children have a double dose of their father's DNA in the same chromosomal region implicated in Prader-Willi syndrome, with no maternal contribution. The mother's gene must be present for health.

People usually learn their chromosomal makeup only when something goes wrong—when they have a family history of reproductive problems, exposure to a toxin, cancer, or symptoms of a known chromosomal disease. While researchers analyze human genome sequences, chromosome studies will continue to be part of health care.

Key Concepts Questions 13.5

*a. What is uniparental disomy?

b. What are two ways that uniparental disomy can arise?

c. What are two ways that uniparental disomy can cause disease?

Study Guide

Summary

13.1 Portrait of a Chromosome

1. **Cytogenetics** is the study of chromosome variations and their effects on phenotypes.
2. **Heterochromatin** stains darkly and harbors many DNA repeats. **Euchromatin** is light staining and contains many protein-encoding genes.
3. A chromosome consists of DNA and proteins. Essential parts are the telomeres, centromeres, and origin of replication sites.
4. Centromeres include DNA repeats and proteins that enable the cell to divide.
5. Subtelomeres have telomere-like repeats that gradually change inward toward the centromere, as protein-encoding genes predominate.
6. Chromosomes are distinguishable by size, centromere position, satellites, and staining patterns. They are imaged and displayed in charts called karyotypes.
7. A **metacentric** chromosome has two fairly equal arms. A **submetacentric** chromosome has a long arm and a short arm. An **acrocentric** chromosome's centromere is near a tip, so that it has one long arm and one very short arm.

13.2 Detecting Chromosomes

8. Amniocentesis and chorionic villus sampling (CVS) are older techniques that culture chromosomes from fetal cells in the amniotic fluid or chorionic villi, respectively.
9. Staining chromosomes is less discriminating than fluorescence *in situ* hybridization (FISH). Ideograms

display chromosome bands and digital karyotypes show chromosome parts.
10. Cell-free fetal DNA analysis detects overrepresentation of DNA pieces from extra chromosomes.
11. Chromosomal shorthand indicates chromosome number, sex chromosome constitution, and type of abnormality.

13.3 Atypical Chromosome Number

12. A **euploid** somatic human cell has twenty-two pairs of autosomes and one pair of sex chromosomes.
13. Polyploid cells have extra chromosome sets.
14. **Aneuploids** have extra or missing chromosomes. **Trisomies** (an extra chromosome) are less harmful than **monosomies** (absence of a chromosome), and sex chromosome aneuploidy is less severe than autosomal aneuploidy. **Nondisjunction** is uneven distribution of chromosomes in meiosis. It causes aneuploidy. Most autosomal aneuploids cease developing as embryos.

13.4 Atypical Chromosome Structure

15. Deletions and **duplications** result from crossing over after pairing errors in synapsis. Crossing over in an inversion heterozygote can also generate deletions and duplications. **Microdeletions** and **microduplications** explain many diseases.
16. In a **Robertsonian translocation,** the short arms of two acrocentric chromosomes break, leaving sticky ends on the long arms that join to form an unusual, large chromosome.
17. In a **reciprocal translocation,** two nonhomologous chromosomes exchange parts.
18. An insertional translocation places a DNA sequence from one chromosome into a nonhomologous chromosome.
19. A **translocation carrier** may have an associated phenotype and produces some unbalanced gametes.
20. A heterozygote for an inversion may have reproductive problems if a crossover occurs between the inverted region and the noninverted homolog, generating deletions and duplications. A **paracentric inversion** does not include the centromere; a **pericentric inversion** does.
21. **Isochromosomes** repeat one chromosome arm but delete the other. They form when the centromere divides in the wrong plane in meiosis. Ring chromosomes form when telomeres are removed, leaving sticky ends that adhere. **Chromothripsis** shatters chromosomes.

13.5 Uniparental Disomy—A Double Dose from One Parent

22. In **uniparental disomy (UPD)**, two copies of a chromosome or part of one are inherited from one parent, and not from the other. It can result from nondisjunction in both gametes, or from a trisomic cell that loses a chromosome.
23. Uniparental disomy causes symptoms if it creates a homozygous recessive state associated with an illness, or if it affects an imprinted gene.

Key Terms

acrocentric (ăk′rō-sĕn′trĭk) A chromosome in which the centromere is near one end.

aneuploid (ăn′yū-ploid′) A cell with one or more extra or missing chromosomes.

chromothripsis A rare event that shatters chromosomes.

comparative genomic hybridization (CGH) A technique using fluorescent labels to detect small copy number variants.

cytogenetics (sītō-jə-nĕt′ĭks) Matching phenotypes to detectable chromosomal abnormalities.

duplication An extra copy of a DNA sequence, usually caused by misaligned pairing in meiosis.

euchromatin (yōō-krō′mə-tĭn′) Parts of chromosomes that do not stain and that contain active genes.

euploid (yōō′-ploid) A somatic cell with the normal number of chromosomes for the species.

heterochromatin (hĕtə-rō-krō′mətĭn) Dark-staining chromosome parts that have few protein-encoding genes.

isochromosome A chromosome that has two copies of one arm but none of the other, as a result of cell division along the wrong plane.

metacentric (mĕtə-sĕn′trĭk) A chromosome with a centromere that divides it into two arms of approximately equal length.

microdeletion A very small chromosomal deletion (missing part); a type of copy number variant.

microduplication A very small second copy of a DNA sequence; a type of copy number variant.

monosomy (mŏn′ō-sō′mē) A cell missing a chromosome; in humans, a cell with 45 chromosomes.

nondisjunction (nŏndĭs-jŭngk′shən) Unequal distribution of chromosomes during meiosis.

paracentric inversion (pər-ē sĕn′-trĭk) A chromosomal inversion in which the inverted section does not include the centromere.

pericentric inversion (pər-ē sĕn-trik) A chromosomal inversion in which the inverted part includes the centromere.

reciprocal translocation A chromosome in which two nonhomologous chromosomes exchange parts, conserving genetic balance but rearranging genes.

Robertsonian translocation A chromosome in which two short arms of nonhomologous chromosomes break and the long arms fuse, forming one unusual, large chromosome.

submetacentric (sŭb mĕt-ə-sĕn′trĭk) A chromosome in which the centromere position establishes a long arm and a short arm.

translocation carrier An individual with a large, translocated chromosome who may not have any symptoms. The person often has the usual amount of genetic material, but it is rearranged.

translocation Exchange of genetic material between nonhomologous chromosomes or a piece of one chromosome being moved to another.

trisomy A human cell with one extra chromosome, for a total of 47.

uniparental disomy (UPD) (yû-ni-pə′rent-əl dīsō mē) Inheriting two chromosomes or chromosome segments from one parent.

Review Questions

1. Identify the structures and/or DNA sequences that must be present for a chromosome to carry information and withstand the forces of cell division.

2. How does the DNA sequence change with distance from the telomere?

3. Distinguish among a euploid, aneuploid, and polyploid cell.

4. How are amniocentesis and chorionic villus sampling (CVS) alike yet different?

5. Explain how analysis of cell-free fetal DNA in the maternal circulation can provide information about a trisomy.

6. What is an ideogram?

7. What happens during meiosis to produce
   a. an aneuploid fertilized ovum?
   b. a polyploid fertilized ovum?
   c. increased risk of trisomy 21 in the offspring of a woman over age 40 at the time of conception?
   d. recurrent spontaneous abortions for a couple in which the man has a pericentric inversion?
   e. several children with Down syndrome in a family where one parent is a translocation carrier?

8. A human liver has patches of octaploid cells—they have eight sets of chromosomes. Explain how this condition might arise.

9. List the gender and possible phenotype of individuals with the following chromosomes:
   a. 47,XXX
   b. 45,X
   c. 47,XX, trisomy 21
   d. 48,XXYY

10. Explain the difference between a triploid and a trisomic cell.

11. About 80 percent of cases of Edwards syndrome are caused by trisomy 18, another 10 percent are caused by mosaic trisomy 18, and 10 percent are attributed to translocation. Distinguish among these three chromosome abnormalities.

12. List three types of atypical chromosomes that can cause duplications and/or deletions, and explain how they do so.

13. Distinguish among three types of translocations.

14. Why would having the same inversion on both members of a homologous chromosome pair not lead to unbalanced gametes, as having only one inverted chromosome would?

15. How many chromosomes does a person have who has 47,XYY syndrome and also trisomy 21?

16. Explain why a female cannot have 47,XXY syndrome and a male cannot have 45,X syndrome.

17. List three causes of 45,X syndrome.

18. Explain how the sequence of genes on an isochromosome differs from normal.

19. Describe how a ring chromosome forms.

20. Explain how uniparental disomy does not appear to follow Mendel's first law (segregation).

Applied Questions

1. The chapter opener describes chromothripsis, in which one or a few chromosomes shatter. Why might whole genome sequencing of white blood cells not reveal chromothripsis?

2. What does the chromosomal constitution 46,XX,del(5)(p15) mean?

3. In a college genetics laboratory course, a healthy student constructs a karyotype from a cell from inside her cheek. She finds only one chromosome 3 and one chromosome 21, plus two unusual chromosomes that do not seem to have matching partners.
   a. What type of chromosomal abnormality does she have?
   b. Why doesn't she have any symptoms?
   c. Would you expect any of her relatives to have particular medical problems?

4. A fetus ceases developing in the uterus. Several of its cells are karyotyped. Approximately 75 percent of the cells are diploid and 25 percent are tetraploid (four copies of each chromosome). What happened, and when in development did it probably occur?

5. A couple has a son diagnosed with 47,XXY syndrome. Explain how the son's chromosome constitution could have arisen from either parent.

6. 22q11.2 deletion syndrome, also called DiGeorge syndrome, causes atypical parathyroid glands, a heart defect, and an underdeveloped thymus gland. About 85 percent of patients have a microdeletion of part of chromosome 22. A girl, her mother, and a maternal aunt have very mild DiGeorge syndrome. They all have a reciprocal translocation of chromosomes 22 and 2.
   a. How can a microdeletion and a translocation cause the same symptoms?
   b. Why were the people with the translocation less severely affected than the people with the microdeletion?
   c. What other problems might arise in the family with the translocation?

7. From 2 to 6 percent of people with autism have an extra chromosome that consists of two long arms of chromosome 15. It includes two copies of the chromosome 15 centromere. Two normal copies of the chromosome are also present. What type of chromosome abnormality in a gamete can lead to this karyotype, which is called isodicentric 15?

Case Studies and Research Results

1. An ultrasound of a pregnant woman detects a fetus and a similarly sized and shaped structure that has disorganized remnants of facial features at one end. Amniocentesis reveals that the fetus is 46,XX, but cells of the other structure are 47,XX,trisomy 2. No cases of trisomy 2 infants have ever been reported. However, individuals who are mosaics for trisomy 2 have a collection of defects, including a rotated and underdeveloped small intestine, a small head, a hole in the diaphragm, and seizures.

    a. How do the chromosomes of cells from the fetus and the other structure differ?
    b. What is the process that occurred during meiosis to yield the abnormal structure?
    c. Why are people with a complete extra chromosome 2 not seen, even though people with an extra chromosome 21 can live many years?

2. Two sets of parents who have children with Down syndrome meet at a clinic. The Phelpses know that their son has trisomy 21. The Watkinses have two affected children, and Mrs. Watkins has had two spontaneous abortions. Why should the Watkinses be more concerned about future reproductive problems than the Phelpses? How are the offspring of the two families different, even though they have the same symptoms?

3. Determining the number of chromosomes in a human cell took many years, because it was so difficult to actually see them. Estimates of the human chromosome number ranged from 30 to 80. In 1923, Theophilus Painter published sketches of human chromosomes from three patients at a Texas state mental hospital. The patients had been castrated in an attempt to control their violent behavior, and Painter prepared chromosomes from cells collected from the removed testicular tissue. He could not at first tell whether the cells had 46 or 48 chromosomes, but finally decided that he saw 48. For a time, the diploid human chromosome number was thought to be 48. Why might Painter have come to this conclusion, based on the population that he studied?

4. The genomes of 4 of 291 people with intellectual disability have a microdeletion in chromosome 17q21.3. The children have large noses, delayed speech, and mild intellectual disability. Each has a parent with an inversion in the same part of chromosome 17.

    a. Which arm of chromosome 17 is implicated in this syndrome?
    b. How can an inversion in a parent's chromosome cause a deletion in a child's chromosome?
    c. What other type of chromosome abnormality might occur in these children's siblings?

5. A child has severe intellectual disability, developmental delay, autism, and unusual facial features. His mother had a learning disability and slightly delayed development. Genome sequencing as well as karyotyping reveal that the child has excess material from chromosomes 9 and 14 and too little from chromosome 10. The mother has all of the genetic material, but she has four "derivative" chromosomes—they have seemingly mixed and matched their parts. These chromosomes consist of pieces of 9, 10, and 14; 9, 10, and 16; 10 and 16; and 9 and 4.

    a. What type of chromosomal event has likely occurred?
    b. How can the nature of the chromosomal event explain the mother's learning disability and developmental delay, but good health otherwise and a full genome?

6. In a family, three adult siblings had 44 chromosomes. Each had two chromosomes consisting of chromosomes 13 and 14.

    a. What type of chromosome abnormality do the siblings have?
    b. Describe the chromosomes of the parents of the three siblings.
    c. Explain why the siblings all have fertility problems.

7. A young child is diagnosed with giant axonal neuropathy (see Clinical Connection 2.2), based on clinical findings as well as genetic testing. The disease is autosomal recessive. Genetic testing of the parents reveals that the mother is a carrier, but the father is wild type for the gigaxonin gene. How did the child's genotype arise? Two answers are possible.

CHAPTER 14

Genomes

What can we learn from our genome sequences?

Learning Objectives

14.1 From Genetics to Genomics
1. Explain how linkage studies, positional cloning experiments, and gene mapping led to the idea to sequence human genomes.
2. Distinguish between the two approaches used to sequence the first human genomes.
3. Discuss the types of information that the first sequenced human genomes provided.

14.2 Analysis of Human Genome Content
4. Define *reference genome*.
5. Explain why genome sequencing requires multiple copies of a genome.
6. Identify limitations of exome and genome sequencing.
7. Discuss practical aspects of implementing exome and genome sequencing in health care.

14.3 A Genomic View Expands Knowledge
8. Explain how a genome-level view of chromosomes reveals details of crossing over.
9. Discuss the possible consequences of sequencing the genomes of newborns.

The BIG Picture

Two decades ago, the first human genomes were sequenced. Today millions of people have had their genomes sequenced. Will you discover the information in yours? Getting the most useful information as possible from our genome sequences will depend on many people permitting their information to be collected and analyzed, even if it remains anonymous. Genomics may impact the future for many of us.

Sequencing the Genomes of the Deceased

Genomes transcend time. DNA sequences provide clues to health, from a cell plucked from the earliest of embryos to the centenarian that embryo might become, to DNA from mummies and ancient humans. Would knowing genome sequences of deceased individuals during their lifetimes have changed their health care? To answer that question, the Genomic Postmortem Research Project of the Marshfield Clinic Research Foundation in Wisconsin is sequencing the genomes of 300 deceased individuals and comparing the information to data in the electronic medical records.

The investigators are searching through more than 27 million gene variants, compiled from databases and recommendations from genetics organizations, that could have been used to prevent, detect, and treat specific diseases. Gene variants are classified as pathogenic, likely pathogenic, "variant of uncertain significance," likely benign, and benign, as described in the Bioethics boxed reading in chapter 12. The study revealed many examples of how genome sequencing might have helped people:

- A woman who died from breast and ovarian cancer at age 59 had no family history of cancer and had never had a mammogram. She had a rare pathogenic variant of *BRCA1*. Had the mutation been detected early, the cancers could likely have been prevented or treated.

- A man with deep vein thrombosis and tangles of abnormal blood vessels in his kidneys, brain, and lungs had hereditary hemorrhagic telangiectasia type 1. If the mutation had been detected, the man could have taken either an old drug (thalidomide) or a newer one (Avastin) to prevent dangerous blood clots.

- Of the 300 deceased individuals, 281 had gene variants that affected their response to drugs, including antidepressants, cancer drugs, painkillers, antipsychotics, blood thinners, and cholesterol-lowering drugs. The information could have guided physicians to initially prescribe the drugs most likely to work.

14.1 From Genetics to Genomics

Genetics is a young science, genomics younger still. As one field has evolved and led to the other, driven by clever technologies, milestones have come at oddly regular intervals. A century after Gregor Mendel announced and published his findings, the genetic code was deciphered; a century after his laws were rediscovered, the first human genomes were sequenced.

Geneticist H. Winkler coined the term *genome* in 1920. A hybrid of "gene" and "chromosome," genome then denoted a complete set of chromosomes and genes. The modern definition refers to all the DNA in a haploid set of chromosomes. The term *genomics*, credited to T. H. Roderick in 1986, indicates the study of genomes. Thoughts of sequencing genomes echoed through much of the twentieth century, as researchers described the units of inheritance from several different perspectives, as this book has done so far.

As the twenty-first century dawned, people began having their genomes sequenced. First came two superstars in genetics, J. Craig Venter and James Watson. For a time papers were published adding different ethnic groups to the roster of the sequenced. Then a few wealthy actors, musicians, politicians, scientists, and entrepreneurs funded their personal genome sequencing, interested in diseases in their families or themselves. Curious journalists did and still do have their genomes sequenced and then report on their collection of gene variants in books, blogs, podcasts, TED talks, and articles. People long dead have had their genomes sequenced, including Egyptian mummies, Ötzi the Ice Man, and Neanderthals and Denisovans (ancient humans described in chapter 17).

At the same time that a few prominent and representative individuals were having their genomes sequenced, the first large-scale genome sequencing research projects began, with an ultimate goal of incorporating genome information into electronic health records and routine health care. Figure 1.2 shows how the cost of genome sequencing has dropped, from an initial $1 billion to under $1,000, not including the all-important interpretation. More than 100 million people have already had their genomes sequenced.

The idea of sequencing a human genome was once nearly unfathomable in its complexity; soon it will likely be routine, and is being done on the earliest embryos to the oldest old. Genome sequence information can already be stored on smartphones. Our genomes are nearly identical, yet with enough variability—we differ at millions of single-base sites—to make life and society interesting.

Beginnings in Linkage and Positional Cloning Studies

Sequencing what was thought of as "the" human genome unofficially began in the 1980s with deciphering signposts along the chromosomes. Many of the initial steps and tools grew from existing technology. Linkage maps and studies of rare families that had chromosome abnormalities and specific syndromes enabled researchers to assign a few genes to their chromosomes. Automated DNA sequencing took genetic analysis to a new level—information in the form of a living language.

The evolution of increasingly detailed genetic maps is similar to zooming in on a geographical satellite map (**figure 14.1**). A cytogenetic map (of a chromosome) is like a map of California within a map of the United States, highlighting only the largest cities. A linkage map is like a map that depicts the smaller cities and large towns, and a physical map is similar to a geographical map indicating all towns in an area. Finally, a sequence map is the equivalent of a Google map showing all of the buildings in a specific town.

Before the first human genomes were sequenced, researchers matched single genes to specific diseases using an approach called positional cloning. The technique began with examining a particular phenotype corresponding to a Mendelian disease in large families. The phenotype was easily matched to a chromosome segment if all the affected individuals shared a chromosome abnormality that relatives without the phenotype did not have. But abnormal chromosomes are rare. Another way that positional cloning located medically important genes on chromosomes used linkage maps that showed parts of a chromosome shared by only the individuals in a family who had the same syndrome. Then researchers isolated pieces of the implicated chromosome, identified short DNA sequences corresponding to the region of interest, and overlapped the pieces, to gradually find the gene behind a phenotype.

Positional cloning was an indirect method that narrowed a gene's chromosomal locus without necessarily knowing the protein product. Throughout the 1980s and 1990s, the method enabled researchers to discover the genes behind Duchenne muscular dystrophy, cystic fibrosis, Huntington disease, and many other single-gene conditions. Positional cloning was slow. It took a decade to go from discovery of a linked genetic marker for Huntington disease to discovery of the gene.

Sequencing the First Human Genomes

The idea to sequence the first human genome occurred to many researchers at about the same time, but they had different goals. It was first brought up at a meeting held by the

Figure 14.1 Different levels of genetic maps are like zooming up the magnification on a geographical map. (a) A cytogenetic map, based on associations between chromosome aberrations and syndromes, can distinguish DNA sequences at least 5,000 kilobases (kb) apart. (b) A linkage map derived from recombination data distinguishes genes hundreds of kilobases apart (cM = centimorgans). (c) A physical map constructed from overlapped DNA pieces distinguishes genes tens of kilobases apart. (d) A sequence map depicts the order of nucleotide bases.

Department of Energy (DOE) in 1984 to discuss the long-term population genetic effects of exposure to low-level radiation. In 1985, researchers convening at the University of California, Santa Cruz, called for an institute to sequence the human genome, because sequencing of viral genomes had shown that it could be done. The next year, virologist Renato Dulbecco proposed that the key to understanding the origin of cancer lay in knowing the human genome sequence. Later that year, scientists packed a room at the Cold Spring Harbor Laboratory on New York's Long Island to discuss the feasibility of a project to sequence the human genome. At first those against the project outnumbered those for it five to one. The major fear was the shifting of goals of life science research from inquiry-based experimentation to amassing huge amounts of data—ironic considering the importance of bioinformatics today.

A furious debate ensued. Detractors claimed that the project would be more gruntwork than a creative intellectual endeavor, comparing it to conquering Mt. Everest just because it is there. Practical benefits would be far in the future. Some researchers feared that such a "big science" project would divert government funds from basic research and developing treatments for the still-new AIDS epidemic. Finally, the National Academy of Sciences convened a committee representing both sides to debate the feasibility, risks, and benefits of the project. The naysayers were swayed to the other side. In 1988, Congress authorized the National Institutes of Health (NIH) and the DOE to fund the $3 billion, 15-year Human Genome Project, which began in 1990 with James Watson at the helm. The project set aside 3 percent of its budget for the Ethical, Legal and Social Implications (ELSI) Research Program. It has helped ensure that genetic information is not used to discriminate. Eventually, an international consortium as well as a private company, Celera Genomics, sequenced the first human genomes. The groups worked separately, and one effort finished a few months before the other, but the accomplishment is referred to historically as "the human genome project."

A series of technological improvements sped the genome project. In 1991, a shortcut called expressed sequence tag (EST) technology enabled researchers to quickly pick out genes most likely to be implicated in disease. This was a foreshadowing of future efforts to focus on the exome, the part that encodes protein and is responsible for 85 percent of single-gene diseases. ESTs are short DNA molecules, called complementary or cDNAs, that are made from the mRNAs in a cell type that is

Figure 14.2 Deriving a DNA sequence. Automated DNA sequencers first determine the sequences of short pieces of DNA, or sometimes of just the ends of short pieces. Then algorithms search for overlaps. By overlapping the pieces, the software derives the overall DNA sequence.

abnormal in a particular illness, such as an airway lining cell in cystic fibrosis. ESTs therefore represent gene expression. Also in 1991, researchers began using devices called DNA microarrays (or chips) to display short DNA molecules that bind their DNA or mRNA complements. Microarray technology became important in DNA sequencing (tiling arrays) as well as in assessing gene expression (expression arrays).

Computer algorithms were developed to assemble many short pieces of DNA with overlapping end sequences into longer sequences (**figure 14.2**). When the project began, researchers cut several genomes' worth of DNA into overlapping pieces of about 40,000 bases (40 kilobases), then randomly cut the pieces into small fragments. The greater the number of overlaps, the more complete the final assembled sequence. The sites of overlap had to be unique sequences, found in only one place in the genome. Overlaps of repeated sequences found in several places in the genome could lead to more than one derived overall sequence—a little like searching a document for the word "that" versus searching for an unusual word, such as "dandelion." Searching for "dandelion" is more likely to lead to a specific part of a document, whereas "that" may occur in several places—just like repeats in a genome.

The use of unique sequences is why the human genome project did not uncover copy number variants. For example, the sequence CTACTACTA would appear only as CTA. Researchers did not at first appreciate the fact that repeats are a different form of information and source of variation than DNA base sequences. A balance was necessary between using DNA pieces large enough to be unique, but not so large that the sequencing would take a very long time.

Two general approaches were used to build the long DNA sequences to initially derive the sequence of the human genome (**figure 14.3**). The "clone-by-clone" technique the U.S. government-funded group used aligned DNA pieces one chromosome at a time. The "whole-genome shotgun" approach Celera Genomics used shattered entire genomes, then used an algorithm to identify and align overlaps in a continuous sequence. Whole-genome shotgun sequencing can be compared to cutting the binding off a large book, throwing it into the air and freeing every page, and reassembling the dispersed pages in order. A "clone-by-clone" dismantling of the book would divide it into bound chapters. Whole-genome shotgunning is faster, but it misses some sections (particularly repeats) that the clone-by-clone method detects.

Technical advances continued. In 1995, DNA sequencing was automated, and software was developed that could rapidly locate the unique sequence overlaps among many small pieces of DNA and assemble them, eliminating the need to gather large guidepost pieces. In 1999, the race to sequence the human genome became intensely competitive. The battling factions finally called a truce. On June 26, 2000, J. Craig Venter from Celera Genomics and Francis Collins, representing the International Consortium, flanked President Clinton in the White House Rose Garden to unveil the "first draft" of the human genome sequence. The milestone capped a decade-long project involving thousands of researchers, culminating a century of discovery. The historic June 26 date came about because it was the only opening on the White House calendar! In other words, the work was monumental; its announcement, somewhat staged. **Figure 14.4** is a conceptual overview of genome sequencing.

Types of Information in Human Genomes

The human genome sequenced by the Public Consortium was actually a composite of the genomes of different individuals. The first two genomes from specific individuals to be sequenced and the findings published, of genome research pioneers J. Craig Venter and James Watson, yielded few surprises. Instead, they showed that we had greatly underestimated genetic variation by focusing only on the DNA base sequence. The numbers of copies of short sequences—copy number variants, or CNVs—contribute significantly to genetic variation, too.

Venter learned that he has gene variants associated with increased risk of Alzheimer disease and cardiovascular disease. His genome sequence confirmed that he has genotypes that cause or are associated with dry earwax, blue eyes, lactose intolerance, a preference for activities in the evening, and tendencies toward

Figure 14.3 **Two routes to sequencing the human genome.** **(a)** The International Consortium began with known chromosomal sites and overlapped large pieces, called contigs, that were reconstructed from many small, overlapping pieces. STS stands for "sequence tagged site," which refers to specific known parts of chromosomes. A BAC (bacterial artificial chromosome) is a cloning vector that uses bacterial DNA to carry DNA pieces (see section 21.2). Restriction sites are sequences where specific enzymes recognize and cut the DNA. **(b)** Celera Genomics shotgunned several copies of a genome into small pieces, overlapped them to form scaffolds, and then assigned scaffolds to known chromosomal sites. They used some Consortium data.

antisocial behavior, novelty seeking, and substance abuse. He metabolizes caffeine fast, which he also knew.

James Watson, according to his genome sequence, carries a dozen rare recessive diseases that would affect glycogen storage, vision, and DNA repair if homozygous, and he is at elevated risk for 20 other diseases. He elected not to learn his status for the *ApoE4* gene variant that increases risk of Alzheimer disease, which a grandparent had, but people inferred the result from the surrounding published parts of his genome (a deduction called imputation that compares the surrounding sequence to that of

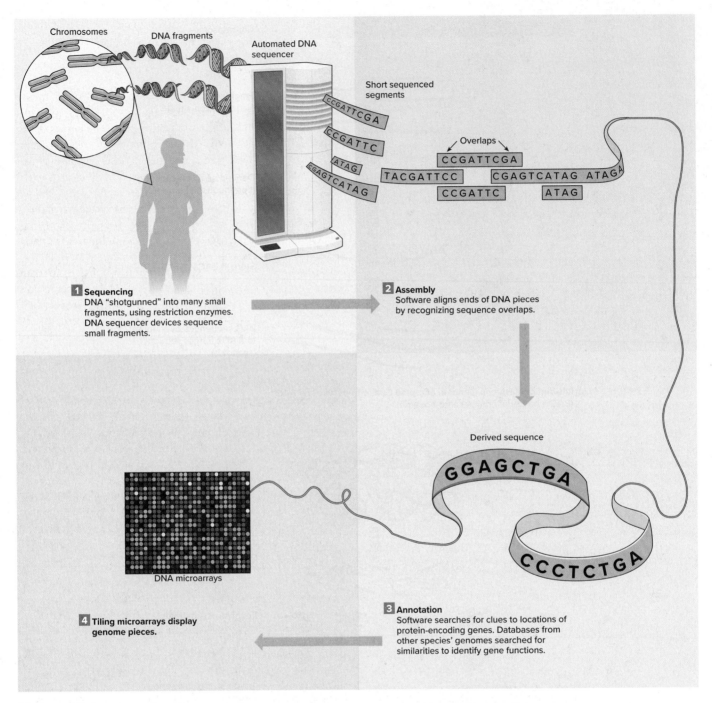

Figure 14.4 Sequencing genomes. The first-generation, whole-genome "shotgun" approach to genome sequencing overlapped DNA pieces cut from several copies of a genome, then assembled the overall sequence. The next-generation genome sequencing techniques used today are based on microfluidics and nanomaterials and are much faster.

other people's genomes). He is a slow metabolizer of beta blockers and antipsychotic medications, indicating that he could overdose on normal weight-based dosages of these medications.

The third person to have his genome sequenced was called, simply, "YH." He is Han Chinese, an East Asian population that accounts for 30 percent of modern humanity. He has no inherited diseases in his family, but his genome includes 116 gene variants that cause recessive diseases, as well as many risk alleles. He shares with J. Craig Venter a tendency to tobacco addiction and high-risk alleles for Alzheimer disease.

An overall comparison of the first three genome sequences of individuals provided an initial peek at our variation. Each man has about 1.2 million SNPs, but a unique collection. Each has only 0.20 to 0.23 percent of SNPs that are nonsynonymous, meaning that they alter an encoded amino acid, and the men share only 37 percent of these more meaningful SNPs. The math indicates, therefore, that about 0.07 percent of our SNPs may affect our phenotypes.

After the first three human genomes were sequenced, others from different population groups were sequenced. Then came the genomes of people who could afford the initially high cost: journalists who were paid to write about the experience and celebrities. Reasons varied, as they still do. The late Steve Jobs (founder of Apple) and late journalist Christopher Hitchens had their cancer genomes sequenced to guide drug choices. In the long run, the genome information did not save their lives. Scholar, documentarian, and host of *Finding Your Roots* Henry Louis Gates Jr. had his genome sequenced to trace his African origins. An actress and a rock star reportedly did it to better understand mental illness in their families.

One geneticist had his genome sequenced to serve as a control for a project to sequence the genomes of all the citizens of Qatar. He discovered that he has gene variants for baldness (which he knew from looking in the mirror), a recessive disease that affects children, Viking ancestors, and most important, a blood clotting disease that explained why he bleeds profusely when injured. Despite learning interesting information, the researcher voiced fears: his family learning things they didn't want to know and even someone using his DNA sequence to frame him for a crime.

Key Concepts Questions 14.1

*a. Explain how linkage relationships were used to match genes to chromosomes.

*b. What were the initial goals of sequencing a human genome?

c. What are the general steps in sequencing a genome?

d. Distinguish between the clone-by-clone and whole-genome shotgunning approaches to sequence the human genome.

14.2 Analysis of Human Genome Content

Before the first drafts of the human genome were completed, researchers were already beginning the analysis stage by identifying sites of variation and discovering functions of individual genes. Even after hundreds of thousands of human genomes had been sequenced, each new one still revealed nearly 9,000 novel gene variants! To ease sequence comparisons and interpretations, researchers derive a **reference genome** sequence, which is a digital DNA sequence assembled from the most common base at each point in many sequenced genomes. It is haploid (one copy). Comparing reference genome sequences for different types of organisms and identifying DNA sequences that they share provides views of evolution, discussed in chapter 17.

Improving Speed and Coverage

Sequencing the first human genomes took 6 to 8 years; today it can be done in hours. Improvements in sequencing technology enabled researchers to work with many more copies of an individual's genome, which is termed *coverage*. Recall from section 9.4 that DNA pieces cut from several genomes must be sequenced and overlapped to derive the overall sequence. Because some pieces are lost, the more copies of a genome used, the more likely the overlapping will pick up every base.

At least 28 human genome copies are necessary to ensure that most sequences are represented in the final derived sequence. A genome with 40-fold coverage, for example, means each site in the genome is read on average 40 times. High coverage is needed to detect a rare DNA sequence, such as the genome sequencing that helped to diagnose the child described in Clinical Connection 1.1. Even today, genomes are not completely sequenced, because the technology is not perfect. The first human genome sequences published, in 2001, had about 150,000 small gaps; current versions have only a few hundred tiny missing pieces.

Sequencing genomes provides much more information than sequencing exomes, which are the exon (coding) parts of protein-encoding genes. Knowing the sequences surrounding the exons can be used to detect "structural variants" such as inversions and reciprocal translocations (see figures 13.15 and 13.18), which flip or move DNA but do not alter the base sequence. Exome sequencing arrays must be designed to distinguish between pairs of SNPs on the same homologous chromosomes (in *cis* configuration) or on different homologs (in *trans* configuration). The *cis* configuration indicates linkage, which is important in predicting transmission of a genotype to offspring.

The Ongoing Goal: Annotation

Just as a book written in a foreign language is meaningless unless translated, knowing the sequence of a human genome is not useful unless we know what the information means. "Annotation" in linguistics means "a note of explanation or comment added to a text or diagram." In genomics, annotations are descriptions of what genes do, and what the significance of a particular gene variant is likely to be. Researchers are accomplishing this enormous task by meticulously consulting the published scientific literature, DNA sequence databases for every identified gene, and patterns of SNPs in noncoding regions that might be associated with specific disease risks. Genetic counselors and people with doctorates in genetics are doing the annotation, and are called curators, variant scientists, or curation specialists.

Annotation of a gene variant might include:

- the normal function of the gene;
- mode of inheritance;
- genotype (heterozygote, homozygote, compound heterozygote);

- frequency of a variant in a particular population; and
- classification as benign, likely benign, variant of uncertain significance, likely pathogenic, or pathogenic.

Knowing the frequency of a gene variant in a population is important in determining the five classifications for logical reasons. A variant that is common—which means that many people live with it—is less likely to cause a serious illness than one that is rare. If a third of a population has a gene variant that is associated with hypertension, for example, that gene probably contributes only slightly to the overall risk for this complex trait. Otherwise a third of the population would have severe hypertension. However, the reverse is not true—a rare variant can be harmless, and just unusual.

To be most valuable, an annotated genome sequence is compared to as much health and family history information as a person can provide. The old-fashioned pedigree is still an invaluable tool and the starting point for many investigations, for this is the information that most people already know—who has what in the family. Genome annotations are also including microbiome data (see section 2.5), because the genes of the organisms that live in and on us can affect expression of our own genes.

Limitations of Genome Sequencing

Genome sequencing does not provide a complete picture of health. Limitations are technical and conceptual.

In a technical sense, genome sequencing cannot detect repeats without additional types of tests because having multiples of a sequence does not alter the sequence. Such repeats include copy number variants, triplet repeat mutations, and even the complete extra chromosomes of trisomies, as well as deletions. Genome sequencing is also blind to uniparental disomy, the rare situation in which a child inherits two alleles of a gene from one parent (see figure 13.23) rather than one from each. The two copies are identical in DNA sequence, and so "count" only once in the overall sequence. In addition, exons that are buried within highly repeated or extensive introns may not be detected. Genome sequencing also does not include mitochondrial genes.

In a conceptual sense, genome information must be interpreted to be useful. This means not just identifying the protein that a gene encodes but deciphering all interactions of genes and the gene networks that they form. Imagine a novel that is "read" one letter at a time, so that it is a string of thousands of copies of 26 letters, rather than a nuanced, coherent story with clues and connections conveyed in words. So, too, is a sequenced but unannotated human genome not informative.

Gene interactions are intricate and complex. Figure 11.4 illustrates just a small sampling of diseases that share genes that have altered expression. One gene's activity affecting the expression of another can explain why siblings with the same single-gene disease suffer to a different extent. For example, a child with severe spinal muscular atrophy (SMA), in which an abnormal protein shortens axons of motor neurons, may have a brother who also inherits the disease but has a milder case thanks to inheriting a variant of a second gene that extends axons. Computational tools and machine learning are used to sort out networks of interacting genes, sometimes called connectomes or interactomes. What were once regarded as simple epistatic interactions—gene affecting gene—may in reality be the tip of an iceberg of complex networks of genes that influence each other.

Finally, epigenetic changes induced by environmental factors provide a layer of information that must be applied to DNA information. These are the influences that are perhaps the most important, because we can act on many of them.

Practical Medical Matters

Until the acceleration of gene discoveries in the 1990s and the sequencing of the first human genomes in the early 2000s, genetics as a medical specialty was a small field, with a few knowledgeable physicians helping families with rare diseases. With the introduction of consumer genetic testing in 2008, the possibility of testing genes not only for disease-causing mutations, but for variants that indicate only risk, was suddenly available to anyone—without requiring medical expertise.

Consumer DNA test offerings are abundant, but for health conditions they are not nearly as complete as the types of tests that a medical professional orders. A consumer genetic test for breast cancer, for example, might test for the three most common mutations in *BRCA1* and *BRCA2*, and not the dozens of other mutations in those genes, nor variants of other genes that elevate breast cancer risk. A breast cancer gene test panel that a health care provider orders, in contrast, might test for variants or even complete sequences of more than 100 genes.

Genetic and genomic testing as part of health care must meet certain practical criteria. The most important requirement for a regulated DNA test or treatment is clinical utility. Does benefit outweigh risk? Is it as accurate or effective as an existing, approved test or treatment? Will it help people who cannot use the existing test or treatment? Efficacy must be demonstrated. For example, molecular evidence may indicate that people with a particular genotype might respond better to a certain drug than people with different genotypes. A clinical trial must evaluate the drug in both groups of people to demonstrate that prescribing-by-genotype—pharmacogenetics—is helpful on the whole-person level.

A DNA test result alone is not sufficient to diagnose a disease, but may support a clinical diagnosis based on symptoms and the results of other types of tests. For example, a person might be a heterozygote for familial hypercholesterolemia, but is not diagnosed with the condition unless the serum cholesterol level becomes elevated or a cardiovascular condition develops (see figure 5.2). However, knowing that a mutation is present can motivate a person to seek further testing, as is the case for breast cancer.

The uncertainty in genetic and genomic testing that makes further diagnostic testing necessary arises from the complications of Mendel's laws discussed in chapter 5. A DNA test alone is not sufficient for diagnosis because of

- incomplete penetrance (genotype does not always foretell phenotype),
- variable expressivity (different severities in different individuals),
- epistasis (gene-gene interactions),
- genetic heterogeneity (mutation in more than one gene causing a phenotype), and
- environmental influences (epigenetics).

How will electronic medical records handle the nuances of genomic data? Will records include entire sequences of the DNA bases A, T, C, and G, or just the diagnostic report that includes gene variants classified as pathogenic? How will the records embrace future discoveries that impact the stored data or diagnoses? How will diagnostic codes work?

While these matters are under discussion, the medical profession has had to catch up to both the profusion of new genetic tests and genomic technologies, and the fact that many patients are knowledgeable about DNA—in general and often their own. Genomics is becoming incorporated into more medical specialties, and it already is part of oncology and pediatrics. Medical students analyze their own genomes, and physicians are attending continuing medical education programs to learn genomics. Clinical geneticists, genetic counselors, and molecular pathologists are the specialists who are leading the way in the new genomic medicine.

While many researchers are busy annotating human genomes, which are highly complex, others are using chemical and microbiological methods to synthesize simple genomes. The effort is part of a field called synthetic biology. **Bioethics** discusses synthetic genomes.

Key Concepts Questions 14.2

*a. What is a reference genome?

*b. Why are many copies of a genome sequenced?

*c. What types of information might be part of a genome annotation?

d. What are limitations of genome sequencing?

e. What are the potential benefits and risks of incorporating genetic and genomic information into health care?

Bioethics

Should We Create Genomes?

A genome is actually a chemical, and so chemists can synthesize one, then transfer it into a receptive cell whose own genome has been removed. Researchers introduced the first synthetic genome in 2010, and have been updating it in new, smaller versions ever since. They began with the tiny genome of *Mycoplasma genitalium*, the smallest organism able to reproduce, and whittled it down. The resulting synthetic "hypothetical minimal genome" is about 531,000 DNA base pairs organized into 473 genes. It is called JCVI-syn3.0, or syn3.0 for short. The researchers even stitched their names into the genome using DNA triplets that they assigned to correspond to letters of the alphabet, functioning like a watermark on documents to distinguish synthetic life from the natural kind. Syn3.0 "lives" as the genetic headquarters of two species of *Mycoplasma*.

As a short cut, the researchers tested eighths of the *Mycoplasma* genome at a time, using DNA-cutting enzymes to destroy one gene at a time. If a cell can't survive with a specific gene harpooned, then that gene is deemed essential for life. Genes are classified as essential, not essential, or quasi-essential (required for the organism to grow well enough to study, but not to live for very long).

The synthetic genome in its host cell is somewhat coddled, given all the small molecules it could possibly require in the laboratory, compared to the natural niche of *Mycoplasma* in goats. The researchers chose *Mycoplasma* as the model because the natural hosts supply nearly all nutrients, enabling the microorganisms to survive with minimal genomes.

Nearly half of the genes in syn3.0 are involved in making proteins, with many contributing to plasma membrane structure and function. Most intriguing and unexpected was that the functions of nearly a third of the genes aren't known. Many genes, however, are highly conserved (found in other organisms), suggesting that they are essential.

Creating a synthetic genome may provide clues to how life began, as complex collections of self-replicating and changeable chemicals coalesced and linked. Perhaps they used clays or minerals as templates, then knit themselves fatty protective coverings. In a practical sense, the ability to design a simple microorganism may ultimately offer new energy sources and ways to develop new drugs and vaccines.

Questions for Discussion

1. Is a synthetic genome really creating life, or is it more like a hermit crab taking up residence in an abandoned shell?
2. Do we know enough to use synthetic life technology to create cells that can improve the world?
3. How might the ability to create genomes be misused?
4. Should the synthetic life research community police itself, or should a regulatory agency or organization control it?

14.3 A Genomic View Expands Knowledge

The ability to sequence human genomes enables researchers to look backward and add details to what we have known about our genetics, and also to look forward to how the expression of gene variants identified at birth will affect health over a lifetime. Different investigations have sequenced genomes to differing degrees. *Exome sequencing* refers to the exons (the protein-encoding parts), *genomic sequencing* to more than an exome but less than a full genome, and *whole genome sequencing* to entire genomes derived from many overlapped copies. Research at the genome level also includes genome-wide association studies, which consult sets of single nucleotide polymorphisms (SNPs) dispersed among the chromosomes (see figures 7.11 and 7.12).

Here are two examples of applications of genome information.

A New View of Crossing Over

Genome sequencing has enabled researchers to look closely at exactly what happens to the region of a pair of chromosomes that cross over during prophase of the first meiotic division, as oocytes and sperm form. Figures 3.5 and 5.11 depict crossing over between homologous chromosomes with a color change of blue to pink. Genome sequencing not only confirms that crossing over occurs, but also reveals instability in the neighboring regions along the involved chromosomes.

Researchers analyzed genome sequences from 155,250 people going back several generations from a database that includes health information from most of the population of Iceland. The country has been reproductively isolated for many years, which is why genetic studies have been able to identify many of their disease-causing gene variants. The Icelandic maps of genetic markers were crucial in establishing the first human reference genomes in the early 2000s, and remain important today.

The Icelandic genomes enabled researchers an unprecedented view of two processes that foster human genetic diversity: crossing over and generation of *de novo* (new) mutations. By comparing DNA sequences at various points along the chromosomes among many people, the researchers uncovered the sites of 4.5 million crossover events that recombined parental chromosomes. They also identified more than 200,000 *de novo* mutations. Epigenetic changes, such as the patterns with which histone proteins bind to DNA, expose the areas of chromosomes where crossing over occurs (see figure 11.6).

The findings reveal that both crossing over and generation of *de novo* mutations are not as random as had been thought: at least 35 genes affect crossover frequency and location. In addition, the two processes—crossing over and *de novo* mutations—are connected: mutations are more than 50 times more likely to arise at the sites of crossovers than elsewhere in the genome (**figure 14.5**). The sexes differ, which reflects epigenetics. More crossovers occur within one kilobase (1,000 bases) of the crossover site in both sexes, but female genomes

Figure 14.5 Whole genome sequencing reveals that new point mutations are more likely to arise within 1,000 bases (1 Kb) of sites of crossing over in males and females. Another mutational "hot spot" is 40 Kb away from the crossover, but only in the genomes of females.

continue to cross over at higher frequencies than elsewhere up to 40 kilobases from the crossover site. The frequency of crossing over increases with maternal age. However, the genomes of males are more likely to have the *de novo* mutations that are a major source of single-gene diseases in children. These findings confirm the observations that sperm tend to introduce single-gene mutations whereas oocytes introduce chromosome imbalances, as well as the association of advanced maternal age with chromosomal anomalies in offspring (see figure 13.6).

One way to interpret these findings is to take a broad, evolutionary view. The diversity that our genomes naturally generate promotes success of humans as a species, but at the expense of recessive alleles that cause genetic diseases when paired in an individual.

Sequencing the Genomes of Newborns

Knowing the genome sequence of an individual from birth can have medical advantages. These may include earlier diagnosis of childhood diseases, identifying conditions that might be prevented with early intervention, and providing a backdrop of information for the electronic health record that will follow a person for life. However, obtaining genome information from birth also raises issues of use or abuse in a variety of scenarios.

To explore the feasibility, value, and possible risks of newborn genome sequencing, the U.S. government funded a pilot study called BabySeq. Two Boston-area hospitals enrolled 127 healthy newborns and 32 babies in the neonatal intensive care unit (NICU). Half of the children in each group had their genomes sequenced. Those who didn't have their genomes sequenced gave a family history and had standard newborn screening for a few dozen genetic diseases from a blood sample taken from the heel.

The researchers and the parents couldn't tell which babies had their genomes sequenced and which had not.

The results attest to the potential value of newborn genome sequencing. The intervention detected childhood-onset diseases that hadn't yet caused symptoms in 15 (9.4 percent) of the babies. Ten of them were from the healthy cohort and five from the NICU group, but none of the mutations were the reasons that the children were in the NICU. Of the 159 infants, 140 were found to have at least one recessive allele that would cause a disease if present in two copies. This type of information would alert individuals, albeit very early, to have prenatal testing when considering parenthood.

Key Concepts 14.3

a. Explain what genome-level information reveals about the effects of crossing over.

b. List possible benefits and risks of newborn genome sequencing.

Study Guide

Summary

14.1 From Genetics to Genomics

1. Genetic maps increase in detail and resolution, from cytogenetic and linkage maps to physical and sequence maps.
2. Positional cloning was used to discover individual genes by beginning with a phenotype and gradually identifying a causative gene, localizing it to part of a chromosome.
3. The Human Genome Project began in 1990 under the direction of the DOE and NIH. Technological advances sped the sequencing.
4. Several copies of a genome were cut and the pieces sequenced, overlapped, and aligned to derive the continuous sequence. For the human genome, the International Consortium used a chromosome-by-chromosome approach and Celera Genomics used whole-genome shotgunning.
5. Sequencing of the first human genomes revealed that we vary very little. Analysis of genome sequences can confirm family history, detect disease-associated recessive alleles and gene variants that contribute to disease risk, and predict drug responses.

14.2 Analysis of Human Genome Content

6. After sequencing genomes became possible, attention turned to refining the process, cataloging human variation, and discovering gene functions.

7. The cost of sequencing decreased while the speed increased, as researchers annotated genes with information on gene function, mode of inheritance, genotype, and frequency of variants.
8. Genome sequencing does not detect copy number variants, mitochondrial DNA, uniparental disomy, or gene-gene and gene-environment interactions.
9. Genome information used in medical tests and treatments must be validated for clinical utility (be novel or at least as useful as an existing test or treatment).
10. The nuances of Mendel's laws discussed in section 5.1 affect the interpretation of genome sequence data.

14.3 A Genomic View Expands Knowledge

11. Genome sequencing provided information on how crossing over and de *novo* mutations are connected.
12. Sequencing the genomes of newborns can have repercussions on health over a lifetime while presenting privacy concerns.

Key Term

reference genome A digital nucleic acid sequence derived from many sequenced genomes that establishes a representative genome sequence for a species.

Review Questions

1. Explain why the phrase "the human genome project" is inaccurate.
2. How did family linkage patterns and chromosomal aberrations lay the groundwork for sequencing the first human genomes?
3. Describe how the four levels of genetic maps differ, and what the newer types of maps depict.
4. Explain how positional cloning was used to identify disease-causing genes.

5. Why did researchers initially disagree over whether to sequence human genomes? How have attitudes toward similar "big data" projects changed?

6. Why is it important to use many copies of a genome when deriving the sequence?

7. Explain why a gene variant that is present at a high frequency in a population is less likely to be harmful than a rare variant in that population.

8. Why is it helpful to include microbiome information in a genome annotation document?

9. List three types of DNA sequence variation that genome sequencing cannot detect.

10. List three criteria for genome information to have clinical utility.

11. Give examples of types of information that people can learn from knowing their genome sequences.

12. Explain what genome sequencing revealed about the consequences of crossing over during meiosis.

13. Explain the design of the BabySeq project.

14. Why is it possible to synthesize a genome?

Applied Questions

1. Suggest a way that the Genomic Postmortem Research Project could help living people.

2. What are uses and limitations of making a clinical decision based on a reference human genome sequence?

3. How should genetic or genomic information be incorporated into clinical diagnosis? Discuss the significance of the differences between Mendelian and complex traits and between the effects of penetrance and epistasis.

4. Compare and contrast DNA testing to an X-ray or other standard medical diagnostic test.

5. Some researchers are suggesting that to cut costs and make genome technology available to more people, testing should only look for genes most likely to be mutant in a particular person. What is an advantage and a disadvantage of this strategy?

6. How does the experiment that Bateson and Punnett conducted on genetic linkage (see figure 5.10) differ from analyzing the genome sequence near the sites of crossing over?

7. What difficulties would be encountered in trying to synthesize a human genome?

Case Studies and Research Results

1. The Exome Aggregation Consortium (ExAC) sequenced the exomes of 60,706 people with diverse geographic ancestries, coming from Latino, European, African, South Asian, and East Asian populations. The consortium identified about 7.4 million gene variants. If a human genome is 3.2 billion bases, with the exome 1.5 percent of that, explain how so many variants are possible.

2. Some researchers have claimed that healthy lifestyle habits (not smoking, following a healthy diet, exercising regularly, having regular wellness exams, and diagnosing and treating common diseases early) would save more lives than sequencing everyone's genome and predicting diseases from the information. Suggest an experiment to evaluate these two approaches to staying healthy.

3. Sequencing the genomes of 2,636 people in Iceland to an average depth of 20-fold revealed 20 million SNPs, 1.5 million "indels" (insertions and/or deletions), and more homozygosity than seen in many other populations.

   a. What does "depth" refer to, and why is it important to the accuracy of findings?

   b. How do SNPs differ from indels?

   c. What is the significance of the increased homozygosity seen among the Icelanders compared to other human populations?

4. If newborn genomic sequencing is voluntary and parents opt out, and then a child develops a disease that could have been prevented with treatment before symptoms began, should they be liable for costs of care that could possibly have been avoided or minimized?

PART 4 Population Genetics

CHAPTER 15

Constant Allele Frequencies

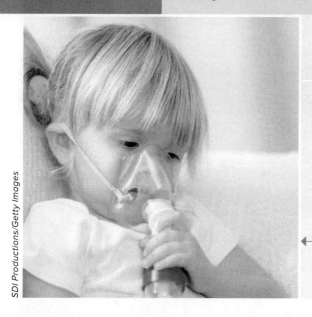

The fact that genetic diseases such as cystic fibrosis occur with different frequencies in different populations impacts the utility of genetic testing.

Learning Outcomes

15.1 Population Genetics Underlies Evolution
1. State the unit of information of genetics at the population level.
2. Define *gene pool*.
3. List the five processes that cause microevolutionary change.
4. State the consequence of macroevolutionary change.

15.2 Constant Allele Frequencies
5. State the genotypes represented in each part of the Hardy-Weinberg equation.
6. Explain the conditions necessary for Hardy-Weinberg equilibrium.

15.3 Applying Hardy-Weinberg Equilibrium
7. Explain how Hardy-Weinberg equilibrium uses population incidence statistics to predict the probability of a particular phenotype.

The BIG Picture

Human genetics at the population level considers allele frequencies. Some parts of the genome that have changed over time enable us to trace our origins, migrations, and relationships.

Tailoring Genetic Tests to Population Groups

About three-quarters of people who have cystic fibrosis (CF) worldwide have at least one copy of the *F508del-CFTR* mutation and 70 percent of Europeans with CF have two copies. But the mutation accounts for only 12 to 31 percent of CF alleles in some Asian nations, and none in Korea, Japan, Thailand, and Vietnam. The Chinese have a *CFTR* mutation seen nowhere else. Tests that cover only a few mutations, or even a few dozen, are not helpful for all people. More than 1,700 mutations have been identified in the gene, *CFTR*. Figures 1.5 and 1.6 describe CF.

People of Asian ancestry tend to have more severe lung disease when they have CF compared to white Europeans because they have poorer lung function overall; have less salty sweat; much better pancreatic function; and fewer cases of CF-associated male infertility. It isn't surprising that CF cases among Asians can be missed, or misdiagnosed as tuberculosis, when tests are geared to European white populations. Because selecting patients for the newer targeted treatments is based on genotype, it is important to tailor genetic tests to populations.

One way to make *CFTR* genetic test panels more inclusive is to use a bioinformatics approach to predict all of the ways that the gene's 188,702 DNA bases can vary, and which of these theoretical alleles can cause the symptoms. A bioinformatics strategy also predicts which mutations complement (cancel each other's effects) by affecting parts of the encoded protein that do not interact. This information is useful to identify partners who are each carriers, but whose offspring would not develop severe, if any, symptoms of CF if they inherit each parent's mutant allele.

Perhaps in the future DNA ancestry information might be paired with panels of single-gene tests that check for the alleles most likely to be present in a particular human population.

15.1 Population Genetics Underlies Evolution

The language of genetics at the family and individual levels is written in DNA sequences. However, at the population level, the language of genetics is allele (gene variant) frequencies. Genetics at the population level goes beyond science, embracing information from history, anthropology, human behavior, and sociology. Population genetics enables us to trace our beginnings, understand our diversity today, and imagine the future.

A biological **population** is any group of members of the same species in a given geographical area that can mate and produce fertile offspring. A population in a sociological sense may be more restrictive, such as an ethnic group or economic stratum. **Population genetics** is a branch of genetics that considers all the alleles of all the genes in a biological population, which constitute the gene pool. The "pool" refers to the collection of gametes in the population; an offspring represents two gametes from the pool. Alleles can move between populations when individuals migrate and mate. This movement, termed **gene flow,** underlies evolution, which is explored in the next two chapters.

Thinking about genes at the population level begins by considering frequencies—that is, how often a particular gene variant occurs in a particular population. Such frequencies can be calculated for alleles, genotypes, or phenotypes, and may include single-base mutations or numbers of short, repeated DNA sequences. For example, an allele frequency for the cystic fibrosis gene (*CFTR*) might be the number of *F508del-CFTR* alleles among the residents of San Francisco. *F508del-CFTR* is the most common allele that, when homozygous, causes the disease globally and in predominantly white populations. The allele frequency is calculated from the two *F508del-CFTR* alleles in each person with CF, plus alleles carried in heterozygotes, considered as a proportion of all alleles for that gene in the gene pool of San Francisco. The genotype frequencies are the proportions of heterozygotes and the two types of homozygotes in the population. Finally, a phenotypic frequency is simply the percentage of people in the population who have

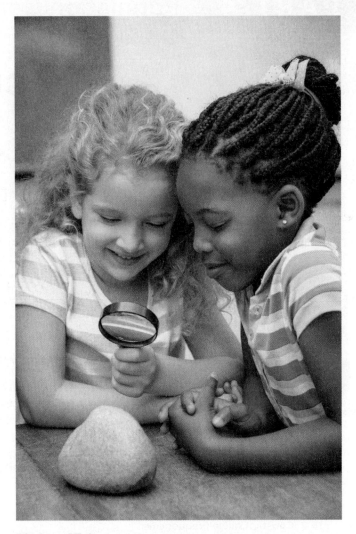

Figure 15.1 Different populations have different frequencies of single-gene diseases. These girls belong to the same social groups, but have different risks for certain inherited diseases because their ancestral groups differ. *Wavebreakmedia /iStock/Getty Images*

CF (or who do not). The situation becomes more complex with multiple alleles for a single gene, because many more phenotypes and genotypes are possible.

Phenotypic frequencies are determined empirically—that is, by observing how common a condition or trait is in a population. Inherited diseases occur with different frequencies in different population groups. The two friends in **figure 15.1** belong to some of the same groups—they are classmates and live in the same town. However, the fact that their ancestors came from different parts of the world means that each child has a different probability of being a carrier for certain inherited diseases that are more prevalent among some groups. The Caucasian child has a higher risk of carrying a mutant cystic fibrosis allele than does the African-American youngster, who has a higher risk of carrying the allele of the beta globin gene that causes sickle cell disease. The elevated gene frequencies arise from marriage

Table 15.1	Frequency of PKU in Various Populations
Population	Frequency of PKU
Chinese	1/16,000
Irish, Scottish, Yemenite Jews	1/5,000
Japanese	1/119,000
Swedes	1/30,000
Turks	1/2,600
U.S. Caucasians	1/10,000

Figure 15.2 Microevolution and macroevolution. Microevolutionary changes, such as variants of the *FOXP2* gene, enabled humans to speak. Larger-scale changes separate us from our closest primate relatives. CAROLINE SEIDEL/DPA/AFP/Getty Images

and migration patterns of ancestors; it isn't that certain genes prefer people of certain backgrounds. A Caucasian child can have sickle cell disease, and an African-American child can have cystic fibrosis.

Genetic counselors use phenotype frequency to estimate the risk that an inherited disease will occur in an individual when there is no known family history of the illness. For example, **Table 15.1** shows disease incidence for phenylketonuria (PKU), an inborn error of metabolism that causes intellectual disability unless the person follows a special, low-protein diet from birth. Note how the frequency differs in different populations. A person from Turkey without a family history of PKU would have a higher risk of having an affected child than a person from Japan.

On a broader level, shifting allele frequencies in populations indicate small steps of genetic change, called **microevolution**. These incremental changes alter genotype frequencies and underlie evolution. Genotype frequencies rarely stay constant. They can change under any of the following conditions:

1. Individuals of one genotype are more likely to reproduce with each other than with individuals of other genotypes (*nonrandom mating*).
2. Individuals move between populations (*migration*).
3. Random sampling of gametes alters allele frequencies (*genetic drift*).
4. New alleles arise (*mutation*).
5. People with a particular genotype are more likely to produce viable, fertile offspring under a specific environmental condition than individuals with other genotypes (*natural selection*).

All of these conditions, except mutation, are quite common. Therefore, genetic equilibrium—when allele frequencies are *not* changing—is rare. Given our tendency to pick our partners and move about, microevolution is not only possible, but also nearly unavoidable. (Chapter 16 considers these factors in depth.) Yet it is important to understand what is required for genetic equilibrium—unchanging allele frequencies—to appreciate the consequences of when it does not happen.

When sufficient microevolutionary changes accumulate to keep two fertile organisms of opposite sexes from producing fertile offspring together, a new species forms. Changes that are great enough to result in speciation are termed **macroevolution** (**figure 15.2**). Speciation can occur through many small changes over time, and/or a few changes that greatly affect the phenotype.

Key Concepts Questions 15.1

*a. What is a biological population?
*b. Define *gene pool*.
*c. Distinguish microevolution from macroevolution.
d. List the five factors that can change genotype frequencies in populations.

15.2 Constant Allele Frequencies

Before examining the abundant genetic evidence for evolution, consider the interesting, but unusual, situation in which frequencies for certain alleles stay constant. This is a condition called **Hardy-Weinberg equilibrium**.

Hardy-Weinberg Equilibrium

In 1908, Cambridge University mathematician Godfrey Harold Hardy (1877–1947) and Wilhelm Weinberg (1862–1937), a German physician interested in genetics, independently used algebra to explain how allele frequencies can be used to predict phenotypic and genotypic frequencies in populations of diploid, sexually reproducing organisms.

Hardy unintentionally cofounded the field of population genetics with a simple letter published in the journal *Science*. The letter began with a curious mix of modesty and condescension:

> *I am reluctant to intrude in a discussion concerning matters of which I have no expert knowledge, and I should have expected the very simple point which I wish to make to have been familiar to biologists.*

Hardy continued to explain how mathematically inept biologists had incorrectly deduced from Mendel's work that dominant traits would increase in populations while recessive traits would become rarer. At first glance, this seems logical. However, the expectation is untrue because recessive alleles enter populations by mutation or migration and are maintained in heterozygotes—they aren't rare. Recessive alleles also become more common when they confer a reproductive advantage, thanks to natural selection.

Hardy and Weinberg disproved the assumption that dominant traits increase while recessive traits decrease using algebra. The expression of population genetics in algebraic terms begins with the simple equation

$$p + q = 1.0$$

where p represents the frequency of all dominant alleles for a gene and q represents the frequency of all recessive alleles. The expression "$p + q = 1.0$" means that all the dominant alleles and all the recessive alleles comprise all the alleles for that gene in a population.

Next, Hardy and Weinberg described the possible genotypes for a gene with two alleles using the binomial expansion

$$p^2 + 2pq + q^2 = 1.0$$

In this equation, p^2 represents the proportion of homozygous dominant individuals, q^2 represents the proportion of homozygous recessive individuals, and $2pq$ represents the proportion of heterozygotes (**figure 15.3**). The letter p designates the frequency of a dominant allele, and q is the frequency of

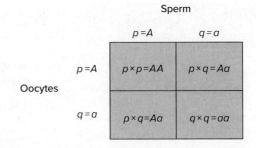

Figure 15.4 Source of the Hardy-Weinberg equation. A variation on a Punnett square reveals how random mating in a population in which gene *A* has two alleles—*A* and *a*—generates genotypes *aa*, *AA*, and *Aa*, in the relationship $p^2 + 2pq + q^2$.

a recessive allele. **Figure 15.4** shows how the binomial expansion is derived from allele frequencies. Note that the derivation is conceptually the same as tracing alleles in a monohybrid cross, in which the heterozygote forms in two ways: recessive allele from the mother and dominant allele from the father, or vice versa (see figure 4.4).

The binomial expansion used to describe genes in populations became known as the Hardy-Weinberg equation. It can be applied to reveal the changes in allele frequency that underlie evolution. If the proportion of genotypes remains the same from generation to generation, as the equation indicates, then that gene is not evolving (changing). This situation, Hardy-Weinberg equilibrium, is theoretical. It happens only if the population is large and undisturbed. That is, its members mate at random, do not migrate, and there is no genetic drift, mutation, or natural selection.

Hardy-Weinberg equilibrium is rare for protein-encoding genes that affect the phenotype, because an organism's appearance and health affect its ability to reproduce. Harmful allele combinations are weeded out of the population while helpful ones are passed on. Hardy-Weinberg equilibrium *is* seen in many DNA sequences that do not affect the phenotype. However, if a gene in Hardy-Weinberg equilibrium is closely linked on its chromosome to a gene that is subject to natural selection, it may be pulled from equilibrium by being inherited along with the selected gene. This bystander effect is linkage disequilibrium, introduced in chapter 5.

Solving a Problem Using the Hardy-Weinberg Equation

To understand Hardy-Weinberg equilibrium, it helps to follow the frequency of two alleles of a gene from one generation to the next. Mendel's laws underlie such calculations.

Consider an autosomal recessive trait: a middle finger that is shorter than the second and fourth fingers. If we know the frequencies of the dominant and recessive alleles, then we can calculate the frequencies of the genotypes and phenotypes and trace the trait through the next generation. The dominant allele *D* confers normal-length fingers; the recessive allele *d* confers a short middle finger (**figure 15.5**). We can deduce the frequencies of the dominant and recessive alleles by observing

Figure 15.3 The Hardy-Weinberg equation in English.

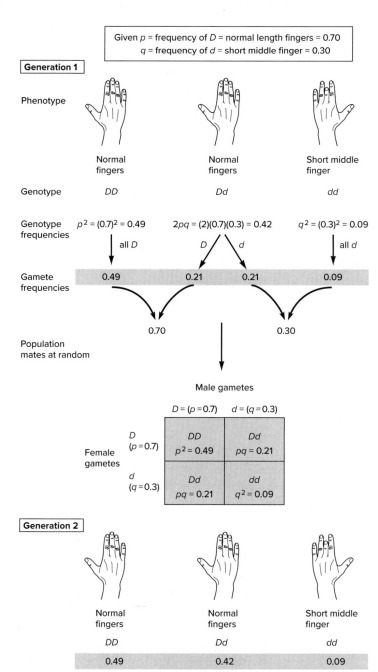

Figure 15.5 Hardy-Weinberg equilibrium. In Hardy-Weinberg equilibrium, allele frequencies remain constant from one generation to the next.

Homozygous recessive = dd
= $0.3 \times 0.3 = 0.09$
= 9 percent of individuals in generation 1

Heterozygous = $Dd + dD$
= $2pq = (0.7)(0.3) + (0.3)(0.7) = 0.42$
= 42 percent of individuals in generation 1

The proportion of homozygous individuals is calculated by multiplying the allele frequency for the recessive or dominant allele by itself. The heterozygous calculation is $2pq$ because D can combine with d in two ways—a D sperm with a d egg, and a d sperm with a D egg.

In this population, 9 percent of the individuals have a short middle finger. Now jump ahead a few generations, and assume that people choose mates irrespective of finger length. This means that each genotype of a female (DD, Dd, or dd) is equally likely to mate with each of the three types of males (DD, Dd, or dd), and vice versa. **Table 15.2** multiplies the genotype frequencies for each possible mating, which leads to offspring in the familiar proportions of 49 percent DD, 42 percent Dd, and 9 percent dd. This gene, therefore, is in Hardy-Weinberg equilibrium—the allele and genotype frequencies do not change from one generation to the next.

Key Concepts Questions 15.2

*a. What does the Hardy-Weinberg equation describe?
b. Explain the components of the Hardy-Weinberg equation.
c. Why aren't all genes under Hardy-Weinberg equilibrium?

15.3 Applying Hardy-Weinberg Equilibrium

A young woman pregnant for the first time learns that her health care providers offer all patients a blood test to see if they are carriers for cystic fibrosis. The test detects more than 100 of the most common CF mutations. The woman had never thought about CF. She and her partner are from the same population (Caucasians of European descent) and neither has a family history of the disease. They would not face the limitation of genetic testing described in the chapter opener. What is her risk of being a carrier? If it is not high, perhaps she will not take the test.

The Hardy-Weinberg equation can answer the patient's question by determining the probability that she and her partner are carriers for CF. If they are carriers, then Mendel's first law predicts that the risk that an offspring inherits the disease is 1 in 4, or 25 percent. The couple might then opt for prenatal testing.

To derive carrier risks, the Hardy-Weinberg equation is applied to population statistics on genetic disease incidence (frequency). To determine allele frequencies for autosomal

the frequency of homozygous recessives, because this phenotype—short finger—reflects only one genotype. If 9 of 100 individuals in a population have short fingers—genotype dd—the frequency is 9/100, or 0.09. Since dd equals q^2, then q equals 0.3. Since $p + q = 1.0$, knowing that q is 0.3 tells us that p is 0.7.

Next, we can calculate the proportions of the three genotypes that arise when gametes combine at random:

Homozygous dominant = DD
= $0.7 \times 0.7 = 0.49$
= 49 percent of individuals in generation 1

Table 15.2 Hardy-Weinberg Equilibrium—When Allele Frequencies Stay Constant

POSSIBLE MATINGS			FREQUENCY OF OFFSPRING GENOTYPES		
Male	Female	Proportion in Population	DD	Dd	dd
0.49 DD	0.49 DD	0.2401 (DD × DD)	0.2401		
0.49 DD	0.42 Dd	0.2058 (DD × Dd)	0.1029	0.1029	
0.49 DD	0.09 dd	0.0441 (DD × dd)		0.0441	
0.42 Dd	0.49 DD	0.2058 (Dd × DD)	0.1029	0.1029	
0.42 Dd	0.42 Dd	0.1764 (Dd × Dd)	0.0441	0.0882	0.0441
0.42 Dd	0.09 dd	0.0378 (Dd × dd)		0.0189	0.0189
0.09 dd	0.49 DD	0.0441 (dd × DD)		0.0441	
0.09 dd	0.42 Dd	0.0378 (dd × Dd)		0.0189	0.0189
0.09 dd	0.09 dd	0.0081 (dd × dd)			0.0081
		Resulting offspring frequencies:	0.49	0.42	0.09
			DD	Dd	dd

recessively inherited characteristics, we need to know the frequency of one genotype in the population. This is typically the homozygous recessive class, because its phenotype indicates its genotype. Recall from chapter 4 that Mendel used homozygous recessive pea plants (*tt*) in his test crosses to distinguish heterozygotes (*Tt*) from homozygous dominant (*TT*) individuals.

The incidence (frequency) of an autosomal recessive disease in a population is used to help calculate the risk that a particular person is a heterozygote. Returning to the example of CF, the incidence of the disease, and therefore also of carriers, may vary greatly in different populations (**table 15.3**).

CF affects 1 in 2,000 Caucasian newborns. Therefore, the homozygous recessive frequency—*cc* if *c* represents the disease-causing allele—is 1/2,000, or 0.0005 in the population. This equals q^2. The square root of q^2 is about 0.022, which equals the frequency of the *c* allele. If *q* equals 0.022, then *p*, or $1 - q$, equals 0.978. Carrier frequency is equal to $2pq$, which equals (2)(0.978)(0.022), or 0.043—about 1 in 23. **Figure 15.6** summarizes these calculations.

Because there is no CF in the woman's family, her risk of having an affected child, based on population statistics, is low. The chance of *each* potential parent being a carrier is about 4.3 percent, or 1 in 23. The chance that *both* are carriers is 1/23 multiplied by 1/23—or 1 in 529—because the probability that two independent events will occur equals the product of the probability that each event will happen alone (the product rule). However, if they *are* both carriers, each of their children would face a 1 in 4 chance of inheriting the illness, based on Mendel's first law (gene segregation). Therefore, the risk that these two unrelated Caucasian individuals with no family history of CF will have an affected child is 1/4 × 1/23 × 1/23, or 1 in 2,116 (**figure 15.7**). The woman takes the test but is much less worried than she was when she first learned it was an option.

Table 15.3 Carrier Frequency for Cystic Fibrosis

Population Group	Carrier Frequency
African Americans	1 in 66
Asian Americans	1 in 150
Caucasians of European descent	1 in 23
Hispanic Americans	1 in 46

Cystic Fibrosis

incidence (autosomal recessive class) = 1/2,000 = 0.0005

$\therefore q^2 = 0.0005$

$\therefore q = \sqrt{0.0005} = 0.022$

$\therefore p = 1 - q = 1 - 0.022 = 0.978$

$\therefore$ carrier frequency = $2pq$ = (2)(0.978)(0.022) = 0.043 = 1/23

Figure 15.6 Calculating the carrier frequency given population incidence: autosomal recessive.

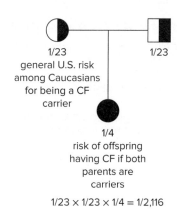

Figure 15.7 The risk of a child being born with cystic fibrosis if the parents are unrelated and do not have any relatives with the illness is calculated by multiplying the probability that both parents are carriers (1/23 × 1/23) by the probability that a child inherits the mutant allele from each (1/4).

For X-linked traits, different predictions of allele frequencies apply to males and females. For a female, who can be homozygous recessive, homozygous dominant, or a heterozygote, the standard Hardy-Weinberg equation, $p^2 + 2pq + q^2 = 1.0$, applies. However, in males, the allele frequency is the phenotypic frequency, because a male who inherits an X-linked recessive allele exhibits it in his phenotype.

The incidence of X-linked hemophilia A, for example, is 1 in 10,000 male (X^hY) births. Therefore, q (the frequency of the h allele) equals 0.0001. Using the formula $p + q = 1$, the frequency of the wild type allele is 0.9999. The incidence of carriers (X^HX^h), who are all female, equals $2pq$, or (2)(0.0001)(0.9999), which equals 0.00019; this is 0.0002, or 0.02 percent, which equals about 1 in 5,000. The incidence of a female having hemophilia A (X^hX^h) is q^2, or $(0.0001)^2$. This is about 1 in 100 million. **Figure 15.8** summarizes these calculations.

Neat allele frequencies such as 0.6 and 0.4, or 0.7 and 0.3, are unusual. In actuality, many single-gene diseases are so rare that the q component of the Hardy-Weinberg equation contributes little. Because this means that the value of p approaches 1, the carrier frequency, $2pq$, is close to $2q$. Thus, the carrier frequency is approximately twice the frequency of the rare, disease-causing allele.

Consider Tay-Sachs disease, which occurs in 1 in 3,600 Ashkenazim (Jewish people of Eastern European descent). This means that q^2 equals 1/3,600, or about 0.0003. The square root, q, equals 0.017. The frequency of the dominant allele (p) is then $1 - 0.017$, or 0.983. What is the likelihood that an Ashkenazim carries Tay-Sachs disease? It is $2pq$, or $(2)(0.983)(0.017)$, or 0.033. This is close to double the frequency of the mutant allele (q), 0.017. Modifications of the Hardy-Weinberg equation are used to analyze genes that have more than two alleles.

Hemophilia A

incidence = 1/10,000 male births = 0.0001

∴ $q = 0.0001$

∴ $p = 1 - q$ $1 - 0.0001 = 0.9999$

∴ carrier frequency (females) = $2pq$ = (2) (0.9999) (0.0001) = 0.00019 = about 1/5,000

∴ affected females = q^2 = (0.0001) (0.0001) = 1/100 million

Figure 15.8 Calculating the carrier frequency given population incidence: X-linked recessive.

Key Concepts Questions 15.3

a. How are allele frequencies in populations inferred?

b. How does the Hardy-Weinberg equation apply to X-linked traits?

*c. What is the approximate carrier frequency for rare diseases?

Study Guide

Summary

15.1 Population Genetics Underlies Evolution

1. A **population** is a group of interbreeding members of the same species in a particular area. All the alleles of all their genes constitute the population's gene pool.

2. **Population genetics** considers allele, genotype, and phenotype frequencies to understand the small changes of **microevolution**. Phenotypic frequencies can be determined empirically, then used in algebraic expressions to derive other frequencies. Changes great enough to cause speciation are termed **macroevolution**.

3. Genotype frequencies change if migration, nonrandom mating, genetic drift, mutations, or natural selection operate. Alleles can move between populations when individuals migrate, a process called **gene flow**.

15.2 Constant Allele Frequencies

4. In **Hardy-Weinberg equilibrium**, allele frequencies are not changing. Hardy and Weinberg proposed an algebraic equation to explain the constancy of allele frequencies. The equation shows why dominant traits do not increase and recessive traits do not decrease in populations.

5. A gene in Hardy-Weinberg equilibrium tightly linked to a gene subject to natural selection, called linkage disequilibrium, may disturb the equilibrium.
6. The Hardy-Weinberg equation is a binomial expansion used to represent genotypes in a population. In Hardy-Weinberg equilibrium, gamete frequencies do not change as they recombine in the next generation. Evolution is not occurring.
7. When $p^2 + 2pq + q^2 = 1.0$ represents a gene with one dominant and one recessive allele, p^2 corresponds to the frequency of homozygous dominant individuals; $2pq$ is the frequency of heterozygotes; and q^2 represents the frequency of the homozygous recessive class. The frequency of the dominant allele is p, and of the recessive allele, q.

15.3 Applying Hardy-Weinberg Equilibrium

8. If we know either p or q, we can calculate genotype frequencies, such as carrier risks. Often such information comes from knowing the value of q^2, which represents the frequency of homozygous recessive individuals in a population.
9. For X-linked recessive traits, the mutant allele frequency for males equals the trait frequency.
10. For rare diseases or traits, the value of p approaches 1, so the carrier frequency ($2pq$) is approximately twice the frequency of the rare trait (q).

Key Terms

gene flow Movement of alleles between populations when individuals migrate and mate.

Hardy-Weinberg equilibrium An idealized state in which allele frequencies in a population do not change from generation to generation.

macroevolution Genetic change sufficient to form a new species.

microevolution Change of allele frequency in a population.

population A group of interbreeding individuals.

population genetics The study of allele frequencies in biological populations.

Review Questions

1. Define *gene pool*.
2. Define *gene flow*.
3. How can evolution occur at both microscopic and macroscopic levels?
4. Why are Hardy-Weinberg calculations more complicated if a gene has many alleles that affect the phenotype?
5. Explain the differences among an allele frequency, a phenotypic frequency, and a genotypic frequency.
6. What are the conditions under which Hardy-Weinberg equilibrium cannot be met?
7. Why is knowing the incidence of a homozygous recessive condition in a population important in deriving allele frequencies?
8. How is the Hardy-Weinberg equation used to predict the recurrence of X-linked recessive traits?

Applied Questions

1. Tay-Sachs disease was once much more prevalent among people of Ashkenazi Jewish descent than in other population groups. Since widespread screening of Jewish populations for carrier status began in the 1970s, nearly all new cases are in other populations. Explain how the carrier screening altered the mutant allele frequency.
2. "We like him, he seems to have a terrific gene pool," say the parents upon meeting their daughter's boyfriend. Why doesn't their statement make sense?
3. Two couples want to know their risk of conceiving a child with cystic fibrosis. In one couple, neither partner has a family history of the disease; in the other, one partner knows he is a carrier. How do their risks differ?
4. How does calculation of allele frequencies differ for an X-linked trait or disease compared to one that is autosomal recessive?
5. Profiling of Y chromosome DNA implicated Thomas Jefferson in fathering children with Sally Hemings, his slave, discussed in chapter 1. What might have been a problem with the conclusion?
6. Torsion dystonia is a movement disease that affects 1 in 1,000 Jewish people of Eastern European descent (Ashkenazim). What is the carrier frequency in this population?

7. Maple syrup urine disease (MSUD) (see Clinical Connection 2.1) is autosomal recessive and causes intellectual and physical disability, difficulty feeding, and a sweet odor to urine. In Costa Rica, 1 in 8,000 newborns inherits the condition. What is the carrier frequency of MSUD in this population?

8. Ability to taste phenylthiocarbamide (PTC) is mostly determined by the gene *PTC*. The letters *T* and *t* are used here to simplify analysis. *TT* individuals taste a strong, bitter taste; *Tt* people experience a slightly bitter taste; *tt* individuals taste nothing.

    A fifth-grade class of 20 students tastes PTC that has been applied to small pieces of paper, rating the experience as "very yucky" (*TT*), "I can taste it" (*Tt*), and "I can't taste it" (*tt*). For homework, the students test their parents, with these results:

    - Of six *TT* students, four have two *TT* parents and two have one parent who is *TT* and one parent who is *Tt*.
    - Of four students who are *Tt*, two have two parents who are *Tt*, and two have one parent who is *TT* and one parent who is *tt*.
    - Of the 10 students who can't taste PTC, four have two parents who also are *tt*, but four students have one parent who is *Tt* and one who is *tt*. The remaining two students have two *Tt* parents.

    Calculate the frequencies of the *T* and *t* alleles in the two generations. Is Hardy-Weinberg equilibrium maintained?

Case Studies and Research Results

1. An extra row of eyelashes is an autosomal recessive trait seen in 900 of the 10,000 residents of an island in the South Pacific. Greta knows that she is a heterozygote for this gene because her eyelashes are normal but she has an affected parent. She wants to have children with a homozygous dominant man so that the trait will not affect her offspring. What is the probability that a person with normal eyelashes in this population is a homozygote for the wild-type allele of this gene?

2. Glutaric aciduria type 1 was the first disease investigated at the Clinic for Special Children, which was founded in 1989 in the heart of the Old Order Amish and Mennonite community in Lancaster, Pennsylvania (see Clinical Connection 16.1). The disease causes severe movement problems that lead to paralysis, brain damage, and early death. In this population, 0.25 percent of newborns have the disease. Researchers at the clinic developed a special formula with the levels of amino acids tailored to counter the metabolic abnormality. Children who use the formula can avoid the symptoms.

    a. What percentage of this population are carriers for glutaric aciduria type 1?
    b. What effect might the treatment have on the mutant allele frequency in this population in the future?

CHAPTER 16

Changing Allele Frequencies

The ability to digest lactose (milk sugar) became more prevalent in human populations after agriculture introduced dairy foods, providing an example of evolution in action.

Learning Outcomes

16.1 Population Matters: Steel Syndrome in East Harlem
1. Explain how DNA variants that are inherited together in members of a family indicate their shared genetic ancestry.
2. Discuss how understanding population substructure can aid in health care.

16.2 Nonrandom Mating
3. Explain how nonrandom mating changes allele frequencies in populations.

16.3 Migration
4. Explain how migration changes allele frequencies in populations.

16.4 Genetic Drift
5. Explain how the random fluctuations of genetic drift affect genetic diversity.
6. Discuss how founder effects and population bottlenecks amplify genetic drift.

16.5 Mutation
7. Discuss how mutation affects population genetic structure.

16.6 Natural Selection
8. Provide examples of negative, positive, and artificial selection.
9. Explain how balanced polymorphism maintains certain diseases in populations.

16.7 Eugenics
10. Explain how eugenics attempts to alter allele frequencies.

The BIG Picture

Several forces mold the genetic composition of populations and, over time, drive evolution. The forces of evolutionary change are nonrandom mating, migration, genetic drift, mutation, and natural selection.

The Evolution of Lactose Tolerance

For millions of people who have lactose (milk sugar) intolerance, dairy food causes cramps, bloating, gas, and diarrhea. They no longer produce lactase, which is an enzyme made in early childhood that breaks down the milk sugar lactose into more easily digested sugars. But people who have lactose intolerance may represent the "normal," or wild type condition. Only 35 percent of people in the world can digest lactose into adulthood. People who *can* digest dairy foods have lactase persistence. One gene controls the ability to digest milk sugar.

Clues in DNA suggest that agriculture drove the differences in our abilities to digest lactose. As dairy farming spread around the world, from 5,000 to 10,000 years ago, people who had gene variants enabling them to digest milk into adulthood had an advantage. They could eat a greater variety of the now more plentiful foods, were healthier, and had more children. Over time, populations that consumed dairy foods had more people with lactase persistence. In contrast, in populations with few or no dairy foods, lactose intolerance was not a problem, and so gene variants impairing the ability to digest lactose remained prevalent because they caused no harm.

Analysis of SNP patterns in DNA extracted from skull bones from individuals who lived in what is now Turkey from 2450 to 2140 B.C.E.

reveal the time when lactase persistence began to become more prevalent. Today, populations vary in the percentage of individuals who are lactose intolerant:

- European Americans 21%
- Hispanics 51%
- African Americans 75%
- Native Americans 79%
- Asian Americans 90%

16.1 Population Matters: Steel Syndrome in East Harlem

Many aspects of modern life alter allele (gene variant) frequencies in populations. As a result, many genes are not in Hardy-Weinberg equilibrium, which is unchanging allele frequencies from generation to generation. The ever-present and interacting forces of nonrandom mating, migration, genetic drift, mutation, and natural selection shape populations at the allele level. Changing allele frequencies can change genotype frequencies, which in turn can change phenotype frequencies.

Human behavior shapes genetic "substructures" within populations in terms of the genotypes of people who have children together. Religious restrictions and personal preferences guide our choices of mates. Wars and persecution diminish certain populations. Economic and political systems enable some groups to have more children than others. We travel, shuttling genes in and out of populations. Natural disasters and new diseases shrink populations, which then resurge, at the expense of genetic diversity. These influences, plus mutation and a reshuffling of genes at each generation, make gene pools quite fluid.

Researchers can trace the genetic ancestry of the individuals who make up a population through linked alleles and SNPs that are passed, as haplotypes, from generation to generation (see figure 5.17 and section 5.3). These DNA sequences, linked on the same copy of a chromosome and transmitted together through generations, are said to be **identical by descent**. When both chromosome copies have the same identical by descent DNA sequence, the phenomenon is termed a **run of homozygosity**. It indicates that two individuals shared a recent ancestor, such as a common great-great-grandmother.

Figure 16.1 illustrates identical by descent alleles and runs of homozygosity in the inbred family of Charles Darwin. Ironically, he studied inbreeding in plants. Charles Darwin and Emma Wedgwood were first cousins, sharing grandparents who were third cousins. They married and had 10 children. Their daughter Mary Eleanor died on her 23rd day, perhaps due to inheriting runs of homozygosity—identical DNA sequences on both members of a pair of autosomes transmitted from both sides of the family. Two other children who died of infection may have had an inherited immune deficiency that made them more susceptible. It is difficult to discern effects of inbreeding from a time when death in early childhood was common.

The more recent the shared ancestor, the longer the runs of homozygosity. Because of the tendency of some stretches of DNA to remain together during meiosis (linkage disequilibrium), even if two people haven't shared an ancestor for 10 generations, runs of homozygosity may reach 4 million DNA bases. These areas of sameness within genomes are what one researcher calls "hidden patterns of relatedness" that can be used to reconstruct demography and ancestry, extending family trees back in time, perhaps farther than diaries and memories can. A person who knows nothing about her or his ancestors can find a matching population group, or relatives, using DNA data, as many people are learning through consumer DNA ancestry testing. In a practical sense, merging genetic information, genealogical records, and electronic health records can lead to more precise diagnoses.

Understanding the genetic substructure of populations, the groups within the larger groups, especially in cities where people are from many diverse backgrounds, can better target health care. This is the case for Steel syndrome. It is an inherited bone disease that is rare in the general population, but clustered among the 8,000 people of Puerto Rican ancestry living in the New York City community of East Harlem. Without knowing that a patient is from this group, a physician might misdiagnose the condition and possibly perform hip surgery that can worsen symptoms.

Steel syndrome causes joint pain, hip dislocation, pinching of the spinal cord in the neck, short stature, and a large head with a long, oval face, broad nose, small low ears rotated slightly backward, and a prominent forehead. Researchers at the Icahn School of Medicine at Mt. Sinai in New York City analyzed 600,000 SNPs in 11,000 DNA samples from a biobank. They identified a genetic signature of Puerto Rican descent so distinctive that they could even tell people who knew only that they were Hispanic/Latino that their ancestors came from the island.

The researchers noticed that several individuals of Puerto Rican ancestry in East Harlem who were short also had joint pain, had undergone joint replacements at young ages, and had pinched necks. Plus, the unusual symptoms clearly ran in families as an autosomal recessive trait. At about the same time, another group of researchers reported a disease of cartilage and bone stemming from a specific point mutation in a collagen gene (*COL27A1*): Steel syndrome, named for the orthopedist who first described it in 1993 in 23 children from Puerto Rico. (See table 12.1 for other medical conditions resulting from abnormal collagen.) Other symptoms include soft bone tips, dislocated elbows and hips, fused finger and toe bones, and scoliosis. Most importantly, Dr. Steel advised against hip surgery because the abnormality differs from common reasons for the surgery, such as a sports injury or accident. Steel syndrome, extremely rare among Hispanics as a whole but more common among the Puerto Rican population, provides a compelling example of the clinical importance of recognizing where a person lives and his or her genetic ancestry.

Figure 16.1 Inbreeding in the family of Charles Darwin. Identical by descent sequences combined to form runs of homozygosity that perhaps account for the deaths of three of Charles and Emma Darwin's children at early ages. A run of homozygosity is actually thousands to millions of DNA bases long.

Key Concepts Questions 16.1

*a. Define *identical by descent*.

*b. Define *run of homozygosity*.

c. Explain how considering population genetic substructure can help in providing health care.

16.2 Nonrandom Mating

In the theoretical state of Hardy-Weinberg equilibrium, individuals of all genotypes are equally likely to successfully mate and to choose partners at random. In reality we may choose partners for many reasons: physical appearance, ethnic background, intelligence, earning potential, and shared interests, to name a few. Nonrandom mating is a major factor in changing allele frequencies in human populations. **Figure 16.2a** depicts nonrandom mating. The figure uses colored pie charts to represent the five forces of evolutionary

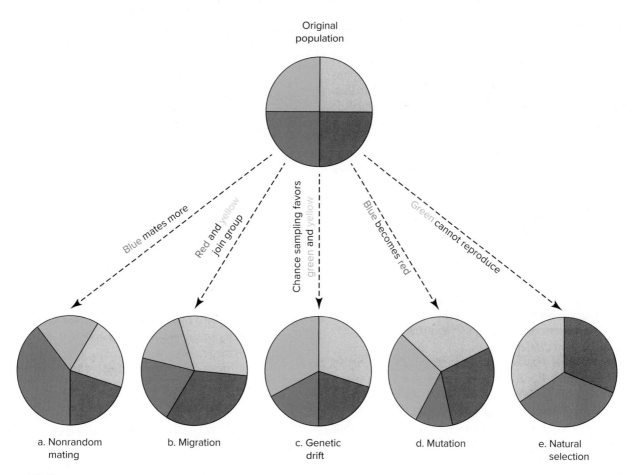

Figure 16.2 **The five sources of evolutionary change at the population level.** Nonrandom mating **(a)** increases the numbers of individuals who have the genotype that has more offspring and decreases the numbers of those who have few or no offspring. Migration **(b)** removes some genotypes from the population more than others, altering the distribution of alleles. Genetic drift **(c)** randomly samples genotypes passed to the next generation, altering allele frequencies. Mutation **(d)** can change one allele into another. Natural selection **(e)** favors some genotypes over others, depending on which are best suited to a particular environment.

change at the population level. Return to the figure throughout the chapter.

Nonrandom mating affects a population when certain individuals contribute more to the next generation than others. In agriculture, using semen from one prize bull to inseminate thousands of cows illustrates extreme and intentional nonrandom mating. A similar situation in people arises when sperm from one donor is used to inseminate many women, and half-siblings discover one another years later.

High prevalence of an otherwise rare inherited condition can be due to nonrandom mating. For example, a form of albinism is uncommon in the general U.S. population, but it affects 1 in 200 Hopi Indians who live in Arizona. The reason for the trait's prevalence is cultural—men with albinism often stay back and help the women, rather than risk severe sunburn in the fields with the other men. They contribute more children to the population because they have more contact with the women.

The events of history echo in nonrandom mating patterns. When a group of people is subservient to another, genes tend to "flow" from one group to the other as the males of the ruling class have children with females of the underclass—often forcibly. Historical records and DNA sequences show such directional gene flow. Section 17.3 offers the compelling example of slave owners in the Deep South of the United States prior to the Civil War marking the genomes of the descendants of many African American women through rape.

Traits may mix randomly in the next generation if we are unaware of them or do not consider them in choosing partners. In populations where AIDS is extremely rare or nonexistent, for example, the two mutations that render a person resistant to HIV infection are in Hardy-Weinberg equilibrium. This situation would change, over time, if HIV began to infect the population, because the people with these mutations would become more likely to survive to produce offspring—and some of them would perpetuate the protective mutation. Natural selection would intervene, ultimately altering allele frequencies. Natural selection is discussed in section 16.6.

Many blood types are in Hardy-Weinberg equilibrium because we do not choose partners by blood type. Yet sometimes the opposite situation occurs. People with mutations in the same gene meet when their families participate in programs for people with the associated disease. For example, more than two-thirds of relatives visiting a camp for children with cystic fibrosis are likely to be carriers, compared to the 1 in 23 or fewer in more diverse population groups.

People try to avoid genetic disease by controlling mate choice and reproduction. In a program that began in New York City called Dor Yeshorim, for example, young people take tests for selected severe recessive diseases that are much more common among Jewish people of eastern European descent (Ashkenazim) or Spanish or Portuguese descent (Sephardim). Results are stored in a confidential database. Two people wishing to have children together can find out if they are carriers for the same disease. If so, they may elect not to have children. Thousands of people have been tested, and the program led to the near-disappearance of Tay-Sachs disease among Ashkenazim. The very few cases each year are usually in non-Jews who have not taken carrier tests.

A population that practices consanguinity has highly nonrandom mating. Recall from chapter 4 that in a consanguineous relationship, "blood" relatives have children together. On the family level, this practice increases the likelihood that harmful recessive alleles from shared ancestors will combine and pass to offspring, causing disease. The birth defect rate in offspring of blood relatives is 2.5 times the normal rate of about 3 percent. On a population level, consanguinity decreases genetic diversity. The proportion of homozygotes rises as the proportion of heterozygotes falls, and runs of homozygosity gather in genomes.

Consanguinity is common. Ten percent of the world's human population are offspring of people who are second cousins or closer, adding up to more than 700 million individuals. Some populations encourage marriage between cousins, which increases the incidence of certain recessive diseases. In parts of the Middle East, Africa, and India, 20 to 60 percent of marriages are between cousins or between uncles and nieces. Worldwide, about 960 million married couples are related, and know of their relationship. Also contributing to nonrandom mating is **endogamy,** which is marriage within a community. In an endogamous society, spouses may be distantly related and be unaware of the connection. Even if people think they are having children with people who aren't their relatives, they may be mistaken if a population is small and endogamous.

Key Concepts Questions 16.2

a. Explain why human mating is usually not random.

*b. Define *endogamy*.

c. Discuss the effects of consanguinity and endogamy on population genetic structure.

16.3 Migration

Large cities, with their pockets of ethnicity, defy Hardy-Weinberg equilibrium by their very existence. New York City exemplifies the contributions that waves of immigration bring to a population's complex and dynamic substructure. The original Dutch settlers of the 1600s had different alleles than those in today's metropolis of English, Irish, Slavs, Africans, Hispanics, Italians, Asians, and many others. Today the New York City neighborhood of Elmhurst, Queens, is the most genetically diverse community in the world!

Migration can result in **admixture,** which is the introduction of new gene variants into one population from another, followed by reproduction between newcomers and residents. The migration depicted in **figure 16.2***b* brings in additional alleles that are already present. Clues to past migrations lie in historical documents as well as in differing allele frequencies in regions defined by geographical or language barriers. A migration caused by a group of people forced to leave their homeland is a common feature of history and is called a *diaspora*, which means "to scatter."

Allele frequencies often reflect who ruled whom. For example, the frequency of ABO blood types today is very similar in northern Africa, the Near East, and southern Spain, all areas where Arabs ruled until 1492. Uneven distribution of allele frequencies can reveal when and where nomadic peoples stopped. For instance, in the eighteenth century, European Caucasians called trekboers migrated to the Cape area of South Africa. The men stayed and had children with the native women of the Nama tribe. The mixed society remained fairly isolated, leading to the distinctive allele frequencies found in the present-day people of color of the area.

In a situation termed a **cline,** allele frequencies change from one neighboring population to another. A cline develops as immigrants introduce alleles and emigrants remove them. Clines may be gradual, if nothing blocks migration paths, or abrupt if barriers block gene flow. Barriers include geography and language.

Geographical formations such as mountains and bodies of water may block migration, maintaining population differences in allele frequencies on either side of the barrier. Consider agriculture, which arose in several places at different times. Agriculture spread faster from east to west in Eurasia because latitude did not change much, and there is less evidence of gene flow in this relatively steady environment. Even if people migrated, they tended to have the same gene variants as the people in their new homes. In the Americas, the situation was different, due to geography, and agriculture spread north to south. Hours of daylight and climate that differed with longitude presented different environmental challenges, which are reflected in greater genetic change from north to south than from east to west.

Language differences may isolate alleles if people who cannot communicate do not have children together. For example, in Italy, a nation rich in family history records with several language variants, blood types tend to cluster with linguistic group more than with geographical area. Perhaps differences in language prevent people from socializing, keeping alleles within groups.

Allele frequencies up and down the lush strip of fertile land that hugs the Nile River illustrate the concept of clines. Researchers found a gradual change in mitochondrial DNA sequences in 224 people who live on either side of the Nile, an area settled 15,000 years ago. The farther apart two individuals live along the Nile, the less alike their mtDNA sequence. The story told in mDNA is consistent with evidence from mummies and historical records that indicate the area was once home to kingdoms separated by wars and language differences. If the region had been one large interacting settlement, then the mtDNA sequences would have been more evenly distributed.

Key Concepts Questions 16.3

a. How does migration alter allele frequencies?
*b. What are clines?
c. How do geography and language affect clines?

16.4 Genetic Drift

Genetic drift refers to fluctuations in allele frequencies from generation to generation that happen by chance, to gametes. It is a characteristic of all populations. The changes in allele frequencies that cause genetic drift occur at random and are unpredictable.

The effects of genetic drift are accelerated when the population becomes very small, and sampling changes allele frequencies. Populations may shrink, amplifying genetic drift, in several circumstances: migration, a natural disaster or geographical barrier that isolates small groups, or as the consequences of human behavior (**figure 16.2c**). Members of a small ethnic community within a larger population might have children only among themselves, keeping certain alleles more prevalent within the smaller group. **Clinical Connection 16.1** discusses genetic diseases among the Old Order Amish and Mennonite populations of North America, in which genetic drift has isolated and amplified disease-causing alleles brought from Europe more than 300 years ago.

Two factors that can accelerate genetic drift are a **founder effect** and a **population bottleneck,** which both decrease population size. A founder effect results when some individuals leave a larger group or become reproductively isolated from them. In contrast, a population bottleneck is a large decrease in the size of an original population. **Figure 16.3** illustrates the distinction between a founder effect and a population bottleneck.

The Founder Effect

In a founder effect, a small group leaves a population to start a new settlement, and the new colony has different allele

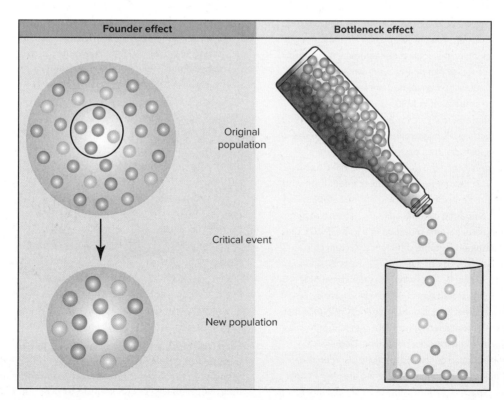

Figure 16.3 Types of genetic drift. A founder effect randomly samples a few individuals from a population to give rise to a new population. A population bottleneck occurs when an event slashes population size and it resurges from the survivors.

Clinical Connection 16.1

The Founder Effect and "Plain" Populations

The Old Order Amish and Mennonite people, called the Plain Populations, arrived in North America from Switzerland in the early eighteenth century, to escape religious persecution. The earliest immigrants settled in Pennsylvania, and additional migrations from Europe established small farming communities in Ohio. The Plain people spread across the Midwest and into Canada, isolating themselves and maintaining simpler ways of living. Today more than a million of their descendants live in hundreds of communities in South, Central, and North America. The genetic traits that are overrepresented in these populations originated in Europe and are still there today.

Genetic drift plus a powerful founder effect, nonrandom mating, and migration contributed to the higher prevalence of certain inherited diseases among these people. Some families went to Children's Hospital of Philadelphia for diagnosis and treatment. Geneticists helped the families and investigated the causes of their inherited diseases, and so the Plain People have contributed greatly to our modern knowledge of genetic disease.

Statistics reveal founder effects in the Plain Populations. In Lancaster County, Pennsylvania, for example, maple syrup urine disease (MSUD; see Clinical Connection 2.1) affects 1 in 400 newborns, but affects only 1 in 225,000 newborns in the general population. Some conditions are rare variants of more common diseases, such as "Amish cerebral palsy." A physician from Philadelphia, Holmes Morton, discovered that the disease was an inborn error of metabolism called glutaric acidemia type 1, and was not due to lack of oxygen at birth, as others had thought. He visited farms, tracking cases against genealogical records and drawing pedigrees. He found that nearly every family with MSUD descended from a couple who transmitted the recessive alleles. They had come to the settlement in 1730.

Dr. Morton founded the Clinic for Special Children in 1989, on a cornfield in Strasburg, Pennsylvania. The goal was to use genetic tools to diagnose children early, when they were still healthy enough to treat. Since the late 1990s, the clinic has used genomic technology. For example, for an Amish infant born with thick skin, no hair, and a quickly developing life-threatening bacterial infection, comparing genetic markers to those of her seven healthy siblings revealed a mutation in a gene, *RAG1*, that caused a severe combined immune deficiency. A stem cell transplant from a sister saved her life.

Research on the Plain People helps the wider community, too. For example, several related Amish children who have autism and seizures led researchers to mutations in a gene, *CNTNAP2*, that causes some cases of unexplained autism, seizures, schizophrenia, and language problems in the broader population. Depression, bipolar disorder, and attention deficit hyperactivity disorder also affect many populations but may stand out among the Plain People.

The Clinic for Special Children's approach of catching genetic disease early and treating symptoms as they arise has worked. Physicians there now treat nearly half of the 110 genetic diseases that they detect and recognize conditions that are lethal, saving children from pointless and painful treatments that physicians less familiar with the diseases unique to the Plain communities might provide. The Amish and Mennonites not only have at least 100 rare diseases, but also have their own variants of the more common single-gene diseases, including cystic fibrosis, phenylketonuria, fragile X syndrome, and clotting and immune disorders.

Modern genetic technology is also coming to the Plain Populations. A clinical trial of gene therapy is testing a lasting, one-time treatment for Crigler-Najjar syndrome. This was the first genetic disease among the Amish that Dr. Morton saw, as he drove around and noticed the eerie blue lights in farmhouse bedroom windows where children were being treated with "bili lights" (**figure 16A**). The ultraviolet radiation in the lights breaks down the buildup of a substance called bilirubin that causes jaundice. A single injection that delivers wild type copies of the mutant gene could provide a one-time treatment. Chapter 22 discusses gene therapy.

Questions for Discussion

1. Public perception is that diseases of the Amish are unique to them. Explain why this is not true.
2. How have certain single-gene diseases become much more prevalent among the Plain People than in other groups?
3. Because the Amish do not permit prenatal diagnosis, inborn errors of metabolism are detected with tests on cord blood that midwives collect. Explain how such analysis can lead to diagnosis of a single-gene disease, even if the child does not have symptoms.

Figure 16A **"Bili lights" are used to treat the jaundice of Crigler-Najjar syndrome,** one of several Mendelian diseases that are more common among the Plain Populations, which include the Old Order Amish and Mennonites. *Li_Al/Shutterstock*

frequencies than the original population (**table 16.1**). Geneticists recognize a founder effect in a community known from local history to have descended from a few founders who brought in inherited traits and illnesses that are rare elsewhere.

A founder effect is easiest to trace when historical or genealogical records are available. This is the case for the 2.5 million Afrikaners of South Africa, who descend from a small group of Dutch, French, and German immigrants who had very large families. In the nineteenth century, some Afrikaners migrated northeast to the Transvaal Province, where they lived in isolation until the Boer War in 1902 introduced better transportation. Today, 30,000 Afrikaners who have an abnormal hemoglobin condition called porphyria variegata descended from one couple who came from the Netherlands in 1688! Their many children also had large families, passing on and amplifying the dominant mutation.

Another type of evidence for a founder effect is when all individuals in a population with a certain illness have the same mutation, which present-day patients inherited from shared ancestors. The people of Puerto Rican ancestry living in East Harlem, described in section 16.1, all have the same point mutation causing Steel syndrome. In contrast, a population with several mutations that cause the same disease is more likely to have picked up those variants from unrelated people joining the group. This is the situation for breast cancer caused by mutation in the *BRCA1* gene. Affected individuals of Ashkenazi Jewish ancestry share a specific 3-base-pair deletion, whereas families from the Ivory Coast in Africa, the Bahamas, and the southeastern United States share a 10-base-pair deletion, probably inherited from West Africans ancestral to all three modern groups. Slaves brought the disease to the United States and the Bahamas between 1619 and 1808, while relatives remaining in Africa have perpetuated the mutant allele there.

Population Bottlenecks

A population bottleneck occurs when many members of a group die and only a few are left to replenish the numbers. The new population has only those alleles in the small group that survived the catastrophe. An allele in the remnant population might become more common in the replenished population than it was in the original larger group. Therefore, the new population has a much more restricted gene pool than the larger ancestral population, with some variants amplified, others diminished.

Population bottlenecks can occur when people (or other animals) colonize islands. An extreme example is seen among the Pingelapese people of the eastern Caroline Islands in Micronesia. Four to 10 percent of the people are born with "Pingelapese blindness." This is an autosomal recessive condition that includes colorblindness, nearsightedness, and cataracts. It is also called achromatopsia. Elsewhere, only 1 in 20,000 to 50,000 people inherits the condition. Nearly 30 percent of the Pingelapese are carriers. The prevalence of the blindness among the Pingelapese stems from a typhoon in 1780 that killed all but 9 males and 10 females who founded the present population. This severe population bottleneck, plus geographic and cultural isolation, increased the frequency of the blindness mutation as the population resurged.

A more widespread series of population bottlenecks occurred as a consequence of the early human expansion from Africa, discussed in chapter 17. As numbers dwindled during the journeys and then were replenished as people settled down, mating among relatives led, over time, to an increase in homozygous recessive genotypes compared to the larger and more genetically diverse ancestral populations that remained in Africa. The evidence can be seen in runs of homozygosity in modern human genomes.

Human-wrought disasters that kill many people can cause population bottlenecks that greatly alter the gene pools of a persecuted group. In contrast, a natural disaster such as a tsunami or earthquake kills anyone in its path. The Ashkenazi Jewish population underwent a series of population bottlenecks that researchers are now illuminating by applying computational tools to genomic information.

The story that is emerging is that about 30 to 35 generations ago, ancestors of Ashkenazim admixed with populations from the Middle East and southern Europe. Then during the Crusades, from 1095 though 1291, communities of Jewish people in Europe were wiped out because of religious persecution. The fourteenth century saw sweeping evictions of Jewish people from western Europe, and by the fifteenth century, very few were left. Historical records indicate that many of the survivors migrated to eastern Europe.

When researchers analyzed genomewide SNPs from the genomes of 128 modern Ashkenazi individuals to follow identical by descent DNA sequences echoing up from long-ago

Table 16.1 Founder Populations

Population	Number of Founders	Number of Generations	Population Size Today
Costa Rica	4,000	12	2,500,000
Finland	500	80–100	5,000,000
Hutterites	80	14	36,000
Japan	1,000	80–100	120,000,000
Iceland	25,000	40	300,000
Newfoundland	25,000	16	500,000
Quebec	2,500	12–16	6,000,000
Sardinia	500	400	1,660,000

Table 16.2 Autosomal Recessive Genetic Diseases More Prevalent Among Ashkenazi Jewish Populations

Disease	Gene	Signs and Symptoms (Phenotype)	Carrier Frequency
Bloom syndrome	RECQL3	Sun sensitivity, short stature, poor immunity, impaired fertility, increased cancer risk	1/110
Breast cancer	BRCA1 BRCA2	Malignant breast tumor	3/100
Canavan disease	ASPA	Brain degeneration, seizures, developmental delay, early death	1/40
Familial dysautonomia	IKAKAP	No tears, cold hands and feet, skin blotching, drooling, difficulty swallowing, excess sweating	1/32
Gaucher disease	GBA	Enlarged liver and spleen, bone degeneration, nervous system impairment	1/12
Niemann-Pick disease type A	SMPD1	Lipid accumulation in cells, particularly in the brain; intellectual and physical disability, death by age 3	1/90
Tay-Sachs disease	HEXA	Brain degeneration causing intellectual disability, paralysis, blindness, death by age 4	1/26
Fanconi anemia type C	FANCA	Deficiencies of all blood cell types, poor growth, increased cancer risk	1/89

shared ancestry, they deduced a population bottleneck so severe, 30 or so generations ago during the Crusades, that the most likely explanation is that during the late-medieval period, only 350 Ashkenazim existed. Today's Ashkenazi gene pool formed when the Jews chased out of western European during the Middle Ages admixed in eastern Europe, forming the modern Ashkenazi gene pool.

Among Jewish people, nonrandom mating and population bottlenecks changed allele frequencies and contributed to the nearly 10 times higher incidence of dozens of inherited diseases compared to other populations. "Jewish genetic disease" tests are not meant to stereotype or discriminate, but are based on a genetic fact of life—some illnesses are more common in certain populations, due to human behavior. However, DNA itself does not discriminate. The "Jewish" mutations can arise anew in anyone. **Table 16.2** describes some inherited diseases that are more common among Ashkenazi Jewish populations.

Key Concepts Questions 16.4

*a. Define *genetic drift*.
b. Explain how founder effects and population bottlenecks amplify the effects of genetic drift.

16.5 Mutation

A continual source of genetic variation in populations is mutation, which changes one allele into another (**figure 16.2d**). Genetic variability also arises from crossing over and independent assortment during meiosis, but these events recombine existing traits rather than introduce new ones.

If a DNA base change occurs in the part of a gene that encodes part of a protein necessary for the protein's function, then an altered trait may result. Another way that genetic change can occur from generation to generation is in the numbers of repeats of copy number variants (CNVs), which function as alleles.

Natural selection, discussed in section 16.6, eliminates alleles that adversely affect reproduction. Yet harmful recessive alleles are maintained in heterozygotes and are reintroduced by new mutation. Therefore, all populations have some alleles that would be harmful if homozygous. The collection of deleterious recessive alleles in a population is called its **genetic load**.

Human behavior and the events of history can influence the diversity of mutations in a population. Consider phenylketonuria (PKU), an autosomal recessive condition that causes intellectual disability unless a special diet followed from birth restricts one amino acid type. PKU mutations worldwide are highly diverse, indicating that the disease has arisen more than once. Conversely, a distinct mutation found in many groups of people is probably more ancient, having occurred before those groups separated and perpetuated it.

Mutations found only in a small geographical region are more likely to be of recent origin, perhaps set apart due to genetic drift. These mutations have had less time to spread. For example, people from Turkey, Norway, Jewish people from Yemen living in Israel, and French-Canadians have their "own" PKU alleles. The Yemeni mutation is so distinctive—a large deletion, compared to point mutations in other populations—that researchers used court and religious records to trace the mutation from families leaving the capital of Yemen north to three towns, and then to Israel. They tested for the telltale deletion in the grandparents of the 22 modern Yemeni Jewish families with PKU in Israel. The earliest court and religious records identify two families with PKU in San'a, the

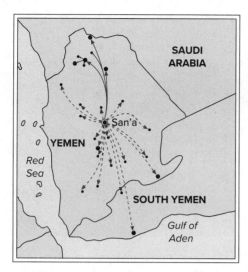

Figure 16.4 A telltale mutation for PKU. The deletion mutation in Israeli Yemeni Jewish families probably arose in San'a, Yemin, in the mid-eighteenth century. The allele spread northward as families moved from San'a in 1809 (solid arrows) and subsequently spread to other regions (broken arrows), and eventually Israel. (Source: Data from Smadar Avigad, et al., A single origin of phenylketonuria in Yemenite Jews, Nature 344:170, March 8, 1990.)

capital of Yemen. By 1809, religious persecution and hard economic times led nine families carrying the distinctive mutation to migrate north and settle in three towns, then spread farther (**figure 16.4**).

The contribution of mutation in countering Hardy-Weinberg equilibrium is small compared to the influence of migration and nonrandom mating, because mutation is rare. The spontaneous mutation rate is about 170 bases per haploid genome in each gamete. Each of us probably has at least five "lethal equivalents"—alleles or allele combinations that if homozygous would kill us or make us too sick to have children. Natural selection has the greatest influence of all the factors on allele frequencies.

> **Key Concepts Questions 16.5**
>
> a. Explain how mutation increases genetic variation in a way that crossing over and independent assortment do not.
>
> b. How does mutation alter allele frequencies?
>
> *c. Define *genetic load*.

16.6 Natural Selection

Environmental change can alter allele frequencies when individuals with certain phenotypes are more likely to survive and reproduce than others. This differential survival to reproduce guided by environmental change is **natural selection** (see **figure 16.2***e*).

The chapter opener chronicles natural selection acting on gene variants that enable adults to digest the sugar lactose.

Another illustration of natural selection related to a dietary pattern involves copy number variants (CNVs), in contrast to the point mutations that cause most instances of lactase persistence. In populations that follow high-carbohydrate diets, most people have more copies of the gene that encodes salivary amylase, the digestive enzyme that begins to break down carbohydrates in the mouth. Members of populations that follow low-carbohydrate diets have fewer copies of the gene and presumably less of the enzyme.

In natural selection, reproductive success is all important, because this is what transmits favorable alleles and weeds out the unfavorable ones, ultimately impacting population structure and therefore driving microevolution (see figure 15.2). In the common phrase used synonymously with natural selection—"survival of the fittest"—"fit" refers to reproductive success, not to physical prowess or intelligence (unless those traits lead directly to reproductive success).

Negative and Positive Selection

Natural selection acts negatively if a trait diminishes in a population because it adversely affects reproduction. Natural selection acts positively if a trait becomes more common in a population because it enhances the chances of reproductive success. However, "negative" and "positive" selection are not value judgments—they simply refer to a trait's decreasing in a population or increasing. In Darwin's time, natural selection was thought to be primarily negative. Today, the ability to sequence DNA has revealed "signatures" of positive selection in the human genome.

One sign of positive selection is a gene in humans that has a counterpart in other primates, but in humans there is at least one distinctive difference in the amino acid sequence, not just the DNA sequence. A change in the DNA sequence that does *not* substitute an amino acid does not change the protein, and therefore has no effect on the phenotype. A change that doesn't alter the phenotype isn't subject to natural selection.

The allele for lactase persistence illustrates positive selection. The allele is part of a large haplotype (genes linked on a chromosome) that shows very little sequence diversity among individuals. The persistence of the DNA sequence of an allele that has greatly increased in frequency in the population over a very short time—since agriculture began—indicates positive selection. Section 17.4 describes traits that were positively selected during recent human evolution.

A dramatic example of positive selection in humans is seen among native people ("highlanders") who live more than two miles above sea level on the Tibetan plateau (**figure 16.5**). Thanks to a "high-altitude adaptation," they thrive in the oxygen-poor air that makes others very ill and not likely to reproduce. Even the Han Chinese, who are a related population that lives in the lowlands, cannot survive up in the mountains for long without developing the headache, dizziness, ringing in the ears, heart palpitations, breathlessness, fatigue, sleep disturbance, lack of appetite, and confusion of chronic mountain sickness. If they ascend to great heights,

Figure 16.5 Positive selection enables the Sherpa to thrive at elevations that make others very sick. The Sherpa have migrated from their native Tibet to Nepal over the past 300 to 400 years. Many of them work as mountaineering guides, especially for climbers ascending Mt. Everest. They are short, strong, and hardy. A variant of the hypoxia inducible factor 2 (*EPAS1*) gene is one of several alleles responsible for their astonishing adaptation to low oxygen conditions. *GOODSHOOT/Alamy*

Figure 16.6 Dogs small and large. It is hard to believe that the diminutive Chihuahua and the Great Dane are members of the same species. Thanks to artificial selection and intense inbreeding, body size in domesticated dogs varies more than it does in any other terrestrial mammal. *Brand X Pictures/Getty Images*

their bloodstreams become crammed with extra red blood cells that extract maximal oxygen from the thin air, raising the risks of heart attack and stroke.

In people other than the native Tibetans exposed to low oxygen conditions, a protein called hypoxia inducible factor 2 signals the kidneys to release the hormone EPO (erythropoietin). EPO stimulates the bone marrow to make more red blood cells, increasing hemoglobin levels in the blood. About 90 percent of native Tibetans have a deletion that removes part of the gene *EPAS1* that encodes hypoxia inducible factor 2. The deletion is extremely rare or not present at all in other populations. The Tibetans who have the deletion also share a set of five SNPs in *EPAS1*. The highlanders' version of the protein enables them to thrive in the thin air. The Tibetans have resided in the area only about 10,000 years. Their ability to survive with less oxygen is considered an example of rapid and recent evolution.

Whole-genome sequencing has revealed other genes that affect adaptation to thin air in other populations. People adapted to high altitudes in the Andes, for example, do not have the Tibetan genetic protection, but have three gene variants that enable their cardiovascular systems to withstand the increase in the number of red blood cells.

Dog Breeding Illustrates Artificial Selection

Natural selection acts on preexisting genetic variants. It is uncontrolled and largely unpredictable. In contrast, artificial selection is controlled breeding to perpetuate individuals with a particular phenotype, such as a crop plant or purebred dog. Darwin's idea of natural selection grew from his observations of artificially selected pigeons.

We have created our pets by controlling their evolution (**figure 16.6**). Pets arose from initial domestication of wild species, followed by artificial selection. Cats were domesticated in the Near East when agriculture began, about 10,000 years ago. They descended from one of five subspecies of wildcats.

Dogs were thought to have been domesticated one time in Southeast Asia about 30,000 years ago, from gray wolves. However, genome sequencing of several dozen ancient dogs supports a hypothesis of two original ancestral wolf populations in eastern and western Eurasia giving rise to two dog populations, with the eastern dogs following people to Europe between 6,400 and 14,000 years ago. Most of today's dog breeds were artificially selected in the nineteenth century, but some, such as the Australian bulldog and the silken windhound, were bred in the 1990s.

The intense inbreeding required to artificially select distinctive breeds lowers their genetic diversity, resulting in more than 300 inherited diseases, from bladder stones in Dalmatians to hip problems in St. Bernards. Yet the differences among dog breeds for some traits stem from only a few genes. For example, variants in only three genes account for 95 percent of the variability in the texture and length of canine coats.

The recent history of dog domestication explains why only a few genes define breeds. We select oddities and quirks in dogs, and these are often the consequence of a mutation in a single gene, rather than the collection of variants in many genes that may underlie a complex trait that hasn't been artificially selected. In nature, that mutant may not have survived; we intentionally breed to select it. For example, ear floppiness and dwarfism, both seen in basset hounds, are each the result of a mutation in a single gene.

Analysis of the genomes of domesticated dogs and their wild relatives confirmed what we know from history—that our pets are no longer like wolves. Blocks of neighboring SNPs on chromosomes distinguish modern breeds. The fact that breeds share few SNP blocks indicates that we have selected away many ancestral DNA sequences. In addition, runs of homozygosity are clear signs of intense inbreeding. The fact that individual domesticated breeds differ from each other less than wild dogs differ from each other reflects the intense population bottlenecks that accompanied dog breeding, severely narrowing the breeds' gene pools.

Antibiotic Resistance Illustrates Natural Selection

Antibiotic drugs are used to treat bacterial infections. Many antibiotic drugs are no longer effective because bacteria become genetically equipped to survive in the presence of a particular drug. More than 2 million antibiotic-resistant infections occur in the United States each year, causing 23,000 deaths, according to the Centers for Disease Control and Prevention. Antibiotic resistance arises from the interplay between mutation and natural selection.

Genetic variants arise in populations of pathogenic bacteria in our bodies, some of which enable the microorganisms to survive in the presence of a specific antibiotic drug. When an infected person takes the drug, symptoms abate as sensitive bacteria die. However, resistant mutants thrive and reproduce, taking over the niche the antibiotic-sensitive bacteria vacated (**figure 16.7**). Soon, enough antibiotic-resistant bacteria accumulate to once again cause symptoms as the immune system responds to the returning infection. Usually antibiotic resistance genes already exist in the bacterial populations, and exposure to the drug *selects* the resistant bacteria. However, some antibiotics actually *induce* mutation.

Resistant bacteria circumvent antibiotic actions in several ways. Penicillin kills bacteria by destabilizing their cell walls. Resistant bacteria produce enzyme variants that dismantle penicillin, or have altered constituents of cell walls that the drug cannot bind. Erythromycin, streptomycin, tetracycline, and gentamicin are antibiotic drugs that kill bacteria by attacking their ribosomes, which are different from ribosomes in a human. The drugs cannot bind the ribosomes of resistant bacteria.

Antibiotic resistance can begin in either of two ways. Bacterial DNA can mutate, passing the resistance from one bacterial generation to the next by cell division. Or, groups of resistance genes are transmitted on mobile pieces of DNA called transposons, which move from bacterium to bacterium as part of DNA circles called plasmids. Bacteria usually pass transposons to similar bacteria, but in the unnatural environment of a hospital, genes may flit to any bacterium, quickly passing drug resistance among bacterial species. This is what has happened with infection by the common bacterium *Staphylococcus aureus*.

S. aureus is normally present in low numbers in the nose and on the skin, but in high numbers it causes boils, food poisoning, toxic shock syndrome, pneumonia, and wound infections. *S. aureus* became resistant to penicillin soon after the drug was introduced in the 1940s. A related penicillin, methicillin, worked for a time, but in 2000 resistant bacterial strains appeared, called MRSA, for methicillin-resistant *S. aureus*. By 2014, MRSA had become a global concern, accounting for 90 percent of *S. aureus* infections in some nations. Another antibiotic, vancomycin, works against some cases of MRSA infection, but *S. aureus* is becoming resistant to it, too. Some plasmids harbor resistance genes to both drugs.

In the rise of MRSA infection, natural selection benefits the pathogen, not us. That is, bacteria that can resist the drugs that we use to fight them will survive and reproduce, ensuring that *S. aureus* infection continues.

Balanced Polymorphism

If natural selection eliminates individuals with detrimental phenotypes from a population, then how do harmful mutant alleles remain in a gene pool? Harmful recessive alleles are replaced in two ways: by new mutation and by persistence in heterozygotes.

A recessive condition can remain prevalent if the heterozygote enjoys a health advantage that affects reproduction, such as resisting an infectious disease or surviving an environmental threat. This "heterozygous advantage" that maintains a recessive, disease-causing allele in a population is called **balanced polymorphism**. Recall that *polymorphism*

Figure 16.7 **Antibiotic resistance.** Resistance to an antibiotic drug develops when a mutation in a bacterial cell gives the cell the ability to survive and reproduce in the presence of the drug. Over time, bacteria harboring the mutation dominate the population. This means return or resurgence of the infection.

Table 16.3 Balanced Polymorphism

Disease 1 (inherited, carrier)	Protects against → Disease 2	Because → Mechanism
Sickle cell disease	Malaria	Atypical red blood cells cannot retain parasites
G6PD deficiency	Malaria	Parasite cannot reproduce in atypical red blood cells
Phenylketonuria	Fungal infection in fetuses	Elevated phenylalanine inactivates fungal toxin
Prion protein mutation	Transmissible spongiform encephalopathy	Prion protein cannot misfold in presence of infectious prion protein
Cystic fibrosis	Diarrheal disease (cholera, typhoid fever)	Fewer chloride channels in intestinal cells prevent water loss
Smith-Lemli-Opitz syndrome	Cardiovascular disease	Lowered serum cholesterol
Kidney disease (apolipoprotein L1 variants G1 and G2)	African sleeping sickness	APO L1 protein cuts up protozoan parasite

means "variant"; the effect is *balanced* because the protective effect of the noninherited condition counters the negative effect of the deleterious allele in two copies, maintaining its frequency in the population. Balanced polymorphism is a type of balancing selection, which more generally refers to maintaining heterozygotes in a population. A few examples follow, and these and others are summarized in **table 16.3**.

Sickle Cell Disease and Malaria

Sickle cell disease is autosomal recessive and causes anemia, joint pain, a swollen spleen, and frequent, severe infections (see figure 12.3). It is the classic example of balanced polymorphism: carriers are resistant to malaria or develop very mild cases.

Malaria is an infection by the parasite *Plasmodium falciparum* or related species that causes cycles of chills and fever. The parasite spends the first stage of its life cycle in the salivary glands of the mosquito *Anopheles gambiae*. When an infected female mosquito draws blood from a human, malaria parasites enter red blood cells, which transport the parasites to the liver. The red blood cells burst, releasing parasites throughout the body.

In sickle cell disease, the blood becomes unwelcoming to the malaria parasite in several ways. The blood is thicker than normal, which may hamper the parasite's ability to infect. The bent shape of the cells prevents parasites from growing and also blocks the parasites from exiting red blood cells, keeping them from reaching other types of host cells and spreading the infection. In addition, many red blood cells burst too soon, expelling the parasites. A sickle cell disease carrier's blood has enough atypical cells to hamper the actions of the parasite, but usually not enough of those cells to block circulation. Two other globin abnormalities that block malaria parasites are alpha thalassemia and hemoglobin C (see table 12.5).

A clue to the protective effect of being a carrier for sickle cell disease came from striking differences in the incidence of the two diseases in different parts of the world (**figure 16.8**). In the United States, 8 percent of African Americans are sickle cell carriers, whereas in parts of Africa, up to 45 to 50 percent are carriers. Although Africans had known about a painful disease

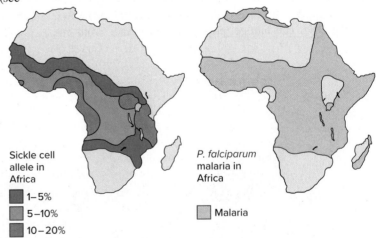

Figure 16.8 Geography reveals balanced polymorphism. Comparing the distribution of people with malaria and people with sickle cell disease in Africa reveals balanced polymorphism. Carriers for sickle cell disease are resistant to malaria because changes in their blood do not impair health, but inhibit the malaria parasite.

that shortened life, the sickled cells weren't reported in a medical journal until 1910 (see section 12.2). In 1949, British geneticist Anthony Allison found that the frequency of sickle cell carriers in tropical Africa was higher in regions where malaria rages all year long. Blood tests from children hospitalized with malaria showed that nearly all were homozygous for the wild type sickle cell allele. The few sickle cell carriers among them had the mildest cases of malaria. Was malaria enabling the sickle cell allele to persist by felling people who did not inherit it? The fact that sickle cell disease is rarer where malaria is rare supports the idea that sickle cell heterozygosity protects against the infection.

Further evidence of a sickle cell carrier's advantage in a malaria-ridden environment is the fact that the rise of sickle cell disease parallels the cultivation of crops that provide breeding grounds for *Anopheles* mosquitoes. About 1000 B.C.E., Malayo-Polynesian sailors from Southeast Asia traveled in canoes to East Africa, bringing new crops of bananas, yams, taros, and coconuts. When the jungle was cleared to grow these crops, mosquitoes occupied the open space. The insects offered a habitat for part of the life cycle of the malaria parasite.

The sickle cell allele may have been brought to Africa by people migrating from southern Arabia and India, or it may have arisen directly by mutation in East Africa. However it happened, people who inherited one copy of the sickle cell allele survived or never contracted malaria—the essence of natural selection. These healthy carriers had more children and passed the protective allele to approximately half of them. Gradually, the frequency of the sickle cell allele in East Africa rose from 0.1 percent to 45 percent in 35 generations. Carriers paid the price for this genetic protection, however, whenever two of them produced a child with sickle cell disease.

A cycle set in. Settlements with large numbers of sickle cell carriers escaped malaria. The people were strong enough to clear even more land to grow food, and support the disease-bearing mosquitoes. Today, however, in African nations that control malaria well, the frequency of the sickle cell allele has decreased. The selective pressure is off. A Glimpse of History discusses a time when malaria was common in parts of the United States.

Cystic Fibrosis and Diarrheal Disease

Balanced polymorphism may explain why CF is so common—its cellular defect protects against certain diarrheal illnesses. Diarrheal disease epidemics have left their mark on many human populations, and continue to be a major killer in the developing world.

Severe diarrhea rapidly dehydrates the body and leads to shock, kidney and heart failure, and death in days. In cholera, bacteria produce a toxin that opens chloride channels in cells of the small intestine. As salt (NaCl) leaves the intestinal cells, water rushes out, producing diarrhea. The CFTR protein does just the opposite, closing chloride channels and trapping salt and water in cells, which dries out mucus and other secretions (see figure 1.6). A person with CF is very unlikely to contract cholera, because the toxin cannot open the chloride channels in the small intestine cells.

A GLIMPSE OF HISTORY

For three centuries, malaria plagued the United States, as human activities opened niches for the mosquitoes that carried the parasites that cause the disease. Starting with Christopher Columbus, European explorers brought malaria, but in mild forms that did not spread. It wasn't until parasite-carrying mosquitoes arrived with slaves from western Africa that malaria became a deadly infectious disease in the United States.

By 1776, swarms of mosquitoes carried malaria from Georgia up through Pennsylvania, as pioneers took the disease westward. Clearing forests and grasslands and digging canals opened up vast new environments for the mosquitoes. Some Africans were protected because they were heterozygous (carriers) for sickle cell disease, but Caucasians were vulnerable, and Native Americans the most vulnerable of all. Settlers from France and Spain brought malaria to Louisiana and Texas, respectively. By 1850, summertime epidemics of malaria stretched from Florida up through the middle of New England and even into the valleys of California. By the 1920s, public health workers began to use dynamite to dig drainage ditches to remove the mosquitoes' habitat (**figure 16B**).

By the beginning of World War II, malaria in the United States was largely confined to the South, but watery conditions around military training grounds again spread the disease. In 1942, the Office of Malaria Control in War Areas began a huge effort to eradicate the disease. Nearly 5 million homes were sprayed with DDT, and residents used mosquito netting at night, as is common today in parts of Africa. By 1951, the disease was gone. The Office of Malaria Control in War Areas eventually became the Centers for Disease Control and Prevention. Today a thousand or so cases are reported each year in the country, nearly all from travelers to areas where malaria is endemic. Had malaria not been stopped in the United States, natural selection might have favored carriers of sickle cell disease.

Figure 16B Stopping the spread of malaria. In the 1920s, public health workers blew up tree stumps to drain the standing water in which mosquito larvae thrive. *Centers for Disease Control and Prevention (CDC)*

CF carriers enjoy the mixed blessing of balanced polymorphism. They do not have enough abnormal chloride channels to cause the labored breathing and clogged pancreas of CF, but they have enough of a defect to block the cholera toxin. During cholera epidemics throughout history, individuals carrying mutant CF alleles had a selective advantage, and they disproportionately transmitted those alleles to future generations.

Because CF arose in western Europe and cholera originated in Africa, an initial increase in CF heterozygosity may have been a response to a different diarrheal infection—typhoid fever. The causative bacterium, *Salmonella typhi*, rather than producing a toxin, enters cells lining the small intestine—but only if CFTR channels are present. The cells of people with severe CF manufacture CFTR proteins that never reach the cell surface, and therefore bacteria cannot get in. Cells of CF carriers admit some bacteria. Protection against infections that produce diarrhea may therefore have kept CF in populations.

Key Concepts Questions 16.6

*a. Define *natural selection*.

*b. Distinguish positive from negative natural selection.

*c. What is artificial selection?

*d. Give two reasons why disease-causing alleles do not vanish from populations.

e. Give an example of how balanced polymorphism maintains an inherited disease in a population.

16.7 Eugenics

We usually think of artificial selection in the context of Darwin's pet pigeons or purebred cats, dogs, and horses. We also practice artificial selection through control of our own reproduction, through individual choices as well as at the societal level. If this behavior alters allele frequencies in subsequent generations, then it is affecting microevolution.

People can, unknowingly, control the genes in their offspring with reproductive choices. They do this by seeking mates with certain characteristics, and by choosing egg or sperm donors with particular traits for use in assisted reproductive technologies (see chapter 23). On a societal level, **eugenics** refers to programs or policies that control human reproduction with the intent of changing the genetic structure of the population. The goal is quite different from that of medical genetics, which is to relieve suffering (see **Bioethics**). Eugenics works in two directions. Positive eugenics creates incentives for reproduction among those considered superior; negative eugenics interferes with reproduction of those judged inferior. Obviously, eugenic measures are highly subjective.

The word *eugenics* was coined in 1883 by Sir Francis Galton to mean "good in birth." He defined eugenics as "the science of improvement of the human race germplasm through better breeding." One vocal supporter of the eugenics movement was Sir Ronald Aylmer Fisher. In 1930, he tried to apply the principles of population genetics to human society, writing that those at the top of a society tend to be "genetically infertile," producing fewer children than the less-affluent classes. This difference in family size, he claimed, was the reason why civilizations ultimately topple. He offered several practical suggestions to remedy this situation, including state monetary gifts to high-income families for each child born to them.

Early in the twentieth century, eugenics focused on maintaining what was thought to be purity. Ironically, one prominent geneticist, Luther Burbank, realized the value of genetic diversity at the beginning of a eugenic effort. Burbank, known for selecting interesting plants and crossing them to breed plants with useful characteristics, in 1906 applied his ideas to people. In a book called *The Training of the Human Plant*, he encouraged immigration to the United States so that advantageous combinations of traits would appear as the new Americans interbred. Burbank's plan ran into problems, however, at the selection stage, which allowed only those with "desirable" traits to reproduce. Who decides which traits are desirable?

On the East Coast of the United States, Charles Davenport led the eugenics movement. In 1910, he established the Eugenics Record Office at Cold Spring Harbor, New York. There he headed a massive effort to compile data from institutions, prisons, circuses, and general society. He attributed nearly every trait rather simplistically to a single gene, such as "feeblemindedness," "criminality," "promiscuity," and "social dependency." A famous case was that of 17-year-old Carrie Buck. In 1927, she was tried as a criminal in her hometown of Charlottesville for having a mother who lived in an asylum for the feebleminded and having a daughter out-of-wedlock (following rape) who was also deemed feebleminded, as was Carrie herself, although she was a B student in school. Ruled Sir Oliver Wendell Holmes Jr., "*three generations of imbeciles are enough.*" Carrie Buck was the first person sterilized to prevent having another "socially inadequate offspring." A crude pedigree drawn at the time showed Carrie Buck and her "inherited trait" of feeblemindedness.

Other nations practiced eugenics. From 1934 until 1976, the Swedish government forced certain individuals to be sterilized as part of a "scientific and modern way of changing society for the better," according to one historian. At first, only people with mental illness were sterilized, but poor, single mothers were later included. The women's movement in the 1970s pushed for an end to forced sterilizations.

In 1994, China passed the Maternal and Infant Health Care Law. It proposed "ensuring the quality of the newborn population" and forbade procreation between two people if physical exams showed "genetic disease of a serious nature" that included intellectual disability, mental illness, and seizures, conditions that are ill-defined in the law and, if inherited, are typically complex and neither dominant nor recessive with carriers. Physical exams would actually miss carriers of the same genetic disease.

Bioethics

Designer Babies: Is Prenatal Genetic Testing Eugenic?

Genetics is sometimes compared to eugenics because DNA-based technologies may affect reproductive choices and can influence which alleles are passed to the next generation. However, medical genetics and eugenics differ in their intent. Eugenics aims to allow only people with certain "valuable" genotypes to reproduce, for the supposed benefit of the population as a whole. The goal of medical genetics, in contrast, is to prevent and alleviate suffering in individuals and families. But the once-clear line between eugenics and genetics is starting to blur as access to prenatal DNA testing and the scope of testing broaden.

For decades prenatal genetic testing focused on detecting only the most common aneuploids—extra or missing chromosomes—or single-gene diseases known to occur in a family. Today it is possible to reconstruct a fetal genome sequence, or selected parts of it, such as mutations that cause single-gene diseases, from several sources:

- Cells of an early embryo (for use in preimplantation genetic diagnosis; see figure 23.6)
- Fetal cells in sampled amniotic fluid
- Cells of the chorionic villi
- Cell-free fetal DNA in the maternal bloodstream

The ability to scrutinize fetal genome sequences could provide information that could theoretically allow a quality control of sorts in terms of which fetuses, with which characteristics, complete prenatal development and are born. Such positive and negative artificial selection could eventually have effects at the societal level.

It is also possible to intervene with a genetic change just as a sperm fertilizes an oocyte. If the introduced DNA integrates into a chromosome, this action would alter the germline—that is, be transmissible to the next generation.

Figure 16C takes a simple view of a complex idea—altering the frequency of inherited traits in a future human population.

Questions for Discussion

1. Is the decreasing incidence of newborns with trisomy 21 Down syndrome a sign of eugenics (see Bioethics in chapter 13)?

Figure 16C **Designer babies.** Will widespread use of genetic technologies to create or select children with characteristics that parents desire have eugenic effects? *Finn Brandt/Vetta/Getty Images*

2. Is genetic manipulation to enhance an individual a eugenic measure?
3. The first researcher to claim to have edited the genomes of human fertilized ova introduced a mutation that blocks HIV from entering cells, as a test case to see if the technology works. Explain how this intervention differs from selecting cells from an embryo or fetus to identify a specific genotype.
4. Do you think that eugenics should be distinguished from medical genetics based on intent, or can widespread genetic testing to prevent disease have an effect on the population that is essentially eugenic?

Another guise of eugenics is war and aggression, if the fighting groups differ genetically. Throughout history, war and conflict have altered gene pools, sometimes dramatically. These effects are eugenic when they take the form of rape of women of one group by men from another, with the intent of "diluting" the genes of the rape victims. In recent years in Rwanda, Congo, and Darfur, the conquerors claimed that their intent was to diminish the genetic contributions of their victims and spread their own genes. That statement is a working definition of eugenics.

Humanity arose in Africa some 200,000 years ago. As pockets of peoples spread across the globe while many remained in Africa, our behaviors and intelligence introduced culture. Chapter 17 explores some of our journeys, through clues to the past found in the sequences of our DNA.

Key Concepts Questions 16.7

*a. What is eugenics?
*b. How do eugenics and medical genetics differ?
*c. Distinguish between positive and negative eugenics.

Study Guide

Summary

16.1 Population Matters: Steel Syndrome in East Harlem

1. The conditions that counter Hardy-Weinberg equilibrium—nonrandom mating, migration, genetic drift, mutation, and natural selection—create population substructures that can be valuable to consider in health care.
2. Researchers use haplotypes of **identical by descent** gene variants or SNPs to trace ancestry.
3. **Runs of homozygosity** (homologs with the same identical by descent haplotypes) indicate that two individuals shared a recent ancestor. **Endogamy** is marriage within a community where people may be related without knowing so.
4. Recognizing the higher prevalence of Steel syndrome among people of Puerto Rican descent living in East Harlem compared to the broader Hispanic population illustrates the value of recognizing a population genetic substructure in making accurate diagnoses and choosing optimal treatments.

16.2 Nonrandom Mating

5. Hardy-Weinberg equilibrium assumes all individuals mate with the same frequency and choose mates without regard to phenotype. This rarely happens. We choose mates based on certain characteristics, and some people have many more children than others.
6. DNA sequences that do not cause a phenotype important in mate selection or reproduction may be in Hardy-Weinberg equilibrium.
7. Consanguinity increases the proportion of homozygotes in a population, which may lead to increased incidence of recessive illnesses or traits.

16.3 Migration

8. Migration can lead to **admixture,** which is the introduction of gene variants into a population.
9. **Clines** are changes in allele frequencies from one area to another. They reflect geographical barriers or linguistic differences and may be abrupt or gradual.

16.4 Genetic Drift

10. **Genetic drift** is the random fluctuation of allele frequencies from generation to generation. New allele frequencies may result from chance sampling.
11. A **founder effect** occurs when a few individuals found a settlement and their alleles form a new gene pool, amplifying their alleles and eliminating others.
12. A **population bottleneck** is a narrowing of genetic diversity that occurs after many members of a population die and the few survivors rebuild the gene pool.
13. Founder effects and population bottlenecks can amplify the impact of genetic drift.

16.5 Mutation

14. Mutation continually introduces new alleles into populations. It can occur as a consequence of DNA replication errors.
15. The **genetic load** is the collection of deleterious recessive alleles in a population.

16.6 Natural Selection

16. Environmental conditions influence allele frequencies via **natural selection**. Alleles that do not enable an individual to reproduce in a particular environment are selected against and diminish in the population, unless conditions change. Beneficial alleles are retained.
17. In **balanced polymorphism,** the frequencies of some deleterious alleles are maintained when heterozygotes have a reproductive advantage under certain conditions.

16.7 Eugenics

18. **Eugenics** is the control of individual reproduction to serve a societal goal.
19. Positive eugenics encourages those deemed superior to reproduce. Negative eugenics restricts reproduction of those considered inferior.
20. Some aspects of genetic technology affect reproductive choices and allele frequencies, but the goal is to alleviate or prevent suffering, not to change society.

Key Terms

admixture New combination of gene variants that arises when individuals from two previously distinct populations have children together.

balanced polymorphism (pŏl′ē-môr′fizəm) Maintenance of a harmful recessive allele in a population because the heterozygote has a reproductive advantage.

cline (klīn) An allele frequency that changes from one geographical area to a neighboring one.

endogamy Marrying within a population group.

eugenics (yōō-jĕn′ĭks) The control of individual reproductive choices to achieve a societal goal.

founder effect An accelerator of genetic drift in which a few individuals found a new settlement, perpetuating a subset of alleles from the original population.

genetic drift Changes in allele frequencies from generation to generation in a population, due to chance.

genetic load The collection of deleterious recessive alleles in a population.

identical by descent Linked DNA sequences that are transmitted together over generations.
natural selection Differential survival and reproduction of individuals with particular phenotypes in particular environments, which may alter allele frequencies in subsequent generations.
population bottleneck A large decrease in population size resulting from an event that kills many members of a population, followed by restoration of population numbers; it accelerates genetic drift.
run of homozygosity Regions of the genome in which contiguous SNPs (single nucleotide polymorphisms) are homozygous, indicating a shared ancestor with another person with the same pattern.

Review Questions

1. What does a "run of homozygosity" indicate?
2. Give examples of how nonrandom mating, migration, a population bottleneck, and a mutation can each alter allele frequencies from Hardy-Weinberg equilibrium.
3. How might a mutant allele that causes an inherited illness in homozygotes persist in a population?
4. Why can increasing homozygosity in a population be detrimental?
5. How does a knowledge of history, sociology, and anthropology help geneticists interpret allele frequency data?
6. Distinguish how a founder effect differs from a population bottleneck.
7. Give examples of negative and positive selection acting on the same population.
8. What do artificial selection and natural selection have in common? How do they differ?
9. Name an inherited disease allele that protects against an infectious illness.
10. Explain how genetic drift, nonrandom mating, and natural selection may interact.
11. What effect, if any, might climate change have on the carrier frequency for sickle cell disease?
12. Cite three examples of eugenic actions or policies.

Applied Questions

1. Why do more Asian Americans have lactose intolerance than European Americans?
2. Is Dor Yeshorim, the program of screening Jewish populations for carriers of single-gene diseases more prevalent among them, an example of positive or negative natural selection?
3. What effect might dating apps have on the nonrandom mating aspect of population genetic change?
4. Before the year 1500, medieval Gaelic society in Ireland isolated itself from the rest of Europe, physically and culturally. Men in the group, called "descendants of Niall," inherited Y chromosomes from a single ancestor. In the society, men took several partners, and sons born out of wedlock were fully accepted. Today, in a corner of northwest Ireland, one in five men has the "descendant of Niall" Y chromosome. In all of Ireland, the percentage of Y chromosomes with the Niall signature is 8.2 percent. In western Scotland, where the Celtic language is similar to Gaelic, 7.3 percent of the males have the Niall Y. In the United States, among those of European descent, it is 2 percent. Worldwide, the Niall Y chromosome makes up only 0.13 percent of the total. What concept from the chapter do these findings illustrate?
5. Indicate whether each of the following situations, examples, or findings illustrates nonrandom mating, migration, genetic drift, mutation, natural selection, or a combination of these factors.
   a. The "granny hypothesis" holds that grandparents who do not develop dementia contribute to the intellectual development of their grandchildren, increasing the likelihood of the grandchildren surviving to reproduce. The evidence for this effect is that variants of 11 genes that lower the risk of vascular (blood-vessel-associated) dementia are more prevalent among the aged.
   b. Ten castaways are shipwrecked on an island. The first mate has blue eyes, the others brown. A coconut falls by chance on the first mate, killing him.
   c. The Old Order Amish of Lancaster, Pennsylvania, have more cases of polydactyly (extra fingers and toes) than the rest of the world combined. All affected individuals descend from the same person, in whom the dominant mutation originated.
   d. The *CCR5* mutation discussed in the chapter 19 opener that keeps HIV out of human cells also blocks infection by the bacterium that causes plague. Seven centuries ago, in Europe, the "Black Death" plague epidemic increased the protective allele in the population. Today the mutation makes 3 million people in the United States and the United Kingdom resistant to HIV infection.

e. African Americans are more likely to develop a form of end-stage kidney disease associated with elevated blood pressure than Europeans. Two variants in a gene on chromosome 22, called *ApoL1*, cause the kidney disease. The encoded protein is secreted into the blood, but only the forms in the African Americans who have the kidney disease also kill the parasites that cause African sleeping sickness. The mutations persist because they protect against African sleeping sickness.

6. Which principles discussed in this chapter do the following science fiction plots illustrate? Some plots may illustrate more than one concept.

   a. In the novel and film *SevenEves,* the moon shatters into seven pieces. The Earth's human population has two years to prepare for the pieces to smash into the surface and make the planet uninhabitable for centuries. Some people already living on huge space stations survive, and others are selected to join them. Everyone else dies under the barrage of moon junk and intense heat. Above, on the "Cloud Ark," the human species dwindles, but eventually resurges from seven surviving women, with help from assisted reproductive technologies to make babies. Five thousand years after the moon blows up, the human population, ready to inhabit a healed Earth, has resurged to 3 billion.

   b. In *The Time Machine,* set in the distant future on Earth, one group of people is forced to live on the surface while another group is forced to live in caves. Over many generations, the groups diverge. The Morlocks that live below ground have dark skin, dark hair, and behave aggressively. The Eloi that live above ground are blond, fair-skinned, and meek.

   c. In *Children of the Damned,* all of the women in a small town are suddenly impregnated by genetically identical beings from another planet.

   d. In Dean Koontz's novel *The Taking,* giant mutant fungi kill nearly everyone on Earth, sparing only young children and the few adults who protect them. The human race must reestablish itself from the survivors.

   e. In the television series *Wayward Pines,* most present-day humans die in the aftermath of environmental collapse triggered by climate change. The survivors, over the ensuing centuries, become increasingly ferocious, although they maintain human family structures.

7. By which mechanisms discussed in this chapter do the following situations alter Hardy-Weinberg equilibrium?

   a. In ovalocytosis, a protein that anchors the red blood cell plasma membrane to the cytoplasm is abnormal, making the membrane so rigid that parasites that cause malaria cannot enter.

   b. In the mid-eighteenth century, a multi-toed male cat from England crossed the sea and settled in Boston, where he left behind many kittens, about half of whom also had extra digits. People loved the odd felines and bred them. Today, in Boston and nearby regions, multi-toed cats are much more common than in other parts of the United States.

   c. Many slaves in the United States arrived in groups from Nigeria, which is an area in Africa with many ethnic subgroups. They landed at a few sites and settled on widely dispersed plantations. Once emancipated, former slaves in the South were free to travel and disperse.

   d. About 300,000 people in the United States have Alzheimer disease caused by a mutation in the presenilin-2 gene. They all belong to five families that migrated from two small villages in Germany to Russia in the 1760s and then to the United States from 1870 through 1920.

8. Zoroastrianism is an ancient religion that arose from a tribe in Iran (Persia) around 1200 B.C.E. who followed the prophet Zoroaster. When Islam began to spread in the year 767 C.E., a small group of Zoroastrians, facing religious persecution, sailed to the state of Gujarat, in India. Male Persians started families with Indian women. Today only about 61,000 Zoroastrians live in India, with 40,000 more in southern Pakistan. They have also settled in the Middle East, Canada, North America, Britain, and Australasia. The Indian population of Zoroastrians has declined by 12 percent every decade, while the population of all of India has surged by 21 percent. Researchers compared areas of the genomes of Zoroastrians in India today to the same genome regions among their neighbors.

   a. At what point did admixture occur in the history of the Zoroastrians?
   b. Name two types of genetic evidence that would reflect the social isolation of the Zoroastrians in India.
   c. Genetic studies of the Indian Zoroastrians show signs of consanguinity and endogamy. How do these two characteristics of populations differ?

9. Between 7,000 and 5,000 years ago, the diversity of DNA sequences of human Y chromosomes plummeted. Researchers reached this conclusion by extrapolating from contemporary DNA sequence comparisons using mutation rates of specific genes as a molecular clock. They also considered geographical and historical information. Two undergraduates at Stanford University applied computational analysis to deduce what might have happened to men during that time to make the Y chromosomes of their descendants so alike. They began with the fact that groups in Africa, Europe, and Asia at that time were patrilineal—the male ancestral line determined membership, and within families, the Y chromosomes were remarkably uniform. The computer simulations that diminished each group did not lead to the present-day uniformity, but simulations that removed entire groups did. The students concluded that war could have accounted for the modern similarity of many Y chromosomes. Which of the five forces driving microevolutionary change does this scenario illustrate?

Case Studies and Research Results

1. When 22 people in their fifties and sixties discovered through DNA ancestry testing that they were half-siblings, they traced their conception to a fertility doctor who had used his own sperm to inseminate their mothers. The doctor's clinic had been shut down in 2009, and a sample of his semen had been saved but never tested because only a few offspring had been identified. When the 22 offspring filed a lawsuit in 2018, a Dutch court supported their case, claiming that the doctor had told the mothers that he was using donor sperm. He had used his own sperm for at least 200 women, over many years. The man's DNA was tested from the sample and confirmed what the ancestry testing had revealed, as did the DNA of his one acknowledged son.

   a. Which of the five factors that influence microevolutionary change described in the chapter does this case illustrate?
   b. What factors should be considered in setting a limit on the number of times a donor's sperm can be used?
   c. Is the use of donor sperm eugenic? Cite a reason for your answer.

2. Section 10.4 introduced the disease kuru, contracted from exposure to prions in a cannibalism ritual in the 1950s in Papua New Guinea. A study of the prion protein gene among 30 elderly Fore women who had eaten contaminated brains revealed that 23 of them were heterozygotes, with a valine encoded at amino acid position 129 from one chromosome 20 and a methionine from the homolog. Decades later, in the United Kingdom, people who died of mad cow disease were all homozygotes. Their prion proteins had only methionine at position 129. What chapter concept do these studies, considered together, illustrate?

3. Lana seemed to be a healthy newborn. She reached developmental milestones ahead of schedule, trying to lift her head up at only 3 weeks. But then she rapidly lost skills. Her head flopped, she stopped trying to turn over, and her arms and legs became spastic. When she no longer made eye contact, her anxious parents took her to the pediatrician. The doctor referred the family to a pediatric neurologist, who was puzzled. "She has all the symptoms of Canavan disease, but that can't be. She's not Jewish." Explain how the neurologist was incorrect.

CHAPTER 17

Human Ancestry and Evolution

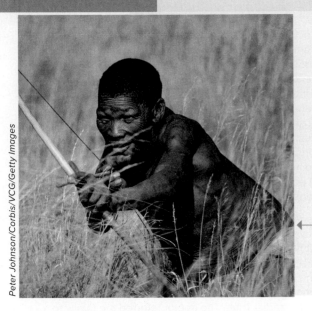

The genomes of the Khoisan, the modern people whose roots in terms of genetic variation go back the farthest, are diverse. Their gene variants reflect their long adaptation to a hunter-gatherer lifestyle. They live in the Kalahari Desert in Botswana and Namibia.

Learning Outcomes

17.1 Human Origins
1. State how DNA sequences indicate ancestry.
2. Explain why we do not have DNA evidence from *Australopithecus* and *Homo* species that dwelled in Africa.
3. Describe what might have happened to Neanderthals and Denisovans, according to evidence in modern human genomes.
4. Explain how DNA analysis of human remains reveals ancient migration patterns.

17.2 Methods to Study Molecular Evolution
5. Explain how DNA and protein sequences and chromosome banding patterns hold clues to evolution.
6. Describe how a "molecular clock" estimates the passage of time.
7. Explain how mitochondrial DNA and Y chromosome sequences are used to track human ancestry.
8. Explain how haplotypes provide clues to ancient migrations.

17.3 The Peopling of the Planet
9. Explain the meaning of mitochondrial Eve.
10. Describe the expansion of groups of people out of Africa.

17.4 What Makes Us Human?
11. Compare human genomes to those of other primates.
12. List genes that distinguish us from our closest relatives.

The BIG Picture

Our genes and genomes hold clues to our deep past and our present diversity. Ancestors of anatomically modern humans (us) interbred with archaic humans, who died out. How will our species continue to evolve?

Indigenous Peoples

It is difficult to imagine a group of people today who do not have cell phones, internet access, or other signs of modern human existence, but about 5,000 such groups are scattered over the planet. An indigenous group of people can trace its ancestry back farther in a geographical region than any other group and has retained its unique culture, social organization, and/or language. Isolation from colonists also separated the gene pools of indigenous peoples, preserving signs of adaptations to past ways of life. Indigenous groups are also called "native" or "first" peoples.

Today less than 5 percent of the world population is indigenous, accounting for about 370 million individuals. They live in 90 nations in groups that range from just a few dozen people to sizable portions of a country's population. Some people live in distinct tribes yet go to school and work and dress just like anyone else. But some indigenous tribes are not very different in lifestyle from their hunter-gatherer ancestors.

Indigenous peoples living today include the Khoisan and Pygmies of Africa, the Etas of Japan, the Hill People of New Guinea, and the Brazilian Arawete, who number only 130 individuals. A fascinating look into the past comes from the genomes of the Khoisan, the modern people whose roots go back the farthest. These hunter-gatherers live today, as they have for millennia, in the Kalahari Desert in southern Africa. They are also known as "San" or "Bushmen," and their language uses "click" sounds.

Researchers compared the complete genome sequences of a Khoisan man named !Gubi to that of Archbishop Desmond Tutu, a well-known South African civil rights activist and a member of the majority Bantu group, as well as to partial genome sequences of three other Khoisan who live near each other. The results indicate that the great genetic diversity from which humanity sprung in Africa persists in the Khoisan today. Their genomes are as different from each other as a modern European genome is from that of a modern Asian genome. The four Khoisan genomes and Desmond Tutu's differ from each other at more than a million places. Comparing the Khoisan genomes to those of later arriving groups may indicate adaptations to an agricultural way of life among more recent Africans.

Khoisan gene variants reflect their lifestyle in the desert. A variant of the actinin-3 muscle gene promotes sprinting over distance running. A gene variant that encodes a chloride channel conserves water. The "bitter taste" gene variant would have enabled a hunter-gatherer to avoid poison and perhaps locate medicinal plants. The Khoisan are also defined by what is *not* in their genomes. They do not have a gene variant that in other populations protects against malaria. Selection would have ignored it in the dry climate where malaria-bearing mosquitoes cannot live.

17.1 Human Origins

Who are we, and where did we come from? We have sparse evidence of our beginnings—pieces of a puzzle in time, some out of sequence, many missing. Traditionally, paleontologists (scientists who study evidence of ancient life) have consulted the record in the Earth's rocks—fossils—to glimpse the ancestors of *Homo sapiens sapiens*, our species. Researchers assign approximate ages to fossils by observing which rock layers fossils are in, and by extrapolating the passage of time from the ratios of certain radioactive chemicals in surrounding rock. But harsh environments destroy most remains of past life.

The DNA sequences of modern organisms provide clues to the past because they change (mutate) over time. Mutation rate provides a "molecular clock." Another source of genetic information over time is the frequencies of gene variants (alleles) that change in populations due to the forces of nonrandom mating, migration, genetic drift, and natural selection, the topics of chapter 16.

The premise behind DNA sequence comparisons is that the more closely related two species are, the more similar their sequences. Similar DNA sequences are more likely to come from individuals or species sharing ancestors than from the exact same set of spontaneous mutations occurring by chance. Rarely, DNA from ancient specimens can add to what we know from DNA sequences of modern organisms. However, the heat of equatorial areas in Africa destroyed much DNA from the past.

Treelike diagrams are used to depict evolutionary relationships, based on fossil evidence and/or inferred from DNA sequence similarities. Typically, the oldest species form the trunk of the tree, with more recent species diverging like branches. The branchpoints represent the most recent common (shared) ancestor. Humans and chimps diverged from a shared ancestor; humans didn't evolve directly from chimps. Similarly, two second cousins share great-grandparents, but one cousin did not descend from the other.

This chapter explores human origins and considers how genetic and genomic evidence adds to our view of our evolution. It concludes with a look at more recent events in our ancestry and a consideration of which characteristics distinguish our species.

Our Place in the Primate Family Tree

A species includes organisms that can successfully produce healthy offspring only among themselves. *Homo sapiens sapiens* ("the wise human") probably first appeared about 200,000 years ago.

We and animals ancestral only to us are members of a taxonomic (biological classification) "tribe" called the Hominini that diverged from an ancestor we shared with other African apes (gorillas, chimpanzees, and bonobos) about 6 million years ago, in Africa. **Figure 17.1** shows our place among our closest primate relatives.

We know of at least three types of Hominini (also called hominins) who lived shortly (in geological time) after the split from our ancestor shared with chimpanzees: *Ardipithecus kadabba* from Ethiopia, *Sahelanthropus tchadensis* from Chad, and *Orrorin tugenensis* from Kenya. Fossil evidence for them is scant. Another hominin, *Ardipithecus ramidus*, lived more recently, about 4.4 million years ago. "Ardi" was discovered in the Afar rift valley of Ethiopia in 1994, an area whose rocks hold fossilized remains of human ancestors going back nearly 6 million years. Ardi was taller than more ancient hominins and was partly able to walk on two legs (bipedal), perhaps while on the forest floor, but could also use all four limbs to climb trees. Ardi likely ate plants.

Fossil evidence is more complete for our ancestors who lived 2 to 4 million years ago. They walked fully upright and conquered vast new habitats on the plains. Several species of a hominin called *Australopithecus* lived at this time, probably following a hunter-gatherer lifestyle. More than one species could coexist because the animals lived in small, widely separated groups that probably never came into contact. The australopithecines were gradually replaced with members of our own genus, *Homo*. The following sections introduce a few of these ancestors known from rare fossil remains and what our computer modeling and imaginations can fill in. DNA evidence becomes important when we reach the archaic humans.

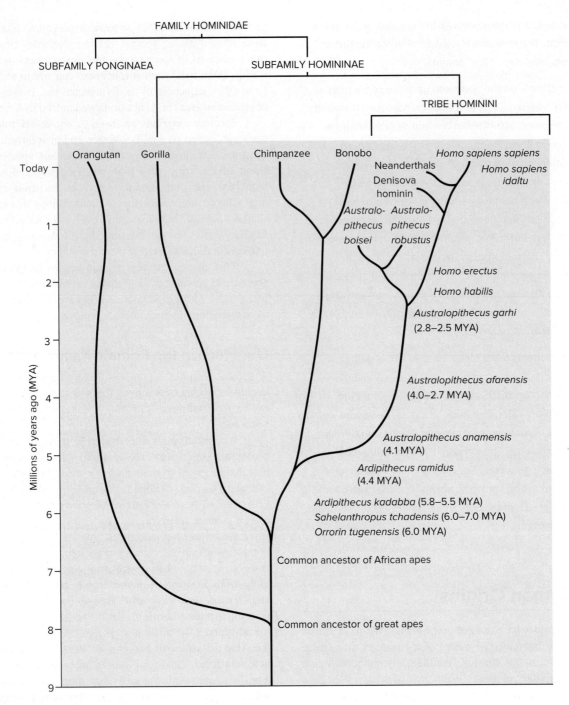

Figure 17.1 Our place on the human family tree. Our most recent ancestors, the archaic humans, are just below us at the upper right of the tree.

Australopithecus

The australopithecines had a mix of apelike and humanlike characteristics. They had flat skull bases, as do all modern primates except humans. Their legs were shorter than ours but their arms were longer. The angle of preserved bones from the pelvis and discovery of *Australopithecus* fossils with those of grazing animals suggest that this ape-human had left the forest for grasslands. Paleontologists have pieced together what little we know about the australopithecines from fossil finds. These hominins lived too long ago to have left DNA evidence.

The oldest species of australopithecine known, *Australopithecus anamensis*, lived about 4.1 million years ago. Another species, *Australopithecus afarensis*, left a partial skeleton, named Lucy, who lived for 20 years about 3.6 million years ago in the Afar river basin of Ethiopia. Lucy was about 4 feet tall, but partial skeletons from slightly older males of her species stood about 5 feet 3 inches tall and were more humanlike than chimplike. Stone tools found with other members of *A. afarensis* from 3.4 million years ago show distinctive cut marks that indicate these hominins sliced meat from bones and removed, and presumably ate, the marrow. Their jaws were strong enough to crack nuts.

Other fossils offer more clues to australopithecine life. Two parallel paths of footprints, preserved in volcanic ash in the Laetoli area of Tanzania, are contemporary with Lucy. A large and small individual walked close together, with a third following the steps of the larger animal in front. A family? In another fossil find, a 3-year-old girl from 3.3 million years ago, the "Dikika infant," apparently died in a flash flood, perhaps while trying to catch fish in rivulets of the Awash river delta near where Lucy lived 300,000 years earlier.

Toward the end of the australopithecine reign, *Australopithecus garhi* may have coexisted with the earliest members of *Homo*. *A. garhi* fossils from the Afar region date from about 2.5 million years ago. Remains of an antelope found near the australopithecine fossils suggest butchering. The ends of the long antelope bones had been cleanly cut with tools, the marrow removed, meat stripped, and the tongue cleanly sliced off. *A. garhi* stood about 4.5 feet tall, and like the Dikika infant and Lucy, the long legs were like those of a human, but the long arms were more like those of an ape. The small cranium and large teeth hinted at apelike ancestors.

Early *Homo*

By 2.3 million years ago, *Australopithecus* coexisted with hominins of genus *Homo*. We do not know how *Homo* replaced *Australopithecus*, or even whether there was just one or several species of early *Homo*. Some australopithecines were "dead ends" that died off. Based on older fossil evidence paleontologists have described several *Homo* species, but more recent fossil finds that reveal variability in bone structure may lead to revision in the classification to a single but variable type of *Homo*.

An early possible species was *Homo habilis*, who cared intensively for its young in caves. *Habilis* means handy, and this primate was the first to use tools for tasks more challenging than stripping meat from bones. *H. habilis* may have descended from hominins who ate a more varied diet than other ape-humans, enabling them to live in a wider range of habitats.

H. habilis coexisted with and was followed by *Homo erectus*. One of the first *H. erectus* individuals described, "Nariokotome Boy," influenced descriptions of the species as tall and thin for two decades, until discovery of more specimens revealed that *H. erectus* matured faster than we do, but their heights and builds varied. Another famed *H. erectus* fossil, named "Daka" for the place where he was found in the Afar region, is from an individual who lived about a million years ago. He had a shallow forehead, massive brow ridges, a brain about a third smaller than ours, and strong, thick legs. Daka lived on a grassland, with elephants, wildebeests, hippos, antelopes, many types of pigs, and giant hyenas. **Figure 17.2** depicts a partial skeleton of an *H. erectus* individual who lived about 1.8 million years ago in the country of Georgia.

H. erectus left fossil evidence of cooperation, social organization, tools, and use of fire. Fossilized teeth and jaws suggest that they ate meat. The distribution of fossils indicates that they lived in families of male-female pairs (most

Figure 17.2 Reconstructing *Homo erectus*. Paleontologists assembled fossilized bones from several *Homo erectus* individuals from the Dmanisi site in the country of Georgia to try to understand what this hominin looked like and how variable it was. *H. erectus* lived about 1.8 million years ago.
Danita Delimont/Alamy Stock Photo

primates have harems). The male hunted, and the female nurtured the young. They were the first to have an angled skull base that enabled them to produce a greater range of sounds, making speech possible. *H. erectus* fossils have been found in China, Java, Africa, Eurasia, and Southeast Asia, indicating that these animals could migrate farther than earlier primates.

A dwarfed variant of *H. erectus* is known from jaw and tooth fragments from three individuals discovered on the island of Flores in Indonesia. The trio were dated back to about 700,000 years ago. Remains of another individual from the island made headlines when they were discovered in 2003. The remains were initially given too recent a date based on erroneously analyzing charcoal, but were then dated to between 100,000 and 50,000 years ago. She was initially given her own species designation, *H. floresiensis*, or "the hobbit" after the fictional character, but many researchers consider her an offshoot of *H. erectus* with a small body.

Archaic Humans

DNA evidence can flesh out fossil evidence for more recent members of *Homo*. *Homo erectus* may have lived until as recently as 35,000 years ago. Meanwhile, the branching of evolution led to hominins that were more like us and that we call the "archaic humans." We know of two types of archaic humans: the Neanderthals (*Homo sapiens neanderthalensis*) and the Denisovans (*Homo sapiens denisova*) (**figure 17.3**). In contrast to archaic humans, we are "anatomically modern humans."

As some of our ancestors left Africa and expanded beyond that continent, reproductively successful populations would eventually have begun to overlap geographically. If populations interbred, then gene flow would occur, leaving evidence in preserved ancient DNA as well as in the genomes of modern peoples. Both have happened. We probably do not yet know anything about some of the peoples who migrated out of Africa over the past tens of thousands of years. The following sections describe the Neanderthals and the Denisovans. These archaic humans may have branched from a shared ancestor with us as far back as 600,000 years ago or even earlier.

The Neanderthals

The ancestors of the Neanderthals left Africa in separate migrations, beginning about 400,000 years ago. They headed toward Europe and west Asia, but did not establish themselves in these places long enough to have left evidence (as far as we know) until about 230,000 years ago. Their gene pools had time to diverge from the ancestral African genomes, perhaps as natural selection favored adaptations to cooler environments.

Anatomically modern humans first appeared about 200,000 years ago, as a patchwork of peoples in Africa, Europe, and Asia. The earlier migrations from Africa of Neanderthal ancestors explains why modern Africans do not have uniquely Neanderthal DNA sequence variants in their genomes, but Europeans and Asians, whose ancestors met up with the Neanderthals much later, do.

Between 100,000 and 50,000 years ago, our ancestors left Africa and encountered Neanderthals in the Middle East en route to Eurasia. The two types of hominins interbred—a discovery deduced from shared DNA sequence variants. When researchers first began to sequence mitochondrial DNA from Neanderthal bones, they concluded that interbreeding was highly unlikely. Then sequencing entire Neanderthal genomes changed the long-held idea of our separateness.

Neanderthals may have lived as recently as 30,000 years ago, in warm caves in Gibraltar when northern Europe was under ice. They might have inhabited these caves since 100,000 years ago. The fossil record indicates that the Neanderthals vanished, but their gene variants persist in modern humans, suggesting strongly that they may have been assimilated as different groups interbred and Neanderthal fertility declined. Without a time machine, we may never know exactly what happened (see **A Glimpse of History**).

The genome sequences of modern humans and Neanderthals are 99.7 percent identical; the Neanderthal contribution to our ancestry is small. Researchers describe the contribution as the proportion of a person's ancestors who were Neanderthal, going back a certain number of generations or years, not the proportion of the genome. Return to figure 1.9, which shows that 50 percent of a person's ancestry is from the mother and 50 percent from the father. Each grandparent contributes 25 percent, and each great-grandparent contributes 12.5 percent, and so on. Going back 100,000 years ago for Asians, approximately 1.8 percent of ancestors were Neanderthals, and for Europeans the proportion is about 1.5 percent.

Modern humans who lived longer ago have greater Neanderthal ancestry. For example, DNA sequencing of a jaw bone from a modern human who lived about 40,000 years ago in Romania, named "Oase1," revealed 6 to 9 percent Neanderthal ancestors who lived about 100,000 years ago, considerably greater than for us today. The long length of the Neanderthal portion of Oase1's genome suggests that the sexual encounter with an archaic partner may have happened just four to six generations earlier!

Computer simulations that use mutation rates and population data suggest that in a population living at about Oase1's time or slightly earlier, modern humans outnumbered Neanderthals 10 to 1, and they lived as one interbreeding group. Genome sequencing reveals that Neanderthal gene variants in modern human genomes aren't evenly dispersed among the chromosomes. The apparent nonrandomness suggests that natural selection was at play once breeding with modern humans began. That is, when Neanderthals lived in tiny, isolated groups, inbreeding was common, and genetic diversity as well as fertility fell. In small populations, genetic drift would have had a large effect. But once Neanderthals became part of a larger population, natural selection would have acted both negatively to remove harmful gene variants and positively to retain helpful ones.

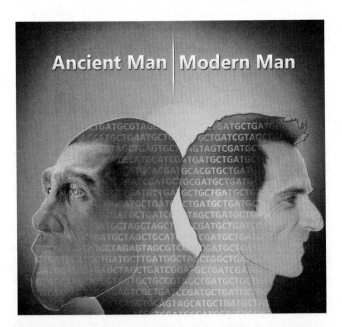

Figure 17.3 Archaic humans. Several types of archaic humans likely coexisted. They probably looked more similar to than different from us. *National Human Genome Research Institute*

A GLIMPSE OF HISTORY

The first bones discovered from Neanderthals came from a limestone cave in Neander Valley, Germany, when quarry workers blasted them out on a summer day in 1856. Assembling the scattered skeletal remains revealed large, brutish individuals with jutting faces and brows and giant jaws and teeth above small chins. Half a century later, discovery of bones from the "Old Man" of La Chapelle-aux-Saints in France led to the depiction of Neanderthals as primitive and slow-witted, stooped perhaps due to arthritis. Another find, of 10 Neanderthals from Shanidar Cave in Iraq whose bones were excavated from 1957 through 1961, included tools and remains of flowers. An old man from that discovery, who researchers named "Nandy," died of what was then very old age—40 to 50 years. Preserved dental plaque on the teeth held starch grains that had been cooked.

The first samples of Neanderthal DNA came from the accidental discovery by cave explorers of bone fragments about 700 feet in from the entrance of a cave called El Sidrón in northwestern Spain, on the westernmost part of the Iberian peninsula. The environment had been warm enough to sustain a Neanderthal community in caves cool enough to preserve their remains, including their DNA. Excavation of the El Sidrón cave began in 2000, and by 2005, researchers had sequenced the first Neanderthal genes from mitochondrial DNA, which is more likely to survive harsh environmental conditions than DNA from cell nuclei.

Neanderthal DNA from a hip bone of a 38,000-year-old female found in Vindija Cave in Croatia was used to reconstruct the first full genome sequence, published in 2010 from researchers at the Max Planck Institute for Evolutionary Anthropology in Leipzig, Germany. They used dental drills to delicately remove pulverized "dust" from bones from three individuals to keep the bones intact for future studies. DNA sequencing in a "clean room" prevented the investigators' DNA contaminating the samples, and other techniques avoided sequencing bacterial or fungal DNA. In 2014 researchers published a second Neanderthal genome sequence from a female who lived 50,000 to 100,000 years ago in the isolated Altai mountains of Siberia. The runs of homozygosity (see figure 16.1) in the woman's genome indicate intensive inbreeding—her parents were siblings.

Figure 17.4 **A Neanderthal phenotype.** Mutations found in Neanderthal DNA suggest that some of them may have had pale skin and red hair. Others had darker hair and skin. *Knut Finstermeier, Max Planck Institute for Evolutionary Anthropology*

The term **introgression** refers to entry of a specific gene variant into a genome from an individual of a different species or subspecies. It is a more expansive form of admixture, which is when individuals from previously separate populations interbreed.

Which genes from Neanderthals persist in modern human genomes? Some Neanderthals had variants of the *MC1R* gene that confers freckles, pale skin, and reddish hair (**figure 17.4**). Before genome sequencing, they were thought to have all had dark hair and eyes. Researchers can propose hypotheses based on genome differences. For example, the Neanderthal Y chromosome is about twice as old as the oldest known modern Y chromosome (from the Mbo people of the Bantu tribe in Cameroon) and has distinct variants in several genes that affect the immune system. Perhaps the immune systems of modern women impregnated with Neanderthal sperm attacked the embryos, leading to miscarriage, which might explain why so few of our gene variants are from Neanderthals. Births of healthy children from matings between Neanderthals and our ancestors must have been rare events.

Neanderthal gene variants persisting in some modern human genomes form about 12,000 haplotypes (sets of linked genes). Identifying the differences in the modern human and Neanderthal genomes enables researchers to reconstruct scenes from prehistory and to understand some modern diseases. That is, variants of several genes that were likely adaptive in Neanderthals may cause disease in modern humans. To investigate this hypothesis, researchers consulted a database called eMERGE (Electronic Medical Records and Genomics) that lists disease diagnostic billing codes and DNA sequence data. They looked for

6,000 Neanderthal haplotypes among the 28,416 people of European descent represented in the database, to see if any diagnoses tracked with specific Neanderthal gene variants. **Table 17.1** lists Neanderthal gene variants identified from this clever study that might affect our health.

One consumer DNA testing company looks for whether customers have zero, one, or two Neanderthal SNPs (single nucleotide polymorphisms) at 1,436 places in the genome. The maximum score possible is 2,872 for having both Neanderthal variants at all 1,436 sites. The company's metric is "you have more Neanderthal variants than __ % of tested people." A few thousand variants is an extremely small portion of a genome. That is, the Neanderthal metric is more entertaining than it is informative or useful.

Gene flow apparently also occurred from anatomically modern humans to Neanderthals. Genome comparisons indicate that modern humans other than our ancestors left Africa about 125,000 to 100,000 years ago, and mated with the isolated Altai Neanderthals in Siberia—but not with the European Neanderthals. Introgression of modern human DNA into Neanderthal genomes might also have occurred in Cyprus, Egypt, Iraq, Israel, Jordan, Lebanon, Syria, and Turkey and the Persian gulf area as Neanderthals moved east to Europe during the interglacial period about 100,000 years ago. Pure Neanderthals may have become extinct because they were about 40 percent less reproductively fit as modern humans.

The Denisovans

The Denisovan story began in 2010, with a description based on DNA extracted from a pinky finger bone discovered in Denisova Cave in the Altai Mountains of southern Siberia. Researchers at first thought that the bone was from a Neanderthal, because Neanderthals had occupied the cave as well. However, sequencing of the mitochondrial DNA in finger bone cells showed that the digit was neither Neanderthal nor modern human. By 2012, researchers had sequenced the genomes of cells from two molars from other Denisovans discovered in different layers of rock in the cave wall, indicating that the teeth had been left at different times. These Denisovan genomes were quite different from each other, suggesting that the cave residents might have lived long enough and left enough offspring to foster some genetic diversity.

In 2018, researchers sequenced the genome of "Denisova 11," whose gene variants revealed her to have been the daughter of a Neanderthal mother and a Denisovan father. Her mitochondrial genome also indicated a Neanderthal mother.

The Denisovan girl or young woman who left the pinky bone, nicknamed "Denise," lived between 32,000 to 50,000 years ago, according to fossil as well as DNA evidence. The genome sequence reveals that Denise had dark skin and brown eyes and hair. Denisovan gene variants in modern populations

Table 17.1	Neanderthal Genes in Modern European Genomes	
Phenotype	**Genes**	**Hypothesis**
Faster blood clotting	SELP (P-selectin)	Benefited Neanderthals in healing from injuries and surviving childbirth. In longer-living modern humans, faster clotting might cause stroke and heart attack.
Innate immunity	TLR1, 6, 10 (Toll-like receptors)	Receptor variants alert immune system to respond to microbes and parasites present 100,000–550,000 years ago. Today they may increase risk of allergies, autoimmunity, and inflammatory diseases.
Tobacco use disorder	SLC6A11 (solute carrier family 6 member 11)	Impaired ability of neurons to take up gamma-aminobutyric acid, a neurotransmitter, from synapses in brain's addiction center.
Thiamine (vitamin B1) transport	SLC35F3 (solute carrier family 35 member F3)	Protein transports vitamin to intestines to metabolize complex carbohydrates that were common in Neanderthal diet, but comprise far less of modern refined-carbohydrate diets.
Urinary problems (infection, incontinence, bladder pain)	STIM1 (stromal interaction molecule 1)	Neanderthal variant not expressed well in brain parts that control bladder function.
Actinic keratosis	Circadian clock genes	Neanderthal adaptations to natural patterns of daylight at higher latitudes predispose modern humans to these precancerous skin lesions.
Depression	Circadian clock genes	Gene variants that cause depression in modern humans are closely linked to circadian rhythm genes that had different variants in Neanderthals.

suggest that these archaic humans lived throughout Asia. Today the people most closely related to the Denisovans are from Papau New Guinea, Melanesia, Oceania, and Polynesia. Some aboriginal Australians have about 3.5 percent of their ancestors who were Denisovans going back about 100,000 years. People who live on Northern Island in Melanesia, in Papua New Guinea, are the closest living relatives of the Denisovans.

The Denisovan cave was also home to the Altai Neanderthals, and likely others. The extensive runs of homozygosity that pepper the Altai Neanderthal genomes indicate siblings having children together, perhaps a consequence of their isolation in an extreme environment. Comparing genomes of the archaic humans to our own indicates that the Denisovans and Neanderthals were more closely related to each other than either was to us and that Neanderthal genes flowed into older Denisovan genomes.

As complicated as this emerging picture is, we do not yet have all of the puzzle pieces—the Denisovan genome clearly indicates admixture with at least one other, yet unknown, type of archaic human. Genes also entered Denisovan and Neanderthal genomes from our very distant ancestors. Because we have such scant evidence, we cannot really know what happened as humanity sorted itself out. Several small populations of archaic humans probably coexisted and eventually mixed for at least 100,000 years before we modern humans, retaining some archaic genes, emerged and persisted. A few ancient gene variants remain in some of us.

Figure 17.5 attempts to capture some of this complex back-and-forth gene flow. A caveat: what we know of the Neanderthals and Denisovans is subject to change with new discoveries and DNA sequences!

Modern Humans

Cave art from about 14,000 years ago indicates that by that time, our ancestors had developed fine-hand coordination and could use symbols. These were milestones in cultural evolution. By 10,000 years ago, people had expanded from the Middle East across Europe, bringing agricultural practices.

Early Farmers

DNA evidence has sketched one scene from the spread of agriculture. Researchers sequenced mitochondrial DNA haplotypes from 21 bodies found in a graveyard in Germany, about 100 miles south of Berlin, and compared them to DNA from 36 modern Eurasian populations. The bodies were from about 7,100 years ago. The comparisons yielded a clear cline, with genetic similarities indicating a long-ago migration of early farmers from Turkey, Syria, Iraq, and other Near Eastern cultures westward from the Balkans, north along the Danube into central Europe—not just the spread of their agricultural techniques by word of mouth. The migration took centuries. According to the DNA, the farmers arrived in Europe and encountered hunter-gatherers descended from the original population from 40,000 B.C.E., and the two groups of people interbred. **Figure 17.6** shows what the ancient farmers from 7,100 years ago might have looked like.

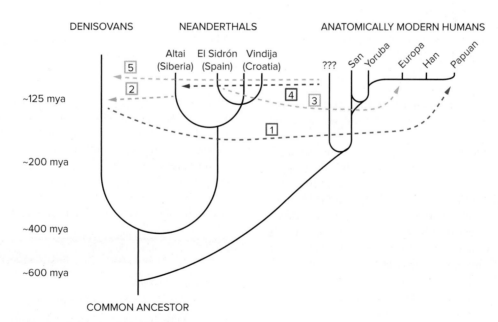

Figure 17.5 Gene flow among archaic and anatomically modern humans. DNA sequence similarities indicate that genes flowed (individuals mated) from Denisovans to some groups of anatomically modern humans (1), from Siberian Neanderthals to Denisovans (2), from Spanish Neanderthals to Europeans (3), and from unknown anatomically modern humans to Siberian Neanderthals (4) and to Denisovans (5). (The abbreviation mya stands for "millions of years ago.") *Source: Adapted from the US National Library of Medicine National Institutes of Health.*

Figure 17.6 Farmers from 7,100 years ago probably looked like modern people. *State Office for Heritage Management and Archaeology Saxony-Anhalt, Karol Schauer*

Ötzi the Ice Man

In 1991, hikers in the Ötztaler Alps of northern Italy discovered an ancient man frozen in the ice (**figure 17.7**). Named "Ötzi the Ice Man," he was on a mountain more than 10,000 feet high 5,200 years ago when he perished. He wore furry leggings, leather suspenders, a loincloth, fanny pack, bearskin cap and cape, and sandal-like snowshoes. He had stained his skin to fashion tattoos, and indentations in his ears suggest that he might have worn earrings. He carried mushrooms that had antibiotic properties. Berries found with him place the season as late summer or early fall. His last meal was ibex and venison.

Ötzi died following a fight. He had a knife in one hand, cuts and bruises, and an arrowhead embedded in his left shoulder that nicked a vital artery. The wound bore blood from two other individuals and his cape had the blood of a third person. Moss found on his body may have been used as wound dressings. When X-rays indicated swelling on the right side of Ötzi's brain, researchers delicately removed two tiny samples of the tissue and analyzed the proteins present. Several were related to blood clotting. Whether Ötzi fell and hit his head or was attacked remains a mystery. However he perished, Ötzi landed in a ditch, where snow covered him and preserved his body, including his DNA.

Ötzi's genome held clues to his health and ancestry. He was lactose intolerant and had type O blood, and had he not been buried in snow immediately following a fight, he might have eventually died of heart disease. Ötzi shared ancestors

Figure 17.7 A 5,200-year-old man. Hikers discovered Ötzi the Ice Man in the Ötztaler Alps of northern Italy in 1991. *AFP/Stringer/Getty Images*

with men living today in Sardinia and Corsica in the Mediterranean area, according to SNP patterns of his Y chromosome. However, Ötzi's mitochondrial genome, indicating maternal ancestry, apparently died out about 5,000 years ago. One interpretation is that some of his male ancestors had left the isolated Alps, migrating toward the Mediterranean sea, while the women who had his mitochondrial genome perished in the local, very harsh environment.

Key Concepts Questions 17.1

*a. Define *hominin*.

b. Describe the australopithecines and early *Homo* species.

c. Describe how researchers compare archaic to modern humans.

d. What has DNA evidence revealed about early farmers and the 5,200-year-old ice man?

17.2 Methods to Study Molecular Evolution

The study of archaic humans and genetic genealogy to investigate present-day relationships use some of the same techniques to analyze and compare DNA sequences. The techniques are introduced here in the context of human evolution and in chapter 18 for deducing how people are related.

Molecules of DNA and protein change in sequence over time as mutations occur and are perpetuated. Knowing the mutation rates for specific genes provides a way to measure the passage of time using a sequence-based "molecular clock." The more alike a gene or protein sequence is between two species or two individuals of the same species, the more closely related the two are presumed to be—that is, the more recently they shared an ancestor. The field of **molecular evolution** compares genomes, DNA or protein sequences, and chromosome banding patterns to deduce how species or individuals are related.

Comparing Genes and Genomes

DNA sequences that encode protein—exons—are often very similar among closely related species, which means that they inherited the gene from a shared ancestor. A change in that gene would not persist in a population unless it provided a reproductive advantage (positive natural selection). At the same time, negative natural selection would have weeded out proteins that did not promote survival to reproduce. DNA sequences that do not affect the phenotype can vary because natural selection does not act on them.

Similar DNA or amino acid sequences found in different species are said to be "highly conserved." Such similarities are the basis of animal models of human disease. **Figure 17.8** compares part of the dystrophin gene, which is mutant in Duchenne muscular dystrophy, among wild type humans, mice, and dogs. The mouse does not become very ill so is not a good model, but "golden retriever muscular dystrophy" (GRMD) is strikingly like the disease in boys. The dogs lose weight rapidly from when they are born and usually do not survive beyond 1 year because their breathing muscles fail. The GRMD mutation is in a splice site that removes an exon in the giant gene, depicted as the blue G in the figure, and as a result the dogs cannot make dystrophin protein. German shorthair pointers with the disease have a full gene deletion, cocker spaniels have a 4-base deletion, Tibetan terriers have a several-exon-long deletion, and Labrador retrievers have an insertion into an intron.

DNA sequences that are similar in closely related species but that do not encode protein often control transcription or translation, and so are also vital and therefore subject to natural selection. In contrast, some genome regions that vary widely among species do not affect the phenotype, and are therefore not subject to natural selection. Within a protein-encoding gene, the exons tend to be highly conserved, but the introns, which are removed from the corresponding mRNA, are not.

Comparing Chromosomes

Before gene and genome sequencing became possible, researchers considered similarities in chromosome banding patterns to assess evolutionary relatedness. Humans share 99 percent of chromosome bands with chimpanzees, but only 35 percent with cats and 7 percent with mice.

Chromosome banding patterns are like puzzle pieces. If both copies of human chromosome 2 were broken in half, we would have 48 chromosomes, as the other three species of apes do, instead of 46. Human chromosome 2 therefore arose from a fusion event. The banding pattern of chromosome 1 in humans, chimps, gorillas, and orangutans matches that of two small chromosomes in the African green monkey, suggesting that this monkey was ancestral to the other primates and its two small chromosomes fused. Most of the karyotype differences among primates are translocations.

We can also compare chromosome patterns between species that are not as closely related as we are to other primates. All mammals, for example, have identically banded X chromosomes. One section of human chromosome 1 that is alike in humans, apes, and monkeys is remarkably similar to parts of chromosomes in cats and mice. A human shares chromosomal segments with a horse, but less with the aardvark, which is the most primitive placental mammal.

The information from comparing chromosome bands using stains is not as precise as information from DNA sequences because a chromosome band can contain some different genes from those within a band at a corresponding locus in another species' genome. In contrast, DNA probes used in a FISH analysis

Species	Part of dystrophin gene	# differences
Human wild type	T A T G T G T A T G T G T T T T A G G C C A G A C C T A T T T G A C T G	
Mouse wild type	T G T A T G T G T T T G T T T C A G G C C C G A C C T G T T T G A T T G G	8
Dog wild type	T A T G T G T G T G T G T T T C A G G C C A G A C C T G T T T G A T T G G	4
Golden retriever	T A T G T G T G T G T G T T T C G G G C C A G A C C T G T T T G A T T G G	5

Figure 17.8 The dystrophin gene. The gene that, when mutant, causes Duchenne muscular dystrophy is highly conserved among 48 mammalian species. Different dog breeds have specific mutations in the gene. Golden retrievers suffer severe illness. *Source: Adapted from Current Genomics, Aug 2013, 14(5):330-342, PubMed Central.*

highlight specific genes (see figure 13.7). FISH can indicate direct correspondence of gene order, or **synteny,** between species, which indicates close evolutionary relationships. For example, versions of eleven genes are closely linked on human chromosome 21, mouse chromosome 16, and chromosome U10 in cows.

Comparing Proteins

Many different types of organisms use the same proteins, with only slight variations in amino acid sequence. The similarity of protein sequences is compelling evidence for descent from shared ancestors—that is, evolution. Many proteins in humans and chimps are alike in 99 percent of their amino acids, and several are identical. When analyzing a human gene's function, researchers routinely consult databases of known genes in many other types of organisms.

Cytochrome c is one of the most ancient and well-studied proteins. It helps to extract energy from nutrients, in the mitochondria. The more closely related two species are, the more alike their cytochrome c amino acid sequence. Human cytochrome c is more like that of the horse than it is that of a kangaroo, which is more distantly related. The human protein is identical to chimpanzee cytochrome c. The homeotic genes, discussed in Clinical Connection 3.1, are another class of genes that has changed little across evolutionary time.

The similarity of proteins among different species has medical applications. Consider insulin, which enables cells to take up glucose from the blood. Inability to make insulin causes type 1 diabetes mellitus.

Insulin is two linked peptide chains, one 21 amino acids and the other 30 amino acids long. Human insulin differs from pig and dog insulin by just one amino acid, from cow insulin by three, and from cat insulin by four. The source of insulin is important, because even though the amino acid sequences are very similar among species, the molecules differ in how quickly the insulin is absorbed into cells, how soon after absorption the level peaks, and the duration of the effect—all important factors when trying to stabilize blood glucose levels. Cats, for example, metabolize insulin twice as fast as dogs or people because their metabolisms are much faster.

Even though insulin is highly conserved, the source is important when it is used as a drug to treat diabetes. Until 1978, when human insulin became available using recombinant DNA technology (see section 21.2), people with diabetes used insulin from cows or from pigs.

Molecular Clocks

A clock measures the passage of time as its hands move through a certain degree of a circle in a specific and constant interval—a second, a minute, or an hour. In the same way, an informational molecule can be used as a "molecular clock" if its building blocks are replaced at a known and constant rate.

The similarity of nuclear DNA sequences in different species can be used to estimate the time when the organisms diverged from a common ancestor, if the rate of base substitution mutation is known. For example, many nuclear genes studied in humans and chimpanzees differ in 5 percent of their bases, and substitutions occur at a rate of 1 percent per million years. Therefore, 5 million years have presumably passed since the two species diverged. Mitochondrial DNA (mtDNA) is also tracked to estimate time of divergence from shared ancestors.

Time scales based on fossil evidence and molecular clocks can be superimposed on evolutionary tree diagrams constructed from DNA or protein sequence data. However, evolutionary trees can become complex when data can be arranged into different tree configurations. A tree for 17 species of mammals, for example, can be constructed in 10,395 ways! With new sequence information, tree possibilities change.

Researchers use computational methods to select out of many possibilities an evolutionary tree that is likely to best represent what really happened. An algorithm connects all evolutionary tree sequence data using the fewest possible number of mutational events to account for observed DNA base sequence differences. Because mutations are rare events, the tree that requires the fewest mutations is most likely to represent reality. Statistical methods can account for different mutation rates for different genes and for different positions of mutations within genes, and can impute estimates from incomplete data.

Using Mitochondrial and Y Chromosome DNA Sequences

To track ancient human migration patterns, researchers use the types of genetic markers that are also used in genetic genealogy:

- single nucleotide polymorphisms (SNPs)
- **short tandem repeats (STRs)** that are 2 to 10 bases long
- other copy number variants (CNVs)

Following markers in mitochondrial DNA provides information on the female lineage, and markers on the Y chromosome, the male lineage, but as **figure 17.9** illustrates, these ancestors are only a few of many. Markers on the autosomes can fill in other ancestors and are called "ancestry informative markers." The contribution of an ancestral DNA sequence decreases as the number of generations increases, which is why Neanderthal DNA is only a small percentage of the genomes of modern Europeans.

MtDNA is ideal for monitoring recent events because it mutates faster than DNA in the nucleus. Its sequences change by 2 to 3 percent per million years. Mutations accumulate faster because mtDNA has no DNA repair. Another advantage of typing mtDNA is that it is more abundant than nuclear DNA because mitochondria have several copies of it, and a cell has many mitochondria. It is also more able to withstand harsh environments than nuclear DNA. These are the reasons why the first studies of DNA from archaic humans examined mtDNA.

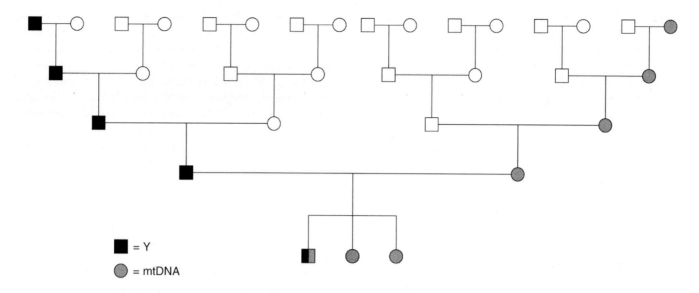

Figure 17.9 Haplogroups. Y chromosome and mtDNA sequences that define haplogroups represent only some of a person's ancestors. Considering autosomal sequences can capture the contributions of other relatives, represented as the unfilled symbols in the interior of this pedigree. The symbol for the man in the most recent generation has two shades that represent his Y chromosome DNA (black) and mitochondrial DNA (gray).

For many years, researchers compared certain rapidly mutating DNA sequences in mitochondria, but today they compare complete mitochondrial genomes, or "mitogenomes." A study of mitogenomes of the Ashkenazi Jews, for example, showed that many maternal lineages came, not from the Near East, as had long been thought from the more limited mtDNA analysis, but from many places in Europe. A picture is emerging of Ashkenazi men traveling in Europe, where they married local women, who converted to Judaism.

Most of the Y chromosome DNA sequence offers the advantage of not recombining. Crossing over, which it could only do with an X chromosome because there is no second Y, would break the linkage from the past generation and therefore make tracing relationships impossible.

Long sets of SNPs, such as those along mitochondrial and Y chromosome DNA, define stretches termed **haplogroups**. (The haplo*types* used to describe linkage in chapter 5 are shorter DNA sequences.) Haplogroups track with geographic events, such as leaving Africa or becoming isolated behind a mountain range, rather than with racial or ethnic groups, which are social designations and not biological ones.

Y haplogroups are classified from "A" through "T," with several subgroups, called subhaplogroups or subclades, indicated by alternating letters and numbers. Haplogroups and their subgroups also describe mtDNA. Populations can be classified into both mtDNA and Y chromosome groups, indicating the sources of female and male lineages, respectively. The subgroups indicate distinguishing mutations.

Shared haplogroups reflect ancient "deep" ancestry, not relatives from a few generations back. Haplogroups might indicate origin in a major part of the world—sub-Saharan Africa, for example—or point to a major population group or subgroup, such as "African American" or "African American with roots in Nigeria." Figure 18.10 depicts maternal (mtDNA) haplogroups and figure 18.11 shows a tree of Y chromosome subhaplogroups.

Admixture

The first molecular evolution studies used Y chromosome and mitochondrial DNA sequences because their gender specificity made them easiest to track. Because we now know of millions of SNPs across the genome, and can sequence genomes, we can fill in more of the blanks in the history of human migrations, by tracing transmission of haplogroups and haplotypes to reveal admixture. Nearly all modern human populations are admixed, to differing extents.

We can trace the occurrence of admixture by following how the pattern of haplotypes changes from one generation to the next if a crossover occurs. The crossover breaks up the linkage of a haplotype or haplogroup, and the admixed generation as a result has shorter chunks of ancestral DNA sequences. If we know the crossover rate for a particular gene that is involved in the exchange of genetic material, then we can extrapolate back to when admixture occurred, assuming a generation time, such as 25 years. **Figure 17.10** illustrates schematically the break up of haplotypes over generations.

Computational tools use changing haplotype data to reconstruct what one researcher calls "snapshots in time" as far back as 2 million years ago, by tracking admixture. One study, for example, looked at 2,400 35,000-base segments of the genome, with a generation time of 25 years, to estimate that an ethnic group called the Yoruba left their native southwestern Nigeria in 1550 and in 1790, times already known from history to correspond to the start of the slave trade and its peak.

Figure 17.10 In admixture, crossing over recombines segments of linked genes representing ancestral populations. The shorter the haplotype blocks, the more generations have passed.

Key Concepts Questions 17.2

a. What is the basis of comparing DNA sequences, protein sequences, chromosome banding patterns, and genome sequences to learn about human origins and evolution?

*b. What is the basis of a molecular clock?

c. How are mitochondrial DNA and Y chromosome sequences used to trace human lineages?

*d. Define *haplogroup*.

e. How does analyzing admixture provide information about the past?

17.3 The Peopling of the Planet

Fossil evidence and extrapolating and inferring relationships from DNA sequence data provide clues to the major movements that peopled the planet. **Figure 17.11** summarizes these migrations. Analyzing the evidence is like reading chapters from different parts of a novel and trying to understand the whole story. Three such chapters in the saga of modern human origins stand out: our beginnings 200,000 years ago, expansion from Africa, and populating the New World. A more recently written chapter is the genetic evidence for the "great migration" of African Americans.

Mitochondrial Eve

Theoretically, if a sequence of mtDNA could have mutated to yield the mtDNA sequences in modern humans, then that ancestral sequence may represent a very early human or humanlike female—a mitochondrial "Eve," or metaphorical first woman. **Figure 17.12** shows how one maternal line may have persisted.

When might this theoretical "first" woman, the most recent female ancestor common to us all, have lived? Researchers in the mid-1980s compared mtDNA sequences for protein-encoding as well as noncoding DNA regions in a variety of people, including Africans, African Americans, Europeans, New Guineans, and Australians. They deduced that the hypothesized ancestral woman lived about 200,000 years ago, in Africa. More recent analysis of mtDNA from 600 living East Africans estimated 170,000 years ago for the beginning of the modern human line. One way to reach this time estimate is by comparing how much the mtDNA sequence differs among modern humans to how much it differs between humans and chimps. The differences in mtDNA sequences among contemporary humans are 1/25 as many as the differences between humans and chimps. The two species diverged about 5 million years ago, according to extrapolation from fossil and molecular evidence. Multiplying 1/25 by 5 million gives a value of 200,000 years ago, assuming that the mtDNA mutation rate is constant over time.

Where did Eve live? The locations of fossil evidence support an African origin, and Charles Darwin suggested it, too. In addition, studies comparing mitochondrial and nuclear DNA sequences among modern populations consistently find that Africans have the most numerous and diverse mutations. For this to be so, African populations must have existed in place for longer than other populations, because it takes time for mutations to accumulate. In many evolutionary trees constructed by computer analysis, the individuals whose DNA sequences form the bases are from Africa. All modern human populations have at least part of an ancestral African genome, plus mutations that occurred after their ancestors left Africa.

Expansion from Africa

Data from mtDNA, Y chromosome DNA, and markers on the autosomes indicate that the peopling of the world took place as a series of founder effects as groups left Africa. Perhaps the migrations happened when the Sahara desert periodically grew wet enough for travel. A large expansion out of Africa occurred about 50,000 years ago, but estimates range widely. These migrations yielded "chains of colonies" that may have overlapped and merged when neighbors met and genes flowed from one region to another.

Geographical and climatic barriers periodically shrank human populations, while natural selection and genetic drift narrowed the African gene pool. At the same time, new mutations among the groups that left Africa established the haplogroups that we use to trace non-African ancestry today. Human evolution continued in Africa as groups left and populated other parts of the world. Global climate change, such as ice ages, forced people into habitable areas, called refugia.

The descendants of some people who left Africa returned. For example, analysis of 2.5 million SNPs across the genomes of people from Chad indicate a Y haplogroup entering African genomes from Eurasians 4750 to 7200 years ago. DNA analysis is helping to fill in the blanks of what we know about human migration both out of and into Africa.

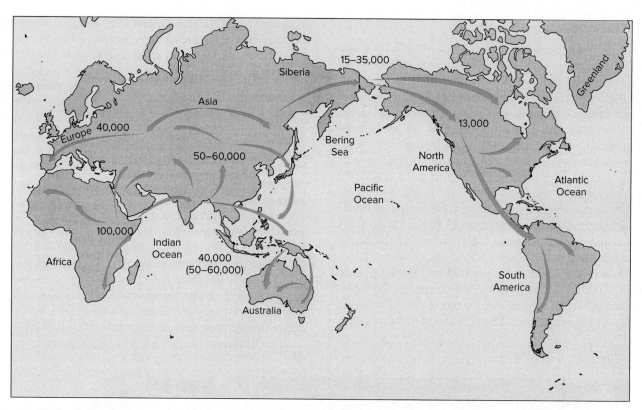

Figure 17.11 The peopling of the planet. This map depicts major migratory paths based on DNA haplogroup information, mostly from mtDNA and the Y chromosome. Fossil and DNA evidence suggest that humanity arose in East Africa, but a southern African origin is possible, too. The Americas were populated relatively rapidly, beginning about 13,000 years ago. Other routes to the Americas may have been possible. These routes are hypotheses, based on available evidence. *Source: Adapted from Hua Liu, Frank Prugnolle, Andrea Manica and Francois Balloux, "A Geographically Explicit Genetic Model of Worldwide Human-Settlement History," The American Journal of Human Genetics, Vol. 79: 2, 2006.*

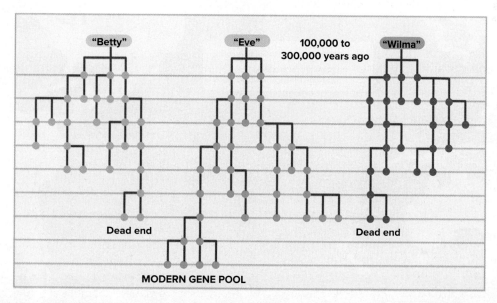

Figure 17.12 Mitochondrial Eve. According to the mitochondrial Eve hypothesis, modern mtDNA retains some sequences from a figurative first woman, "Eve," who lived in Africa 300,000 to 100,000 years ago. In this schematic, the lines represent generations, and the circles, females. Lineages cease whenever a woman does not have a daughter to pass on the mtDNA.

Populating the New World

People spread across Eurasia and elsewhere by 40,000 years ago, lastly through southern Siberia and Mongolia. By 20,000 years ago, humanity was everywhere except the Americas and Antarctica (**figure 17.13**). From Siberia, people could cross the Bering Land Bridge, which emerged between Siberia and Alaska during times when the glaciers had retreated and winds from the southwest blew snow away. The land bridge (also called an ice-free corridor) was about 620 miles (about 1,000 kilometers) wide, for a total area of about 620,000 square miles (1,600,000 square kilometers). The areas for several hundred miles on either side of the bridge, and the bridge itself, are called Beringia. The land bridge was present from 25,000 to 15,000 years ago, according to most evidence. Today it is beneath the Bering and Chukchi Seas.

Mutations occurred in the population of Beringia. Then between 23,000 and 19,000 years ago, a severe population bottleneck killed off all but only about 1,000 people who survived the journey over the bridge from Siberia. Some of them continued southward along the Pacific coastline, bringing some of the mutations that distinguished them from European ancestors.

As the ice age ended about 18,000 years ago, the small founding populations in the Americas began a period of rapid expansion that lasted about 3,000 years. The burst of reproduction amplified alleles that had survived while at the same time alleles unique to people who had perished vanished from the population. The people spread south and east through the Americas as the first Native Americans, a process that may have taken about 2,000 years. Early on Native Americans split into a southern group and a northern group, called ancestral A and B lineages, respectively.

Although DNA evidence points to a single large migration across Beringia, other types of evidence suggest entry into the northern United States and Canada through "vegetated corridors" that opened up between the retreating ice sheets. Evidence for the coastal route is in human DNA. Evidence for an interior route includes signs of salmon fishing, charcoal in remains of hearths, and marks on bones of butchering horses, bison, and camels. Carbon-14 dates these animals to about 13,000 to 15,000 years ago.

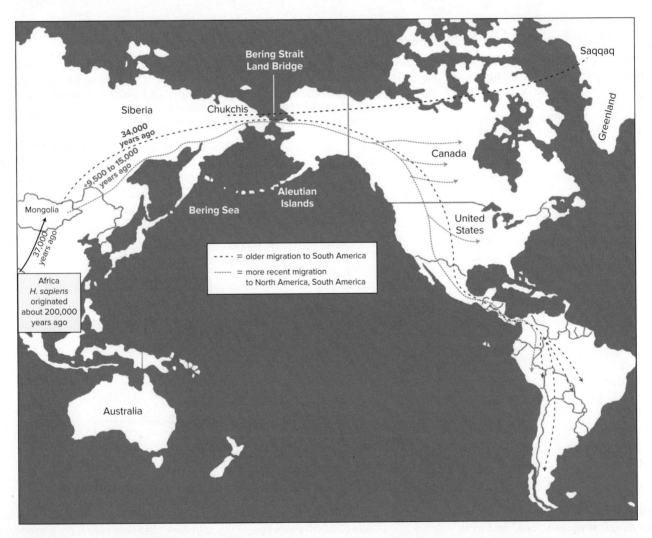

Figure 17.13 Tracing Native American origins. Analyses of mitochondrial DNA and Y chromosome DNA sequences reveal that the ancestors of Native Americans came from Mongolia and Siberia.

Today, Native American genomes reflect the long-ago trek across the land bridge: five major mtDNA haplogroups (A, B, C, D, and X) and two Y chromosome haplogroups (C and Q), plus a few rare haplogroups. These markers are seen in all present-day indigenous populations in southern Siberia and some in northern Siberia, indicating a single gene pool traveling in a single migration. However, a complete genome sequence from a boy who lived 24,000 years ago in Siberia had gene variants from the European edge of Asia. Today about a third of Native American genomes come from western Eurasia and two-thirds from eastern Asia. Some ancestors of Native Americans did not survive to contribute to modern gene pools. Genetic diversity among Native American populations has decreased over time for a number of reasons, including genetic drift, infection, and genocide.

DNA sequence information extrapolated back from present-day Native American populations is consistent with molecular clock data from ancient DNA. In addition, Native American populations have an STR marker and some mtDNA haplogroups that are not seen in eastern Siberian peoples, indicating mutations that happened after crossing the Bering Strait.

A comparison of 678 autosomal STR markers from 29 Native American populations and 49 other indigenous groups worldwide found that Native Americans are very different from other populations, yet are very much like each other—as might be expected from a multigenerational journey southward by small but hardy groups, along the coastline of the Americas.

By 14,000 to 13,000 years ago, Native Americans had arrived inland, according to evidence in fossilized excrement. DNA from the skeleton of an adolescent girl discovered underwater in the Yucatan peninsula dates from 12,000 years ago and includes sequences from Asia and from the Americas.

It seems that with every new DNA discovery, we rewrite prehistory or add details. This was the case for the discovery in Greenland of a 4,000-year-old tuft of hair from a male Paleo-Eskimo from the first group of settlers, the Saqqaq. The DNA from the hair indicated that the man had coarse, dark hair and skin, brown eyes, shovel-shaped teeth, dry earwax, and, if he had lived longer, would have become bald. He had type A-positive blood, he was at increased risk for high blood pressure and diabetes, he had a high tolerance for alcohol, and he may have become addicted to nicotine had it been possible to cultivate tobacco in the frozen wasteland. The most revealing information concerned the Saqqaq Eskimo's origins. His mitochondrial DNA was distinct from that of either Native Americans or modern Eskimos, yet was very similar to that of the Chukchis people of Siberia. Apparently, the ancestors of the Greenland Eskimos crossed a land bridge about 5,400 years ago, separate from the crossings that founded the Native Americans and modern Eskimos, and left no present-day descendants. So some chapters from the history of human existence have been erased.

Chapter 18 returns to the techniques used to trace deep human ancestry in forensic and consumer applications.

The Great Migration of African Americans

DNA evidence has helped to follow more recent migrations, such as the "Great Migration" from 1910 through 1970, when millions of African Americans left the South in search of economic opportunity. They traveled, mostly by rail, to cities in the northeastern, midwestern, and western parts of the United States. The genetic evidence supports historical evidence of their journeys. African American genomes include significant gene variants from Europeans, with a directional gene flow from male to female, reflecting widespread rape of slaves.

Researchers analyzed more than half a million SNPs among 3,726 African Americans living today in all 50 states. They tracked the proportions of gene variants from African, European, and Native American ancestors, extrapolating backward using haplotype data. They inferred that:

- Overall 82.1 percent of the ancestors of the average modern African American were African, 16.7 percent were European, and 1.2 percent were Native American. The proportions vary slightly geographically, with average African ancestry 83 percent in the South, 80 percent in the North, and 79 percent in the West.
- Native American DNA sequences entered in the seventeenth century, mostly from females marrying African slaves.
- European admixture occurred in two phases, around 1740 and then 1863. Admixture declined as the Civil War ended.
- Similar levels of African ancestry for all modern age groups suggest little genomic input from Europeans from 1930 through 1960.
- African Americans with a greater proportion of European genes were more likely to have left the South during the first wave of migration, perhaps because some could pass for white, or because owners helped them. Genetic data are consistent with historical data that indicate earlier migrants had lighter skin.
- When the researchers compared all possible pairs of the 3,726 individuals for identical by descent genome regions and geographical census data, the map of the United States that emerged nearly exactly parallels the rail lines that freed slaves and their descendants used during the Great Migration, including lines up the Atlantic coast and the Illinois Central line to Chicago (**figure 17.14**).

Key Concepts Questions 17.3

*a. Who does "mitochondrial Eve" represent?

b. Describe the main events of how people likely populated the world.

*c. How did people reach the Americas?

d. How similar are Native American genomes to the genomes of other populations?

e. What does DNA evidence indicate about the great migration of African Americans?

Figure 17.14 SNP genotyping reconstructed migration routes from the southern United States to cities in the north and west. The genetic data directly parallel the railroad routes that the people used to escape. This family traveled from the South to Chicago in 1919. *Chicago History Museum/Archive Photos/Getty Images*

17.4 What Makes Us Human?

We can investigate the traits and abilities that distinguish us from chimpanzees and bonobos, our closest relatives, because they live today. For the archaic humans that we know about only from bits of bones, genome sequencing has provided clues to some of their traits, suggesting abilities gained as the archaic humans interbred with our ancestors.

We assess similarity between ourselves and chimps and bonobos by DNA sequence, numbers of copies of sequences, or sequences not in our genomes. We share about 98.7 percent sequence similarity with our closest primate relatives, but we also differ in the number of copies of certain DNA sequences. These include insertions and deletions, which are collectively called **indels**. Considering indels, our degree of genome similarity to chimps and bonobos is only about 96.6 percent. The degree of similarity may even be as low as 94 percent if sequences not in the human genome are considered. That is, what *isn't* present defines us as well as what *is* present.

Evolution of Human Genomes

Comparisons of human genome sequences to those of other species are interesting and humbling. For example, our genome is not that different from that of a pufferfish, with added introns and repeated sequences. Our exomes are remarkably similar for animals that live in such different environments.

Overall, the human genome has a more complex organization of the same basic parts that make up the fruit fly and roundworm genomes. For example, the human genome has 30 copies of the gene that encodes fibroblast growth factor, compared to two copies in the fly and worm genomes. This growth factor is important for the development of highly complex organs.

Genome studies indicate that over a great span of evolutionary time, genes and gene pieces provided vertebrates, including humans, with certain defining characteristics:

- complex neural networks
- blood clotting pathways
- adaptive immunity (see section 19.2)
- refined apoptosis (cell death)
- greater control of transcription
- complex development
- more intricate signaling within and among cells

Comparing the human genome to itself provides clues to evolution, too. The many duplicated genes and chromosome segments in the human genome suggest that it doubled, at least once, since diverging from the genome of a vertebrate ancestor about 500 million years ago. Either the human genome doubled twice, followed by loss of some genes, or one doubling was followed by additional duplication of certain DNA sequences.

The extensive duplication within the human genome is what distinguishes our genome from those of other primates. Some of the doublings are vast. Half of chromosome 20 repeats, rearranged, on chromosome 18. Much of chromosome 2's short arm reappears as almost three-quarters of chromosome 14, and a block on its long arm is echoed on chromosome 12. The gene-packed yet tiny chromosome 22 includes eight huge duplications and several gene families. Repeated DNA sequences in the human genome may provide raw material for future evolution. A second copy of a DNA sequence can mutate, allowing a cell to experiment with a new function while the old one carries on, a little like trying a new car before selling the old one. More often, the twin gene mutates into a silenced pseudogene, leaving a ghost of the gene behind as a similar but untranslated DNA sequence.

A duplication can be located near the original DNA sequence it was copied from, or away from it. A sequence repeated right next to itself is called a tandem duplication, and it usually results from mispairing during DNA replication. A copy of a gene on a different chromosome may arise when messenger RNA is copied (reverse transcribed) into DNA, which then inserts elsewhere among the chromosomes.

Duplication of an entire genome results in polyploidy, discussed in section 13.3. It is common in plants and some insects, but not vertebrates. If a polyploid event was followed by loss of some genes and duplication of others, the result would look much like the modern human genome (**figure 17.15**). The remnants of such an ancient whole-genome

Figure 17.15 Evolution of the human genome. The many duplicated DNA sequences in the human genome suggest a complete duplication followed by other chromosome-level events.

duplication would have become further obscured with time, as inversions and translocations altered the ancestral DNA landscape.

Genes That Help to Define Us

Comparison of the chimpanzee and human genomes has revealed "human accelerated regions." These are highly conserved sequences that show signs of positive selection in humans, such as an amino acid change seen in all human groups but not in the chimp or orangutan versions of the same gene. (Recall from section 16.6 that positive selection maintains a beneficial trait.) Signs of positive selection in the human genome can complement views of our ancestry from fossils.

Relatively few DNA sequence differences distinguish us from the Denisovans and Neanderthals: 31,000 single nucleotide changes (SNPs and mutations), and fewer than 125 small losses or gains of DNA bases (indels). Only 96 of the DNA differences in the human genome alter amino acids in proteins (are nonsynonymous) and can therefore be subject to natural selection. Subtracting Denisovan DNA sequences that are conserved in ape and monkey genomes, indicating their antiquity, reduces the list of distinctive human mutations to 23. Eight of them have to do with brain function and development, neural connectivity and synapses, and two of these mutations are implicated in autism. About 3,000 sequence differences are in parts of the human genome that regulate gene expression, so just a few distinctions can nonetheless have large effects.

Traits that may be uniquely human include spoken language, abstract reasoning ability, highly opposable thumbs, and larger frontal lobes of the brain. We can't know whether the Denisovans and Neanderthals were able to speak and the extent of their abilities to think. Following are examples of traits determined by single genes that help to distinguish us from our closest relatives (**figure 17.16**).

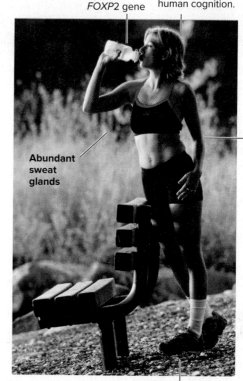

Figure 17.16 What makes us human? A collection of traits distinguishes us from our closest primate relatives.
Ingram Publishing

Speech

A family living in London whose members have unintelligible speech led to the discovery of a single gene (*FOXP2*) that controls speaking ability. Modern humans and Neanderthals have the same two *FOXP2* variants that are not in the chimpanzee genome.

Different Developmental Timetables

More primitive primates lack or have very little fetal hemoglobin (see figure 11.2). In more recently evolved and more complex primates, fetal hemoglobin extends the fetal period, allowing time for substantial brain growth and development before birth. With larger brains came greater skills. Other single genes make possible the longer childhood and adolescence of humans compared to chimps.

Bare Skin

Nearly all mammals have thick, abundant hairs that insulate, protect, and are important in social displays and species recognition. Chimpanzees and gorillas express a keratin gene whose counterpart in humans is not expressed.

Our naked skins make sense in terms of evolution in a few ways. Relatively hairless skin enables us to sweat. When our forebears moved onto the plains as the forests shrank, about 2.5 to 3 million years ago, individuals with less hair and more abundant sweat glands could travel farther in search of food, or to avoid becoming food. Natural selection favored them. Today, our skin is peppered with sweat glands that pump out quarts of sweat a day, compared to our furred fellow mammals, who have sparser and less efficient sweat glands. Lack of hair may have also enabled our ancestors to shed skin parasites such as lice. Human skin is thin yet strong, thanks to a unique combination of keratin proteins that fill the flattened skin cells.

Walking

Walking requires the ability to place one foot in front of the other. People who have Joubert syndrome can't do this. Nerve cell fibers cannot cross from their origin on one side of the brain to the other, so moving just one arm or leg is impossible; both move at once. The part of the brain that controls posture, balance, and coordination is compromised. The gene that causes Joubert syndrome, *AHI1*, is identical in all modern human groups examined, but has different alleles in chimps, gorillas, and orangutans. Perhaps in the lineage leading to humans, the gene came to control walking by making it possible to place one foot in front of the other.

Running

Homo erectus was the first hominin to be able to run long distances, thanks to specific anatomical adaptations. The nuchal ligament that connects the skull to the neck became more highly developed in *H. erectus*, keeping the head in place while running. Leg muscles were more highly developed than those of chimps or australopithecines, acting as springs. *H. erectus* originated large buttocks, whose muscles contract during running. All three of these structures are not merely the result of being able to walk, but enabled early *Homo*, and us, to run. This skill would have helped our ancestors to escape predators, find food, and locate new homes.

A Big Brain

Only a few genes may control the difference in brain size between us and chimps. About 2.4 million years ago, a gene called *MYH16* underwent a nonsense mutation, which prevented production of a type of muscle protein called a myosin. The mutation is seen in all modern human populations, but not in other primates. Without this type of myosin, jaws developed less, which allowed expansion of the bony plates of the skull, permitting greater brain growth. Researchers nicknamed the mutation "room for thought." Fossil evidence indicates that the switch from "big jaw, small brain" to "small jaw, big brain" happened when *Homo* gradually replaced *Australopithecus*, about 2 million years ago. The genetic analysis is new, but the idea isn't. Charles Darwin wrote in 1871 that different-sized jaw muscles were at the root of the distinction between apes and humans.

The observation that humans and chimps look and behave differently but are genetically similar may have more to do with gene expression than genome sequence. For example, a study comparing gene expression in the liver and brain found many more differences between the two species in the brain than in the liver, suggesting that our mental capabilities have extended beyond those of chimps more than have the functions of our livers. The brain, therefore, is a big part of what makes us human.

A larger brain presumably led to greater cognitive skills. In this area, gene copy number may be more important than the nature of the genes. At least 134 genes are present in more copies in the human genome than in genomes of apes. Many of these genes affect brain structure or function. Some of the genes foster long-term memory, and others, when mutant, cause intellectual disability or impair language skills. Single genes implicated in fueling human brain growth control the migration of nerve cells in the front of the fetal brain.

Recent mutations may have propelled our brain power, too. A mutation in a gene that encodes an enzyme called a fatty acid desaturase enables the digestive system to use long-chain fatty acids, which are found in fish and shellfish and are useful in brain development. The mutation occurred in Africa from 180,000 to 80,000 years ago, when some of our ancestors lived near the lakes of Central Africa. Positive selection greatly increased the prevalence of the mutation, but at a cost—the encoded enzyme also causes inflammation and may be responsible for the increased prevalence of hypertension, stroke, type 2 diabetes, and coronary heart disease among African Americans.

Sense of Smell

Our chemical senses—smell and taste—have decreased as our reliance on them has decreased. The sense of smell derives from a 1-inch-square patch of tissue high in the nose that consists of 12 million cells that bear odorant receptor (OR) proteins. (In contrast, a bloodhound has 4 billion such cells!) Molecules from a smelly substance bind to combinations of these receptors, which then signal the brain in a way that creates the perception of the associated odor.

The 906 human odorant receptor genes are clustered on chromosome 17. More than half of them are pseudogenes, similar to active "smell" genes but riddled with nonsense mutations. Perhaps these genes are remnants of a distant past, when we depended more on our chemical senses for survival, and natural selection has silenced them. At the same time, natural selection has acted positively in a way that has retained functioning OR genes. While the pseudogenes contain many diverse SNPs, the functional OR genes are remarkably alike in sequence among individuals, indicating a successful function. In addition, the nucleotide differences that persist among the retained OR genes alter the encoded amino acids, suggesting that natural selection favored these sequences.

The Future of Humanity?

How will evolution mold humanity in the future? Perhaps the most profound characteristic that distinguishes us from archaic humans is our ability to alter the environment on a large scale, which has led to our domination of the planet. We exist in varieties of skin color, body size, intellectual ability, and hundreds of other observable and measurable traits **(figure 17.17)**. Yet at the same time, people from different populations are meeting and mixing their genomes at unprecedented levels, thanks to transportation and communication over vast distances. The planet is far different from the days of the australopithecines, who dwelled in geographical areas so dispersed that they existed as an overlapping series of distinct species, or even the Neanderthals and Denisovans who might never have ventured far from their caves. Today, phenotypes persist that in millennia past would have disappeared due to negative natural selection, thanks to health care and technology.

What will humans be like 10,000 years from now? Will admixture increase so that we all have genome parts that were once associated with specific population groups? Will new technologies and medical care improve our genomes, or allow gene variants that would have been lethal in the past to persist and be passed to future generations? Will we manipulate our genomes to direct our own future evolution? Understanding our origins poses many questions about our futures.

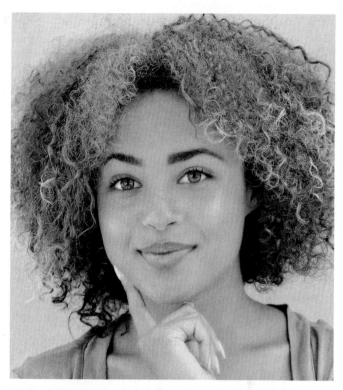

Figure 17.17 Many people have combinations of characteristics that reflect ancestors from different populations. Note this woman's curly hair, caramel-colored skin, and light eyes. *Geri Lavrov/Getty Images*

Key Concepts Questions 17.4

a. How does the human genome differ from the genomes of other animals, in a general sense?

*b. What are some genes that provide traits that are uniquely human?

c. How do we influence our own future evolution?

Study Guide

Summary

17.1 Human Origins

1. DNA provides information on evolution because sequence similarities indicate descent from common (shared) ancestors. Changing allele frequencies indicate events at a population level.

2. Evolutionary tree diagrams depict descent from shared ancestors.

3. Humans are hominins, a taxonomic group that includes our ancestors who lived after the split from other African apes about 6 million years ago.

4. The australopithecines preceded and then coexisted with early *Homo* species, who lived in caves, had strong family units, and used tools, then lived in societies and used fire.

5. Neanderthals and Denisovans were archaic humans. A small percentage of their genomes persists in the genomes of some modern human populations, indicating **introgression** (introduction of gene variants from one species or subspecies to another) and admixture (presence of DNA variants from more than one species or subspecies).

6. The spread of agriculture is inferred from changed allele frequencies from ancient migrations.

7. The DNA of Ötzi the Ice Man indicates that his maternal ancestors died out but his paternal ancestors escaped the harsh Alps and survived.

17.2 Methods to Study Molecular Evolution

8. Molecular evolution considers differences in genomes, chromosome banding patterns, protein sequences, and DNA base sequences, using mutation rates to estimate species relatedness.
9. Genes in the same order on chromosomes in different species show **synteny**.
10. For a highly conserved gene, the DNA sequence is similar or identical in different species, indicating importance and shared ancestry.
11. Evolutionary tree diagrams represent gene sequence information from several species, using molecular clocks based on mutation rates.
12. Molecular clocks based on mtDNA date recent events through the maternal line because this DNA mutates faster than nuclear DNA. Y chromosomes are used to trace paternal lineage. Markers (SNPs, **short tandem repeats,** and CNVs) in mtDNA, Y chromosome DNA, and autosomal DNA are used to study human origins and expansions. Many linked markers inherited together form **haplogroups**. Haplotypes are shorter than haplogroups.
13. Changing haplotype patterns indicate admixture, which, using crossover rates and generation times, estimates when populations mixed.

17.3 The Peopling of the Planet

14. The rate of mtDNA mutation and current mtDNA diversity can be extrapolated to hypothesize that a theoretical first woman lived in Africa about 200,000 years ago. *Homo sapiens* began to leave Africa about 56,000 years ago.
15. A series of migrations, creating founder effects, peopled the planet, with genetic diversity decreasing from that of the ancestral African population, but with new mutations occurring.
16. After the last ice age, people crossed the Bering Strait from Siberia, occupying the Americas from about 14,000 years ago.

17.4 What Makes Us Human?

17. Humans and chimps share 98.7 percent of their protein-encoding gene sequences. **Indels,** introns, and repeats create genome differences between humans and chimps. The two species also differ in gene expression.
18. The human genome shows many signs of past duplication.
19. Humans differ from other apes in adaptations that allowed greater brain growth, less hairy skins and the advantages that they provide, and the abilities to speak, walk, and run.
20. We affect our own evolution by treating disease, by the partners we choose to have children with, and by altering the environment.

Key Terms

haplogroup (hăp′ lō-groōp) An extensive set of SNP markers linked on a chromosome that are inherited together and therefore can be used to trace deep ancestry.

indels Insertions and deletions of DNA in a genome.

introgression Entry of a specific gene variant into a genome from an individual of a different species or subspecies.

short tandem repeats (STRs) Repeats of 2 to 10 DNA bases that are compared in DNA profiling.

synteny (sĭn′tə-nē) Correspondence of gene order on the same chromosome in two or more species.

Review Questions

1. What types of evidence are used to construct evolutionary tree diagrams?
2. Explain how the evolutionary tree diagram in figure 17.1 indicates that we did not descend directly from chimpanzees.
3. Why were there several species of *Australopithecus* but possibly only one of early *Homo* in Africa?
4. What is a limitation of relying on fossil evidence in describing extinct species?
5. Why are Neanderthal DNA sequences in the genomes of modern Europeans but not in those of modern Africans?
6. Why is the gene flow among Neanderthals, Denisovans, and our ancestors so complicated?
7. What do we know about ancient farmers and Ötzi the Ice Man from their DNA?
8. State two assumptions behind molecular evolution studies.
9. Which of the forces discussed in chapter 16 explains why some DNA sequences vary greatly among species, while others are highly conserved?
10. Why are exons highly conserved but introns not?
11. Describe the type of information that Y chromosome and mitochondrial DNA sequences provide, and explain why it is incomplete.
12. How can haplotypes provide information on genetic admixture and ancient migrations?
13. Name a human population that has not admixed with others.
14. Describe how people expanded out of Africa, while evolution continued in Africa.
15. Provide an example of how changing geography or climate affected human evolution.
16. Name three traits that are uniquely human.

Applied Questions

1. Explain what comparing genomes from indigenous peoples to other modern peoples can reveal about evolution.

2. "I'm 4 percent Neanderthal!" claimed the politician. Explain how this is not likely and how she may be misinterpreting findings from a consumer DNA test.

3. A man with white skin checks the box on forms about personal information for "African American," claiming that we are all, if we go back far enough in our family trees, from Africa. Is he correct?

4. What do drug developers need to consider to create a product based on or using a protein from a species other than humans?

5. In Central Africa, the Mbuti Pygmies are hunter-gatherers who live amid agricultural communities of peoples called the Alur, Hema, and Nande. Researchers compared autosomal STRs, mitochondrial DNA, and Y chromosome DNA haplogroups among these people. The pygmies had the most diverse Y chromosomes, including about a third of the sequence that was the same as those among the agricultural groups, who had greater mtDNA diversity than the pygmies. None of the agricultural males had Y chromosome DNA sequences that matched those of the pygmies. Create a narrative of gene flow to explain these findings.

6. Explain why analyzing mtDNA or Y chromosome DNA cannot provide a complete picture of a person's ancestry.

7. Explain how a female can trace her paternal lineage if she doesn't have a Y chromosome.

8. Name a technology that is contributing to human admixture.

9. Humans have chins, but other primates, including *Australopithecus* and Neanderthals, do or did not. Propose a function that might explain the positive selection of chins in anatomically modern humans.

Forensics Focus

1. The Millers raise chickens in their backyard. On their block are six pet dogs, all mixed breeds. One day, Mr. Miller discovers that his chickens are dead, and amid the feathers and signs of struggle are clumps of dog hair. He collects the hair and sends it to a dog ancestry testing company, for clues to which of his canine neighbors is the culprit. What will the company test?

2. In many African cultures, "family" is not dictated by genetics, but by who cares for whom. Any adult can be "mother" or "father" to any child. Since the early 1990s, many parts of East Africa have been under civil war. Thousands of Africans have asked to be admitted to the United States to join relatives. The "family reunification program" enables parents, children, and spouses of U.S. citizens to come to the United States. In 2008, addressing rumors that many people were lying that they were related to people in the United States, the State Department began to ask refugees from Kenya to voluntarily provide a DNA sample, to be compared to that of the U.S. citizen claimed to be a relative. When the DNA testing turned up many cases of people claiming to be family who were not blood relatives, the resettlement program was temporarily stopped.

   a. How should immigration policy address cultural definitions, such as that of "family"?
   b. How should DNA be collected?
   c. What part of the genomes should be compared to get the most useful information?

Case Studies and Research Results

1. Isolated indigenous people, like those described in the chapter opener, often resent scientists who come great distances to collect their DNA for "helicopter research," which is defined as "a clinical or epidemiological research project conducted by foreign scientists who use local contacts to gain access to a population group and obtain samples." An alternative is "community-based participatory research," which involves the people being studied in the investigation. Suggest such a project based on the information in this chapter.

2. The 4,000-year-old Saqqaq Paleo-Eskimo from Greenland is known only from a tuft of hair. How did researchers learn about other characteristics?

3. Researchers compared 4 million SNPs across the genomes of 107 people living today in six African nations and in three African-descendant populations in South America. A group called the Noir Marron (also called Maroon) who live in French Guiana have 98 percent African ancestry, the highest proportion known for African descendants. The ancestors of the Noir Marron escaped as slaves from Suriname in the sixteenth and early seventeenth centuries. The percent African ancestry among African Brazilians in the study was 71 percent and among African Colombians 75 percent. The Noir Marron genomes have several large runs of homozygosity not seen in other populations. Suggest a scenario of human behavior that could account for these findings.

4. *Glossogenetics* is the field of study that associates language with patterns of genetic variation. Explain the finding that languages in a particular geographic area track with Y chromosome sequences, but not with mtDNA sequences.

CHAPTER 18

The Genetics of Identity

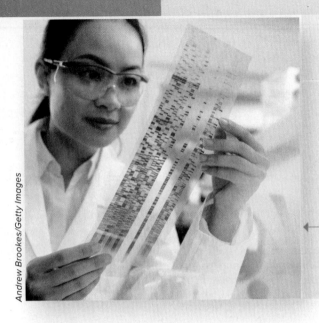

A forensic scientist consults a DNA profile. The black bars represent short tandem repeats that form patterns used to exclude suspects in a crime. Figure 18.6 shows an automated approach that uses electrophoresis of the STR DNA fragments.

Andrew Brookes/Getty Images

Learning Outcomes

18.1 Genetics as an Informational Science
1. Explain how DNA sequences can be used to distinguish individuals.
2. Describe the types of DNA sequences that serve as markers.

18.2 Forensic DNA Typing
3. Define allele as used in DNA profiling.
4. Distinguish a heterozygote from a homozygote for an allele that is a DNA repeat.
5. Explain how the 20 standard forensic markers are used with population statistics to distinguish individuals.
6. Discuss circumstances in which forensic DNA typing is useful.

18.3 Genetic Genealogy
7. How are mitochondrial and Y chromosome haplogroups used to trace deep ancestry?
8. Explain how geographic origins and ethnicity estimates are derived.
9. How do ranges of shared centimorgans indicate familial relationships?

18.4 Forensic STRs Meet Genetic Genealogy SNPs
10. How can investigators combine crime scene evidence with DNA information in genealogy databases to identify suspects?

The BIG Picture

Short tandem repeats (STRs) and single nucleotide polymorphisms (SNPs) are highly variable parts of human genomes that are compared to identify individuals. The approaches have applications in forensic science and in genetic genealogy.

Postconviction DNA Testing

Forensic DNA typing has led to the arrests of many criminals, but it has also led to exonerations. Josiah Sutton had served 4.5 years of a 25-year sentence for rape when he was exonerated, thanks to the Innocence Project. This nonprofit legal clinic and public policy organization, created in 1992, has used DNA retesting to free hundreds of wrongfully convicted prisoners. Sutton became a suspect after a woman in Houston identified him and a friend 5 days after she had been raped, threatened with a gun, and left in a field. The two young men supplied saliva and blood samples, from which DNA profiles were done and compared to DNA profiles from semen found in the victim and in her car. At the trial, a crime lab employee testified that the probability that Sutton's DNA matched that of the evidence by chance was 1 in 694,000, leading to a conviction. Jurors ignored the fact that Sutton's physical description did not match the victim's description of her assailant.

In 1999, the forensic investigators looked at only 7 of the 13 STR markers that were typically compared to generate a DNA profile, then called a DNA fingerprint. Doing the test correctly revised the statistics dramatically: Sutton's pattern was shared not with 1 in 694,000 black men, as had originally been claimed, but with 1 in 16. That is, his DNA profile was too common to point only to him.

While in jail, Sutton read about DNA profiling and requested independent testing, but was refused. Then journalists investigating the Houston crime laboratory learned of his case and alerted the Innocence Project. Retesting the DNA evidence led to freeing Sutton.

18.1 Genetics as an Informational Science

Population genetics was once an academic discipline that was more theoretical than practical. Today, it has become an informational science. Although our genomes are more than 99 percent alike in sequence, the differences are sufficient to distinguish and even to identify people. Only 30 to 80 genome sites, functioning as genetic markers, need to be considered to uniquely describe each person. This is because our genomes can vary many more ways than there are people on the planet. Genome sequencing can be used to distinguish individuals, but this level of detail is not necessary to do so. Recall from chapter 7 that a genetic marker is a DNA sequence that is associated with, and used to represent, a gene or trait of interest.

The types of genetic markers used to distinguish individuals and to identify relatives do not encode protein and therefore do not affect the phenotype. If they did, then two people might have a marker in common because it is part of or near a gene that natural selection has favored, rather than having been inherited from a shared ancestor. Instead, markers used to establish identity or reveal ancestry are not closely linked to protein-encoding genes and are in Hardy-Weinberg equilibrium, discussed in chapter 15.

Assessing variation in genetic marker DNA sequences can be used to identify individuals if three conditions are met:

1. The frequencies of markers in specific populations are known.
2. Markers at several sites in the genome are considered at the same time.
3. The product rule is applied.

The informational nature of the genetic material has been recognized since the beginning of the field. Gregor Mendel knew that information that conferred visible traits was transferred from one generation to the next: his *elementen*. The discovery of DNA's four-letter alphabet in the 1950s suggested how the molecule could copy itself and in the 1960s how the sequence of DNA nucleotide bases could direct cells to synthesize the proteins that provide traits.

In the 1980s and 1990s, researchers developed several types of genetic markers. These short DNA sequences were first used to map genes to chromosomes, and then to provide landmarks for sequencing genomes. Practically, genetic markers were adapted for use in forensic science because they can distinguish individuals, and in health care to indicate disease-causing mutations to which they were linked. Genetic markers could also provide signals to track human origins, migrations, and genetic diversity. Commonly used markers are single nucleotide polymorphisms (SNPs) and copy number variants (CNVs), especially short tandem repeats (STRs) (see table 7.5).

The genetic technologies developed over the past four decades are converging into a new genetics of identity. Investigators are overlapping the repeated DNA sequences used in forensics (section 18.2) and the SNPs used in genetic genealogy (section 18.3) to solve crimes (section 18.4). The numbers of repeats at certain sites in genomes and long stretches of SNPs linked into haplotypes are considered alleles (gene variants), even though they do not encode protein like conventional genes.

The new genetics of identity builds on concepts discussed in previous chapters: genetic linkage in chapter 5, genome-wide association studies using SNPs in chapter 7, and evolution in chapters 15 through 17. The new field of genetic identity is both practical and disturbing, and touches on history, politics, law, health care, bioethics, social work, psychology, sociology, psychiatry, and definitions of self and family.

Key Concepts Questions 18.1

a. Why is it not necessary to compare entire genome sequences to distinguish people?

b. Explain why DNA sequences that do not encode proteins are valuable in distinguishing individuals or telling whether they are related.

*c. List two types of genetic markers.

18.2 Forensic DNA Typing

In DNA profiling, the number of copies of a repeated sequence is considered to be an allele because it is information—just not protein-encoding information. A person is classified as a heterozygote or a homozygote based on the number of copies of the same repeat at the same chromosomal locus on the two homologs. A homozygote has the same number of copies of a repeat on both homologs, such as individual 2 in **figure 18.1**. A heterozygote has two different numbers of copies of a repeat, such as the other two individuals in the figure. The copy numbers are distributed in the next generation according to Mendel's law of segregation. A child of individual 1 and individual 2 in figure 18.1, for example, could have either the combination of the parental copy numbers: two repeats and three repeats, or four repeats and three repeats.

DNA profiling was pioneered on detecting copy number variants of very short repeats and using them to identify or distinguish individuals. In general, the technique calculates the probability that certain combinations of repeat numbers

Figure 18.1 DNA profiling detects differing numbers of repeats at specific chromosomal loci. Individuals 1 and 3 are heterozygotes for the number of copies of a 5-base sequence at a particular chromosomal locus. Individual 2 is a homozygote, with the same number of repeats on the two copies of the chromosome. (Repeat number is considered an allele.)

will be in two DNA sources by chance. For example, if a DNA profile of skin cells taken from under the fingernails of an assault victim matches the profile from a suspect's hair, and the likelihood is very low that those two samples would match by chance, that is strong evidence of guilt rather than a coincidental similarity. DNA evidence is more often valuable in excluding a suspect, and should be considered along with other types of evidence.

An Evolving Technology

Obtaining a DNA profile is a molecular technique, but interpretation requires statistical analysis of population genetic data. The methods for generating a DNA profile grew out of genetic marker testing that identified DNA sequences closely linked to disease-causing genes. Genetic markers for disease were used in the 1980s, before we knew the sequences of human disease-causing genes and could detect them directly. Sir Alec Jeffreys at Leicester University in the United Kingdom was the first investigator to use DNA typing in forensic cases. He followed DNA sequences called **variable number of tandem repeats**—VNTRs—that are each 10 to 80 bases long (**table 18.1** and **A Glimpse of History**).

DNA for profiling can come from any cell with a nucleus, including hair, skin, secretions, white blood cells, and cells scraped from the inside of the cheek. Sir Jeffreys cut DNA with naturally occurring, scissorlike enzymes, called **restriction enzymes,** that cut at specific sequences. He then separated the DNA pieces using a technique called agarose gel electrophoresis. The DNA fragments carry a negative electrical charge because of the phosphate groups. When placed in an electrical field on a gel strip, the DNA pieces move toward the positively charged end by size. The shorter the piece, the faster it travels. When the different-sized fragments stop moving, a pattern of smears remains, such as the gel sheet that the woman in the chapter opener photo is scrutinizing. A person heterozygous for a particular VNTR—meaning a different number of copies of the repeat on each of the two homologs—would have a DNA profile with two bands at that site. A homozygote would have just one band, because his or her DNA would have only one size piece corresponding to that part of the genome. One of these early DNA profiles was done to test whether Dolly the cloned sheep actually developed from the claimed breast cell of a ewe. She did.

A GLIMPSE OF HISTORY

The criminal case that made Sir Alec Jeffreys—and what was then called DNA fingerprinting—famous was that of Colin Pitchfork, a 27-year-old baker and father who raped and strangled two 15-year-old girls in the Leicester countryside. Other early cases proved that a boy was the son of a British citizen so that he could enter England and freed a man unjustly imprisoned for raping two schoolgirls.

The first exoneration thanks to DNA profiling came in 1993, in the United States. Kirk Bloodsworth, while on death row for the 1985 conviction of raping and murdering a 9-year-old, read about Colin Pitchfork and requested DNA testing of a stain on the victim's underwear. Bloodsworth's DNA and the DNA on the evidence did not match. The real criminal had lived a floor below Mr. Bloodsworth in prison!

Researchers began to replace VNTRs with short tandem repeats (STRs), because shorter DNA molecules were more likely to persist in a violent situation, such as an explosion, fire, or natural disaster (see section 12.5). Today, the Federal Bureau of Investigation (FBI) in the United States tracks 20 STRs for DNA profiling, which generates 40 data points, because a person has two copies of each STR. The FBI system, called CODIS for Combined DNA Index System, shares DNA profiles among local, state, and federal crime laboratories. It began in 1990 with 13 "core loci" STRs, but added 7 more in 2017 (**figure 18.2**). The power of DNA profiling is that the number of repeats of each of the 20 STRs occurs with certain probabilities in

Table 18.1	Characteristics of Repeats Used in DNA Profiling			
Type	Repeat Length	Distribution	Example	Fragment Sizes
VNTRs (minisatellites)	10–80 bases	Not uniform	TTCGGGTTGT	50–1,500 bases
STRs (microsatellites)	2–10 bases	More uniform	ACTT	50–500 bases

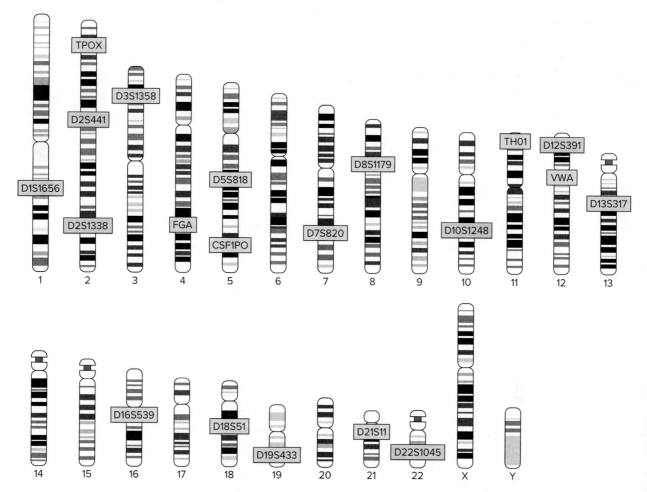

Figure 18.2 **The CODIS core loci are short tandem repeats (STRs) that provide a DNA profile.** Twenty short tandem repeat (STR) sites in the genome are compared to rule out suspects in crimes. The yellow labels indicate the 13 original CODIS sites and the green labels indicate the 7 loci added in 2017. The STR in figure 18.7 is the one on the short arm of chromosome 11.

specific populations. The probability that any two unrelated individuals have the same 40 CODIS markers (20 pairs) by chance far exceeds the global human population number.

STRs range in size from 2 to 10 bases. Figure 7.10 depicts a seven-base STR. Most of the STRs used for forensic applications are four bases long. **Figures 18.3, 18.4,** and **18.5** depict the patterns of shared STRs among a forensic match (suspect's DNA to crime scene evidence), full siblings, and half siblings. Although the relationships are obvious at a glance, the probabilities of the STR repeat numbers in the relevant populations quantify the conclusions.

DNA profiling also includes the genotype for a gene called amelogenin to determine the gender of the person who left the DNA sample. Amelogenin is present on both the X and Y chromosomes but is six bases shorter on the X chromosome. Therefore, a long amelogenin allele indicates a male.

Today a technique called capillary electrophoresis is used to separate DNA pieces (**figure 18.6**). Fluorescently labeled short pieces of DNA that correspond to parts of the genome including the CODIS STRs are applied to sample DNA, where they bind to their complements. Then the polymerase chain reaction (PCR; see figure 9.19) copies the selected sequences, amplifying all the CODIS markers simultaneously. Instead of the slab of gel used when the technology was developed, the DNA pieces travel through hair-thin tubes, called capillaries. A laser shone through a window excites the fluorescent dyes that tag each DNA piece, and the strength of the fluorescence is measured and displayed as a peak on a readout. A single tall peak for a particular STR indicates that the person is a homozygote—for example, six copies of the repeat on each homolog. Two peaks of lesser intensity indicate that the person is a heterozygote, such as having six copies on one homolog but eight on its mate. DNA can now be typed at the scene of a crime in just minutes using a portable device.

The FBI maintains a National DNA Index (NDIS). It has DNA from more than 14 million convicted offenders, more than 3 million arrested individuals, and 0.9 million samples collected from crime scenes but never associated with an individual. The FBI claims that NDIS has aided nearly half a million investigations.

If DNA is so degraded that even STRs are destroyed, mitochondrial DNA (mtDNA) is often used instead. Two regions of repeats, called hypervariable regions I and II, are

Black = shared markers

STR	Suspect		Crime Scene Evidence	
	All markers shared (40/40)			
	Allele Sizes (# of repeats)			
D1S1656	14	14	14	14
TPOX	11	13	11	13
D2S441	10	14	10	14
D2S1338	22	24	22	24
D3S1358	16	16	16	16
FGA	20	24	20	24
D5S818	11	13	11	13
CSF1PO	10	10	10	10
D7S820	8	9	8	9
D8S1179	13	15	13	15
D10S1248	13	13	13	13
THO1	7	8	7	8
D12S391	12	12	12	12
VWA	15	17	15	17
D13S317	12	12	12	12
D16S539	8	14	8	14
D18S51	9	12	9	12
D19S433	14	14	14	14
D21S11	28	30	28	30
D22S1045	16	16	16	16

Figure 18.3 The suspect and crime scene DNA share all 40 markers.

STR	Sibling #1		Sibling #2	
	Half shared (20/40 per person)			
	Allele Sizes			
D1S1656	14	14	14	14
TPOX	11	13	9	13
D2S441	10	14	10	14
D2S1338	22	28	22	24
D3S1358	16	16	18	18
FGA	20	18	20	24
D5S818	11	13	11	13
CSF1PO	10	10	13	13
D7S820	7	10	8	9
D8S1179	13	17	13	15
D10S1248	13	13	17	17
THO1	7	9	7	8
D12S391	12	12	15	15
VWA	15	17	15	19
D13S317	18	18	18	18
D16S539	8	14	10	14
D18S51	9	12	9	12
D19S433	14	14	11	11
D21S11	28	30	29	30
D22S1045	18	18	18	18

Figure 18.4 Two full siblings share 20 of the 40 markers, indicated in red.

assessed because they are distinct in different populations. Because a single cell can yield hundreds or thousands of copies of the mitochondrial genome, even extremely small forensic samples contain this DNA. MtDNA analysis was critical in analyzing evidence from the September 11, 2001 terrorist attacks, most of which was extremely degraded.

Using Population Statistics to Interpret DNA Profiles

In forensics in general, the more clues, the better. The power of DNA profiling comes from tracking repeats on several chromosomes. The numbers of copies of a repeat are assigned probabilities (likelihood of being present) based on their observed frequencies in the population from which the source of the DNA comes. Considering repeats on different chromosomes makes it possible to use the product rule to calculate the probabilities of particular combinations of repeat numbers occurring in a population, based on Mendel's law of independent assortment.

The Hardy-Weinberg equation ($p^2 + 2pq + q^2 = 1$; see figure 15.3) and the product rule are applied to derive the statistics that back up a DNA profile, using the following steps:

1. The pattern of peaks on the electropherograms indicates whether an individual is a homozygote or a heterozygote for each repeat.
2. Calculate genotype frequencies using parts of the Hardy-Weinberg equation: p^2 and q^2 denote each of the two homozygotes for a two-allele repeat, and $2pq$ represents the heterozygote.
3. Multiply the frequencies to reveal the probability that this combination of repeat copy numbers would occur in a particular population.
4. If the genotype combination is rare in the population the suspect comes from, and the genotype is in the suspect's DNA and in crime scene evidence, such as a rape victim's body or stolen property, guilt is highly likely.

STR	Half-sibling #1		Half-sibling #2	
	Quarter shared (10/40 per person)			
	Allele Sizes			
D1S1656	14	15	12	12
TPOX	11	13	9	13
D2S441	10	12	10	14
D2S1338	21	23	22	24
D3S1358	16	18	17	19
FGA	20	18	20	24
D5S818	9	13	11	13
CSF1PO	10	11	8	10
D7S820	6	11	8	9
D8S1179	13	17	13	15
D10S1248	14	14	14	14
THO1	7	8	7	8
D12S391	12	12	15	15
VWA	13	16	15	17
D13S317	11	15	10	12
D16S539	8	12	10	14
D18S51	9	10	11	12
D19S433	12	14	13	17
D21S11	28	31	29	30
D22S1045	17	17	16	16

Figure 18.5 Two half-siblings share 10 of the 40 markers, indicated in black.

Figure 18.7 summarizes the procedure for considering multiple STRs.

For the sequences used in DNA profiling, Hardy-Weinberg equilibrium is assumed. When it doesn't apply, problems can arise. For example, the requirement of nonrandom mating for Hardy-Weinberg equilibrium wouldn't be met in a community with a few very large families where distant relatives might inadvertently marry each other—a situation in many small towns. Cousins, for example, might share parts of a DNA profile. In one case, a young man was convicted of rape based on a DNA profile—part of which he shared with his father, the actual rapist. Considering a larger number of repeat sites can minimize such complications. If more repeat sites had been considered in the rape case, chances are that they would have revealed a polymorphism that the son had inherited from his mother, but that the guilty father lacked. Presence of the polymorphism would have indicated that the son was not guilty, but a close male relative might be.

The accuracy and meaning of a DNA profile depend upon the population that is the source of the allele frequencies. If populations are too broadly defined, then allele frequencies are typically low, leading to very low estimates of the likelihood that a suspect matches evidence based on chance. Not recognizing population genetic substructure was why Steel syndrome was lumped in with more common orthopedic conditions when the broad population of Hispanics living in East Harlem was considered. Its distinction became apparent when only people of Puerto Rican descent were considered (see section 16.1).

The first DNA profiling databases relied on populations that weren't realistic enough to yield valid statistics. They worked with just three categories: Caucasian, black, or Hispanic. People from Poland, Greece, or Sweden were all considered Caucasian, and a dark-skinned person from Jamaica and one from Somalia would be considered blacks. Perhaps the most incongruous of all were the Hispanics. Cubans and Puerto Ricans are part African, whereas people from Mexico and Guatemala have mostly Native American gene variants. Spanish and Argentinians have neither black African nor Native American genetic backgrounds. Native Americans and Asians were left out altogether. Analysis of databases for Caucasians, blacks, and Hispanics revealed significantly more homozygous recessives for certain polymorphic genes than the Hardy-Weinberg equation would predict, confirming that allele frequencies were not in equilibrium.

More restrictive ethnic databases are required to apply allele frequencies to interpret DNA profiles. A frequency of 1 in 1,000 for an allele in all Caucasians may actually be much higher or lower in a subgroup whose members intermarry. However, even narrowly defined ethnic databases may be insufficient to interpret DNA profiles from people of mixed heritage, such as someone whose mother was Scottish/Ethiopian and whose father was Greek/Thai. As it becomes easier for people from different parts of the world to meet, human population substructure will grow more complex, and require DNA profiling to consider more markers.

Using DNA Profiling to Identify Victims

DNA profiling was first used in criminal cases and to identify human remains from plane crashes. Then terrorist attacks and natural disaster applications brought the technology to a new level.

Identifying World Trade Center Victims

In late September 2001, a company that provides breast cancer tests received three unusual types of DNA samples:

- evidence from the World Trade Center in New York City;
- cheek brush scrapings from relatives of people missing from the site; and
- "reference samples" from the victims' toothbrushes, razors, and hairbrushes.

Technologists analyzed the DNA for copy numbers of the 13 then–standard STRs and the sex chromosomes (the amelogenin gene). STR analysis yielded results on pieces of soft

Figure 18.6 Several steps identify STRs. In this example, the STR TH01 consists of TCAT repeats on chromosome 11. **(a)** Short pieces of DNA called primers that are complementary to regions near the STR are added to the DNA sample, each bound to a fluorescent dye. **(b)** The primers direct the polymerase chain reaction to make many copies of the STR. **(c)** The amplified DNA pieces—in this case either six or eight copies of TCAT—enter a gel electrophoresis device. Here, the smaller pieces move toward the positive charge faster than the larger pieces. Laser excitation of the dye bound to the primers emits a signal that is converted into an image called an electropherogram. Two peaks indicate a heterozygote and a single, larger peak a homozygote—for this STR.

tissue, but bone bits that persisted despite the ongoing fire at the site required hardier mitochondrial DNA analysis. If the DNA pattern in crime scene evidence matched DNA from a victim's toothbrush, identification was fairly certain. Forensic investigators used the DNA results to match family members to victims. DNA profiling provides much more reliable information on identity than traditional forensic identifiers such as dental patterns, scars, and fingerprints, and objects found with the evidence, such as jewelry.

Identifying Natural Disaster Victims

Different types of disasters present different challenges for DNA profiling. New York City workers searched rubble in a limited area for remains, but the approximately 250,000 bodies strewn about by the Indian Ocean tsunami in 2004 were everywhere. Disaster workers had to exhume bodies that had been buried quickly to stem the spread of infectious disease. Remains that were accessible after the waves hit quickly decayed in the hot, wet climate. These conditions, combined with the lack of roads and labs, led to 75 percent of the bodies being identified by standard dental record analysis, and 10 percent from fingerprints. Fewer than half of 1 percent of the victims were identified by their DNA.

Forensic scientists had learned from 9/11 the importance of matching victim DNA to that of relatives, to avoid errors when two people matched at several genome sites by chance. In New York City, many of those relatives were from nearby neighborhoods; in the 2004 tsunami, 12 countries were directly affected and victims came from 30 countries. Entire families were washed away, leaving few or no relatives to provide DNA, even if everyday evidence such as toothbrushes had remained.

Confirming Genocide

DNA profiling can provide evidence of mass murder. During four days in July 1995, Bosnian Serbs killed 8,100 Muslim men and boys in the Srebrenica genocide of Bosnia and Herzegovina. The dead were initially dumped into mass graves but then were dug up, broken apart, and widely dispersed in a cover-up operation. Many of 32,000 other bodies had been buried deeply in clay-rich soil, where the lack of oxygen turned the soft tissue clinging to bones into a soaplike substance called *grave wax*.

Figure 18.7 To solve a crime. A dead man was found who had bits of skin and blood beneath a fingernail. The evidence was sent to a forensics lab, where the patterns of five STRs in the blood from beneath the fingernail was compared to patterns for the same STRs in blood from the victim and blood from a suspect. The pattern for the crime scene evidence matched that for the suspect. Allele frequencies from the man's ethnic group used in the Hardy-Weinberg equation yielded the probability that his DNA matched that of the skin and blood under the murdered man's fingernail by chance. This is a hypothetical case. Using the full set of 20 STRs yields much more significant probabilities. The numbers under the peaks are the numbers of copies of the STR. The decimals are allele frequencies. © Andrey Kuzmin/Alamy Stock Photo

Conclusion: The probability that another person in the suspect's population group has the same pattern of these alleles is approximately 1 in 3,226.

An investigation by the International Commission on Missing Persons (ICMP), which formed in 2000, used DNA profiling to identify more than 7,000 of the victims of the July 1995 Srebrenica massacre. Researchers collected blood samples from more than 85,000 relatives in search of loved ones and matched the samples to the DNA profiles of the remains, using STRs. At the ICMP laboratory in northern Sarajevo, investigators cleaned and pulverized bone samples, then extracted the DNA and tested it for the forensic STRs. The DNA profiles provided a great deal more information than had classical evidence, such as clothing found with bodies and dental patterns—enough to convict war criminal Radovan Karadzic in 2016, after a six-year trial. DNA profiling has identified more than 70 percent of the 40,000 people missing in total after the conflict. In addition, by noting where each bone fragment was found, researchers could reconstruct the movements to hide the bodies. DNA profiling is also being used on war crime victims from Syria, Iraq, and Libya.

Reuniting Families

Genetics intersects politics when forensic DNA typing is used to reunite family members separated under circumstances such as war and immigration. A precedent is the Grandmothers of the Plaza de Mayo project, which began in 1977 in Argentina. In 1976, the eight-year "Dirty War" had begun, when the Argentinean military and police abducted and sold or gave away at least 210 *niños desaparecidos* ("desperate children"). Relatives of missing children started the Grandmothers project to use conventional evidence, such as documents, to reunite families (**figure 18.8**). Grandmothers were the focus because so many of the mothers had died.

By 1984, when conventional evidence hadn't reunited many families, the American Association for the Advancement of Science sent a team of geneticists to Buenos Aires to assist the Grandmothers project. The group included Mary-Claire King, not yet famous for discovering the *BRCA* breast cancer genes.

Figure 18.8 Forensic DNA typing helped the Grandmothers of the Plaza de Mayo reunite families after war. *Alejandro Pagni/AFP/Getty Images*

Because DNA fingerprinting was too new to use, the researchers relied on the older way to match people according to variability of the six human leukocyte antigen (HLA) genes (see section 19.1).

The HLA proteins dot blood cells and for many years were used to type organs for transplant. To improve the test, Dr. King added analysis of a 1200-base-long DNA sequence in the mitochondrial genome that is highly variable and does not encode protein. Like all mtDNA, it is passed from mothers only. Tracking female relatives was practical, because documentation and conversation tended to focus on women. In addition, mtDNA is more abundant and hardier than the DNA from the nucleus that yields the STRs used in forensic testing. The mitochondrial approach worked so well that Argentina established a voluntary DNA databank.

The Grandmothers project solved 50 cases: 44 of the children had been "gifts" to military leaders. Adoptive parents of the other 6 children were likely unaware of the kidnapping. Children raised by adults who *had* known of the kidnapping learned that their adoptive parents knew that the biological parents had been murdered.

DNA testing is still used to reunite families. Testing for the 20 CODIS markers in saliva or cheek samples is used in immigration cases in the United States when other evidence of relationships is not available. The DNA is analyzed at labs accredited by the American Association of Blood Banks, which include academic centers, the American Red Cross, and forensics/legal companies. The tests are fast, statistically powerful enough to provide a nearly 100 percent "probability of relationship" in matching offspring to parents, but aren't as useful beyond identifying parents or full siblings because the percentage of the genome shared drops (see table 7.3).

Forensic DNA Phenotyping

Comparing DNA repeats that do not encode protein is an unbiased way to reveal genetic relationships. However, protein-encoding genes can also be used to solve crimes, particularly those in which body parts with defining features, such as heads, are missing. That was the case for Swedish journalist Kim Wall.

In August 2017, Wall had gone on a private submarine with its famous inventor. A few days after friends noticed her missing, a headless torso washed up on an island near Copenhagen. Then a leg appeared, then the other leg, and finally a severely damaged head. DNA from the body parts matched DNA from blood on the submarine.

Although the inventor claimed innocence in dismembering the journalist, a **forensic DNA phenotyping** test said otherwise. The then-new test considered SNP markers to identify variations of eye, hair, and skin color of the victim, and predicted probabilities using allele frequencies in 59 global populations (**figure 18.9**). The findings matched the phenotype of Kim Wall—a redhead with blue eyes, freckles, and pale skin. The technique can also analyze the shapes, sizes, and distances between various body parts, such as facial features. The second question in Case Studies and Research Results in chapter 1 is about an artist who collects DNA in public places and uses a forensic DNA phenotyping tool to reconstruct faces.

Limitations of Forensic DNA Evidence

DNA profiling is based on population genetics, but it requires logic and consideration of details from a crime scene to avoid false accusations, such as how the DNA got to where it was found. Conventional sources, such as fingerprints, fibers, scratches, and footprints, also provide context to interpret DNA evidence. Wrong conclusions can be reached in a few ways.

Background DNA refers to picking up traces of DNA left at a place that later became a crime scene. This was the case for Amanda Knox, a young woman from the United States who was imprisoned for 5 years in Italy for allegedly stabbing her roommate. Traces of Knox's DNA were found on the murder weapon, which was a knife, but it was there because she had used it previously to cut bread. Starch grains were indeed found on the knife. Amanda Knox was eventually exonerated of killing Meredith Kercher.

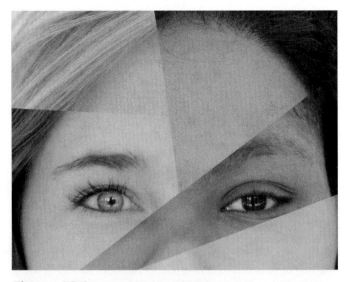

Figure 18.9 Forensic phenotyping. SNPs corresponding to genes that affect visible traits are used to identify body parts. *Susan Walsh/Purdue School of Science*

Secondary transfer DNA, a form of *touch DNA*, can contaminate a forensic sample when the criminal carries another person's DNA, such as from a handshake. (*Primary transfer* is when the suspect's DNA is on an object, such as a glove or a weapon.) In one complicated case of secondary transfer, a professor was accused of killing his wife. When he and his wife had shared a towel earlier on the day of the murder, his skin cells had been transferred to her face. Later, the murderer, wearing gloves, touched her face, picked up the husband's cells, and unknowingly transferred them to his weapon. Reconstructing this scenario helped to exonerate the husband.

Rarely, samples become cross-contaminated in the laboratory. This happened to a man in the United Kingdom arrested for spitting in public. He was accused of a rape and served 5 months in jail, until the error was discovered. The same disposable plastic plate had been used for the man's saliva sample and then for semen from the rape victim. Although three months earlier mobile phone records showed that the man, or at least his phone, had been far from the crime scene, it was not until the DNA mix-up was revealed that he was exonerated.

Rarely, an innocent person's DNA will by chance match several STRs in DNA from a crime scene. A few notable cases like this have led to very elderly and unwell people in nursing homes being accused of heinous crimes, until a more likely candidate or explanation arose. **Bioethics** considers whether DNA collected to investigate one case should be used to solve a past case.

Genetic Privacy

The ease of assigning highly individualized genetic name tags is helpful in criminal investigations, but it poses privacy issues. Consider a *DNA dragnet,* which is a forensic approach that compiles DNA profiles of all residents of a town where a violent crime is unsolved. Sir Jeffreys conducted some of the first DNA dragnets in the late 1980s. The largest to date occurred in 1998 in Germany, where more than 16,000 men had their DNA profiled in a search for the man who raped and murdered an 11-year-old. The dragnet indeed caught the killer.

A controversial application of DNA profiling is a familial DNA search, which is based on the fact that close relatives share large portions of their genomes. DNA from a crime scene is compared to DNA in databases from convicted felons. If nearly half of the CODIS sites match, then a first-degree relative of the convict becomes a suspect. This might be a son, brother, or the father of a man in prison, for example.

Bioethics

Should DNA Collected Today Be Used to Solve a Past Crime?

In 2003, a masked man broke into a woman's home in Salisbury, Maryland, and raped her. She couldn't describe her assailant, but forensic investigators collected and stored the man's DNA. The crime remained unsolved—until 2009.

Alonzo Jay King was arrested for assault in 2009 in Wicomico County, Maryland, for "menacing a group of people with a shotgun." State law required that he provide a DNA sample on a cheek swab, because the crime was considered violent. Maryland law also holds that taking a DNA sample and entering the information in a database is legal as long as the suspect has been arraigned, but the sample must be destroyed if the suspect is not convicted.

When FBI forensic scientists ran King's DNA profile against a database of half a million unidentified DNA samples taken from crime scenes (part of CODIS), the sample matched DNA on the underwear of the victim from the 2003 unsolved rape case. When a second sample from King matched, too, he was brought before a grand jury, indicted, then tried and convicted of the older crime and sentenced to life in prison without parole. However, the conviction was based on the DNA from the scene of the more recent crime. A court of appeals ruled that taking the DNA in 2009 had been an unlawful seizure and unreasonable search and set the conviction aside.

The case went to the Supreme Court, which ruled in 2013 that taking DNA from King was like taking fingerprints or a photograph—a noninvasive way to establish or confirm identity. The court deemed DNA profiling a "legitimate police booking procedure." The five judges in the assenting opinion argued for the accuracy of DNA profiling compared to other forensic methods and the fact that the parts of the genome that are compared have nothing to do with either traits or health, so privacy was not at issue. The dissenting opinion of four judges focused more on common sense—DNA wasn't needed to identify King in 2009. Everyone knew who he was. What wasn't known, and the DNA revealed, was that he'd committed the *earlier* crime. Whether or not DNA collected at one time for one crime can be used to convict an individual for a *different* crime is the bioethical issue.

Questions for Discussion

1. What should be criteria for requiring a suspect to provide a DNA sample to law enforcement officials?
2. Why would a DNA profile not reveal anything about the arrested individual's health?
3. Do you agree with the majority opinion or the dissenting opinion of the Court? Cite a reason for your answer.
4. How is a DNA profile like a fingerprint and unlike one?
5. Why is destroying DNA evidence futile?
6. How do you think the rape victim might have felt about the use of DNA testing in this case?

In the case of California's "Grim Sleeper," who killed nine women and one girl over more than two decades, the suspect was finally arrested when DNA found on a victim closely matched that of a young man in prison for trafficking weapons. The young man did not have brothers, was too young to have a son old enough to be the Grim Sleeper, but did have a father who could have committed the crimes. After an intense investigation, Los Angeles police collected DNA from the father, a garbage truck driver named Lonnie David Franklin Jr., from a discarded pizza slice. A search of his home turned up 180 photographs of potential victims, which were posted on the police department's website so that the public could help in identifying them. He was convicted in 2016. Yet for every criminal that familial DNA searches identify, some innocent people are accused, based on the chance of sharing CODIS markers with convicted felons.

Privacy issues related to DNA sequences were first raised in the context of DNA dragnets and familial searches of convicted felon databases. In the realm of medical research, investigations attempted to *de-identify* patient specimens by removing personal information, such as birthdays and hometowns, from test results and diagnoses, and instead linking them using codes. This was deemed necessary because of the idea of *genetic exceptionalism*, which designates genetic information as different from other types of medical information, requiring it to be more difficult to access. However, the issue of genetic privacy has exploded as millions of people have voluntarily uploaded their DNA information, obtained from consumer genetic testing companies, into public databases. We may be looking at a future when DNA information is no longer private.

Key Concepts Questions 18.2

a. How does the concept of Hardy-Weinberg equilibrium enable interpretation of DNA profiles?

*b. Distinguish between VNTRs and STRs.

c. How do the terms *heterozygote* and *homozygote* apply to repeated DNA sequences?

d. How is probability used to interpret DNA profiles?

e. List applications of forensic DNA typing.

f. What are limitations of forensic DNA typing?

g. How does forensic DNA typing raise privacy issues?

18.3 Genetic Genealogy

People can trace their ancestry using DNA data at three levels:

1. The major mitochondrial DNA and Y chromosome haplogroups
2. Geographic origins and ethnicity estimates
3. Shared haplotypes of autosomal SNPs

Mitochondrial and Y Chromosome Haplogroups

Chapter 17 introduced mitochondrial DNA and Y chromosome haplogroups, which are used to trace two of the many lineages that contribute to a person's genome (see figure 17.9). Haplo*groups* estimate ancient migration patterns and go back thousands of years, revealing clues to deep ancestry. In contrast, haplo*types* go back only hundreds of years. Figure 17.10 shows how admixture and crossing over mix up and break apart haplotypes over generations.

At least 78 haplogroups are recognized. The most ancient maternal (mtDNA) haplogroup, L, represents a single woman who was the common ancestor of all members of the group, mitochondrial Eve, from Africa. Descendants of six of its subhaplogroups remained in Africa, but descendants of subhaplogroup L3 left the continent, branching into haplogroups M and N and then others. Figure 17.12 traces the spread of Eve's lineages.

Figure 18.10 and **table 18.2** describe the ancestral maternal migration pattern of an Ashkenazi Jewish person (the author). The full haplogroup "address" is K1a1b1a. It is found today almost entirely within the Jewish community and accounts for a fifth of their maternal lineages. Haplogroup K1a1b1a leads back to a woman who lived about 4,500 years ago in what is now central and eastern Europe and was ancestral to the small group of Ashkenazi founders in eastern Europe from about 2,500 years ago.

The 20 Y chromosome haplogroups are designated A through T, with the most ancient, A, representing the "Adam" in parallel to mitochondrial Eve. The Y haplogroups include many subhaplogroups. **Figure 18.11** shows the origin of Y haplogroup E1b1a, which includes African Americans and remains widespread in Africa. Note how many lineages have died out. **Table 18.3** lists selected Y haplogroups.

In 2014, the company providing DNA ancestry tests the longest, AncestryDNA, discontinued tracking mitochondrial and Y chromosome haplogroups, because they represent only two of hundreds or thousands of lineages that contribute to a person's genetic makeup. In their place came profiles based on hundreds of thousands of SNPs spanning the autosomes.

Geographic Origins and Ethnicity Estimates

In figure 1.1, a woman holding the collection device that she used to send her saliva to a consumer genetic testing company declares, "I'm 42 percent Italian, 24 percent Ethiopian, 19 percent Swedish, and 14 percent Scottish!" Some companies return pie charts of geographic origins, others polygons superimposed on geographic maps (**figure 18.12**). These depictions are called ethnicity estimates. They are derived from databases of individuals who know their family histories back at least five generations to people who lived in the same place and had children only with others in their ethnic group. This type of information comes from the documents, photographs, and family stories of traditional genealogy.

Figure 18.10 A representative maternal haplogroup, **K1a1b1a.** The arrows trace migration out of Africa, to eastern Europe and beyond.

way to reveal peoples' origins—hence the term *genetic genealogy*.

Ancestry findings require some approximations and estimates. AncestryDNA began with associating SNP profiles with reported family origins to establish 26 *reference populations*. These are overlapping geographic regions where the population has a relatively distinctive genetic profile, based on an overall 3,000 samples (**table 18.4**). The representation of nations is biased toward Europe, but this may change as more people are tested. The sample numbers in the table are updated every few months.

To derive an ethnicity estimate, an algorithm systematically compares a customer's SNP profile to 25 of the 26 reference populations at a time, omitting a different population at each iteration. If the "left out" population at each iteration does not alter the distribution of the profile in the other 25 populations, then the profile is not in the omitted population. This strategy identifies the populations in which the profile is found. After the geography is thus narrowed down, the SNP pattern frequencies in the customer's populations are used to determine the relative proportions of ancestry from the different parts of the world. The sizes of the overlapping sections of polygons or of pie slices approximate the degree of ancestry from specific regions.

Another type of information, profiles of about 750,000 SNPs covering the autosomes, comes from databases of genetic diversity that were established at about the same time as the publication of the first human genome sequences in 2002. Because some of the SNP databases didn't consider geography, a melding of traditional genealogy and extensive SNP maps became a

Complex computational analyses lie behind ethnicity estimates. For a hypothetical simplified example of the nature of the data, imagine a SNP that is 2000 bases from one tip of chromosome 7. In an Irish population 90 percent of the people might be A/A (2 adenines) and 10 percent A/G, but in

Table 18.2	A Representative Maternal Lineage	
Haplogroup	**Years Ago**	**Migration Pattern**
L	180,000	Within Africa.
L3	65,000	Eastern Africa. Some stayed, some crossed Red Sea into Arabian Peninsula.
N	59,000	Arrived in southwest Asia from Red Sea or Sinai Peninsula.
R	57,000	Southwest Asia. Descendants lived among members of haplogroup N, then branched.
U	47,000	Earliest branch from R. Descendants moved into Europe, Asia, and back to Africa.
K	27,000	Branch of haplogroup U8, in Middle East. Descendants moved throughout east central Asia, Europe, north Africa, and Arabian peninsula.

Figure 18.11 **A representative paternal haplogroup E1b1a, which includes African Americans and Africans.** The generations depicted are actually many generations. Additional letter and number designations with each step indicate new mutations.

Table 18.3	Selected Y Haplogroups
Haplogroup	**Dispersal**
B	Ancient African lineage, highest frequency among Pygmies
C3	From southeast or central Asia to northern Asia and Americas
D	Japan
G	Dispersed from India or Pakistan into central Asia, Europe, and Middle East
H	India, Sri Lanka, Pakistan
I, R1b	Europe
M, S	Papua New Guinea
Q	Links Asia to Americas (Q3 is the only lineage among Native Americans, marked by one mutation that originated during the migration through Beringia).

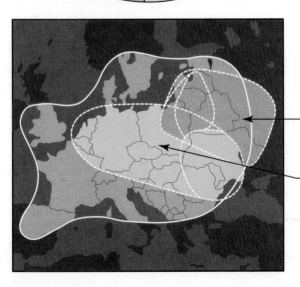

18.12 Ethnicity estimates. (a) This pie chart indicates an individual of mostly western European ancestry. (b) The polygons superimposed on a geographical map indicate a person of Ashkenazi Jewish ancestry.

340 Part 4 *Population Genetics*

Table 18.4 Distribution of 3,000 SNP Profiles Across 26 Reference Populations

Continent	Region	# Samples/3,000
Africa		464 total
	Southeastern Bantu	18
	North	26
	Southcentral hunter gatherers	35
	Benin/Togo	60
	Cameroon/Congo	115
	Ivory Coast/Ghana	99
	Mali	16
	Nigeria	67
	Senegal	28
Europe		1579 total
	Great Britain	111
	Ireland	138
	Eastern	432
	Iberian Peninsula	81
	European Jewish	189
	Scandinavia	232
	Italy/Greece	171
	West	166
	Finland/Northwest Russia	59
Asia		780 total
	Near East	141
	Caucasus	58
	Central	26
	East	394
	South	161
Oceania		36 total
	Melanesia	18
	Polynesia	18
Native American		131

Autosomal Haplotypes and Shared Centimorgans

Mitochondrial DNA and Y chromosome haplogroups represent only a few ancestral lineages. Ethnicity estimates provide broad brushstrokes of where ancestors may have originated. A third type of ancestry information that is more specific comes from the extent of shared autosomal regions between pairs of individuals, measured in centimorgans (cM). The greater the number of centimorgans shared, the closer the relationship.

The algorithms used to compare SNP profiles among people who submit DNA samples to testing companies are trained on many pairs of individuals who know that they are related in a certain way, and assign possible relationships based on shared identical by descent autosomal haplotypes. The calculations compare every submitted DNA sample to every other one, representing many millions of individuals and pairings. The matching chromosomal regions are much larger than the theoretical identical by descent segments used to trace consanguinity in Charles Darwin's family depicted in figure 16.1.

Table 5.2 lists the ranges of shared centimorgans corresponding to degree of relatedness, such as siblings, parent-to-child, and cousins, and table 7.3 lists coefficients of relationship, which estimate the percentage of a genome shared. The centimorgan ranges may be confusing when, for example, two people try to discover whether they are first cousins who share the upper limit of cM or half-siblings who share the lower limit. The ranges are a function of how the genetic material is distributed during meiosis. The longer the chromosome segments, the more closely related two individuals are assumed to be, because it takes time for crossovers at each meiosis to break the linkage (see figure 17.10). The number of autosomal segments over which the SNP haplotypes are distributed is considered, too. The ratio of the number of centimorgans shared to the number of segments falls as relatives become more distant (**figure 18.13**).

For populations that have been through bottlenecks, like the Ashkenazi Jewish people and French-Canadians, the DNA data predicting distant cousin relationships are not as reliable as those predicting closer relationships. This is because of the high degree of endogamy, which is evident as runs of homozygosity (identical by descent sequences on both chromosomes of a pair). That is, many people in such groups are distant cousins. For example, a person related to another individual as a third cousin on one side of a tangled family tree and also as a second cousin on the other side of the tree can have enough shared centimorgans to fall into the category of first cousins.

The 750,000 SNPs consulted to identify relatives from shared autosomal haplotypes were chosen because their patterns vary greatly in populations. The variability is why the SNPs can represent a genome that is so much larger, 3.2 billion bases. The 750,000 SNPs standing in for a genome are equivalent to 24 words in a typical 300-page, 100,000-word book! That leaves room for coincidence—people may share SNP patterns and DNA sequences by chance.

a Nigerian population, perhaps 70 percent are A/A, 15 percent A/G, 10 percent T/A, and 5 percent A/T. The algorithms do not recognize extremely rare SNPs, which is a limitation of this approach. One explanation for people receiving ancestry results that do not include a population mentioned by relatives is that the group is not included in the reference populations. It may be a subset of people hidden within the population substructure.

Figure 18.13 Centimorgans shared across chromosomal segments indicate how two people may be related.

Autosomal matching can lead to unsettling discoveries. A person who shares 3,500 centimorgans with an unknown man, yet none with the man who raised her as a father, learns that the circumstances of her conception may not have been what she had thought. Finding one half-sibling could mean that her mother had an affair or was sexually assaulted. Finding many half-siblings suggests a sperm donor. Thousands of women with infertility conceived using donor sperm, usually from medical students, in the 1950s and 1960s. Later, donor sperm more often came from sperm banks that promised anonymity. Egg donors were also promised anonymity. People who discover from consumer DNA ancestry testing that they are NPE—not parent expected—are helping each other in online support groups. The intersection of genetic information, psychology, and sociology can be upsetting. Chapter 23 examines assisted reproduction.

Key Concepts Questions 18.3

a. What are the three levels at which people use DNA data to trace their ancestry?

*b. Explain how mitochondrial DNA and Y chromosome haplogroups are used to trace lineages.

c. Why do mitochondrial DNA and Y chromosome haplogroups provide limited information?

*d. Define *ethnicity estimate*.

*e. What is the measurement used to estimate how closely two people may be related?

18.4 Forensic STRs Meet Genetic Genealogy SNPs

The FBI uses patterns of alleles of 20 short tandem repeats (STRs) to compare crime scene evidence to a suspect's DNA. Genetic genealogy companies compare patterns of many single nucleotide polymorphisms (SNPs) to connect relatives who have submitted their DNA for testing. In a powerful combination of these two approaches, forensic investigators are consulting SNPs surrounding STRs to identify people who have voluntarily uploaded their DNA sequence information to a database who may be relatives of criminals who left DNA at crime scenes—even if there are no suspects. The availability of data from the consumer testing companies meant that investigators no longer had to rely only on DNA databases of convicted felons.

Solving the Golden State Killer case in April 2018 opened the door to using STRs and SNPs to crack cold cases, and others soon followed, now at the rate of several cases each week. **Bioethics** in chapter 1 introduced the case in the context of genetic privacy. We return to it to illustrate the combination of DNA testing and traditional sleuthing that is being used to eliminate and find suspects.

The Idea to Cross-Reference Information

In 2012, graduate student Melissa Gymrek at the Whitehead Institute had a daring idea: could she use online resources to crack the anonymity promised to people who had submitted their DNA to the U.S. government's 1000 Genomes Project? The project had begun in 2008 to collect human genome sequences to study inherited variation. Part of the informed consent form read, ". . . *it will be hard for anyone to find out anything about you personally from any of this research.*" Gymrek had her doubts. If she could identify people, then she and her professor Yaniv Erlich could warn government officials that the supposedly anonymous database was not private.

Gymrek checked public genealogy databases for surnames corresponding to specific Y chromosome STR haplotypes that were identified in the 1,000 Genomes Project. Further clues came from basic public information such as state of residence and birth year. Consulting family websites and Facebook pages confirmed when she had correctly identified an individual from DNA. Gymrek and her colleagues were able to track women, who of course do not have Y chromosomes to check, by cross-referencing to other DNA sequences of cells that the 1000 Genomes Project had deposited in cell banks.

When the researchers had easily identified 50 people, they contacted officials at the National Institutes of Health. The agency took swift measures to de-identify data, such as by encrypting birth year. *Science* magazine published their paper, "Identifying personal genomes by surname inference," to begin a discussion and possibly prevent further exposure of personal genetic information. The protection promised in 2008 when the project started was impossible by 2012. Gymrek is today at the forefront of using DNA information in forensics.

Soon, researchers began combining STR and SNP maps to extend the marked regions of chromosomes. In 2017 Noah Rosenberg of Stanford University and colleagues looked at two datasets for 872 people: 642,563 SNPs and 13 STRs. Using both sources of information enabled identification of 98 percent of the people. When the researchers used 20 STRs, combined with the SNP maps, they could identify *all* of the people. **Figure 18.14** is a simplified look at how STR and SNP patterns in DNA can impinge upon one another, providing enough information to distinguish individuals.

Finding a Killer

He was known first as the East Area Rapist, then the Visalia Ransacker, and finally the Golden State Killer. Joseph James DeAngelo committed 12 murders, more than 50 rapes, and more than 100 robberies. His reign of terror lasted from 1976 through 1986, in California.

Lead investigator and chief of forensics Paul Holes worked the case from Sacramento for years. He had a long-standing interest in using DNA evidence. In 2002, he learned about a kidnapping case that it had solved. He remembered the case in 2012 when he became interested in identifying suspects using Y chromosome STRs. In 2018, Holes used the technology in an East Area Rapist case, after he found crime scene evidence stored in a freezer from a 1980 rape and murder that had been untouched in previous investigations and yielded abundant semen for analysis. The DNA was extracted and mapped for the SNPs that DNA ancestry companies use, a form of information that could be uploaded to GEDmatch, the voluntary database for ancestry DNA results.

Figure 18.14 Combining SNPs and STRs provides a powerful forensic tool.

Holes used a pseudonym on GEDmatch, not expecting much. So he was stunned when the profile brought up 25 third cousins to the man who had left his DNA at the long-ago crime scene. Sketching family trees, Holes deduced that the killer and a third cousin shared a great-great-great grandparent, building backward in a process called triangulation. The effort led back to the 1800s, when people had very large families. The researchers constructed 25 possible family trees involving more than 1,000 individuals! Then the detective work really began.

Holes and his team collected and compared data for four months, working around the clock, looking through newspaper clippings, census statistics, licenses, 2.7 million entries in police databases, and gravesite locators, eliminating male relatives whose ages and geographical locations didn't fit the times of the crimes. Further DNA testing cleared some suspects. Finally, only two cousins matched the age and geographic profile of the killer. Even further DNA testing discounted one man, leaving Joseph James DeAngelo.

At first the investigators thought it couldn't be DeAngelo, because he had been a police officer, but then they realized that this experience could have helped him while providing cover. Neighbors confirmed the man's strange behavior over the years. Conventional surveillance began.

The investigators needed a DNA sample from their suspect. In March 2018 they got one from swabbing a door handle that DeAngelo touched, but it had a mixture of DNA from several people. Still, the door handle DNA yielded the same GEDmatch profile as the crime scene evidence. That meant the killer's DNA was in the mix. Finally the investigators obtained a purer DNA sample from a facial tissue plucked from DeAngelo's trash. Although the arresting officers prepared themselves for a violent confrontation, the suspect went into custody easily.

Comparing DNA sequences that do not encode proteins but that vary, in repeat patterns or SNP haplotypes, has proven effective in distinguishing individuals. Applications in forensics and in genetic genealogy have shown that knowing a genetic profile of one person can lead to identifying others, using other types of data. Just as the forensic team in Sacramento was able to deduce that Joseph James DeAngelo was the Golden State Killer without his having submitted his DNA to the genealogy database, so too can consumers who take ancestry DNA tests use social media to identify relatives. What kinds of limits will we need to set, if any, on the uses of genetic information?

Key Concepts Questions 18.4

*a. What are the two types of information that investigators used to identify the Golden State Killer?

b. How was it possible to identify people from a supposedly anonymous database?

c. What was the role of GEDmatch in identifying the Golden State Killer?

Study Guide

Summary

18.1 Genetics as an Informational Science

1. Only 30 to 80 sites in genomes need be compared to uniquely identify individuals.
2. Genetic markers used to identify people are in Hardy-Weinberg equilibrium and not subject to natural selection, enabling the assumption that identical by descent DNA sequences that people have in common are due to shared ancestry.
3. **Restriction enzymes** cut DNA at specific sequences and are used to generate pieces of a chromosome containing markers. The most commonly used genetic markers are SNPs and CNVs, particularly STRs.

18.2 Forensic DNA Typing

4. DNA profiling uses numbers of STRs as alleles, with the same number on both homologs considered a homozygote and different numbers a heterozygote.
5. DNA profiling applies statistics (allele frequencies in populations and the product rule) to DNA sequence information from any cell type that has a nucleus.
6. Forensic DNA profiles are based on 20 STRs, mtDNA sequences, and amelogenin gene size to determine sex.
7. Originally DNA profiling separated DNA fragments by charge and compared patterns, on gels, to deduce relationships. STRs replaced **variable number of tandem repeats** (VNTRs) because they withstand harsher conditions. Capillary electrophoresis replaced gels.
8. DNA profiling is used to identify victims of terrorist attacks, natural disasters, genocide, and to reunite families and solve crimes.
9. **Forensic DNA phenotyping** consults SNP markers to detect variants of hair, skin, and eye color to identify the people that body parts came from.
10. Limitations of forensic DNA evidence include background DNA, secondary transfer DNA, cross-contamination, and chance.
11. DNA dragnets and familial searches are effective but may compromise privacy by identifying people other than a suspect.

18.3 Genetic Genealogy

12. Mitochondrial DNA and Y chromosome haplogroups and subhaplogroups are consulted to trace maternal and paternal deep ancestry, which are only two of many lineages.
13. Ethnicity estimates are derived by comparing DNA from consumers to extensive databases that match patterns of SNPs to populations that have been in the same place for many generations with no admixture. The estimate is derived using population frequencies of SNP patterns.
14. Autosomal haplotypes from many SNPs are used to match individuals who send their DNA to consumer DNA testing companies. Ranges of shared centimorgans indicate specific possible familial relationships.

18.4 Forensic STRs Meet Genetic Genealogy SNPs

15. It was possible to identify people in a supposedly private government DNA database by matching Y haplogroups to surnames on social media sites.
16. SNP and STR genotypes are merged to identify people.
17. The Golden State Killer was captured by identifying a third cousin on GEDmatch, setting a precedent for using consumer DNA data in forensics and raising privacy issues.

Key Terms

forensic DNA phenotyping Considering SNP markers to identify variations of eye, hair, and skin color and population frequencies, to identify remains from body parts.

variable number of tandem repeats Repeats that are 10 to 80 DNA bases long, used in early DNA profiling technology.

restriction enzyme An enzyme, typically from a bacterium, that cuts DNA at a specific sequence. Used in early DNA profiling, to create recombinant DNA molecules, and in genome editing.

Review Questions

1. Why is it not necessary to compare whole genome sequences to identify individuals?
2. Why are DNA sequences that are used to distinguish individuals chosen because they are in Hardy-Weinberg equilibrium?
3. What are the two types of genetic markers that are commonly used today in forensics?
4. Why did STRs replace VNTRs in DNA profiling?
5. Why are specific population databases needed to interpret the likelihood that a DNA profile comes from the suspect?
6. Why is a probability value assigned to a CNV? What is the source of the probabilities?
7. Name a circumstance, from the textbook or from the news, in which analysis of DNA repeats in mitochondrial DNA might be valuable.

8. Explain how the information provided by forensic DNA phenotyping and determining haplogroups differs.

9. Discuss three ways that forensic DNA typing can lead to erroneous conclusions.

10. Distinguish among DNA dragnets, familial searches, and the strategy used to capture the Golden State Killer.

11. What are the three levels of DNA information about ancestry?

12. Distinguish a haplogroup from a haplotype. What type of information does each provide?

13. How is figure 18.11 similar to a pedigree and how is it different?

14. How are ethnicity estimates derived?

15. Cite two limitations of DNA ancestry testing.

16. Why is the concept of identical by descent important in ancestry testing used to find relatives?

17. Explain how SNP and STR profiles are used to find suspects in a crime.

Applied Questions

1. Discuss how DNA profiling can lead to conviction as well as exoneration.

2. Explain why all the genetic markers are shared in figure 18.3 but not in figures 18.4 and 18.5. How would analysis of the 20 STRs be used to rule out paternity?

3. MeowPlex is a test that detects 11 four-base STRs on nine autosomes plus the SRY gene of domestic cats, pumas, and ocelots. Describe a scenario where this test might be useful.

4. Describe a situation in the news that uses a DNA test to identify an individual.

5. Should DNA information relevant to health be considered differently than results from other types of tests, such as cholesterol measurement or blood cell counts?

6. What can consumer DNA ancestry testing companies do to increase the diversity of their reference databases? Why is this important?

7. Consult figure 16.1. What is the range of centimorgans that Charles Robert Darwin shared with Josiah Wedgewood II? With his wife Emma?

8. How can social media be used to identify people whose DNA had been stored in a database?

9. A young woman takes a DNA ancestry test and learns that she shares 100 to 150 centimorgans with eight people, 300 to 420 with five people, 690 cM with two others whom she recognizes, and 2600 and 2750 with two others, whom she also recognizes. How are these people related to her?

10. A man who sent his DNA into a testing company is shocked to discover, when his results arrive, eight individuals with whom he shares more than 1500 centimorgans. Suggest an explanation for these findings.

11. Consumer DNA ancestry testing companies and Facebook have pages for people who belong to the same mtDNA or Y chromosome haplogroups. Would you join such a group, and if so, why? What information would you seek, and what information would you post about yourself?

12. Uploading ancestry DNA data to GEDmatch is voluntary. Since the capture of the Golden State Killer, consumer DNA testing companies have added disclaimers warning users that the FBI can access DNA data without a warrant. People who took tests before the Golden State Killer case did not agree to the FBI accessing their data. What might be the reasons for their concern?

Forensics Focus

1. A DNA profiling method that is simplified so that it can be used in the field uses only three STRs. How should the three be selected to best distinguish individuals?

2. The Indian Ocean tsunami of 2004 gave forensic investigators the idea to form and implement a Global DNA Response Team to collect samples after a natural disaster. What rules should such a program have?

3. Irene is 80 years old and lives alone with her black cat Moe. One day when she is taking out the garbage, a man jumps out from behind a garbage can and shoves her. As she struggles to get up, he runs inside, pushes open her door, enters, and grabs her purse, which is next to a slumbering Moe on the kitchen table. Moe, sensing something is wrong or perhaps just upset that his nap has been interrupted, scratches the man, who yelps and forcefully flings the cat against a wall, yanking out one of the animal's claws in the process.

   Meanwhile, outside, Irene is back on her feet, trying to get her bearings when the fleeing thief knocks her down again. This time she blacks out, and the man escapes. A few minutes later a neighbor finds her and calls the police. At the crime scene, they collect a drop of blood on the table, a curly blond hair, a few straight black hairs, several gray hairs, and a cat's claw with human skin under it. The police send Irene to the hospital, Moe to the veterinary clinic, and the claw, blood, and hairs to the state forensics lab. The police then thoroughly search the immediate area and neighborhood, but do not find any suspects. When Irene regains consciousness, she remembers nothing about her attacker.

a. List DNA tests that could help identify the perpetrator.
b. A familial DNA search closely matches two of the forensic samples to a 58-year-old man in prison for murder. One sample matches at 12 CODIS sites and the other at 10 sites. The investigators find that the convict has four sons and four nephews. What should the police do with this information?
c. How might the investigators proceed if consulting the convicted felon database yields no hits? That is, what is another source of DNA information?

4. DNA dragnets have been so successful that some people have suggested storing DNA samples of everyone at birth, so that a DNA profile could be obtained from anyone at any time. Do you think that this is a good idea or not? Cite reasons for your answer.

5. Rufus the cat was discovered in a trash can by his owners, his body covered in cuts and bite marks and bits of gray fur clinging to his claws—gray fur that looked a lot like the coat of Killer, the huge hound next door. Fearful that Killer might attack their other felines, Rufus's distraught owners brought his body to a vet, demanding forensic analysis. The vet suggested that the hair might have come from a squirrel, but agreed to send samples to a genetic testing lab. Identify the samples that the vet might have sent, and what information each could contribute to the case.

6. In a crime in Israel, a man knocked a woman unconscious and raped her. He didn't leave any hairs at the crime scene, but he left eyeglasses with unusual frames, and an optician helped police locate him. The man also left a half-eaten lollipop at the scene. DNA from blood taken from the suspect matched DNA from cheek-lining cells collected from the base of the telltale lollipop at four STRs on different chromosomes. Allele frequencies from the man's ethnic group in Israel are listed beside the profile pattern shown here:

STR1	Homozygote	Allele frequency 0.2
STR2	Allele 1 = 0.3	
	Allele 2 = 0.7	
STR3	Homozygote	Allele frequency = 0.1
STR4	Allele 1 = 0.4	
	Allele 2 = 0.2	

a. How would the electropherogram for STRs 1 and 3 look different than that for STRs 2 and 4?
b. What is the probability that the suspect's DNA matches that of the lollipop rapist by chance? (Do the calculation.)
c. The man's population group has a high level of endogamy. How does this affect the usefulness of the DNA profile? (P.S.—He was so frightened by the DNA analysis that he confessed!)

Case Studies and Research Results

1. From 2007 until 2018, 13 disembodied feet washed up on shores in British Columbia, 12 in running shoes, and 1 in hiking boots. Name a type of DNA test discussed in the chapter that could be used to match the feet to missing persons.

2. Cab driver David Butler was accused of having murdered a sex worker in London because his DNA was found under her fingernails. Investigators found him quickly because the UK has a national DNA database for all citizens. But Butler was innocent. His cells were on the corpse because he had a medical condition that causes flaky skin and some of it had gotten on the bills he gave her in change. Which limitation of DNA profiling does this case illustrate?

3. Of the 26 reference populations use to trace genetic ancestry based on 3,000 DNA samples, 1,579 came from Europe, 464 from Africa, 780 from Asia, 36 from Oceania, and 131 from Native Americans. Explain how this representation of geographic regions can lead to a customer receiving a false negative finding (not belonging to a group expected from family discussions).

PART 5　Immunity and Cancer

CHAPTER 19

Genetics of Immunity

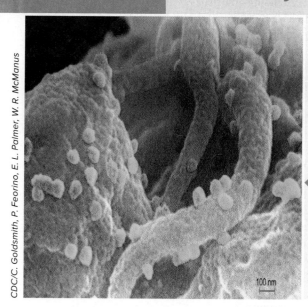

These HIV particles (green) are attached to a white blood cell (pink). In people with two CCR5 mutations, the white blood cells do not make one of two types of receptors that HIV requires to enter. These people cannot become infected. Treatments attempt to mimic having this protective genotype. Perhaps the same approach of mimicking a resistance mutation will be helpful in treating COVID-19, the pandemic disease caused by infection with a novel coronavirus.

Learning Outcomes

19.1 The Importance of Cell Surfaces
1. List the components of the immune system.
2. Describe the basis of blood groups.
3. Explain what human leukocyte antigens are and what they indicate about health.

19.2 The Human Immune System
4. Distinguish among physical barriers, innate immunity, and adaptive immunity.
5. Distinguish between the humoral and cellular immune responses.

19.3 Abnormal Immunity
6. Discuss conditions that result when the immune system is underactive, overactive, and misdirected.

19.4 The Immune Response During Pregnancy
7. Describe the immune system challenge that arises during a pregnancy.
8. Which type of immune system cell controls interactions between the maternal and fetal immune systems?
9. Explain how retained fetal cells were discovered.
10. Discuss how an Rh incompatibility arises.

19.5 Altering Immunity
11. Describe how medical technologies boost or suppress immunity to prevent or treat disease.

12. Explain the requirements for the body to accept an organ from another person.

19.6 Using Genomics to Fight Infection
13. Discuss how we can use knowledge of the genome sequences of pathogens.

The BIG Picture

The immune system enables us to share the planet with other organisms, while it ignores the members of the microbiome that normally live in and on our bodies. Genes control the immune response. We can alter immunity to prevent and treat several types of diseases.

Mimicking a Mutation to Protect Against HIV

In 2008, a 40-year-old man received a stem cell transplant to treat leukemia at a hospital in Berlin. He had been HIV positive for at least a decade, and had taken anti-HIV drugs for 4 years. Leukemia was his first HIV-related illness. Known at first as "the Berlin patient," Timothy Brown became the star of a groundbreaking experiment.

Brown's stem cell donor was both a tissue match and one of the 0.5 percent of Caucasians who are genetically resistant to HIV infection. As a homozygote for the gene *CCR5 delta 32,* the donor did not have a type of receptor, called CCR5, that HIV must bind to enter a cell (figure 19.11 shows this receptor).

The transplanted stem cells eventually replaced Brown's blood and bone marrow with HIV-resistant cells, and all signs of the infection vanished, as did his leukemia. As Brown made headlines, researchers were already planning to mimic the natural mutation that protects against HIV.

In 2017, another patient like Brown had his HIV treated with a stem cell transplant from a *CCR5* delta 32 homozygote. Eighteen months later, the virus was still undetectable. Another approach being tested edits out the *CCR5* genes, discussed in chapters 21 and 22.

19.1 The Importance of Cell Surfaces

We share the planet with plants, microbes, fungi, and other animals, but we can become ill when some of them, or their parts, enter our bodies, causing an infection. The immune system protects us against infectious illnesses. It also detects and destroys cancer cells, the topic of chapter 20. The human immune system is a mobile army of about 2 trillion cells, the biochemicals they release, and the organs where they are produced and stored. Genes control many aspects of immune system function, including susceptibility to infection.

Protection against infection is based on the ability of the immune system to recognize "foreign" or "nonself" cell surfaces that are not part of the body. These include surfaces of certain microorganisms such as bacteria and yeasts; nonliving "infectious agents" such as viruses and prions; and tumor cells and transplanted cells. A stimulated immune system launches a highly coordinated attack that includes general and highly specific responses. Organisms or infectious agents that cause disease are called pathogens. **Clinical Connection 19.1** highlights one commonly encountered type of pathogen—viruses. **Figure 19.1** shows another, a bacterium. A bacterium is a cell; a virus is simpler than a cell.

Figure 19.1 A bacterial pathogen. *Escherichia coli* is a normal resident of the human small intestine, but under certain conditions can produce a toxin that causes severe diarrhea (food poisoning) and can damage the kidneys.
CDC/Janice Haney Carr

Understanding how genes control immunity makes it possible to enhance or redirect the system's ability to fight disease. Mutations can impair immune function, causing immune deficiencies, autoimmune diseases, allergies, and cancer. Genes affect immunity by conferring susceptibilities or resistances to certain infectious diseases. Like other inherited characteristics, degree of immune protection varies from person to person. One individual may suffer frequent respiratory infections, yet another is rarely ill. A study of the immune systems of people who survived the 1918 influenza pandemic revealed that many decades later, they can still rapidly destroy an influenza virus. Yet 50 to 100 million people died of that disease, many in just days.

A few types of genes encode proteins that powerfully affect immunity. **Antibodies** are proteins that directly attack foreign antigens, which are any molecules that can elicit an immune response. Most antigens are proteins or carbohydrates. Genes also specify the cell surface antigens that mark the body's cells as "self."

Blood Groups

Some of the antigens that dot our cell surfaces determine blood types. Figures 5.3 and 5.4 describe the familiar ABO blood types. Humans have 36 major blood types based on protein and carbohydrate antigens on the surfaces of red blood cells. Each blood type includes subtypes, generating hundreds of ways that the topographies of our red blood cells differ from individual to individual. **Table 19.1** lists a few blood groups.

For blood transfusions, blood is typed and matched from donor to recipient. For more than a century, an approach called *serology* typed blood according to red blood cell antigens. A newer way to type blood is to identify the *instructions* for the cell-surface antigens—that is, the genes that encode these proteins. This approach, termed *genotyping*, uses a tiny device that detects 100 distinct DNA "signatures" for blood types. Genotyping is especially useful for people who have a chronic disease that requires multiple transfusions, such as leukemia or

Table 19.1	Blood Groups
Blood Group	**Description**
MN	Codominant alleles *M*, *N*, and *S* determine six genotypes and phenotypes. The antigens bind two glycoproteins.
Lewis	Allele *Le* encodes fucosyltransferase (FUT3), which adds an antigen to the sugar fucose, which the product of the *H* gene places on red blood cells. *H* gene expression is necessary for the ABO phenotype (see figure 5.3). People with *LeLe* or *Lele* have the Lewis antigen on red blood cells and in saliva. People of genotype *lele* do not.
Secretor	People with the *Se* allele secrete A, B, and H antigens into body fluids.

Clinical Connection 19.1

Viruses

A viral infection can bring much misery, but many of the aches and pains are actions of the human immune system.

A virus is a single or double strand of RNA or DNA wrapped in proteins, some with attached sugars, and in some types, an outer lipid bilayer. Section 11.4 discusses how viruses enter and use a host cell's energy resources, protein synthetic machinery, and secretion pathway to reproduce. A virus may have only a few protein-encoding genes, but many copies of the same protein can assemble to form an intricate covering, like the panes of glass in a greenhouse or the crown-like coverings of spike-shaped proteins of a coronavirus. Ebola viruses, for example, produce only seven types of proteins, but these form a structure that can reduce a human body to little more than a bag of blood and decomposed tissue in days. In contrast, the smallpox virus codes for more than 100 different types of proteins.

Many DNA viruses reproduce by inserting their DNA into the host cell's DNA (see figure 19.12). For some RNA viruses, like HIV, the viral enzyme reverse transcriptase copies the viral RNA into DNA, which can then insert into a chromosome. For certain other viruses, including SARS-CoV-2 (figure 9A), the cause of COVID-19, the infected cell responds to the viral RNA as if it is messenger RNA, translating it into protein. For both DNA and RNA viruses, activated viral genes direct the host cell to produce viral proteins, which assemble into viral particles. The cell bursts, releasing many new viruses.

Figure 19A SARS-CoV-2, the virus that causes COVID-19. The triplets of spike proteins that festoon the virus and give it the name coronavirus bind to protein receptors, called ACE2, on cells in the human respiratory tract, causing severe lung disease in some people.
CDC/ Alissa Eckert, MS; Dan Higgins, MAMS

Diverse viruses infect all types of organisms. Viral genetic material cannot repair itself, so the mutation rate may be high. This is one reason we cannot develop an effective vaccine against the common cold, or norovirus, and why new influenza vaccines must be developed each year.

Influenza viruses have RNA as their genetic material and are of three strains: A, B, and C. Influenza A is the most common and comes from birds, sometimes passing through pigs. Each viral strain has subtypes, based on two types of glycoproteins on their surfaces. Specifically, influenza A has many copies of any of 16 variants of a large surface glycoprotein called a hemagglutinin (HA), and many copies of any of nine variants of another surface glycoprotein called a neuraminidase (NA). The 1918 Spanish influenza and the swine influenza of 2009 were "H1N1," whereas the bird influenza of 2004 was "H5N1." Traditional vaccines consist of two types of influenza A and one type of the less common influenza B. A "universal" vaccine that targets parts that all influenza viruses share, such as a stalk-like structure to which hemagglutinin attaches, is being developed.

Some viral diseases have been in existence for centuries, such as influenza, but new ones can arise at any time, as the world learned in 2020. Late the previous year, a novel coronavirus emerged in China, causing severe respiratory symptoms. Because no one had immunity to the new virus and it spread easily, from person-to-person in everyday activities and due to travel, the initial epidemics quickly became a global phenomenon, a pandemic. Six coronaviruses had been known to infect humans, causing "colds" or the respiratory diseases SARS (severe acute respiratory syndrome) and MERS (middle east respiratory syndrome). Like many viruses, SARS-CoV-2 comes from bats, which mount a weak immune response that is enough to keep the animals healthy but that also allows the viruses to flourish. An infectious disease that "jumps" from one animal species to another, like from bats to humans, is termed a zoonosis.

Questions for Discussion

1. How does the structure of a virus differ from that of a cell?
2. How does SARS-CoV-2 cause a human cell to produce more viruses?
3. Why is the mutation rate in viruses high?

sickle cell disease. They produce so many antibodies against so many types of donor blood that it is often difficult to determine their blood types by serology.

The Major Histocompatibility Complex

Many proteins on our cell surfaces are encoded by genes that are part of a 6-million-base-long DNA sequence on the short arm of chromosome 6 called the major histocompatibility complex (MHC). This region includes more than 200 genes, and confers about 50 percent of the genetic influence on immunity. MHC genes are classified into three groups based on their functions.

MHC class III genes encode proteins that are in plasma (the liquid portion of blood) and provide nonspecific immune functions. MHC classes I and II genes encode the **human leukocyte antigens (HLAs)**, so-named because they were first studied in leukocytes (white blood cells). The HLA proteins link to sugars, forming branchlike glycoproteins that extend from cell surfaces.

The proteins that the class I and II HLA genes encode differ in the types of immune system cells they alert. Some HLA glycoproteins bind bacterial and viral proteins, displaying them like badges, which alerts other immune system cells. This action, called antigen processing, is often the first step in an immune response. The cell that displays the foreign antigen is called an **antigen-presenting cell**. **Figure 19.2** shows how a large cell called a **macrophage** displays bacterial antigens. Certain white blood cells called T cells (or T lymphocytes) are also antigen-presenting cells. **Dendritic cells** are antigen-presenting cells found in places where the body contacts the environment, such as in the skin and in the linings of the respiratory and digestive tracts. Dendritic cells signal T cells, initiating an immune response. Dendritic cells and macrophages are known as *sentinel cells* because they alert other components of the immune system to the presence of a pathogen.

A person's HLA "type" identifies all of his or her cells as "self," or as belonging to that individual. The three class I HLA genes encode proteins that vary greatly and are found on all types of cells that have nuclei. These proteins display peptides from within the cells that may come from pathogens. The six class II genes encode proteins found mostly on antigen-presenting cells. In addition to the common HLA markers are more specific markers that distinguish particular tissue types.

An individual's overall HLA type is based on six major HLA genes. So variable are these genes that only 2 in every 20,000 unrelated people match for the six major HLA genes by chance. When transplant physicians attempt to match donor tissue to a potential recipient, they determine how alike the two individuals are for the variants of these six genes. Usually at least four of the genes must match for a transplant to have a reasonable chance of success. HLA genotyping has become very complex because hundreds of alleles are now known.

More than 100 diseases are strongly associated with inheriting particular HLA types. This is the case for ankylosing spondylitis, which inflames and deforms vertebrae. A person with either of two subtypes of an HLA called B27 is 100 times as likely to develop the condition as someone who does not have either form of the antigen. HLA-associated risks are not absolute. More than 90 percent of people who suffer from ankylosing spondylitis have the B27 antigen, which occurs in only 5 percent of the general population.

Figure 19.2 Macrophages are antigen-presenting cells. A macrophage engulfs a bacterium, then displays foreign antigens on its surface, which are held in place by major histocompatibility complex (MHC) self proteins. This "antigen presentation" sets into motion many immune reactions, including binding to a lymphocyte. *CNRI/Science Source*

However, 10 percent of people who have ankylosing spondylitis do *not* have the B27 antigen, and some people who have the antigen never develop the disease.

> **Key Concepts Questions 19.1**
>
> *a. What is the basis of an immune response?
> b. Name the major components of the immune system.
> c. How are genes part of an immune response?
> *d. Distinguish between viruses and bacteria.
> *e. What determines blood types and HLA types?

19.2 The Human Immune System

The human immune system is a network of vessels called lymphatics that transport lymph fluid to bean-shaped structures called lymph nodes. Lymph fluid forms as the liquid portion of blood exits through the walls of the smallest blood vessels and bathes cells throughout the body, as tissue fluid. The tissue fluid is termed *lymph* once it is in lymphatic vessels. Lymph fluid carries white blood cells called lymphocytes and the wandering, scavenging macrophages that capture and degrade bacteria, viruses, and cellular debris. **Figure 19.2** compares a macrophage to a lymphocyte, and figure 2.1 shows a macrophage engulfing bacteria. **B cells** and **T cells** are the two major types of lymphocytes. The spleen and thymus gland are also part of the immune system (**figure 19.3**).

The genetic connection to immunity is the proteins required to carry out an immune response. The immune response attacks pathogens, cancer cells, and transplanted cells with two lines of defense—an immediate generalized **innate immune response** and a more specific, slower **adaptive immune response**. These defenses act after various physical barriers block pathogens. **Figure 19.4** summarizes the basic components of the immune system, discussed in the following sections. ("Immune response" is used synonymously with "immunity.")

Physical Barriers and Innate Immunity

Several familiar structures and fluids keep pathogens from entering the body in the innate immune response: unbroken skin, mucous membranes such as the lining inside the mouth, earwax, and waving cilia that push debris and pathogens up and out of the respiratory tract. In addition, most microbes that reach the stomach perish in a vat of churning acid, while others that survive longer are typically flushed out in diarrhea. These physical barriers are nonspecific. That is, they keep out anything foreign, not just particular pathogens.

If a pathogen breaches physical barriers, then innate immunity provides a rapid, broad defense. The term *innate* means that these general defenses are ready to fight infection all the time, without requiring specific stimulation. Many of the aches and pains we experience from an infection are actually due to the innate immune response, not directly to the actions of the pathogens.

Once inside the body, the pathogen encounters the sentinel cells (macrophages and dendritic cells) that are festooned with proteins, called **toll-like receptors**, that span the plasma membrane. Toll-like receptors bind to broad classes of pathogens that have some of the same molecules on their surfaces, which activates the innate and adaptive immune responses through cascades of signals. As a result, macrophages degrade bacteria while displaying some bacterial antigens like flags to further signal the immune response. Virally infected cells may produce

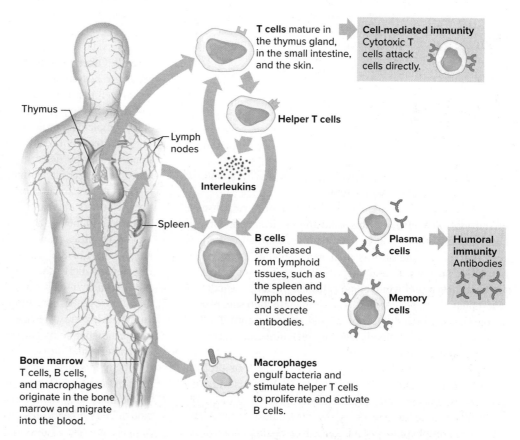

Figure 19.3 Cells of the immune system. T cells, B cells, and macrophages build an overall immune response. All three types of cells originate in the bone marrow and circulate in the blood.

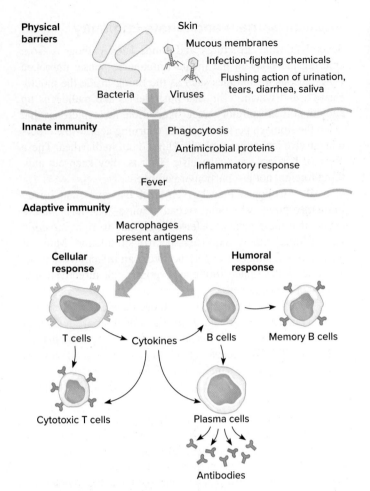

Figure 19.4 Levels of immune protection. Pathogens must breach physical barriers, then encounter nonspecific cells and molecules of the innate immune response. If this is ineffective, the adaptive immune response begins: Antigen-presenting cells stimulate T cells to produce cytokines, which activate B cells to divide and differentiate into plasma cells, which secrete antibodies. Once activated, some B cells "remember" the pathogen, allowing faster responses to subsequent encounters.

Table 19.2		Types of Proteins in Innate Immunity
Protein Class	**Protein**	**Function**
Complement		Puncture bacteria, dismantle viruses, activate mast cells to release histamine, attract phagocytes
Collectins		Recognize how surfaces of bacteria, yeasts, and some viruses differ from the surfaces of human cells
Cytokines	Colony-stimulating factors	Stimulate bone marrow cells to produce lymphocytes
	Interferons	Block viral replication, stimulate macrophages to engulf viruses, stimulate B cells to produce antibodies, attack cancer cells
	Interleukins	Control lymphocyte differentiation and growth, cause fever that accompanies bacterial infection
	Tumor necrosis factor	Stops tumor growth, releases growth factors, stimulates lymphocyte differentiation, dismantles bacterial toxins

virus-fighting biochemicals (interferons) or die by apoptosis, which limits spread of the infection.

A central part of the innate immune response is **inflammation,** which creates a hostile environment for certain types of pathogens at an injury site. Inflammation sends in cells that engulf and destroy pathogens. Such cells are generally called phagocytes, and their engulfing action is phagocytosis. Certain types of white blood cells and the large, wandering macrophages are phagocytes. Also at the infection site, plasma accumulates, which dilutes toxins and brings in antimicrobial chemicals. Increased blood flow with inflammation warms the area, turning it swollen and red.

In addition to inflammation, three classes of proteins participate in innate immunity (**table 19.2**). These are the complement system, collectins, and cytokines. Mutations in the genes that encode these proteins lower resistance to infection.

The **complement** system consists of plasma proteins that assist, or complement, several other defenses. Some complement proteins puncture bacterial plasma membranes, bursting the cells. Other complement proteins dismantle viruses or trigger release of histamine from **mast cells,** another type of immune system cell that is involved in allergies. Histamine dilates blood vessels, increasing fluid flow to the infected or injured area. Still other complement proteins attract phagocytes to an injury site.

Collectins broadly protect against bacteria, yeasts, and some viruses by detecting slight differences in their surfaces from the surfaces of human cells. Groups of human collectins respond to the surfaces of different pathogens, such as the distinctive sugars on yeast, the linked sugars and lipids of certain bacteria, and the surface features of some viruses.

Cytokines play roles in both innate and adaptive immunity. As part of the innate immune response, cytokines called **interferons** alert other components of the immune system to the presence of cells infected with viruses. These cells are then destroyed, which limits the spread of the viral infection. **Interleukins** are cytokines that cause fever, temporarily triggering a higher body temperature that directly kills some infecting bacteria and viruses. Fever also counters microbial growth indirectly, because higher body temperature reduces the iron level in the blood. Because bacteria and fungi require more iron as the body temperature rises, they cannot survive in a fever-ridden body. Phagocytes also attack more vigorously when the temperature rises. Tumor necrosis factor is another type of cytokine that activates other protective biochemicals, destroys certain bacterial toxins, and attacks cancer cells.

Adaptive Immunity

Adaptive immunity must be stimulated into action. It may take days to respond, compared to minutes for innate immunity. Adaptive immunity is highly specific and directed.

B cells and T cells carry out the adaptive immune response, which has two components. In the **humoral immune response,** B cells produce antibodies after activation by T cells. ("Humor" means fluid; antibodies are carried in fluids.) In the **cellular immune response,** T cells produce cytokines and activate other cells. B and T cells differentiate in the bone marrow and migrate to the lymph nodes, spleen, and thymus gland, as well as circulate in the blood and tissue fluid.

The adaptive arm of the immune system has three basic characteristics. It is *diverse,* vanquishing many types of pathogens. It is *specific,* distinguishing the cells and molecules that cause disease from those that are harmless. The immune system also *remembers,* responding faster to a subsequent encounter with a foreign antigen than it did the first time. The first assault initiates a **primary immune response**. The second assault, based on the system's "memory," is a **secondary immune response**. The immune system's memory is why we get some infections, such as mononucleosis, only once. However, upper respiratory infections and influenza recur because the causative viruses mutate, presenting a slightly different face to our immune systems each season.

The Humoral Immune Response—B Cells and Antibodies

An antibody response begins when an antigen-presenting macrophage activates a T cell. The T cell in turn contacts a B cell that has surface receptors that can bind the type of foreign antigen the macrophage presents. The immune system has so many B cells, each with different combinations of surface antigens, that one or more B cells corresponding to a particular foreign antigen is nearly always available. Turnover of B cells is high. Each day, millions of B cells perish in the lymph nodes and spleen, while millions more form in the bone marrow, each with a unique combination of surface molecules.

Once an activated T cell finds a B cell match, it releases cytokines that stimulate the B cell to divide. When the B cell divides, it yields two types of daughter cells (**figure 19.5**). The first, **plasma cells,** are antibody factories, each secreting 1,000 to 2,000 identical antibodies per second into the bloodstream. Plasma cells live only days. These cells provide the primary immune response. Plasma cells derived from different B cells secrete different antibodies. Each type of antibody corresponds to a specific part of the pathogen, like touching a person on different parts of the body. This multipronged attack is called a polyclonal antibody response (**figure 19.6**). The second type of B cell daughter cells, **memory cells,** are far fewer and usually dormant. They respond to the foreign antigen faster and with more force should it appear again. This is a secondary immune response.

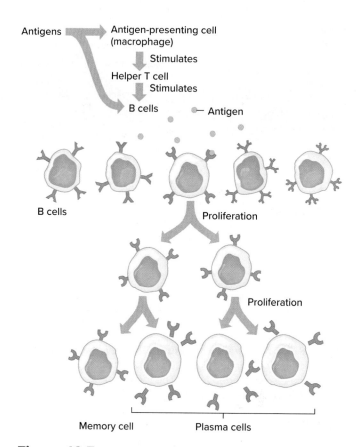

Figure 19.5 Production of antibodies. In the humoral immune response, B cells proliferate and mature into antibody-secreting plasma cells. Only the B cell that binds the antigen proliferates; its daughter cells may develop into rare memory cells or the more abundant plasma cells.

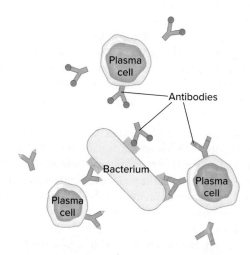

Figure 19.6 An immune response recognizes many targets. A humoral immune response is polyclonal, which means that different plasma cells produce antibody proteins that recognize and bind to different features of a foreign cell's surface.

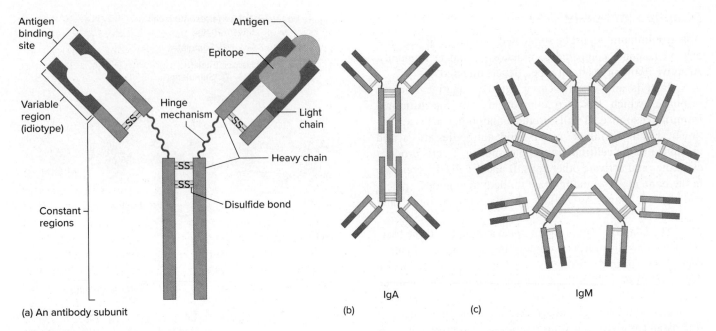

Figure 19.7 **Antibody structure.** The simplest antibody molecule **(a)** consists of four polypeptide chains, two heavy and two light, joined by disulfide bonds between pairs of sulfur atoms. Part of each polypeptide chain has a constant sequence of amino acids, and the remainder varies. The tops of the Y-shaped molecules form antigen binding sites. **(b)** IgA consists of two Y-shaped subunits, and **(c)** IgM consists of five subunits.

An antibody molecule is built of several polypeptides and is therefore encoded by several genes. The simplest type of antibody molecule is four polypeptide chains connected by disulfide (sulfur-sulfur) bonds, forming a shape like the letter Y (**figure 19.7**). A large antibody molecule might consist of three, four, or five such Y-shaped subunits.

In a Y-shaped antibody subunit, the two longer polypeptides are called **heavy chains** and the other two **light chains**. The lower portion of each chain is an amino acid sequence that is similar in all antibody molecules, even in different species. These areas are called constant regions, and they provide the *activity* of the antibody. The amino acid sequences of the upper portions of each polypeptide chain, the variable regions, can differ greatly among antibodies. These parts provide the *specificities* of antibodies to their corresponding antigens.

Antibodies can bind certain antigens because of the three-dimensional shapes of the tips of the variable regions. These specialized ends are **antigen binding sites,** and the parts that actually contact the antigen are called **idiotypes**. The parts of the antigens that idiotypes bind are **epitopes**. An antibody contorts, forming a pocket around the antigen.

Antibodies have several functions. Antibody-antigen binding may inactivate a pathogen or neutralize the toxin it produces. Antibodies can clump pathogens, making them more visible to macrophages, which then destroy them. Antibodies also activate complement, extending the innate immune response. In some situations, the antibody response can be harmful.

Antibodies are of five major types, distinguished by where they act and what they do (**table 19.3**). (Antibodies are also called immunoglobulins, abbreviated *Ig.*) Different antibody types predominate in different stages of an infection.

Table 19.3	Types of Antibodies	
Type*	Location	Functions
IgA	Milk, saliva, urine, and tears; respiratory and digestive secretions	Protects against pathogens at points of entry into body
IgD	On B cells in blood	Stimulates B cells to make other types of antibodies, particularly in infants
IgE	In secretions with IgA and in mast cells in tissues	Acts as receptor for antigens that cause mast cells to secrete allergy mediators
IgG	Blood plasma and tissue fluid; passes to fetus	Protects against bacteria, viruses, and toxins, especially in secondary immune response
IgM	Blood plasma	Fights bacteria in primary immune response; includes anti-A and anti-B antibodies of ABO blood groups

*The letters *A, D, E, G,* and *M* refer to the specific conformation of heavy chains characteristic of each class of antibody.

How can a human body manufacture seemingly limitless varieties of antibodies from the information in a limited number of antibody genes? This great diversity is possible because parts of different antibody genes combine. During the early development of B cells, sections of their antibody genes move to other chromosomal locations, creating new genetic instructions for antibodies.

The assembly of antibody molecules is like putting together many different outfits from the contents of a closet containing 200 pairs of pants, a drawer containing 15 different shirts, and four belts. Specifically, each variable region of a heavy chain and a light chain consists of three sections, called V (for variable), D (for diversity), and J (for joining). The *V, D,* and *J* genes—several of each—for the heavy chains are on chromosome 14, and the corresponding genes for the light chains are on chromosomes 2 and 22. *C* (constant) genes encode the constant regions of each heavy and light chain. Control sequences oversee the mixing and matching of the *V, D,* and *J* genes. A promoter sequence precedes the *V* genes and an enhancer sequence precedes the *C* genes. **Figure 19.8** shows how the genetic instructions for the antibody parts are combined in different ways to encode the heavy and light polypeptide chains.

Enzymes cut and paste the pieces of antibody gene parts. The number of combinations of parts that can be used to build antibodies is so great that virtually any antigen that a person with a healthy immune system might encounter will elicit an immune response.

The Cellular Immune Response—T Cells and Cytokines

T cells provide the cellular immune response. It is called "cellular" because the T cells themselves travel to where they act, unlike B cells, which secrete antibodies into the bloodstream. T cells descend from stem cells in the bone marrow, then travel to the thymus gland ("T" refers to thymus). As the immature T cells, called thymocytes, migrate toward the interior of the thymus, they display diverse cell surface receptors. Then selection happens. As the wandering thymocytes touch lining cells in the gland that are studded with self antigens, thymocytes that do not attack the lining cells begin maturing into T cells, whereas those that harm the lining cells die by apoptosis—in great numbers. Gradually, T cells-to-be that recognize self cells persist, while those that harm body cells are destroyed.

Several types of T cells are distinguished by the types and patterns of receptors on their surfaces, and by their functions. **Helper T cells** have many functions: They recognize foreign antigens on macrophages, stimulate B cells to produce antibodies, secrete cytokines, and activate another type of T cell called a **cytotoxic T cell** (also called a killer T cell). **Regulatory T cells** (T regs) help to suppress an immune response when it is no longer required. The cytokines that helper T cells secrete include interleukins, interferons, tumor necrosis factor, and colony-stimulating factors, which stimulate white blood cells in bone marrow to mature. Cytokines interact with and signal each other, sometimes in complex cascades.

Distinctive surfaces distinguish subsets of helper T cells. Certain antigens called cluster-of-differentiation antigens, or CD antigens, enable T cells to recognize foreign antigens displayed on macrophages. One such cell type, called a CD4 helper T cell, is an early target of HIV. The novel coronavirus SARS-CoV-2 alters the balance of helper T cell subsets in a way that unleashes a combination of cytokines that can destroy organs.

Cytotoxic T cells do not have CD4 receptors but have CD8 receptors. These cells attack virally infected and cancerous cells by attaching to them and releasing chemicals. They do this by linking two surface peptides to form structures called T cell receptors that bind foreign antigens. When a cytotoxic T cell encounters a nonself cell—a cancer cell, for example—the T cell receptors draw the two cells into physical contact. The T cell then releases a protein called perforin, which pierces the cancer cell's plasma membrane, killing it (**figure 19.9**). Cytotoxic T cell receptors also attract cells that are covered with certain viruses, destroying the cells before the viruses on them can enter,

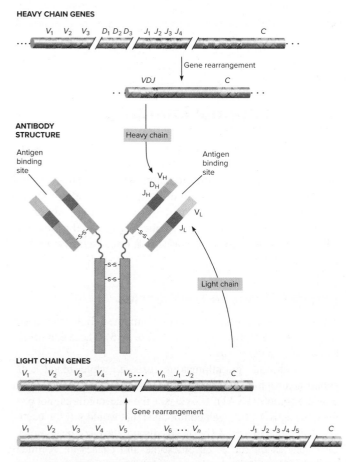

Figure 19.8 Antibody diversity. The human immune system can produce antibodies to millions of possible antigens because each polypeptide is encoded by more than one gene. That is, the many components of antibodies can combine in many ways.

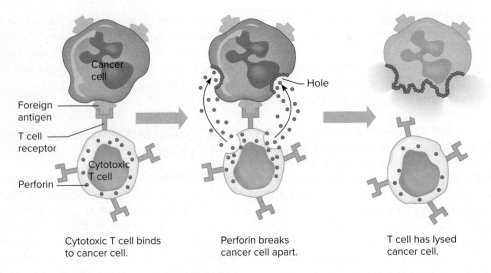

Figure 19.9 **Death of a cancer cell.** A cytotoxic T cell binds to a cancer cell and injects perforin, a protein that pierces (lyses) the cancer cell's plasma membrane. The cancer cell dies, leaving debris that macrophages clear away.

replicate, and spread the infection. Cytotoxic T cells continually monitor body cells, recognizing and eliminating virally infected cells and tumor cells. **Table 19.4** lists and defines types of immune system cells.

Table 19.4	Types of Immune System Cells
Cell Type	**Function**
Macrophage	Presents antigens
	Performs phagocytosis
Dendritic cell	Presents antigens
Mast cell	Releases histamine in inflammation
	Releases allergy mediators
B cell	Matures into antibody-producing plasma cell or into memory cell
T cell	
Helper	Recognizes nonself antigens presented on macrophages
	Stimulates B cells to produce antibodies
	Secretes cytokines
	Activates cytotoxic T cells
Cytotoxic	Attacks cancer cells and cells infected with viruses upon recognizing antigens
Regulatory	Suppresses immune response
Neutrophil	Attacks bacteria
Natural killer	Attacks cancer cells and cells infected with viruses without recognizing antigens; activates other white blood cells

Key Concepts Questions 19.2

a. What are the three major parts of the immune response?

*b. Distinguish innate immunity from adaptive immunity.

c. Describe antibody structure and function.

d. What do the three types of T cells do?

19.3 Abnormal Immunity

The immune system continually adapts to environmental change. Because the immune response is so diverse, its breakdown affects health in many ways. Immune system malfunction may be inherited or acquired, and immunity may be too weak, too strong, or misdirected. Abnormal immune responses may be polygenic, with several genes contributing to susceptibility to infection, or caused by mutation in a single gene.

Inherited Immune Deficiencies

The more than 20 types of inherited immune deficiencies affect innate and/or adaptive immunity (**table 19.5**). These conditions can arise in several ways.

In chronic granulomatous disease, white blood cells called neutrophils engulf bacteria, but, due to deficiency of an enzyme complex (NADPH oxidase), the neutrophils cannot produce the activated oxygen compounds that would kill the bacteria. Mutations in any of five genes disrupt the enzyme complex. One gene is on the X chromosome and causes an X-linked recessive form of the condition, and the other four genes are on autosomes, causing autosomal recessive forms. Antibiotic drugs and gamma interferon (a cytokine) are used to prevent bacterial infections in these patients, and a bone marrow or an umbilical cord stem cell transplant can cure the disease.

Table 19.5	Inherited Immune Deficiencies	
Disease	**Inheritance**	**Defect**
Chronic granulomatous disease	ar, AD, Xlr	Abnormal neutrophils cannot kill engulfed bacteria
Immune defect due to absence of thymus	ar	No thymus, no T cells
Neutrophil immunodeficiency syndrome	ar	Deficiencies of T cells, B cells, and neutrophils
Severe combined immune deficiency (SCID)		
Adenosine deaminase deficiency	ar	No T or B cells
Adenosine deaminase deficiency with sensitivity to ionizing radiation	ar	No T, B, or natural killer cells
IL-2 receptor mutation	Xlr	No T, B, or natural killer cells
X-linked lymphoproliferative disease	Xlr	Absence of protein that enables T cells to bind B cells
X1	Xlr	Abnormal interleukin-2

Note: ar = autosomal recessive; AD = autosomal dominant; Xlr = X-linked recessive.

Mutations in genes that encode cytokines or T cell receptors impair cellular immunity, which primarily targets viruses and cancer cells. Because T cells activate the B cells that manufacture antibodies, abnormal cellular immunity (T cell function) disrupts humoral immunity (B cell function). Mutations in the genes that encode antibody segments, which control how the segments join, or that direct maturation of B cells, mostly impair immunity against bacterial infection. Inherited immune deficiency can also result from defective B cells, which usually increases vulnerability to certain bacterial infections.

Severe combined immune deficiencies (SCIDs) are called "combined" because they affect both humoral and cellular immunity. About half of SCID cases are X-linked. In a less severe form, the individual lacks B cells but has some T cells. Before antibiotic drugs became available, children with this form of SCID died before age 10 of overwhelming bacterial infection. In a more severe form of X-linked SCID, absence of B and T cells causes death by 18 months of age, usually of severe and diverse infections.

A young man named David Vetter taught the world about the difficulty of life without immunity years before AIDS arrived. David had an X-linked recessive form of SCID, called SCID-X1, that caused him to be born without a thymus gland. His T cells could not mature and activate B cells, leaving him defenseless in a germ-filled world. Born in Texas in 1971, David spent his short life in a large vinyl "bubble" enclosure, awaiting a treatment that never came (**figure 19.10**). As he reached adolescence, David wanted to leave his bubble. He did, and received a bone marrow transplant from his sister. Sadly, her bone marrow contained Epstein-Barr virus. Her healthy immune system could fight the virus so she had no symptoms, but the virus caused lymphoma, a cancer of the immune system, in David. He died in just a few weeks. Today gene therapy can treat SCID-X1 as well as another form of SCID called adenosine deaminase deficiency (see figure 22.7).

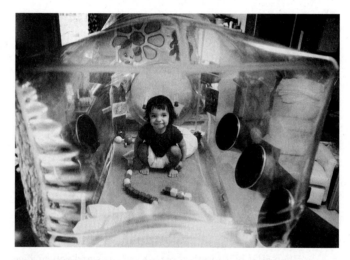

Figure 19.10 **David Vetter, the original "bubble boy," had severe combined immune deficiency (SCID) X1.** Because his T cells could not mature, he was virtually defenseless against infection. Gene therapy can now treat this disease.
Bettman/Contributor/Getty Images

Acquired Immune Deficiency Syndrome

AIDS is not inherited, but acquired by infection with HIV, a virus that gradually shuts down the immune system. The effect of HIV on a human body seems out of proportion to its simple structure. Its genome is a millionth the size of ours, and its nine genes, consisting of about 9,000 RNA bases, encode only 15 proteins! But HIV affects more than 200 human proteins as it invades the immune system.

HIV infection begins as the virus enters macrophages, impairing this first line of defense. In these cells, the virus adheres with its surface protein, called gp120, to two coreceptors on the host cell surface, CD4 and CCR5 (**figure 19.11**). (CCR5 is the glycoprotein altered by mutation that was

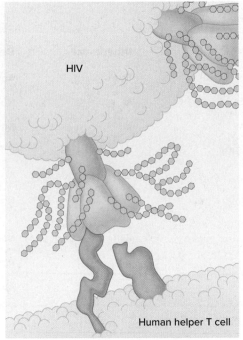

Figure 19.11 HIV binds to a helper T cell. (a) The part of HIV that binds to helper T cells is gp120 (gp stands for glycoprotein). **(b)** The carbohydrate chains that shield the protein part of gp120 move aside as they approach the cell surface, allowing the viral molecule to bind to a CD4 receptor. After also binding to the CCR5 receptor, the viral envelope fuses with the plasma membrane and HIV enters. (The size of HIV is greatly exaggerated.) The man described in the chapter opener received stem cells from a donor whose cells do not have CCR5 co-receptors.

are useless against the next. For several years, the bone marrow produces 2 billion new T and B cells a day. A million to a billion new HIV particles bud daily from infected cells.

So genetically diverse is the population of HIV in a human host that, within days of the initial infection, viral variants arise that resist the drugs used to treat AIDS. HIV's changeable nature is why combining drugs that work in different ways is the most effective strategy to slow the disease into a chronic, lifelong, but treatable illness, instead of a killer (**table 19.6**). The several classes of anti-HIV drugs work at different points of discussed in the chapter opener.) Another glycoprotein, gp41, anchors gp120 molecules into the viral envelope.

When HIV binds both coreceptors, virus and cell surface contort in a way that enables the virus to enter the cell. Once inside, reverse transcriptase copies the viral RNA into DNA, which replicates to form a DNA double helix. The viral DNA then enters the nucleus and inserts into a chromosome. As the viral genes are transcribed and translated, the cell fills with viral pieces, which are assembled into complete new viral particles that eventually bud from the cell (**figure 19.12**).

HIV also enters helper T cells, and once the cells start to die at a high rate, bacterial infections begin, because B cells aren't activated to produce antibodies. Much later in infection, HIV variants arise that can bind to a receptor called CXCR4 on cytotoxic T cells, killing them, thereby leaving the body vulnerable to viral infections and cancer.

Not only does HIV replicate quickly, but it changes quickly and can hide. The virus mutates easily because it cannot repair replication errors, which are frequent—1 per every 5,000 or so bases—because of the "sloppiness" in copying viral RNA into DNA. The immune system cannot keep up; antibodies against one viral variant

Figure 19.12 How HIV infects. HIV integrates into the host chromosome, then commandeers transcription and translation, ultimately producing more virus particles.

Table 19.6 Anti-HIV Drugs

Drug Type	Mechanism
Reverse transcriptase inhibitor	Blocks copying of viral RNA into DNA
Protease inhibitor	Blocks shortening of certain viral proteins
Fusion inhibitor	Blocks ability of HIV to bind a cell
Entry inhibitor	Blocks ability of HIV to enter a cell

infection: blocking binding or entry of the virus into T cells, replicating viral genetic material, and processing viral proteins. The reverse transcriptase inhibitors are also used to prevent HIV infection in people at high risk, which is a strategy called pre-exposure prophylaxis (PrEP).

Clues to developing new drugs to treat HIV infection come from people at high risk who resist infection. Researchers identified variants of four receptors or the molecules that bind to them that block HIV from entering cells by looking at the DNA of people who had unprotected sex with many partners but who never became infected. Some of them were homozygous recessive for the 32-base deletion in the *CCR5* gene described in the chapter opener. The abnormal CCR5 coreceptors are too stunted to reach the infected cell's surface, so HIV has nowhere to dock. Heterozygotes, with one copy of the deletion, can become infected, but they remain healthy for several years longer than people who do not have the deletion. Curiously, the same mutation may have enabled people to survive various plagues in Europe during the Middle Ages. Apparently, more than one pathogen uses the CCR5 entryway into human cells.

Immunological Amnesia

A recently discovered form of immune deficiency is the disappearance of antibodies against the pathogens that caused past infections in an individual after the person contracts measles. It is as if the immune response "forgets" what it has already encountered.

Measles is an infectious disease of childhood that causes cough, high and prolonged fever, and a characteristic rash. Pneumonia and encephalitis (brain inflammation) are complications.

Vaccines against measles, given since the 1960s, have largely vanquished the once-common disease, but in recent years, as some people do not allow their children to be vaccinated, cases of measles have appeared. A study of children in the Netherlands before and after they contracted measles revealed a broad loss of antibody protection. Such "immunological amnesia" could make affected individuals more susceptible to other infectious diseases, such as influenza and tuberculosis.

Autoimmunity

In **autoimmunity**, the immune system produces antibodies, called **autoantibodies**, that attack the body's own cells. About 5 percent of the population has an autoimmune disorder. The signs and symptoms correspond to the cell types under attack (**table 19.7**).

Most autoimmune conditions are not single-gene diseases, but are polygenic. Dozens of genes are each associated with more than one autoimmune disorder, and an autoimmune disorder may result from the actions of variants of several genes that each contributes to susceptibility, perhaps in the presence of a specific environmental trigger such as a food. For example, the digestive condition Crohn's disease is associated with 32 genome regions and a few genes that have variants that increase risk.

Table 19.7 Autoimmune Disorders

Disorder	Symptoms	Autoantibodies Against
Diabetes mellitus (type 1)	Thirst, hunger, weakness, weight loss	Pancreatic beta cells
Graves disease	Restlessness, weight loss, irritability, increased heart rate and blood pressure	Thyroid gland cells
Hemolytic anemia	Fatigue, weakness	Red blood cells
Multiple sclerosis	Weakness, poor coordination, failing vision, disturbed speech	Myelin in the white matter of the central nervous system
Myasthenia gravis	Muscle weakness	Neurotransmitter receptors on skeletal muscle cells
Rheumatic fever	Weakness, shortness of breath	Heart valve cells
Rheumatoid arthritis	Joint pain and deformity	Cells lining joints
Systemic lupus erythematosus	Red facial rash, fever, weakness, joint pain	Connective tissue
Ulcerative colitis	Lower abdominal pain	Colon cells

Although most autoimmune conditions are polygenic, the rarer autoimmune conditions that are caused by a mutation in a single gene can reveal the cellular basis of symptoms. This is the case for autoimmune polyendocrinopathy syndrome, which is caused by dominant mutations in a gene on chromosome 21, *AIRE*. Symptoms include persistent *Candida* infection (a fungus) and impaired functioning of the thyroid and/or adrenal glands. The gene encodes a protein, AIRE, that normally enables T and B cells to distinguish self from "non-self." When the gene is mutant, T cells developing in the thymus are not destroyed if they attack self, and instead are released, causing symptoms if they enter the circulation. At the same time, too few regulatory T cells enable B cells to persist that produce autoantibodies, and these attack self cells, too. The errant T and B cells harm endocrine glands in a sequence. Within a family, affected individuals may appear to have different disorders if different glands are affected, when they actually share an inherited inability to educate both T and B cells to recognize self.

Identifying the genes behind autoimmune disorders is complicated. Some of the more common autoimmune disorders may arise in several ways when parts of the immune response overact. This is the case for systemic lupus erythematosus, better known as lupus. Variants of at least 10 genes can predispose a person to develop lupus. The condition affects 1.5 million people in the United States and more than 5 million worldwide.

In lupus, a butterfly-shaped rash appears on the cheeks. The condition also produces autoantibodies that affect the connective tissue of the kidneys, joints, lungs, brain, spinal cord, and the heart and blood vessels. A person may need dialysis when the kidneys are involved, blood pressure medication to counter increasing pressure in the lungs, and drugs to minimize buildup of fatty deposits on interior artery walls. Lupus can also cause strokes, memory loss, fever, seizures, headache, and psychosis.

Lupus involves several aspects of the immune response, including cell surface characteristics, secretion of interferons, production of autoantibodies, activation of B and T cells, antigen presentation, adhesion of immune system cells to blood vessel linings, inflammation, removal of complexes of immune cells and foreign antigens, and cytokine production.

In general, autoimmunity may arise in several ways:

- A virus replicating in a cell incorporates proteins from the cell's surface onto its own. When the immune system "learns" the surface of the virus to destroy it, it also learns to attack human cells that normally bear the protein.
- Some cells that should have died in the thymus escape the massive die-off, persisting to attack self tissue later on.
- A nonself antigen coincidentally resembles a self antigen, and the immune system attacks both. In rheumatic fever, for example, antigens on heart valve cells resemble those on *Streptococcus* bacteria; antibodies produced to fight a strep throat also attack the heart valve cells.
- If X inactivation is skewed, a female may have a few cells that express the X chromosome genes of one parent (see figure 6.9). The immune system may respond to these cells as foreign if they have surface antigens that are not also on the majority of cells. Skewed X inactivation may explain why some autoimmune diseases are much more common in females.

Section 19.4 highlights a special situation in which two immune systems must coexist—pregnancy.

Allergies

An **allergy** is an immune system response to a substance, called an **allergen,** that does not actually present a threat. Many allergens are particles small enough to be carried in the air and enter a person's respiratory tract. The size of the allergen may determine the type of allergy. For example, grass pollen is large and remains in the upper respiratory tract, where it causes hay fever. But allergens from house dust mites, cat dander, and cockroaches are small enough to infiltrate the lungs, triggering asthma. Asthma is a chronic disease in which contractions of the airways, inflammation, and accumulation of mucus block air flow.

Both humoral and cellular immune responses take part in an allergic reaction (**figure 19.13**). Antibodies of class IgE bind to mast cells, sending signals that open the mast cells, which release allergy mediators such as histamine and heparin. Allergy mediators cause inflammation, with symptoms that may include runny eyes from hay fever, narrowed airways from asthma, rashes, or the overwhelming body-wide allergic

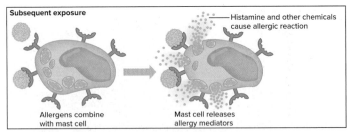

Figure 19.13 Allergy. In an allergic reaction, an allergen such as pollen activates B cells, which divide and give rise to antibody-secreting plasma cells. The antibodies attach to mast cells. When the person encounters the allergen again, the allergen particles combine with the antibodies on the mast cells, which then burst, releasing the chemicals that cause itchy eyes and a runny nose.

reaction called anaphylactic shock. Allergens also activate a class of helper T cells that produce cytokines.

A mutation in one gene may cause several common allergies. For example, half the normal amount of a skin protein called filaggrin can lead to development of atopic dermatitis (a type of eczema), asthma, peanut allergy, and hay fever. Filaggrin is a gigantic protein that binds to the keratin proteins that form most of the outermost skin layer, the epidermis. Filaggrin normally breaks down, releasing amino acids that rise to the top of the skin and keep it moisturized. The epidermis forms a barrier that keeps out irritants, pathogens, and allergens.

People with the rare inherited disease ichthyosis vulgaris have two mutations in the gene for filaggrin, and experience severe skin flaking. Many more individuals—possibly 1 in 10 of us—have a mutation that causes ichthyosis that is so mild that we just treat it with skin lotion. When researchers realized that people with either form of ichthyosis very often also have the itchy red inflamed skin of atopic dermatitis, they realized there could be a connection—when the epidermis is cracked, allergens enter and reach deeper skin layers, where they activate dendritic cells to signal inflammation. Atopic dermatitis results. The dendritic cells also activate immune memory, so that when years later the person inhales the same allergens that once crossed the broken skin, an immune response ensues in the airways, causing a form of asthma due to a skin barrier deficiency. Perhaps treating breaks in the skin early in life can prevent atopic dermatitis, asthma, and other allergies.

> ### Key Concepts Questions 19.3
>
> a. What are some causes of inherited immune deficiencies?
>
> *b. Describe the effects of HIV on the human immune system.
>
> *c. Explain how the immune system malfunctions in autoimmune disorders.
>
> *d. Explain how the immune system malfunctions in allergic reactions.

19.4 The Immune Response During Pregnancy

What tempers the immune system of a pregnant woman to accept cells from her fetus? Half of a fetal genome comes from the father, and so fetal cell surfaces likely include some antigens from him that would be "foreign" to the mother-to-be. Similarly, some of her antigens might be foreign to the fetus. Yet pregnant woman and fetus routinely swap cells. Most women retain fetal cells after pregnancy. The presence of cell populations from more than one individual in one body is called microchimerism ("little mosaic"). We do not completely understand how immune tolerance evolved between pregnant woman and fetus, but there is evidence for immunological "crosstalk" between the two.

T Regs

Regulatory T cells enable pregnant woman and fetus to tolerate differences in cell surface antigens. This is why samples of lymph nodes from fetuses have up to 1 percent maternal cells. The woman's cells stimulate the fetal immune system to produce T regs, which dampen the fetal immune response. The maternal immune system similarly produces T regs that inhibit response to fetal cells. In one experiment, fetal lymph node samples did not react against cells from the pregnant woman unless the fetal regulatory T cells were removed. Children retain these cells for several years.

Systemic Sclerosis

The persistence of fetal cells in a woman's circulation long after the pregnancy was discovered in women who have a condition called systemic sclerosis (also called scleroderma). The skin hardens into an armorlike texture, described by one patient as "the body turning to stone." Symptoms usually begin in middle age and the condition affects mostly women. It was long thought to be autoimmune, but discovery of Y chromosomes in skin cells from patients who are mothers of sons revealed a very different source of the illness—lingering cells from a fetus. Cells from long-ago female fetuses can presumably have the same effect but cannot be distinguished from cells of the mother on the basis of a sex chromosome check.

The degree of genetic difference between a mother and a son may play a role in development of systemic sclerosis. Mothers who have the condition tend to have cell surfaces that are more similar to those of their sons than mothers who do not have the condition. Perhaps the similarity of cell surfaces enabled the fetal cells to escape destruction by the mother's immune system.

Rh Incompatibility

A potential blood type incompatibility provides evidence of the communication between the immune systems of a pregnant woman and a fetus. "Rh" stands for the rhesus factor, discovered in rhesus monkeys. A person is Rh^+ if red blood cells have a surface molecule called the RhD antigen. Rh type is important when an Rh^+ man and an Rh^- woman conceive a child who is Rh^+ (**figure 19.14**). The woman's immune system manufactures antibodies against the few fetal cells that enter her bloodstream. Not enough antibodies form to harm the fetus that sets off the reaction, but the number of antibodies continues to increase. If she carries another Rh^+ fetus, the antibodies can attack the fetal blood supply, causing potentially fatal hemolytic disease of the fetus and newborn. It can be treated at birth with a transfusion of Rh^- blood.

Fortunately, natural and medical protections make this complication rare today. If a woman's ABO blood type is O and the fetus is A or B, her anti-A or anti-B antibodies attack the fetal cells in her circulation before her immune system produces anti-Rh antibodies. A pregnant woman who potentially has an incompatibility with the fetus can take RhoGAM, which is antibody against the Rh antigen. RhoGAM shields fetal cells so that they do not stimulate the maternal immune system to manufacture the harmful antibodies. When the woman becomes

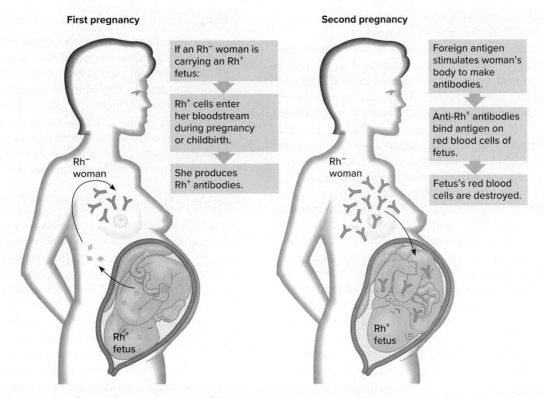

Figure 19.14 Rh incompatibility. Fetal cells entering the pregnant woman's bloodstream can stimulate her immune system to make anti-Rh antibodies, if the fetus is Rh⁺ and she is Rh⁻. A drug called RhoGAM prevents attacks on subsequent fetuses.

pregnant again, fetal DNA in her circulation can be tested to see if it is Rh⁻ or Rh⁺. If the second fetus is Rh⁻, the woman does not need RhoGAM. A first Rh⁺ fetus developing in an Rh⁻ woman can be affected if the woman's blood has been exposed to Rh⁺ cells in amniocentesis, a blood transfusion, an ectopic (tubal) pregnancy, a miscarriage, or pregnancy termination.

Key Concepts Questions 19.4

a. Explain why a woman's immune system might attack a fetus.

*b. What component of the immune system enables a fetus to grow in a woman's body without her immune system attacking?

c. Explain how systemic sclerosis may result from retained fetal cells in the woman's body.

*d. How does Rh incompatibility between pregnant woman and fetus develop?

19.5 Altering Immunity

Medical technologies can alter or augment immune system functions. Vaccines trick the immune system into acting early, preventing infection. Antibiotic drugs, which are derived from organisms such as fungi and soil bacteria, have been used for decades to assist an immune response. Cytokines and altered antibodies are used as drugs. Transplants require suppression of the immune system so that the body will accept a nonself replacement part.

Vaccines

A **vaccine** is an inactive or partial form of a pathogen that stimulates the immune system to alert B cells to produce antibodies. People have thought about developing ways to prevent infectious diseases for a millennium. At first they used inoculation, which is a technique that exposes a person to a live pathogen from an infected person, such as to pus from a sore (see **A Glimpse of History**). When an inoculated person then encounters the natural pathogen, a secondary immune response ensues, even before symptoms arise.

A vaccine differs from an inoculation in that it uses a related pathogen that causes milder symptoms, or a weakened form of the pathogen. Vaccines consisting of entire viruses or bacteria can, rarely, cause illness if they mutate to a pathogenic form. This was a risk of the smallpox vaccine. A safer vaccine uses only the part of the pathogen's surface that elicits an immune response, such as the "spike" protein of a coronavirus. Vaccines against different illnesses can be combined into one injection, or the genes encoding antigens from several pathogens can be inserted into a harmless virus and delivered as a "super vaccine."

Individuals are vaccinated, but the overall effect is at the population level, so it is a public health matter. A substantial proportion of a population must be vaccinated to control an infectious disease. This establishes **herd immunity**, which means that vaccinating enough people in a population stops the spread of infection as pathogens enter protected human bodies (**figure 19.15**). If the population includes

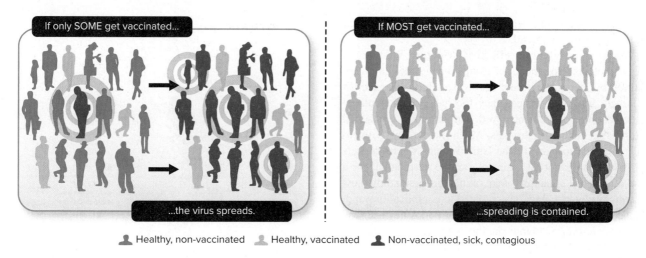

Figure 19.15 **Herd immunity.** If many people in a population are vaccinated against a specific pathogen, then spread of the associated disease will stop.

A GLIMPSE OF HISTORY

The beginnings of what would become vaccine technology date back to the eleventh century in China. Because people saw that those who recovered from smallpox never got it again, they crushed scabs from pox into a powder that they inhaled or rubbed into pricked skin, a strategy called inoculation, which uses the whole, live pathogen. In 1796, the wife of a British ambassador to Turkey witnessed the Chinese method of inoculation and mentioned it to English country physician Edward Jenner. Intrigued, Jenner was inoculated the Chinese way, and then thought of a different approach, vaccination.

It was widely known that people who milked cows contracted a mild illness called cowpox but did not get smallpox. The cows became ill from infected horses. Because the virus seemed to jump from one species to another, Jenner wondered whether exposing a healthy person to cowpox lesions might protect against smallpox. A slightly different virus causes cowpox, but Jenner's approach worked, leading to development of the first vaccine (the word comes from the Latin *vacca,* for "cow"). It was a vaccine and not an inoculation because the preparation was not made from smallpox virus.

Jenner tried his first vaccine on a volunteer, 8-year-old James Phipps. Jenner dipped a needle in pus oozing from a cowpox sore on a milkmaid named Sarah Nelmes, then scratched the boy's arm with it. He then exposed the boy to people with smallpox. Young James never became ill. Eventually, improved versions of Jenner's smallpox vaccine eradicated a disease that once killed millions (**figure 19B**). By the 1970s, vaccination became unnecessary. Several nations have resumed smallpox vaccination in case the virus is ever used as a bioweapon. Smallpox has not naturally infected a human since 1977. Because many doctors are unfamiliar with smallpox and people are no longer vaccinated, an outbreak could be a major health disaster.

Figure 19B **Smallpox.** This woman had such a severe case of smallpox that the lesions joined. *Source: CDC/ Carl Flint; Armed Forces Institute of Pathology*

unvaccinated individuals who come into contact with each other, the disease can spread.

Vaccines are highly effective. George Washington famously protected regiments of troops during the American Revolution by inoculating them against smallpox. The British were largely immune because they too had been inoculated (vaccines hadn't been invented yet), but smallpox devastated the vulnerable Native Americans. Much more recently, measles has returned to parts of the United States where pockets of individuals have not let their children be vaccinated, for a variety of reasons including fear or not understanding how herd immunity works. Measles can lead to complications and, rarely, death. The World Health Organization agency uses the term *vaccine hesitancy* to indicate the "reluctance or refusal to vaccinate despite the availability" of vaccines. In some nations, wars or natural disasters have prevented widespread vaccination and, as a result, some infectious diseases have returned. One danger of refusing vaccination is that an infectious disease such as influenza that is mild or harmless to most people can kill a person who has a compromised immune system.

Most vaccines are injections. New delivery methods include nasal sprays (influenza vaccine) and genetically modified fruits and vegetables. A banana as a vaccine makes sense in theory, but in practice it is difficult to obtain a uniform product. Edible plants are grown from cells that are given genes from pathogens that encode the antigens that evoke an immune response. When the plant vaccine is eaten, the foreign antigens stimulate phagocytes beneath the small intestinal lining to "present" the antigens to nearby T cells. From here, the antigens go to the bloodstream, where they stimulate B cells to divide to yield plasma cells that produce antibodies that coat the small intestinal lining, protecting against pathogens in food. Potatoes and tomatoes have also been genetically modified to function as vaccines. Edible vaccines are still experimental.

Immunotherapy

Immunotherapy amplifies or redirects the immune response. It originated in the nineteenth century to treat disease. Today, a few immunotherapies are in use, with more in clinical trials.

Monoclonal Antibodies Boost Humoral Immunity

When a B cell recognizes a single foreign antigen, it manufactures a single, or monoclonal, type of antibody. A large amount of a single antibody type could target a particular pathogen or cancer cell because of the antibody's great specificity. In contrast, a natural antibody response is polyclonal, in which antibodies against a pathogen have different structures that correspond to different antigens that are part of the pathogen's surface.

In 1975, British researchers Cesar Milstein and George Köhler devised monoclonal antibody (MAb) technology, which mass-produces a single B cell, preserving its specificity and amplifying its antibody type. First, they injected a mouse with a sheep's red blood cells (**figure 19.16**). They then isolated a single B cell from the mouse's spleen and fused it with

Figure 19.16 Monoclonal antibody technology. Monoclonal antibodies are pure preparations of a single antibody type that recognize a single antigen type. They are useful in diagnosing and treating disease because of their specificity. *Source: CDC/ Carl Flint; Armed Forces Institute of Pathology*

a cancerous white blood cell from a mouse. The fused cell, called a hybridoma, had a valuable pair of talents. Like the B cell, it produced large numbers of a single antibody type. Like the cancer cell, it divided continuously.

MAbs are used in basic research, veterinary and human health care, agriculture, forestry, and forensics. They are used to diagnose everything from strep throat to turf grass disease. In a home pregnancy test, a woman places drops of her urine onto a paper strip containing a MAb that binds hCG, the "pregnancy" hormone. The color changes if the MAb binds its target. In cancer diagnosis, if a MAb attached to a fluorescent dye and injected into a patient or applied to a sample of tissue or body fluid binds its target—an antigen found mostly or only on cancer cells—fluorescence indicates disease. MAbs linked to radioactive isotopes or to drugs deliver treatment to cancer cells. The MAb drug trastuzumab (Herceptin) blocks receptors on certain breast cancer cell surfaces, preventing them from receiving signals to divide. MAbs used in humans are made to match human antibodies, because the original mouse-derived products caused allergic reactions.

More than 30 MAb-based treatments are available for human conditions. In addition to cancer, MAbs are used to treat rheumatoid arthritis, multiple sclerosis, cardiovascular disease, lupus, inflammatory bowel disease, and psoriasis.

Cytokines Boost Cellular Immunity

Cytokines, as coordinators of immunity, are used to treat a variety of conditions. However, it has been difficult to develop these body chemicals into drugs because they act only for short periods. They must be delivered precisely where they are needed, or overdose or side effects can occur.

Interferons (IFs) were the first cytokines tested on a large scale, and today are used to treat cancer, genital warts, multiple sclerosis, and some other conditions. Interleukin-2 (IL-2) is a cytokine that is administered intravenously to treat kidney cancer recurrence. Colony-stimulating factors, which cause immature white blood cells to mature and differentiate, are used to boost white blood cell levels in people with suppressed immune systems, such as individuals with AIDS or those receiving cancer chemotherapy. Treatment with these factors enables a patient to withstand higher doses of a conventional drug.

Because excess of another cytokine, tumor necrosis factor (TNF), underlies some diseases, blocking its activity treats some conditions. One drug consists of part of a receptor for TNF. Taking the drug prevents TNF from binding to cells that line joints, relieving rheumatoid arthritis pain. Excess TNF in rheumatoid arthritis prevents cells lining the joints from secreting lubricating fluid.

Transplants

When a car breaks down, replacing the damaged part often fixes the trouble. The same is sometimes true for the human body. The heart, kidneys, liver, lungs, corneas, pancreas, skin, and bone marrow are routinely transplanted, and sometimes several organs at a time. Organs may come from deceased donors or from live donors if they can survive without the parts, such as a kidney or part of a liver.

Although transplant medicine had a shaky start, many problems have been solved. Today, thousands of transplants are performed annually and recipients gain years of life. The challenge to successful transplantation lies in genetics, because individual inherited differences in cell surfaces determine whether a particular recipient will accept tissue from a particular donor.

Transplant Types

Transplants are classified by the relationship of donor to recipient (**figure 19.17**).

An autograft transfers tissue from one part of a person's body to another. A skin graft taken from the thigh to replace burned skin on the chest, or a leg vein that replaces a coronary artery, are autografts. The immune system does not reject the graft because the tissue is self. Technically, an autograft is not a transplant because it involves only one person.

An isograft is a transplant of tissue from a monozygotic twin. Because the twins are genetically identical, the recipient's immune system does not reject the transplant. Ovary isografts are used to treat infertility.

An allograft comes from an individual who is not genetically identical to the recipient, but is a member of the same species. A kidney transplant from an unrelated donor is an allograft. A xenograft transplants tissue from one species to another. (See **Bioethics**.)

Rejection Reactions—or Acceptance

The immune system recognizes most donor tissue as nonself and launches a tissue rejection reaction in which T cells, antibodies, and activated complement destroy the foreign tissue. The greater the difference between recipient and donor cell surfaces, the more rapid and severe the rejection. An extreme example is the hyperacute rejection reaction against tissue transplanted from another species. Donor tissue from another type of animal is usually destroyed in minutes as blood vessels blacken and cut off the blood supply.

Physicians use several approaches to limit rejection to help a transplant recipient survive. These include closely matching the HLA types of donor and recipient and stripping donor tissue of antigens. Gene expression microarrays can be used to better match donors to recipients.

Immunosuppressant drugs inhibit production of the antibodies and T cells that attack transplanted tissue. If recipients receive an infusion of bone marrow stem cells from the organ donors, they may only need immunosuppressant

Figure 19.17 Transplant types. An autograft is a transfer of tissue within an individual. An isograft is between identical twins. An allograft is between members of the same species, and a xenograft is between members of different species.

Bioethics

Replacement Parts

In 1902, a German medical journal reported an astonishing experiment. A physician, Emmerich Ullman, had attached the blood vessels of a patient dying of kidney failure to a pig's kidney set up by her bedside. The patient's immune system rejected the attachment almost immediately.

Nearly a century later, in 1997, a similar experiment took place. Robert Pennington, a 19-year-old suffering from acute liver failure and desperately needing a transplant, survived for 6.5 hours with his blood circulating outside of his body through a living liver removed from a 15-week-old, 118-pound pig named Sweetie Pie. The pig liver served as a bridge until a human liver became available. But Sweetie Pie was no ordinary pig. She had been genetically modified and bred so that her cells displayed a human protein that controlled rejection of tissue transplanted from an animal of another species. Because of this bit of added humanity, plus immunosuppressant drugs, Pennington's body tolerated the pig liver's help for the few crucial hours. Baboons have also been organ donors (**figure 19C**).

Successful xenotransplants would help alleviate the organ shortage. A possible danger of xenotransplants is that people may acquire viruses from the donor organs. Viruses can "jump" species, and the outcome in the new host is unpredictable.

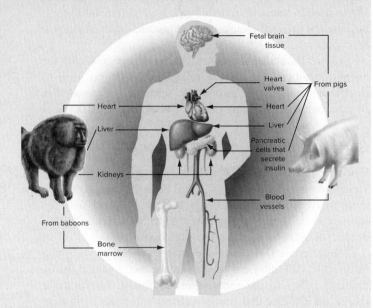

Figure 19C Baboons and pigs can provide tissues and organs for transplant into humans.

For example, a pig virus called PERV—for "porcine endogenous retrovirus"—can infect human cells in culture. However, several dozen patients who received transplants of pig tissue did not show evidence of PERV years later. That study, though, looked only at blood. We still do not know what effect pig viruses can have on a human body. Because many viral infections take years to cause symptoms, introducing a new infectious disease in the future could be the trade-off for using xenotransplants to solve the current organ shortage.

A xenotransplant may seem an extreme measure. Other methods are in development to replace our parts. In one approach, human pluripotent stem cells are incorporated into early embryos of pigs that have a mutation that prevents a specific organ from developing. The human cells divide and develop into the missing pig part. Best of all is to fashion missing parts from a person's own cells, so that the immune system is not alerted. Organoids are organ-like structures grown in laboratory glassware from a patient's cells. They are used widely in research, but may in the future provide replacement parts (**Table 19A**). Another source of replacement parts is to keep body parts of deceased individuals alive in laboratory culture. Yet another technology that may one day provide organs for transplant is 3D bioprinting, which fashions a tissue or part of an organ layer-by-layer using "inks" that consist of cells.

Questions for Discussion

1. Discuss the pros and cons of the following systems for rationing human organs:
   a. first come, first served
   b. closeness of match of cell surface antigens

Table 19A	Human Organoids
Organoid	**Use**
Mini-kidneys	Investigate how polycystic kidney disease progresses
Bladder	Tests treatments for bladder cancer
Esophagus	To investigate cancer, birth defects, gastric reflux
Small intestine (duodenum)	To study nutrient transport, digestion, intestinal disorders, infections, and microbiome interactions
Brain (cortex)	Revealed how Zika virus causes microcephaly; repurposed an HIV drug to treat a genetic brain disease; traced the origin of autism
Blood-brain barrier	To study drug entry into the brain
Retina	To design gene therapies to treat eye diseases; to watch how color vision develops to treat colorblindness and macular degeneration
Liver	To study regeneration, detoxification, metabolism, and clotting

c. ability to pay
d. the importance of the recipient
e. the youngest
f. the most severely ill, who will soon die without the transplant
g. the least severely ill, who are strong enough to survive a transplant
h. people who are not responsible for their condition, such as a nonsmoker with hereditary emphysema versus a person who has emphysema caused by smoking

2. Choose a book or film with a transplant plot and discuss the source of the transplants from the points of view of the donor, the recipient, the families of both, and the government.
3. Discuss the issues that people might find disturbing about growing human organs in pigs.
4. What needs to be tested before human organoids can be used clinically?

drugs for a short time. This is because the bone marrow stem cells help the recipient's body accept the transplanted tissue. Gene expression profiling can identify transplant recipients unlikely to reject their new organs, sparing some people from the need to take immunosuppressants, which have side effects. Yet another approach to increasing the likelihood that a transplant will not be rejected is to use drugs that decrease the activity of T regs.

Rejection is not the only problem that can arise from an organ transplant. Graft-versus-host disease can develop when bone marrow transplants are used to correct certain blood deficiencies and cancers. The transplanted bone marrow, which is actually part of the donor's immune system, attacks the recipient—its new body—as foreign. Symptoms include rash, abdominal pain, nausea, vomiting, hair loss, and jaundice.

Sometimes a problem arises if a bone marrow transplant to treat cancer is too closely matched to the recipient. If the cancer returns with the same cell surfaces as it had earlier, the patient's new bone marrow is so similar to the old marrow that it is equally unable to fight the cancer. The best tissue for transplant may be a compromise: different enough to control the cancer, but not so different that rejection occurs.

Key Concepts Questions 19.5

*a. Explain how a vaccine protects against an infectious disease.
*b. Define *herd immunity*.
c. How are monoclonal antibodies used clinically?
d. How are cytokines used clinically?
*e. Describe the types of transplants.
*f. What happens in organ rejection?

19.6 Using Genomics to Fight Infection

Immunity against infectious disease arises from interactions of two genomes—ours and the pathogen's. Human genome information is revealing how the immune system halts infectious disease. Information from pathogen genomes reveals how they make us sick.

Researchers use genomic information to better understand how an infection affects the body and how infections spread, causing outbreaks and epidemics. Genomic information can inspire new treatment approaches. The DNA sequence for *Streptococcus pneumoniae*, for example, revealed genetic instructions for a huge protein that enables the bacterium to adhere to human cells. Potential drugs could dismantle this adhesion protein. Knowing genome sequences of pathogens can also help to halt the spread of the diseases they cause, as the following two examples illustrate.

Reverse Vaccinology

Older vaccines consisted of parts of pathogens that were detected using standard microbiological approaches. In a strategy called reverse vaccinology, researchers consult genome sequence information from pathogens to identify genes that encode "hidden" antigens that might serve as the basis for a vaccine. If the pathogen proteins that correspond to the genes induce immunity in experimental animals such as mice, the preparation is tested in humans.

The first reverse vaccine targeted meningococcus B, a bacterial infection that inflames the membranes around the brain and/or spinal cord. It can be fatal, or cause deafness or brain damage. Using part of the bacterium as a vaccine is difficult, because some molecules on the surface resemble human proteins so much that a vaccine might direct an immune response against the body. Proteins unique to the bacterium are too varied to easily use as the basis for a vaccine.

To develop a vaccine against meningococcus B, researchers used bioinformatics (computer analysis of DNA sequences) to identify hundreds of bacterial antigens, and tested the antigens in mice to see if they would stimulate an immune response and thereby prevent the infection. A few antigens that were highly specific to the bacterium and protected mice became the basis of the vaccine that is now used for humans. Vaccines against severe acute respiratory syndrome (SARS) and several varieties of influenza were also developed using information from pathogen genomes. Reverse vaccines do not consist of DNA, but rather are based on finding protein antigens that elicit a human immune response using pathogen genome information.

Genome Sequencing to Track Outbreaks

DNA is a changeable molecule, and so over time, even over short periods, mutation alters genome sequences. Epidemiologists can follow the rapidly changing genome sequences in pathogens taken from body fluids or on objects to identify the source of an outbreak and trace the spread of infectious disease. This approach is an example of a strategy called metagenomics, which sequences random pieces of DNA, or entire genomes, that are present in a particular environment, such as a drop of lake water, a subway car seat surface, or a blood or stool sample. Following are two examples of how a genome sequencing approach helped protect public health.

Hospital-Acquired Pneumonia

Infection by the drug-resistant bacterium *Klebsiella pneumoniae* killed six patients at the National Institutes of Health's (NIH) Clinical Center in Bethesda, Maryland, and caused life-threatening pneumonia in five others. *K. pneumoniae* is normally found on the skin and in the mouth and intestines, but certain strains, when inhaled, can severely damage the lungs. Researchers from the nearby National Human Genome Research Institute teamed pathogen genome sequencing with classic epidemiological sleuthing to quickly identify the source of the outbreak at the clinical center, halting the spread and saving lives.

K. pneumoniae lung infections kill half of the people they infect and are resistant to many antibiotic drugs. In addition, the bacteria exist as variants that are so alike genetically that standard microbiological techniques to identify pathogenic variants miss subtle genetic changes. Genome sequencing, however, checks every nucleotide. When researchers sequenced bacterial genomes from the first ("index") case in the outbreak, a 43-year-old woman who recovered, they discovered distinguishing single nucleotide polymorphisms (SNPs) at 41 sites in the 6-million-base genome of *K. pneumoniae*. The bacteria were evolving—mutating—so rapidly that secretions sampled from different parts of the patient's body yielded *K. pneumoniae* with slightly different SNP patterns.

Using an algorithm to compare bacterial sequences, the researchers found that some of the sick people shared SNP patterns with bacteria from the lungs and groin of the index patient, and some infected individuals had SNP patterns like those from the first patient's throat. One patient got the infection from a contaminated ventilator, which provided a clue to transmission of the infection in the hospital.

To see who might be infected but not yet show symptoms, the researchers next sequenced the genomes of all 1,115 patients in the hospital at the time. This part of the investigation identified people who carried the bacteria without developing symptoms and had spread the infection to others, who did get sick. Identifying all of these people, and discovering objects that they had touched, enabled the investigators to control the outbreak.

Using pathogen genome sequencing revealed two new facts about the infection: *K. pneumoniae* survives on inanimate objects and can be transmitted by carriers who never become ill. Comparing pathogen genomes may help to prevent some of the 1.7 million hospital-acquired infections that occur in the United States each year, killing nearly 100,000 people.

In another study, researchers used genome sequencing to link fecal bacteria in municipal wastewater treatment plants to bloodstream infections acquired in nearby hospitals. *Enterococcus faecium* bacteria resistant to the antibiotic vancomycin were behind the outbreaks.

Toxic *Escherichia coli*

Pathogen genome sequencing helped epidemiologists control a more widespread outbreak than that of pneumonia at the NIH Clinical Center. It involved *E. coli,* a bacterium that is a normal part of the human intestinal microbiome (see figure 19.1). A strain of the bacterium, called "STEC" for Shiga-toxigenic *Escherichia coli,* produces a toxin that causes severe diarrhea and, in some individuals, a form of kidney failure that can be lethal.

An outbreak in Germany sickened nearly 4,000 people and resulted in 52 fatalities. The specific strain of *E. coli* had never been seen in an epidemic. Identifying the pathogen was a life-and-death race against time, and standard microbiological culturing techniques were too slow. So researchers conducted a metagenomic analysis on stool samples from people who suffered from diarrhea during the outbreak. Each sample contained DNA from thousands of microorganisms, as well as a great deal of DNA from the patient.

Of the 45 samples that the researchers analyzed, 40 included *E. coli* and 27 of those 40 carried Shiga toxin genes. The investigators then used traditional epidemiological methods to determine that the affected individuals had all eaten raw bean sprouts in a particular town in Germany. The tainted sprouts were traced to a farm where runoff from cattle feces had contaminated them. Five of the patients' stool samples had genomes from other types of bacteria, including *Salmonella, Campylobacter,* and *Clostridium* species. The infection does not respond to antibiotics, and the only treatment is to keep the patient hydrated.

Epidemiologists have used genomic sleuthing to contain other outbreaks. Because it is now possible to sequence a microbial genome rapidly, a health provider can much more accurately diagnose an infection, guiding treatment choice. As researchers continue to identify variants of human immune system genes that protect against infection as well as gene variants in the pathogens that provoke the immune response, we will be better able to prevent, contain, and treat infectious diseases.

Key Concepts Questions 19.6

*a. Explain how reverse vaccinology uses pathogen genome information.

b. How are pathogen genome sequences used to identify the source of an outbreak of an infectious disease and to trace its spread?

Study Guide

Summary

19.1 The Importance of Cell Surfaces

1. The cells and biochemicals of the immune system distinguish self from nonself, protecting the body against infections and cancer.
2. Genes encode immune system proteins and may confer susceptibilities to certain infectious diseases.
3. An antigen is a molecule that elicits an immune response. Patterns of cell surface proteins and glycoproteins determine blood types. **Human leukocyte antigens** (HLAs) are cell surface antigens that bind foreign antigens that **antigen-presenting cells** display to the immune system. **Dendritic cells** and **macrophages** are antigen-presenting cells.

19.2 The Human Immune System

4. If a pathogen breaches the body's physical barriers, **toll-like receptors** on macrophages and dendritic cells bind proteins that several types of pathogens share, triggering the **innate immune response**, which produces the redness and swelling of **inflammation**, plus **complement, collectins,** and **cytokines** such as **interferons** and **interleukins**. The innate immune response is broad and general.
5. The **adaptive immune response** is slower and specific, and it remembers antigens it has responded to.
6. The **humoral immune response** begins when macrophages display foreign antigens near HLAs. This activates **T cells,** which activate **B cells,** which give rise to **plasma cells** that secrete specific **antibodies**. Some B cells give rise to **memory cells**. The **primary immune response** is the first reaction to encountering a nonself antigen, and the **secondary immune response** is a reaction to subsequent encounters.
7. An antibody subunit is Y-shaped and has four polypeptide chains, two **heavy chains** and two **light chains**. Each antibody molecule has regions of constant amino acid sequence and regions of variable sequence.
8. The tips of the Y of each antibody subunit form **antigen binding sites,** which include the more specific **idiotypes** that bind foreign antigens at their **epitopes**.
9. Antibodies bind antigens to form immune complexes large enough for other immune system components to detect and destroy. Antibody genes are rearranged during early B cell development, providing instructions to produce a great variety of antibodies.
10. T cells carry out the **cellular immune response**. Their precursors are selected in the thymus to recognize self cells. **Helper T cells** secrete cytokines that activate other T cells and B cells. A helper T cell's CD4 antigen binds macrophages that present foreign antigens. **Cytotoxic T cells** release biochemicals that kill bacteria and destroy cells covered with viruses. **Regulatory T cells** dampen an immune response.

19.3 Abnormal Immunity

11. Mutations in antibody or cytokine genes, or in genes encoding T cell receptors, cause inherited immune deficiencies. Severe combined immune deficiencies (SCIDs) affect both branches of the immune system.
12. HIV binds to the coreceptors CD4 and CCR5 on macrophages and helper T cells, and, later in infection, triggers apoptosis of cytotoxic T cells. As HIV replicates, it mutates, evading immune system attack. Falling CD4 helper T cell numbers allow opportunistic infections and cancers to flourish. People who cannot produce a complete CCR5 protein resist HIV infection.
13. In *immunological amnesia*, the infectious disease measles causes a loss of the ability to manufacture antibodies against pathogens to which a person has previously been exposed.
14. In **autoimmunity**, the body manufactures **autoantibodies** against its own cells.
15. In people who are susceptible to **allergies, allergens** stimulate IgE antibodies to bind to **mast cells,** which causes the cells to release allergy mediators. Certain helper T cells release selected cytokines.

19.4 The Immune Response During Pregnancy

16. Pregnant women and fetuses swap cells, but regulatory T cells provide immune tolerance in both directions.
17. A study of women with systemic sclerosis revealed fetal cells retained for many years in the woman's body.
18. In Rh incompatibility, an Rh⁻ pregnant woman develops antibodies against an Rh⁺ fetus.

19.5 Altering Immunity

19. A **vaccine** presents a disabled pathogen, or part of one, to elicit a primary immune response. Many individuals in a population must be vaccinated against a particular pathogen to establish **herd immunity,** which stops transmission of the associated infectious disease.
20. Immunotherapy enhances or redirects immune function. Monoclonal antibodies are useful in diagnosing and treating some diseases because of their abundance and specificity. Cytokines are used to treat various conditions.
21. Transplant types include autografts (within oneself), isografts (between identical twins), allografts (within a species), and xenografts (between species). A tissue rejection reaction occurs if donor tissue is too unlike recipient tissue.

19.6 Using Genomics to Fight Infection

22. Knowing the genome sequence of a pathogen provides more information than can be obtained from traditional microbiological classification.

23. Reverse vaccinology uses pathogen genome sequence information to identify antigens that can form the basis of an effective vaccine.
24. Comparing SNP patterns among samples of bacteria taken from patients in an outbreak or epidemic can reveal the spread of the infection, including inanimate objects and carriers. It can also identify bacterial strains that produce toxins. Tracking outbreaks is an application of metagenomics, which is sequencing all the DNA in a particular environment.

Key Terms

adaptive immune response A slow, specific type of immune response following exposure to a foreign antigen.

allergen (al′er-jen) A substance that can provoke an allergic response.

allergy (al′er-jē) A response of the immune system to a substance that is not dangerous.

antibody (ăn′tē-bŏd′ē) A multisubunit protein, produced by B cells, that binds a specific foreign antigen, alerting the immune system or destroying the pathogen that bears the antigen.

antigen binding sites (ăn′tē-jən) Specialized ends of antibody chains.

antigen-presenting cell (ăn′tē-jən) A cell displaying a foreign antigen.

autoantibodies (ô′tō-ăn′tē-bŏdēz) Antibodies that attack the body's own cells.

autoimmunity (ô′tō-ĭ-myōō′nĭ-tē) An immune system attack against the body's own cells.

B cell A type of lymphocyte that secretes antibody proteins in response to nonself antigens displayed on other immune system cells.

cellular immune response (ĭ-myōōn′) Process in which T cells release cytokines that stimulate and coordinate an immune response.

collectins (ko-lek′tinz) Immune system molecules that detect viruses, bacteria, and yeasts.

complement Plasma proteins that have a variety of immune functions.

cytokine (sītō-kīn′) A biochemical that a T cell secretes that controls immune function.

cytotoxic T cells A class of T cells that attack cancer cells and self cells infected with viruses.

dendritic cells Immune system cells that reside in places where the body contacts the environment, serving as antigen-presenting cells to stimulate an immune response.

epitope (ep′ĭ-tōp) Part of an antigen that an idiotype of an antibody binds.

heavy chains The two longer polypeptide chains of an antibody subunit.

helper T cells A class of T cells that recognize nonself antigens on macrophages, stimulate B cells to produce antibodies, secrete cytokines, and activate cytotoxic T cells.

herd immunity The protection from a certain infection that arises when a large majority of a population has been vaccinated, blocking spread of the disease.

human leukocyte antigens (HLAs) (lōōkəsīt′ ăn′tĭ-jənz) Cell surface proteins that are important in immune system function and are encoded by closely linked genes on the short arm of chromosome 6.

humoral immune response (yōō′ mər-əl) Process in which B cells secrete antibodies into the bloodstream.

idiotype (id′ēo-tīp) Part of an antibody molecule that binds an antigen.

inflammation Part of the innate immune response that causes an infected or injured area to swell with fluid, turn red, and attract phagocytes.

innate immune response (ĭ-myōō′n) Components of the immune response that are present at birth and do not require exposure to an environmental stimulus.

interferon (in″tər-fēr′on) A type of cytokine.

interleukin (in″tər-loo′kin) A type of cytokine.

light chains The two shorter polypeptide chains of an antibody subunit.

macrophage A large, wandering cell that engulfs and digests debris, including cells, and presents nonself antigens on its surface, alerting other immune system cells.

mast cells Cells that release histamine, which dilates blood vessels near infected or injured areas. Misdirected mast cell activity causes allergic reactions.

memory cells B or T cell descendants that carry out a secondary immune response.

plasma cell A cell descended from a B cell that produces abundant antibodies of a single type.

primary immune response The immune system's response to initial encounter with a nonself antigen.

regulatory T cells A class of T cells that help to suppress an immune response when it is no longer required.

secondary immune response Immune system activation in response to a second or subsequent encounter with a foreign antigen.

T cell A type of lymphocyte that produces cytokines and coordinates the immune response.

toll-like receptors Proteins embedded in the plasma membranes of immune system "sentinel" cells (macrophages and dendritic cells) that bind proteins common to many pathogens, stimulating innate and adaptive immune responses.

vaccine (vak-sē′n) An inactive or partial form of a pathogen that stimulates antibody production.

Review Questions

1. Match the cell type to the type of biochemical it produces.

   1. mast cell
   2. T cell
   3. plasma cell
   4. macrophage and dendritic cell
   5. all cells with nuclei

   a. antibodies
   b. class II human leukocyte antigens
   c. interleukin
   d. histamine
   e. interferon
   f. heparin
   g. tumor necrosis factor
   h. class I human leukocyte antigens
   i. toll-like receptors

2. What does "nonself" mean? Give an example of a nonself cell in your own body.
3. Distinguish between viruses and bacteria.
4. What is the physical basis of a blood type?
5. Distinguish between using serology or genotyping to type blood.
6. Explain why an HLA-disease association is not a diagnosis.
7. Explain how mucus, tears, cilia, and earwax are part of the immune response.
8. Distinguish between the following:
   a. a T cell and a B cell
   b. innate and adaptive immunity
   c. a primary and a secondary immune response
   d. a cellular and a humoral immune response
   e. an autoimmune condition and an allergy
   f. an inherited and an acquired immune deficiency
9. Which components of the human immune response explain why we experience the same symptoms of an upper respiratory infection (a "cold") when many different types of viruses can cause these conditions?
10. State the function of each of the following immune system biochemicals:
    a. complement proteins
    b. collectins
    c. antibodies
    d. cytokines
    e. filaggrin
11. What does HIV do to the human immune system?
12. Cite three reasons why developing a vaccine against HIV infection has been challenging.
13. What would be the consequences of lacking each of the following?
    a. helper T cells
    b. cytotoxic T cells
    c. B cells
    d. macrophages
14. Explain how the immune system can respond to millions of different nonself antigens using only a few hundred antibody genes.
15. How are SCID and AIDS similar and different?
16. What part do antibodies play in allergic reactions and in autoimmune diseases?
17. What do a plasma cell and a memory cell descended from the same B cell have in common? How do they differ?
18. Why is a deficiency of T cells more dangerous than a deficiency of B cells?
19. Cite two explanations for why autoimmune diseases are more common in females.
20. How do each of the following illnesses disturb immunity?
    a. graft-versus-host disease
    b. SCID
    c. systemic sclerosis
    d. AIDS
    e. atopic dermatitis
21. Explain why a maternal immune system might attack a fetus.
22. What component of the immune system enables a fetus to develop and grow without immune attack from the pregnant woman?
23. Explain how systemic sclerosis may result from retained fetal cells in the maternal body.
24. What is Rh incompatibility?
25. Why is a polyclonal antibody response valuable in the body, but a monoclonal antibody valuable as a diagnostic tool?
26. State how each of the following alters immune system functions:
    a. a vaccine
    b. an antibiotic drug
    c. a cytokine-based drug
    d. an antihistamine drug
    e. a transplant
27. Explain how a "reverse" vaccine is similar to but also different from a traditional vaccine.
28. How can knowing the genome sequence of a pathogen be useful in fighting an outbreak of infectious disease?

Applied Questions

1. Explain how the famous case of Timothy Brown is leading to development of a vaccine against HIV.
2. "Winter vomiting disease," a form of gastroenteritis sometimes called "stomach flu," is caused by a virus called norovirus. It makes a person miserable for a day or two. Why do some people get the illness every year?
3. Children who have Rasmussen's encephalitis have 100 or more seizures a day because they make antibodies

that attack brain cell receptors that normally bind neurotransmitters. Is this condition most likely an inherited immune deficiency, an autoimmune disease, or an allergy? State a reason for your answer.

4. In people with a certain HLA genotype, a protein in the joints resembles an antigen on the bacterium that causes Lyme disease. This infection is transmitted in a tick bite and causes flulike symptoms soon and joint pain (arthritis) a few weeks or months later. When the individuals with the HLA genotype become infected, their immune systems attack the bacteria and their joints. Explain why antibiotics treat the early phase of the disease, but not the arthritis.

5. A person exposed for the first time to Coxsackie virus develops a painful sore throat. How is the immune system alerted to the exposure to the virus? When the person encounters the virus again, why don't symptoms develop?

6. A young woman who has aplastic anemia will soon die as her lymphocyte levels drop sharply. What type of cytokine might help her?

7. In 2019 the World Health Organization declared people who refuse to have their children vaccinated a "top threat to global health," marking the first time that vaccines have made the agency's list of the top 10 biggest threats to global health. Some young adults whose parents refused to have them vaccinated have had it done on their own. Do you think that a government should restrict activities of unvaccinated children to protect the public health?

8. Some bacteria that cause urinary tract infection are resistant to many antibiotic drugs. A patient may try two or even three drugs before a health care provider will order a test to identify the pathogen. Suggest a technology that could increase the probability of initially using a drug to which the bacteria will be sensitive.

Case Studies and Research Results

1. Researchers infected human neural progenitor cells with Zika virus and carried out three types of experiments.
   - Genome editing technology removed one human gene at a time from the neural progenitor cells and identified nine genes that the virus uses to infect human cells. The proteins that the genes encode participate in how the endoplasmic reticulum (ER) processes viral proteins.
   - "Transcriptomics" catalogued the mRNAs that the infected cells produce as infection ensues and compared the mRNAs to those made in brain cells from children who died from microcephaly due to genetic syndromes. Many of the gene variants were overexpressed in both the neural progenitor cells infected with Zika virus in the laboratory culture and from the children who had microcephaly.
   - Transcriptomics revealed activation of genes encoding toll-like receptors, interleukins, tumor necrosis factor, CXCR4, an interferon, and a colony-stimulating factor, in the neural progenitor cells.

   a. One of the roles of the human ER in Zika virus infection is to cut two viral proteins in half, which activates them. What would you have to know in order to target this activity in developing a drug? (See chapter 2.)
   b. How might the transcriptomics findings for microcephaly be used to determine the critical period for Zika virus infection? (See chapter 3.)
   c. Does Zika virus infection activate innate immunity, adaptive immunity, or both, and how do you know?
   d. Which of the experimental results explains why adults who develop symptoms of Zika virus infection experience fever, joint pain, a rash, and conjunctivitis?

2. State whether each of the following situations involves an autograft, an isograft, an allograft, or a xenograft.
   a. A man donates part of his liver to his daughter, who has a liver damaged by cystic fibrosis.
   b. A woman with infertility receives an ovary transplant from her identical twin.
   c. A man receives a heart valve from a pig.
   d. A woman who has had a breast removed has a new breast built using her fatty thigh tissue.

3. Mark and Louise are planning to have their first child, but they are concerned because they think that they have an Rh incompatibility. He is Rh⁻ and she is Rh⁺. Will there be a problem? Why or why not?

4. Twenty-four children and teens at a summer camp for people with cystic fibrosis contract severe lung infections from the multidrug-resistant bacterium *Mycobacterium abscessus*. Describe a technology, based on genetics, that infectious disease experts might use to discover exactly how the infection spread at the camp.

5. In a Dutch family, a father, son, and daughter frequently develop fungal infections (candidiasis) of the mouth, throat, and skin of the feet. Each of the three family members has at least one autoimmune condition. The mother is unaffected. Investigation of the immune responses of the father and children found that they do not make sufficient helper T cells, nor do they produce enough interferon. They also make autoantibodies against two types of interleukins.
   a. What is the likely mode (or modes) of inheritance of the underlying immune system dysfunction that affects the father and two children?
   b. Explain how the immune systems of the father and children are abnormal.
   c. How can autoimmunity indirectly increase the risk of infection?

6. Until a few years ago, a commonly stated reason to refuse to have children vaccinated was a link to autism that appeared in a paper published in *The Lancet* in 1998, in which English physician Andrew Wakefield described "a pervasive developmental disorder" in 12 children. The paper not only had no controls and a tiny sample size, but was retracted when it emerged that attorneys representing allegedly harmed children were paying Dr. Wakefield to make his claims. Investigate how social media has contributed to persistence of the idea that vaccination causes autism.

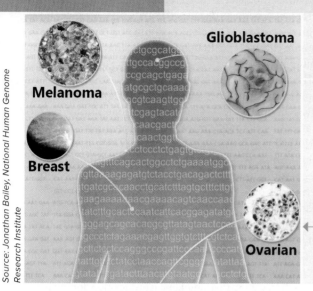

CHAPTER 20

Cancer Genetics and Genomics

Decisions on how to treat cancer are increasingly considering both location (body part) and mutation (genetic changes that occur as a cancer starts, progresses, and responds to treatment).

Learning Outcomes

20.1 Cancer Is an Abnormal Growth That Invades and Spreads
1. List the characteristics of a cancerous tumor.
2. Explain how loss of cell cycle control causes cancer.
3. Explain how most cancers are not inherited, but are genetic.

20.2 Cancer at the Cellular Level
4. Describe cancer cells.
5. Explain how cancer cells can arise from stem cells.

20.3 Cancer Genes and Genomes
6. Distinguish between driver and passenger mutations in cancer.
7. Discuss how mutations in several genes drive cancer.
8. Describe what can happen to chromosomes in cancer cells.
9. Explain how mutations in oncogenes and tumor suppressor genes cause or increase susceptibility to cancer.
10. Discuss environmental factors that increase risk of or cause cancer.

20.4 Diagnosing and Treating Cancer
11. Explain how cancer diagnosis and treatment have become more specific and increasingly based on genetics and genomics.

The BIG Picture

A complex chain reaction of changes at the gene, chromosome, and genome levels causes and propels the common and diverse collection of diseases that we call cancer.

Treating Cancer by Location and Mutation

In the past, we have classified the more than 100 types of cancer according to the cell type, tissue type, or body part initially affected. Treatment was to remove the tumor and/or use drugs that kill all rapidly dividing cells. Today, cancers are being increasingly classified by, and treatment decisions based on, mutations in cancer cells.

Attacking cancer based on mutations is particularly helpful for forms of the disease that are so rare that there is little research and no specific treatments. This is the case for Erdheim-Chester disease, a cancer of macrophages that causes widespread organ failure and bone pain. Only about 500 cases have been identified since the 1930s. About half of patients with the cancer have a mutation, V600E, in a gene called *BRAF*. These patients are benefitting from a targeted drug (Zelboraf, or vemurafenib) developed to treat the skin cancer melanoma, which is also caused by the V600E mutation.

When the head physician at a major cancer center tested the drug on Erdheim-Chester patients with the V600E mutation, results were astonishing. "The first patient we treated had been on the way to hospice. She couldn't walk. Two weeks after joining the clinical trial she came back, walking and feeling fine." She and others have recovered from a cancer that had no treatment—by considering the underlying mutation.

Choosing a drug based on one mutation oversimplifies the genetic complexity of cancer cells, which may have dozens of mutations. The

National Cancer Institute's Molecular Analysis for Therapy Choice (MATCH) program is assigning thousands of cancer patients to receive combinations of targeted drugs tailored to the subsets of 143 well-studied mutations in their cancer cells.

20.1 Cancer Is an Abnormal Growth That Invades and Spreads

Cancer is a type of disease in which certain cells become able to divide more often, leading to an abnormal growth (a tumor) or disruption of the proportion of blood cell types (a "liquid" tumor). One in three people develop cancer. A person may learn of cancer after noticing symptoms and reporting them to a health care provider, who then orders diagnostic tests. A change in bowel habits, lower abdominal discomfort, blood-tinged stools, and fatigue are warning signs of colon cancer, for example. Or, a cancer diagnosis may follow a routine screening test, if the test reveals elevated levels of a biomarker in the blood or shows a growth on a scan. Sometimes cancer diagnosis is secondary, such as discovering a shadow on an X ray taken to diagnose pneumonia. Then cells sampled from the suspicious area may reveal a tumor.

However a cancer is detected, chances are it has been present for years, perhaps decades (**figure 20.1**). A cancer takes time to grow because it is the culmination of a series of genetic and genomic changes—mutations—that enable certain cells to divide more frequently than normal, forming a new growth that eventually crowds healthy tissue.

Eradicating cancer is challenging because of its great diversity. Although a few "driver" mutations initiate a cancer by affecting a small set of cellular processes, this can happen in so many ways that the cancers of no two patients are exactly alike. Nor are multiple tumors within the same body necessarily genetically alike, or even the cells within a single tumor.

Cancer results from a derangement of the cell cycle. It may be set into motion by inherited cancer susceptibility (risk) genes in all cells, or, more commonly, by environmental exposure to factors such as ultraviolet radiation in sunlight or toxins in cigarette smoke that cause somatic mutations in the affected cells. The observation that smokers who develop lung cancer have 10 times as many mutations in their cancer cells as nonsmokers who develop lung cancer indicates the powerful role of the environment in causing cancer.

Researchers have sequenced the genomes of thousands of tumors to better understand the genetic changes that cause and accompany a cancer. Sequences of mutations that occur

Figure 20.1 Levels of cancer. It takes years for a cancer to produce symptoms, as more and more cells lose their specializations and divide more frequently than the cells from which they descend. Mutations—from single-base changes to large-scale chromosomal upheavals—drive the disease, accompanied by changes in gene expression, some of which are responses to environmentally induced epigenetic effects.

over time in somatic cells, the shattering of chromosomes, and changes in gene expression underlie the progression of cancer as it spreads. Mutations may affect the expression of other genes, but so may epigenetic influences, such as DNA methylation and chromatin remodeling (see section 11.2). One researcher calls the accumulating DNA changes that lie behind cancer "genomic scars."

This chapter explores cancer at the cellular, genetic, and genomic levels. Figure 20.1 summarizes the major concepts. The depiction can be read from either direction, but the central portion—the cellular level—is the way of looking at cancer that may ultimately prove the most useful. All of the genes in which mutations increase susceptibility to, or cause, cancer affect three basic cellular pathways: cell fate, cell survival, and genome maintenance. Cell fate refers to differentiation (specialization). Cell survival refers to oxygen availability and preventing apoptosis. Genome maintenance refers to the abilities to survive in the presence of reactive oxygen species and toxins, to repair DNA, to maintain chromosome integrity and structure, and to correctly splice mRNA molecules.

Cancer is a complication of being a many-celled organism. Our specialized cells must follow a schedule of mitosis—the cell cycle—so that organs and other body parts grow appropriately during childhood, stay a particular size and shape throughout adulthood, and repair damage by replacing tissue. If a cell in solid tissue escapes normal controls on its division rate, its daughter cells form a tumor. In the blood, cancer cells divide more frequently than others, taking over the population of blood cells.

A tumor is benign if it grows in place but does not spread into, or "invade," surrounding tissue. A tumor is cancerous, or malignant, if it infiltrates nearby tissue. Pieces of a malignant tumor can enter the bloodstream or lymphatic vessels and travel to other areas of the body, where the cancer cells "seed" the formation of new tumors. The process of spreading is termed **metastasis,** which means "not standing still." Metastasis may be challenging to treat because the new growth may be in an inaccessible part of the body, or genetically different enough from the original, or primary, tumor that drugs that were effective early in the illness no longer work. However, several newer drugs treat metastatic cancer, and they are discussed in section 20.4. Metastases are difficult to detect. If a few sites of metastases appear on a medical scan, there may actually be dozens of growths in the body.

An early hint at the genetic nature of cancer was the observation that most substances known to be **carcinogens** (causing cancer) are also mutagens (damaging DNA). Researchers first discovered genes that could cause cancer in humans in 1976. These genes were versions of genes from certain viruses that had been associated with tumors in birds in studies dating from the mid-nineteenth century. In the 1980s and 1990s, research to identify cancer-causing genes began with rare families that had many young members who had the same type of cancer and specific unusual chromosomes. Finding genes in the affected chromosome regions whose protein products could alter cell cycle control led to the discovery of more than 100 **oncogenes**. An oncogene is a gene that causes cancer when it is expressed when it normally wouldn't be in healthy cells, or is overexpressed. In a sense, oncogenes are inappropriately activated.

Family studies also identified more than 30 **tumor suppressor genes,** which cause cancer when they are deleted or inactivated. The normal function of a tumor suppressor gene is to keep the cell cycle running at the appropriate rate for a particular cell type under particular conditions.

Most mutations that cause cancer are in oncogenes or tumor suppressor genes. The effects of mutations in oncogenes are typically dominant because just one mutation can increase the cell division rate, and the effects of tumor suppressor genes are recessive because if one copy is deleted, the other is still present and protective. A third category of cancer genes includes mismatch mutations in DNA repair genes (see section 12.7) that allow other mutations to persist. When these other mutations activate oncogenes or inactivate tumor suppressor genes, cancer can result. Most DNA repair disorders are inherited in a single-gene fashion, and are quite rare. They typically cause diverse and widespread tumors, often beginning at a young age. We return to the genes behind cancer in section 20.3.

Loss of Cell Cycle Control

The most fundamental characteristic of cancer is the underlying disruption of the cell cycle. **Figure 20.2** repeats the cell cycle shown in figure 2.14. Cancer begins when a cell divides more frequently, or more times, than the noncancerous cell it descended from (**figure 20.3**). Mitosis in a cancer cell is like a runaway train, racing along without signals and controls.

The timing, rate, and number of mitoses a cell undergoes depend on protein growth factors and signaling molecules from outside the cell, and on transcription factors from within. Because these biochemicals are under genetic control, so is the cell cycle. Cancer cells probably arise often because mitoses are so frequent that an occasional cell escapes control. However, the immune system destroys most cancer cells after recognizing tumor-specific antigens on their surfaces.

The discovery of the checkpoints that control the cell cycle revealed how cancer can begin. A mutation in a gene that normally halts or slows the cell cycle can lift the constraint, leading to inappropriate mitosis. Failure to pause long enough to repair DNA can allow a mutation in an oncogene or tumor suppressor gene to persist.

Loss of control over telomere length may also contribute to cancer by affecting the cell cycle. Recall that telomeres, or chromosome tips, protect chromosomes from breaking (see figure 2.15). Human telomeres consist of the DNA sequence TTAGGG repeated thousands of times. The repeats are normally lost from the telomere ends as a cell matures, from 15 to 40 nucleotides per cell division. The more specialized a cell, the shorter its telomeres. The chromosomes in skin, nerve, and muscle cells, for example, have short telomeres. Chromosomes

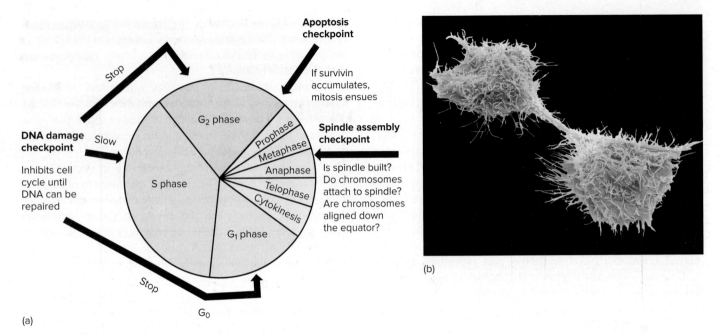

Figure 20.2 Cell cycle checkpoints. (a) Checkpoints ensure that mitotic events occur in the correct sequence. Many types of cancer result from faulty checkpoints. **(b)** These fibrosarcoma cancer cells descend from connective tissue cells (fibroblasts) in bone. The photo captures them in telophase of mitosis. *(b): Steve Gschmeissner/SPL/Getty Images*

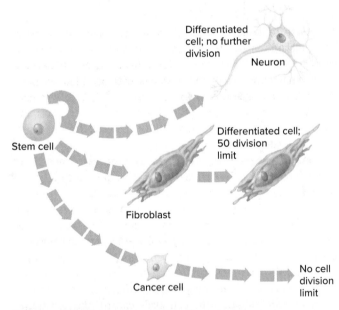

Figure 20.3 Cancer sends a cell down a pathway of unrestricted cell division. Cells may be terminally differentiated and no longer divide, such as a neuron, or differentiated yet still capable of limited cell division, such as a fibroblast (connective tissue cell). Cancer cells either lose specializations or never specialize; they divide unceasingly. (Arrows represent some cell divisions; not all daughter cells are shown.)

in a sperm cell or oocyte, however, have long telomeres. This makes sense—as the precursors of a new organism, gametes must retain the capacity to divide many times.

Gametes keep their telomeres long using an enzyme, telomerase, that consists of RNA and protein. Part of the RNA—the sequence AAUCCC—is a template for the 6-DNA-base repeat TTAGGG that builds telomeres. Telomerase moves down the DNA like a zipper, adding six "teeth" (bases) at a time. Mutation in the gene that encodes telomerase, called *TERT,* causes some cancers.

In normal, specialized cells, telomerase is turned off and telomeres shrink, signaling a halt to cell division when they reach a certain size. In cancer cells, telomerase is turned back on. Telomeres extend, and this releases the normal brake on rapid cell division. As daughter cells of the original abnormal cell continue to divide uncontrollably, a tumor forms, grows, and may spread. Usually the longer the telomeres in cancer cells, the more advanced the disease. However, turning on telomerase production in a cell is not sufficient in itself to cause cancer. Many other things must go wrong for cancer to begin.

Cancer cells can divide continuously if given sufficient nutrients and space. Cervical cancer cells of a woman named Henrietta Lacks, who died in 1951, vividly illustrate the hardiness of these cells. Her cells persist today as standard cultures in many research laboratories. These "HeLa" cells divide so vigorously that when they contaminate cultures of other cells, they soon take over.

Cells vary greatly in their capacity to divide. Cancer cells divide more frequently or more times than the cells from which they arise. Yet even the fastest-dividing cancer cells, which complete mitosis every 18 to 24 hours, do not divide as often as some cells in a normal human embryo do. A tumor grows more slowly at first because fewer cells divide. By the time the tumor is the size of a pea—when it is usually detectable—billions of cells are actively dividing. A cancerous tumor eventually grows faster than surrounding tissue because a greater proportion of its cells are dividing.

Inherited versus Sporadic Cancer

Cancer is *genetic*, because it is caused by changes in DNA, but it is not usually *inherited*. Only about 10 percent of cases result from inheriting a cancer susceptibility allele from a parent. The inherited allele is a germline mutation, meaning that it is present in every cell of the individual, including the gametes. Cancer develops when a second mutation occurs in the other allele in a somatic cell in the affected body part (**figure 20.4**).

The other 90 percent of cancers are sporadic (an isolated event, not part of a familial pattern). Sporadic cancers are caused by somatic mutations, which affect only non-sex cells (see figure 12.2). A sporadic cancer may result from a single dominant mutation or from two recessive mutations in copies of the same gene. The cell loses control of its cell cycle, and accelerated division of its daughter cells forms the tumor. Eventually, tumor cells may contain dozens of mutations that are not in neighboring, healthy cells.

Germline cancers are rare, but they have high penetrance and tend to strike earlier in life than do sporadic cancers. Germline mutations may explain why some heavy smokers develop lung cancer, but many do not; the unlucky ones may have inherited a susceptibility allele that put them halfway to cancer even before they began smoking. Years of exposure to the carcinogens in smoke eventually cause a mutation in a tumor suppressor gene or oncogene of a lung cell, giving it a proliferative advantage. Without the susceptibility gene, two recessive somatic mutations are necessary to trigger the cancer. This, too, can be the result of an environmental insult, but it takes longer for two events to occur than one.

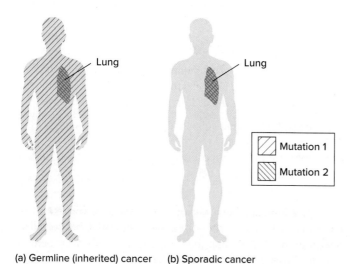

(a) Germline (inherited) cancer (b) Sporadic cancer

Figure 20.4 Germline versus sporadic cancer. (a) In germline cancer, every cell has one gene variant that increases cancer susceptibility and a somatic mutation occurs in cells of the affected tissue. This type of predisposition to cancer is inherited as a single-gene trait and is due to an initial germline mutation. **(b)** A sporadic cancer forms when a dominant mutation occurs in a somatic cell or two recessive mutations occur in the same gene on homologous chromosomes in a somatic cell. An environmental factor can cause the somatic mutations of cancer.

Key Concepts Questions 20.1

*a. Distinguish between a benign and a cancerous tumor.

b. What is the relationship between genes and cancer?

c. What are the three cellular processes that cancer disrupts?

*d. Define *oncogene*.

*e. Define *tumor suppressor*.

f. Distinguish between inherited and sporadic cancers.

20.2 Cancer at the Cellular Level

Cancer begins at the genetic and cellular levels. If not halted by the immune system or treatment, it may spread through tissues and eventually take over organs and organ systems.

Characteristics of Cancer Cells

A cancer cell looks different from a normal cell (**figure 20.5**). Some cancer cells are rounder than the cells they descend from because they do not adhere to surrounding normal cells as strongly as other cells do. Because the plasma membrane is more fluid, different substances cross it. A cancer cell's surface may sport different antigens than are on other cells or different numbers of the antigens that are also on normal cells. The "prostate specific antigen" (PSA) blood test that indicates increased risk of prostate cancer, for example, detects elevated levels of this protein that may come from cancer cell surfaces.

When a cancer cell divides, both daughter cells are cancerous, because they inherit the altered cell cycle control. Therefore, cancer is said to be *heritable* because it is passed from parent cell to daughter cell. A cancer is also *transplantable*, because a cancer cell injected into a healthy animal of the same species will proliferate and form a solid tumor or a blood cancer.

A cancer cell is **dedifferentiated**, which means that it is less specialized than the normal cell types near it that it might have descended from. A skin cancer cell, for example, is rounder and softer than the flattened, scaly, healthy skin cells above it in the epidermis, and is more like a stem cell in both appearance and division rate.

Cancer cell growth is unusual. Normal cells in a container divide to form a single layer; cancer cells pile up on one another. In an organism, this pileup would produce a tumor. Cancer cells that grow all over one another are said to lack contact inhibition—they do not stop dividing when they crowd other cells.

Cancer cells have surface structures that enable them to squeeze into any space, a property called *invasiveness*. They anchor themselves to tissue boundaries, called basement membranes, where they secrete enzymes that cut paths through

The changes that craft a cancer cell from a healthy cell, and the proliferation of cancer cells and eventual invasion and metastasis, take time. Pancreatic cancer, for example, begins 10 to 15 years before the first abdominal pain, then progresses rapidly if not treated. Smoking-induced lung cancer begins with irritation of the lining tissue in respiratory tubes and may not produce symptoms for two decades. Cancer cells on the move eventually reach the bloodstream or lymphatic vessels, which take them to other body parts. This is metastasis. However, cancer cells can remain dormant, not dividing or moving, for years or even decades. This is why doctors typically no longer declare cancer "cured" if five years have passed without detection of recurrence. "No evidence of disease" is the preferred designation.

Once a tumor has grown to the size of a pinhead, interior cancer cells respond to the oxygen-poor environment by secreting a protein called vascular endothelial growth factor (VEGF). It stimulates nearby capillaries (the tiniest blood vessels) to sprout new branches that extend toward the tumor, bringing in oxygen and nutrients and removing wastes. This growth of new capillary extensions to bring in a blood supply is called **angiogenesis,** and it is critical to a cancer's growth and spread. Capillaries may snake into and out of the tumor. Cancer cells wrap around the blood vessels and creep out upon this scaffolding, invading nearby tissue. Cancer cells may also secrete hormones that encourage their own growth, even if the cells they descend from do not produce these hormones.

When cancer cells move to a new body part, the DNA of these secondary tumor cells often mutates, and chromosomes may break or rearrange. Many cancer cells are aneuploid (with missing or extra chromosomes). The metastasized cancer becomes a new genetic entity that may resist treatments that were effective against most cells of the original tumor.

Origins of Cancer Cells

Factors that influence whether or not cancer develops include how specialized the initial cell is and the location of that cell in the tissue. Cancer can begin at a cellular level in at least four ways:

- activation of stem cells that produce cancer cells
- dedifferentiation
- increase in the proportion of a tissue that consists of stem or progenitor cells
- faulty tissue repair

Dedifferentiation is not an all-or-none phenomenon. Most cancer cells are more specialized than stem cells, but considerably less specialized than the differentiated cells near them in a tissue. From which does the cancer cell arise, the stem cell or the specialized cell? Either direction is possible. A cancer cell may descend from a stem cell that yields slightly differentiated daughter cells that retain the capacity to self-renew, or a cancer cell may arise from a specialized cell that loses some of its features and becomes able to divide. Certain stem cells, called **cancer stem cells,** veer from normal development and produce both cancer cells and abnormal specialized cells. Cancer stem cells are found in cancers of the brain, blood, and epithelium (particularly in the breast, colon, and prostate).

Figure 20.5 Cancer cells stand out. (a) A cutaneous melanoma is a cancer of the pigment-producing cells (melanocytes) in the skin. This cancer may have any or all of five characteristics, abbreviated abcde: it is <u>a</u>symmetric, has <u>b</u>orders that are irregular, <u>c</u>olor variations, a <u>d</u>iameter of more than 5 millimeters, and <u>e</u>levation. **(b)** These cutaneous melanoma cells stain orange. The different staining characteristics of cancer cells are due to differences in gene expression patterns between the normal and cancerous states. *(a): McGraw-Hill Education; (b): Nancy Kedersha/Science Source*

healthy tissue. Unlike a benign tumor, an invasive malignant tumor grows irregularly, sending tentacles in all directions. A cancer cell can move. Mutations affect the cytoskeleton (see figure 2.8), breaking down actin microfilaments and releasing actin molecules that migrate to the cell surface, moving the cell from where it was anchored in the tissue.

Figure 20.6 illustrates how cancer stem cells may cause brain tumors. In figure 20.6a, as stem cells give rise to progenitors and then differentiated cells (neurons, astrocytes, and oligodendrocytes), a cell surface molecule called CD133 is normally lost (designated by CD133⁻) at the late progenitor stage. In contrast, in figure 20.6b, cancer cells retain the molecule (designated by CD133⁺), giving rise to a tumor. Some progenitor cells that descend from a cancer stem cell can relentlessly divide, and they ultimately accumulate, forming a brain tumor.

Cancer may also begin when cells lose some of their distinguishing characteristics as mutations occur when they divide. Or, cells on the road to cancer may begin to express "stemness" genes that override signals to remain specialized (**figure 20.7**).

Figure 20.6 Cancer stem cells. (a) In the developing healthy brain, stem cells self-renew and give rise to early progenitor cells, which divide to yield late progenitor cells. These late progenitor cells lose the CD133 cell surface marker, and they divide to give rise to daughter cells that specialize as neurons or one of two types of supportive cells, astrocytes or oligodendrocytes. **(b)** A cancer stem cell can divide to self-renew and give rise directly to a cancer (tumor) cell, which can also spawn abnormal daughter cells (top row). Or, an early progenitor cell can give rise to normal differentiated cells (middle row) or a tumor, which may spawn some abnormal daughter cells (bottom row). Note that only the differentiated cells have lost the marker.

Chapter 20 *Cancer Genetics and Genomics*

Figure 20.7 Dedifferentiation reverses specialization. Mutations in a differentiated cell could reactivate latent "stemness" genes, giving the cell greater capacity to divide while removing some specialization. *Source: Adapted from "Neurobiology: At the root of brain cancer," Michael F. Clarke, Nature 432:281–282. Macmillan Publishers Ltd, November 18, 2004.*

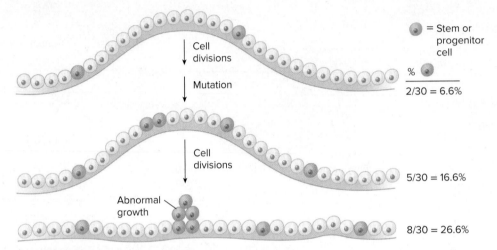

Figure 20.8 Shifting the balance in a tissue toward cells that divide. If a mutation renders a differentiated cell able to divide to yield other cells that frequently divide, then over time these cells may take over, forming an abnormal growth.

Figure 20.9 Too much repair may trigger tumor formation. If epithelium is occasionally damaged, resting stem cells can become activated and divide to fill in the tissue. If injury is chronic, the persistent activation of stem cells to renew the tissue can veer out of control, fueling an abnormal growth.

Another possible origin of cancer may be a loss of balance at the tissue level in favor of cells that can divide continually or frequently—like a population growing faster if more of its members are of reproductive age. Consider a tissue that is 5 percent stem cells, 10 percent progenitors, and 85 percent differentiated cells. If a mutation, over time, shifts the balance in a way that creates a greater proportion of stem and progenitor cells, the extra cells pile up, and a tumor forms (**figure 20.8**).

Uncontrolled tissue repair may also cause cancer (**figure 20.9**). If too many cells divide to fill in the space left by injured tissue, and those cells keep dividing, an abnormal growth may result.

With so many millions of cells undergoing so many error-prone DNA replications, and so many ways that cancer can arise, it isn't surprising that cancer is so common. Yet most of the time, the immune system destroys a cancer before it progresses very far.

> **Key Concepts Questions 20.2**
>
> a. What are the characteristics of cancer cells?
>
> *b. What is a cancer stem cell?
>
> c. Explain how altering the balance between stem and progenitor cells and differentiated cells, or excess tissue repair, can cause cancer.

20.3 Cancer Genes and Genomes

Researchers have sequenced the genomes of cells from thousands of tumors. As a result, a more comprehensive view is emerging of how mutations cause cancer and keep changing the course of the disease. An average cancer cell has 267 somatic mutations and chromosome alterations. Algorithms based on genomic profiles of cancer cells are used to predict drug responses.

Driver and Passenger Mutations

A driver of a vehicle takes it to the destination; a passenger goes along for the ride. In cancer genetics, a **driver mutation** provides the selective growth advantage to a cell that defines the cancerous state. A **passenger mutation** occurs in a cancer cell, but does not cause or propel the cancer's growth or spread. Drivers can be oncogenes or tumor suppressor genes, and may be generated from rearranged chromosomes.

Only about 200 genes are known to have driver mutations, but thousands of genes can harbor passenger mutations, which occur in cancerous as well as noncancerous cells. About 98 percent of the mutations in cancer cells are passengers, just along for the ride. Looked at another way, about 2 percent of our genes are involved in the cell cycle or DNA repair and can therefore be implicated in some way in cancer.

Tumors vary greatly in the numbers of driver and passenger mutations. A cancer generally has four or five driver mutations. The number of passenger mutations increases with age. For a 40-year-old and an 80-year-old with cancer in the same type of tissue, the older person's tumor cells will have many more passenger mutations than the younger person's tumor cells. This makes sense. The passage of time brings DNA replication errors and environmental exposures.

The effect of driver mutations is cumulative. One model describes the accumulation of mutations as a cancer forms and progresses as "three strikes" that correspond to three stages: breakthrough, expansion, and invasion (**figure 20.10**).

The initial mutation ("strike") enables a normal epithelial (lining) cell, for example, to divide slightly faster than cells near it. In this way, a clone (a group of cells that descend from one cell) of faster-dividing cells gradually forms. The second strike occurs when a second driver mutation boosts the division rate in the already mutation-bearing cells, and their proportion within the tissue increases. The cells form a tumor that is benign (not yet cancerous). Even if each driver mutation boosts the cell division rate by only 0.4 percent, and if cells divide only once or twice a week, in several years the tumor will grow to a size that a person might be able to feel, and consist of billions of cells. In the expansion stage, cancer cells acquire the ability to withstand low levels of nutrients, oxygen, and growth factors, which may happen in interior portions of a solid tumor.

In the third strike, invasion, additional driver mutations representing several pathways send tentacles of the tumor into surrounding tissue. **Table 20.1** lists the sequences of mutational events behind four types of cancer. Metastasis is not given its own "strike" stage because mutations that enable a tumor to metastasize may have been present from the beginning of the disease.

Researchers use several techniques that compare tumors to deduce the mutational steps that drive the disease. One approach is to count and compare mutations in tumor cells from people at different stages of the same type of cancer, as classified by body part. The older the tumor, the more genetic changes have accumulated. Therefore, a mutation present in all stages among several individuals' tumors acts early in the disease process, whereas a mutation seen only in the tumor cells of sicker people acts later. Looked at another way, when a cancer is considered over time in the same individual, even an advanced, widely spread cancer retains the initial driver mutations.

The evolution of a cancer is similar to the evolution of species in that genetic changes accumulate over time. Just as linear changes may represent only part of an evolutionary tree diagram, linear progressions of mutations may be too simple a way to view the development of cancer. In many tumors, cell lineages branch when they acquire new mutations, which may accelerate or accompany metastasis (**figure 20.11**).

The following subsections look at specific oncogenes and tumor suppressor genes. Chromosome abnormalities can cause these mutations. The chromosomes in cancer cells may be abnormal in number and/or structure. They may bear translocations, inversions, or have extra or missing pieces. A translocation that joins parts of nonhomologous chromosomes can hike expression of a gene enough to turn it into an oncogene. Duplications can increase the number of copies of a particular oncogene from 2—one on each of a pair of homologs—to up to 100. A deletion may remove a tumor suppressor gene. A one-time event called chromothripsis shatters several chromosomes and may kill the cell—or trigger

Figure 20.10 The "three strikes" of cancer.

Table 20.1	Three Strikes Drive Cancer			
	Melanoma	**Pancreatic Cancer**	**Cervical Cancer**	**Colorectal Cancer**
Breakthrough	BRAF	KRAS	TP53, RB	APC
Expansion	TERT	CDK2NA	PIK3CA	KRAS
Invasion	CDK2NA	SMAD4	MAPK1	SMAD4
	TP53	TP53	STK11	TP53
	PIK3CA		FBXW7	PIK3CA
				FBXW7

cancer. The opener of chapter 13 discusses chromothripsis. Point mutations can also cause cancer.

A Closer Look at Oncogenes

Genes that normally trigger cell division are called **proto-oncogenes**. They are active where and when high rates of cell division are necessary, such as in a wound or in an embryo. When proto-oncogenes are transcribed and translated too rapidly or frequently, or perhaps at the wrong time in development or place in the body, they function as oncogenes (*onco* means "cancer"). Usually oncogene activation is associated with a point mutation or a chromosomal translocation or inversion that places the gene next to another that is more highly expressed (transcribed). Oncogene activation causes a gain-of-function that is dominant. A mutation in just one copy of the

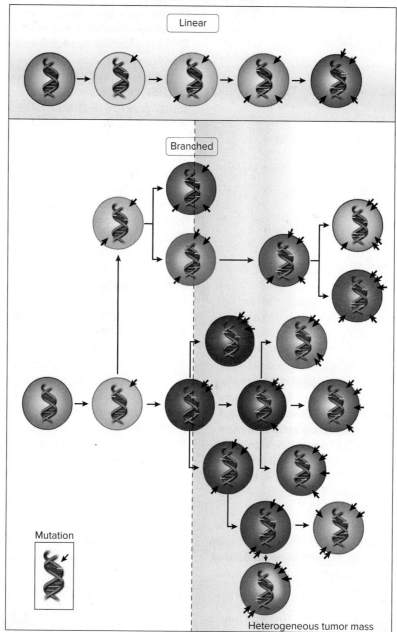

Figure 20.11 **The evolution of a cancer.** Genetic changes in cancer may occur in a linear and/or branching pattern. A heterogeneous tumor mass evolves as new mutations arise, while old ones remain. An initially linear cancer may subsequently branch. Table 20.1 also shows how mutations accumulate as cancer progresses.

gene raises cancer risk. In contrast, a tumor suppressor gene mutation is usually a deletion that causes a loss-of-function and is recessive (see figure 4.8).

Proto-oncogenes can also become oncogenes by being physically next to highly transcribed genes. Three examples of genes that can activate proto-oncogenes are a viral gene, a gene encoding a hormone, and parts of antibody genes.

A virus infecting a cell may insert DNA next to a proto-oncogene. When the viral DNA is rapidly transcribed, the adjacent proto-oncogene (now an oncogene) is also rapidly transcribed. Increased production of the oncogene's encoded protein then switches on genes that promote mitosis, triggering the cascade of changes that leads to cancer (**table 20.2**). Viral damage to a human genome may be catastrophic, activating and amplifying oncogenes as well as inverting and translocating chromosomes.

Another way that a proto-oncogene may be activated is when it is moved next to a gene that is normally very actively transcribed. This happens when an inversion on chromosome 11 places a proto-oncogene next to a DNA sequence that controls transcription of the parathyroid hormone gene. When the gland synthesizes the hormone, the

Table 20.2	Viruses That Can Cause Cancer
Virus	**Cancers**
Epstein-Barr virus	Lymphomas, stomach, throat
Hepatitis C	Liver
Herpesvirus 8	Kaposi sarcoma
Human immunodeficiency virus (HIV)	Kaposi sarcoma, cervical, B cell lymphoma
Human papilloma virus (HPV)	Cervix, throat, anus, vagina, penis

oncogene is expressed, too. Cells in the gland divide, forming a tumor.

Antibody genes are among the most highly transcribed, so it isn't surprising that a translocation or inversion that places a proto-oncogene next to an antibody gene causes cancer. Cervical cancer and anal cancer following human papillomavirus infection may begin when proto-oncogenes are mistakenly activated with antibody genes. Similarly, in Burkitt lymphoma, a cancer common in Africa, a large tumor develops from lymph glands near the jaw. Epstein-Barr virus stimulates specific chromosome movements in maturing B cells to assemble antibodies against the virus, and a translocation places a proto-oncogene on chromosome 8 next to an antibody gene on chromosome 14. The oncogene is overexpressed, and the cell division rate increases. Tumor cells of Burkitt lymphoma patients have the translocation (**figure 20.12**).

A proto-oncogene may not only move next to another gene, but also be transcribed and translated with it as if they are one gene. The double gene product, called a **fusion protein,** activates or lifts control of cell division. For example, in acute promyelocytic leukemia, a translocation between chromosomes 15 and 17 brings together a gene coding for the retinoic acid cell surface receptor and an oncogene called *myl*. The fusion protein functions as a transcription factor, which, when overexpressed, causes cancer.

The nature of the retinoic acid receptor/*myl* fusion protein explains why some patients who receive retinoid (vitamin A–based) drugs to treat acne recover from their cancers. Their immature cancer cells, stuck in an early stage of development where they divide frequently, suddenly differentiate, mature, and die. Perhaps the cancer-causing fusion protein prevents affected white blood cells from getting enough retinoids to specialize, locking them in an embryonic-like, rapidly dividing state. Supplying extra retinoids allows the cells to resume their normal developmental pathway. **A Glimpse of History** describes the fusion protein that led to development of one of the first cancer drugs targeted to cells with specific mutations.

Another way that an oncogene can cause cancer is by excessive response to a growth factor. In about 25 percent of women with breast cancer, affected cells have 1 to 2 million copies of a cell surface protein called HER2 that is the product of an oncogene. The normal number of these proteins on a cell is 20,000 to 100,000.

The HER2 proteins are receptors for epidermal growth factor. In breast cells, the receptors traverse the plasma membrane, extending outside the cell into the extracellular matrix while dipping into the cytoplasm. They function as a tyrosine kinase, as is the case for the leukemia described in A Glimpse of History. When the growth factor binds to the tyrosine (an amino acid) of the receptor, the tyrosine picks up a phosphate group, which signals the cell to activate transcription of genes whose protein products stimulate cell division. In HER2 breast cancer, too many tyrosine kinase receptors send too many signals to divide.

HER2 breast cancer usually strikes early in adulthood and spreads quickly. However, a monoclonal antibody-based drug binds to the receptors, blocking the signals to divide. The drug works when the extra receptors arise from multiple

Figure 20.12 **A translocation that causes cancer.** **(a)** In Burkitt lymphoma, a proto-oncogene on chromosome 8 moves to chromosome 14, next to a highly expressed antibody gene. Overexpression of the translocated proto-oncogene, now an oncogene, triggers the molecular and cellular changes of cancer, leading to explosive numbers of B cells. **(b)** Burkitt lymphoma typically affects the jaw. *(a): Source: Adapted from "Tissue repair and stem cell renewal in carcinogenesis," Philip A. Beachy, Sunil S. Karhadkar, and David M. Berman, Nature 432:324–331. Macmillan Publishers Ltd, 2004; (b): Mark Boulton/Alamy Stock Photo*

A GLIMPSE OF HISTORY

On August 13, 1958, two men entered hospitals in Philadelphia and reported weeks of fatigue. Each had a very high white blood cell count and was diagnosed with chronic myelogenous leukemia (CML). Too many immature white blood cells were crowding the healthy cells. The men's blood samples eventually fell into the hands of pathologist Peter Nowell and cytogeneticist David Hungerford. They had developed ways to stimulate white blood cells to divide in culture, and they probed the chromosomes of both leukemic and normal-appearing white blood cells in the two tired men and five others with CML.

Nowell and Hungerford discovered a small, unusual chromosome that was only in the leukemic cells, the first chromosome abnormality linked to cancer. Later, it would be dubbed "the Philadelphia chromosome" (Ph[1]). By 1972, new chromosome stains that distinguished AT-rich from GC-rich regions revealed that Ph[1] is the result of a reciprocal translocation (see figure 13.19). By 1984, researchers had homed in on the two genes juxtaposed in the translocation between chromosomes 9 and 22. The gene from chromosome 9 is called the Abelson oncogene (*abl*), and the other gene, from chromosome 22, is called the breakpoint cluster region (*bcr*). Two different fusion genes form. The *bcr-abl* fusion gene is part of the Philadelphia chromosome, and it causes CML. The other fusion gene does not affect health.

The discovery that the fusion protein, called the BCR-ABL oncoprotein, is a form of the enzyme tyrosine kinase (the normal product of the *abl* gene) led directly to development of the first kinase inhibitor drug. A kinase is an enzyme that transfers a phosphate group (PO_4) to another molecule, and is a key part of signal transduction pathways. The cancer-causing form of tyrosine kinase is active for too long, which sends signals into the cell, stimulating it to divide, too many times.

Through the 1980s, drug developers tested more than 400 small molecules in search of one that would block the activity of the errant tyrosine kinase, without derailing other important enzymes. They found what would become the first kinase inhibitor in 1992. The drug nestles into the pocket on the tyrosine kinase that must bind ATP to stimulate cell division. With ATP binding blocked, cancer cells do not receive the message to divide, and they cease doing so. After passing safety tests, the drug worked so well that it set a new record for drug approval—10 weeks. However, after several years of using the original kinase inhibitor, resistant cancer cells may appear. Other drugs continue targeted treatment for these patients.

---

copies of the gene, rather than from extra transcription of a single *HER2* gene.

A Closer Look at Tumor Suppressor Genes

Some cancers result from loss or silencing of a tumor suppressor gene that normally inhibits expression of genes involved in all of the activities that turn a cell cancerous. Cancer can result when a tumor suppressor gene is deleted or if the promoter region binds too many methyl (CH_3) groups, which blocks transcription. Recall that a mutation in a tumor suppressor gene is recessive. One copy of the mutation may be inherited, and the second copy mutates in the cell type that becomes cancerous.

Wilms' tumor is an example of a cancer that develops from loss of tumor suppression. A gene is deleted that normally halts mitosis in the rapidly developing kidney tubules in the fetus. As a result, an affected child's kidney retains pockets of cells dividing as frequently as if they were still in the fetus, eventually forming a tumor. Following are descriptions of specific tumor suppressor genes.

Retinoblastoma

Retinoblastoma (RB) is a rare childhood eye tumor (**figure 20.13**). About 1 in 20,000 infants develops RB, and half of them have inherited susceptibility to the disease. They have one germline mutant allele for the *RB1* gene in each of their cells, and then cancer develops in a somatic cell where the second copy of the *RB1* gene mutates. Therefore, inherited retinoblastoma

Figure 20.13 Retinoblastoma type 1 is due to a mutation in a tumor suppressor gene. The retinal tumor appears opaque in photographs because it reflects light. Fortunately nearly all cases of retinoblastoma are treatable today with surgery and/or chemotherapy. *National Cancer Institute/Science Source*

requires two point mutations or deletions, one germline and one somatic. In some sporadic (noninherited) cases, two somatic mutations occur in the *RB1* gene, one on each copy of chromosome 13. Either way, the cancer usually starts in a cone cell of the retina, which provides color vision.

Study of retinoblastoma inspired the "two-hit" hypothesis of cancer causation—that two mutations (germline and somatic or two somatic) are required to cause a cancer related to tumor suppressor deletion or malfunction. (A second form of retinoblastoma is caused by mutation in an oncogene, *MYCN*.)

The discovery that many children with RB have deletions in the same region of the long arm of chromosome 13 led researchers to the *RB1* gene and its protein product, which linked the cancer to control of the cell cycle. The RB1 protein normally binds transcription factors so that they cannot activate genes that carry out mitosis, halting the cell cycle at G_1. When the *RB1* gene is mutant or missing, the hold on the transcription factor is released, and cell division ensues.

For many years, the only treatment for retinoblastoma was removal of the affected eye. Today, children with an affected parent or sibling, who have a 50 percent chance of having inherited the mutant *RB1* gene, can be monitored from birth so that noninvasive treatment (chemotherapy) can begin early. Full recovery is common.

Mutations in the *RB1* gene cause other cancers. Some children successfully treated for retinoblastoma develop bone cancer as teens or bladder cancer as adults. Mutant *RB1* genes have been found in the cells of patients with breast, lung, or prostate cancers, or acute myeloid leukemia, who never had the eye tumors. Expression of the same mutation in different tissues may cause these cancers.

p53 Normally Prevents Many Cancers

A single gene that causes a variety of cancers when mutant is *p53*. Recall from chapter 12 that the p53 protein transcription factor "decides" whether a cell repairs DNA replication errors or dies by apoptosis. If a cell loses a *p53* gene, or if the gene mutates and malfunctions, a cell with damaged DNA is permitted to divide, and cancer may be the result.

More than half of human cancers arise from a point mutation or deletion in *p53*. This may be because p53 protein is a genetic mediator between environmental insults and development of cancer (**figure 20.14**). A type of skin cancer, for example, is caused by a *p53* mutation in skin cells damaged by an excessive inflammatory response that can result from repeated sunburns. That is, *p53* may be the link between sun exposure and skin cancer.

In most *p53*-related cancers, mutations occur only in somatic cells. However, in the germline condition Li-Fraumeni syndrome, family members who inherit a mutation in *p53* have a very high risk of developing cancer—50 percent do so by age 30, and 90 percent by age 70. A somatic *p53* mutation in the affected tissue results in cancer because a germline mutation in the gene is already present.

Stomach Cancer Lifts Cellular Adhesion

E-cadherin normally acts as a cellular adhesion protein found in tissue linings, but it is also a tumor suppressor because when its gene is deleted, cancer results. This was the case for the Bradfield family. Golda Bradfield died of stomach cancer in 1960. By the time some of her grown children developed the cancer too, the grandchildren began to realize that their family had a terrible legacy. Genetic testing revealed familial diffuse gastric cancer, caused by an "exon skipping" missense mutation in the E-cadherin gene that deletes an entire exon as the mRNA is transcribed from the gene (see section 12.5).

Golda's grandchildren had genetic tests. Eleven of them had inherited the mutant gene, but scans of their stomachs did not show any tumors. Still, they all had their stomachs removed

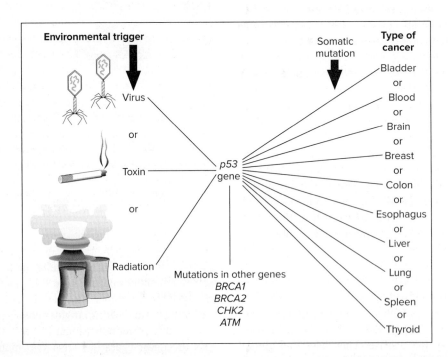

Figure 20.14 *p53* **cancers reflect environmental insults.** The environment triggers mutations or changes in gene expression that lead to cancer. The *p53* gene is a mediator—"the guardian of the genome." The protein products of many genes interact with p53 protein. (*p53* is also called *TP53*, for tumor protein.)

to prevent the cancer. Most of them already had hundreds of tumors, too tiny to have been seen on medical scans. The cousins without stomachs are doing well. Like people who have their stomachs surgically shrunk to lose weight, the cousins avoid hard-to-digest foods and eat a little at a time, throughout the day. The inconvenience, they say, is a fair trade for eliminating the fear of their grandmother's, parents', and aunts' and uncles' fates—stomach cancer.

BRCA1/BRCA2 Mutations Disrupt Repair

Breast cancer that runs in families may be due to inheriting a germline mutation and then having a somatic mutation occur in a breast cell (a familial form), or two somatic mutations affecting the same breast cell (a sporadic form), as figure 20.4 depicts for lung cancer. However, breast cancer is so common that a family with many affected members may actually have multiple sporadic cases, rather than an inherited form of the disease causing them all.

Only about 5 percent of breast cancers are familial, caused by mutations in any of at least 20 genes. Most of the genes associated with susceptibility to familial breast cancer encode proteins that interact in ways that enable DNA to survive damage. If DNA cannot be repaired, mutations that directly cause cancer can accumulate and persist.

The two major breast cancer susceptibility genes are *BRCA1* and *BRCA2*. Together they account for 15 to 20 percent of the 5 percent of cases that are familial. A *BRCA* mutation is inherited in an autosomal dominant manner, with incomplete penetrance because it increases susceptibility, rather than directly causing cancer.

Inheriting a mutation in *BRCA1*, which stands for "breast cancer predisposition gene 1," greatly increases the lifetime risk of inheriting breast and/or ovarian cancer. This risk, however, varies in different population groups because of the modifying effects of other genes.

The most common *BRCA1* mutation deletes two adjacent DNA bases, altering the reading frame and shortening the protein. Thousands of mutations in the gene are known, most of them rare. Cancer risk depends on the site of the mutation. For *BRCA1*, mutations in the ends of the gene are more likely to increase risk of breast cancer and mutations in the middle part of the gene are more likely to increase risk of ovarian cancer.

BRCA1 encodes a protein that interacts with many other proteins that counter DNA damage in several ways. One important form of protection is the mending of areas of the genome where both DNA strands are broken at the same site. These double-stranded breaks are particularly dangerous because they cut the chromosomes all the way through, making rearrangements such as deletions and translocations possible. Imagine an earthquake cutting across both sides of a highway rather than just one. **Figure 20.15** depicts the central role of *BRCA1* in protecting DNA.

BRCA1 mutations have different incidences in different populations. Only 1 in 833 people in the general U.S. population has a *BRCA1* mutation. That figure is more than 1 in 50 among Ashkenazi Jewish people, due to population bottlenecks

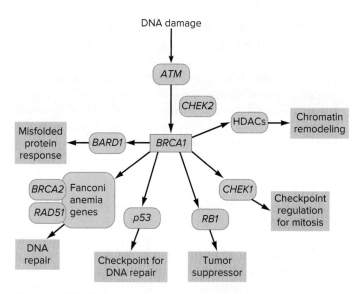

Figure 20.15 **The *BRCA1* gene controls many cellular defense mechanisms.** *BRCA1* functions as a genetic crossroads in handling DNA damage. Inheriting a *BRCA1* mutation increases susceptibility to cancers of the breast, ovary, cervix, uterine tube, uterus, peritoneum, pancreas, and prostate gland. Mutations in any of the genes with which *BRCA1* interacts increases cancer susceptibility too. (HDACs are histone deacetylases.)

and nonrandom mating (endogamy). The *BRCA1* gene was initially discovered in Ashkenazi families in which several members developed the associated breast cancer at very young ages. In this population, a woman who inherits a *BRCA1* mutation has up to an 87 percent risk of developing breast cancer over her lifetime (five times the general population risk), and a 50 percent risk of developing ovarian cancer. For Ashkenazi families without a strong clinical history of breast cancer, the risks are 65 percent for breast cancer and 39 percent for ovarian cancer.

Some women who learn that they have inherited an allele predisposing to breast or ovarian cancer have their breasts and/or ovaries removed. This action may be more compelling for an Ashkenazi woman facing an 87 percent lifetime risk of breast or ovarian cancer than for a woman in a population group with a much lower risk. The general population risk of developing a *BRCA1* cancer if a person inherits a mutation is only about 10 percent, based on empirical (observational) evidence.

BRCA2 breast cancer is also more common among the Ashkenazim. Women in this group who inherit a mutation in *BRCA2* face a 45 percent lifetime risk of developing breast cancer and an 11 percent risk of developing ovarian cancer. Men have a 6 percent lifetime risk of developing breast cancer, which is 100 times the risk for men in the general population. Inheriting a *BRCA2* mutation also increases the risk of developing cancers of the colon, kidney, prostate, pancreas, gallbladder, skin, or stomach.

Tests for any cancer susceptibility genes can reveal a "variant of uncertain significance" (see **Bioethics** in section 12.7). This means that the gene sequence is not wild type, but the identified variant has not been associated with increased cancer susceptibility or any other phenotype.

Environmental Causes of Cancer

Environmental factors contribute to cancer by mutating or altering the expression of genes that control the cell cycle, apoptosis, and DNA repair. Inheriting a susceptibility gene places a person farther along a particular road to cancer, but cancer can happen in somatic cells in anyone. It is more practical, for now, to identify environmental cancer triggers and develop ways to control them or limit our exposure to them, than to alter genes that cause or predispose a person to develop cancer.

Looking at cancer at a population level reveals the interactions of genes and the environment. For example, researchers examined samples of non-Hodgkin's lymphoma tumors from 172 farmers, 65 of whom had a specific chromosomal translocation. The 65 farmers had greater exposure to insecticides, herbicides, fungicides, and fumigants, compared to the farmers with lymphoma who did not have the translocation. These findings associate exposure to the chemicals with developing the cancer. Some subtypes of non-Hodgkin's lymphoma are more prevalent among members of the military exposed to "burn pits" of debris in Iraq and Afghanistan, and among first responders to the terrorist attacks of 9/11.

Determining precisely how an environmental factor such as diet affects cancer risk can be complicated. Consider the cruciferous vegetables, such as broccoli and brussels sprouts, which are associated with decreased risk of colon cancer. These vegetables release chemicals that activate enzymes that detoxify carcinogenic products of cooked meat, called heterocyclic aromatic amines. With a vegetable-poor, meaty diet, these amines accumulate and elevate cancer risk. They cross the lining of the digestive tract and circulate to the liver, where enzymes metabolize them into compounds that cause driver mutations for colon cancer. This is one reason why a plant-based diet is healthier than eating red meat.

Both environmental and genetic factors contribute to cancer risk. Consider melanoma, a type of skin cancer (see figure 20.5). Sun exposure elevates risk, but certain variants of the melanocortin-1-receptor (*MC1R*) gene, known for imparting red hair, fair skin, and freckles, double the risk of developing melanoma, even if a person avoids intense sunlight (see the opener to chapter 5).

Key Concepts Questions 20.3

*a. Define *driver mutation*.

*b. Define *passenger mutation*.

c. Explain how cancer is a multistep process.

d. Discuss the role of chromosomes in causing cancer.

*e. Define *oncogene*.

*f. Define *tumor suppressor gene*.

g. How does the environment contribute to causing cancer?

20.4 Diagnosing and Treating Cancer

The experience of having cancer has changed markedly in recent years, as both diagnosis and treatment have evolved from general measures to highly specific strategies to classify and control the disease. New approaches go beyond identifying oncogenes and tumor suppressor mutations.

The ability to sequence cancer exomes and genomes is providing a new level of clinical specificity. For example, if several members of a family have cancer of the same body part, yet they all test negative for known cancer gene driver mutations, then sequencing their exomes can reveal other genes that are mutant in the cancer cells and have a function that could disrupt the cell cycle or DNA repair. If researchers can then find the same mutations in other families with the same type of cancer based on body part, then a new entry can be added to the list of cancer driver genes, and perhaps a new drug target identified—or even a discovery that an existing drug might work.

Diagnosing Cancer

For a long time, cancer was diagnosed only after a person reported symptoms to a physician, such as abdominal bloating, a lump, bruising, a scaly patch of skin, or fatigue. Follow-up tests then might include scans to distinguish a solid mass from a cyst, then biopsies to check for cancer cells in a tissue sample, using special stains to highlight the cancer cells, as in figure 20.5, or noting cell shape changes and loss of specialized characteristics. In contrast to such a diagnostic test that follows reporting of symptoms, a screening test for cancer is done on groups of people who are at elevated risk, but do not have symptoms. An example of a cancer screening test is to test for *BRCA1* or *BRCA2* mutations in an Ashkenazi woman who has young relatives who have or had breast or ovarian cancer, but who does not have symptoms or signs of the diseases herself. Sometimes large populations are screened if the test is economical, such as screening men over age 50 for prostate specific antigen, which is a protein biomarker that could indicate prostate cancer.

Detecting mutations is one genetic approach to identifying risk of and diagnosing cancer. Another is **gene expression profiling,** which provides a view of the proteins that a cancer cell makes at a certain point in time and the relative abundance of the proteins. A striking example is mixed lineage leukemia (MLL) (**figure 20.16**). The discovery of this type of blood cancer came from the observation that 10 percent of children with acute lymphoblastic leukemia (ALL) do not respond to the chemotherapy that helps the other 90 percent, although all the patients have the same symptoms of fever, fatigue, and bruising. The leukemic cells all look alike under a microscope. However, gene expression analysis using DNA microarrays (chips) revealed that in the 10 percent of patients who did not respond to standard drugs, the cancerous cells make too little of about 1,000 proteins and too much of 200 others, compared to protein production in children who did respond

Figure 20.16 **Cancer cells that look alike may be genetically distinct.** These leukemias—ALL, MLL, and AML—differ in gene expression patterns. The columns of squares represent DNA from tumor samples and the rows compare the activities of particular genes. Red tones indicate higher-than-normal expression and blue tones show lower-than-normal expression. The different patterns indicate distinct cancers, although the cells look alike under a microscope.

to the standard "cocktail" of chemotherapy drugs. Those 10 percent, who actually have MLL and not ALL, respond to different drugs.

Treating Cancer

The traditional ways of treating cancer are broad. The oldest approach, surgery, is still effective and often the first treatment, if it removes a primary tumor before it has invaded healthy tissue and spread through the bloodstream. The other two standard approaches, chemotherapy and radiation, hit all rapidly dividing cells, causing adverse effects when they harm healthy cells.

Newer ways of treating cancer are targeted to specific mutations in cancer cells. These treatments disrupt the cell cycle and/or alter the immune response to more readily destroy cancer cells. Some newer treatments are used with more traditional ones, such as drugs that block hormones from stimulating breast cancer cells, and some are first-line treatments.

Drugs that interfere with the increased mitotic rate of cancer cells target the enzymes that propel the cell cycle—kinases and cyclins. Recall that a kinase adds a phosphate group to certain proteins, which alters the proteins' functions in ways that propel the cell cycle. Kinases bind to three amino acids in particular—tyrosine, serine, and threonine. The kinase inhibitors are drugs that bind to any of these amino acids in kinases. For example, drugs called mTOR inhibitors bind to a serine/threonine protein kinase and drugs called CDK4/6 inhibitors block another type of kinase called a cyclin-dependent kinase. CDK4/6 inhibitors are used to treat breast cancer.

Several approaches harness parts of the immune response to build up the body's defense against cancer. Flu-like symptoms may result from the release of cytokines that is part of activating the immune response. For many patients, drugs that boost the immune response extend the time when the cancer does not progress, but a few patients achieve long-lasting responses.

Drugs called **immune checkpoint inhibitors** block a stranglehold that cancer cells can have on certain T cells, freeing the T cells to attack cancer cells. Just as the cell cycle has checkpoints in which groups of interacting proteins control the sequence of events, the immune response has checkpoints in which groups of molecules control the degree to which classes of T cells are activated or suppressed. A T cell response might be normally suppressed to prevent its attacking body (self) cells, such as in the thymus. Cancer cells can intervene in the immune checkpoints in ways that block and disable T cells, taking advantage of the body's system to control the immune response.

Figure 20.17 depicts a monoclonal-antibody-based immune checkpoint inhibitor drug that interferes with the binding between the T-cell receptor PD-1 and its ligand (binding partner) PD-L1 on a cancer cell surface. Cancer cells may have many PD-L1s, enabling them to disable many T cells. Other drugs block PD-L1. In either approach, freeing the PD-1 enables the T cell to escape the inactivation by the cancer cell. The immune checkpoint inhibitors are being used to treat several types of cancers. Patients are selected based on the cancer subtype and mutations and the types of drugs already taken. For some cancers, such as melanoma, an immune checkpoint inhibitor may be the first drug used.

A strategy called **chimeric antigen receptor T cell therapy (CAR T)** improves upon the ability of the immune system to recognize and destroy cancer cells (**figure 20.18**). It creates DNA instructions for a hybrid protein antigen on T cells that does not occur in nature. The protein, called a chimeric antigen receptor, or CAR, is part T-cell receptor and part antibody. (Figure 19.4 shows T-cell receptors and

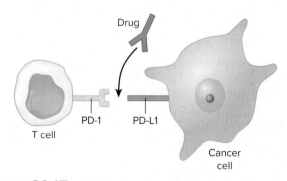

Figure 20.17 **A PD-L1 immune checkpoint inhibitor drug is a monoclonal antibody that frees T cells from the cancer cells that bind them.** Immune checkpoint inhibitors are effective for several types of cancers. (PD stands for programmed cell death protein 1 and PD-L1 stands for programmed cell death protein 1 ligand.)

Figure 20.18 Chimeric antigen receptor (CAR) T cell therapy arms a patient's T cells to attack cancer cells. **(a)** A blood sample is taken from a patient and the T cells are separated. **(b)** In laboratory glassware, the patient's T cells are exposed to a gene modified to include the genetic instructions for both a T-cell receptor and an antibody. **(c)** Millions of copies of the modified T cells are cultured in the dish and then **(d)** transferred to the patient. The modified cells circulate and produce the hybrid protein that consists of both the T-cell receptor **(e)** and the antibody. The receptor guides the T cell to a cancer cell, where the antibody part binds, which alerts the immune system to send perforins that shatter the cancer cell and granulysins that dismantle the cancer cell from within. *(c): 2017 Terese Winslow LLC U.S. Govt. has certain rights*

antibodies.) The receptor part is engineered to guide the T cell to a cancer cell, like a drone. The antibody part then binds the target, alerting the immune system to respond and kill the cancer cell using perforins to break the plasma membrane and proteins called granulysins that enter the cancer cells and destroy them.

The first patient to receive CAR-based treatment was a 65-year-old man with leukemia who had exhausted all treatments. He recovered after receiving CARs made from his own cells. In 2012, a 6-year-old with another form of leukemia and who had also tried everything possible received CAR treatment using her own cells. She had been near death when treated that April, but was back in school by September, her hair regrown and energy back. Many of these therapies are in clinical trials for a variety of cancers, and a few have been approved for use, with a patient's own cells. Research continues on creating CAR T cells from donors.

Genome editing tools (see section 21.4) are used to create the chimeric antigen receptors. Typically a patient receives a billion or more altered T cells.

Living with cancer entails not just diagnosis and treatment, but also the possibility of recurrence of the disease. This is why patients are checked frequently, even if they feel well, and even if years have elapsed since the first illness. **Clinical Connection 20.1** discusses how liquid biopsy detects cancer recurrence with a blood test.

Continuing analysis of cancer cell genomes combined with the ongoing annotation of human genomes (discovering what genes do) will tell us more about exactly what happens when a cell becomes malignant. But conquering cancer may be an elusive goal. The DNA of cancer cells mutates in ways that enable cells to pump out any drug sent into them. Cancer cells have redundant pathways, so that if a drug shuts down angiogenesis or invasiveness, the cell completes the task another way. A more realistic goal than eradicating cancer may be to kill enough cancer cells, and sufficiently slow their spread, so that it takes the remainder of a human lifetime for tumors to grow back enough to harm health. In this way, cancer can become a chronic, manageable condition. As long as our cells divide, we are at risk of developing cancer.

> **Key Concepts Questions 20.4**
>
> a. How can exome sequencing help to identify a previously unknown cancer gene in a family?
>
> *b. Contrast traditional approaches to diagnosing and treating cancer to newer methods.
>
> c. Explain how testing for mutations and for gene expression patterns can help a physician to select a treatment for a patient.
>
> *d. Define kinase inhibitor.
>
> e. Explain how immune checkpoint inhibitors work.
>
> f. Discuss how chimeric antigen receptor T cell therapy treats a cancer.

Clinical Connection 20.1

A Liquid Biopsy Monitors Cancer Recurrence and Response to Treatment

The fact that cancerous tumors shed DNA into the bloodstream can be used to monitor disease and perhaps even to detect disease in people who do not have symptoms (**figure 20A**). Checking DNA pieces in the blood plasma for oncogene or tumor suppressor mutations is termed a *liquid biopsy*. The entire genome sequence of a cancer can be deduced by overlapping DNA pieces from the bloodstream. The simple blood test of a liquid biopsy is much less painful and invasive than a traditional biopsy, which samples cancer cells from a solid tumor, such as in a breast or the liver, or from a lining such as the skin.

The DNA detected in a liquid biopsy is called circulating tumor DNA, or ctDNA. It is similar to the cell-free fetal DNA that is detected in pregnant women (see figure 13.8). Such prenatal tests sometimes detect cancer in an unsuspecting pregnant woman.

The first uses of liquid biopsy were in people who already had cancer, and required monitoring. If a liquid biopsy soon after surgery to remove a tumor does not have ctDNA, but 2 years later does, then the cancer likely has recurred, and may have new mutations. Liquid biopsy is also useful for monitoring response to treatment. If a drug is working, the level of ctDNA will decrease. If a cancer has become resistant to a drug, the level of ctDNA will increase.

Using tumor DNA as a biomarker is more specific than using a protein biomarker because a protein biomarker may also be present on healthy cells, such as elevated prostate-specific antigen. Circulating tumor DNA can also be collected from urine (bladder cancer), sputum (lung cancer), and feces (colorectal cancer). A liquid biopsy is particularly useful when tissue is difficult to obtain or not enough cells are sampled, or when cancer has spread and the initial site is unknown.

More controversial than detection of ctDNA for people who already have cancer is its use to screen high-risk populations, such as people who smoke (for lung cancer) or people with family histories of specific cancers. A finding of ctDNA in a person who does not have cancer symptoms can warn the individual to seek additional types of tests that might diagnose cancer very early.

Liquid biopsy might be used someday on everyone, to detect cancers that are entirely unsuspected. For example, ovarian cancer is sometimes diagnosed at a very late stage because the symptoms of bloating and fatigue are vague and

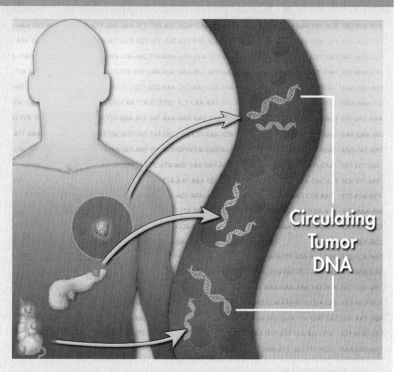

Figure 20A Liquid biopsy. Detecting tumor cell DNA in the blood is a less invasive way to monitor cancer recurrence than a traditional surgical biopsy. *Source: National Human Genome Research Institute.*

common, and may be attributed to other factors, such as a change in diet or exercise routine. The danger of using liquid biopsy on people of average risk, however, is a false positive finding—an oncogene or tumor suppressor variant that doesn't cause cancer in everyone, due to effects of other genes. Liquid biopsies are being done in clinical trials on thousands of healthy people at low or average risk of cancer to assess the predictive value of the technology.

Questions for Discussion

1. What would you want to know before you have a liquid biopsy?
2. What are the advantages and possible disadvantages of a liquid biopsy?
3. Explain why the results of a liquid biopsy would likely change as a cancer is monitored over many years.
4. When in the course of checking someone for cancer do you think a liquid biopsy should be offered?

Study Guide

Summary

20.1 Cancer Is an Abnormal Growth That Invades and Spreads

1. Cancer is a genetically driven loss of cell cycle control, creating a population of highly proliferative cells that invades surrounding tissue.
2. Spread of a cancer beyond the organ of origin is called **metastasis**.
3. A **carcinogen** is a chemical that causes cancer, often by causing a mutation. Cancer-causing mutations (activated **oncogenes,** inactivated **tumor suppressor genes,** and mutant DNA repair genes) affect cell fate, cell survival, and genome maintenance. They affect transcription factors, cell cycle checkpoint proteins, growth factors, repair proteins, or telomerase.
4. Sporadic cancers result from two recessive somatic mutations in the two copies of a gene. They are more common than inherited cancers that are caused by germline mutations that confer susceptibility in all cells along with somatic mutations in affected tissue.

20.2 Cancer at the Cellular Level

5. A tumor cell divides more frequently or more times than cells surrounding it, has an altered surface, loses the specializations of the cell type it arose from (**dedifferentiates**), and produces daughter cells like itself.
6. A malignant tumor infiltrates tissues and can metastasize by attaching to basement membranes and secreting enzymes that penetrate tissues and open a route to the bloodstream. **Angiogenesis** establishes a blood supply for a tumor. A cancer cell can travel, establishing secondary tumors.
7. Cell specialization and position within a tissue affect whether cancer begins and persists.
8. **Cancer stem cells** can divide to yield cancer cells and abnormally differentiated cells.
9. A cell that dedifferentiates and/or expresses "stemness" genes can begin a cancer.
10. A mutation that enables a cell to divide continually can alter the percentages of cells in a tissue that can divide, resulting in an abnormal growth.
11. Chronic repair of tissue damage can provoke stem cells into producing an abnormal growth.

20.3 Cancer Genes and Genomes

12. A **driver mutation** provides a selective growth advantage to a cell. A **passenger mutation** occurs in a cancer cell but does not cause or contribute to the disease. Passengers greatly outnumber drivers.
13. Driver mutations hit with three strikes: breakthrough (initial mutation in one cell), expansion (second mutation hikes division rate), and invasion. Mutations that will cause metastasis may be present early.
14. Sequences of driver mutations may be linear or branched.
15. Chromosomes in cancer cells may be highly abnormal (translocations, inversions, deletions, duplications, and chromothripsis).
16. Cancer is often the result when **proto-oncogenes** become oncogenes or when tumor suppressor genes are deleted or silenced. Mutations in DNA repair genes cause cancer by enabling mutations to persist.
17. Proto-oncogenes normally promote controlled cell growth, but are overexpressed because of a point mutation, placement next to a highly expressed gene, or transcription and translation with another gene, producing a **fusion protein**. Oncogenes may also be overexpressed growth factor receptor genes.
18. A tumor suppressor gene normally enables a cell to respond to factors that limit its division.
19. Environmental factors contribute to cancer by mutating genes and altering gene expression.

20.4 Diagnosing and Treating Cancer

20. Traditional cancer treatments are surgery, radiation, and chemotherapy. Newer approaches based on identifying mutations and **gene expression profiling** are used to subtype cancers and better target treatments.
21. **Kinase inhibitors** disrupt the cell cycle of cancer cells.
22. **Immune checkpoint inhibitors** stop cancer cells from blocking T cell activity.
23. **Chimeric antigen receptor T cell therapy** creates a cell-surface protein that is part T-cell receptor and part antibody. It functions like a drone to highlight cancer to the immune system.

Key Terms

angiogenesis (an″je-o-jen′ə-sis) Extension of blood vessels, such as toward a tumor.

cancer stem cells Stem cells that divide and yield cancer cells and abnormal specialized cells.

carcinogen (kar-sĭnə-jən) A substance that causes cancer.

chimeric antigen receptor T cell therapy (CAR T) (ki-meer-ik ăn′tē-jən re″sep′tor) A cancer treatment that combines DNA instructions for a T-cell receptor with those for an antibody, creating a cell surface molecule that guides T cells to cancer cells, alerting the immune system and eliciting a response.

dedifferentiated A cell less specialized than the cell it descended from. A characteristic of a cancer cell.

driver mutation A mutation that provides the selective growth advantage of a cancer cell.

fusion protein A protein that forms from translation of transcripts from two genes.

gene expression profiling Use of DNA microarrays to detect the types and amounts of cDNAs reverse transcribed from the mRNAs in a particular cell source.

immune checkpoint inhibitors Drugs that bind receptors in ways that block the bonding of cancer cells to T cells that suppress them, unleashing the T cells to attack the cancer cells.

kinase inhibitors Drugs that block kinases, which are enzymes that are part of signal transduction pathways, interfering with the cell cycle in cancer cells.

metastasis (mĕtə-stā′-sĭs) Spread of cancer from its site of origin to other parts of the body.

oncogene (ŏn′kə-jēn) A gene that normally controls the cell cycle, but causes cancer when overexpressed.

passenger mutation A mutation in a cancer cell that does not provide a selective growth advantage. It doesn't cause cancer but accompanies it.

proto-oncogene (prōtō-ŏn′kə-jēn) A gene that normally triggers cell division, but when overexpressed causes cancer.

tumor suppressor gene (tōōmər səprĕs′ər jēn) A recessive gene whose normal function is to limit the number of divisions a cell undergoes.

Review Questions

1. Distinguish between a benign and a malignant tumor.
2. Explain how a sporadic cancer differs from a Mendelian disease in terms of predicting risk of occurrence in the relatives of a patient.
3. Name the three basic cellular pathways that cancer disrupts.
4. Why is metastasis dangerous?
5. Explain the connection between cancer and control of the cell cycle.
6. List four characteristics of cancer cells.
7. Describe four ways that cancer can originate at the cell or tissue level.
8. Distinguish between a driver mutation and a passenger mutation.
9. Describe two abnormalities of chromosomes in cancer.
10. Explain how comparing mutations in cells from the same cancer type at different stages can reveal the sequence of genetic changes that drive the cancer.
11. How can classifying cancer genomically differ from classifying cancer histologically or by initial affected body part?
12. Distinguish between a proto-oncogene and an oncogene.
13. Describe two events that can activate an oncogene.
14. Explain why an oncogene is associated with a gain-of-function and a mutation in a tumor suppressor gene is associated with a loss-of-function.
15. What do retinoblastoma type 1, a *p53*-related cancer, inherited stomach cancer, and *BRCA1* breast cancer have in common?
16. Explain how a virus can cause cancer, and provide an example.
17. What is the role of deficient DNA repair in causing cancer?
18. Describe the three "strikes" of the origin and progression of cancer.
19. Distinguish between mutations and altered gene expression in cancer cells.
20. Explain why not all cancers affecting the same cell type respond the same way to a particular drug.
21. Describe an alternative to performing surgery to determine whether a metastatic tumor is responding to a new drug.
22. Describe how kinase inhibitors, immune system checkpoint inhibitors, and CAR T cell therapy work.

Applied Questions

1. A monoclonal antibody–based drug binds a receptor present in many copies on the cells of certain aggressive and early-onset breast cancers. (Monoclonal antibodies are discussed in section 19.4.)
    a. How does this treatment approach differ from standard chemotherapy?
    b. The drug is now used to treat tumors of the stomach and esophagus. Explain how the same drug can be used for cancers that affect different body parts.
2. Breast cancer can develop from inheriting a germline mutation and then undergoing a second mutation in a breast cell, or from two recessive mutations in a breast cell, one in each copy of a tumor suppressor gene. Cite another type of cancer, discussed in the chapter, that can arise in these two ways.
3. Why is a "cocktail" of several drugs likely to be more effective at slowing the course of a cancer than using a single, powerful drug?
4. von Hippel-Lindau syndrome is inherited. The responsible mutation lifts control over the transcription of certain genes, which, when overexpressed, cause tumors to form in the kidneys, adrenal glands, and blood vessels.

Is the von Hippel-Lindau gene an oncogene or a tumor suppressor gene? Cite a reason for your answer.

5. A new cancer drug derived from a black sponge binds to the mitotic spindle. What information is needed to predict whether the drug will cause the same side effects as conventional chemotherapy drugs?

6. A tumor is removed from a mouse and broken up into cells. Each cell is injected into a different mouse. Although all the mice used in the experiment are genetically identical and raised in the same environment, the animals develop cancers with different rates of metastasis. Some mice die quickly, some linger, and others recover. What do these results indicate about the characteristics of the original tumor cells?

7. State a genetic factor and an environmental factor that increase risk of developing melanoma.

8. Select a gene from table 20.1 and describe its general function as shown in figure 20.15.

9. If cancer genome sequencing from liquid biopsy becomes a clinical tool, why might it be useful to repeat the sequencing every few years, rather than basing treatment decisions for many years on an initial cancer genome sequence?

Case Studies and Research Results

1. More males than females develop a brain cancer called *glioblastoma*, and the males do not survive as long as the females. A study revealed that the tumor cells of males with glioblastoma have different gene expression profiles than those of females. Explain how this distinction differs from having different driver mutations.

2. A woman is diagnosed with invasive breast cancer, which means that it has grown outside the milk ducts. The cancer cells are tested for a set of messenger RNAs that are correlated to how likely certain chemotherapy drugs are to be effective. One company calls the results an *oncotype,* and the name has come to generally describe this type of test. Which technology discussed in the chapter does it illustrate?

3. People with certain types of prostate or breast cancer can opt to do *active surveillance,* which consists of frequent monitoring with scans and biomarker measurements, rather than undergoing chemotherapy, surgery, or radiation soon after diagnosis. What type of information would be helpful in deciding whether to follow active surveillance?

4. Most cases of retinoblastoma (RB) are due to a germline mutation in the tumor suppressor gene *RB1* and typically affect both eyes. Researchers discovered extra copies of an oncogene, called *MYCN,* in children who had a tumor in one eye, but had no affected relatives, and who did not have mutations in *RB1*. *MYCN* sometimes becomes overexpressed in later stages of RB, but in the newly recognized type of disease, called MYCN RB, the oncogene is the initial driver mutation. The mutation is somatic. That is, the *MYCN* mutation originated in an eye cell of the affected child. Which form of retinoblastoma, the one due to the oncogene or the tumor suppressor, is likely to recur in a sibling of the affected child? Explain your answer.

5. The genomes of four patients with acute myeloid leukemia are sequenced and mutations in the following genes noted:

    patient 1   *IDH1* and *NPM1*
    patient 2   just *IDH1*
    patient 3   *IDH1, NPM1,* and *IDH2*
    patient 4   *IDH1, NPM1, IDH2,* and *FLT3*

    Explain how these patients can have the same diagnosis, yet mutations in different genes.

6. Which technology explained in the chapter could be used to investigate different activities in cells at the edges of a tumor compared to cells in the interior?

7. A polygenic risk score for breast cancer consists of common variants of 80 genes that collectively increase the risk of developing the cancer by 14 percent. Do you think that this test is useful? Cite a reason for your answer.

PART 6 Genetic Technology

CHAPTER 21
DNA Technologies

Ingram Publishing/Alamy

A brand of veggie burger has the taste and texture of meat. It includes a protein from soy that is similar to the hemoglobin found in blood. For ease of manufacture, the soy protein is made in yeast cells. The veggie burger that tastes and feels impossibly like a hamburger is the result of biotechnology.

Learning Outcomes

21.1 Patenting DNA
1. State the criteria for a patentable invention.
2. Discuss the history of patenting organisms and DNA.

21.2 Modifying DNA
3. Distinguish between recombinant DNA and transgenic organisms.
4. Describe applications of recombinant DNA technology.

21.3 Monitoring Gene Function
5. Explain how a DNA microarray is used to monitor gene expression.

21.4 Gene Silencing and Genome Editing
6. Describe ways to decrease expression of a specific gene.
7. Explain how genes and genomes are edited.

The BIG Picture

Ancient biotechnologies used selective breeding to give us bakeries and breweries, foods and medicines. Modern biotechnologies manipulate DNA to give us new ways to study, monitor, and treat disease, and alter foods and the environment.

Engineering a Better Veggie Burger

Some veggie burgers have colorful chunks of corn, beans, and carrots, while others more closely mimic a beef-based burger in texture, taste, and a quality called *mouthfeel*. Much of the sensual similarity to beef is due to heme, a molecule that is packed into muscle cells of cattle but found in all species, from bacteria to beans to buffalo.

Heme consists of iron within a small organic (carbon-containing) heme group, which in turn lies within a wound-up alpha or beta globin chain (see figure 11.1). Globin chains differ by species. In mammals, the protein in the muscles that becomes meat is myoglobin, with a single globin chain. In blood it is hemoglobin, with four globins.

The trick to creating a plant-based burger that mimics meat is finding a plant species whose hemoglobin imparts the unique savory flavor of beef. One such patty uses the heme-surrounding protein from soybeans, called leghemoglobin. But soybean root nodules do not provide enough leghemoglobin; biotechnology does.

Researchers inserted the soybean leghemoglobin gene into yeast cells and mass-produced the needed protein. They fermented the leghemoglobin with "flavor-precursor molecules" including vitamins, amino acids, sugars, lactic acid, and plant oils, then mixed in wheat and soy protein. Human taste testers and analytical chemistry tools evaluated the release of the meaty aromas when the mixture was cooked. Because the soybean leghemoglobin made in yeast is identical to the protein in the plants, the product does not require labeling in the United States as a "genetically modified organism," or GMO.

This burger eliminates the requirement to kill animals. As a plant-based food it is healthier to eat than red meat. It is also more environmentally friendly. Compared to beef, the burger has an 89 percent smaller carbon footprint and uses 87 percent less water and 96 percent less land. It cuts water contamination by 92 percent. The product is used in soup and stew bases, in snack cubes, and in powders.

Figure 21.1 **The universality of the genetic code makes biotechnology possible.** All the cells of the greenish mice contain the gene encoding a jellyfish's green fluorescent protein (GFP). Researchers use GFP to mark genes of interest. The GFP mice glow less greenly as they mature and more hair covers the skin. The nongreen mice are not genetically modified. *Eye of Science/Science Source*

21.1 Patenting DNA

DNA is the language of life, the instruction manual for keeping an organism alive. Yet we also use DNA for other purposes. Manipulating DNA is part of **biotechnology,** which is the use or alteration of cells or biological molecules for specific applications, including products and processes. Biotechnology is an ancient art as well as a modern science, and is familiar as well as futuristic. Using yeast to ferment fruit into wine is a biotechnology, as is extracting and using biochemicals from organisms.

The terminology for biotechnology can be confusing. The popular terms *genetic engineering* and *genetic modification* refer broadly to any biotechnology that manipulates DNA. It includes altering the DNA of an organism to suppress or enhance the activities of its own genes, as well as combining the genetic material of different species. Multicellular organisms that harbor DNA from other species are termed **transgenic** and their DNA is called **recombinant DNA.** Atlantic salmon that grow faster on less food than unaltered salmon, due to genes from Chinook salmon and a gene from an eel-like ocean pout, is an example of a transgenic organism. Recombinant DNA is also in single-celled organisms and in isolated cells growing in culture.

Creating transgenic organisms is possible because all life uses the same genetic code—that is, the same DNA triplets encode the same amino acids (**figure 21.1**). Mixing DNA from different species may seem unnatural, but in fact DNA moves and mixes between species in nature—bacteria do it, and it is why we have viral DNA sequences in our chromosomes. But human-directed genetic modification is intended to give organisms traits they would not have naturally, such as goats that produce spider silk, tomatoes that grow in salt water, and bacteria that synthesize human insulin.

What Is Patentable?

The invention of transgenic organisms raises legal questions, because the design of combinations of DNA sequences not seen in nature may be considered intellectual property and, therefore, patentable. To qualify for patent protection, a transgenic organism, as any other invention, must be new, useful, and not obvious to an expert in the field. A corn plant that manufactures a protein naturally found in green beans but not in corn, making the corn more nutritious as a food for people, is an example of a patentable transgenic organism.

A DNA sequence might be patentable if it is part of a medical device used to diagnose an inherited or infectious disease. DNA-based tests, for example, can identify specific mutations that cause cystic fibrosis. Another test can distinguish among bacterial pathogens and identify strains that are resistant to specific antibiotic drugs. DNA is also patentable as a research tool, as are algorithms used to extract information from DNA sequences and databases built of DNA sequences. The technologies used to manipulate DNA are also patentable, such as the machine learning described in chapter 18 used in ancestry testing and the polymerase chain reaction and DNA sequencing methods (see sections 9.3 and 9.4).

The holder of a patent controls how the invention can be used for 20 years following the date of issue. Patent law has had to keep up with modern biotechnology. In the 1980s, when sequencing a gene was painstakingly slow, only a few genes were patented. In the mid-1990s, with faster sequencing technology and shortcuts to finding the protein-encoding parts of the genome, the U.S. National Institutes of Health and biotech companies began seeking patent protection for thousands of short DNA sequences, even if their functions weren't known. Because of the flood of applications, the

U.S. Patent and Trademark Office tightened requirements for usefulness and revoked patents on DNA sequences alone. The invention must be useful as a tool for research or as a novel or improved product, such as a diagnostic test or a drug. In the United States, more than one in five human genes is patented in some way, and only a few gene patents have been challenged.

Patent Disputes

DNA patenting became highly controversial in 2009, when several groups, including the American Civil Liberties Union and the Public Patent Foundation, challenged patents on two breast cancer genes (*BRCA1* and *BRCA2*) held by biotechnology company Myriad Genetics and the University of Utah. The company did not license the full gene sequences to other companies, so some patients were forced to take Myriad's test for increased familial breast cancer risk, costing them more than $3,000 to sequence the entire genes. (Companies could market tests for the most common mutations in the genes, though.) The patents discouraged research and prevented patients from getting second opinions.

In 2010, a federal judge in the United States ruled that seven patents on the *BRCA* genes were "improperly granted" because they are based on a "law of nature." In 2011, the court invalidated the patents on the two genes, but a federal appeals court overruled that action, claiming that an isolated gene is not the same as a gene in a cell, which is part of a chromosome.

In 2013 the U.S. Supreme Court ruled that "a naturally occurring DNA segment is a product of nature and not patent eligible merely because it has been isolated." The court did allow patenting of **complementary DNA (cDNA),** which is synthesized in a laboratory using an enzyme (reverse transcriptase) that makes a DNA molecule that is complementary in sequence to a specific mRNA. The cDNA represents only the exons of a gene because the introns are spliced out during transcription into mRNA. A cDNA is considered not to be a product of nature because its exact sequence is not in the genome of an organism.

Soon after the controversy over patenting breast cancer genes began to die down, another issue arose: ownership of the gene editing technology CRISPR-Cas9. The technology adapts a natural bacterial defense against viral infection to remove, replace, or add genes in the genomes of other types of organisms. The University of California, Berkeley, published on the technique first, but the Broad Institute filed the patent in an expedited program and was granted the initial patent. The argument was that the original Berkeley patent application did not specify cell types, whereas the Broad patent specifically mentioned use to edit genes of eukaryotic organisms—including humans and other animals (such as mice used in research, livestock, and pets) as well as plants. While the two sides were arguing, other researchers filed patents using CRISPR-Cas9 in specific ways, such as creating crops with pesticide resistance, treating genetic diseases, engineering chimeric antigen receptors (see figure 20.18), and creating induced pluripotent stem cells (see section 2.4). CRISPR-Cas 9 is discussed further in sections 21.4 and 22.4.

Table 21.1 highlights a few biotechnology inventions.

Key Concepts Questions 21.1

a. How do modern applications of biotechnology differ from ancient applications?

*b. What are the requirements for a patented invention?

*c. Why did the U.S. Supreme Court rule that a gene's DNA sequence is not patentable?

d. What were the bases of the controversies over patenting the *BRCA* genes and CRISPR-Cas9?

21.2 Modifying DNA

Recombinant DNA technology adds genes from one type of organism to the genome of another. It was the first gene modification biotechnology, and was initially done with bacteria to make them produce peptides and proteins useful as drugs. When bacteria bearing recombinant DNA divide, they yield many copies of the "foreign" DNA, and under proper

Table 21.1 Key Biotechnology Inventions

Invention	Function in Nature	Function as Technology
CAR-T (figure 20.18)	Combines T-cell receptor binding with antibody anti-cancer response	Directs antibody to cancer cells
CRISPR-Cas9 (figure 21.9)	Cuts across two DNA strands in bacteria to halt viral infection	Molecular scissors to remove, replace, or add genes to selected sites
Green fluorescent protein (figure 21.1)	Jellyfish gene provides bioluminescence, repelling predators	Attached to selected genes causes encoded proteins to glow, serving as "reporter"
Monoclonal antibodies (figure 19.16)	A natural antibody response is polyclonal (several antibody types)	An antibody response directed against a specific antigen

conditions they produce many copies of the protein that the foreign DNA specifies. Recombinant DNA technology is also known as gene cloning. "Cloning" in this context refers to making many copies of a specific DNA sequence.

Constructing and Selecting Recombinant DNA Molecules

In February 1975, molecular biologists convened at Asilomar, on California's Monterey Peninsula, to discuss the safety and implications of combining genes of two species. They discussed restricting the types of organisms and viruses used and ways to prevent escape of a resulting organism from the laboratory, including "physical containment," such as using specialized hoods and airflow systems, and "biological containment," such as weakening organisms so that they would die outside the laboratory.

Recombinant DNA technology turned out to be safer than expected, and it spread rapidly to industry in diverse ways. **Table 21.2** lists some of the drugs produced using recombinant DNA technology. Recombinant DNA research initially focused on providing direct gene products such as peptides and proteins, including insulin, growth hormone, and clotting factors. The technology targets carbohydrates and lipids through the genes that encode enzymes required to synthesize them.

Manufacturing recombinant DNA molecules requires three components:

1. **Restriction enzymes,** which are enzymes that cut DNA at specific sequences
2. **Cloning vectors,** which are pieces of DNA used to deliver specific DNA sequences to cells
3. Recipient cells, such as bacteria or cultured single cells

Selected DNA is inserted into vectors, and then the loaded vectors are delivered to cells. The cells then use the added genetic instructions to manufacture the desired protein.

Hundreds of types of restriction enzymes are naturally found in bacteria, where they cut DNA of infecting viruses, protecting the bacteria. Methyl (CH_3) groups shield the bacterium's own DNA from its restriction enzymes. Some of the enzymes cut DNA at sequences of four, five, or six bases that are symmetrical in a specific way—the recognized sequence reads the same, from the 5′ to 3′ direction, on both strands of the DNA. For example, the restriction enzyme EcoR1, whose actions are depicted in **figure 21.2,** cuts at GAATTC. The complementary sequence on the other strand is CTTAAG, which, read backwards, is

Table 21.2	Drugs Produced in Single Cells Using Recombinant DNA Technology
Drug	**Use**
Atrial natriuretic peptide	Dilates blood vessels, promotes urination
Colony-stimulating factors	Help restore bone marrow after marrow transplant; restore blood cells following cancer chemotherapy
Deoxyribonuclease (DNase)	Thins secretions in lungs of people with cystic fibrosis
Epidermal growth factor	Accelerates healing of wounds and burns; treats gastric ulcers
Erythropoietin (EPO)	Stimulates production of red blood cells in cancer patients
Factor VIII	Promotes blood clotting in treatment of hemophilia A
Glucocerebrosidase	Corrects enzyme deficiency in Gaucher disease
Human growth hormone	Promotes growth of muscle and bone in people with very short stature due to hormone deficiency
Insulin	Allows cells to take up glucose in treatment of type 1 diabetes
Interferons	Treat genital warts, hairy cell leukemia, hepatitis B and C, Kaposi sarcoma, multiple sclerosis
Interleukin-2	Treats kidney cancer recurrence
Lung surfactant protein	Helps lung alveoli to inflate in infants with respiratory distress syndrome
Renin inhibitor	Lowers blood pressure
Somatostatin	Decreases growth in muscle and bone in pituitary gigantism
Superoxide dismutase	Prevents further damage to heart muscle after heart attack
Thrombin	Stops postsurgical bleeding
Tissue plasminogen activator	Dissolves blood clots in treatment of heart attack, stroke, and pulmonary embolism

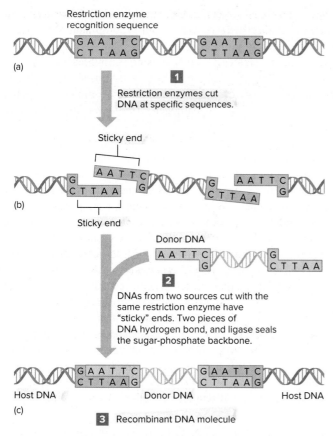

Figure 21.2 Recombining DNA uses restriction enzymes to insert a foreign DNA sequence. These enzymes can be used as molecular scissors because they cut DNA from any source at the same base sequence.

GAATTC. (Try this with other sequences to see that it rarely works this way.) In the English language, this type of symmetry is called a palindrome, referring to a sequence of letters that reads the same in both directions, such as "Madam, I'm Adam." Unlike the language palindrome, however, palindromic sequences in DNA are on complementary strands.

When some restriction enzymes cut, they leave single-stranded extensions, creating overhangs. These are called "sticky ends" because they are complementary to each other, forming hydrogen bonds as their bases pair. Restriction enzymes work as molecular scissors in creating recombinant DNA molecules because they cut at the same sequence in DNA from any source. That is, the same sticky ends result from the same restriction enzyme, whether the DNA is from a moose or a maple.

Another natural "tool" used in recombinant DNA technology is a cloning vector. This structure carries DNA from the cells of one species into the cells of another. A vector can be any piece of DNA into which other DNA can insert. A commonly used vector is a **plasmid**, which is a small circle of double-stranded DNA that exists naturally in some bacteria, yeasts, plant cells, and cells of other types of organisms. Viruses that infect bacteria, called bacteriophages, are another type of vector. Disabled retroviruses are used as vectors, too, as are DNA sequences from bacteria and yeast called artificial chromosomes.

The gene to be delivered must be short enough to insert into the vector. Gene size is measured in kilobases (kb), which are thousands of bases. Cloning vectors can hold up to about 2 million DNA bases.

To create a recombinant DNA molecule, a restriction enzyme cuts DNA from a donor cell at sequences that bracket the gene of interest (**figure 21.3**). The enzyme leaves

Figure 21.3 Recombinant DNA. DNA isolated from a donor cell and a plasmid from a bacterium are cut with the same restriction enzyme and mixed. Sticky ends from the donor DNA hydrogen bond with sticky ends of the plasmid DNA, forming recombinant DNA molecules. When such a modified plasmid is introduced into a bacterium, it is mass produced as the bacterium divides. *(5): Maximilian Stock Ltd/Science Source; (6): CDC/Jim Gathany*

Chapter 21 *DNA Technologies* **399**

single-stranded ends on the cut DNA, each bearing a characteristic base sequence. Next, a plasmid is isolated and cut with the same restriction enzyme used to cut the donor DNA, generating the same complementary single-stranded base sequences at the ends of each. When the cut plasmid and the donor DNA are mixed, the single-stranded sticky ends of some plasmids base pair with the sticky ends of the donor DNA. The result is a recombinant DNA molecule, such as a plasmid carrying the human insulin gene. The plasmid and its human gene can now be transferred into a cell, such as a bacterium or a white blood cell.

Cells bearing the gene must then be separated from cells that lack plasmids or that have taken up empty plasmids. Researchers have ways of marking and separating the useful cells.

When cells containing the recombinant plasmid divide, so does the plasmid. The enzymes, ribosomes, energy molecules, and factors necessary for protein synthesis transcribe and translate the plasmid DNA and its foreign gene, producing the desired protein. Then the protein is separated, collected, purified, and packaged to create a product, such as a new drug.

Products from Recombinant DNA Technology

In basic research, recombinant DNA technology is used to isolate genes from complex organisms and observe their functions on the molecular level. Recombinant DNA is also used to mass-produce protein-based drugs and vaccines.

Drugs manufactured using recombinant DNA technology are pure, and are the human version of the protein. Before the technology, human growth hormone came from cadavers, follicle-stimulating hormone came from the urine of postmenopausal women, and clotting factors were pooled from hundreds or thousands of donors. These sources introduced risk of infection, especially after HIV and hepatitis C became more widespread.

The first drug manufactured using recombinant DNA technology was insulin, which is produced in bacterial cells (*E. coli*). Before 1982, people with type 1 diabetes mellitus obtained the insulin that they injected daily from pancreases removed from cattle in slaughterhouses. Cattle insulin is so similar to the human peptide, differing in only 2 of its 51 amino acids, that most people with diabetes could use it. However, about 1 in 20 patients is allergic to cow insulin because of the slight chemical difference. Before recombinant DNA technology, the allergic patients had to use expensive combinations of insulin from other animals or from human cadavers.

Insulin is a peptide and is therefore straightforward to mass-produce in bacteria. Some drugs, however, require that sugars be attached to peptides or proteins, or that the proteins fold in specific, intricate ways to function. These molecules must be produced in eukaryotic cells that readily carry out these modifications. Yeast cells, Chinese hamster ovary cells, insect cells, and even carrot and tobacco cells have been used to produce human proteins. **Bioethics** considers another drug derived from recombinant DNA technology, erythropoietin (EPO).

Safer vaccines are created using recombinant DNA technology. Recall from section 19.5 that a vaccine is a live or killed pathogen or part of one that stimulates an immune response upon encountering the pathogen. Traditional influenza vaccine, for example, is cultured with egg components, to which some people are allergic, and is a time-consuming process. An alternative vaccine consists of the genes that encode the hemagglutinin proteins from two influenza A strains and one influenza B strain (see Clinical Connection 19.1). The influenza virus genes are introduced aboard a virus that naturally infects the cells of armyworms. The vaccine avoids eggs, and can be manufactured faster than conventional vaccines. One type of vaccine against SARS-CoV-2, the coronavirus infection that emerged in 2019, is also made in this system.

A vaccine that protects against malaria is based on altering a bacterium (*Pantoea agglomerans*) that normally inhabits the mosquito gut. Mosquitoes transmit *Plasmodium falciparum,* the parasite that causes malaria. The bacteria are given genes from *E. coli* bacteria that enable them to produce proteins that tear apart the insect's intestines. Using recombinant bacteria is easier than genetically modifying mosquitoes to prevent malaria, discussed in section 21.4.

Transgenic Organisms

Eukaryotic cells growing in culture are generally better at producing human proteins than are prokaryotic cells such as bacteria. An even more efficient way to express some recombinant genes is in a body fluid of a transgenic animal. Secretion of a human protein from an intact non-human animal more closely mimics the environment in the human body than do cultured cells. The genetic change must be introduced into a fertilized ovum so that it is present in every cell of the transgenic organism.

Transgenic non-human mammals can express human genes in their milk, including genes that encode clotting factors, clot busters, and the connective tissue protein collagen. Human antibodies produced in rabbit and cow milk, for example, are used to treat cancers. Recall from figure 19.8 that antibodies are assembled from the products of several genes. Researchers attach the appropriate human antibody genes to promoters for milk proteins. (Promoters are the short sequences at the starts of genes that control transcription rates.) These promoters normally oversee production of abundant milk proteins. The mammary gland cells of transgenic animals can assemble antibody parts to secrete the final molecules—just as if they were being produced in a plasma cell in the human immune system. **Table 21.3** lists some drugs made in transgenic animals.

Several techniques are used to insert DNA into animal cells to create transgenic animals. Chemicals and brief jolts of electricity open transient holes in plasma membranes that admit "naked" DNA, or a gunlike device is used to shoot tiny metal

Bioethics

EPO: Built-in Blood Booster or Performance-Enhancing Drug?

"Cycle, run, and swim longer and faster than anyone else!" proclaims a website selling a product that supposedly boosts levels of erythropoietin. "EPO" is a glycoprotein hormone that the kidneys produce in response to low levels of oxygen in the blood. The hormone moves in the bloodstream to the bone marrow, where it binds receptors on cells that give rise to red blood cell progenitors. Soon, more red blood cells enter the circulation, carrying more oxygen to the tissues (**figure 21A**). Enhanced stamina results.

The value of EPO as a drug became evident after the 1961 invention of hemodialysis to treat kidney failure. Dialysis removes EPO from the blood, causing severe anemia. But boosting EPO levels proved difficult because there isn't enough in human plasma to pool from donors. Instead, in the 1970s, the U.S. government obtained EPO from South American farmers with hookworm infections and Japanese aplastic anemia patients, who secrete abundant EPO into urine. But when the AIDS epidemic came, human body fluids were no longer safe to use. Recombinant DNA technology solved the EPO shortage, and today it is used to treat anemia in dialysis and AIDS patients and in people receiving cancer chemotherapy to avoid the need for transfusions.

EPO's ability to increase the oxygen-carrying capacity of blood under low oxygen conditions is why athletes train at high altitudes to increase endurance. Since the early 1990s, athletes have abused EPO to reproduce this effect, at great risk. EPO thickens the blood, raising the risk of a blockage that can cause a heart attack or stroke. Olympic athletes now take urine tests that detect recombinant EPO, which has a slightly different configuration of sugars than the natural form.

People with any of four types of familial erythrocytosis get the effects of extra EPO naturally. In one autosomal dominant form, the receptor for EPO is active for too long, leading to large and abundant red blood cells. A member of a family from Finland with this condition won several Olympic medals from 1960 to 1972 for skiing thanks to his inborn ability. Researchers traced the family's mutation to a single-base change in one ancestor who lived in the 1850s, in the EPO receptor gene *EPOR*. A gain-of-function dominant mutation in the EPO gene in another family is a frameshift (see table 12.4). Instead of disabling the gene, the mutation enables different mRNAs to be transcribed, which increases the level of the hormone. Autosomal recessive forms of the condition also increase the level of EPO in the bloodstream.

Erythrocytosis may cause dizziness, headaches, nosebleeds, and shortness of breath, or have no symptoms. However, it increases the risk of stroke and heart attack from blocked circulation.

Questions for Discussion

1. Should using a substance made naturally in the body be considered performance enhancement?
2. Should tests be developed to identify athletes whose genes, anatomy, or physiology give them a competitive advantage? Why or why not?
3. When developing drugs that use recombinant DNA technology, should researchers consider how the product could be abused?

Figure 21A At least two genes control EPO secretion. Certain variants of these genes increase the number of red blood cells, increasing endurance but also raising risk of heart attack and stroke. *Image Source/Getty Images*

Table 21.3	Drugs Made in Transgenic Animals	
Animal	**Product**	**Used to Treat**
Chicken	Recombinant enzyme made with egg white protein (albumin)	Lysosomal acid lipase deficiency
Cow	Antibodies in plasma	Infectious disease treatment
Goat	Antithrombin (clotting factor) in milk	Prevents blood clots
Rabbit	Recombinant C1 esterase inhibitor in milk	Hereditary angioedema

particles coated with DNA inside cells. DNA may also cross the plasma membrane in tiny fatty bubbles called liposomes. For non-animal cells, DNA must pass through cell walls as well as plasma membranes. Plant cells can be genetically modified using a natural delivery system, the bacterium *Agrobacterium tumefaciens*. It causes tumor-like bumps on trees, but can be attached to genes of interest that are then delivered to crop plants.

Getting foreign DNA into a fertilized ovum is the first step in creating a transgenic animal. The recombinant DNA must enter the nucleus, replicate with the cell's own DNA as part of a chromosome, and be transmitted to daughter cells when the cell divides. Finally, an animal must develop from the fertilized ovum in a surrogate mother. If the trait is dominant, the transgenic animal must express it in the appropriate tissues at the right time in development to be useful. If the trait is recessive, crosses between heterozygotes may be necessary to yield homozygotes that express the trait. Then the organisms must pass the characteristic on to the next generation.

Animal Models

Not many drugs are made from herds of transgenic farm animals, because they are difficult to maintain. Transgenic animals are more useful as models of human disease (**figure 21.4**). Inserting the mutant human beta globin gene that causes sickle cell disease into mice, for example, results in a mouse model of the disease. Drug candidates can be tested on these animal models and abandoned before being tested in humans if they cause significant side effects. A more practical example of a transgenic animal that models humans was Sweetie Pie, the pig described in Bioethics in chapter 19. Her liver kept a young man alive until he could receive a transplant.

Transgenic animal models have limitations. Before the more precise gene editing became possible, researchers could not control where a transgene inserted into a genome, and how many copies inserted, so that different animals could have different numbers of the gene of interest. The level of gene expression necessary for a phenotype to emerge may also differ in the model and humans. This was the case for a mouse model of familial Alzheimer disease. The transgene has the exact same DNA sequence that disrupts amyloid precursor protein in a Swedish family with the condition, but apparently did nothing to the mice—until researchers increased transcription rate 10-fold. Only then did the telltale plaques and tangles, and neuron cell death, appear in the mouse brains.

Animal models might not mimic the human condition exactly because of differences in rates of development. For example, for some inherited diseases that do not cause symptoms until adulthood in humans, mice simply do not live long enough to have the associated phenotype. For this reason, a transgenic monkey is a more accurate model of Huntington disease than a mouse, because as a primate the monkey is much more similar to a human in life span, metabolism, reproduction, behavior, and cognition.

Genetically Modified Foods

Traditional agriculture is the controlled breeding of plants and animals to select individuals with certain combinations of inherited traits that are useful to us, such as seedless fruits and lean meat. Agriculture is a form of genetic intervention based on phenotype, such as taste or appearance, and is imprecise, affecting many genes. Agriculture selects from preexisting variants, rather than altering them at the DNA level. In contrast, DNA-based techniques manipulate one or a few specific genes at a time. Organisms altered to have genes from other species or to over- or underexpress their own genes are broadly termed "genetically modified" organisms, or GMOs. An organism given genes from another species is transgenic. The products of technologies that block gene expression, such as RNA interference, or that remove a gene, such as some applications of gene editing, discussed in section 21.4, are not considered to be genetically modified.

Crops are genetically modified to protect them against specific pests, weeds, natural diseases, and toxins such as pesticides. Golden rice is a GM crop that, when invented in 1999, had two genes from daffodils and one from bacteria, which enabled the rice plants to produce more beta-carotene, a precursor to vitamin A. The rice was intended to improve human nutrition in the many nations where rice is a staple and vitamin A deficiency widespread and is being used in several nations now. Other genetic modifications in rice and other crop plants are still being investigated to boost vitamin A intake.

Genetic modification can introduce interesting traits from unusual source species:

- Lettuce plants produce more vitamin C when they are transgenic for a gene from rats.
- Apples resist fire blight, a bacterial infection, thanks to a transgene from a moth.
- Potatoes given genes from moths and bees resist potato blight, a fungal infection.
- Grapes with a silkworm gene resist Pierce's disease, a bacterial infection.

Genetically modified foods have been controversial. Some nations outlaw GM foods, but people in the United States have been eating them since the mid-1990s, and (apparently) safely. A government report found that more than 60 percent of processed foods in supermarkets contain at least one GM ingredient. From 92 to 94 percent of corn, cotton, and soybean crops in the United States are genetically modified.

Figure 21.4 Animal models mimic human disease.
Transgenic mice that have a mutation that causes Huntington disease in humans are tested for coordination on a rotating drum with a grooved surface. *Courtesy MRC Harwell*

People object to GM foods for a variety of reasons. Officials in France and Austria have called such crops "not natural," "corrupt," and "heretical." An enraged consumer declared on a TV news program, "I will not eat food that contains DNA!" GM crops may have inherited traits not normal to them, but DNA is DNA. All foods contain DNA.

Other reasons for objecting to GM foods are practical. A food that contains a nutrient not normally in it—like a protein from peanuts in corn—could harm a person with peanut allergy who eats it, unless the product is labeled. Some companies genetically modify plants in ways that require farmers to buy certain herbicides or purchase new seeds each season.

Field tests may not adequately predict the effects of GM plants on ecosystems. Buffer zones of non-GM plants are routinely planted around fields of GM varieties, but wind pollination can take GM plants far. A sampling of 400 canola plants growing along roads in North Dakota found that 86 percent of them had a gene indicating that they descended from GM plants.

Overreliance on GM crops may lead to genetic uniformity, which is just as dangerous as traditional monoculture. Genetic sameness creates vulnerability in a population if environmental conditions change and no variants are present to survive natural selection. Some GM organisms, such as fish that grow to twice normal size or can survive at temperature extremes, can disrupt natural ecosystems. **Table 21.4** describes several crops that are altered using transgenic technology or two technologies described in section 21.4, gene silencing and genome editing.

Bioremediation

Recombinant DNA technology and transgenic organisms provide processes as well as products. In **bioremediation**, bacteria or plants with the ability to detoxify certain pollutants are released or grown in a particular area. Natural selection has sculpted such organisms, perhaps as adaptations that render them unpalatable to predators. Bioremediation uses genes that enable an organism to metabolize a substance that, to another species, is a toxin. The technology uses unaltered organisms, and also transfers "detox" genes to other species so that the protein products can more easily penetrate a polluted area.

Nature offers many organisms with interesting metabolic characteristics that might benefit other species. A type of tree that grows in a tropical rainforest on an island near Australia, for example, accumulates so much nickel from soil that slashing its bark releases a bright green latex ooze. Genes from this tree can be used to clean up nickel-contaminated soil. Transgenic microorganisms that make proteins that detoxify contaminants are sent into plants whose roots distribute the detox proteins in the soil. In this way, transgenic yellow poplar trees can thrive in mercury-tainted soil thanks to a bacterial gene that encodes an enzyme that converts a highly toxic form of mercury to a less toxic gas. The tree's leaves release the gas.

Cleaning up munitions dumps from wars is another use of bioremediation, such as deploying bacteria that normally break down trinitrotoluene, or TNT, the major ingredient in dynamite and land mines. The enzyme that provides this ability is linked to the *GFP* gene from jellyfish (see figure 21.1). Bacteria spread in a contaminated area glow near land

Table 21.4 Engineered Crops

Plant	Altered Trait	Method
Alfalfa	Decreased lignin extends feeding season for cattle	RNAi suppresses enzyme
Apple	Less browning	Transgene from one apple species suppresses expression of browning enzyme genes in another apple species
Cassava	Increased iron and zinc	Transgenes from mustard weed block transporter proteins, increasing levels of these minerals
False flax	Seeds make more omega-3 fatty acids, normally eaten in fish oil	Transgenes for omega-3 fatty acid production from alga, moss, and water mold
Potato	Less bruising/browning, less acrylamide (carcinogen)	RNAi blocks synthesis of enzyme (polyphenol oxidase)
Rice	Resistance to drought-related stress	Transgene from mustard weed encodes enzyme that enables a sugar to accumulate in cells, retaining water during drought
Soybean	Oil without trans fats, increased oleic acid, extended shelf life	TALENS (gene editing)

mines, revealing the locations more clearly than a metal detector could. Once the land mines are removed, the bacteria die as their food vanishes.

Bioremediation is also used to decrease the amount of phosphorus in pig manure, which presents a serious environmental problem. The animals do not make an enzyme to extract the mineral nutrient phosphorus from a compound called phytate in cereal grains in their feed, so they are given dietary phosphorus supplements. As a result, their manure is full of phosphorus. The element washes into natural waters, contributing to fish kills, oxygen depletion in aquatic ecosystems, algal blooms, and the greenhouse effect. Fortunately, biotechnology has solved the "pig poop" problem.

In the past, pig raisers tried to keep their animals healthy and the environment clean by feeding animal byproducts from which the pigs can extract more phosphorus and by giving supplements of the enzyme phytase, which liberates phosphorus from phytate. But consuming animal byproducts can introduce prion diseases, and giving phytase before each meal is costly. A phytase transgenic pig, however, secretes bacterial phytase in its saliva, which enables the animal to excrete low-phosphorus manure.

A transgenic organism has a gene in each of its cells from an organism of another species. The genetically modified pig, called the Enviropig, has a phytase gene from the bacterium *E. coli* in each cell, as well as a promoter DNA sequence from a mouse that controls secretion of phytase from the salivary glands. Enviropig's manure has 75 percent less phosphorus than unaltered pig excrement.

Key Concepts Questions 21.2

*a. Explain the steps of recombinant DNA technology.
b. Discuss applications of recombinant DNA technology.
c. List examples of useful transgenic plants and animals.

21.3 Monitoring Gene Function

We usually can do little about the gene variants that we inherit. Gene expression, in contrast, is where we can control our environment. Monitoring gene expression requires detecting the mRNAs in particular cells under particular conditions. To do this, devices called gene expression **DNA microarrays,** or gene chips, detect and display the mRNAs in a cell. The creativity of the technique lies in choosing the types of cells to interrogate. Figure 20.16 depicts use of the technology to distinguish types of leukemia.

Evaluating a spinal cord injury illustrates the basic steps in creating a DNA microarray to assess gene expression. Researchers knew that in the hours after such a devastating injury, immune system cells and inflammatory biochemicals flood the affected area. Gene expression profiling revealed how fast healing begins.

A DNA microarray is a piece of glass or plastic that is about 1.5 centimeters square—smaller than a postage stamp. Many small pieces of DNA (oligonucleotides) of known sequence are attached to one surface, in a grid pattern. The researcher records the position of each DNA piece in the grid. In many applications, a sample from an abnormal situation (such as disease, injury, or environmental exposure) is compared to a normal control. **Figure 21.5** compares cerebrospinal fluid (CSF; the liquid that bathes the spinal cord) from an injured person (sample A) to fluid from a healthy person (sample B). Messenger RNAs are extracted from the samples and complementary DNAs (cDNAs) are made. Recall from section 21.1 that a cDNA represents the exons of a gene.

The cDNAs from the injury sample are labeled with a red fluorescent dye and the cDNAs from the control sample are labeled with a green fluorescent dye, then they are applied to the microarray. The device might display genes that are likely to be involved in a spinal cord injury, or the entire human exome or genome, to be certain all genes are probed. Considering so many DNA sequences allows for surprises, avoiding the assumption that we know what to look for.

DNA that binds to cDNAs on the grid fluoresce, and a laser scanner detects and converts the results to a colored image. Each position on the microarray can bind DNA pieces from both samples (injured and healthy), either sample, or neither. The scanner also detects fluorescence intensities, which provides information on how strongly the gene is expressed (how much mRNA is in the sample).

A computer algorithm interprets the pattern of gene expression. For the spinal cord example, the visual data mean the following:

- Red indicates a gene expressed in CSF only when the spinal cord is injured (and presumably leaking inflammatory molecules).
- Green indicates a gene expressed in CSF only when the spinal cord is intact.
- Yellow indicates positions where both red- and green-bound dyes fluoresce, representing genes that are expressed whether or not the spinal cord has been injured.
- Black, or a lack of fluorescence, corresponds to DNA sequences that are not expressed in CSF because they do not show up in either sample.

The color and intensity pattern of the microarray provides a glimpse of gene expression following spinal cord injury. The technique is even more powerful when it is repeated at different times after injury. When researchers did exactly that on injured rats, they discovered genes expressed just after the injury whose participation they never suspected. Their microarrays, summarized in **table 21.5,** revealed waves of expression of genes involved in healing. Analysis on the first day indicated activation of the same suite of genes whose protein products heal injury to the deep layer of skin—a total surprise that suggested new points for drugs to intervene.

Figure 21.5 A DNA microarray experiment reveals gene expression in response to spinal cord injury. In this example, the red label represents DNA from a patient with a spinal cord injury and the green label represents control DNA from a healthy person. DNA targets on the microarray that bind red but not green can reveal new points of intervention for drugs.

Table 21.5	Gene Expression Profiling Chronicles Repair After Spinal Cord Injury
Time After Injury (rats)	Type of Increased Gene Expression
Day 1	Protective genes to preserve remaining tissue
Day 3	Growth, repair, cell division
Day 10	Repair of connective tissues
	Angiogenesis
Days 30–90	Blood vessels mature
	New type of connective tissue associated with healing

Key Concepts Questions 21.3

*a. What do DNA microarrays detect?

b. List the steps of a DNA microarray experiment.

c. How can microarrays that track gene expression be used to reveal the genes that participate in a process?

21.4 Gene Silencing and Genome Editing

Since the 1960s, when the flow of genetic information from DNA to RNA to protein was discovered and described, researchers have been trying to manipulate the process to diminish ("knock down") or silence ("knock out") the expression of specific genes. **Gene silencing** techniques block the synthesis of, or degrade, mRNA. Genome editing techniques create double-stranded breaks in the DNA double helix, enabling insertion of a desired DNA sequence or removal of a sequence. Different approaches to gene silencing and genome editing have had varying degrees of success. **Figure 21.6** indicates where these techniques intervene in the flow of genetic information. All of these biotechnologies are based on complementary base pairing and some borrow directly from natural DNA repair pathways.

Gene Silencing

Antisense technology is a form of gene silencing that blocks expression of a gene by introducing RNA that is complementary to the gene's mRNA transcript. The introduced RNA, called antisense RNA, binds to the mRNA, preventing its translation into protein. An early application of antisense technology was a tomato intended to stay fresh longer because the antisense RNA blocked the activity of a ripening enzyme. Ripening slowed, but the tomatoes still softened and tasted terrible.

Antisense technology is used to produce a drug that treats spinal muscular atrophy (SMA). This autosomal recessive disease damages the motor neurons in the spinal cord, producing extreme weakness and "floppiness," eventually impairing breathing. Most babies with SMA do not live beyond 2 years. It is the most prevalent genetic cause of infant mortality, affecting 1 in 10,000 newborns in the United States. One in 50 people are carriers.

A genetic quirk makes an antisense approach to treating SMA possible. In people with the disease, the causative gene, "survival motor neuron 1" (*SMN1*), is nonfunctional, but nearby on chromosome 5 is a second gene, *SMN2*. In most people, healthy or with SMA, the two copies of *SMN2* have a mutation that blocks transcription. However, an antisense molecule similar to DNA can patch the mutation, enabling *SMN2* to be transcribed into mRNA and translated into protein. The drug is a series of spinal cord injections of the antisense molecule that stimulates synthesis of the normally silenced SMN2 protein. Treated children reach developmental milestones that they otherwise wouldn't, even walking. Their progress is being monitored.

A second variation of antisense technology uses synthetic molecules called **morpholinos** that consist of sequences of 25 DNA bases attached to organic groups that are similar to, but not exactly the same as, the sugar-phosphate backbone of DNA. Morpholinos can block splice-site mutations that would otherwise delete entire exons. For example, a morpholino-based drug targets the dystrophin gene, restoring the function of a skipped exon and enabling muscle cells to produce some dystrophin. The drug has been approved to treat Duchenne muscular dystrophy, but is still being tested to see if it can restore enough dystrophin to provide sustained improvement in mobility for affected boys.

Figure 21.6 Gene silencing and genome editing. Gene silencing techniques prevent mRNA from being translated into protein. Genome editing techniques enable researchers to remove double-stranded segments of DNA, so that they can induce deletions or swap in genes.

Another approach to gene silencing uses **ribozymes**. These are RNA molecules that are part of ribosomes (the organelles on which translation occurs) that have catalytic activity, like enzymes. Ribozymes fit the shapes of certain RNA molecules. Because ribozymes cut RNA, they can be used to destroy RNAs from pathogens, such as HIV. Figure 13.13 illustrates yet another way to turn off gene expression—using the XIST long noncoding RNA that normally shuts off one X chromosome in the cells of female mammals to silence the extra chromosome of trisomy 21 Down syndrome.

RNA interference (RNAi) is a gene silencing technique based on the fact that RNA molecules can fold into short, double-stranded regions where the base sequence is complementary. **Figure 21.7** shows how this bonding produces a hairpin shape. Messenger RNAs naturally fold into shapes that facilitate their translation into protein. This reliance on conformation is why application of an antisense sequence and formation of RNA hairpins can disrupt protein synthesis. Short, double-stranded RNAs introduced into cells can have great effects.

The basis of RNAi technology is that short, double-stranded RNAs sent into cells separate into single strands and one binds its complement in mRNA, preventing its translation. The small RNAs that carry out RNA interference are called small interfering RNAs (siRNAs).

Several proteins and protein complexes orchestrate RNAi, and are also part of microRNA function (**figure 21.8** and see section 11.2). First, an enzyme called Dicer cuts long, double-stranded RNAs into pieces 21 to 24 nucleotides long. These pieces contact a group of proteins that form an RNA-induced silencing complex, or RISC. One strand of the double-stranded short RNA, called the guide strand, adheres. Now, as part of RISC, the guide strand finds its complementary RNA and binds. Then another part of RISC, a protein called argonaute, degrades the targeted RNA, preventing its translation into protein. SiRNAs act in the nucleus, too, where they alter methylation and the binding of histones to certain genes.

RNA interference can be used to create vaccines by knocking down expression of key genes in viruses that cause diseases; treat cancer by silencing oncogenes; and engineer a better-tasting decaf coffee by silencing an enzyme required for caffeine synthesis.

Figure 21.8 **RNA interference.** Dicer cuts double-stranded parts of RNA molecules, which then associate with RNA-induced silencing complexes (RISCs). The RNA is open, revealing single strands that locate and bind specific mRNAs. Nucleases then break down the targeted mRNAs, preventing their translation into protein.

Another agricultural application of RNAi is to engineer crops to make RNAi molecules that a specific pest species eats. The RNA molecules then kill or sterilize the pest. RNAi is being applied to the Colorado potato beetle, for example. Crops treated with RNAi offer several advantages. They rely less on toxic pesticides, and one manipulation lasts the plant's lifetime, unlike the repeated spraying of an insecticide. Only one pest species is affected. RNA cannot harm other species, as can a protein.

The first approved drug based on RNAi technology treats hereditary transthyretin-mediated amyloidosis with an intravenous infusion every three weeks. The condition causes tingling, burning sensations, poor balance, urinary tract infections, lightheadedness, gastrointestinal effects, and heart arrhythmias that can be fatal. The protein that is affected, transthyretin, is made in the liver, where it normally forms quartets that carry vitamin A. A mutation changes the shape of transthyretin so that it aggregates into spikes that accumulate in the affected body parts.

Genome Editing

Genome editing, like recombinant DNA technology, uses restriction endonucleases to cut and paste DNA molecules in patterns that might not exist in nature. However, the two technologies differ in that genome editing precisely cuts, whereas recombinant DNA technology inserts DNA randomly. In recombinant DNA technology, the enzymes generate offset ends that are "sticky" so that pieces connect by base-pairing. In genome editing, the enzymes generate precise breaks across both strands of the DNA double helix that are aligned, not offset. Then natural DNA repair systems

Figure 21.7 **Hairpins.** RNA hydrogen bonds with itself, forming hairpin loops. RNAi uses similar molecules to "knock down" expression of specific genes by binding their mRNAs. Translation cannot proceed if the loops or antisense molecules prevent an mRNA from folding normally.

mend the breaks, removing or even replacing or adding DNA nucleotides to the targeted area. Understanding natural DNA repair pathways inspired the development of gene and genome editing in the late 1980s.

Genome editing can be done on somatic cells or on cells of the germline (developing oocytes and sperm). In somatic applications, the change must affect all or enough cells to alter the phenotype, such as treating symptoms of a genetic disease (see section 22.3). Germline genome editing introduces the genetic change to every cell in an organism because it is carried out at the beginning of development.

The three major genome editing techniques are usually referred to by their easier-to-remember abbreviations and acronyms (**table 21.6**). Although ZFNs and TALENs were invented several years before CRISPR-Cas9, they use proteins and can only cut one gene at a time. CRISPR-Cas9 uses RNA and is cheaper and easier to use, and can remove, replace, or add more than one gene at a time. All three genome editing techniques are more precise than the gene therapies described in chapter 22 that deliver genes but do not replace the existing mutant genes.

The restriction enzymes that are used in genome editing function like scissors directed to cut DNA at a specific point. A "DNA binding" domain leads the enzyme to a specific short DNA sequence, while the endonuclease part severs both strands of the double helix at the same point.

Zinc finger nuclease (ZFN) technology uses protein motifs (parts of proteins that have characteristic shapes) called zinc fingers that consist of a beta-pleated sheet linked to an alpha helix by a zinc atom (see figure 10.19*b*). Different zinc fingers bind different three-base DNA sequences. If zinc fingers bind, then another nuclease (FokI) cuts the DNA. Zinc finger action is a little like folding a small loop of yarn and snipping it across both strands with one cut.

In **transcription-activator-like effector nuclease (TALEN) technology,** a restriction enzyme from a bacterium (*Xanthomonas*) that infects plants cuts DNA on both strands. TALENs also work on animal cells. Consider spider silk, which is a strong, stretchy material used in military and biomedical applications. TALENs introduce a gene from a golden orb-web spider into the silkworm spider, enabling the recipient to produce a more elastic silk in greater amounts.

Clustered regularly interspaced short palindromic repeats (CRISPRs) are short sequences of DNA that include several repeats. CRISPRs are natural components of the genomes of certain bacteria, where they provide an action similar to an immune response. Specifically, CRISPRs enable bacteria to deploy a restriction enzyme called Cas9 ("CRISPR-associated protein 9") to recognize and cut out DNA sequences from infecting viruses that have inserted into the bacterial genome (other enzymes can be used). The CRISPR DNA sequences along with the viral sequences are transcribed into CRISPR RNAs. These RNAs then serve as "guides" that bring Cas9 to other sites where viral DNA has integrated into the bacterial genome (**figure 21.9**). Cas9 then cuts both strands. The CRISPR RNAs remain to eject DNA from future viral infections.

The beauty of CRISPR is its great versatility—researchers can design and synthesize guide RNAs to direct which DNA sequences are removed, and even add sequences to replace the ones snipped out. It works in any species. While bacteria might remove infecting viral DNA, a human cell might be programmed to replace a mutant gene with the wild type variant. Treating genetic disease and creating animal models are applications of CRISPR. **Table 21.7** lists some innovative uses.

Table 21.6	Genome Editing Techniques
Abbreviation	**Full Name**
CRISPR-Cas9	Clustered regularly interspaced short palindromic repeats-CRISPR-associated protein 9
TALENs	Transcription-activator-like effector nucleases
ZFNs	Zinc finger nucleases

Figure 21.9 CRISPR-Cas9. In the genome editing tool CRISPR-Cas9, a synthetic guide RNA directs the restriction endonuclease Cas9 (purple) to a selected target sequence of DNA. The enzyme makes a double-stranded cut, and the target sequence can be swapped in. Source: Adapted from Chemical & Engineering News 93(26), 20–21.

Table 21.7 Applications of CRISPR Genome Editing

Application	Example
GMO containment	Deleting genes that encode hormones that catfish require to reproduce, so that catfish with other genetic modifications can be raised on farms where the hormones are added to the water, but not survive in the wild.
Bringing back extinct species	Replace genes with counterparts from extinct relatives, such as mammoth genes in an elephant.
Limiting spread of infectious diseases	Introduce genes into disease vectors such as mosquitoes and ticks that make them infertile.
Avoiding allergy to vaccines	Remove the gene that encodes albumin, the egg white protein, which makes some people react to vaccines grown in hen eggs.
Adding genes to resist disease	Swapping in a *CCR5* mutation to provide resistance to HIV (see the chapter opener of chapter 19).
Creating organ donors	Introduce human genes into pig genomes to create cell surfaces that the human immune system will not reject.
Adding traits to show animals	Introduce genes for valued traits (such as fur color, body size, and stamina) into pets or show animals in just one or two generations.

Gene Drives

Genome editing techniques are powerful, and therefore controversial. One major concern is use of CRISPR to reduce populations of, disarm, or eliminate species that are vectors for human pathogens. Top targets are mosquitoes, particularly *Aedes aegypti*, which spreads Zika virus disease, dengue fever, and yellow fever, and the *Anopheles* species that transmit malaria.

The application of genome editing to kill, alter, or render infertile a pathogen is called a **gene drive**. It is based on a natural form of DNA repair, called homing, which removes one of a pair of alleles of a selected gene and replaces it with another copy of the remaining allele. Homing genes are found in certain single-celled organisms. The proteins that they encode recognize certain DNA sequences of 15 to 30 bases, cut them out, and replace them with a copy of the DNA sequence on the other strand. In a gene drive, homing proteins are incorporated into CRISPR-Cas9 to delete and then replace a selected gene.

A gene drive counters Mendel's first law, of segregation. **Figure 21.10** compares Mendelian inheritance of a single gene, in which each parent transmits alleles with a frequency of 50 percent, to a gene drive, in which the parent whose allele is copied essentially transmits it with a frequency near 100 percent. Gene drives are an incredibly fast way to alter a wild population. However, a change to the genetic structure of a population resulting from a gene drive may not be permanent because of evolution. New mutations may occur and offer a selective advantage, similar to the way that antibiotic resistance arises.

Gene drives were first demonstrated on laboratory fruit flies, then applied to the mosquitoes that transmit malaria. Developing mosquito oocytes are given the homing gene; guide RNAs corresponding to three genes required for female fertility, which bear mutations; and a gene variant giving the altered insects a reproductive advantage. The oocytes are fertilized with sperm wild type for the three genes, and the insects that result are all heterozygotes for the three genes. But when the heterozygotes mate, homing happens and suddenly the next generation of mosquitoes are homozygotes for the three fertility mutations. Population size plummets to below the level necessary to sustain the ability to transmit the protozoan that causes malaria.

Malaria kills more than 1 million people a year, and dengue, yellow fever, and Zika virus sicken and kill many, too. However, drastic alterations to mosquito populations could have unintended consequences on ecosystems by removing a component of food webs, opening up new ecological niches or harming populations of other species.

Editing the Human Germline

Altering the germline in humans is even more controversial than doing so in insects. Although many researchers have agreed not to attempt to edit the human germline, attempts have been made to do so.

Early experiments were done on human embryos destined not to survive because their cells had extra nuclei. Then experiments introduced CRISPR at the moment of fertilization, creating germline-edited embryos because the change was transmitted to every cell. One such experiment corrected a dominant mutation that causes heart disease, and another claimed to have created twin girls whose genomes were edited to remove the *CCR5* gene that encodes one of the co-receptors to which HIV binds to enter human cells. Genome sequencing reportedly revealed that the genetic change had indeed been made and that CRISPR had not introduced any unplanned insertions or deletions.

Discussion, at times heated, continues among developmental biologists, physicians, regulatory authorities, policy experts, and

Figure 21.10 **A gene drive.** Using a type of natural DNA repair called homing, a gene drive counters Mendel's law of segregation by cutting out an allele and replacing it with a copy of the remaining allele. Incorporating gene variants that confer infertility or interfere with the relationship between an infectious disease vector and a pathogen that it carries can wipe out the vector or destroy its ability to transmit the infectious disease. Source: Adapted from Dr. Kevin Esvelt, Wyss Institute at Harvard

bioethicists about whether and how research that uses genome editing should be carried out, and how the technology could ultimately be used in the clinic. Genetics organizations do not object to the use of CRISPR on gametes and embryos if appropriate permissions are granted, but object to use of the technology that leads to a human pregnancy. They advise that future clinical applications of human germline genome editing should only proceed if the outcome would serve a medical need, if it is ethically justified, and if the public is informed. Chapter 22 returns to the use of genome editing to diagnose and treat disease.

Key Concepts Questions 21.4

a. How do antisense technologies, ribozymes, and RNA interference silence gene expression?

b. Explain how the three genome editing techniques work.

c. What are some applications of genome editing?

*d. Why is genome editing controversial?

Study Guide

Summary

21.1 Patenting DNA

1. **Biotechnology** alters cells or biochemicals to provide a product or process. It extracts natural products, alters an organism's DNA, or combines DNA from different species.

2. A **transgenic organism** is multicellular and has DNA from a different species in all its cells. **Recombinant DNA** comes from more than one type of organism. Both are possible because of the universality of the genetic code.

3. Patented DNA must be useful, novel, and nonobvious. A gene sequence by itself can no longer receive patent protection in the United States.

4. Patent battles arose over breast cancer genes and gene editing.

21.2 Modifying DNA

5. **Recombinant DNA technology** mass-produces proteins in bacteria or other single cells. Begun hesitantly in 1975, the technology has matured into a valuable method to produce proteins that are used as drugs.

6. To construct a recombinant DNA molecule, **restriction enzymes** cut the gene of interest and a **cloning vector** (such as a **plasmid**) at a short palindromic sequence,

creating complementary "sticky ends." The DNAs are mixed and vectors that pick up foreign DNA selected. Useful proteins are isolated and purified.

7. Heterozygotes for a transgene are bred to yield homozygotes. Transgenic plants are altered to be more nutritious, easier to grow, and/or to resist pests and toxins. Transgenic animals are used to model human disease and to produce protein-based drugs in body fluids. **Bioremediation** provides traits that have environmental effects.

21.3 Monitoring Gene Function

8. **DNA microarrays** hold DNA pieces to which fluorescently labeled **complementary DNA (cDNA)** probes from samples are applied. They are used in **gene expression profiling**.

21.4 Gene Silencing and Genome Editing

9. **Gene silencing** uses **antisense technology, ribozymes, morpholinos,** and **RNA interference (RNAi)** to block translation of mRNAs.
10. **Genome editing** uses enzymes to create double-stranded breaks in DNA, enabling researchers to add, replace, or delete specific genes. **ZFNs, TALENs,** and **CRISPR-Cas9** are used to edit genomes.
11. Genome editing can correct a genetic defect, create an animal model, or introduce a **gene drive** to change or eradicate a species, among other applications.
12. Germline editing of the human genome is controversial.

Key Terms

antisense technology A form of gene silencing that blocks expression of a gene using an RNA or similar molecule that is complementary to the gene's mRNA transcript.

bioremediation Use of plants or microorganisms or their molecules to detoxify environmental pollutants.

biotechnology The alteration of cells or biochemicals for a specific application.

cloning vector A piece of DNA used to transfer DNA from a cell of one organism into the cell of another.

clustered regularly interspaced short palindromic repeats (CRISPR) A genome editing tool that uses bacterial DNA and RNA to direct double-stranded cuts in target DNA.

complementary DNA (cDNA) A DNA molecule synthesized from the exon (protein-encoding) sequences of a gene, omitting the control sequences and introns.

DNA microarray Target genes embedded in or attached to a glass or plastic chip to which labeled cDNA pieces from a clinical sample bind and fluoresce. Microarrays show patterns of gene expression and gene variants present.

gene drive Genome editing to eliminate transmission of a selected gene variant, being developed to eliminate certain pathogens.

gene silencing Blocking transcription of or degrading mRNA, preventing translation into protein.

morpholinos Synthetic molecules that resemble 25 linked DNA nucleotides that block splice-site mutations, restoring deleted exons and function in some genes.

plasmid (plăz′mĭd) A small circle of double-stranded DNA found in some bacteria, yeasts, and plant cells. Used as a vector in recombinant DNA technology.

recombinant DNA DNA that includes sequences from more than one species.

recombinant DNA technology Adding genes from one type of organism to the cells of another.

restriction enzyme An enzyme, typically from a bacterium, that cuts DNA at a specific sequence. Used to create recombinant DNA molecules and in genome editing.

ribozyme An RNA that is the catalytic part of a ribosome.

RNA interference (RNAi) Introduction of a small interfering RNA molecule that binds to and prevents translation of a specific mRNA.

transcription-activator-like effector nuclease (TALEN) technology A genome editing method that uses a bacterial restriction enzyme to cut DNA across both strands.

transgenic organism (trăns-jĕn′ĭk) An individual with a genetic modification, typically introduction of a gene from another species, in every cell.

zinc finger nuclease (ZFN) technology A genome editing technique that uses protein motifs called zinc fingers to bind specific DNA triplets, enabling a nuclease to cut the DNA across both strands.

Review Questions

1. Cite three examples of a DNA sequence that meets requirements for patentability.
2. Why is a DNA sequence by itself not patentable, but a DNA sequence that is used to diagnose a disease in a medical test is patentable?
3. Add an entry to table 21.1 based on chapter material.
4. Describe the functions of each of the following tools used in biotechnology:
    a. restriction enzymes
    b. cloning vectors
    c. DNA microarrays
    d. short nucleic acid molecules with specific sequences
    e. CRISPR sequences
5. Why would recombinant DNA technology and creation of transgenic organisms be restricted or impossible if the genetic code were not universal?
6. What is an advantage of a drug produced using recombinant DNA technology compared to one extracted from natural sources?
7. How are transgenic animals better models of human disease than animal models whose DNA is unaltered? What are limitations of transgenic animal models?

8. How is genetic modification of a crop usually more precise and predictable than using conventional breeding to create a new plant variety?

9. Explain the advantages of using a DNA microarray that covers the exome or genome rather than selected genes whose protein products are known to take part in the disease process being investigated.

10. Describe how a technology to silence a gene or edit a genome can be used to treat a disease.

11. How does the action of the type of enzyme used in genome editing with CRISPR-Cas9 differ from similar enzymes used in recombinant DNA technology?

12. Explain how a gene drive counters Mendel's first law.

13. Discuss why germline genome editing in humans is more difficult and more controversial than somatic cell genome editing in humans.

Applied Questions

1. Do you think that the veggie burger that has the texture of meat because of a soy heme protein made in yeast cells should be considered genetically modified? Cite a reason for your answer.

2. Do you agree with the Supreme Court that a gene is not patentable but a cDNA is? Cite a reason for your answer.

3. What is the basis of patent protection for chimeric antigen receptor (CAR) T cell therapy to treat cancer, shown in figure 20.18?

4. Genetic modification endows organisms with novel abilities. From the following three lists (choose one item from each list), devise an experiment to produce a particular protein, and suggest its use.

Organism	Biological Fluid	Protein Product
Pig	Milk	Human beta globin chains
Cow	Semen	Human collagen
Goat	Milk	Human EPO
Chicken	Egg white	Human tPA
Aspen tree	Sap	Human interferon
Silkworm	Blood plasma	Jellyfish GFP
Rabbit	Honey	Human clotting factor
Mouse	Saliva	Human alpha-1-antitrypsin

5. Collagen is a connective tissue protein that is used in skincare products, shampoos, desserts, and artificial skin. For many years, it was obtained from the hooves and hides of cows collected from slaughterhouses. Human collagen can be manufactured in transgenic mice. Describe the advantages of the mouse source of collagen.

6. A human oncogene called *ras* is inserted into mice, creating transgenic animals that develop a variety of tumors. Why are mouse cells able to transcribe and translate human genes?

7. In a DNA microarray experiment, researchers attach certain DNA pieces to the grid. For example, to study an injury, genes known to be involved in the inflammatory response might be attached. What is a limitation of this approach?

Case Studies and Research Results

1. A rat has, in each of its cells, the human gene that encodes amyloid precursor protein, and is therefore an animal model of Alzheimer disease. What type of biotechnology was used to create the rat?

2. To investigate causes of acne, researchers used DNA microarrays that cover the entire human genome. Samples came from facial skin of people with flawless complexions and from people with severe acne. In the simplified portion of a DNA microarray shown, one sample is labeled green and comes from healthy skin; a second sample is labeled red and represents skin with acne. Sites on the microarray where both probes bind fluoresce yellow. The genes are indicated by letter and number.

   a. Which genes are expressed in skin whether or not a person has acne?
   b. Which genes are expressed only when acne develops?
   c. List three DNA pieces that correspond to genes that are not expressed in skin.

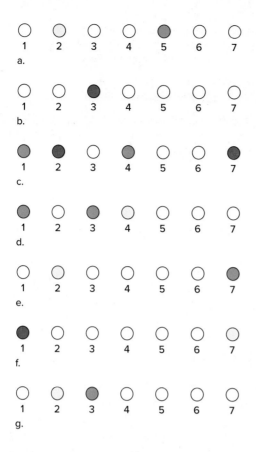

d. How would you use microarrays to trace changes in gene expression as acne begins and worsens?

e. Design a microarray experiment to explore gene expression in response to sunburn.

3. Most viral infections of the lower respiratory tract (the lungs and airways) in babies are caused by rhinovirus (a common cold virus), influenza virus, or respiratory syncytial virus (RSV). Little is known about RSV, and treatment is nonspecific—just supportive. A study showed that white blood cells from babies infected with RSV express 2,000 genes at significantly different levels than white blood cells from healthy infants.

   a. What technology discussed in the chapter could be used to assess the differences among the three types of infections?
   b. What further information would reveal more about RSV infection?
   c. How would you develop a diagnostic test to distinguish RSV infection from rhinovirus and flu infections?
   d. Design an experiment to enable physicians to determine which babies with RSV infection will develop symptoms severe enough to require supplemental oxygen and hospitalization.

4. Crohn's disease is inflammation in the intestines due to an overactive immune response. At the molecular level, the disease is caused by high levels of a protein called SMAD7, which blocks signals from another protein, transforming growth factor B1 (TGFB1). To restore functioning of TGFB1, researchers developed a drug that complementary base-pairs to SMAD7 mRNA, allowing the TGFB1 signal to increase enough to improve symptoms. Which biotechnology described in the chapter does this research apply?

5. Describe a technology that could eliminate the deer ticks that transmit the bacteria that cause Lyme disease. Discuss the risks and benefits of doing this.

CHAPTER 22

Genetic Testing and Treatment

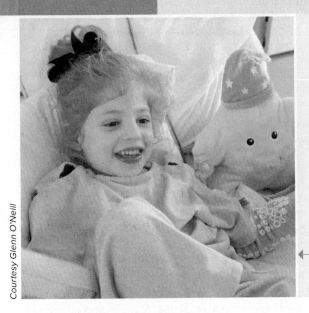

Eliza O'Neill had gene therapy to treat Sanfilippo syndrome type A, a lysosomal storage disease.

Learning Outcomes

22.1 Genetic Counseling
1. Describe the services that a genetic counselor provides.

22.2 Genetic Testing
2. Describe types of genetic tests that are done at different stages of human prenatal development and life.
3. Discuss the benefits and limitations of direct-to-consumer genetic testing.
4. Explain how pharmacogenetics and pharmacogenomics personalize drug treatments.

22.3 Treating Genetic Disease
5. Describe three approaches to correcting inborn errors of metabolism.
6. Discuss how gene therapy adds a functional gene to correct or counter symptoms of a single-gene disease.

22.4 CRISPR-Cas9 in Diagnosis and Treatment
7. Explain how CRISPR-Cas9 genome editing may be used to help diagnose and develop treatments for single-gene diseases.

The BIG Picture

DNA-based tests to predict and help to diagnose disease are becoming more common as researchers identify the functions and variants of more genes, and develop faster ways to detect mutations. Proteins are manipulated to treat certain inborn errors of metabolism. Gene therapy, and gene and genome editing are types of interventions that add, delete, or replace genes to correct the faulty instructions behind specific diseases.

Eliza Gets Gene Therapy

Eliza O'Neill's first symptoms were not very alarming or unusual—slight developmental delay, hyperactivity, recurrent ear infections, and not interacting much with the other children at preschool. After an autism diagnosis didn't describe all of the symptoms, Eliza's pediatrician recommended an MRI. The scan revealed fluid at the back of Eliza's brain and flattened vertebrae in her neck. These findings led to additional tests. A urine test showed the telltale buildup of a sugar, heparan sulfate, that was caused by deficiency of a lysosomal enzyme (see figure 2.5). Then blood tests to detect the enzyme deficiency and identify the mutant gene led to the diagnosis: mucopolysaccharidosis (MPS) type IIIA, also called Sanfilippo syndrome type A. It progresses to intellectual disability, loss of mobility, insomnia, seizures, loss of hearing and vision, dementia, and early death.

When Eliza was diagnosed at age 3½, her parents formed the Cure Sanfilippo Foundation and began researching options. They soon learned that a clinical trial to test a gene therapy for the disease was already being planned at a major children's hospital. Eliza was worsening, losing speech and becoming more hyperactive. Shortly after her sixth birthday, she entered the gene therapy trial. A trillion viruses, each bearing a wild type copy of the gene that encodes the missing enzyme, entered her body through a vein in her hand. The viruses were modified to cross the blood-brain barrier and, if all went well, to allow newly formed enzyme to melt away the buildup of heparan sulfate that caused Eliza's symptoms.

Within a few months, biochemical tests on the first treated children showed a lessening of the biochemical buildup. However, it is too soon to tell the extent to which gene therapy is able to slow progression of this disease.

22.1 Genetic Counseling

Over the past few decades, the field of human genetics has evolved from an academic life science, to a narrow medical specialty, to a growing part of many clinical fields, to a source of personal information that ordinary people can easily access. This chapter presents the types of genetics services that a health care consumer might encounter: genetic counseling, genetic testing, and protein and gene-based therapies.

Genetic tests provide views of our genomes at several levels. They may detect single-base or copy number variants in individual genes, display or identify abnormal chromosomes or chromosome numbers, assess variability across the genome, or determine the entire protein-encoding part of the genome (the exome) or the whole genome. The U.S. National Institutes of Health maintains information about more than 75,000 DNA-based tests in the Genetic Testing Registry (www.ncbi.nlm.nih.gov/gtr/). Health care consumers, physicians, and researchers use the registry to find laboratories offering specific tests. Such tests identify genotypes that cause, contribute to, or raise the risk of developing a specific disease.

A **genetic counselor** is a health care professional with a master's degree who can help patients and their families understand the inheritance pattern of a specific medical condition, evaluate risk, and navigate the path of genetic testing (**figure 22.1**). Genetic counseling addresses medical, psychological, sociological, and ethical issues, and a genetic counselor has medical, scientific, and communication skills. A counselor can recommend and interpret a DNA test; explain uncertainties; and suggest ways to cope with anxiety, fear, or guilt associated with taking genetic tests—or not taking them.

Reasons to seek genetic counseling:

- Family history of abnormal chromosomes
- Elevated risk of single-gene disease
- Family history of complex disease
- Family history of cancer

Genetic counseling sessions:

- Family history
- Pedigree construction
- Information provided on specific diseases, modes of inheritance, tests to identify at-risk family members
- Testing arranged, discussion of results and treatments
- Links to support groups, appropriate services, clinical trials
- Follow-up contact

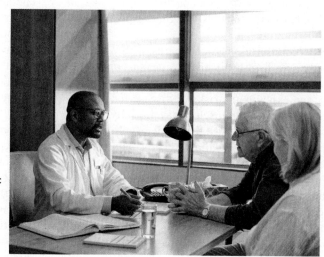

Figure 22.1 The genetic counseling process. *Morsa Images/Getty Images*

Genetic counseling began in pediatrics and prenatal care. In 1947, geneticist Sheldon Reed coined the term "genetic counseling" for the advice he gave to physician colleagues on how to explain heredity to patients with single-gene diseases. In 1971, the first class of specially trained genetic counselors graduated from Sarah Lawrence College, in Bronxville, New York. Today many countries have genetic counseling training programs.

Genetic counseling has branched into such specialties as cancer, cardiovascular disease, neurology, hematology, and ophthalmology. A genetic counselor can explain the difference between susceptibility from a gene variant that contributes a small degree to risk and a single-gene diagnostic test with a higher risk based on Mendelian inheritance. Education and public policy may be part of the job. Genetic counselors in New York, for example, hold "DNA days" to educate state legislators. Genetic counselors are integral parts of research teams conducting exome and genome sequencing projects, searching the scientific literature and disease databases to assign functions to genes.

Genetic counselors work with patients as parts of health care teams, typically at medical centers. A consultation may entail a single visit to explore a test result, such as finding that a pregnant woman is a carrier for spinal muscular atrophy, or a several-month-long relationship as the counselor guides a decision to take (or not take) a test for an adult-onset condition, such as Huntington disease or susceptibility to *BRCA* breast cancer.

The information that a genetic counselor imparts is similar to the topics covered in this book, but personalized and applied to a specific disease or other phenotype or concern. A counselor might explain Mendel's laws but substitute a family's condition for pea color. Or, a counselor might explain how an inherited susceptibility can combine with a controllable environmental factor, such as smoking or poor diet, to affect health.

A genetic counseling session begins with a discussion of the family's health history. Using an online tool or pencil and paper, the counselor constructs a pedigree, then deduces and explains the risks of disease for particular family members. The counselor may initially present possibilities and defer discussion of specific risks and options until test results are available. The counselor also explains which second-degree relatives—aunts, uncles, nieces, nephews, and cousins—might benefit from being informed about a test result. The genetic counselor provides detailed information on the condition and refers the family to appropriate medical specialists, including therapists, and to support groups. If a couple wants to have a biological child who does not have the illness, a discussion of assisted reproductive technologies (see chapter 23) might be advised.

A large part of the genetic counselor's job is to determine when specific biochemical, gene, chromosome, or exome tests are warranted and to arrange for people to take the tests. The counselor then interprets test results and helps the patient or family choose among medical options, in consultation with other medical specialists. Until genetic tests for more conditions became available fairly recently, people most often sought genetic counseling for either prenatal diagnosis or a disease in the family.

Prenatal genetic counseling typically presents population (empiric) and family-based risks, explains tests, and discusses whether the benefits of testing outweigh the risks. The couple, or woman, decides whether noninvasive prenatal testing of cell-free fetal DNA, amniocentesis, chorionic villus sampling, maternal serum screening, ultrasound scans, or no testing is the best option (see section 13.2). Part of a prenatal genetic counseling session is to explain that tests that rule out some conditions do not guarantee a healthy baby. If a test reveals that a fetus has a serious medical condition, the genetic counselor discusses potential outcomes, treatment plans, and the possible option of ending the pregnancy.

Genetic counseling when an inherited disease is in a family is more tailored to the individual situation than general prenatal counseling. For recessive diseases, the affected individual is usually a child. The illness is often a surprise, and recognition of a problem may take months and a diagnosis, years. However, family exome and genome analysis (see Clinical Connection 1.1 and figure 4.18), combined with rigorous mutation database searching, can now diagnose in minutes what once took years, in some cases.

Communicating the risk concerning other children in the family, future or living, may be difficult. People may think that if one child in a family has an autosomal recessive condition, then the next three will be healthy. Actually, each child has a one in four chance of inheriting the illness. Counseling for subsequent pregnancies requires great sensitivity. Some people opt not to terminate a pregnancy when it is discovered that the fetus has a condition that already affects another of their children, because they perceive that choice as a rejection of the child. Yet some people will see ending the pregnancy as the best option. The severity of an illness and available treatments may be important in how people feel about options. Genetic counselors must respect these feelings and tailor the discussion accordingly, while still presenting all the facts and choices.

Genetic counseling for adult-onset diseases does not have the problem of potential parents making important decisions for existing or future children, but presents the conflicting feelings of people choosing whether or not to find out if a disease is likely in their future. Often, they have seen loved ones suffer with the illness. This is the case for Huntington disease (see the opener to chapter 4). Predictive tests are introducing a new type of patient, the "genetically unwell" or those in a "premanifest" state—people with mutant genes but no symptoms (yet). Such a disease-associated genotype indicates elevated risk, but is not a medical diagnosis, which is based on symptoms and results of other types of tests.

When genetic counseling began in the 1970s, it was "nondirective," meaning that the counselor presented options but did not offer an opinion or suggest a course of action. That approach is changing as the field moves from analyzing hard-to-treat, rare single-gene diseases to considering inherited susceptibilities to more common illnesses that are more treatable, and for which lifestyle changes might realistically alter the outcome. A more recent description of a genetic counseling session is "shared deliberation and decision making between the counselor and the client."

Genetic counselors regularly communicate with physicians and other health care professionals. They are important parts of teams at molecular diagnostic testing laboratories, where they guide physicians in ordering and interpreting tests. Before a doctor orders a test, the counselor helps to interpret risks from the patient's pedigree, discusses the pros and cons of the appropriate test, and raises ethical issues that might arise when other family members are considered. While the test is under way, the genetic counselor ensures that time constraints are respected, such as an advancing pregnancy, and updates the physician. Once test results are in, the counselor may request a repeat test if the findings are inconsistent with the patient's symptoms, interpret the results, suggest additional tests, and alert the physician if the patient may be a candidate for participating in a clinical trial or able to use a new treatment.

The United States has about 5,000 genetic counselors. Most practice in urban areas, but telehealth services are bringing genetic counseling to all areas. Physicians, nurses, social workers, and PhD geneticists may also provide counseling. Dietitians, physical therapists, psychologists, and speech-language pathologists may discuss genetics with their patients, although they may not be specifically trained to do so. Genetic testing companies may connect patients with genetic counselors for phone, online, or email consultations. Genetic counselors also work for testing companies as curators, interpreting risks of specific gene variants (see Bioethics in chapter 12). Some states require that genetic counselors be licensed, requiring specific training.

As genetic testing becomes more commonplace, the need for genetic counselors and other genetics-savvy professionals to help individuals and families best use the new information will increase. Medical schools offer courses in which students interpret their own genetic tests and exome or genome sequences. Physicians attend 2-day workshops to have their genomes sequenced and interpreted, and are given iPads loaded with their results. The goal of this continuing medical education is to learn how to incorporate genetic and genomic testing into clinical practice.

Key Concepts Questions 22.1

*a. What services and types of information does a genetic counselor provide?

b. To what issues must a genetic counselor be sensitive?

c. Where do genetic counselors work?

d. How is the medical field preparing for increased use of genetic and genomic testing?

22.2 Genetic Testing

Genetic tests are administered at all stages of human existence, and for a variety of reasons (**table 22.1**). Identifying mutations can help in diagnosis and choosing treatments, such as when a person only has some symptoms that are part of a recognized syndrome. Results of a genetic test are unlike the results of a cholesterol check or an X-ray because they can have effects beyond the individual, to family members who share genotypes that affect health.

As the pace of exome and genome sequencing accelerates, and such testing becomes more widely available, it may become more economical to obtain all the information and parse it for results relevant to an individual, than to do single-gene tests. Cost is based on several factors, including the effort of sequencing and interpreting information. Obtaining a "raw sequence" of a genome may cost under $1,000, yet an analysis of every known variant of the *BRCA1* and *BRCA2* genes costs more than $3,400, for example.

The following subsections consider clinical genetic tests according to a developmental time frame, and then take a closer look at three general types of genetic tests.

Genetic Testing Through the Human Life Cycle

Some prenatal genetic tests have been in use for decades; others are new or in development.

Preconception and Prenatal Testing

Parents-to-be wonder what their child will look like and be like. An algorithm for "gamete donor selection based on genetic calculations" predicts the results of meiosis in the gametes of an individual and analyzes how two people's possible gametes might combine to produce a child with a particular combination of genotypes, interpreted as possible traits. For example, a couple might ask to have their DNA analyzed to predict the likelihood of their having a child with red hair, green eyes, and low risk of *BRCA* breast cancer, familial Alzheimer disease, and cystic fibrosis. The algorithm refers to a "hypothetical offspring" rather than a "designer baby" because testing an oocyte or sperm for certain inherited traits would destroy it.

Although we still can't order up a particular baby, it has been possible for many years to collect and separate sperm into fractions that are enriched for cells with X or Y chromosomes, to attempt to increase the probability of conceiving a girl or boy, respectively. Using a technique called flow cytometry, sperm are labeled with fluorescent markers. X-bearing sperm cells glow more intensely than Y-bearing sperm because the X chromosome is so much larger. Flow cytometry assigns each type of sperm a positive or negative charge, and uses the distinction to separate and collect them. The approach is not as specific as selecting a single sperm, but can increase the probability of having, for example, a daughter in a family that has an X-linked condition, rather than a son who would face a 50:50 chance of inheriting the illness.

Table 22.1 Genetic Testing

Test	Description	Chapter
PRENATAL		
Sperm selection	Enriches for X- or Y-bearing sperm, then intrauterine insemination or *in vitro* fertilization	6, 23
Preimplantation genetic diagnosis and sequencing	Sequences genes, exomes, and genomes and examines chromosomes of 1 cell of 8-celled embryo conceived *in vitro*; remaining embryo develops for a few days and is then implanted in the uterus	23
Chorionic villus sampling	Tests DNA and chromosomes of chorionic villus cell	13
Rescue karyotyping	Detects small deletions and duplications in archived cells from past unexplained pregnancy losses	13
Noninvasive prenatal diagnosis	Tests cell-free fetal (placental) DNA	13
Maternal serum markers	Measure levels of biomarkers in pregnant woman's blood	13
Amniocentesis	Tests DNA and chromosomes of amniocytes	13
NEWBORNS		
Screening	Screen metabolites and DNA in heelstick blood sample for 50-plus actionable conditions	22
Genome sequencing	Sequence genomes of many newborns to assess clinical value	22
CHILDREN		
Chromosomal microarray analysis	Detects small deletions, duplications, and other copy number variants associated with certain phenotypes	8
Exome sequencing	Diagnoses unrecognized syndromes or atypical cases; family comparisons distinguish *de novo* from inherited mutations in children	1, 4, 8
ADULTS		
Dor Yeshorim program	Preconception carrier tests for genetic diseases more prevalent among people of Jewish ancestry	16
Sickle cell disease	Tests athletes to identify carriers at risk for symptoms	11, 12
Comprehensive carrier testing	Preconception carrier tests for many single-gene diseases	22
Population carrier screen	Tests for heterozygotes for diseases more prevalent in certain population groups	16
Ancestry testing	Y chromosome and mitochondrial DNA sequences identify paternal and maternal lineages; autosomal markers identify relatives	18
Forensics testing	Copy numbers of short tandem repeats (STRs) in crime scene or disaster evidence	14
Military	Identify remains; risk for depression, PTSD; rapid infection diagnosis	22
Susceptibility	*BRCA* cancers, Alzheimer disease	5, 20
Pharmacogenetics	Drug efficacy, adverse effects, and dose	22
Genome-wide association studies	Identify genes contributing small degrees to a phenotype	7
Paternity	Half of a child's genome comes from the father	15
Centenarians	Identify gene variants that extremely old people share	3
POSTHUMOUS		
Disease diagnosis	Identify mutations in cells of deceased individuals	2, 14
History	Identify remains	5, 15
Human origins	Compare genomes of modern and archaic humans	17

During the first few days following conception, when the embryo consists of only a few cells, sampling one of them can reveal mutations as well as presence of an X or Y chromosome. Then the rest of the embryo can be placed in the uterus to continue development. This technique, called **preimplantation genetic diagnosis,** is discussed further in chapter 23. Genetic testing techniques used later in pregnancy are discussed in chapter 13. They are chorionic villus sampling, amniocentesis, maternal serum markers, and noninvasive prenatal testing (which tests cell-free fetal DNA).

Another type of chromosome test, called rescue karyotyping, is done on tissue samples that were stored when a woman's uterus was scraped following a miscarriage. Finding certain chromosome anomalies, such as translocations or inversions, can indicate increased risk for future pregnancies. More often, rescue karyotyping alleviates concern by revealing an aneuploid (extra or missing chromosome) embryo or fetus, which is not likely to repeat.

Genetic Testing of Newborns

Newborn screening identifies infants who are at high risk of developing certain inherited diseases that are considered "actionable"—that is, parents and health care professionals can provide treatments and services to improve the quality of life for the child. Testing a few drops of a newborn's blood for metabolites (biochemicals that indicate an inborn error of metabolism) has been routine for decades (**figure 22.2**). For most conditions, an analytical chemistry technique called mass spectrometry detects abnormal metabolites, but DNA testing has been added. The number of conditions screened for in newborns has steadily grown, from the first one tested for, phenylketonuria (PKU), to more than 50 conditions today.

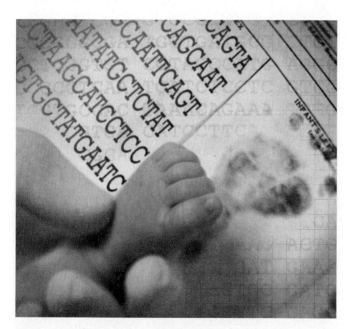

Figure 22.2 Newborn screening. In addition to taking a newborn baby's footprint, a few drops of blood are sampled with a heel prick for metabolic and genetic testing.

Source: National Human Genome Research Institute

A genetic screening test that indicates a newborn is at increased risk for a specific condition is followed up with diagnostic tests that look for evidence of the disease. Early accurate diagnosis of a genetic disease can lead to treatment and it indicates risks to future pregnancies. The downside of newborn screening is that follow-up tests may *not* show an abnormality, and the family may then experience anxiety that has been called "patients in waiting." The genotype associated with an illness is present, but the phenotype isn't—yet. The condition may actually never manifest, perhaps because of effects of other genes. Studies have found that this incomplete knowledge may interfere with parent-child bonding and cause anxiety for years.

The challenges of newborn screening for genetic disease will be amplified with the much more complete information that genome sequencing provides. Section 14.3 considers the benefits and implications of sequencing the genomes of newborns.

Genetic Testing of Children

Tests on children range from analyzing single genes corresponding to a known set of symptoms, to chromosome tests, to exome and genome sequencing.

A **chromosomal microarray analysis** detects tiny deletions and duplications that are associated with autism spectrum disorder, developmental delay, intellectual disability, behavioral problems, and other phenotypes. If a parent has the deletion or duplication and is healthy, then it is considered harmless. Finding a deletion or duplication that a parent does not also have might be comforting because it may explain the child's problems, although it wouldn't necessarily change treatment.

Increasingly, exome or genome sequencing has been used to speed the diagnosis of children whose symptoms do not match known syndromes, such as the child described in Clinical Connection 1.1. Such diagnoses can help to tailor a child's treatment and lead to development of genetic tests that can help other children.

Exome and genome studies are revealing that some children could not be diagnosed because they actually have *two* genetic diseases. For example, a 9-year-old boy was very weak and had episodes when he would stop breathing. He also had droopy eyelids, difficulty feeding, increased respiratory secretions, and at 8 months developed an enlarged heart. His sister had died at 20 months when she stopped breathing when she had a fever. The boy's exome sequence showed that he had congenital myasthenic syndrome, in which a fever causes too-rapid breakdown of a neurotransmitter. A drug prevented the boy from dying during a fever, as his sister had. His exome sequence also revealed mutations in a second gene, *ABCC9*, which enlarged his heart.

Genetic testing of children for diseases that will not cause symptoms for many years is controversial. For Huntington disease, it is generally agreed not to do a presymptomatic test in anyone under age 18, because it can be so upsetting. (Juvenile HD, discussed in the chapter 4 opener, is an exception.) However, parents send in their children's DNA to consumer genetics companies that test for health-related gene variants.

Genetic Testing of Adults

Like children, adults take single-gene tests as part of diagnostic workups based on symptoms or other test results. Adults also take genetic tests to detect increased risk of developing certain cancers. Chapter 18 discusses use of adult DNA in ancestry and forensics applications. In the military, genetic testing is used to identify remains; to detect a disease that might put a soldier at risk, such as sickle cell disease or a susceptibility to develop depression; and to diagnose a communicable disease on the battlefield.

Many adults begin to think about genetic testing when they are considering having children, and wonder what traits and illnesses they might pass on. Tests can identify carriers of more than 1,000 recessive disorders. If a woman is a carrier of a specific disease, then her partner can be tested and, if he is a carrier too, genetic counseling provided. Diseases for which tests are offered affect children, have a carrier frequency of at least 1 in 100, cause physical or cognitive impairment requiring treatment, and can be diagnosed prenatally. **Bioethics** considers a problem that can arise with exome and genome sequencing: detecting and reporting unexpected results, called secondary (or unsolicited) findings, about conditions that were not the reason for the sequencing.

Genetic testing isn't only for the young. We have a lot to learn from the genomes of those who have survived past age 100 with good health. Genetic testing after death is informative, too. The opener to chapter 14 discusses a project to sequence the genomes of the deceased to refine the diagnoses of what had killed them, exploring whether the genetic knowledge might have altered their treatment. The opener to chapter 2 describes a girl with Rett syndrome diagnosed years after her death from the DNA in a saved baby tooth.

A Closer Look at Three Types of Genetic Tests

The genetic tests just discussed are either part of research protocols or are regulated by government agencies. Health care professionals recommend genetic tests to identify people at

---

Bioethics

Secondary Findings: Does DNA Sequencing Provide Too Much Information?

In medical practice, an "incidentaloma" occurs when a diagnostic work-up for one condition discovers another—such as an X-ray to rule out pneumonia revealing lung cancer. A genetic "incidental" (or "secondary") finding arises when sequencing a person's exome or genome to discover a mutation that accounts for one set of symptoms identifies a mutation that indicates a second condition that may or may not already have produced symptoms. The term "secondary finding" is used because "incidental" implies that the information is not useful.

Secondary genetic findings were first identified in children having their exomes sequenced to explain unfamiliar combinations of developmental delay, intellectual disability or other neurological symptoms, and/or birth defects. An early case illustrated how lifesaving unexpected information can be.

A 2-year-old had severe feeding problems, seizures, failure to thrive, developmental delay, and intellectual disability. Doctors had ruled out infection, Angelman syndrome, Rett syndrome, and mitochondrial disease. Family exome sequencing (see figure 4.18) revealed an autosomal dominant mutation that originated in the boy, in a gene, *SYNGAP1*, which affects synapse formation. That explained the developmental delay, intellectual disability, and seizures. But the exome sequence also revealed a mutation in the connective tissue protein fibrillin, which causes Marfan syndrome, for which the boy had no symptoms. An ultrasound of his heart revealed an enlarged aortic root. A burst aorta can be the first, and deadly, symptom of the syndrome. Drug treatment saved the boy's life.

As more people have their exomes and genomes sequenced, secondary findings are inevitable, because we all have mutations. How should a clinician determine which results to report to a patient? Should all of the information in an exome or genome be deciphered?

The American College of Medical Genetics and Genomics recommends reporting 58 conditions that are prevalent, caused by mutation of a single gene, and that are "actionable" (treatment possible). The list includes cancers, connective tissue diseases, and malignant hyperthermia, in which exposure to a certain anesthetic can be lethal. Discussion of what a patient wants to know comes *before* the sequencing. If a patient does not want to know about a particular condition, it isn't included in the exome or genome analysis.

Questions for Discussion

1. What factors should a physician take into account when discussing with a patient which genetic findings to report?
2. Should everyone have their exome or genome sequenced and the information entered into a database, even if they don't want to know the results? The reason for such a database would be to speed gene discovery and development of new drugs and diagnostics.
3. Many people have taken genetic tests for mutations in *APOE4*. Certain variants greatly increase the risk of developing Alzheimer disease. A more recently discovered gene has variants that protect against the disease. How can clinicians prevent harm from the incompleteness of genetic information?
4. How can an overextended health care system that can barely handle people who are ill right now provide for people who have taken genetic tests that indicate they may become sick in the future?

increased risk of developing an illness or to help diagnose a disease. For-profit companies offer other types of DNA-based tests, such as ancestry tests.

In the United States, the Clinical Laboratory Improvement Amendments, or CLIA, control genetic testing of body materials for the prevention, diagnosis, or monitoring response to treatment of a disease or health impairment. The CLIA regulations, instituted in 1988, added "specialty areas" in 1992 to cover very complex tests, such as those involving immunology or toxicology. These tests did not include genetic testing, which at the time was limited. Since then, the genetics community has asked that genetic tests be included as a specialty area, to no avail. State regulations can override CLIA, but only if they are equally or more stringent. Therefore, the regulation of clinical genetic tests remains unclear. Availability of exome and genome sequencing will complicate the regulation of testing because of the increased amount and depth of information that sequencing provides.

Tests that offer genetic information, but are not intended to be used to diagnose a disease, do not come under the CLIA regulations. The distinction between information and diagnosis is often based on careful wording. Some company websites use vague phrases such as "genetic risk factors" and a genetic variant "linked to" or "associated with" a particular trait, talent, or health condition. Following are three examples of DNA tests that provide information, but not a diagnosis, and are marketed directly to consumers: tests for inborn athletic ability, nutrigenetics testing, and pharmacogenetic and pharmacogenomic tests.

Inborn Athletic Ability

Several dozen companies worldwide offer consumer genetic tests for athletic ability. Customers mail in saliva or cheekbrush scrapings that yield DNA and receive lists of variants of specific genes a few weeks later. The genes have general physiological functions that are related to fitness and athletic ability, perhaps studied at the population level. **Table 22.2** lists some of the genes that have variants that could impact such characteristics as muscle response to exercise, metabolism, tendency to gain weight on specific diets, endurance, strength, and susceptibility to specific injuries.

Two genes have gotten the most attention in predicting athletic ability. The angiotensin-1-converting enzyme gene (*ACE*) has an "insertion/deletion polymorphism," in which 250 base pairs are either present or not. The "I" (or insertion) allele is more common among endurance athletes, such as people who do triathlons, as well as among elite mountaineers and in populations who live at very high altitudes. The "D" (or deletion) allele, in contrast, is more prevalent among elite swimmers than other groups and is associated with increased power and strength with training.

The second commonly discussed "athletic" gene is *ACTN3* (alpha-actinin 3), which encodes a protein that binds actin, a component of the cytoskeleton in every cell and also one of the two major proteins of muscle. One genotype is more common among elite sprinters and another among endurance athletes. The opener to chapter 7 discusses this gene.

Table 22.2 Genes Associated with Athletic Characteristics

Gene Abbreviation	Encoded Protein	Protein Function
ADRBR	Beta-2 adrenergic receptor	Regulates blood glucose level, vasodilation, fat utilization
APOA2	Apolipoprotein A2	Weight gain with high saturated fat diet
COL1A1	Type 1 collagen alpha-1	Increased need for hydration, decreased response to endurance activity
COL5A1	Type 5 collagen alpha-1	Susceptibility to tendon injury
FABP2	Fatty acid binding protein 2	Increases fatty acid uptake in small intestine, increasing blood levels of triglycerides, LDL, and total cholesterol and lowering HDL
GDF5	Growth differentiation factor 5	Increases risk of tendon disease and osteoarthritis
GST	Glutathione S transferase	Detoxification, antioxidant activity, decreased free radicals
IL6	Interleukin-6	Immune response and inflammation in injury repair
PPAR	Peroxisome proliferator-activated receptor gamma	Fat storage and glucose metabolism
TRHR	Thyrotropin-releasing hormone receptor	Controls metabolic rate
VDR	Vitamin D receptor	Degree of muscle growth and bone density increase with strength training
VEGF	Vascular endothelial growth factor	Increased blood vessel extensions with exercise

Some parents are testing their young children for gene variants associated with athletic ability and using the results to decide whether the child should pursue a sport that entails sprinting or endurance, for example. Geneticists warn that DNA-based tests for athletic ability are too simplistic. However, company websites have caveats and stress that gene action combines with environmental factors to influence athletic ability.

Nutrigenetics Testing

"Nutrigenetics" websites offer consumer genetic tests along with general questionnaires about diet, exercise, and lifestyle habits. The companies return supposedly personalized profiles with dietary suggestions, often with an offer of a costly package of exact supplements that an individual purportedly needs to prevent realization of his or her genetic fate.

After the media spread the word of these services a few years ago, the U.S. Government Accountability Office tested the tests. An investigator took two DNA samples—one from a 9-month-old girl and the other from a 48-year-old man—and created 14 lifestyle/dietary profiles for these "fictitious consumers"—12 for the female, 2 for the male. The samples were sent to four nutrigenetics companies, none of which asked for a health history. Here is an example of the information sent to the companies:

- The DNA from the man was submitted as being from a 32-year-old male, 150 pounds, 5'9", who smokes, rarely exercises, drinks coffee, and takes vitamin supplements.
- The DNA from the baby girl was submitted as being from a 33-year-old woman, 185 pounds, 5'5", who smokes, drinks a lot of coffee, doesn't exercise, and eats a lot of dairy, grains, and fats.
- The same baby girl DNA was also submitted as that of a 59-year-old man, 140 pounds, 5'7", who exercises, never smoked, takes vitamins, hates coffee, and eats a lot of protein and fried foods.

The elevated risks found for the three individuals were exactly the same: osteoporosis, hypertension, type 2 diabetes, and heart disease. One company offered the appropriate multivitamin supplements for $1,200, which the investigation found to be worth about $35. Recommendations stated the obvious, such as advising a smoker to quit. The advice tracked with the fictional lifestyle/diet information, and *not* genetics. Concluded the study: "Although these recommendations may be beneficial to consumers in that they constitute common sense health and dietary guidance, DNA analysis is not needed to generate this advice." Some of the suggestions could even be dangerous, such as vitamin excesses in people with certain medical conditions. These types of companies have flourished since the 2008 government study.

Matching Patient to Drug

People may react differently to the same dose of the same drug, even accounting for weight, gender, and age, because we differ in the rates at which our bodies respond to and metabolize drugs. Genetic tests can identify these differences. A **pharmacogenetic** test detects a variant of a single gene that affects drug metabolism, and a **pharmacogenomic** test detects variants of multiple genes or gene expression patterns that affect drug metabolism. Pharmaceutical and biotechnology companies use these tests in developing drugs, and health care providers are increasingly using these tests in prescribing drugs.

Genetic testing to guide drug selection offers several advantages:

- Identifying patients likely to suffer an adverse reaction to a drug
- Selecting the drug most likely to be effective
- Monitoring response to drug treatment
- Predicting the course of the illness (prognosis)

Many pharmacogenetic tests detect specific variants of genes that encode enzymes, called cytochromes, that metabolize toxins, including toxic byproducts of normal metabolism as well as drugs. The P450 cytochrome enzymes are especially diverse, encoded by 57 genes with many variants. P450 genotyping can be used to predict how quickly a person metabolizes a particular drug. A "rapid metabolizer" may dismantle a drug so fast that a higher dose is required, whereas a "poor metabolizer" may break down a drug so slowly that effects linger and may become toxic. **Table 22.3** lists examples of the effects of certain gene variants on drug metabolism.

One of the first drugs to be described using pharmacogenetics was the blood thinner warfarin (also known as Coumadin). This drug has a small range of concentration in which it keeps blood at a healthy consistency, but people can vary up to 10-fold in the dose required. Too little drug allows dangerous clotting; too much causes dangerous bleeding. In the past, physicians would give an initial standard dose, based on a patient's age, gender, health status, weight, and ethnicity, then monitor the patient for a few weeks to check for too much clotting or bleeding, tweaking the dose until it was about right. This general approach led to hospitalization for abnormal bleeding in 43,000 of the 2 million people prescribed the drug each year.

A "pharmacogenetic algorithm" is now used to prescribe warfarin. It considers two genes: two variants of *CYP2C9* and one variant of *VKORC1*, which are associated with increased sensitivity to the drug. People with these gene variants require lower doses of warfarin. The genetic test for warfarin response is especially helpful for the 50 percent of patients who fall at the extremes of the range of drug concentration that is effective. However, considering clinical information remains important in determining the dose of warfarin in particular, and other drugs in general, because variants of genes other than the ones that are tested for can affect how an individual human body metabolizes a drug.

The ability to choose the most effective drug for an individual can prevent suffering by not wasting time taking drugs that do not work or that have serious adverse effects. People with cancer or depression can greatly benefit from a

Table 22.3 Pharmacogenetics

Gene Abbreviation	Encoded Protein	Genotype	Phenotype
CYP2D6	Cytochrome P450	2D6	Poor metabolizer; increased sensitivity to certain beta blocker heart drugs
HTR2A	Serotonin receptor	G allele	Strong response to SSRI drugs to treat depression and anxiety
OPRM1	Opioid receptor	A118G (asp to asn)	Requires higher doses of opiates for pain relief
SLCO1B1	Solute carrier organic anion transporter	T/C (val/ala)	4.5-fold increased risk of muscle pain from statin drugs
		C/C (ala/ala)	17-fold increased risk of muscle pain from statin drugs
VKORC1	Vitamin K epoxide reductase complex subunit 1	1639G>A	Increased sensitivity to warfarin (a blood thinner)

pharmacogenetic approach. For many years cancer drugs targeted all cells with fast cell cycles, causing hair loss, digestive distress, and bone marrow suppression. Newer targeted drugs are based on genotypes (see section 20.4) and for some patients may be the best initial treatment.

In the past, health care providers chose anti-depressant drugs based mostly on patient self-assessments, rather than on an analysis of how an individual metabolizes a particular drug. A more recent study analyzed more than 2,000 people taking a certain anti-depressant for variants of the gene that encodes the P450 cytochrome CYP2C19. A third of the participants, who had not been satisfied with their response to the drug, had gene variants that caused them to metabolize the drug more rapidly or more slowly than people who responded to it. Many of these people had already switched drugs, but changing the dose of the original drug enabled them to go back on it.

Key Concepts Questions 22.2

*a. On what biological process is genotyping gametes to predict characteristics of possible future children based?

*b. Which genetic tests are used on embryos and fetuses?

c. In what circumstances are genetic tests done on children?

d. Why might adults take genetic tests?

e. What are limitations of consumer genetic tests for athletic ability and nutrition?

f. Explain how pharmacogenetics and pharmacogenomics help health care providers select the best drugs for patients.

22.3 Treating Genetic Disease

Tests for genetic diseases greatly outnumber treatments, because treatments are challenging to develop. They must correct the abnormality in enough of the appropriate cells and tissues to prevent or minimize symptoms, while not harming other parts of the body. Treatments for single-gene diseases have evolved through several stages, in parallel to development of new technologies:

- Removing an affected body part
- Replacing an affected body part or biochemical with material from a donor
- Delivering pure, human proteins derived from recombinant DNA technology
- Refolding correctly a misfolded protein
- Blocking gene expression (gene silencing)
- Using **gene therapy** to add wild type alleles without removing mutant alleles
- Using **gene editing** and genome editing to replace, delete, or add alleles

Protein-Based Therapies

Preventing a disease phenotype may be as straightforward as adding digestive enzymes to applesauce for a child with cystic fibrosis or giving a clotting factor to a boy with hemophilia. Inborn errors of metabolism are particularly treatable when the biochemical pathways are well understood.

Lysosomal storage diseases are a well-studied subgroup of inborn errors of metabolism. The little girl described in the chapter opener has such a disease, Sanfilippo syndrome type A. In the disease, a deficient or abnormal enzyme leads to buildup of the substrate (the molecule that the enzyme acts on)

Table 22.4 Lysosomal Storage Disease Treatments

Treatment	Mechanism
Enzyme replacement and substitution therapy	Recombinant human enzyme infused to compensate for deficient or absent enzyme
Substrate reduction therapy	Oral drug that reduces level of substrate so enzyme can function more effectively
Pharmacological chaperone therapy	Oral drug that binds to patient's misfolded protein, restoring function

and deficit of the breakdown product of the substrate. Recall from figure 2.5 that a lysosome is an organelle that dismantles debris. It houses more than 50 types of enzymes, and each breaks down a specific molecule.

Treatment of another lysosomal storage disease, type 1 Gaucher disease, illustrates three general approaches to counteracting an inborn error of metabolism that affects an enzyme, summarized in **table 22.4** and **figure 22.3**. In type 1 Gaucher disease, the enzyme glucocerebrosidase is deficient or absent. As the substrate builds up because there is little or no enzyme to break it down, lysosomes swell, ultimately bursting cells. Symptoms include an enlarged liver and spleen, bone pain, and deficiencies of blood cells. Too few red blood cells cause the fatigue of anemia; too few platelets cause easy bruising and bleeding; and too few white blood cells increase the risk of infection. The disease is variable in age of onset, severity of symptoms, and rate of progression.

Early treatments for Gaucher disease corrected the affected body parts: removing the spleen, replacing joints, transfusing blood, or transplanting bone marrow. In 1991, **enzyme replacement therapy** became available, which supplies the missing or deficient enzyme, in this case recombinant glucocerebrosidase. It is effective but costs about $550,000 a year and the recombinant enzyme takes several hours to infuse, twice a month. Only a few diseases have enzyme replacement therapies.

A variation of enzyme replacement therapy is enzyme substitution therapy, which provides an enzyme other than the one that is abnormal in a specific inborn error of metabolism. For example, the enzyme deficiency in phenylketonuria (PKU) causes the amino acid phenylalanine to build up in the blood, to toxic levels. For many years newborn screening has identified babies with PKU, and the symptoms, including intellectual disability, largely prevented by following a diet so limited in the amino acid that it is called a "medical food." The protocol is difficult to follow and cognitive deficiencies may persist. Supplying the enzyme that is deficient is challenging because it is highly unstable and requires a second molecule to function. Researchers used recombinant DNA technology to mass-produce a different enzyme that lowers phenylalanine levels—in plants! Used as a drug, this nonhuman enzyme works so well that adult patients report lifting of the "brain fog" to which they'd become accustomed.

In 2003 came a different approach: **substrate reduction therapy**. A drug taken by mouth decreases the amount of the substrate, the molecule on which the deficient enzyme acts. In a third approach called **pharmacological chaperone therapy**, an oral drug binds a patient's misfolded enzyme, stabilizing it sufficiently to allow some function.

Less costly than developing new drugs is to "repurpose" drugs already available to treat other conditions. For example, the chapter 3 opener describes a failed cancer drug that helps children who have a form of progeria, the accelerated aging condition.

Gene Therapy

Gene therapy delivers working copies of genes to specific cell types or body parts. A common delivery method uses modified viruses called vectors. More than 2,000 clinical trials of gene therapies have been conducted worldwide since 1990. The first approval of a gene therapy, in Europe, was in 2012 for a rare enzyme deficiency that causes pancreatitis but it was discontinued due to the high cost and low demand. The first gene therapy in the United States approved for marketing, in 2017, treats a form of inherited blindness.

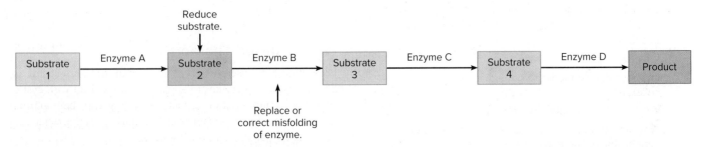

Figure 22.3 Counteracting a metabolic abnormality. If an enzyme is deficient or its activity blocked, the substrate builds up and the pathway's product is deficient. Enzyme replacement or enzyme substitution therapy delivers an absent or deficient enzyme or one that can substitute. Substrate reduction therapy decreases the amount of substrate and pharmacological chaperone therapy corrects misfolded proteins

Adding functional genes to treat an inherited disease may provide a longer-lasting effect than treating symptoms or supplying a protein. A gene therapy may require a single or a few treatments, whereas a traditional drug or even enzyme replacement, substrate reduction, or pharmacological chaperone therapy may have to be administered frequently for the patient's entire life.

Types and Targets of Gene Therapy

Gene therapy strategies vary in the way that healing genes are delivered and where they are sent. **Germline gene therapy** alters the DNA of a gamete or fertilized ovum, so that all cells of the individual have the change and the correction is heritable, passing to offspring. The transgenic plants and animals discussed in chapter 21 had germline genetic manipulations, but many individuals, organizations, and governments object to altering the human germline (gametes or fertilized ova), as discussed in section 21.4.

Somatic gene therapy corrects only the cells that an illness affects or the cells that divide and give rise to them. It is nonheritable: A recipient does *not* pass the genetic correction to offspring. Clearing lungs congested from cystic fibrosis with a nasal spray containing functional *CFTR* genes is an example of somatic gene therapy.

Gene therapy strategies vary in invasiveness (**figure 22.4**). In *ex vivo* **gene therapy** ("outside the body"), cells are altered outside the body and then placed in the body, such as through an intravenous infusion or into the spinal fluid. In the more invasive *in vivo* **gene therapy** ("in the living body"), the gene and its vector are introduced directly into the body, such as through a catheter inserted and snaked to the appropriate organ. There, the vector must enter the targeted cells, the human DNA be transcribed into mRNA, the mRNA translated into protein, and then the protein must function sufficiently to halt progression of the disease or improve symptoms.

Researchers obtain therapeutic genes using the polymerase chain reaction, recombinant DNA technologies, and other techniques that cut DNA. In the future, researchers—and, someday, clinicians—may deliver synthetic genes. The origin should not make a difference—DNA is DNA and the genetic code universal.

Delivery of Gene Therapy

The next step in gene therapy, gene transfer, usually delivers the healing DNA with other DNA that is mobile. To create viral vectors, researchers remove the genes that cause symptoms or alert the immune system to infection and add the corrective gene.

Different viral vectors are useful for different treatments. A certain virus may transfer its cargo with great efficiency to a specific cell type, but carry only a short DNA sequence. Another virus might carry a large piece of DNA but enter many cell types, causing "off-target" side effects. Even if a viral vector goes where intended, it must enter enough cells to alleviate symptoms or slow or halt progression of the disease. Finally, a viral vector must not integrate into a gene that harms the patient, such as an oncogene or tumor suppressor gene, which could cause cancer. Some introduced genes do not insert into chromosomes at all, but remain outside, still within the nucleus, in a circle of DNA called an **episome**. If the cell type divides, then the episomes are lost and the effect of the gene therapy wanes. However, gene therapy via episomes in nondividing cell types can have a lasting effect.

Some gene therapies use viruses that normally infect the targeted cells. For example, a herpes simplex virus delivers the gene encoding a pain-relieving peptide to nerve endings in skin. Researchers team parts of viruses to target a certain cell type. Adeno-associated virus (AAV), for example, infects many cell types, but adding a promoter from a parvovirus gene restricts it to red blood cell progenitors in bone marrow. Different subtypes of AAV home to different body parts. AAV9 can cross the blood-brain barrier and is given by intravenous infusion, replacing earlier versions that required delivery into the brain through catheters. The girl described in the chapter opener received gene therapy via AAV9 vectors. Fatty structures called liposomes are also used as vectors. **Figure 22.5** shows four targets of somatic gene therapies.

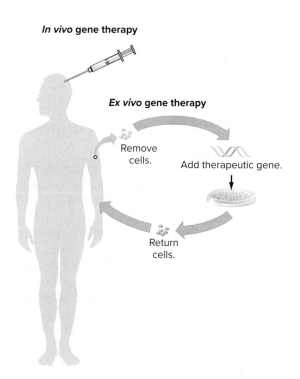

Figure 22.4 Gene therapy invasiveness. Therapeutic genes are delivered to cells removed from the body that are then returned (*ex vivo* gene therapy) or delivered directly to an interior body part (*in vivo* gene therapy).

The first gene therapy efforts were aimed at well-studied inherited diseases, even though they are rare. Knowing the biochemical pathway that is disrupted enabled researchers to know where to intervene. **Figure 22.6** illustrates such a biochemical pathway for adenosine deaminase (ADA) deficiency, a form of severe combined immune deficiency that was the first disease for which gene therapy was tested (see table 19.5 and **Clinical Connection 22.1**). Today, about 8 percent of gene therapy clinical trials are for single-gene diseases. About 65 percent target cancers, and about 8 percent focus on cardiovascular disease.

Some of the new treatments for genetic diseases have enabled children to live so much longer that new symptoms can emerge that had never been seen before because no one with the disease lived long enough to develop them. Investigators are alert to this possibility and are developing ways to treat new symptoms. **Table 22.5** describes four examples of gene therapy.

Figure 22.6 ADA deficiency. Absence of the enzyme adenosine deaminase (ADA) causes deoxy ATP to build up, which destroys T cells, which then cannot stimulate B cells to secrete antibodies. The result is severe combined immune deficiency (SCID). © Anna Powers

Endothelium. The tile-like endothelium that forms capillaries can be genetically altered to secrete proteins into the circulation.

Muscle. Immature muscle cells (myoblasts) given healthy dystrophin genes may treat muscular dystrophy.

Liver. To treat certain inborn errors of metabolism, only 5 percent of the liver's 10 trillion cells need to be genetically altered.

Lungs. Gene therapy can reach damaged lungs through an aerosol spray. Enough cells would have to be reached to treat hereditary emphysema (alpha-1-antitrypsin deficiency) or cystic fibrosis.

Figure 22.5 Some sites of gene therapy. MedicalRF.com/Getty Images

Key Concepts Questions 22.3

a. Explain how the three types of protein-based therapies work.
b. Distinguish between somatic and germline gene therapies.
*c. What is the role of viruses in gene therapy?

22.4 CRISPR-Cas9 in Diagnosis and Treatment

The genome editing tool CRISPR-Cas9 may one day replace gene therapy because it is more precise (see figure 21.9). CRISPR-Cas9 not only delivers a gene to a specific part of the genome, but can replace a gene, remove a gene, or add a gene, and do so in multiple places or insert, correct, or delete multiple genes. In contrast, genes introduced through older gene therapy methods may integrate into chromosomes at random or remain outside the chromosomes as parts of episomes. **Figure 22.7** compares gene therapy to gene editing. Although genome editing using CRISPR-Cas9 is being developed rapidly, it is largely a research tool, with applications in diagnosis and treatment just beginning.

An Animal Model Identifies a Family's Mutation

CRISPR-Cas9 genome editing can be combined with genome sequencing to engineer an animal model of a family's inherited disease, to refine a diagnosis and reveal how the symptoms arise.

Table 22.5	Gene Therapies			
Disease	**Inheritance**	**Phenotype**	**Delivered Gene***	**Effect**
Hemophilia B	xlr	Impaired clotting	Factor IX *(F9)*, IV	Less need for clotting factor
Myotubular myopathy	xlr	Skeletal muscle cells do not fully differentiate, death by teens	Myotubularin 1 *(MTM1)*, IV	Better breathing, developmental milestones reached
Recessive dystrophic epidermolysis bullosa	ar	Skin blisters and peels due to absence of collagen fibrils that connect epidermis to dermis	Collagen *(COL7A1)*, skin grafts	Better wound healing
Spinal muscular atrophy	ar	Extreme weakness, impaired breathing, death in childhood	Survival motor neuron 1 *(SMN1)*, IV	Motor functions restored

xlr = X-linked recessive
ar = autosomal recessive, IV = intravenous
* via viral vector

Figure 22.7 A comparison of gene therapy and gene editing. Gene therapy *adds* a correct DNA sequence, which may insert into a chromosome at random or remain outside it in an episome. Gene editing *corrects* a DNA sequence by adding, deleting, or replacing a base or bases at the site of the mutation. The left side illustrates gene therapy to treat retinal dystrophy due to two mutations in the *RPE65* gene on chromosome 1. The treated retinal pigment epithelium (RPE) cells do not divide, so even though the wild type gene does not insert into a chromosome, its effect persists. The right side depicts the precise correction possible using gene editing. The rods provide black and white vision; cones, not shown, provide color vision and are affected too. Not drawn to scale.

This is the case for a family with thoracic aortic aneurysm and dissection (TAAD), in which the wall of the aorta—the largest artery in the body, leading from the heart—stretches, thins, balloons out, and may burst, causing sudden death. If detected early, the ballooning, called an aneurysm, can be repaired. TAAD runs in families and is part of several genetic syndromes that affect connective tissue.

A 35-year-old man came to the attention of researchers at a genome sequencing center. The man had experienced sudden severe chest pain at age 19. When scans revealed that his aorta was on the verge of rupturing, he underwent surgery to replace the blood vessel. The man's mother had also had an aortic repair, and two cousins had aortic aneurysms. Everyone in the family had tests for all mutations known to cause aortic aneurysm, and all were negative. So the researchers sequenced the genomes of the two cousins and identified seven candidate genes. Segregation analysis, based on Mendel's first law, tracked which mutations occurred only in the affected family members and never in the healthy ones. Only one gene fit those criteria, and it encodes an enzyme called lysyl oxidase (LOX).

Could mutations in *LOX* cause the symptoms in this family? LOX is necessary for the connective tissue proteins collagen and elastin to link, which they do when they form the walls of arteries. To test the hypothesis that *LOX* mutations cause TAAD, the researchers used CRISPR-Cas9 to replace certain DNA nucleotides in the mouse version of the gene to recreate the mutant DNA sequence in the family. Mice whose cells had been given two copies of the human mutant *LOX* gene using CRISPR died just after birth as their aortas burst. Mice with one copy—similar to the family members—survived but with fragmented and fragile aortic walls. (Many genetic diseases that are autosomal dominant in humans are autosomal recessive in mice.)

Clinical Connection 22.1

The Rocky History of Gene Therapy

Gene therapy has had its ups and downs, as does any new medical technology.

Initial Success

In the late 1980s, Ashanthi ("Ashi") DeSilva's parents did not think their little girl would survive her near-continual coughs and colds. She could walk only a few steps before becoming winded. Ashi had severe combined immune deficiency due to ADA deficiency (see **figure 22A**). At age 2, she began enzyme replacement therapy, but within 2 years it stopped working. She would likely die of infection. Then her physician heard about a clinical trial of gene therapy that would give her white blood cells (T cells) functional ADA genes. On September 14, 1990, at the National Institutes of Health in Bethesda, Maryland, Ashi received an intravenous infusion of her own (*ex vivo*) corrected white blood cells. She did well. Today a longer-lasting form of the gene therapy alters stem cells instead of the more specialized T cells.

Setbacks

In September 1999, 18-year-old Jesse Gelsinger died, 4 days after receiving gene therapy, from an overwhelming immune response to the viruses used to introduce the genes. He had ornithine transcarbamylase (OTC) deficiency and could not make a liver enzyme required to break down dietary proteins. In the disease, the nitrogen released from digestion of proteins combines with hydrogen to form ammonia (NH_3), instead of being excreted in urine. The ammonia rapidly accumulates in the bloodstream and travels to the brain, usually causing irreversible coma within 72 hours of birth. Half of affected babies die within a month and another quarter by age 5. The survivors can control their symptoms by following a very-low-protein diet and taking drugs that bind ammonia.

Jesse was diagnosed at age 2. He had a mild case because he was a mosaic—some of his cells could produce the enzyme. At age 10, Jesse's diet lapsed and he became very ill. Years later, after he went into a coma in December 1998 from missing a few days of medications and recovered, he volunteered for a gene therapy trial his doctor had mentioned. On September 13, 1999, a trillion adenoviruses carried functional human *OTC* genes through a tube into Jesse's liver. But the viruses entered not only the hepatocytes as intended, but also the macrophages that alert the immune system. Jesse's organ systems began to fail. Four days later he was brain dead, and the devastated medical team stood by as his father turned off life support.

Also in 1999, researchers in France gave gene therapy for X-linked severe combined immune deficiency (SCID X1) to two babies (see figure 19.10). In SCID X1, T cells do not have certain cytokine receptors, which prevents the immune system from recognizing infection. Most children die in infancy. *Ex vivo* gene therapy removes

Figure 22A Gene therapy introduces a gene that can compensate for, but not replace, a mutant allele.
Source: Jane Ades/NHGRI

a boy's T cell progenitor cells from bone marrow, gives them wild type alleles, and infuses the corrected cells back. The researchers used a retrovirus, which only enters dividing cells. If the healing viruses could infect the T cell progenitor cells, the corrected cells would differentiate into mature T cells that could alert the immune system. However, by 2002, one boy in the trial developed leukemia, followed by several others. The retroviruses had inserted into part of a chromosome that harbors a proto-oncogene, turning it into an active oncogene. Use of a "self-inactivating" vector has made this gene therapy, which is effective against SCID X1, much safer.

More Success

A dramatic gene therapy success story is the treatment of a form of hereditary blindness called retinal dystrophy, due to mutation in both copies of the *RPE65* gene.

At the backs of the eyes, cells that comprise a thin layer called the retinal pigment epithelium (RPE) cannot make an enzyme necessary to convert vitamin A to a form that the rods and cones, the cells that transmit light energy to the brain, can use. Gene therapy supplies the gene that encodes the enzyme to cells of the RPE in a brief surgical procedure. Figure 5.5 shows the anatomy of the retina and where genes intervene. The gene therapy restores vision in just days and so far prevents or slows further visual loss. Four days after surgery, a boy had to hide his eyes from the glare of the sun. A girl could see the food on her plate for the first time.

Gene therapy also works for adrenoleukodystrophy (ALD). The X-linked recessive disease affects peroxisomes, the tiny sacs inside cells that house enzymes. They have porthole-like openings formed by a protein called ABCD1. A mutation in *ABCD1* prevents admission of an enzyme needed to process certain fats used to make myelin, the fatty material that insulates neurons, enabling them to send messages. The affected brain cells, called microglia, descend from progenitor cells in the bone marrow.

Gene therapy for ALD alters bone marrow cells outside the body and sends them into the bloodstream, where they go to the brain and give rise to corrected microglia. The viral vector is HIV,

(Continued)

stripped of the genes that cause AIDS. The disabled HIV is an excellent vector because it does not insert into proto-oncogenes, is efficient, can carry large genes, and enters many types of cells. Brain neurons gain enough myelin to enable treated boys to attend school—with only about 15 percent of their microglia corrected! Teaming the gene therapy for ALD with newborn screening may completely prevent symptoms of this otherwise devastating disease.

Questions for Discussion

1. The goal of a clinical trial is to demonstrate safety and efficacy. Patients are selected so that the trial will yield the most useful information to help as many people as possible—not to help a particular individual. However, different clinical trial protocols select participants in different ways. State the risks and benefits, from different perspectives, for each of the following real clinical trial protocols to evaluate gene therapies. What further information do you need to judge whether each procedure is ethical or not?

   a. A clinical trial for metachromatic leukodystrophy enrolls families with more than one affected child. The diagnoses of the older children led to the diagnoses of the younger ones, and the younger children are given the gene therapy. The reasoning is that the treatment can prevent symptoms, whereas it cannot improve or likely reverse symptoms in the older children.
   b. A clinical trial for the gene therapy now approved to treat *RPE65* retinal dystrophy enrolled several families with more than one affected child. The older children were treated first because their vision was worse and therefore they had more to gain. The continuing deterioration in vision was studied in the younger siblings, who received the gene therapy a year later.
   c. A clinical trial for a form of the neurological condition Batten disease treated two young daughters of a famous Hollywood producer shortly after they were diagnosed, partly because the case was publicized in the media, some people have said. Other children wait years to enter clinical trials or die waiting.

A genetic test is now possible for TAAD caused by mutations in *LOX*. Although this is a rare form of the condition, availability of an accurate and precise test can help to diagnose family members who do not yet have symptoms, possibly saving lives.

Somatic Genome Editing Tackles Sickle Cell Disease

The first applications of CRISPR-Cas9 to treat disease are targeting conditions in which the affected cells are easy to reach. Chimeric antigen receptor technology (see section 20.4) at first used ZFNs and TALENs to treat cancer, and is now being done using CRISPR-Cas9 (see table 21.6). Other treatments based on genome editing alter cells that descend from the hematopoietic (blood-forming) stem cells in bone marrow because they can be removed, altered, and placed back in the body (*ex vivo* delivery).

Sickle cell disease is particularly amenable to genome editing. Recall from figure 12.3 that it is caused by a missense mutation in the beta globin gene. Recall from figure 11.2 that the embryo and fetus use different genes to produce embryonic and fetal forms of hemoglobin.

Reactivating the quiescent globin genes can treat sickle cell disease. One approach uses induced pluripotent stem cells (iPS cells; see table 2.3) that are cultured in laboratory glassware from skin fibroblasts of people who have the disease. Chemicals are added to the iPS cells to direct their development into the cell type that precedes mature red blood cells, called reticulocytes. Then CRISPR-Cas9 replaces one sickle cell mutation with a wild type copy of the beta globin gene. The reticulocytes normally extrude their nuclei as they mature into red blood cells to make room for as much hemoglobin as possible. The red blood cells that descend from the gene-edited reticulocytes indeed fill with as much normal hemoglobin as in a mature red blood cell of a heterozygote—that is, enough to not have the disease.

A second approach to using CRISPR-Cas9 to treat sickle cell disease silences a gene that normally switches fetal hemoglobin production off after birth. Researchers introduced small deletions into the switch gene, called *BCL11A*, into hematopoietic stem cells. The red blood cells that descend from the manipulated cells make fetal hemoglobin, which does not sickle and effectively carries oxygen.

A new biotechnology typically takes about 30 years to go from initial concept to application in treating a disease—sometimes even longer. The first clinical trial for a gene therapy was in 1990; the first product was approved at the end of 2017. By comparison, gene editing is in its infancy, genome editing even farther from the clinic. Gene silencing and protein-based therapies are also, slowly, becoming available, one disease at a time. The most common mutation that causes cystic fibrosis, for example, was discovered in 1989. It was exactly 30 years later that a trio of drugs that correct the faulty ion channels became able to help 90 percent of patients (see Clinical Connection 4.1).

Understanding the nature of the genetic material in the 1950s and how genes encode proteins in the 1960s, then unraveling the controls of gene expression, all paved the way for the applications of human genetics that are beginning to improve our lives by providing new types of tests and treatments.

Key Concepts Questions 22.4

*a. Distinguish genome editing from gene therapy.
b. Explain how genome editing can help to identify specific mutations that cause a disease in a family.
c. Describe two strategies that use genome editing to treat sickle cell disease.

Study Guide

Summary

22.1 Genetic Counseling

1. **Genetic counselors** provide information on inheritance patterns for specific medical conditions, disease risks and symptoms, and tests and treatments.
2. Prenatal counseling and counseling a family coping with a particular disease pose different challenges.
3. Genetic counselors interpret genetic tests and assist other health care professionals in incorporating genetic information into their practices.

22.2 Genetic Testing

4. Genetic tests are performed before birth and on newborns to detect certain single-gene diseases, on children to assist in diagnosis, on adults for diagnosis and carrier detection, and on human remains. **Preimplantation genetic diagnosis** examines genes and chromosomes of a cell from a very early embryo, in glassware. A **chromosomal microarray analysis** detects small deletions and copy number variants and is mostly used for children.
5. The Clinical Laboratory Improvement Amendments regulate some genetic tests.
6. Consumer genetics tests may provide inaccurate, inappropriate, or incomplete information on traits such as athletic ability and dietary habits.
7. **Pharmacogenetic** and **pharmacogenomic** tests provide information on how individuals metabolize certain drugs.

22.3 Treating Genetic Disease

8. **Gene therapy** adds a gene, but **gene editing** can more precisely add, remove, or replace a gene.
9. **Enzyme replacement therapy, substrate reduction therapy,** and **pharmacological chaperone therapy** treat biochemical imbalances at the protein level.
10. Drugs may be repurposed to treat genetic diseases.
11. Gene therapy delivers genes to cells and directs production of a needed substance at appropriate times and in appropriate tissues, in therapeutic (not toxic) amounts.
12. **Germline gene therapy** alters gametes or fertilized ova, affects all cells, and is transmitted to future generations. It is not done in humans. **Somatic gene therapy** affects somatic cells and is not passed to offspring.
13. *Ex vivo* gene therapy is applied to cells outside the body that are then implanted or infused into the patient. *In vivo* gene therapy delivers gene-carrying vectors directly into the body.
14. Several types of vectors deliver therapeutic genes. Viruses are most commonly used. Some gene therapies target stem or progenitor cells, because they can divide and some can move. A gene delivered in gene therapy may become part of a chromosome or remain outside the chromosomes in a small circle of DNA called an **episome**.
15. Gene therapies are attempted on diseases with well understood pathological mechanisms.

22.4 CRISPR-Cas9 in Diagnosis and Treatment

16. Genome editing adds, deletes, or replaces genes at specific loci. Gene therapy introduces genes nonspecifically.
17. Genome editing aids diagnosis by creating animal models that have human mutations.
18. CRISPR-Cas9 can correct the beta globin mutation behind sickle cell disease in induced pluripotent stem cell–derived red blood cell precursors, or it can reactivate fetal hemoglobin genes.

Key Terms

chromosomal microarray analysis A technique that detects small deletions and copy number variants in chromosomes.

enzyme replacement therapy Infusion of a recombinant human enzyme to compensate for a deficient or absent enzyme in a single-gene disease.

episome A small circle of DNA in the nucleus, separate from the chromosomes.

ex vivo gene therapy Gene therapy in which a wild type gene is introduced into cells outside of the body and then the manipulated cells are infused into the body.

gene editing Adding, removing, or replacing a gene with a copy with a different DNA sequence, using a tool such as CRISPR.

gene therapy Adding a wild type copy of a gene to a cell in which the gene is mutant.

genetic counselor A clinician who calculates risks of inherited disorders in families, applying the laws of inheritance to pedigrees and interpreting genetic test results.

germline gene therapy Genetic alteration of a gamete or fertilized ovum, which perpetuates the change throughout the organism and transmits it to future generations.

in vivo **gene therapy** Introduction of vectors carrying therapeutic human genes directly into the body part where they will act.

pharmacogenetics (farm a kō jə-net-ĭks) Testing for variants of a gene that affect metabolism of a specific drug.

pharmacogenomics (farm a kō jə-nōm-ĭks) Testing for variants of many genes or gene expression patterns that affect metabolism of a specific drug.

pharmacological chaperone therapy Use of an oral drug that binds to a misfolded enzyme, stabilizing it sufficiently to permit some level of function.

preimplantation genetic diagnosis (PGD) Removing a cell from an early embryo and testing it for a mutation to deduce the genotype of the embryo.

somatic gene therapy (sō-măt' ĭk) Genetic alteration of a specific cell type, which is not transmitted to future generations.

substrate reduction therapy Use of an oral drug that reduces the level of substrate so that an enzyme can function more effectively.

Review Questions

1. How are the services that a genetic counselor provides different from those of a nurse or physician?

2. How are the consequences of a genetic test different from those of a cholesterol check?

3. Compare and contrast the types of information obtained from preconception comprehensive carrier screening, prenatal diagnosis, and newborn screening in terms of the types of actions that can be taken in response to receiving the results.

4. Using information from this or other chapters, or the internet, cite DNA-based tests given to a fetus, a newborn, a young adult, and a middle-aged person.

5. Select a gene from table 22.2 and discuss how its variants might impact the ability to succeed at a particular sport or exercise.

6. Explain how a pharmacogenetic test can improve quality of life for a person with cancer.

7. Distinguish among enzyme replacement or substitution therapy, substrate reduction therapy, and pharmacological chaperone therapy.

8. What factors would a researcher consider in selecting a viral vector for gene therapy? In selecting a disease amenable to gene therapy?

9. How do gene therapy and genome editing differ?

10. Explain how researchers combined genome sequencing, application of Mendel's first law, and CRISPR-Cas9 to diagnose familial thoracic aortic aneurysm and dissection.

11. Explain why blood conditions will likely be among the first diseases to be treated with CRISPR-Cas9 genome editing.

12. Describe two ways that CRISPR-Cas9 genome editing is used experimentally to correct the mutation underlying sickle cell disease.

Applied Questions

1. Suggest a treatment other than gene therapy that might have helped Eliza O'Neill, the child described in the chapter opener.

2. Explain why high penetrance is an important criterion for including a disease in newborn screening.

3. Discuss the challenges that a genetic counselor faces in explaining to parents-to-be a prenatal diagnosis of trisomy 21 in a fetus with no family history, compared to a prenatal diagnosis of translocation Down syndrome in a family with a reproductive history of pregnancy loss and birth defects. (Chapter 13 discusses Down syndrome.)

4. Lysosomal acid lipase deficiency causes liver cells (hepatocytes) to accumulate triglycerides and low density lipoproteins (LDL), but have low levels of high density lipoproteins (HDL). The resulting "fatty liver" hardens, and its functions become impaired. Certain genotypes are lethal by the age of 1 year. The condition is a lysosomal storage disease and is autosomal recessive. The statin drugs widely used to combat this unhealthy lipid profile are ineffective, because the drugs target a different pathway (endogenous cholesterol synthesis) than the one that is affected in lysosomal acid lipase deficiency. Compare how the mechanisms of enzyme replacement therapy and gene therapy to treat this disease would differ.

5. Select a disease described in an earlier chapter and explain how any of the technologies discussed in this chapter might be applied to create a treatment.

Case Studies and Research Results

1. How might a genetic counselor address the concerns of the following patients? (See other chapters for specific information.)

    a. A couple in their early forties is expecting their first child. Cell-free fetal DNA testing indicates that the fetus is XXX (see table 13.4). When they learn the results, the couple fears severe birth defects caused by the extra chromosome.

    b. Two people of normal height have a child with achondroplastic dwarfism, an autosomal dominant trait. Will future children have the condition (see figure 5.1)?

    c. A couple has received results from a consumer genetic testing company. Tests based on genome-wide association studies indicate that they each have inherited susceptibility to asthma as well as gene variants that in some populations are associated with autism. Both also have several gene variants that are found in lung cancers. Each has a few recessive mutant alleles, but not in the same genes. On the basis of these results, they do not think that they are "genetically healthy" enough to have children.

2. Jill and Scott had thought 6-month-old Dana was developing just fine until Scott's sister, a pediatrician, noticed that the baby's abdomen was swollen and hard. Knowing that the underlying enlarged liver and spleen could indicate an inborn error of metabolism, Scott's sister suggested the child undergo several tests.

    Dana had inherited sphingomyelin lipidosis, also known as Niemann-Pick disease type A. Both parents were carriers, but Jill had tested negative when she took a Jewish genetic disease panel during her pregnancy because her mutation was rare and not tested for. Dana was successfully treated with a transplant of umbilical cord blood cells from a donor. She caught up developmentally and became more alert. Monocytes, a type of white blood cell, from the cord blood traveled to her brain and manufactured the deficient enzyme. Dietary or enzyme replacement therapy do not work for this condition because the enzyme cannot cross from the blood to the brain. Monocytes, however, can enter the brain.

    a. Explain the effect of Dana's treatment on her phenotype and genotype.
    b. Why did the transplant have to come from donated cord blood, and not from Dana's own, which had been stored?
    c. What advice might a genetic counselor give this couple if they conceive again?
    d. What is the probability that Dana can pass on the disease to a child if her partner is a carrier?

3. A clinical trial of gene therapy for the clotting disorder hemophilia B introduced the gene for factor IX via AAV into the livers of several men. Researchers halted the trial when they detected the altered gene in semen of treated men. Why did they take this precaution?

4. Eden and Jamal were concerned that their son Perry did not make eye contact or respond as his older siblings had, but the pediatrician assured them that the boy was well, just easily distracted. In school, though, Perry could not focus, and had angry outbursts. Eden, a nurse, learned about chromosomal microarray analysis in a continuing education class. She had Perry tested, which revealed a small duplication. The report stated that the finding was a "variant of uncertain significance" (see chapter 12 Bioethics). The duplication was abnormal, but it hadn't been associated with a syndrome. Eden wanted to include the test results in Perry's school records, in case a situation should arise in which a medical explanation or special services might be helpful, but Jamal feared that the genetic test result might stigmatize the boy. Do you agree with either parent? Cite a reason for your answer.

CHAPTER 23

Reproductive Technologies

In vitro *fertilization coupled with genetic testing of preimplantation embryos enables parents-to-be to select embryos for completing development that have the normal number of chromosomes and have not inherited specific disease-causing gene variants. This is a stylized view of sperm meeting oocyte.*

Learning Outcomes

23.1 Savior Siblings and More
1. Explain how a child can be conceived to provide tissue for an older sibling.
2. Define *assisted reproductive technology (ART)*.

23.2 Infertility and Subfertility
3. Distinguish infertility from subfertility.
4. Describe causes of infertility in the male.
5. Describe causes of infertility in the female.
6. List infertility tests.

23.3 Assisted Reproductive Technologies
7. Describe ARTs that donate sperm, uterus, or oocyte.
8. List the steps of *in vitro* fertilization.
9. Explain how preimplantation genetic diagnosis enables people to select embryos conceived *in vitro* that have normal chromosomes and do not have certain mutations.
10. Explain how testing a polar body can reveal information about a genotype of a fertilized ovum.

23.4 Extra Embryos
11. Discuss uses for extra embryos resulting from ARTs.

The BIG Picture

Assisted reproductive technologies provide intriguing and sometimes complex variations on the processes of conceiving, selecting embryos with or without specific gene variants, and carrying offspring to term. Testing early embryos is evolving from visualizing chromosomes to identifying specific single-gene variants to exome and genome sequencing. Reproductive technologies have also added to our understanding of the events of gamete formation and maturation, and early embryonic development.

Choosing an Embryo

The couple met when they were in their early forties. When they began discussing starting a family, they had genetic testing and discovered that they each are carriers of the same serious neurological condition, Pompe disease, and that the man has a dominant variant of the factor V Leiden gene that explained his family history of deep vein thrombosis and pulmonary embolism from abnormal blood clotting. The couple chooses *in vitro* fertilization (IVF) with preimplantation genetic diagnosis (PGD) to increase the chances that they have a healthy child. They are concerned because of their family medical histories, their own mutations, and the elevated risk of an abnormal number of chromosomes in an oocyte due to the woman's age. It might take much longer to conceive and carry a pregnancy to term the natural way.

For several weeks the woman injects hormones to prepare her body to ovulate. At the right time, a physician retrieves the largest oocytes bulging from an ovary and mixes them with the man's sperm. Twelve fertilized ova result. They divide in the lab dish over the next 5 days, reaching the stage just before they would implant in the uterus. A few cells are sampled from the nine embryos that remain and look robust, and they are tested for chromosome number, Pompe disease, and factor V Leiden.

Four embryos have one extra or missing chromosome and one embryo has two missing chromosomes. They are donated for research. Of the four embryos with the normal 46 chromosomes, two are XX and two are XY. One XX embryo has the factor V Leiden gene variant but is wild type for Pompe disease. One XY embryo does not have the factor V Leiden mutation but has one copy of the Pompe disease mutation. One XX and one XY embryo inherited neither of the parents' mutations.

The couple chooses the XY embryo that inherited neither mutation. They didn't select the XX with the factor V Leiden mutation because it confers inappropriate clotting risk. The couple freezes the XX embryo that does not have the family mutations. Perhaps when the couple is ready to have another child, genome sequencing of embryos will further check that their daughter will be healthy by revealing *de novo* mutations—those that originate in the fertilized ovum and are therefore not inherited.

23.1 Savior Siblings and More

A couple in search of an oocyte donor advertises in a college newspaper seeking an attractive, bright young woman from an athletic family. A cancer patient has ovarian tissue removed and frozen before undergoing treatment. Two years later, a strip of the frozen tissue is thawed and oocytes separated and fertilized in a laboratory dish with her partner's sperm. Several cleavage embryos develop and two are implanted in her uterus. She becomes a mother—of twins. A man paralyzed from the waist down has sperm removed and injected into his partner's oocytes. He, too, becomes a parent when he thought he never would.

Lisa and Jack Nash wanted to have a child for an unusual reason. Their daughter Molly, born on July 4, 1994, had Fanconi anemia. This autosomal recessive condition would destroy her bone marrow and her immunity and likely cause leukemia; the median age at death of patients with this disease is 30. An umbilical cord stem cell transplant from a sibling could cure her, but Molly had no siblings. Nor did her parents wish to have another child who would have a one in four chance of inheriting the disease, as Mendel's first law dictates. Technology offered a way for them to alter the odds.

In late 1999, researchers mixed Jack's sperm with Lisa's oocytes in a laboratory dish. After allowing 15 of the fertilized ova to develop to the 8-cell stage, researchers separated and applied DNA probes to one cell from each embryo (**figure 23.1**). A cell that had wild type Fanconi anemia alleles and that matched Molly's human leukocyte antigen (HLA) type was identified and its 7-celled remainder implanted into Lisa's uterus to develop. This technology, revolutionary at the time, is preimplantation genetic diagnosis (PGD). The result was Adam, born in late summer. His umbilical cord stem cells saved Molly's life. At first the Nashes were sharply criticized for intentionally conceiving a "savior sibling," but as other families followed their example, conceiving and selecting a child to provide cells for a sibling became more accepted.

Figure 23.1 Savior siblings. Preimplantation genetic diagnosis removes one cell from an early embryo and tests it for an inherited disease that is in a particular family. The test cannot be done on a fertilized ovum because the cell would be destroyed. *Science Photo Library/Alamy*

Increased knowledge of how the genomes of two individuals come together and interact has spawned several novel ways to have children. **Assisted reproductive technologies (ARTs)** replace the source of a male or female gamete, aid fertilization, or provide a uterus. These procedures were developed to treat infertility, but are increasingly including genetic testing, too, so that parents-to-be can select offspring-to-be, like the couple in the chapter opener. The PGD that was shockingly new when the Nashes did it is now a routine follow-up to IVF, to increase the chances of selecting a viable embryo for further development.

The U.S. government does not regulate ARTs, but the American Society for Reproductive Medicine provides voluntary guidelines. The United Kingdom has pioneered ARTs and its Human Fertilisation and Embryology Authority has served as a model for government regulation. A great advantage of the British regulation of reproductive health services and technologies is that databases include success rates for the different procedures. Another advantage is that access to reproductive technology is not limited to those who can afford it.

Key Concepts Questions 23.1

a. Explain how preimplantation genetic diagnosis saved Molly Nash.

*b. What are assisted reproductive technologies?

23.2 Infertility and Subfertility

Infertility is the inability to conceive a child after a year of frequent sexual intercourse without the use of contraceptives. It is also called *reproductive system disorder* and affects about 15 percent of couples. Some specialists use the term *subfertility* to distinguish those individuals and couples who can conceive unaided, but for whom this may take longer than average. On a more personal level, infertility can be a seemingly endless monthly cycle of raised hopes and crushing despair.

As a woman ages, fertility naturally declines, due to fewer remaining oocytes, called the ovarian reserve. By the start of menopause at about age 51, only a thousand or so oocytes remain in her ovaries. Spontaneous abortion occurs in 15 percent of women under 35, in 20 to 35 percent of women between ages 35 and 45, and in up to 50 percent of women over 45 who try to become pregnant. The incidence of pregnancy-related problems rises with maternal age, including chromosomal anomalies, fetal deaths, premature births, and low-birth-weight babies. Losing very early embryos may appear to be infertility because the bleeding accompanying the embryo as it leaves the body resembles a heavy menstrual flow. Older men are at increased risk of having children who develop autism or schizophrenia. Sperm motility declines with age of the man, and erectile dysfunction is more likely.

One in six couples has difficulty conceiving or giving birth to children. According to the American Society for Reproductive Medicine, males and females contribute about equally to infertility, with 25 percent of infertile couples having two factors at play. When a physical problem is not obvious, the cause is usually a mutation or chromosomal aberration that impairs fertility in the male. A common combination that accounts for low fertility is a woman with an irregular menstrual cycle and a man with a low sperm count.

Male Infertility

Infertility in the male is easier to detect but sometimes harder to treat than female infertility. Four in 100 men in the general population are infertile, and half of them do not make any sperm, a condition called azoospermia. The general term *male factor infertility* refers to a man who has a low sperm count, many abnormally shaped sperm, or many nonmotile (unable to move) sperm.

Oligospermia, in which a man has difficulty fathering a child because he produces fewer than the average 15 to 200 million sperm cells per milliliter of ejaculate, has several causes. If a low sperm count is due to a hormonal imbalance, administering the appropriate hormones may boost sperm production. Sometimes a man's immune system produces IgA antibodies that cover the sperm and prevent them from binding to oocytes. Male infertility can also be due to a varicose vein in the scrotum. This enlarged vein emits heat near developing sperm, which prevents them from maturing. Surgery can remove a scrotal varicose vein. The heat from using laptop computers for long periods of time can raise scrotal temperature, damaging sperm, as can wearing tight underwear.

Most causes of male infertility are genetic. About a third of infertile men have small deletions of the Y chromosome that remove the only copies of key genes whose products control spermatogenesis. Other genetic causes of male infertility include mutations in genes that encode androgen receptors or protein fertility hormones, or that regulate sperm development or motility (movement). **Clinical Connection 23.1** describes a type of autosomal recessive male infertility that is unusual in that it is not part of a syndrome.

To speed conception, a man with a low sperm count can donate several semen samples over a period of weeks at a fertility clinic. The samples are kept in cold storage, then pooled. Some of the seminal fluid is withdrawn to leave a sperm cell concentrate, which is then placed in the woman's body. Test kits that assess sperm count and motility are available over-the-counter. A smartphone-based semen analyzer uses a microchip sample device and the phone's camera as a microscope to track sperm count, motility, concentration, and velocity. Men who cannot ejaculate due to a spinal cord injury can use a vibrating device applied to the tip of the penis or inserted into the rectum that stimulates release of sperm. The sperm are collected and transferred to the partner using a syringe.

Sperm quality is more important than quantity. Sperm cells that are unable to move or are shaped abnormally cannot reach an oocyte. Inability to move may be due to a hormone imbalance and abnormal shapes may be due to impaired apoptosis (programmed cell death) that normally removes such sperm. The genetic package of an immobile or abnormally shaped sperm cell can be injected into an oocyte and sometimes this leads to fertilization, discussed later in the chapter. However, even sperm that look and move normally may be unable to fertilize an oocyte.

Female Infertility

Abnormalities in any part of the female reproductive system can cause infertility (**figure 23.2**). Many women with subfertility or infertility have irregular menstrual cycles, making it difficult to pinpoint when conception is most likely. In an average menstrual cycle of 28 days, ovulation usually occurs around the 14th day after menstruation begins. At this time a woman is most likely to conceive.

For a woman with regular menstrual cycles who is under 30 years old and not using birth control, pregnancy typically happens within 3 or 4 months. A woman with irregular menstrual periods can tell when she is most fertile by using an ovulation predictor test, which detects a peak in the level of luteinizing hormone that precedes ovulation by a few hours. An older way to detect the onset of ovulation is to record body temperature each morning using a digital thermometer with subdivisions of hundredths of a degree Fahrenheit, which can indicate the 0.4 to 0.6 rise in temperature when ovulation starts. Several smartphone apps track a woman's menstrual cycle, enabling her to predict the time of ovulation. Sperm can survive in a woman's body for up to 5 days but the oocyte is only viable for 24 to 48 hours after ovulation.

Clinical Connection 23.1

The Case of the Round-Headed Sperm

In fewer than a tenth of a percent of men who are infertile, sperm cells do not have the tip, called the acrosome, which contains the enzymes that break through the layers surrounding an oocyte. This condition is called "globozoospermia" (**figure 23A**). An Ashkenazi Jewish family led researchers to a gene that, when mutant, causes an autosomal recessive form of male infertility due to sperm with round rather than oval heads.

The family went to a center for reproductive medicine in the Netherlands. Of the six sons, three were infertile (**figure 23B**). Four daughters were fertile. The affected sons' sperm were misshapen. The mode of inheritance was recessive, because the parents were obviously fertile.

Researchers suspected consanguinity—a shared ancestor would increase the risk of both parents having inherited the same rare autosomal recessive mutation. But the family denied knowing a relative who had married a relative. Reasoning that perhaps DNA testing might reveal consanguinity that the family did not know about, researchers scanned the genomes of all six sons for "runs of homozygosity" that would indicate they had relatives marrying relatives not too far back in the family tree (see section 16.1).

A run of homozygosity in this case was defined as 25 consecutive SNPs that were homozygous. The genomes of all six sons had these telltale regions, evidence that at some point, a cousin had children with a cousin or an aunt/uncle had children with a nephew/niece. One run of homozygosity was seen in all three infertile brothers, but was heterozygous in two of the three fertile brothers. The remaining brother was homozygous wild type for the region.

Next, the researchers scrutinized the part of the long arm of chromosome 3 where the run of homozygosity lay in the three infertile brothers. It houses 50 genes, only one of which is expressed in the testes. This gene encodes "spermatogenesis-associated protein 16" and is called *SPATA16*. It has 11 exons. The mutation in the family is a single-base change, from G to A, at the 848th position in the gene, near the end of exon 4. The mutation affects the splicing out of introns as the gene is transcribed.

The wild type protein product of the *SPATA16* gene is transported from the Golgi apparatus into vesicles that take it to the acrosome as it telescopes out of the front end of a differentiating sperm cell. By attaching the gene for the jellyfish's green fluorescent protein (see figure 21.1) to the wild type *SPATA16* gene in cells growing in culture, researchers visualized the protein being transported to the forming acrosome in immature healthy sperm.

Questions for Discussion

1. What is the evidence that the infertility in the family is inherited as an autosomal recessive trait?
2. What is the evidence for inheritance of the mutation from shared ancestors?
3. Discuss how researchers discovered the gene that is mutant in the family.

Figure 23A **A sperm cell's streamlined form facilitates movement.** A misshapen sperm cannot fertilize an oocyte. *Dr. Tony Brain/Science Source*

Figure 23B **Inheriting infertility.** In a family with autosomal recessive globozoospermia, three of six sons are infertile.

The hormonal imbalance that usually underlies irregular ovulation has various causes. These include a tumor in the ovary or in the pituitary gland in the brain that controls the reproductive system, an underactive thyroid gland, or use of steroid-based drugs such as cortisone. If a nonpregnant woman produces too much prolactin, the hormone that promotes milk production and suppresses ovulation in new mothers, she will not ovulate.

Fertility drugs can stimulate ovulation, but they can also cause women to "superovulate," producing and releasing more than one oocyte each month. A commonly used drug, clomiphene, raises the chance of having twins from 1 to 2 percent to 4 to 6 percent. If a woman's ovaries are completely inactive or absent (due to a birth defect or surgery), she can become pregnant only if she uses a donor oocyte. Some cases of female infertility are due to *reduced ovarian reserve*—too few oocytes. This is typically discovered when the ovaries do not respond to fertility drugs. Signs of reduced ovarian reserve are an ovary with too few follicles (observed on an ultrasound

Figure 23.2 Sites of reproductive problems in the female.

scan) or elevated levels of follicle-stimulating hormone on the third day of the menstrual cycle.

The uterine tubes are a common site of female infertility because fertilization usually occurs in open tubes. Blockage can prevent sperm from reaching the oocyte, or entrap a fertilized ovum, keeping it from descending into the uterus. If an embryo begins developing in a blocked tube and is not removed and continues to enlarge, the tube can burst and the woman can die. This is called a tubal or ectopic pregnancy.

Uterine tubes can also be blocked due to a birth defect or, more likely, from an infection such as pelvic inflammatory disease. A woman may not know she has blocked uterine tubes until she has difficulty conceiving and medical tests uncover the problem. Introducing a dye into the uterine tubes and imaging the area can reveal a blockage. In some cases, surgery can open blocked uterine tubes.

Excess tissue growing in the uterine lining may make it inhospitable to an embryo. This tissue can include benign tumors called fibroids or areas of thickened lining from a condition called endometriosis. The tissue can grow outside of the uterus, too, in the abdominal cavity. In response to the hormonal cues to menstruate, the excess lining bleeds, causing cramps. Endometriosis can hamper conception, but curiously, if a woman with endometriosis conceives, the cramps and bleeding usually disappear after the birth.

Secretions in the vagina and cervix may be hostile to sperm. Cervical mucus that is thick or sticky due to infection can entrap sperm, keeping them from moving far enough to encounter an oocyte. Vaginal secretions may be so acidic or alkaline that they weaken or kill sperm. Douching daily with an acidic solution, such as acetic acid (vinegar) or an alkaline solution, such as bicarbonate, can alter the pH of the vagina so that, in some cases, it is more receptive to sperm cells. Too little mucus can prevent conception, too; this is treated with low daily doses of oral estrogen. Sometimes mucus in a woman's body has antibodies that attack sperm. Infertility may also result if the oocyte does not release sperm-attracting biochemicals.

One reason the incidence of female infertility increases with age is that older women are more likely to produce oocytes that have an abnormal chromosome number, which often causes spontaneous abortion because defects are too severe for development to proceed for long. The cause is usually misaligned spindle fibers when the second meiotic division begins, causing aneuploidy (extra or missing chromosomes). Perhaps the longer exposure of older oocytes to harmful chemicals, viruses, and radiation contributes to the risk of meiotic errors. However, one study of more than 1,300 women correlated increasing exposure to air pollution to decrease in ovarian reserve. The study considered women of all ages, using "geo-localization" to track their exposure levels to pollutants and yearly blood tests to measure anti-Mullerian hormone, which increases with increased ovarian reserve. Women with the highest quartile exposure to the pollution had significantly lower ovarian reserve—and age was not a factor.

Infertility Tests

A number of medical tests can identify causes of infertility. The man is checked first, because it is easier, less costly, and less painful to obtain sperm than oocytes.

Sperm are checked for number (sperm count), ability to swim (motility), and shape (morphology). An ejaculate containing up to 40 percent of unusual forms of sperm is still considered normal, but many more than this can impair fertility. A urologist performs sperm tests. A genetic counselor can evaluate Y chromosome deletions associated with oligospermia. If a cause of infertility in the male is not identified, a gynecologist does scans and other tests on the woman to determine whether all reproductive organs are present and function.

Some cases of subfertility or infertility have no clear explanation. Psychological factors may be at play, or it may be that inability to conceive results from consistently missed timing. Sometimes a subfertile couple adopts a child, only to conceive one of their own shortly thereafter; the causes of infertility may remain a mystery.

> **Key Concepts Questions 23.2**
>
> *a. What are causes of male infertility?
> *b. What are causes of female infertility?
> c. Describe medical tests used to identify the causes of infertility.

23.3 Assisted Reproductive Technologies

Many people with fertility problems who do not choose to adopt children use alternative ways to conceive. Several of the ARTs were developed in nonhuman animals, including mice, cows, and rabbits. In some nations up to 6 percent of births a year result from ART.

ART procedures can be performed on material from the parents-to-be ("nondonor") or from donors, and may be "fresh" (collected just prior to the procedure) or "frozen" (preserved in liquid nitrogen using a rapid procedure called vitrification). Some ARTs cost thousands of dollars and are only covered by health insurance in a few states in the United States.

Donated Sperm—Intrauterine Insemination

The oldest ART is **intrauterine insemination (IUI)**, in which a doctor places donated sperm into a woman's cervix or uterus. (It used to be called artificial insemination.) The success rate is 5 to 15 percent per attempt. The sperm are first washed free of seminal fluid, which can inflame female tissues. A woman might seek IUI using donor sperm if her partner is infertile or has a mutation that the couple wishes to avoid passing to their child. Women also undergo IUI to be a single parent without having sex, or a lesbian couple may use it to have a child.

The first documented IUI in humans was done in 1790. For many years, physicians donated sperm, as did medical students to earn a few extra dollars. It was long a practice shrouded in secrecy. In a famous case from 1884, a doctor at Jefferson Medical College used sperm from the medical student deemed to be the most attractive in his class to inseminate the young wife of an older wealthy man who was infertile due to a sexually transmitted infection. Years later, the doctor informed the husband of what had been done to his wife, who had been anesthetized for the insemination and never learned what had happened. The sperm donor eventually admitted his role in the deception, told in a medical journal article in 1909. In the past few years, after evidence surfaced implicating physicians in "fertility fraud"—inseminating often many patients with their sperm without the patients' knowledge—states have passed legislation to prevent this practice.

Attempts to freeze sperm date to the mid-1800s, but it took until 1953 for the procedure to work well enough for fertilization using frozen sperm to lead to a birth. By the 1970s the first "cryobanks" began to offer frozen (cryopreserved) sperm from anonymous donors. Using frozen donor sperm became more in demand in the 1980s, when the HIV/AIDS epidemic made it safer to use than fresh specimens, and single women and lesbian couples began seeking IUI using donor sperm. More recently, frozen human sperm survived for more than a year aboard the International Space Station, with normal activity upon thawing.

Today sperm banks are still in business, and clients can still select donors with certain characteristics, but the anonymity once promised to donors is no longer possible. Consumer genetic testing has greatly simplified identifying half-siblings who share a sperm (or less commonly an oocyte) donor.

A potential genetic danger of anonymous sperm donation is *inadvertent consanguinity,* when people who are related, such as half-siblings, have children together, not knowing of their relationship. This is more likely to have happened among older individuals who were conceived before consumer DNA testing enabled half-siblings to recognize one another. Consanguinity, knowing or unknowing, raises the risk of transmitting recessive diseases from mutations that genetic relatives inherited from a shared ancestor. Sperm banks now limit the number of donations from one man.

A Donated Uterus—Surrogate Motherhood

A man can be a genetic parent, contributing half of his genetic self in his sperm, but a woman can be both a genetic parent (donating an oocyte) and a gestational parent (loaning her uterus).

If a man produces healthy sperm but his partner's uterus cannot maintain a pregnancy, a surrogate mother may help by being inseminated with the man's sperm. When the child is born, the surrogate mother gives the baby to the couple. In this variation of the technology, the surrogate is both the genetic and the gestational mother. Attorneys usually arrange surrogate relationships. The surrogate mother signs a statement signifying her intent to give up the baby. In some U.S. states, and in some nations, she is paid for her 9-month job, but in the United Kingdom compensation is illegal. Outlawing compensation is to prevent wealthy couples from taking advantage of women who become surrogates to earn money.

A problem with surrogate motherhood is that a woman may not be able to predict her responses to pregnancy and childbirth in a lawyer's office months before she must hand over the baby. When a surrogate mother changes her mind, the results are wrenching for all.

Another type of surrogate mother lends only her uterus, receiving a fertilized ovum conceived from a man and a woman who has healthy ovaries but lacks a functional uterus. This variation is an "embryo transfer to a host uterus," and the pregnant woman is a "gestational-only surrogate mother." She turns the child over to the biological parents. About 1,600 babies are born in the United States to gestational surrogates each year.

In Vitro Fertilization

In *in vitro* fertilization (**IVF**), which means "fertilization in glass," sperm and oocytes join in a laboratory dish. At or before the fifth day of development, an embryo or embryos are transferred to a woman's uterus. If all goes well, an embryo implants into the uterine lining and continues development. **Figure 23.3** reviews these early stages of the embryo.

Louise Joy Brown, the first "test-tube baby," was born in 1978 amid great attention and sharp criticism. A prominent bioethicist said that IVF challenged "the idea of humanness and of our human life and the meaning of our embodiment and our relation to ancestors and descendants." Yet Louise is, despite her unusual beginnings, an ordinary woman. Millions of children have been born following IVF. Along the way, researchers have studied IVF "failures" that revealed how certain mutations interfere in early prenatal development (**table 23.1**).

A woman might undergo IVF if her ovaries and uterus work but her uterine tubes are blocked. After a woman takes hormones for a few weeks, a physician removes several of the largest oocytes from an ovary through the cervix and transfers them to

Table 23.1	Mutations That Interfere with Early Prenatal Development
Gene	**Effect of Mutation**
TUBB8	Block in meiosis I in oocytes due to absent or abnormal tubulin, required for forming meiotic spindle.
ZP3	Oocytes do not have the zona pellucida (see figure 3.15) and degenerate, leaving empty ovarian follicles.
PATL2	Block in oocyte maturation.
DNAH1	Misshapen sperm tails (flagella).
CFAP43	Misshapen sperm tails
CFAP44	Misshapen sperm tails
TLE6	Fertilized ovum does not cleave due to absence of a group of proteins beneath the oocyte plasma membrane.

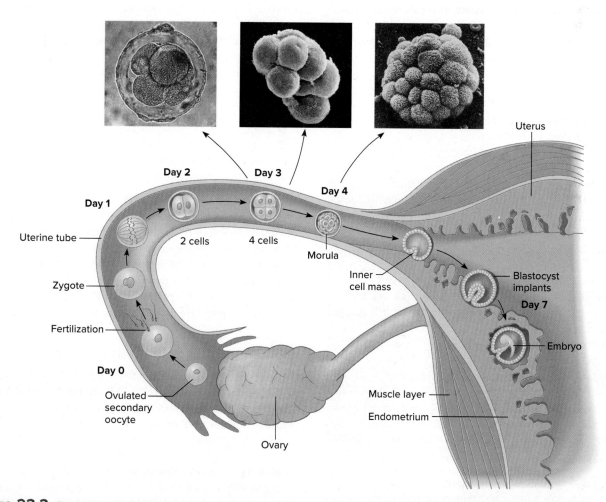

Figure 23.3 Reviewing the earliest stages of prenatal development. The zygote forms in the uterine tube when a sperm nucleus fuses with the nucleus of an oocyte at fertilization. The first divisions, called cleavage, proceed while the zygote moves toward the uterus. By day 7, the blastocyst begins to implant in the uterine lining. *(left): Petit Format/Nestle/Photo Researchers/Science Source; (middle): PROFESSORS P.M. MOTTA and J. VAN BLERKOM/Science Source; (right): PETIT FORMAT/Science Source/Science Source*

a culture dish. If left in the body, only one oocyte would exit the ovary, but in culture, many oocytes can mature sufficiently to be fertilized *in vitro*. Chemicals, sperm, and other cell types similar to those in the female reproductive tract are added to the culture. An acidic solution may be applied to the zona pellucida, which is the layer around the egg, to thin it to ease the sperm's penetration.

IVF may also be used if a female is fertile but has not conceived because her partner has a low sperm count or many abnormal or nonmotile sperm. A technique called **intracytoplasmic sperm injection (ICSI)** has been used as an adjunct to IVF when sperm cannot readily enter an oocyte. It was developed in the 1990s to treat male factor infertility, but is today used on about two-thirds of all IVF cycles, because it increases the chance of fertilization. ICSI sucks a sperm up into a tiny syringe and it is then microinjected into an oocyte. ICSI is equally likely to lead to a live birth whether using fresh or frozen and thawed sperm (**figure 23.4**). ICSI makes fatherhood possible for men who cannot ejaculate, such as those who have suffered spinal cord injuries.

Two to 5 days after sperm wash over the oocytes in the dish, or are injected into them, a cleavage stage embryo or blastocyst is transferred to the uterus. If the hormone human chorionic gonadotropin appears in the woman's blood a few days later, and its level rises, she is pregnant. Transferring a blastocyst (the hollow sphere at day 5) is more likely to result in a successful pregnancy than transferring a cleavage-stage embryo (solid ball of cells), but carries a slightly higher risk of pregnancy complications like the placenta detaching early, and about doubles the chance of having twins if two embryos are transferred.

Embryos resulting from IVF that are not soon implanted in the woman can be frozen for later use. Cryoprotectant chemicals are used to prevent salts from building up or ice crystals from damaging delicate cell parts. Freezing takes a few hours; thawing about a half hour. The "oldest" pregnancy using a frozen embryo occurred 20 years after the freezing.

IVF costs on average $12,000 per cycle. Medications can add $3,000 to $5,000. ICSI adds another $1,500. Children born following IVF have only a slight increase in the rate of birth defects (about 4 to 7 percent) compared to children conceived naturally (about 3 to 5 percent), but the elevation may be due to the medical problems that caused the parents to seek IVF and higher surveillance of IVF pregnancies and outcomes. Children born after IVF on average have slightly higher birth weights, which is thought to be due to the time the early embryos spend in laboratory culture if they were frozen and thawed.

In the past, several embryos were implanted at one time to increase the success rate of IVF, but this led to many multiple births, which are riskier than single births. In some cases, physicians had to remove embryos to make room for others to survive. To avoid the multiples problem, and because IVF has become more successful as techniques have improved, physicians typically transfer only one or two embryos.

The chances of a live birth following IVF vary greatly, depending on certain risk factors that lower the likelihood of success. These include:

- maternal age;
- increased time being infertile;
- number of previous failed IVF attempts;
- use of a woman's own oocytes rather than a donor's; and
- infertility with a known cause.

Several websites assess risks and predict IVF success. In one example, a couple had been infertile for 11 years. They attempted IVF four times that resulted in two inabilities to fertilize and two spontaneous abortions. They had used the woman's oocytes and ICSI because too many sperm were abnormal. The chance of success per IVF attempt for this couple was estimated to be about 8 percent, but if they used a donor oocyte, it doubled. **Table 23.2** lists live birth rates for some ARTs.

Gamete and Zygote Intrafallopian Transfer

IVF may not work because of the artificial environment for fertilization. A procedure called **gamete intrafallopian transfer (GIFT)** improves the setting. (Uterine tubes are also called fallopian tubes.) Fertilization is assisted in GIFT, but it occurs in the woman's body rather than in glassware.

In GIFT, several of a woman's largest oocytes are removed. The man submits a sperm sample, and the most active cells are separated from it. The collected oocytes and sperm are deposited together in the woman's uterine tube, at a site past any obstruction that might otherwise block fertilization. GIFT is about 22 percent successful but depends upon a woman's age.

Figure 23.4 Intracytoplasmic sperm injection. ICSI was developed to enable some infertile men, such as those with spinal cord injuries or certain illnesses, to become fathers. Today, it is a routine accompaniment to IVF. A single sperm cell is injected into the cytoplasm of an oocyte. *Science Photo Library/Getty Images*

Table 23.2 Live Births Following *In Vitro* Fertilization

Procedure	Live Birth Rate
Overall IVF	37%
Maternal age at IVF:	
<35	41-43%
35-37	33-36%
38-40	23-27%
>40	13-18%
Overall ICSI	36%
Transfer of cleavage embryo	29%
Transfer of blastocyst embryo	32-42%
Transfer of frozen embryo	30.5%
Use of frozen oocytes	29.9%

Source: Statistics from the European Society of Human Reproduction and Embryology

In **zygote intrafallopian transfer (ZIFT)**, an IVF ovum is introduced into the woman's uterine tube. Allowing the fertilized ovum to make its own way to the uterus increases the chance that it will implant. ZIFT is also 22 percent successful.

GIFT and ZIFT are done less frequently than IVF. These procedures may not work for women who have scarred uterine tubes. The average cost of GIFT or ZIFT is $15,000 to $20,000.

Bioethics considers the unusual situation of collecting gametes from a person shortly after the person has died.

Oocyte Banking and Donation

Oocytes can be stored, as sperm are, but the procedure is more challenging. Because an oocyte is the largest type of cell, it contains a large volume of water. Freezing can form ice crystals that damage cell parts. Reasons to freeze oocytes include:

- Cancer treatment (radiation or chemotherapy) that can harm oocytes and cause early menopause
- Exposure to toxins or teratogens in the workplace
- Disease of the ovaries
- Premature ovarian failure, which may be due to XO syndrome or fragile X syndrome
- Ovary removal to prevent ovarian cancer due to a mutation, such as in *BRCA1* or *BRCA2*
- Inability to obtain viable sperm on the day of IVF
- Extra oocytes remain after IVF
- Donation
- Delaying motherhood for personal reasons

The American Society for Reproductive Medicine declared oocyte freezing no longer experimental in 2012. Then in 2018 it issued guidelines that introduced the term *planned oocyte cryopreservation* to represent all women wishing to delay parenthood, not only those with cancer for whom the technique was originally developed. Phrases used in the media such as *social egg freezing* were thought to disparage women trying to prevent infertility.

Oocytes are frozen in liquid nitrogen at temperatures below −40°C, when they are at metaphase of the second meiotic division. At this time, the chromosomes are aligned along the spindle, which is sensitive to temperature extremes. If the spindle comes apart as the cell freezes, the oocyte may lose a chromosome, which would devastate development if fertilization is even possible. Another problem with freezing oocytes is retention of a polar body, leading to a diploid oocyte.

To minimize risk of damage to oocytes when freezing, they are first treated with cryoprotectants that remove water and prevent ice crystal formation. When the frozen oocytes are to be used, they are slowly warmed and rehydrated by gradually removing the cryoprotectants. Frozen oocytes are fertilized *in vitro* using ICSI because the outermost layer (the zona pellucida) may be altered. Studies show that frozen and fresh oocytes from the same woman are equally likely to be fertilized, and there is no difference in live birth rate. Thousands of babies have been born following oocyte freezing. However, in 2018 problems with too little liquid nitrogen and temperature fluctuations led to the loss of more than 1,200 frozen oocytes and embryos at facilities in two U.S. cities.

An alternative to freezing oocytes is freezing strips of ovarian tissue that can be stored, thawed, and implanted at various sites, such as under the skin of the forearm or abdomen or in the pelvic cavity near the ovaries. The tissue ovulates and the oocytes are collected and fertilized *in vitro*. The first child resulting from fertilization of an oocyte from implanted ovarian tissue was born in 2004. The mother, age 25, had been diagnosed with advanced Hodgkin's lymphoma. The harsh chemotherapy and radiation cured her cancer, but destroyed her ovaries. Before her cancer treatment, five strips of tissue from her left ovary were frozen. Later, several pieces of ovarian tissue were thawed and implanted in a pocket that surgeons crafted on one of her shriveled ovaries, near the entrance to a uterine tube. Menstrual cycles resumed, and shortly thereafter, the woman became pregnant with her daughter, who is healthy. Freezing ovarian tissue may become routine for cancer patients of childbearing age.

Women who have no oocytes or wish to avoid passing on a mutation can obtain oocytes from donors, who are typically younger women. Some women become oocyte donors when they undergo IVF and have "extras." The potential father's sperm and donor's oocytes are placed in the recipient's uterus or uterine tube, or fertilization occurs in the laboratory and a blastocyst is transferred to the woman's uterus. A program in the United Kingdom funds IVF for women who cannot afford the procedure if they donate their "extras." The higher success rate of using oocytes from younger women confirms that it is the oocyte that age affects, and not the uterine lining.

Bioethics

Removing and Using Gametes after Death

A gamete is a packet containing one copy of a person's genome. If, after death, gametes are collected and combined with an opposite gamete type, the deceased person can become a parent. This "postmortem gamete retrieval" has happened for years, but nearly always for men.

One of the first cases of postmortem sperm removal affected a couple in their early thirties who had delayed becoming parents, confident that their good health would make pregnancy possible later. Then the man suddenly died of an allergic reaction to a drug. His wife knew how much he had wanted to be a father, so she asked the medical examiner to collect her husband's sperm. The sample was sent to a cryobank, where it lay deeply frozen for more than a year. In the summer of 1978, the sperm were defrosted and used to fertilize one of the woman's oocytes. The couple's daughter was the first case of postmortem sperm retrieval in which the father did not actively participate in the decision. Since 1990, U.S. servicemen who feared infertility from exposure to chemical or biological weapons have taken advantage of sperm bank discounts to the military, preserving their sperm before deploying.

Postmortem sperm retrieval raises legal and ethical issues based on timing. In one case, a woman conceived twins using her husband's preserved sperm 16 months after he died of leukemia; he had stated his wishes for her to do so. The Social Security Administration refused to provide survivor benefits to their daughters, claiming that the husband was not a father, but a sperm donor. The Massachusetts Superior Court reversed this decision. In New Jersey, a mother claimed Social Security benefits for twins conceived after her husband's death. An appeals court upheld the denial of benefits, claiming that the children had to have been dependents at the time of their father's death.

Postmortem oocyte retrieval was considered in a case reported in 2010. A 36-year-old woman stood up on a plane following many hours of sleeping in one position, and her heart stopped. By the time a doctor on board restarted it, the woman's brain had been robbed of oxygen for several precious minutes. The plane made an emergency landing and she was taken to a hospital, where she was placed on a respirator. Scans showed that blood clots in her lungs had caused the collapse.

By the fourth day, the woman's brain was dangerously swelling, and by the ninth day, her brain activity was nearly nil, although she could still open her eyes and move spontaneously. Her husband, parents, and in-laws asked that the tubes keeping her alive be withdrawn. Then, they changed their minds, and asked that the breathing tube be reinserted so that oocytes could be retrieved.

The physicians wanted to know one other person's opinion—the patient's. So they consulted with the young woman's gynecologist, who had no record or recollection of the patient stating she wished to have children. The young woman was not completely brain dead, so the decisions would not be the same as for donating an organ after death. If her oocytes were to be used to give her husband a child, IVF and a surrogate mother would be necessary. Before that could happen, though, the woman would have to undergo 2 weeks of hormone treatments to ovulate, when she would have to lie flat, which could kill her. For these practical reasons, and the fact that the woman had never stated that she wished to be a parent, the family elected to turn off life support, and she quickly died.

Like other ARTs, postmortem gamete retrieval is not regulated at the federal level in the United States. Bioethicists have identified situations to avoid:

- Someone other than a spouse wishing to use the gamete
- A too-hasty decision based on grief
- Use of the gamete for monetary gain

Questions for Discussion

1. How does the case of the woman whose oocytes were to be retrieved following her brain death differ from that of a pregnant woman in a coma who is kept alive for weeks so that her baby can be born?
2. The people described in this essay did not have other children. How might the situation differ for a couple who already has children?
3. Do you think that Social Security or another benefit system should cover fetuses, embryos, or gametes?
4. How might postmortem gamete retrieval be abused?

Embryo donation is a variation on oocyte donation. A woman with malfunctioning ovaries but a healthy uterus carries an embryo that results when her partner's sperm is used in intrauterine insemination of a woman who produces healthy oocytes. If the woman conceives, the embryo is gently flushed out of her uterus a week later and inserted through the cervix and into the uterus of the woman with malfunctioning ovaries. The child is genetically that of the man and the woman who carries it for the first week, but is born from the woman who cannot produce healthy oocytes. Embryo donation includes use of IVF "leftovers."

In another technology, cytoplasmic donation, older women have their oocytes injected with cytoplasm from the oocytes of younger women to "rejuvenate" the cells.

Although resulting children conceived through IVF appear to be healthy, they are being monitored for a potential problem—heteroplasmy, or two sources of mitochondria in one cell (see chapter 5 Bioethics). Researchers do not yet know the health consequences, if any, of having mitochondria from the donor cytoplasm plus mitochondria from the recipient's oocyte. These conceptions also have an elevated incidence of XO syndrome, which often causes spontaneous abortion.

Because oocytes are harder to obtain than sperm, oocyte donation technology has lagged behind that of sperm banks, but is catching up. One IVF facility that has run a donor oocyte program since 1988 has a brochure that describes 120 oocyte donors of various ethnic backgrounds, like a catalog of sperm donors. The oocyte donors are young and have undergone extensive medical and genetic tests. Recipients may be up to 55 years of age.

Preimplantation Genetic Diagnosis

Preimplantation genetic diagnosis (PGD) is often teamed with IVF to detect genetic and chromosomal abnormalities *before* pregnancy starts, as described in the chapter opener and in section 23.1. The procedure is called *diagnosis* if it detects a specific gene variant known to be in one or both parents-to-be, and *screening* if it is used to construct a karyotype (that is, check chromosome number), without looking for a specific chromosomal anomaly. The couple selects a very early "preimplantation" embryo that tests show has not inherited a specific detectable genetic condition. "Preimplantation" refers to the fact that the embryo is tested at a stage prior to when it would naturally implant in the uterus. PGD has about a 29 percent success rate, and it adds on average $3,500 to the cost of IVF.

PGD is possible because one cell, or blastomere, can be removed for testing from an embryo consisting of at least 8 cells, and the remaining cells can complete development normally in a uterus, even after being frozen. Before the embryo is transferred into the woman, the sampled single cell is karyotyped, or its DNA is amplified and probed for genes that the parents carry, or the exome or genome is sequenced. Embryos that have normal chromosomes and do not have disease-causing genotypes in the family are selected to complete development or are stored. Testing cells from a day-5 embryo, when it is 80 to 120 cells, is more successful than testing a cell from an 8-celled embryo, as was once done, because several cells can be studied. Testing at the 8-cell stage is called a "blasto*mere* biopsy" and the later one a "blasto*cyst* biopsy" (**figure 23.5**). A blastocyst biopsy tests cells of the trophectoderm, which is the surrounding tissue that does not develop into the embryo. Accuracy in detecting a mutation or abnormal chromosome in PGD is about 97 percent. Errors generally happen when a somatic mutation affects the sampled cell but not the rest of the embryo. Amplification of the selected blastomere DNA may cause such a somatic mutation. A newer technique tests DNA leaked from blastocysts into the culture medium.

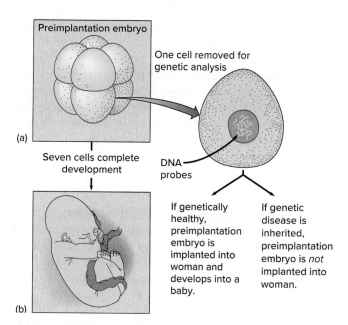

Figure 23.5 Preimplantation genetic diagnosis (PGD) probes disease-causing genes or chromosome aberrations in an 8-celled preimplantation embryo or from a later stage, the blastocyst. (a) A single cell is separated and tested to see if it contains a disease-causing genotype or atypical chromosome. (b) If it doesn't, the remaining seven cells divide a few more times and are transferred to a woman's uterus—typically the oocyte donor—to complete development.

PGD is not new. The first children were born following PGD in 1989. Probes for Y chromosome-specific DNA sequences were used to eliminate males and select females, who could not have inherited the X-linked conditions their mothers carried. The alternative to PGD would have been to face the 25 percent chance of conceiving an affected male.

In 1992, the first child was born who underwent PGD to avoid a specific inherited disease. She was checked as an 8-celled preimplantation embryo to see if she had escaped the severe cystic fibrosis that affected her brother. Since then, PGD has helped to select many children free of single-gene conditions.

PGD is routinely used to screen early embryos derived from IVF for normal chromosome number before implanting them into women, even if there is no family history of single-gene diseases, because the technology can detect *de novo* (new) mutations. In this way, PGD increases the success rate of IVF because it eliminates many embryos that are not likely to complete development. It is also more precise than examining early embryos to see which ones appear most robust, as researchers who worked with human embryos did for years before studies revealed that the best-looking embryos are not necessarily the most genetically healthy ones.

PGD can introduce a bioethical "slippery slope" when it is used for reasons other than to increase the probability that a child is born healthy. Some people may regard gender selection using PGD as a misuse of the technology. Using PGD to avoid disease can be controversial too. In the United Kingdom, where the government regulates reproductive technology, inherited cancer susceptibility is an approved indication for PGD. These cancers do not begin until adulthood, the susceptibility is incompletely penetrant (not everyone who inherits the disease-associated genotype will actually develop cancer), and the cancer may be treatable.

Sequential Polar Body Analysis

A technique called **sequential polar body analysis** may substitute for PGD and provides genetic information even earlier in development. It determines whether a specific mutation is present in a first and a second polar body. The approach is based on the fact that meiosis completes in the female only as a secondary oocyte is fertilized (see figure 3.12).

If a woman is heterozygous for a mutation, then an oocyte will contain the mutation and its associated first polar body will have the wild type allele, or vice versa. This is in accordance with Mendel's first law, gene segregation. A second polar body, which forms at fertilization, genetically matches the fertilized ovum and can be tested (**figure 23.6**).

Because sequential polar body analysis is still experimental, researchers follow it up with PGD to test their predictions and ensure that IVF embryos transferred to the woman's uterus to develop are free of the family's mutation. The idea of probing polar bodies dates from the 1980s, but the technique first led to results in 2011. So far mutations behind more than 150 single-gene diseases have been detected at this initial stage of prenatal development—the very first cell.

Table 23.3 summarizes the ARTs.

Table 23.3	Some Assisted Reproductive Technologies
Technology	**Procedures**
GIFT	Deposits collected oocytes and sperm in uterine tube.
IVF	Mixes sperm and oocytes in a dish. Chemicals simulate intrauterine environment to encourage fertilization.
IUI	Places or injects washed sperm into the cervix or uterus.
ICSI	Injects immature or rare sperm into oocyte, before IVF.
Oocyte freezing	Oocytes retrieved and frozen in liquid nitrogen.
PGD	Analyzes chromosomes and gene variants in early embryos, *in vitro*. Selected embryos are implanted in the uterus and child will be free of tested-for conditions.
Sequential polar body analysis	Genetic testing of polar body attached to just-fertilized ovum enables inference that fertilized ovum is free of a family's mutation.
Surrogate mother	Woman carries a pregnancy for another.
ZIFT	Places IVF ovum in uterine tube.

Key Concepts Questions 23.3

*a. Describe intrauterine insemination.

b. Distinguish between a genetic and gestational surrogate mother and a gestational-only surrogate mother.

*c. List the steps of IVF.

d. Distinguish between GIFT and ZIFT.

e. Explain how PGD optimizes the outcome of IVF.

*f. How can genetic testing of a polar body reveal information about a fertilized ovum?

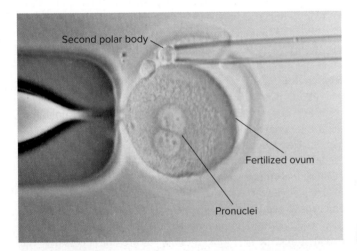

Figure 23.6 Deducing a fertilized ovum's genotype by testing a second polar body. If a mutation is detected in a second polar body, then it can be inferred to also be in the fertilized ovum. *Courtesy Dr. Anver Kuliev*

23.4 Extra Embryos

ARTs can result in "extra" oocytes, fertilized ova, or very early embryos. Frozen embryos can be:

- Stored indefinitely.
- Stored and destroyed after a set time.
- Donated for research.
- Thawed later for use by the biological parents or by other parents
- Discarded.

In the United States, nearly half a million embryos derived from IVF sit in freezers; some have been there for years. Couples who donate their frozen embryos to others can do so anonymously through the reproductive health care facility, but the children can identify their genetic parents years later using consumer DNA testing. Some couples use social media to connect with people who want to use the embryos. One couple who had twins from 2 of 18 IVF embryos chatted online with and selected another couple to receive the remaining embryos. That couple also had twins, and they then shipped the remaining frozen embryos to a third couple, who also had twins. Extra frozen embryos are an excellent resource for same-sex couples to have children, with the help of a surrogate mother if necessary.

An ethical dilemma can arise when embryos already donated are discovered to possibly have a disease-causing mutation. In one case, a boy was diagnosed with a brain tumor shortly after his father was diagnosed with lung cancer, and genetic tests revealed that they both had Li-Fraumeni syndrome, which causes early-onset cancers. Affected families with this rare condition have several members with cancer. The boy had been conceived using IVF and the extra embryos had been donated. Bioethicists concluded that the parents had a "duty to warn" others who had used their embryos that the children might have inherited the mutation. If the adoptive parents are aware of the risk, close monitoring can lead to early detection and more effective treatments of cancer.

Embryos donated for research can reveal surprising findings that may challenge long-held ideas, indicating that we still have much to learn about early human prenatal development. This was the case for a study from Royal Victoria Hospital in Montreal. Researchers examined the chromosomes of sperm from a man with XXY syndrome. Many of the sperm would be expected to have an extra X chromosome, due to nondisjunction (see figure 13.11), which could lead to a preponderance of XXX and XXY offspring. Surprisingly, only 3.9 percent of the man's *sperm* had extra chromosomes, but five of his 10 spare *embryos* had an abnormal X, Y, or chromosome 18. That is, even though most of the man's sperm were normal, his embryos weren't. The source of reproductive problems in XXY syndrome, therefore, might not be in the sperm, but in early embryos—a finding that was previously unknown and not expected, and was only learned because of observing early human embryos develop.

In another study, Australian researchers followed the fates of single blastomeres that had too many or too few chromosomes. Did the abnormal cells preferentially end up in the inner cell mass, which develops into the embryo, or the trophectoderm, which becomes extra-embryonic membranes? The study showed that cells with extra or missing chromosomes become part of the inner cell mass much more frequently than expected by chance. The finding indicates that the ability of a blastomere sampled for PGD to predict health may depend on whether it is fated to be part of the inner cell mass.

ARTs paired with the ability to sequence and interpret human genomes introduce both ownership and parentage issues in an immediate sense, but also broader issues of the evolution and future of the human species. When we can routinely scan the genomes of human gametes, fertilized ova, and embryos, who will decide which traits are worth living with, and which aren't?

ARTs operate on molecules and cells, but affect individuals and families, returning to the levels of genetics presented in chapter 1. Ultimately, by introducing artificial selection and perhaps one day even altering germlines, these interventions and technologies may affect the human gene pool.

Key Concepts Questions 23.4

*a. What can be done with extra fertilized ova and early embryos?

b. What can we learn from using early embryos in research?

Study Guide

Summary

23.1 Savior Siblings and More

1. **Assisted reproductive technologies (ARTs)** replace what is not present or nonfunctional in reproduction, using laboratory procedures and people other than the couple with infertility.

23.2 Infertility and Subfertility

2. **Infertility** is the inability to conceive a child after a year of frequent unprotected intercourse. Subfertile individuals or couples manufacture gametes, but take longer than usual to conceive.

3. Causes of infertility in the male include low sperm count, a malfunctioning immune system, a varicose vein in the scrotum, structural sperm defects, drug exposure, vasectomy reversal, heat to the testes, and abnormal hormone levels. Mutation may impair fertility.

4. Causes of infertility in the female include absent or irregular ovulation, blocked uterine tubes, an inhospitable or abnormally shaped uterus, secretions that are hostile to sperm, or lack of sperm-attracting biochemicals. Early pregnancy loss due to abnormal chromosome number is more common in older women and may appear to be infertility.

23.3 Assisted Reproductive Technologies

5. In **intrauterine insemination (IUI)**, donor sperm are placed into a woman's reproductive tract in a clinical setting.
6. A gestational and genetic surrogate mother provides her oocyte. Then IUI introduces sperm from a man whose partner cannot conceive or carry a fetus. The surrogate provides her uterus for 9 months. A gestational-only surrogate mother receives a fertilized ovum that resulted from *in vitro* **fertilization (IVF)** of a secondary oocyte by a sperm that came from the couple who asked her to carry the fetus. She is not genetically related to the fetus.
7. In IVF, oocytes and sperm join in a dish, fertilized ova divide a few times, and embryos are placed in the woman's body, circumventing blocked tubes or malfunctioning sperm. **Intracytoplasmic sperm injection (ICSI)** introduces immature or nonmotile sperm into oocytes.
8. Embryos can be frozen and thawed and then complete development when placed in a uterus.
9. **Gamete intrafallopian transfer (GIFT)** introduces oocytes and sperm into a uterine tube past a blockage; fertilization occurs in the woman's body. **Zygote intrafallopian transfer (ZIFT)** places an early embryo in a uterine tube.
10. Oocytes can be frozen and stored. In embryo donation, a woman undergoes IUI. A week later, the embryo is washed out of her uterus and put into the reproductive tract of the woman whose partner donated the sperm.
11. **Preimplantation genetic diagnosis (PGD)** checks chromosomes and gene variants in one cell from a cleavage embryo or blastocyst. The remaining embryo is transferred to a uterus if there are no detected abnormal chromosomes or mutations.
12. **Sequential polar body analysis** infers absence of a mutation in a fertilized ovum by checking its associated second polar body.

23.4 Extra Embryos

13. Extra fertilized ova and early embryos generated in IVF are used, donated to couples, stored, donated for research, or discarded.
14. Donated embryos enable researchers to study aspects of early human development that they could not investigate in other ways.

Key Terms

assisted reproductive technologies (ARTs) Procedures that replace a gamete or provide a uterus to help people with infertility have children.

gamete intrafallopian transfer (GIFT) (găm′ēt ĭntrə-fə-lōpē-ən) An infertility treatment in which sperm and oocytes are placed in a woman's uterine tube.

***in vitro* fertilization (IVF)** (ĭn vē′trō) Placing oocytes and sperm in a laboratory dish with appropriate biochemicals so that fertilization occurs, then, after a few cell divisions, transferring the embryo(s) to a woman's uterus.

infertility The inability to conceive a child after a year of frequent unprotected intercourse.

intracytoplasmic sperm injection (ICSI) (ĭn′trə-sītō-plăzmĭk) An infertility treatment that injects a sperm cell nucleus into an oocyte, to overcome a low sperm count or many abnormally shaped or non-motile sperm.

intrauterine insemination (IUI) (ĭn′trə-yōō′tər-ĭn) An infertility treatment that places donor sperm in the cervix or uterus.

preimplantation genetic diagnosis (PGD) (jə-nĕt′ĭk) Removing a cell from a cleavage stage embryo or blastocyst and testing it for a mutation to deduce the genotype of the embryo.

sequential polar body analysis Testing the DNA of a second polar body to infer the genotype of its associated fertilized ovum.

zygote intrafallopian transfer (ZIFT) (zī′ gōt ĭn′trə-fə-lō′ pē-ən) An assisted reproductive technology in which an ovum fertilized *in vitro* is placed in a woman's uterine tube.

Review Questions

1. List three facts about early human prenatal development that were learned using ARTs.
2. Distinguish among azoospermia, oligospermia, and globozoospermia.
3. Why are men typically tested for infertility before women?
4. Cite a situation in which both man and woman contribute to subfertility.
5. What are causes of infertility among older women?
6. Which ARTs might help the following couples? (More than one answer may fit some situations.)
   a. A woman is born without a uterus but manufactures healthy oocytes.
   b. A man has cancer treatments that damage his sperm.

c. A genetic test reveals that a woman will develop Huntington disease. She wants a child but does not want to pass on the disease.

d. Two women wish to have and raise a child together.

e. A man and woman are carriers of sickle cell disease. They do not want to have an affected child or terminate a pregnancy to avoid the birth of an affected child.

f. A woman's uterine tubes are scarred and blocked.

g. A young woman must have radiation and chemotherapy to treat ovarian cancer but wishes to have a child.

7. How does ZIFT differ from GIFT and IVF?

8. Explain how PGD works, and list two events in early prenatal development that might explain cases in which the PGD result is inaccurate.

9. How do each of the following ARTs deviate from the normal biological process?

   a. IVF
   b. GIFT
   c. embryo donation
   d. gestational-only surrogacy
   e. intrauterine insemination
   f. cytoplasmic donation

10. Explain how sequential polar body analysis is based on Mendel's first law.

Applied Questions

1. A company that offers PGD terms an embryo "normal and healthy" if it has 46 chromosomes. Why is this statement misleading?

2. Two famous male actors had twins using a surrogate mother who carried two embryos that had been fertilized *in vitro,* one with one man's sperm and the other with the other man's sperm. In terms of genetics, how closely are the babies, a boy and a girl, related to each other if they have different fathers?

3. ICSI was developed to help men who could not ejaculate but who produce viable sperm to become fathers. Similarly, oocyte freezing was developed to help women undergoing toxic cancer treatments preserve their fertility. Why are both ICSI and oocyte freezing now done on healthy men and women?

4. An embryo bank offers IVF leftover embryos, which would otherwise remain in the deep freeze or be discarded, to people wanting to have children, for $2,500 each. The bank circumvents bioethical concerns by claiming that it sells a service, not embryos. People in favor of the bank claim that purchasing an embryo is not different from paying for sperm or eggs, or an adopted child. Those who object to the bank claim that it makes an embryo a commodity.

   a. Do you think that an embryo bank is a good idea, or is it unethical?
   b. Whose rights are involved in the operation of the embryo bank?
   c. Who should be liable if a child that develops from the embryo has an inherited disease?
   d. Is the bank elitist because the cost is so high?

5. Many people spend thousands of dollars pursuing pregnancy. What is an alternative solution to their quest for parenthood?

6. Big Tom is a bull with valuable genetic traits whose sperm is used to conceive more than 1,000 calves. Which application of an ART in humans parallels this situation?

7. State who the genetic parents are and who the gestational mother is in each of the following cases:

   a. A man was exposed to unknown burning chemicals while serving in the U.S. military in Iraq, and abused drugs for several years before and after that. Now he wants to become a father, but he is concerned that past exposures to toxins have damaged his sperm. His wife undergoes IUI using sperm from the husband's brother, who has led a calmer and healthier life.

   b. A 26-year-old woman has her uterus removed because of cancer. However, her ovaries are intact and her oocytes are healthy. She has oocytes removed and fertilized *in vitro* with her husband's sperm. Two resulting embryos are implanted into the uterus of the woman's best friend.

   c. Max and Tina had a child by IVF and froze three extra embryos. Two are thawed years later and implanted into the uterus of Tina's sister, Karen. Karen's uterus is healthy, but she has ovarian cysts that often prevent her from ovulating.

   d. Forty-year-old Christopher wanted children, but did not want a partner. He donated sperm, which were used for IUI of a woman who carried the fetus to term for a fee, and gave birth to a daughter. Christopher adopted the child.

8. Describe a scenario in which use of an ART and DNA testing can change an individual's view of her or his origins.

Case Studies and Research Results

1. Suggest an experiment to test the effect of a substance or behavior on female or male fertility, using a study design like the one used to assess the effect of air pollution on ovarian reserve.

2. Do you agree or disagree with the conclusion of bioethicists that the family with Li-Fraumeni syndrome has a duty to warn the people who used their extra embryos?

3. Doola is 28 years old and is trying to decide if she and her husband are ready for parenthood when she learns that her 48-year-old mother has Alzheimer disease. The mother's physician tells Doola that because of the early onset, the Alzheimer disease could be inherited through a susceptibility gene. Doola is tested and indeed has the same dominant allele. She wants to have a child right away, so that she can enjoy many years as a mother. Her husband David feels that it wouldn't be fair to have a child knowing that Alzheimer disease likely lies in Doola's future. He suggests adoption.

   a. Which technology might help them avoid passing the Alzheimer disease susceptibility gene variant to their child?
   b. Is Doola correct in assuming that she is destined to develop Alzheimer disease?

4. Two large, well-known companies offered female employees a $20,000 benefit to freeze their eggs. Why might the companies have done this, and why might many people have objected to the offer?

Index

Notes: Page numbers followed by *f* or *t* indicate figures and tables; page numbers followed by *cc* indicate material in Clinical Connections.

A

AAV (adeno-associated virus), 425
Abandoned DNA, 178
ABCC9 gene, 419
Abelson oncogene *(abl)*, 385
Abnormal immunity, 356–361, 369
    AIDS, 97, 357–359, 358*f*, 359*t*, 401
    allergies, 360–361, 360*f*
    autoimmunity, 359–360, 359*t*, 372
    inherited immune deficiencies, 356–357, 357*f*, 357*t*
ABO blood group system, 348
    codominance in, 93–94, 94*f*, 98*t*
    effects of migration on, 290
Accidental exposures to mutagens, 226
Accutane, teratogenicity of, 62, 63
ACE (angiotensin-1-converting enzyme) gene, 421
ACE2 receptor, 349*f*
Acentric chromatids, 258*f*, 260*t*
Acetyl groups (CH_3CO_2), 207
Achondroplasia, 53*t*, 92, 98*t*, 224
Achromatopsia, 293
Acne inversa, 223*t*
Acne medicine, teratogenicity of, 63
Acquired immune deficiency syndrome (AIDS), 357–359, 358*f*, 359*t*
    as phenocopy, 97
    use of EPO to treat, 401
Acrocentric chromosomes, 244, 245*f*, 261
Actinic keratosis, 312*t*
Actinin-3 muscle gene, 307
Actinin 3 protein, 130
ACTN3 (alpha-actinin 3) gene, 421
Acute lymphoblastic leukemia (ALL), 388–389
ADA (adenosine deaminase) deficiency, 357, 357*t*, 426, 426*f*, 428*cc*
Adaptive immunity, 351, 352*f*, 353–356
    cellular immune response, 355–356, 356*f*
    humoral immune response. *See* Humoral immunity
Addictive drugs, 153–155, 155*f*
Adenine (A), 4, 4*f*, 13, 170, 170*f*, 171, 172*f*, 187

Adeno-associated virus (AAV), 425
Adenosine deaminase (ADA) deficiency, 357, 357*t*, 426, 426*f*, 428*cc*
Adenosine triphosphate (ATP), 24, 101, 191
Adenovirus, 176*f*
ADHD (attention deficit hyperactivity disorder), 232
Admixture, 317, 318*f*
Adoption, 9
Adoption studies, 137–138
Adrenoleukodystrophy (ALD), 24, 428*cc*–429*cc*
Adult-onset inherited diseases, 64–65, 416
Adult(s), genetic testing of, 418*t*, 420
Adult stem cell, 36, 36*t*
Aedes aegypti (mosquito), 409
Affected sibling pair study, 141, 141*t*, 145
AFP (alpha fetoprotein), 60
Africa
    indigenous peoples of, 306, 307, 311, 327
    migration from, 318, 319*f*
African-Americans, Great Migration of, 321, 322*f*
Agarose gel electrophoresis, 330
Aging
    genes and longevity, 65–67, 67*t*, 68
    overview, 64
    "premature aging" disorders, 65, 65*t*, 66*f*
Agriculture, 10, 137, 313, 314*f*, 402
AHI1 gene, 324
AIDS. *See* Acquired immune deficiency syndrome (AIDS)
AKT1 gene mutation, 219, 219*f*
Albinism, 76, 77*f*, 85, 85*f*, 289
Alcohol, teratogenicity of, 63, 63*f*, 152
ALD (adrenoleukodystrophy), 428*cc*–429*cc*
Alexandra, Tsarina of Russia, 120
Alexei, Crown Prince of Russia, 120
Alkylating agents, as mutagens, 226
ALL (acute lymphoblastic leukemia), 388–389
Allele(s), 5, 7, 13, 73, 122
    dominant, 7, 14
    lethal. *See* Lethal alleles

    multiple, in cystic fibrosis, 92
    recessive, 7, 14, 294
Allele frequencies, 277–285, 286–302
    calculating, 278
    for *CFTR* gene, 278
    "designer babies," ethics of, 301, 301*f*
    "designer babies," ethics of, 301, 301*f*
    eugenics and, 300–301, 302
    genetic drift and, 291–294, 302
        founder effect. *See* Founder effect
        population bottlenecks, 293–294, 294*t*
    Hardy-Weinberg equilibrium. *See* Hardy-Weinberg equilibrium
    macroevolution, 279, 279*f*, 283
    microevolution, 279, 279*f*, 283
    migration and, 290–291, 302
    mutation and, 294–295, 295*f*, 302
    natural selection and, 295–300, 302
        antibiotic resistance, 297, 297*f*
        artificial selection, 296–297, 296*f*
        balanced polymorphism, 297–300, 298*t*
        negative and positive selection, 295–296, 296*f*
    nonrandom mating and, 288–290, 289*f*, 302
    population genetics, 278–279
    Steel syndrome, 287, 288*f*, 293, 302
Allelic diseases, 223, 223*t*, 238
Allergens, 360
Allergic reaction, 360, 360*f*
Allergies, 360–361, 360*f*, 403
Allison, Anthony, 299
Allograft, 365, 365*f*
Alpha-actinin 3 *(ACTN3)* gene, 421
Alpha fetoprotein (AFP), 60
Alpha (α) globin chains, 203, 204*f*, 221
Alpha globin gene complex, 212
Alpha globin genes, 225, 225*f*, 229
Alpha synuclein, 42
Alpha thalassemia, 225, 298
Alport syndrome, 221*t*
ALS (amyotrophic lateral sclerosis), 36, 197*t*, 212
Altai Neanderthals, 311, 313, 313*f*
Alternative splicing, 189, 199, 210, 210*f*
Alu sequence, 213

I-1

Alzheimer disease, 223*t*
　abnormal imprinting in, 126
　amyloid beta and, 65
　ApoE4 gene and, 269
　Down syndrome and, 249
　early-onset, 206
　familial, 96*cc*, 97*t*, 402
　misfolded proteins in, 197*t*
　single-gene mutations and, 65
　tracing causative mutation, 84–85
Amelogenin gene, 333
Amino acids, 62, 183, 183*f*, 198
Amino acid sequences, 195, 195*f*, 221
Amish, Old Order. *See* Old Order Amish; Plain Populations
Amnesia, immunological, 359
Amniocentesis, 57, 68, 245, 246*f*, 261, 418*t*
Amnion, 57*f*, 59, 59*f*
Amniotic fluid, 56, 57
Amphetamines, 153
Amygdala, 96*cc*
Amyloid beta, 65
Amyloid plaques, 96*cc*
Amyloid precursor protein (APP), 96*cc*, 206
Amyotrophic lateral sclerosis (ALS), 36, 197*t*, 212
Anal cancer, 384
Anandamide, 153
Anaphase, 30–31, 31*f*, 40
Anaphase I, 47*f*, 48
Anaphase II, 48, 48*f*
Anaphylactic shock, 360–361
Ancestry, race and, 134
AncestryDNA, 338
"Ancestry informative markers," 316
Ancestry testing, 418*t*
Anchor proteins, 173, 173*f*
Androgen insensitivity syndrome, 113
Anemia, 232*t*
Aneuploidy, 247–249, 247*t*, 262
　autosomal, 249–252, 249*t*, 250*f*, 251*f*
　　trisomy 13, 249*t*, 251–252
　　trisomy 18, 249*t*, 250–251, 251*f*
　　trisomy 21, 244*f*, 249–250, 249*t*, 250*f*, 251*f*, 252
　in sex chromosomes, 252–254, 252*t*
　　female, 252–253
　　　female infertility due to, 437, 437*f*
　　male, 253–254
Angelman syndrome, 126, 126*f*
Angiogenesis, 378
Angiotensin-1-converting enzyme *(ACE)* gene, 421
Anhidrotic ectodermal dysplasia, 122*f*
Animal(s). *See specific animals and species*
Animal models
　of familial Alzheimer disease, 402
　in genome sequencing, 426–427, 429
　of Huntington disease, 402, 402*f*
　in recombinant DNA technology, 402, 402*f*
Ankylosing spondylitis, 350
Anne, Queen of Romania, 101
Annotation, 271–272
Anopheles gambiae (mosquito), 298
Anorexia nervosa, 147
Antennapedia, 58*cc*
Antibiotic resistance, 297, 297*f*, 368
Antibodies, 348
　autoantibodies, 359
　immunoglobulins, 354*t*, 435
　monoclonal, humoral immunity and, 364–365, 364*f*
　structure of, 354, 354*f*, 355
　types of, 354, 354*f*, 354*t*
"Anticipation," 230
Anticodon, 186, 186*f*, 192, 199
Antigen binding sites, 354, 354*f*
Antigen-presenting cells, 350, 350*f*
Antigen receptors, chimeric (CARs), 429
Antigen receptors T cell therapy, chimeric (CAR T), 389–390, 392
Antigen(s), 93, 107, 348, 350, 360, 365
Anti-HIV drugs, 358–359, 359*t*
Antiparallelism, 170–171, 171*f*, 172*f*, 179
Antisense technology, 406
Anxiety disorders, 150*t*
Aortic aneurysm, 10, 223, 427
APOE (apolipoprotein E), 96*cc*, 97*t*
Apolipoprotein(s), 132*cc*, 132*t*
Apolipoprotein E *(APOE)*, 96*cc*, 97*t*
Apoptosis, 28, 29*f*, 32–33, 33*f*, 39, 435
Apoptosis checkpoint, 31, 31*f*, 376*f*
APP (amyloid precursor protein), 96*cc*, 97*t*, 206
Arawete people (Brazil), 306
Archaea, 18
Archaic humans, 310–313
　Denisovans, 310, 310*f*, 312–313, 323
　Neanderthals, 310–312, 311*f*, 312*t*
"Ardi," 307
Ardipithecus kadabba, 307, 308*f*
Ardipithecus ramidus, 307, 308*f*
Aristotle, 63
Artificial selection (dog breeding), 296–297, 296*f*
Ashkenazi Jewish people, 338
　BRCA1 gene mutation in, 237, 293
　Dor Yeshorim program, 290
　genetic disease among, 294, 294*t*
　infertility among, 436*cc*, 436*f*
　mitogenomes of, 317
　Tay-Sachs disease in, 283, 284
A (acceptor) site, 192
ASPA gene, 294*t*
Asperger, Hans, 161
Asperger syndrome, 159
Assisted reproductive technologies (ARTs), 434, 438–444, 444*t*, 446
　"extra" embryos from, 444–445, 446
　gamete intrafallopian transfer (GIFT), 440–441, 444*t*
　intrauterine insemination (IUI), 438
　oocyte banking and donation, 441–443
　PGD. *See* Preimplantation genetic diagnosis
　postmortem gamete retrieval, 442
　sequential polar body analysis, 444, 444*f*
　surrogate motherhood, 438
　in vitro fertilization (IVF), 439–440, 439*f*, 440*f*
　zygote intrafallopian transfer (ZIFT), 440–441
Association, 141
Asthma, 360
ASXL3 gene, 86
AT (ataxia telangiectasia), 236
Ataxia telangiectasia (AT), 236
Atherosclerosis, 132*cc*
Athletic ability, 130–131, 421–422, 421*t*
Atopic dermatitis, 361
ATP (adenosine triphosphate), 24, 101, 191
ATP7A gene, 223*t*
Attention deficit hyperactivity disorder (ADHD), 232
Auriculocondylar syndrome, 70
Australian Rett Syndrome Study, 17
Australopithecines, 307–309, 308*f*, 324
Australopithecus afarensis, 308, 308*f*
Australopithecus anamensis, 308, 308*f*
Australopithecus garhi, 308*f*, 309
Autism, 126, 159–161, 159*t*, 160*f*, 162*f*, 163, 232
　exome sequencing, 159, 160*t*
　famous people with, 159*t*
　finding fault for, 161
　"multiple unrelated affected subjects," 159, 163
　prevalence, 150*t*
　recessive analysis, 160, 160*f*, 163
　trio analysis, 160, 160*f*, 163
　understanding, 161, 162*f*
Autism spectrum disorders, 150*t*
Autoantibodies, 359–360
Autograft, 365, 365*f*
Autoimmune disorders, 359*t*
Autoimmunity, 359–360, 359*t*, 372

Autophagy, 23, 39
Autosomal aneuploidy
　trisomy 13, 249*t,* 251–252
　trisomy 18, 249*t,* 250–251, 251*f*
　trisomy 21, 244*f,* 249–250, 249*t,* 250*f,* 251*f,* 252
Autosomal dominant disorders, 58*cc,* 72, 87
　Huntington disease. *See* Huntington disease (HD)
　pedigree for, 85, 85*f*
Autosomal dominant inheritance, 77, 77*f,* 77*t,* 87
Autosomal genes, mutation of, 224
Autosomal recessive disorders, 72, 87
　autism in, 160*f*
　of *COL27A1* gene, 287
　cystic fibrosis. *See* Cystic fibrosis (CF)
　pedigree for, 85, 85*f*
　risk of being carrier for, 282, 282*f*
Autosomal recessive inheritance, 77–78, 77*t,* 79*cc*
Autosomes, 6, 13
Avery, Oswald, 167, 167*f,* 169*t*
Azoospermia, 435

B

Baboons, 366, 366*f*
BabySeq, 274
BabySeq project, 11
BAC (bacterial artificial chromosome), 269*f*
Bacteria, 18, 19, 38, 79*cc,* 188. *See also specific bacteria*
Bacterial artificial chromosome (BAC), 269*f*
Bacteriophages, 399
Bainbridge-Ropers syndrome, 11, 86
Balanced polymorphism, 297–300, 298*t*
　cystic fibrosis and diarrheal disease, 299–300
　geographical distribution of, 298*f*
　natural selection and, 297–300, 298*t*
　sickle cell disease and malaria, 298–299, 298*f*
Baldness allele, 122
Bantu people (Africa), 307, 311
Bardet-Biedl syndrome, 27
Barr bodies, 122*f,* 123, 250, 251*f*
Base excision repair, 234, 238
Bateson, William, 74, 101, 103
B cells, 351, 351*f,* 353, 353*f,* 356*t*
BCL11A switch gene, 429
bcr (breakpoint cluster region), 385
BCR-ABL oncoprotein, 385
Beard growth, 121–122

Beaudet, Arthur, 260
Becker muscular dystrophy, 223*t,* 227
Beckwith-Wiedemann syndrome, 125
Behavioral disorders. *See also specific disorders*
　chromosomal, 254
　prevalence of, 150, 150*t*
　PTSD (posttraumatic stress disorder), 148–149
Behavioral genetics, 148–164
　autism, 126, 159–161, 159*t,* 160*f,* 162*f,* 232
　drug addiction, 152–155, 162
　　addictive drugs, 153–155, 155*f*
　　cannabis, 153, 154*cc*
　　substance use disorders, 152–153, 153*f*
　intelligence, 151–152, 152*f,* 162
　mood disorders, 150*t,* 155–156, 162
　　bipolar disorder, 150*t,* 156
　　major depressive disorder, 150*t,* 155, 156*f*
　role of genes in behavior, 149–150, 149*f,* 150*t,* 162
　schizophrenia. *See* Schizophrenia
　sleep, 150–151, 151*f,* 162
Behavioral traits, multifactorial, 131
Benign tumors, 375
Beringia, 320
Bering Land Bridge, 320
"Berlin patient," 347
Beta amyloid protein, 65
Beta (β) globin chains, 203, 204*f*
Beta globin gene complex, 212
Beta globin gene mutation, 220–221, 220*f*
Beta thalassemia, 220–221
Bettelheim, Bruno, 161
Bifidobacterium breve, 38
"Bili lights," 292, 292*f*
Binet, Alfred, 151
Biochemicals, 18–19
Bioethics, 3, 12, 13, 337, 342, 400
　autism, 161
　autism, finding fault for, 161
　bioethics of "infidelity testing," 178
　of cloning, 55
　of "designer babies," 301, 301*f*
　of DNA profiling, 337
　of erythropoietin use, 401, 401*f*
　of gender selection, 443–444
　genetic "incidental" findings, 420
　genetic information, 87
　genomes, creating, 273
　of "infidelity testing," 178
　postmortem gamete retrieval, 442
　of preimplantation genetic diagnosis, 301

　privacy issues in genetic testing, 12–13
　replacement parts, 366–367, 366*f*
　stem cell tourism, 37
　transgender identity, 116
　trisomy 21 Down syndrome, 252
　variant of uncertain significance, 237
　of xenotransplants, 366, 366*f*
Bioinformatics, 339
　strategy, genetic testing, 277–278
Biological membranes, 25–26, 26*f*
　structure, 25–26, 26*f*
Biopsy(ies), 391*cc,* 391*f,* 443, 443*f*
Bioremediation, 403–404
Biotechnology, 396
　inventions, 397*t*
Biotinidase deficiency, 20
Bipolar disorder, 150*t,* 156
Birth defects, 62–64, 62*f,* 68
Bitter taste receptor *(TAW2R16)* gene, 307
Black rhino, 181
Blastocysts, 55, 57*t,* 68, 443, 443*f*
Blastomeres, 53, 68, 443
"Blender experiments," 167, 168*f*
Bleuler, Paul, 161
"Blighted ovum," 52
Blindness, 95, 95*f,* 101, 293
Blood, 132*cc,* 203, 204
Blood cells, 132*cc*
Blood clotting, 312*t*
Blood groups
　ABO system. *See* ABO blood group system
　cell surfaces and, 348, 348*t,* 350
Bloodsworth, Kirk, 330
Blood types
　Bombay phenotype, 94, 98*t*
　Duffy, 104
　in Hardy-Weinberg equilibrium, 290
　law of segregation in, 94*f*
　Lewis or MN blood group, 348*t*
　Rh, 104–105
Bloom syndrome, 294*t*
"Blue people of Troublesome Creek," 217–218
Blue person disease, 217–218
BMI (body mass index), 143, 143*f*
Body mass index (BMI), 143, 143*f*
Body weight, 143–144, 143*f,* 143*t,* 145
Bombay phenotype, 94, 98*t*
Bone marrow, 351*f*
Bone marrow transplants, 36–37
Bouchard, Thomas, 138
Bradfield, Golda, 386

BRAF gene mutation, 373
Brain
  abnormalities due to Zika virus, 64
  effects of drug addiction on, 153
  large size as human trait, 323*f*, 324
Brain tumors, 379, 379*f*
Branched tumor evolution, 383*f*
BRCA1 gene, 294*t*, 397
  genetic test, 272
  mutation of, 129, 237, 293
    breast cancer and, 129, 293, 294*t*, 377*f*, 387, 387*f*
    postmortem diagnosis of, 265
BRCA2 gene, 294*t*, 397
  breast cancer and, 237, 294*t*, 377*f*, 387, 387*f*
  genetic test, 272
  mutation of, 237
Breakpoint cluster region *(bcr)*, 385
Breakthrough stage, in mutation, 381, 382*f*, 382*t*
Breast cancer, 294*t*
  in Ashkenazi Jewish people, 237, 293, 387
  attempts to patent *BRCA1/BRCA2 genes*, 397
  BRCA1, 129, 237, 293, 294*t*, 377*f*, 387, 387*f*, 397
  BRCA2, 237, 294*t*, 377*f*, 387, 387*f*, 397
  HER2 cell surface protein in, 384
  polygenic risk score for, 394
  risk, 146
"Brittle bone disease" (osteogenesis imperfecta), 65, 95, 98*t*
Brown, Louise Joy, 439
Brown, Timothy, 347
Buck, Carrie, 300
Bulbourethral glands, 44*f*, 45
Bunker, Chang, 60
Bunker, Eng, 60
Burbank, Luther, 300
Burkholderia cepacia, 79*cc*
Burkitt lymphoma, 384, 384*f*
"Bushmen" (Africa), 307
"Bystander effect," 226

C

C4 genes, 157–158, 158*f*
CAH (congenital adrenal hyperplasia), 114, 114*f*
Calcium homeostasis modulator 1 *(CALHM1)*, 97*t*
Calico cats, 123, 123*f*
Campylobacter bacteria, 368

Canavan disease, 294*t*
Cancer. *See also specific types of cancer*
  cell cycle, disruption, 33
  at cellular level. *See* Cancer cells
  description of, 374–377, 374*f*, 376*f*, 392
  diagnosis and treatment of, 374*f*, 388–390, 394
    chimeric antigen receptors T cell therapy, 389–390, 392
    gene expression profiling, 388, 389*f*
  drug development, 385
  environmental causes of, 131, 131*f*, 374*f*, 386, 386*f*, 388
  genes and genomes, 381–388, 392
    BRCA1/BRCA2 mutations, 237, 293, 377*f*, 387, 387*f*
    cellular adhesion proteins, 386–387
    driver and passenger mutations, 381–382, 382*f*, 382*t*, 383*f*
    oncogenes, 382–385, 384*f*
    tumor suppressor genes, 385–387
  hereditary nonpolyposis colon cancer, 236
  inherited *v.* sporadic, 377, 377*f*
  loss of cell cycle control in, 375–376, 376*f*
  microRNAs in treatment of, 209
  somatic mutation and, 220
  treatment based on mutation, 373–374
Cancer cells, 377–381, 392
  characteristics of, 377–378
  mutation of, metastasis and, 381, 383*f*
  origins of, 378–381, 379*f*, 380*f*
Cancer stem cells, 378–379, 379*f*
Cannabidiol (CBD), 154
Cannabis (marijuana), 153, 154*cc*, 154*f*
Cannabis sativa, 154*cc*, 154*f*
Capacitation, 53
Capillary electrophoresis, 331, 334*f*
Carbohydrates, 19, 20, 39
Carbons, 170, 171*f*, 183, 183*f*
Carcinogens, 375
Cardiovascular disease
  aortic aneurysm in Marfan syndrome, 10, 223, 427
  Dravet syndrome, 154*cc*
  low birth weight and, 64
  as polygenic trait, 132*cc*, 132*t*
Carrier(s)
  of autosomal recessive disorders, 282, 282*t*
  of sickle cell disease, 80
  of Tay-Sachs disease, 78
  of X-linked traits, 123

Carrier frequency, calculating, 281–282, 282*t*
"Carrier" protein, 25
Carrier screening, 418*t*, 420
CARs (chimeric antigen receptors), 429
CAR T (chimeric antigen receptors T cell therapy), 389–390, 392
Case-control studies, 141, 141*t*
Caseins, 23
Caspases, 32, 33*f*
Cataplexy, 150
Cats, coat colors of, 123, 123*f*
CBD (cannabidiol), 154
CCR5 delta 32 gene, 347–348, 359
CCR5 gene, 348, 409
CCR5 protein, 218
CCR5 receptor, 347–348, 357–358, 358*f*
CD4 T cells, 355, 357, 358*f*
CD133 molecule, 379, 379*f*
cDNA (complementary DNA), 397, 404
Celera Genomics, 267, 268
Celiac disease, 213
Cell(s), 2, 13, 17–40. *See also specific cells and types of cells*
  in cancer. *See* Cancer cells
  cell division and death, 28–33
  components of, 18–28, 39
    biological membranes, 25–26, 26*f*
    chemical constituents, 18–19
    cytoskeleton, 26–28
    organelles, 18, 19, 21–24, 21*f*, 25*f*, 25*t*
  differentiation of cell types, 6
  human microbiome, 38–39
  inner cell mass, 36, 55
  overview of, 18, 18*t*, 19*f*
  stem cells, 33–37
  surfaces of, in immunity, 348–351, 348*f*, 369
    blood groups and, 348, 348*t*, 350
    major histocompatibillity complex and, 350–351, 350*f*
Cell banks, 37
Cell cycle, 28–33, 29*f*, 39
  checkpoints in, 375, 376*f*
  checkpoints of, 31–32, 31*f*
  defined, 28
  disruption of, in cancer, 33, 375–376, 376*f*
  interphase, 28–30, 29*f*
  mitosis, 29*f*, 30–31, 30*f*
Cell death, 32–33, 33*f*
Cell division, 28–33
  apoptosis, 28, 29*f*, 32–33, 33*f*
  cell cycle, 28–33, 29*f*
Cell fate, 374*f*, 375

Cell-free fetal DNA, 57, 246–247, 247f, 262
Cell-free tumor DNA (ctDNA), 391cc
Cell lineages, 33–35, 33f, 34f, 40
Cell-mediated immunity, 351f
Cell membrane. *See* Plasma membrane
Cell specialization, 34, 34f
Cell survival, 374f, 375
Cellular adhesion, 26, 39
Cellular adhesion proteins, 386–387
Cellular aging, 55
Cellular communication, 26
Cellular immunity, 355–356, 356f, 365
CENP-A (centromere protein A), 243–244
Centenarians, 66
Centimorgans, 103, 104t, 108, 341–342, 342f
"Central dogma," 184, 184f
Central precocious puberty, 124
Centrioles, 21f, 30, 40
Centromere(s), 29, 30f, 40, 213t
  in chromosome structure, 243–244, 244f
  position of, 244–245, 245f
Centromere protein A (CENP-A), 243–244
Centrosomes, 30, 40
Cerebrospinal fluid (CSF), 404
Cervical cancer, 382t, 384
Cervical mucus, 437t, 437f
Cervix, 45, 45f
CF. *See* Cystic fibrosis (CF)
CF genotype, 95
CFTR gene. *See* Cystic fibrosis transmembrane regulator *(CFTR)* gene
CFTR protein, 197, 299–300
CGH (comparative genomic hybridization), 255
Chain termination, 177
"Channelopathies," 26
Chaperone proteins, 193f, 196, 199
Charcot-Marie-Tooth disease, 229
Chargaff, Erwin, 168, 169t
Chase, Martha, 167, 168f, 169t
Checkpoints, cell cycle, 31–32, 31f, 375, 376f
Chelators, 221
Chemical constituents, cells, 18–19
Chemicals, as mutagens, 226, 227
Chemical senses, 324
Chernobyl disaster, 226
Chickenpox virus, 64
Children, genetic testing of, 418t, 419
Chimeric antigen receptors (CARs), 429

Chimeric antigen receptors T cell therapy (CAR T), 389–390, 392
Chimpanzees, 8
China, one-child policy, 111–112
Cholera, 298t, 299
Chondrodysplasia, 221
Chorion, 57f, 59, 59f
Chorionic villi, formation, 56
Chorionic villus sampling (CVS), 57, 68, 245, 246f, 261, 418t
Chromatids, 30, 30f, 40
Chromatin, 172–173, 179, 208f
Chromatin loops, 173, 173f
Chromatin remodeling, 207–208, 207f–208f, 214
Chromosomal microarray analysis (CMA), 419, 432
Chromosomal mosaicism, 250
Chromosomal sex, 115t
Chromosomal shorthand, 247, 247f, 247t
Chromosome(s), 5–6, 13, 242–265. *See also* Sex chromosomes; *specific chromosomes*
  abnormalities in cancer cells, 381
  age effects on, 52
  atypical numbers of, 247–254, 247t, 260t, 261
    aneuploidy, 247–249, 247t
    polyploidy, 247, 247f, 247t
  atypical structure of, 254–260, 260t, 262
    deletions and duplications, 254–255, 255f
    inversions, 258–259, 258f, 259f
    isochromosomes and ring chromosomes, 259–260, 260f
    translocation Down syndrome, 255–258
  banding patterns, 315
  centromeres of, 243–244, 244f
  chromothripsis, 242–243, 248f, 261t, 263
  comparison of, 315–316
  detecting, 245–246, 262–263
    by amniocentesis, 245, 246f
    by cell-free fetal DNA testing, 246–247, 247f
    chorionic villus sampling, 245, 246f
    chromosomal shorthand, 247, 247f, 247t
    fluorescence *in situ* hybridization, 245–246, 246f
    indirectly, 245, 246f
  isochromosomes, 259, 259f
  karyotype (chart) of, 244–245, 244f, 245f

  number of, 247
  telomeres of, 243, 243f
Chromosome 7, 247f
Chromosome 21, 244, 244f. *See also* Trisomy 21
Chromosome 22, 244
Chromosome microarray analysis, 418t
Chromothripsis, 242–243, 248f, 261t, 262
Chronic granulomatous disease, 356–357, 357t
Chronic myelogenous leukemia (CML), 385
Cigarettes, teratogenicity of, 63
Cilia, 26–27, 27f
Ciliopathy, 27
Circadian clock genes, 312t
"Circadian pacemaker," 151
Circadian rhythms, 151
cis configuration, 103, 103f, 104
CJD (Creutzfeldt-Jakob disease), 198
Cleavage of zygote, 53, 54f, 57t, 68
Cleft lip, empiric risk of, 135, 135t, 136f
CLIA (Clinical Laboratory Improvement Amendments), 421
Clines, 290–291, 313
Clinical Connection
  cannabis (marijuana), 153, 154cc, 154f
  cystic fibrosis, 79cc
  DNA profiling, forensic applications of, 337
  familial Alzheimer disease, 96cc, 97t
  founder effect in Plain Populations, 292cc, 292f
  fragile X syndrome, 231cc
  globozoospermia, 436cc, 436f
  heart health, 132cc, 132t
  history of gene therapy, 428cc–429cc, 428f
  homeotic mutations, 56, 58cc
  inborn errors of metabolism, 20cc
  inheritance of color vision, 118cc–119cc, 118f
  liquid biopsies, 391cc, 391f
  Robertsonian translocations, 255–257, 256f, 257cc
  viruses, 349cc, 349f
Clinical depression, 155, 156f
Clinical Laboratory Improvement Amendments (CLIA), 421
"Clinically silent" mutations, 232t
Clinic for Special Children, 292cc
Clinton, Bill, 268
Clitoris, 45, 45f
"Clock" genes, 151, 162, 312t

"Clone-by-clone" sequencing technique, 268
Cloning, 36, 55, 266, 267f, 330
Cloning vectors, 398, 399
Clostridium bacteria, 368
Clostridium difficile, 38
Clustered regularly interspaced short palindromic repeats (CRISPRs), 408, 409, 409t, 410
CMA (chromosomal microarray analysis), 419, 432
CML (chronic myelogenous leukemia), 385
CNTNAP2 gene mutation, 292cc
CNVs. *See* Copy number variants (CNVs)
Cocaine, 152, 153
cocaine, teratogenicity of, 63
Coding strand, of DNA, 184, 184f, 199
CODIS (Combined DNA Index System), 330–331, 331f, 337, 338
Codominance, 93–94, 94f, 98t, 107
Codon, 185, 189, 199
Coefficient of relatedness, 137, 137t, 143t, 145
Cohort studies, 141, 141t
Collagen disorders, 221–223, 221t, 222f, 238, 287
Collectins, 352, 352t
College students, genetic testing of, 12
Collins, Francis, 268
Colony-stimulating factors, 352t, 365
Colorblindness, 117, 118cc–119cc
Colorectal cancer, 382t
Columbus, Christopher, 11, 299
Combined DNA Index System (CODIS), 330–331, 331f, 337, 338
"Community-based participatory research," 327
Comparative genomic hybridization (CGH), 255
Complementary base pairing, 171, 172f, 177, 179
  in gene silencing, 406–407, 406f, 407f
  in genome editing, 406, 407–408, 408f, 408t, 409t
Complementary DNA (cDNA), 397, 404
Complement system, 352, 352t
Complete dominance, 92
Complex traits, 6, 7f, 13, 130–147
  athletic ability, 130–131
  body weight, 143–144, 143f, 143t, 145
  defined, 131
  genes and environment in, 131, 131f, 145
  heart health, 131, 132cc, 132f, 132t

investigation of, 135–142, 145
  in adopted individuals, 137–138
  empiric risk, 135–136, 135t, 136f, 145
  GWAS, 139–142, 140f, 140t, 141f, 141t, 142f, 143t
  heritability, 136–137, 136f, 136t, 137f, 137t, 145
  twin studies, 138–139, 138t, 139f
polygenic traits, 133–135, 145
  fingerprint patterns, 133, 133f
  height, 133–134, 134f
  skin color and race, 134–135, 134f
Compound heterozygote, 74, 87, 92
Comprehensive carrier testing, 418t
Concordance, 138, 138t, 142, 143t, 145, 157
Conditional mutation, 233, 238
Conditional probability, 85, 86f
Conformation, 183, 185, 199
Congenital myasthenic syndrome, 419
Congenital skin crease Kunze type, 42
Conjoined (Siamese) twins, 60, 60f
Connective tissues, 18, 18t
Consanguinity, 78, 88, 290
  CYP5R3 gene mutation and, 217–218, 240
  5-alpha reductase deficiency and, 114
  globozoospermia and, 436cc, 436f
  inadvertent, 438
Conservation genetics, 9
Conservative replication, 174
Constant regions, 354, 354f
Contact inhibition, 377
Continuous strand synthesis, 175f
Cooley, Thomas, 220
Cooley's anemia, 220
Copy number variants (CNVs), 137, 232, 268, 329
  dietary patterns and, 295
  as genetic markers, 316
  in GWAS studies, 139, 139t, 140f, 145
  mutation caused by, 230, 232, 238
  repeats of, 294
Cord blood banks, 37
Cord blood transplants, 37
Core microbiome, 38
Corona radiata, 53, 54f
Coronavirus, 349cc, 349f
Correlation, 141
Coumadin (Warfarin), 422, 423t
Coverage, 271
COVID-19, 349cc, 349f
Cowpox, 363
Creutzfeldt-Jakob disease (CJD), 198
Crick, Francis, 166, 169, 169f, 169t, 174, 184, 190, 201

Crigler-Najjar syndrome, 292, 292f
CRISPR-Cas9, 10, 397
  as diagnostic tool, 426–427, 429, 430
  as restriction enzyme, 408, 408f
  in treatment of genetic disease, 426–427, 429
CRISPRs (clustered regularly interspaced short palindromic repeats), 408, 409, 409t, 410
Cristae, 24, 25f
Critical period, 62, 62f, 68
Crohn's disease, 359
Crossing over, 47, 47f, 67
  in admixture, 317, 318f
  frequency of, 104
  linkage and, 103, 103f
  new view of, 274, 274f
Crouzan syndrome, 53t
Crowdfunding, 37
"Crowd surfing," 27
Cryopreservation, 440, 441
Cryptorchidism, 129
CSF (cerebrospinal fluid), 404
CTCF anchor protein, 173, 173f
ctDNA (cell-free tumor DNA), 391cc
CTG repeats, in myotonic dystrophy, 230
CVS (chorionic villus sampling), 57, 68, 245, 261, 418t
CXCR4 gene mutation, 242–243
CXCR4 receptor, 358
Cyanosis, 232
Cyclins, 32
CYP2C9 gene, 422
CYP2D6 gene, 423t
CYP5R3 gene, 217
CYP5R3 gene mutation, 217–218
Cystic fibrosis (CF), 1, 222t
  autosomal recessiveness of, 72, 77, 79cc
  carrier frequency, 282, 282f, 282t, 283f
  classification of, 223
  F508del-CFTR mutation, 277
  gene mutation in, 5, 5f, 6f
  genetic testing, 277–278
  haplotypes in, 107, 107f
  intron mutation in, 228
  multiple alleles, 92, 98t
  PGD used to avoid, 443
  protection against diarrheal diseases, 299–300
  protein misfolding in, 183, 197
  single-gene inheritance, 72, 78, 79cc
Cystic Fibrosis Foundation, 79cc
Cystic fibrosis transmembrane regulator *(CFTR)* gene, 4–5, 5f, 6f, 79cc, 247

allele frequency for, 278
mutation, 222*t*, 277
as somatic gene therapy, 425
Cytochrome b_5 methemoglobin reductase, 217
Cytochrome *c*, 316
Cytochrome P450, 422–423, 423*t*
Cytogenetic map, 266
Cytogenetics, 243, 261
Cytokines
  cellular immunity and, 365
  in immune response, 352, 352*t*, 353, 355
Cytokinesis, 28, 31*f*, 40
Cytoplasm, 19, 21*f*, 39
Cytoplasmic donation, 442
Cytosine (C), 4, 4*f*, 13, 170, 170*f*, 171, 172*f*
Cytoskeleton, 26–27, 26–28, 39
Cytosol, 19
Cytotoxic T cells, 355, 356, 356*f*, 356*t*

D

"Daka" fossil, 309
Dalton, John, 118*cc*
Darwin, Charles, 159*t*, 287, 288*f*, 296, 300, 318, 324
Daughter cells, 35
Davenport, Charles, 300
Deacetylases, 207
DeAngelo, Joseph James, 343
"Death receptor," 32
Dedifferentiation, in cancer cells, 378
Deinococcus radiodurans, 234
Deletion mutation, 229, 238
Deletions, 227*t*, 229
  in chromosomes, 247*t*, 254–255, 255*f*
  indels, 322
  insertion/deletion polymorphism, 421
  in Y chromosome, infertility due to, 435
Dendritic cells, 350, 350*f*, 356*t*, 361
Dengue virus, 349*f*
"Denise" (Denisovan), 312
Denisova Cave (Siberia), 312
Denisovans *(Homo sapiens denisova)*, 310, 310*f*, 312–313, 323
De novo mutations, 72, 88, 274
  deletions and duplications, chromosomal, 255
  inversion, 258–259
  ring chromosomes, 260
  translocation, 257, 258
  transmission of, 86
"Density shift" experiments, 174

Dentin, 211
Dentinogenesis imperfecta, 211, 211*f*
Dentin phosphoprotein (DPP), 211, 211*f*
Dentin sialophosphoprotein (DSPP), 211, 211*f*
Dentin sialoprotein (DSP), 211, 211*f*
Deoxyribonuclease (DNase), 167, 167*f*
Deoxyribonucleic acid. *See* DNA (deoxyribonucleic acid)
Deoxyribose, 167–168, 179
Deoxyribose sugar, 170, 171*f*, 172*f*
Dependence, in drug addiction, 152
Depression, 155, 156*f*, 312*t*
Dermatoglyphics, 133, 133*f*
"Designer babies," bioethics of, 301, 301*f*
DeSilva, Ashanthi ("Ashi"), 428*cc*
"Detox" genes, 403
Deuteranopia, 118*cc*
Developmental timetables, human, 313
DHT (dihydrotestosterone), 113, 114, 114*f*
Diabetes mellitus
  abnormal imprinting in, 126
  as polygenic trait, 131
  type 1, as autoimmune disorder, 359*t*
Diagnosis. *See also specific diseases*
  CRISPR-Cas9 in, 426–427, 429, 430
  exome sequencing in, 421
  genetic counseling and, 417
  genome sequencing in, 421
  infertility tests, 437
  karyotyping in, 245
  PGD. *See* Preimplantation genetic diagnosis (PGD)
  prenatal, noninvasive, 246, 418*t*
Diagnostic odyssey, 10, 11
Diarrheal disease, cystic fibrosis and, 299–300
Dicentric chromatids, 258*f*, 260*t*
Dicephalic twins, 60, 60*f*
Dicer enzyme, 407
Dickinson, Emily, 159*t*
Dietary patterns, 295, 388
Differentiation, 6, 13
DiGeorge syndrome, 58*cc*
Digital karyotype, 263
Dihybrid cross, 81, 82*f*, 88, 102*f*, 103
Dihybrid(s), 81
Dihydrotestosterone (DHT), 113, 114, 114*f*
"Dikika infant," 309
Diploid cells, 18, 39, 45, 48*f*
Discontinuous strand synthesis, 175, 175*f*
Disease. *See also specific diseases and disorders*

adult-onset inherited, 64–65, 416
allelic diseases, 223, 223*t*
diagnosing with genome sequencing, 421
faulty ion channels, 26
genetic. *See* Genetic disease
genome sequencing, 11
identifying with exome sequencing, 10
imprinting diseases, 125–126, 126*f*
inborn errors of metabolism, 20
infectious. *See* Infectious disease(s)
lysosomal storage disease, 23
mitochondrial diseases, 99, 100
precision medicine approach to, 9–10
prion diseases, 197–198, 197*f*, 197*t*, 198*f*
single-gene. *See* Single-gene disorders
stem cells and, 36–37
Diseasome, 205–206, 206*f*
Dispersive replication, 174
Dizygotic (DZ) twins, 57, 68, 138–139, 138*t*, 142
DMD. *See* Duchenne muscular dystrophy (DMD)
DMD gene, 223*t*
DNA (deoxyribonucleic acid), 1, 2–6, 13, 165–181
  background, 336
  A and B forms, 168, 169*f*
  complementary, 397, 404
  fetal ("cell free"), 57
  identification and description of, 166–170, 179
  as hereditary molecule, 166–167, 166*f*, 167*f*
  protein distinguished from, 167, 168*f*
  structure, 167–170, 169*f*, 169*t*
  mitochondrial. *See* Mitochondrial DNA (mtDNA)
  placental, tests of, 245
  recombinant. *See* Recombinant DNA technology
  replication of, 4, 174–177, 179
    polymerase chain reaction, 175–177, 176*f*
    semiconservative, 174, 174*f*
    steps in, 174–175, 174*f*–175*f*, 176*f*
  RNA distinguished from, 168, 184–185, 184*t*, 185*f*
  secondary transfer, 337
  sequencing, 177–179, 177*f*, 178*f*
    chimpanzees, 8
    chromosome structure, 243
    Neanderthals, 311
    premise behind, 307

DNA (deoxyribonucleic acid), (Continued)
  spontaneous mutation in, 224, 224f
  structure of, 170–173, 179
    complementary base pairing, 171, 172f
    discovery of, 167–170, 169f, 169t
    nucleotide building blocks, 170, 170f
    nucleus, configuration in, 172–173, 173f
    polynucleotide chains, 170–171, 171f, 172f
  transcription of, 4
  translation of, 4
DNA amplification, 176, 176f
DNA analysis, 9, 9f
"DNA binding" domain, 408
DNA-cutting enzymes, 273
DNA damage checkpoint, 31, 31f, 376f
"DNA damage response" genes, 233
DNA dragnets, 337
DNA microarrays, 140f, 268, 404, 405f
DNAP (DNA polymerase), 174–175
DNA phenotyping, 336, 336f
DNA polymerase (DNAP), 174–175, 179, 233
DNA primer, 177
DNA profiling, 8–9, 14, 329–338, 330f
  accuracy and meaning in, 333
  bioethics of, 337
  forensic applications of. See Forensic science
  genetic privacy, 337–338
  history of, 330
  interpretation of profiles, 332–333, 335f
  postconviction testing, 328–329
  technology of, 330–332, 330t, 331f, 332f, 333f, 334f
  victim identification, 333–336
DNA repair, 233–236, 238
  BRCA1 mutations, 387
  disorders, 235–236
  longevity and, 66
  types, 234–235, 234f, 235f
DNA repair genes, mismatch mutations in, 375
DNA replication, 4, 13
DNase (deoxyribonuclease), 167, 167f
DNA sequences, 329
  chimpanzees, 8
  "identical by descent," 287
  palindromic, 113, 399
DNA sequencing
  chimpanzees, 8

derivation of, 268, 268f
  of dystrophin gene, 315, 315f
  of Y chromosome, 316–317, 317f
DNA technologies, 395–410, 395–411
  DNA profiling, 330–332, 330t, 331f, 332f, 333f, 334f
  gene expression profiling, 367
    in cancer diagnosis and treatment, 389, 389f
    in spinal cord injury, 404, 405f, 406t
  gene silencing, 406–407, 406f, 407f
  genome editing, 406, 407–408, 408f, 408t, 409t
  patenting issue, 396–397, 396f
  recombinant DNA. See Recombinant DNA technology
DNA template strand, 187, 188f
DNA testing, 272–273
Dogs
  artificial selection of, 296–297, 296f
  cryptorchidism, 129
  muscular dystrophy in, 315, 315f
  study of narcolepsy in, 150–151, 151f
Dolly the cloned sheep, 36, 330
Dominance
  autosomal dominant inheritance, 58cc, 72, 77, 77f, 77t
  codominance, 93–94, 94f, 98t
  complete, 92
  incomplete, 92–93, 93f, 98t
  recessiveness compared, 80
  X-linked dominant inheritance, 120, 120f, 121t
Dominant allele, 7, 14
Dominant toxic gain-of-function, 230
Dominant traits, 73
  autosomal dominant inheritance, 77, 77f, 77t
Donohue, R. P., 104
Donor oocytes, 443
Dopamine, 153, 154, 155f
Dor Yeshorim program, 290, 418t
Double helix, 4, 4f
"Double muscle" gene, 130
Down syndrome, 152
  fingerprint patterns in, 133
  genome editing to treat, 250, 251f
  karyotyping to diagnose, 244, 244f
  maternal age and, 250
  physical problems, 249
  trisomy 21 Down syndrome, 244f, 247t, 249, 251, 251f, 253
DPP (dentin phosphoprotein), 211, 211f
Dravet syndrome, 154cc
Driver mutations, 381–382, 382f, 382t, 383f

Drosophila melanogaster (fruit fly), 103
Drug(s)
  anti-HIV drugs, 358–359, 359t
  development of
    anti-cancer drugs, 373–374, 385
    retinoids, 384
    using recombinant DNA technology, 398t, 400
  fertility drugs, 436
  immunosuppressants, 365, 367
  matching patient to, 422–423, 423t
  metabolism of, race and, 135
  "repurposing," 424
  response to, gene variants and, 271
  teratogenicity of, 63
Drug addiction, 152–155, 162
  addictive drugs, 153–155, 155f
  cannabis, 153, 154cc
  nicotine addiction, 153–154, 155f
  smoking cessation drug, 155f
  substance use disorders, 152–153, 153f
Drug testing, stem cells in, 36
DSP (dentin sialoprotein), 211, 211f
DSPP (dentin sialophosphoprotein), 211, 211f
Duchenne muscular dystrophy (DMD), 18, 205, 406
  DNA sequencing of gene, 315, 315f
  drug treatment of, 228–229
  mutation as cause of, 222t, 223t
  ribozyme treatment of, 407
Ductus deferens, 44, 44f
Duffy (blood type), 104
Dulbecco, Renato, 267
Duplications, 227t, 255, 262
  in chromosomes, 248t, 254–255, 255f
  within human genome, 322
Dutch Famine Study, 209
Dutch Hunger Winter, 64, 209cc
Dwarfism, 223t
Dystrophic epidermolysis bullosa, 221t
Dystrophin, 222t, 227, 406
Dystrophin gene, 188, 315, 315f, 406
DZ (dizygotic) twins, 57, 68, 138–139, 138t, 142

E

Earle, Pliny, 84
Early-onset Alzheimer disease, 206
Eating disorders, prevalence of, 150t
Ebola virus, 349cc
EcoR1 restriction enzyme, 398, 399f
Ectoderm, 56, 57f, 68
Ectopic (tubal) pregnancy, 437, 437f

Edwards syndrome (trisomy 18), 249*t*, 250–251, 251*f*
Ehlers-Danlos syndrome, 221*t*
Einstein, Albert, 159*t*
Ejaculation, 45
Eleanor, Mary, 287
Electronic cigarettes, teratogenicity of, 63
Electronic Medical Records and Genomics (eMERGE) database, 311
Electropherogram, 334*f*
"Elementen," 72
Elephants, ivory poachers and, 165
Elliptocytosis, 104–105
El Sidrón Cave (Spain), 311
ELSI (Ethical, Legal and Social Implications) Research Program, 267
Embryo(s), 53, 68
 choosing, in IVF, 433–434
 development, 60–61, 61*f*
 "extra" disposition of, 444–445, 444*t*, 446
 formation of, 56, 57*f*
 implantation of, 54*f*, 56
 primordial (gastrula), 56, 57*f*, 57*t*
 "three-parent" embryos, 99, 100
Embryo donation, 442, 445
Embryonic induction, 60
Embryonic period, 60–61, 61*f*, 62
Embryonic stem (ES) cells, 35, 36, 36*t*, 40
eMERGE (Electronic Medical Records and Genomics) database, 311
Empiric risk
 of complex traits, 135–136, 135*t*, 136*f*, 143*t*, 145
 of schizophrenia, 157, 157*t*
Endocrine part, pancreas, 205, 205*f*
Endoderm, 56, 57*f*, 68
Endogamy, 290
Endoplasmic reticulum (ER), 21*f*, 23, 39
 rough, 21*f*, 22*f*, 23
 smooth, 21*f*, 22*f*, 23
Endorphins, 153
Endosome, 23, 39
Enhanceosomes, 207
Entry inhibitor, 359*t*
Environment
 bioremediation of, 404
 environmental causes of cancer, 131, 131*f*, 374*f*, 386, 386*f*, 388
 as source of error in GWAS studies, 142
 as source of radiation, 226
Environmental epigenetics hypothesis, 146

Environmental influences, 2
 altering gene expression in space, 182–183
 on genes, 6, 131, 131*f*
 on heritability, 136, 136*f*
 on migraine headache, 131
 on schizophrenia, 156, 157
 toxins, 245
 twin studies of, 138–139, 138*t*, 139*f*
 on weight, 144
Enviropig, 404
Enzyme replacement therapy, 424, 424*t*
Enzymes, 19, 20, 39
 lysosomes, 23
 restriction, 330, 398–399, 399*f*
EPAS1 gene, 296, 296*f*
Epididymis, 44, 44*f*
Epigenetic changes, 56, 68, 123, 203
 centromere replication, 244
 in chromatin remodeling, 208
 in gene expression, 139
 genomic imprinting as, 124
Epigenetic effects in cancer, 374*f*
Epistasis, 94, 97, 107
 in gene expression, 94, 98*t*
 role in heritability, 137
 as source of error in GWAS studies, 142
Epithelial tissues, 18, 18*t*
Epitopes, 354, 354*f*
EPO (erythropoietin), 400, 401, 401*f*
Epsilon (ε) globin chains, 204
Epstein-Barr virus, 357, 384
Equational division, 46, 48*f*, 67
Equator, 30
ER. *See* Endoplasmic reticulum (ER)
Erdheim-Chester disease, 373
Erythrocytosis, familial, 401
Erythropoietin (EPO), 400, 401, 401*f*
Escherichia coli
 development of insulin and, 400
 Shigatoxigenic (STEC), 348*f*, 368
EST (expressed sequence tag), 267–268
Eta people (Japan), 306
Ethical, Legal and Social Implications (ELSI) Research Program, 267
Ethnic groups
 databases, in DNA profiling, 333
Euchromatin, 243, 243*f*, 261
Eugenics, 300–301, 302
 modern genetics compared to, 301, 301*f*
 use of pedigrees in, 84
Eukarya, 18
Eukaryotic cells, 18, 19*f*, 39, 188
 vs. prokaryotic cells, 18
Euploidy, 247, 262

Evolution, 7–8
 of human genome, 322, 323*f*
 at molecular level. *See* Molecular evolution, study of
 population genetics, 278–279, 283
Evolutionary trees, 307, 308*f*, 313*f*, 316
Excision repair, 234, 238
Exocrine/endocrine progenitor cells, 205*f*
Exocrine part, pancreas, 205, 205*f*
Exome, 3, 13
Exome sequencing, 10–11, 88, 271, 274, 418*t*
 autism, 159, 160*f*
 in diagnosing disease, 421
 in pedigree analysis, 85–86, 86*f*
Exons, 188, 199, 210*f*
"Exon skipping" missense mutation, 386
Exon skipping mutation, 228, 238
Exosomes, 23, 39
Expanding mutations, 227*t*
Expanding repeats, 230, 230*f*, 238
Expansion stage, in mutation, 381, 382*f*, 382*t*
Expressed sequence tag (EST), 267–268
Expressivity, 95, 98*t*, 107, 126
Ex vivo gene therapy, 425, 425*f*, 428*cc*
Eye color, single-gene inheritance of, 76, 77*f*, 88, 131
Eye disorders
 colorblindness, 117, 118*cc*–119*cc*
 retinitis pigmentosa, 129
 retinoblastoma, 385, 385*f*, 394
Eyelashes, 285

F

F508del-CFTR mutation, 277, 278
Fabry disease, 123, 129
Factor IX, 222*t*
Factor VIII gene, 230
Factor V Leiden gene, 433–434
Factor X deficiency. *See* Hemophilia B
Factor XI deficiency, 227
Familial advanced sleep phase syndrome, 151, 163
Familial Alzheimer disease, 96*cc*, 97*t*
Familial amyotrophic lateral sclerosis, 197*t*
Familial dysautonomia (FD), 228
Familial erythrocytosis, 401
Familial hypercholesterolemia (FH), 20
 incomplete dominance in, 93–94, 94*f*, 98*t*
 mutation as cause of, 222*t*, 227
Families, 7
FANCA gene, 294*t*

Fanconi anemia, 294t, 434
Farmers, early, 313, 314f
FAS (fetal alcohol syndrome), 63, 63f, 152
Fatal familial insomnia, 151
Fatigue, 24
Fatty acid desaturase enzyme, 324
FBN1 gene, 223t
FD (familial dysautonomia), 228
Fecal transplantation, 38
Female infertility, 435–437, 437f
Female reproductive system, 45, 45f, 67
Ferritin, 196
Fertility drugs, 436
Fertilization, 53, 54f, 57t
Fetal alcohol spectrum disorder, 63
Fetal alcohol syndrome (FAS), 63, 63f, 152
Fetal cells, scleroderma and, 361
Fetal ("cell free") DNA, 246–247, 247f
Fetal hemoglobin, 204, 323
Fetal period, 61, 61f
Fetal ultrasound, 245, 246f
Fetus, 53, 68
Fever, in immune response, 352
FGFR3 gene, 223t
FGFR (fibroblast growth factor receptor) genes, 52–53
FH. *See* Familial hypercholesterolemia (FH)
Fibrillin, 95, 222t, 420
Fibroblast growth factor receptor *(FGFR)* genes, 52–53
Fibroblasts, 36
Fibroid tumors, 437, 437f
Fibrosarcoma, 376f
Filaggrin, 361
Finding Your Roots, 271
Fingerprint patterns, 133, 133f
FISH (fluorescence *in situ* hybridization), 245–246, 246f, 261, 315–316
Fisher, Ronald Aylmer, 300
5-alpha reductase deficiency and, 114
FKBP5 gene, in PTSD, 149
"Flavr Savr" tomato, 10
Fluorescence in situ hybridization (FISH), 315–316
Fluorescence *in situ* hybridization (FISH), 245–246, 246f, 261
FMR1 gene, 231cc
Folic acid, 60
Food(s), 9, 10, 402–403
Forced sterilization, 300
Forensic DNA phenotyping, 336, 336f

Forensic science
  DNA profiling, 337
    postconviction testing, 328–329
  fingerprint analysis, 133
  genetic testing in, 418t
  identifying ivory poachers, 165
  time of death, tracking, 202–203
  use of mtDNA in, 101
Fore people (New Guinea), 197, 197f
Founder effect, 291, 293t
  migration from Africa and, 318
  Plain Populations, 292cc, 292f
FOXP2 gene, 323
Fragile X-associated tremor/ataxia syndrome (FXTAS), 231cc
Fragile X syndrome, 231cc
  intellectual disability and, 152
Frameshift mutation, 190, 199
Frameshift mutations, 227t, 228f, 229
Framingham Heart Study, 141
Franklin, Lonnie David, Jr., 338
Franklin, Rosalind, 168, 169, 169f, 169t
Fraternal (dizygotic) twins, 57, 138–139, 138t, 142
Friedman, Jeffrey, 143
Frontotemporal dementia, 197t
Fruit fly *(Drosophila melanogaster),* 103
Fugate, Elizabeth Smith, 217
Fugate, Martin, 217
Fugate family, 217, 240
Fusion genes, 229, 385
Fusion inhibitor, 359t
Fusion proteins, 384
FXTAS (fragile X-associated tremor/ataxia syndrome), 231cc

G

G_0 phase, 29, 29f, 31f, 39
G_1 phase, 29, 29f, 30, 31f, 39
G_2 phase, 29, 29f, 30, 31f, 39
G6PD. *See* Glucose-6-phosphate dehydrogenase (G6PD)
G6PD deficiency, 233
Gain-of-function mutations, 80, 80f, 218, 227
  dominant toxic, 230
  in oncogene activation, 382
Galton, Francis, 300
Gamete(s), 44, 45, 67
  maturation of, 49–53, 67–68
    meiosis and mutations, 52–53, 53t
    oogenesis, 50–52, 50f–52f
    spermatogenesis, 49–50, 49f, 52f
  postmortem retrieval of, 442

Gamete intrafallopian transfer (GIFT), 440–441, 444t
Gamma (γ) globin chains, 204, 204f
Gamma radiation, 226
GAN (giant axonal neuropathy), 28cc, 183
Garrod, Archibald, 166
Gastrula (primordial embryo), 56, 57f, 57t, 68
Gastrulation, 56, 57f
Gates, Henry Louis, Jr., 271
Gaucher disease, 223t, 229, 294t, 424
GBA gene, 223t, 294t
GEDmatch, 343
Gelsinger, Jesse, 428cc
Gender. *See* Sex; Sex chromosomes
Gender identity, 114–115, 115t
Gender selection, 417, 418t, 443–444
Gene(s), 2, 13, 74, 170. *See also specific genes*
  circadian clock genes, 312t
  environmental influences on, 6, 131, 131f
  expression of. *See* Gene expression
  gene variants, 2, 7, 8, 10
  homeotic, 56, 58cc, 316
  linked. *See* Linkage
  longevity and, 10, 65–67, 67t
  mitochondrial. *See* Mitochondrial genes
  mutation of
    in Ashkenazim, 237, 293
    autosomal genes, 224
    consanguinity and, 217–218
    gain- or loss-of function, 382–383
    mismatch mutations, 375
    pseudogenes and, 229–230, 230f
    single-gene mutations, 44, 65
  Neanderthal gene variant, 312t
  polygenic complex traits, 131
  resistance genes, 297
  role in behavior, 149–150, 149f, 150t
  role in cancer. *See* Cancer
  X-linked and Y-linked, 116
Genealogical records, 293
Gene doping, 215
Gene drives, 409, 410f
Gene expression, 4, 13, 202–216, 214
  altered, in clones, 55
  building tissues and organs, 204–205, 205f
  control of, 206–210, 214
    chromatin remodeling, 207–208, 207f–208f
    microproteins, 207, 210
    microRNAs, 207, 208–209, 208f
    in "diseasome," 205–206, 206f

dominance relationships and, 92–94, 93f, 94f, 98t
Dutch Hunger Winter, 209cc
epigenetic changes in, 139
epistasis, 94, 98t, 107
expressivity, 95, 98t, 107
genetic heterogeneity. See Genetic heterogeneity
globin chain switching, 203–204, 204f
human genome sequence and, 97–98, 98t
lethal allele combinations, 92, 93f, 98t
maximizing genetic information, 210–212, 214
Mendelian ratios, 92–98, 107
multiple alleles, 92, 98t
patterns, in GWAS studies, 139, 139t
penetrance, 95, 98t, 107, 125
phenocopies, 95, 97, 98t, 107
in the placenta, 203
pleiotropy, 95, 97, 98t, 107
response to spinal cord injury, 404, 405f, 406t
in space, altering, 182–183
time of death, 202–203
timing of, 202, 203
Gene expression profiling, 367
in cancer diagnosis and treatment, 389, 389f
in spinal cord injury, 404, 405f, 406t
Gene flow, 278, 312, 313, 313f, 321
Gene pool, 8, 14
Gene silencing, 406–407, 406f, 407f
Gene therapy, 28, 424–426
history of, 428cc–429cc, 428f
for Sanfilippo syndrome, 414, 423
types and targets of, 425, 425f, 426f
Genetically modified foods, 364, 402–403, 403t
Genetically modified organisms (GMOs), 10, 396, 402
Genetic code, 199
deciphering, 189t, 191
degeneracy of, 191
nonoverlapping nature of, 190, 191f
triplet nature of, 190, 190f
universality of, 191
Genetic counseling, 415–417, 415f, 430
evaluating infertility tests, 437
Genetic disease, treatment of, 423–426, 430
CRISPR-Cas9, 426–427, 429
gene therapy. See Gene therapy
Genetic diversity, 46
Genetic drift, 291–294, 302
effect on Plain Populations, 292cc

founder effect, 291, 292cc, 293, 293t
population bottlenecks, 293–294, 294t
Genetic exceptionalism, 338
Genetic genealogy, 2, 13, 338–342
centimorgans, 341–342, 342f
geographic origins and ethnicity estimates, 338–339, 340f, 341t
SNPs, 342–343, 343f
tracking migration patterns, 316–317, 317f
Genetic heterogeneity, 95, 95f, 97
in gene expression, 95, 95f, 96cc, 98t, 107
as source of error in GWAS studies, 142
Genetic information, 87
maximizing, 210–212, 214
Genetic load, 294
Genetic markers, 106, 108, 329
Genetic modification, 10
Genetic privacy, 12–13
Genetic prospecting, 11
"Genetic rescue," 181
Genetics, 2, 13–14
applications of, 8–11, 9f
conservation genetics, 9
DNA analysis, 9, 9f
DNA profiling, 8–9
exome sequencing, 10–11
genetic modification, 10
precision medicine, 9–10
as informational science, 329
levels of, 3–8, 4f
DNA. See DNA (deoxyribonucleic acid)
families, 7
human body, 6–7
populations to evolution, 7–8
Genetic testing, 417, 418t, 419–423, 430
of adults, 418t, 420
bioethics, 12–13
bioinformatics strategy, 277–278
for cardiovascular disease, 132cc
of children, 418t, 419
costs of, 417
cystic fibrosis, 277–278
forensic, 418t
for inborn athletic ability, 421–422, 421t
kinds of tests, 420–423
matching patient and drug, 422–423, 423t
of newborns, 418t, 419, 419f
nutrigenics, 422
posthumous, 418t
preconception, 417, 419

prenatal, 418t, 419
secondary findings in, 420
Genetic Testing Registry, 415
Genetic uniformity, GMOs and, 403
Gene transfer, 425
Gene variants, 2, 7, 8, 10, 91
Genocide, identifying victims of, 334–335
Genome(s), 2–3, 266. See also Human genome
sequencing. See Genome sequencing
Genome editing, 10, 14, 406, 407–408, 408f, 408t, 409t
to create chimeric antigen receptors, 390
for Down syndrome, 250, 251f
immunity and, 358f
Genome maintenance, 374–375, 374f
Genome sequencing, 2–3, 3f, 88, 329
animal models in, 426–427, 429
Bainbridge-Ropers syndrome, 11
"clone-by-clone" sequencing technique, 268, 269f
comparison to other species, 322
costs of, 271
development of, 266–268, 268f–270f
in diagnosing disease, 420
first human genome, 266–268
in hospital-acquired pneumonia, 368
immunity against infection and, 368
limitations of, 272
Neanderthal and modern human compared, 310, 323
newborn, 11, 274–275
in newborns, 418t
in pedigree analysis, 85–86, 86f
posthumous, 265–266
reference genome sequences, 271
toxic *Escherichia coli*, 348f, 368
"whole-genome shotgun" technique, 268, 269f, 270f
Genome-wide association studies (GWAS), 418t
of body weight, 144
of change in intelligence, 151
of complex traits, 139–142, 140f, 140t, 141f, 141t, 142f, 143t, 145
limitations of, 141–142, 142f, 143t
polygenic risk scores, 140–141, 141f
of schizophrenia, 157
single nucleotide polymorphisms in, 139–141, 139t, 140f
study designs in, 141, 141t
Genomic imprinting, 55, 124–125, 124f, 125f, 127
diseases of, 125–126, 126f
imprinting centers, 124

Genomic Postmortem Research Project, 265
Genomics, 3, 13, 266–271
　analysis of genome content, 271–273
　　annotation, 271–272
　　consumer DNA testing, 272
　　limitations of, 272
　　speed and coverage, 271
　development, 266–268, 268f–270f
　　first human genome sequencing, 266–268
　　linkage and positional cloning studies, 266, 267f
　　sequencing genomes of newborns, 274–275
　in fighting infection, 368
　metagenomics, 11–12
　new view of crossing over, 274, 274f
　posthumous genetic testing, 418t
Genomic sequencing, 274
Genotype, 7, 14, 74
Genotyping, 348, 350
Geographical barriers, allele frequencies and, 291
Geographical distribution of balanced polymorphism, 298f
German measles (rubella), 64
Germ cells, 18, 39
Germline, human, editing, 409–410
Germline cancers. *See* Inherited (germline) cancers
Germline cells, 45
Germline gene therapy, 425
Germline genome editing, 408
Germline mutations, 218–219, 238
GFP (green fluorescent protein), 396f, 403
Ghrelin, 143, 144t
Giant axonal neuropathy (GAN), 28cc, 183
Gibberellin, 74
GIFT (gamete intrafallopian transfer), 440–441, 444t
Gigaxonin, 28
Globin chain switching, 203–204, 204f
Globin gene mutations, 232–233, 232t
Globozoospermia, 436cc, 436f
Glossogenetics, 327
Glucocerebrosidase, 424
Glucose-6-phosphate dehydrogenase (G6PD)
　protective against malaria, 298t
Glutamate, 161
Glutamic acid, 220, 220f
Glutaric acidemia type 1, 292cc
Glutaric aciduria type 1, 285

Glycine-proline-modified proline, 221
Glycolipids, 26f
Glycoproteins, 26f
GMOs (genetically modified organisms), 10, 396, 402
Golden retriever muscular dystrophy (GRMD), 315, 315f
Golden rice, as GMO, 402
Golden State Killer case, 343
Golgi apparatus, 21f, 22f, 23, 39
Gonadal mosaicism, 224, 238
Gonadal sex, 115t
Gonad(s), 44, 67
Government Accountability Office, 422
gp120 surface protein, 357–358, 358f
Graft-*versus*-host disease, 367
Grandmothers of the Plaza de Mayo project, 335–336, 336f
Grapheme, 178
Graves disease, 359t
"Great Migration" of 1910-1970, 321, 322f
Green fluorescent protein (GFP), 396f, 403
Griffith, Frederick, 166, 166f, 167, 169t
"Grim Sleeper" case, 338
GRMD (golden retriever muscular dystrophy), 315, 315f
GTP (Guanosine triphosphate), 191
Guanine (G), 4, 4f, 13, 170, 170f, 171, 172f
Guanosine triphosphate (GTP), 191
!Gubi, 307
Guevedoces, 114
Gut microbiome, 38, 144
GWAS. *See* Genome-wide association studies (GWAS)
Gymrek, Melissa, 342

H

H1N1 virus, 349cc
H5N1 virus, 349cc
HA (hemagglutinin), 349cc
Hair color, 91–92, 142, 142f
Haplogroups, 317, 317f, 321, 338, 339f, 339t, 340f, 340t
Haploid cells, 18, 39, 45, 48f
Haplotypes, 107, 107f, 108
Hardy, Godfrey Harold, 280
Hardy-Weinberg equation, 280, 280f, 281, 332, 335f
Hardy-Weinberg equilibrium
　applications of, 281–283, 282f, 282t, 284, 383f
　defined, 279

　DNA profiling and. *See* DNA profiling
　equation, 280, 280f, 281
　nonrandom mating and, 288–290
　solving problems with, 280–281, 281f
hCG (human chorionic gonadotropin), 56
HD. *See* Huntington disease (HD)
HDL (high-density lipoprotein), 66, 132cc
Head shaking, 11
Healing, stem cells in, 35, 35f
Heart health
　as complex trait, 131, 132cc, 132f, 132t
　Dravet syndrome, 154cc
　low birth weight and, 64
Heavy chains, 354, 354f, 355, 355f
Height, as polygenic trait, 133–134, 134f
"HeLa" cells, 376
Helicase, 174, 175f
"Helicopter research," 327
Helix-turn-helix, 186
Helper T cells, 351f, 355, 356t, 358, 358f
Hemagglutinin (HA), 349cc
Heme groups, 204f
Hemings, Sally, 9
Hemizygous males, 117, 127
Hemodialysis, 401
Hemoglobin
　fetal, 204, 323
　formation of, 203–204, 204f
　globin gene mutations, 217–218
　hemoglobin C, malaria and, 298
Hemolytic anemia, 359t
Hemolytic disease of fetus and newborn, 361, 362f
Hemophilia A, 120, 123, 129
Hemophilia B (factor X deficiency), 84
　mutation as cause of, 222t, 224
　pedigree, 117, 119f
　as X-linked recessive trait, 117, 119f
Hensel, Abigail, 60, 60f
Hensel, Brittany, 60, 60f
HER2 cell surface protein, 384
HERC2 gene, 76, 77f
Herceptin (traztuzumab), 365
"Herd immunity," 362, 363f
hereditary hemorrhagic telangiectasia type 1, 266
Hereditary nonpolyposis colon cancer (HNPCC), 236
Heredity, 2, 13
Heritability, 152
　of complex traits, 136–137, 136f, 136t, 137f, 137t, 143t, 145

estimating, 136–137, 137f, 137t
  missing, 136
  of traits in monozygotic twins, 138–139, 138t, 142
Hermaphroditism, 113
Herrick, James, 221
Hershey, Alfred, 167, 168f, 169t
HERVs (human endogenous retroviruses), 212
Heterochromatin, 243, 243f, 244, 261
Heterocyclic aromatic amines, 388
Heterogametic sex, 113, 127
Heteroplasmy, 99–101, 100f, 108, 443
Heterozygote, manifesting, 123, 127
Heterozygous individuals, 74, 87
HEXA gene, 294t
HGPRT enzyme, 124
HGPS (Hutchinson-Gilford progeria syndrome), 43, 53t, 65, 65t, 66f, 223
High altitude adaptation, 295–296, 296f
High-density lipoprotein (HDL), 66, 132cc
Hill people (New Guinea), 306
Hippocampus, in Alzheimer disease, 96cc
Hirschsprung disease, 223t
Histamine, 352, 360, 360f
Histones, 172, 173f, 179
History, DNA analysis and, 9, 9f
Hitchens, Christopher, 271
HIV (human immunodeficiency virus)
  anti-HIV drugs, 357–359, 359t
  birth defects due to, 64
  infection with, 357–359, 358f, 359t
  resistance to
    CCR5 mutation and, 218
  genome editing and, 358f
HLAs (human leukocyte antigens), 336, 350, 365
HNPCC (hereditary nonpolyposis colon cancer), 236
Hodgkin's lymphoma, 441
Holes, Paul, 343
Holmes, Oliver Wendell, Jr., 300
Holstege, Henne
  somatic mutation and, 220
Homeobox, 58cc
Homeotic genes, 56, 58cc, 316
Homeotic mutations, 56, 58cc
Home pregnancy tests, 56, 364
Hominins, 307, 308f
Homocysteine, 62
Homo erectus, 309–310, 309f, 324
Homo floresiensis, 309
Homogametic sex, 113, 127
Homo habilis, 309

Homologous pairs (homologs), 45, 67
Homo sapiens denisova (Denisovans), 310, 310f, 312–313, 323
Homo sapiens neanderthalensis. See Neanderthals (*Homo sapiens neanderthalensis*)
Homo sapiens sapiens, 307, 308f, 324
Homosexuality, 115, 115t, 126
Homo species, 309, 309f
Homozygosity, 287, 288f
Homozygosity in dogs, 297
Homozygosity mapping, 141, 141t, 145
Homozygous individuals, 74, 87
Hormones, 32, 40
  female infertility due to imbalance of, 436
  pregnancy tests and, 56
  sex-influenced traits and, 122
  to treat XO syndrome, 253
  to treat XXY syndrome, 254
Hospital-acquired pneumonia, 368
"Housekeeping" functions, cell, 35
HTR2A gene, pharmacogenetics of, 423t
"Human accelerated regions," 323
Human chorionic gonadotropin (hCG), 56
Human endogenous retroviruses (HERVs), 212
Human exome sequence, 271
Human genome, 1–13. *See also* Genome(s)
  content, analysis of, 271–273, 275
  editing. *See* Genome editing
  evolution of, 322, 323f
  example of, 1–2
  first, sequencing, 266–268, 275
  genetics. *See* Genetics
  genomics. *See* Genomics
  information, types of, 268–271
  noncoding RNAs, 208, 212–213, 213t
  parts of, 214f
  protein not encoded by, 212–214
    noncoding RNAs, 208, 212–213, 213t
    repeats, 213–214, 214f
    viral DNA, 212
  role of genes and genomes in, 2–3
  sequencing. *See* Genome sequencing
  sequencing genomes of deceased, 265–266
Human Genome Project, 267, 268, 275
Human genome sequence, 97–98, 98t, 271
Human germline, editing, 409–410
Human immunodeficiency virus. *See* HIV (human immunodeficiency virus)

Human leukocyte antigens (HLAs), 336, 350, 365
Human microbiome, 38–39, 38f, 40
Human Microbiome Project, 38
Human organoids, 366t
Human papillomavirus, 242, 384
Humans, 306–326. *See also* Molecular evolution, study of
  archaic, 310–313
  defining characteristics of, 322–325, 323f
  future of, 325
  indigenous peoples, 306–307
  migrations of. *See* Migration
  modern, 313–314
  origins of, 307–314, 325
  primates, 307–309, 308f
Humoral immunity
  B cells and antibodies, 351f, 353–355, 353f–355f, 354t
  monoclonal antibodies and, 364–365, 364f
Hungerford, David, 385
Hunter syndrome (mucopolysaccharidosis II), 123
Huntingtin, 222t
Huntington disease (HD), 222t
  animal models of, 402
  autosomal dominance of, 72
  complete penetrance in, 97
  as gain-of-function mutations, 80
  genetic information, 87
  inheritance of, 126
  juvenile, 71–72, 419
  as lethal allele disorder, 92
  misfolded proteins in, 197t
Hutchinson-Gilford progeria syndrome (HGPS), 43, 53t, 65, 65t, 66f, 223
Hydatidiform mole, 125
Hydrogen bonds, 171, 171f, 172f
Hydrophilic end, phospholipid, 25
Hydrophobic end, phospholipid, 25
Hypocretin, 151
Hypomania, 156
"Hypothetical minimal genome," 273
Hypothetical offspring, 417
Hypoxia inducible factor, 296

I

Ichthyosis, 117, 117f
Ichthyosis vulgaris, 361
ICM (inner cell mass), 36, 55, 68
ICMP (International Commission on Missing Persons), 335

ICSI (intracytoplasmic sperm injection), 125, 440, 440f
Identical by descent DNA sequences, 287
Identical twins. *See* identical) twins; Monozygotic (MZ
Ideogram, 247f
Idiotypes, 354, 354f
IFs (interferons), 352, 352t, 365
IgA antibodies, 354t, 435
IgD antibodies, 354t
IgE antibodies, 354t
IGF-2 (insulin-like growth factor 2), 209cc
IKAKAP gene, 294t
IL-2 (interleukin-2), 365
IL-2 receptor mutation, 357t
Immune checkpoint inhibitors, 389, 389f
Immune deficiencies, inherited, 356–357, 357f, 357t
Immune system, 351–356, 351f, 352f, 369
  immune response
    adaptive, 351, 352f, 353–356
    innate, 351–352, 352f, 352t
    during pregnancy, 361–362
    primary or secondary, 353
  of Neanderthals, 311
  physical barriers, 351, 352f
Immunity, 347–370
  abnormal. *See* Abnormal immunity
  adaptive, 351, 352f, 353–356
  altering, 362–367, 369
    immunotherapy, 364–365
    transplants, 365, 367
    with vaccines, 362–364, 363f
  autoimmunity, 359–360, 359t, 372
  genome editing and, 358f
  "herd immunity," 362
  humoral. *See* Humoral immunity
  against infection, 367–368
    genome sequencing and, 368
    reverse vaccinology, 367
  innate, 312t, 351–352, 352f, 352t
  role of cell surfaces in, 348–351, 348f
  role of immune system. *See* Immune system
Immunoglobulins, 354t
Immunological amnesia, 359
Immunosuppressant drugs, 365, 367
Immunotherapy, 364–365
  altering immunity with, 364–365
  cytokines, 365
  monoclonal antibodies, 364–365, 364f
Implantation of embryo, 54f, 56
Imprinting. *See* Genomic imprinting

Inborn errors of metabolism, 20cc
  basis of, 80
  genetic testing for, 419, 419f
  glutaric acidemia type 1, 292cc
  treatment of, 423
Inbreeding, 287, 288f
Inc13 in celiac disease, 213
Incidence, 135, 145
"Incidentaloma," 420
Incomplete dominance, 92–94, 93f, 98t, 107
Incomplete penetrance, 125
Incomplete twins, 60, 60f
Inconclusive pedigree, 85, 85f
Incontinentia pigmenti (IP), 120, 120f
Indels, 323
Independent assortment, 48, 48f, 67, 105f
Indian Ocean tsunami, 334
Indigenous peoples. *See also specific groups*
  genomes of, 306–307
  Native Americans, 320–321, 320f
  Siberian, 321
Individuals with Disability Education Act of 1990, 161
Induced mutation, 225–226
  intentional use of mutagens, 226
Induced pluripotent stem (iPS) cells, 35, 36t, 37, 40, 216
  reprogramming of, 36
  in treating sickle cell disease, 429
Infantile autism, 161
Infectious disease(s)
  DNA analysis and, 9
  immunity against, 368
  during pregnancy, schizophrenia and, 157
  reverse vaccinology and, 367
  viral infections, 64, 408
Infertility, 445
  female, 435–437, 437f
  male, 253–254, 435
  medical tests, 437
Infidelity DNA testing, bioethics of, 178
Inflammation, 352, 360
Influenza, 349cc
Influenza viruses, 176f, 349cc
Informational science, genetics as, 329
Ingram, V. M., 221
Inheritance
  autosomal dominant, 58cc, 72, 77, 77f, 77t
  autosomal recessive, 77–78, 77t
  law of. *See* Mendel's laws of inheritance

  of mitochondrial genes, 98–101, 99f, 100f
  single-gene. *See* Single-gene inheritance
Inherited (germline) cancers, 377, 377f
  retinoblastoma, 385–386, 385f, 394
  sporadic cancer compared, 377, 377f
Inherited colon cancer, 236
Initiation codon, 192, 193f
Innate immunity, 312t, 351–352, 352f, 352t
Inner cell mass (ICM), 36, 55, 68
Innocence Project, 328–329
Insertion/deletion polymorphism, 421
Insertion mutation, 229, 238
Insertion(s), 227t, 229, 322
Insulin, 316, 400
Insulin-like growth factor 2 (IGF-2), 209cc
Intellectual impairment, 63, 155
Intelligence, 151–152, 152f, 162
Intelligence quotient (IQ), 151–152, 152f, 162
Intelligence tests, 151–152
Interferons (IFs), 352, 352t, 365
Interleukin-2 (IL-2), 365
Interleukin(s), 352, 352t
Intermediate filaments, 26, 27–28, 27f, 39
International Commission on Missing Persons (ICMP), 335
International Human Genome Mapping Consortium, 268, 269f
International Society for Stem Cell Research, 37
Interphase, in cell cycle, 28–30, 29f, 39, 46
Intersex, 113–114
Intracytoplasmic sperm injection (ICSI), 125, 440, 440f
Intrauterine growth retardation (IUGR), 64
Intrauterine insemination (IUI), 438, 444t
Introgression, 311
Introns, 188, 199, 211f, 213t
Invasion stage, in mutation, 381, 382f, 382t
Invasiveness, in cancer cells, 377
Inversions, chromosomal, 247t, 255, 255f, 258–259, 258f, 260t
In vitro fertilization (IVF), 439–440, 439f, 440f, 444t
  Beckwith-Wiedemann syndrome and, 125
  choosing embryo from, 433–434

of cryopreserved oocytes, 441
mitochondrial disease and, 100
In vivo gene therapy, 425, 425*f*
Ion channels, 26*f*
IP (incontinentia pigmenti), 120, 120*f*
iPS. *See* Induced pluripotent stem (iPS) cells
IQ (Intelligence quotient), 151–152, 152*f*, 162
Irons, Ernest, 221
Isochromosomes, 259, 259*f*, 262
Isoforms, 189, 199, 210, 210*f*
Isograft, 365, 365*f*
Isotretinoin, teratogenicity of, 62, 63
IUGR (intrauterine growth retardation), 64
IUI (intrauterine insemination), 438, 444*t*
Ivacaftor (Kalydeco), 79*cc*
IVF. *See In vitro* fertilization (IVF)
Ivory poaching, 165

J

Jacobs, Patricia, 255
Jacobs (XYY) syndrome, 252*t*, 254, 255
James, Jesse, 176
Jefferson, Thomas, 9, 9*f*
Jeffreys, Alec, 330, 337
Jenner, Edward, 363
Jobs, Steve, 271
Joubert syndrome, 324
Junctional epidermolysis bullosa, 36–37
Junk DNA, 188
Juvenile Huntington disease, 71–72, 419

K

Kabuki syndrome, 216
Kallmann syndrome, 121
Kalydeco (ivacaftor), 79*cc*
Kanner, Leo, 161
Karadzic, Radovan, 335
Karyotype(s), 6, 6*f*, 13, 244–245, 244*f*, 245*f*
Kelly, Mark, 182
Kelly, Scott, 182
Keratin gene, 323
Kercher, Meredith, 336
Khoisan people (Africa), 306, 307
Kidney cancer, 365
Kinases, 32
King, Alonzo Jay, 337
King, Mary-Claire, 335
Klebsiella pneumoniae, 368
Klinefelter (XXY) syndrome, 252*t*, 253–254, 445
Knockouts, 218

Knox, Amanda, 336
Kuru, 197, 197*f*

L

Lacks, Henrietta, 376
Lactase, 286, 295
Lactase deficiency, 20
Lactobacillus, 38
Lactose tolerance/intolerance, 286–287, 295
Lamin A gene, 223
Language barriers, allele frequencies and, 290
Large intergenic noncoding RNAs, 213*t*
Laron syndrome, 66
Larvae feeding, time of death and, 202–203
Laughing disease, 198
Law of independent assortment, 80
  following multiple genes, 80–81, 81*f*, 82*f*
  linkage and, 102, 102*f*, 105*f*
Law of segregation, 73, 74–75, 75*f*, 75*t*, 87
  in blood typing, 94*f*
  gene drives counter to, 409, 410*f*
  risk of inheriting disease and, 282
  sex ratio and, 115
LD (linkage disequilibrium), 106, 287
LDL (low-density lipoprotein), 23, 132*cc*, 227, 228*f*
LDL receptor, 222*t*, 228*f*
LDL receptor gene, 133, 137
Lee, Pearl, 220
Leigh syndrome, heteroplasmy in, 100
Leptin, 143–144, 144*t*
Leptin receptor, 143, 144*t*
Leptin transporter, 144*t*
Lesch-Nyhan syndrome, 20, 123–124
Lethal alleles, 92, 295
  gene expression of, 92, 93*f*, 98*t*
  polyploids, 247, 247*f*
Leucine zippers, 186
Levene, Phoebus, 167–168, 169*t*
Lewis blood group, 348*t*
Lewy body dementia, 197*t*
Liebenberg syndrome, 58*cc*, 58*f*
Li-Fraumeni syndrome, 386
Ligase, 175, 175*f*, 179
Light chains, 354, 354*f*
Linear tumor evolution, 382*f*
Linkage, 101–107, 102*f*, 108
  genome-wide associations and, 106–107, 107*f*
  in pea plants, discovery of, 101–103, 102*f*, 103*f*

  solving problems in, 104–106, 106*f*
Linkage disequilibrium (LD), 106, 108, 287
Linkage maps, 103–104, 103*f*, 105*f*, 108, 266, 266*f*
Linkage studies, 266, 267*f*
Linker regions, 172
Lipid droplet, 22*f*
Lipids, 19, 20, 39
Lipoprotein lipase, 132*cc*
Liposomes, as vectors, 425
Liquid biopsies, 391*cc*, 391*f*
Logarithm of the odds (LOD) score, 106–107, 108
Longevity, genes and, 10, 65–67, 67*t*
Long noncoding RNAs, 212–213, 214
Long QT syndrome, with deafness, 110
Loss-of-function mutation, 80, 80*f*, 218, 383
Lou Gehrig's disease, 36
Louis XVI, king of France, 101
Louis XVII, Prince Royal of France, 101
Low-density lipoprotein (LDL), 23, 132*cc*, 227, 228*f*
Lowe syndrome, 129
LOX (lysyl oxidase) enzyme, 427, 429
LSD (lysergic acid diethylamide), 153
"Lucy," 308–309
Lung cancer, 131, 131*f*
Lyme disease, 372
Lymphatics, 351, 351*f*
Lymph fluid, 351
Lymph nodes, 351, 351*f*
Lysergic acid diethylamide (LSD), 153
Lysosomal acid lipase deficiency, 431
Lysosomal storage diseases, 23
  lysosomal acid lipase deficiency, 431
  treatment of, 423–424, 424*t*
  X inactivation and, 123
Lysosome, 21*f*, 22*f*, 23, 24*f*, 39
Lysyl oxidase *(LOX)* enzyme, 427

M

Mab (monoclonal antibody) technology, 364, 364*f*
MacLeod, Colin, 167, 167*f*, 169*t*
Macroevolution, 279, 279*f*, 283
Macromolecules, 18–19
Macrophages, 350, 350*f*, 351, 351*f*, 356*t*
"Mad cow disease," 198
Major depressive disorder, 155, 156*f*
Major histocompatibility complex (MHC), 157, 350–351, 350*f*

Malaria
  attempts to eradicate, 299, 299f
  gene drives to destroy vectors, 409
  protection against, 232t
  recombinant vaccines against, 400
  sickle cell disease protective against, 298–299, 298f
Male infertility, 435
Male reproductive system, 44–45, 44f, 67
Malignant tumors, 375
Malnutrition
  birth defects and, 64, 68
  effects on methylation, 209cc
  gut microbiome, 38
  IUGR and, 64
Mania, in bipolar disorder, 156
Manifesting heterozygote, 123, 127
MAOA (monoamine oxidase A), 164
Maple syrup urine disease (MSUD), 20cc, 285, 292cc
Marfan syndrome, 10, 222t, 223t
  aortic aneurysm in, 10, 223
  pleiotropy in, 95, 98t
Marijuana (cannabis), 153, 154cc, 154f
"Massively parallel DNA sequencing," 179
Mast cells, 352, 356t, 360, 360f
MATCH (Molecular Analysis for Therapy Choice), 373
Maternal age
  chromosome abnormalities and, 53
  spontaneous abortion and, 435
  testing cell-free fetal DNA and, 246–247
  trisomy 21 and, 246f, 250
Maternal and Infant Health Care Law of 1994 (China), 300
Maternal immune response, 157
Maternal serum markers, 252, 418t
Matthaei, Heinrich, 191
Mbo people (Cameroon), 311
Mbuti Pygmies, 327
MC1R (melanocortin-1-receptor) gene, 91, 388
McCarty, Maclyn, 167, 167f, 169t
McClintock, Barbara, 213
McCune-Albright syndrome, 240
MCI (mild cognitive impairment), 96cc
MCR1 gene, in Neanderthals, 311, 311f
McWilliams, Millie, 86
MDMA, 164
Measles, 359
MECP2 gene, 17
Meiosis, 29f, 45–48, 46f–48f, 46t, 67
  aneuploidy and polyploidy in, 250

  Mendel's observation of, 74, 75f
  mitosis compared, 46t
  mutations and, 52–53, 53t
  oogenesis, 50–51, 51f
  spermatogenesis, 50, 50f
Meiosis II, 46, 48, 48f
Melanin, 76, 91, 134
Melanocortin-1-receptor *(MC1R)* gene, 91, 388
Melanocortin-4 receptor, 144t
Melanocytes, 91, 134
Melanoma
  cutaneous, 378f
  environmental factors in, 374f, 388
  overexpressed HERVs in, 212
  "three strikes" in, 382t
Melanosomes, 134
MELAS (mitochondrial myopathy encephalopathy lactic acidosis syndrome), 99
Memory cells, 353, 353f
Mendel, Gregor, 6, 72–74, 73f, 77, 78, 80–81, 97, 329
Mendelian diseases. *See* Single-gene disorders
Mendelian ratios, 92–98, 101, 107
Mendelian traits, 6, 7f, 13
Mendel's laws of inheritance, 88, 102, 102f. *See also* Law of independent assortment; Law of segregation
  derivation of, 97–98
  pedigree analysis and, 85, 85f
  seeming exceptions to, 92
Meningococcus B, 367
Menkes ("kinky hair") disease, 223t
Mervar, Jane, 71
Mervar, Karl, 71
Meselson, Matthew, 174
Mesoderm, 56, 57f, 68
Messenger RNA (mRNA), 4, 21, 39, 185, 185t, 199, 397
  analyzing for time of death, 202
  disease caused by, 5, 5f, 6f
  in gene expression profiling, 404
  mitochondrial diseases and, 99
  processing, 188–189, 188f
  in protein synthesis, 23, 191, 192, 192f, 193f
  translation of, 189–190, 189f
Metacentric chromosomes, 244, 245f, 261
Metagenomics, 11–12, 368, 370
Metaphase, 30, 31f, 40
Metaphase I, 47, 47f
Metaphase II, 48, 48f
Metastasis, 375, 381, 383f

Methemoglobin, 217
Methemoglobinemia type I, 217, 240
Methionine, 192, 193f
Methyl (CH_3) groups
  in chromatin remodeling, 207
  effects of starvation on, 209cc
  in genomic imprinting, 124, 127
  in GWAS studies, 139
  in inactivated DNA, 123
  restriction enzymes and, 398
Methylome, 207
MHC (major histocompatibillity complex), 157, 350–351, 350f
Mice, 402, 402f, 427
Michelangelo, 159t
Microbiome, 9, 14
  human, 38–39, 38f, 40
Microcephaly, 64
Microchimerism, 361, 362f
Microdeletions, 255, 262
Microduplications, 255, 262
Microevolution, 279, 279f, 283
Microfilaments, 21f, 26, 27, 27f, 39
Microglia, 158, 158f
Microproteins, 207, 210, 214
MicroRNAs (miRNAs), 207, 208–209, 208f, 214, 407
Microtubules, 21f, 26–27, 27f, 31
Middlesex (novel), 114
Miescher, Friedrich, 166, 169t
Migraine headache, 42, 131
Migration, 318–321, 319f, 325, 326
  from Africa, 318, 319f
  allele frequencies and, 290–291, 302
  effect on Plain Populations, 292cc
  "Great Migration" of 1910-1970, 321, 322f
  mitochondrial Eve, 318, 319f
  into New World, 320–321, 320f
  spread of agriculture and, 313, 314f
  tracking patterns in, 316–317, 317f
Mild cognitive impairment (MCI), 96cc
Milk, 121
Milk proteins, 21
Milk secretion, 21, 22f
Minerals, 20
Minisatellite sequences, 226
Minnesota Twin Family Study, 138
miRNAs (microRNAs), 207, 208–209, 208f, 214, 407
Miscarriage. *See* Spontaneous abortion
Misfolded proteins, 23
Mismatch mutations, in cancer, 375
Mismatch repair, 234–235, 235f, 238
Missense mutations, 221t, 227, 227t, 228f, 238, 386

Missing heritability, 136, 145
Mitochondria, 39
  bioethics of replacing, 100
  cristae, 24, 25f
  energy production by, 24, 25f
Mitochondrial diseases, 99, 110, 159
Mitochondrial DNA (mtDNA), 336
  analysis of, 101
  in clones, 55
  DNA profiling of, 331–332
  genes encoded in, 97, 99f
  in genetic genealogy, 316, 317f, 338
  of mitochondrial "Eve," 318, 319f
  as molecular clock, 316
  Native American, 320–321
  variations in allele frequencies and, 291
Mitochondrial "Eve," 318, 319f
Mitochondrial genes, 108
  heteroplasmy, 99–101, 100f, 108
  inheritance of, 98–101, 99f, 100f
  mitochondrial diseases and, 99
  mtDNA analysis, 101
  mutation of, 224
Mitochondrial myopathy encephalopathy lactic acidosis syndrome (MELAS), 99
Mitochondrion, 21f, 22f
Mitogenomes, 317
Mitosis, 28, 29f, 30–31, 31f, 39
  aneuploidy and polyploidy in, 249
  in cancer cells, 375
  meiosis compared, 46t
  of zygote, 53, 54f
Mitotic spindle, 29–30
Mixed lineage leukemia (MLL), 388
MKRN3 gene, 124
MLL (mixed lineage leukemia), 388
MN blood group, 348t
Modes of inheritance, 72, 87
Molecular Analysis for Therapy Choice (MATCH), 373
Molecular clocks, 307, 315, 316–317
  admixture, occurrence of, 317, 318f
  genetic markers in, 316–317
Molecular evolution, study of, 315–317, 326
  chromosome comparison, 315–316
  gene and genome comparison, 315, 315f
  molecular clocks, 316–317
  protein comparison, 316
Monoamine oxidase A (MAOA), 164
Monoclonal antibodies, 364–365, 364f
Monoclonal antibody (Mab) technology, 364, 364f

Monogenic (single-gene) diseases, 72, 87. *See also* Single-gene inheritance
Monohybrid cross, 73, 74f, 87
Monosomy, 247t, 248, 262
Monozygotic (MZ; identical) twins, 59–60, 59f, 68
  altering gene expression in space, 182–183
  heritability of traits in, 138–139, 138t, 142
  homosexuality in, 115
  ovary isografts from, 365, 365f
  sleep study of, 150
Mood disorders, 155–156
  bipolar disorder, 150t, 156
  major depressive disorder, 150t, 155, 156f
  prevalence of, 150t
Morgan, Thomas Hunt, 103
Mormons, family records of, 84
"Morphogen" proteins, 58cc
Morpholinos, 406
Morton, Holmes, 292cc
Morula, 53, 57t, 68
Mosaicism, 219f
  chromosomal, 245, 250
  gonadal, 224
  microchimerism, 361, 362f
  somatic, 219, 219f
  X inactivation and, 122, 122f
Mosquitoes
  Aedes aegypti, 409
  Anopheles gambiae, 298
  as vectors, destruction of, 409, 410f
  Zika virus and. *See* Zika virus
Motile cilia, 27
Mouthfeel, 395
MPS (mucopolysaccharidosis) type IIIA, 414–415
mRNA. *See* Messenger RNA (mRNA)
MRSA, 297
MSUD (maple syrup urine disease), 20cc, 285, 292cc
mtDNA. *See* Mitochondrial DNA (mtDNA)
MTHFR C677T gene variant, 63
mTOR inhibitors, 389
Mucopolysaccharidosis II (Hunter syndrome), 123
Mucopolysaccharidosis (MPS) type IIIA, 414–415
Müllerian ducts, 112, 112f
Multifactorial traits, 7f
Multigene families, 244
Multiple alleles, in gene expression, 92, 98t
Multiple endocrine neoplasia, 223t

Multiple endocrine neoplasia 2, 53t
Multiple sclerosis
  as autoimmune disorder, 359t
  overexpressed HERVs in, 212
"Multiple unrelated affected subjects" approach, 159, 163
Mundios, Stefan, 58cc
Muscle tissues, 18, 18t
Muscular dystrophy
  Duchenne type. *See* Duchenne muscular dystrophy (DMD)
  in golden retrievers, 315, 315f
Mutagens, 238, 375
  accidental exposures to, 226
  defined, 224
  intentional use of, 226
  natural exposure to, 226–227
Mutant, definition of, 218
Mutant phenotype, 74, 87
Mutate, 218
Mutation(s), 5, 13, 74, 217–240, 238
  allele frequencies and, 294–295, 295f, 302
  allelic diseases, 223, 223t
  of beta globin gene, 220–221, 220f
  birth defects caused by, 62–64
  blue person disease, 217–218
  in cancer, 373–374, 374f, 380, 380f
  causes of, 224–227, 238
    accidental exposures to mutagens, 226
    induced mutation, 225–226
    natural exposure to mutagens, 226–227
    spontaneous mutation, 224–225, 224f–225f
  in clones, 55
  collagen disorders, 221–223, 221t, 222f, 287
  consanguinity and, 217–218
  defined, 218
  disease caused by, 5, 5f, 6f
  DNA repair and
    BRCA1 mutations, 387, 387f
  driver mutations, 381–382, 382f, 382t, 383f
  factors that lessen the effects, 233
  frameshift, 190
  homeotic, 58cc
  impairment of immune function, 348
  induced mutation, 225–226
    intentional use of mutagens, 226
  monogenic, 72
  mutagens
    accidental exposures to, 226
    defined, 224

Mutation(s), *(Continued)*
   intentional use of, 226
   natural exposure to, 226–227
   nature of, 218–220, 238
      germline and somatic, 218–220, 219f
      polymorphisms compared, 218
   passenger mutations, 381–382, 382f, 382t, 383f
   position of, 232–233, 238
   single-gene disorders, 44
   spontaneous mutation, 224–225, 224f
      mutational hot spots, 224–225, 225f
      rates of, 224
   types, 238
      copy number variants, 230, 232
      deletions and insertions, 229
      expanding repeats, 230, 230f
      missense and nonsense, 221t, 227–228, 228f
      point mutations, 227, 228f
      pseudogenes and transposons, 229–230, 229f
      splice-site mutations, 227–229, 228f
Mutational hot spots, 224–225, 225f
Myasthenia gravis, 359t
MYCN oncogene, 394
MYCN oncogene mutation, 385
Mycobacterium abscessus, 79cc
Mycoplasma genitalium, 273
Myelin, 149f, 150
MYH16 gene, 324
myl oncogene, 384
Myoglobin, 196
Myotonic dystrophy, 230, 230f
MZ twins. *See* Monozygotic twins

N

NA (neuraminidase), 349cc
NADPH oxidase, 356
Nagoya Protocol, 11–12
Nail-patella syndrome, 105–106, 106f
"Nandy" (Neanderthal), 311
Narcolepsy, 150–151, 151f
"Nariokotome Boy," 309
NASA (National Aeronautics and Space Administration), 182
Nash, Adam, 434
Nash, Jack, 434
Nash, Lisa, 434
Nash, Molly, 434
Nathans, Jeremy, 118cc
National Aeronautics and Space Administration (NASA), 182

National DNA Index (NDIS), 331
Native Americans, 320–321, 320f
Natural disasters, 334
Natural exposure to mutagens, 226–227
Natural killer cells, 356t
Natural selection, 295–300
   antibiotic resistance, 297, 297f
   artificial selection: dog breeding, 296–297, 296f
   balanced polymorphism, 297–300, 298t
   negative and positive selection, 295–296, 296f
Naura, Western Samoa, obesity in, 144
ncRNAs (noncoding RNAs), 208, 212–213, 213t
NDIS (National DNA Index), 331
Neanderthals *(Homo sapiens neanderthalensis),* 310–312, 311f, 312t
   Altai Neanderthals, 311, 313, 313f
   comparison of modern DNA with, 310, 323
   discovery of remains, 311
   gene variants in, 312t
   "Oasel," 310
Necrosis, 32, 203
Neel, James, 144
Negative selection, 295–296, 296f
Nelmes, Sarah, 363
Neonatal intensive care unit (NICU), 274, 275
Nervous tissues, 18, 18t
Neural plasticity, 149
Neural tube, 60, 61, 68
Neural tube defect (NTD), 60, 135
Neuraminidase (NA), 349cc
Neurexins, 161, 163
Neurofibrillary tangles, 65, 96cc
Neurofibromatosis type 1, 222t, 224
Neurofibromin (NF), 211, 222t
Neurofibromin gene, 211, 211f, 212t
Neuroglia, 149, 162
Neuroligins, 161, 163
Neurons, 149, 162
Neuropeptide Y, 144t
Neurotransmitters, 149, 149f, 155, 162
Neutrophils, 356t
Newborns
   genetic testing, 419, 419f
   genome sequencing, 11, 274–275, 418t
   hemolytic disease in, 361, 362f
   screening
      for fragile X syndrome, 231cc
New World, migration to, 320–321, 320f

Next-generation sequencing (NGS), 177
NF (neurofibromin), 211
NF1 (Neurofibromatosis type 1), 222t, 224
NGS (next-generation sequencing), 177
Niall Y chromosome, 303
Nicholas II, Tsar of Russia, 101
Nicotine, teratogenicity of, 63
Nicotine addiction, 153–154, 155f
Nicotinic receptors, 153, 154, 155f
NICU (neonatal intensive care unit), 274, 275
Niemann-Pick disease type A (sphingomyelin lipidosis), 294t, 432
Nirenberg, Marshall, 191
Nitrogenous bases, 4, 4f, 13, 170, 170f
Noir Marron, 327
Noncoding RNAs (ncRNAs), 208, 212–213, 213t
Nondisjunction, 248, 248f, 250, 252, 252t, 260, 262
Non-Hodgkin's lymphoma, 388
Noninvasive prenatal diagnosis, 246, 418t
Nonrandom mating
   allele frequencies and, 288–290, 289f, 302
   effect on Plain Populations, 292cc
Nonself antigens, 360
Nonsense-mediated decay, 227, 238
Nonsense mutations, 221t, 227, 227t, 228f, 238, 324
Nonsynonymous codons, 191, 199
Noonan syndrome, 126
Norovirus, 371
Notochord, 60
Nowell, Peter, 385
NTD (neural tube defect), 60, 135
Nuchal ligament, in *H. erectus,* 324
Nuclear envelope, 19, 21f, 22f
Nuclear lamina, 19
Nuclear pores, 21f, 22f
Nucleic acids, 19, 20, 39
Nucleolus, 19, 21f, 22f, 39
Nucleosomes, 172, 179
Nucleotide excision repair, 234, 234f, 238
Nucleotide(s), 170, 170f, 179
Nucleus (nuclei), 2, 13, 19, 21f, 22f
   configuration of DNA in, 172–173, 173f
Numerical abnormalities, chromosomes, 247–254, 247t, 260t, 262. *See also* Chromosome(s)
Nutrients, birth defects caused by, 63–64
Nutrigenics testing, 422

O

"Oasel" (Neanderthal), 310
Obesity, 144, 147. *See also* Body weight
OCA2 gene, 76, 77*f*
Odorant receptor (OR) proteins, 324–325
Office of Malaria Control in War Areas, 299
Okazaki fragments, 175, 175*f*
"Old Man" of La Chapelle-aux-Saints, 311
Old Order Amish. *See also* Plain Populations
   "Amish cerebral palsy," 292*cc*
   bipolar disorder in, 156
   family records of, 84
Oligospermia, 435
Oncogenes, 375, 382–385, 384*f*
One-child policy, in China, 111–112
O'Neill, Eliza, 414
100,000 Genomes Project, 10
Oocyte banking and donation, 441–442, 444*t*
Oocyte(s), 44, 67
   banking and donation of, 441–442, 444*t*
   postmortem retrieval of, 442
   primary, 50–51, 51*f*
   secondary, 50–51, 50*f*, 51*f*
Oogenesis, 50–52, 50*f*–52*f*, 68
   vs. spermatogenesis, 52*f*
Oogonium(a), 50, 68
Open reading frame, 191, 199
Opium, 153
OPRM1 gene, pharmacogenetics of, 423*t*
Opsin genes, 118*cc*
Orexin, 151
Organelles, 18, 19, 21–24, 21*f*, 25*f*, 25*t*, 39
   energy production in, 24, 25*f*
   intracellular digestion in, 23–24, 24*f*
   secretion in, 19, 21, 22*f*, 23
   structures and functions of, 19, 21*f*, 25*t*
Organogenesis, 60
Organoids, 36
Organs, 6
Organ systems, 6
Organ transplantation, 366–367, 366*f*
Ornithine transcarbamylase (OTC) deficiency, 428*cc*
OR (odorant receptor) proteins, 324–325
Orrorin tugensis, 307, 308*f*
Osteoarthritis, 221*t*
Osteogenesis imperfecta ("brittle bone disease"), 65, 95, 98*t*
Osteogenesis imperfecta type 1, 221*t*
Osteoporosis, susceptibility to, 6
Ötzi the Ice Man, 266, 314, 314*f*
Ovarian cancer, detecting, 391*cc*
Ovarian tissue, reimplantation of, 441
Ovaries, 45, 45*f*, 67
Ovulation, 51*f*, 52
Ovulation predictor tests, 435
Ovum (ova), injecting foreign DNA into, 402
Oxygen binding, 232*t*

P

p53 gene, 386, 386*f*
Painter, Theophilus, 265
Palindromic DNA sequences, 113, 399
Pancreas
   pancreatic cancer, 378, 382*t*
   pancreatic islets, 205
   structure of, 204–205, 205*f*
Pancreatic islets, 205
Pantoea agglomerans, 400
Paracentric inversion, 258*f*, 259, 262
Parent-of-origin effects, 124–126, 127
   genomic imprinting, 124–125, 124*f*, 125*f*, 127
   imprinting diseases, 125–126, 126*f*
   sperm and egg production timetables, 126, 127
Parkinson disease, 42, 146, 201, 223*t*
   misfolded proteins in, 197*t*
Passenger mutations, 381, 382*f*, 382*t*
Patau syndrome (trisomy 13), 249*t*, 251–252
Patent and Trademark Office, U.S., 397
Patent(s)
   breast cancer genes, 397
   disputes, 397
   DNA technologies, 396–397, 396*f*
   transgenic organisms, 396, 396*f*
Paternal age effects, 52–53, 53*t*, 126
Paternity testing, 418*t*
Patient selection, in GWAS studies, 142
Pattern baldness, 44, 122
Pauling, Linus, 169, 221
PCR (polymerase chain reaction), 175–177, 176*f*, 179, 331
PCSK9 gene, 146
Peanut allergies, 403
Pea plants
   discovery of linkage in, 101–103, 102*f*, 103*f*
   Mendel's experiments with, 72–74, 73*f*, 74*f*
parent-of-origin effects in, 124
Pedigree(s), 7, 7*f*, 14
   analysis of, 83–86, 88
      conditional probability problem, 85, 86*f*
      exome and genome sequencing and, 85–86, 86*f*
      history of, 84–85, 84*f*
      Mendel's laws and, 85, 85*f*
   for autosomal recessive disorders, 85, 85*f*
   components of, 83, 83*f*
   for Hemophilia B, 117, 119*f*
   for heritability of traits, 137, 137*f*
Penetrance
   in gene expression, 95, 97, 98*t*, 107, 125
   incomplete, 125
   of single-gene diseases, 72
Pennington, Robert, 366
Peptide bonds, 192, 193*f*
Peptide(s), 183
Perchloroethylene exposure, 146
Perforin, 355, 356*f*
Pericentric inversion, 259, 259*f*, 262
Period 2 gene, 151
Peripheral neuropathy, 223*t*
Peroxisome, 21*f*, 23–24, 39
PERV (porcine endogenous retrovirus), 366
Pfeiffer syndrome, 53*t*
PGD. *See* Preimplantation genetic diagnosis (PGD)
Ph[1] ("Philadelphia chromosome"), 385
Phagocytes, 352
Phagocytosis, 352
Pharmacogenetics, 9, 14
Pharmacogenetics testing, 418*t*, 422–423, 423*t*
Pharmacogenomics testing, 422
Pharmacological chaperone therapy, 424, 424*t*
Phenocopies, 95, 97, 98*t*, 107
Phenotype(s), 7, 14, 74
   effect of heteroplasmy on, 100
   effect of X inactivation, 123–124, 123*f*
   Neanderthal, 311, 311*f*
   sex and, 113–115, 114*f*
Phenotypic frequency, 278–279. *See also* Allele frequencies
Phenotypic sex, 115*t*
Phenylketonuria (PKU), 298*t*
   frequency of, 279, 279*t*
   genetic testing for, 419, 419*f*
   misfolded proteins in, 197*t*
   mutations of, 294

Phenylthiocarbamide (PTC), 285
"Philadelphia chromosome" (Ph[1]), 385
Phipps, James, 363
Phocomelia, 62
Phospholipids, 25
Physical barriers, in immunity, 351, 352f
Physical maps, 266, 267f
Pigs
   enviropig, 404
   as organ donors, 366, 366f
Pima Indians, obesity among, 144
Pingelapese blindness, 293
piRNA (piwi-interacting RNA), 213t
Pitchfork, Colin, 330
PITX1 gene, 58cc
Piwi-interacting RNA (piRNA), 213t
PKU. *See* Phenylketonuria (PKU)
Placenta
   formation of, 56
   gene expression, 203
Plain Populations
   founder effect in, 292cc, 292f
   genetic diseases among, 291
Plasma cells, 353, 353f
Plasma membrane, 19, 21f, 22f, 39
Plasmids, 297, 399, 399f, 400
Plasmodium falciparum, 298, 298f
Plasmodium falciparum, 400
Plastin 3 gene, 94
Pleiotropy, 95, 97, 98t, 107
Pluripotent stem cells (PSCs)
   induced, 35, 36t, 37, 40, 216, 429
   in xenotransplants, 366
Point mutations, 227, 228f, 238
Polar bodies, 50–51, 50f, 68
Polyclonal antibody response, 353, 353f
Polycystic kidney disease, 65
Polydactyly
   pedigree for, 84f, 85
   penetrance and expressivity in, 95, 98t, 125
Polyendocrinopathy syndrome, 360
Polygenic complex traits, 131, 145
Polygenic risk score for breast cancer, 394
Polygenic risk scores, 140–141, 141f, 145
Polygenic traits, 133–135, 145
   autoimmunity as, 359–360
   fingerprint patterns, 133, 133f
   height, 133–134, 134f
   skin color and race, 134–135, 134f
Polyglutamine diseases, 230
Polymerase chain reaction (PCR), 175–177, 176f, 179, 331
Polymorphisms, 238
   balanced. *See* Balanced polymorphism

   insertion/deletion, 421
   mutations compared, 218
   SNPs. *See* Single nucleotide polymorphisms (SNPs)
Polynucleotide chains, 170–171, 171f, 172f
Polypeptide chains, 192, 193f, 203, 204, 204f, 354, 354f
Polypeptide(s), 183, 192, 199, 203, 204
Polyploid, 46, 67
Polyploidy, 247, 247f, 247t, 322
Pompe disease, 433–434
Population bottlenecks
   in Beringia, 320
   genetic drift and, 293–294, 294t
Population genetics, 8, 283
   constant allele frequencies. *See* Hardy-Weinberg equilibrium
   defined, 278
Population(s)
   defined, 278, 283
   in DNA profiling, 333
   shrinking, allele frequencies and, 291
Porcine endogenous retrovirus (PERV), 366
Porphyria variegata, 293
Positional cloning studie, 266, 267f
Positive selection, 295–296, 296f
Posthumous, genome sequencing, 265–266
Postmortem gamete retrieval, 442
Postmortem interval, 202, 215
Postsynaptic neuron, 149, 149f
Posttranslational modifications, 193
Posttraumatic stress disorder (PTSD), 148–149
Prader-Willi syndrome, 125–126, 126f, 129
Precision medicine, 9–10
Preconception genetic testing, 417, 419
Preeclampsia, 122
Pregnancy
   home testing for, 56
   immune response during, 361–362
      Rh incompatibility, 361–362, 362f
      systemic sclerosis, 361
      T regs, 361
   infection during, schizophrenia and, 157
   microchimerism and, 361, 362f
Preimplantation genetic diagnosis (PGD), 418t, 419
   in assisted reproduction, 443–444, 443f, 444t
   bioethics of, 301

   of Pompe disease, 433–434
"Preimplantation" stage, 53
"Premature aging" disorders, 65, 65t, 66f
Pre-mRNA, 188
"Premutations," 231cc
Prenatal development, 68
   cleavage and implantation, 53 54f, 55–56
   earliest stages of, 439f
   embryo formation, 56, 57f
   embryonic development, 60–61, 61f
   fertilization, 53, 54f
   fetal period, 61, 61f
   formation of supportive structures, 56–57
   genetic counseling during, 416
   homeotic mutations, 58cc, 58f
   multiple fetuses, 57, 59–60, 59f, 60f
   sexual development, 113–114, 114f
   stages and events, 57t
Prenatal genetic testing, 418t, 419
Presenelin 1 *(PSEN1)*, 97t
Presenelin 2 *(PSEN2)*, 97t
Presynaptic neurons, 149, 149f
Prevalence, 135, 145
Primary cilia, 27
Primary germ layers, 56, 57f, 68
Primary immune response, 353
Primary sex ratio, 115
Primary (1°) structure, of proteins, 195, 195f, 199
Primase, 175, 175f
Primates, 307–309, 308f
   apes and monkeys. *See specific animals*
   Australopithecines, 308–309
   chromosome banding patterns in, 315
   early genus homo, 309, 309f
   Neanderthals. *See* Neanderthals
Primitive streak, 60
Primordial embryo (gastrula), 56, 57f, 57t
Prion diseases, 197–198, 197f, 197t, 198f, 199
Prion protein mutation, 298t
Prion(s), 197, 199
Privacy issues, 337–338
   in DNA profiling, 338
   in genetic testing, 12–13
Probiotics, 38
Procollagen, 221–222, 222f
Product rule, 82, 82f, 88, 332
Progenitor cells, 33, 33f, 34–35, 40. *See also* Stem cell
   in brain tumors, 378–379, 379f

exocrine/endocrine, 205f
Progeria Research Foundation, 44
Progeria(s), 65, 66f
  HGPS, 43, 65, 65t
  treatment of, 424
Progerin, 65
Prokaryotic cells, 18, 19f, 39
  vs. eukaryotic cell, 18
Promoters, 186, 187f, 199, 213t, 227, 400
Promoter sequences, 206, 355
Prophase, 30, 30f, 31f, 40
Prophase I, 46–47, 47f
Prophase II, 48, 48f
Prostate cancer, gene expression in, 206, 210
Prostate gland, 44, 44f
Prostate specific antigen (PSA), 210, 211f, 212t
  blood test, 377, 388
Protanopia, 118cc
Protease inhibitor, 359t
Proteasome, 196, 199
Protein(s), 19, 20, 39, 182–201. See also *specific proteins*
  amyloid beta, 65
  cellular adhesion proteins, 386–387
  comparing protein sequences, 316
  in drug addiction, 152–153
  folding, 195–197, 195f, 196f
  fusion proteins, 384
  importance of, 183, 183f, 198–199
  misfolding, 183, 196, 196f, 197t
  not encoded by human genome
    noncoding RNAs (ncRNAs), 208, 212–213, 213t
  oncoproteins, 385
  processing, 195–198, 195f, 199
    diseases of misfolding, 196–197, 197t
    folding and misfolding, 196, 196f
    prion diseases and, 197–198, 197f, 197t, 198f
  scaffold proteins, 172, 173f
  study of, 167, 168f
  Tau protein, 65
  transcription of, 184–189, 184f, 199
    RNA processing, 188–189, 188f
    RNA structure and types, 184–186, 184f–186f, 184t, 185t
    steps in transcription, 186–187, 187f, 188f
    transcription factors, 186
  translation of, 189–194, 189f, 189t, 199
    building a protein, 191–193, 192f, 193f

genetic code, 189–191, 189t, 190f, 191f
Protein coats, 167, 168f
Protein synthesis, endoplasmic reticulum, 23
Proteomics, 205, 214
Proteus syndrome, 219, 219f, 224
Proto-oncogenes, 382–384, 384f
PSA (prostate specific antigen), 210, 211f, 212t
  blood test, 377, 388
PSA-linked molecule (PSA-LM), 210, 212t
PSA-LM (PSA-linked molecule), 210, 212t
PSCs. See Pluripotent stem cells (PSCs)
PSEN1 (presenelin 1), 97t
PSEN1 gene, 223t
PSEN 2 (presenelin 2), 97t
Pseudoautosomal regions of Y chromosome, 113, 113f
Pseudogenes, 212, 213t, 214
  mutation and, 229–230, 229f
  odorant receptors, 324–325
  silences, 322
Pseudohermaphroditism, 114, 115
Pseudomonas, 79cc
P (peptide) site, 192, 193f
Psoriasis treatment, teratogenicity of, 64
PTC (phenylthiocarbamide), 285
PTSD (posttraumatic stress disorder), 148–149
Punnett, R. C., 101, 103
Punnett squares, 75, 76f, 87
  conditional probability problems, 85, 86f
  multiple-gene problem, 81–82, 82f
Purines, 170, 170f, 179
Pygmies (Africa), 306, 327
Pyloric stenosis, empiric risk of, 135
Pyrimidines, 170, 170f, 179
Pythagoras, 233
PYY protein, 144t

Q

Quantitative trait loci (QTLs), 133, 145
Quaternary (4°) structure of proteins, 195f, 196, 199

R

Race, as polygenic trait, 134–135
Race-based prescribing, 135
Radiation, 226
RAG1 gene mutation, 292cc
Ransacker, Visalia, 343

Ransome, Joseph, 118cc
RB1 gene mutation, 385–386
RB1 tumor suppressor gene, 394
Reading frame, 190, 191f, 199, 228
Recessive alleles, 7, 14, 294
Recessive analysis, for autism, 160, 160f, 163
Recessiveness, 80, 81f
Recessive traits, 73
  autosomal recessive inheritance, 77–78, 77t
  as loss-of-function mutation, 80, 80f
  X-linked, 117–120, 117f, 119f, 121t
Recipient cells, 398
Reciprocal translocation, 257–258, 258f, 262
Recombinant DNA technology, 10, 396, 397–404
  animal models, 402, 402f
  bioremediation, 403–404
  constructing molecules, 398–400, 399f
  drugs developed with, 398t, 400
  genetically modified foods, 402–403, 403t
  products from, 400
  transgenic organisms, 400–404
Recombinant progeny, 102
Recombination mapping, 106, 106f, 108
RECQL3 gene, 294t
Red blood cells, 233
Red hair color, 91–92
Red-ragged fibers, 99
Reduced ovarian reserve, 436
Reduction division, 46, 46f, 47f, 67
Reed, Sheldon, 416
Reference genome sequences, 271
Refugia, 318
Regenerative medicine, 37
Regulatory cells (T regs), 355, 361
Rejection reactions, in transplants, 365, 367
Relatives, 7f
Repeats, 213–214, 213t
  expanding repeats, 230, 230f
  mutational hot spots, 224–225, 225f
  STRs, 316, 321, 330–332, 330t, 331f, 332f
  triplet repeat diseases, 230, 231cc
  VNTRs, 330
Replication bubbles, 175, 176f
Replication fork, 174, 175f, 179
Reproduction, eugenics movement and, 300–301
Reproductive success, 295

Index I-21

Reproductive system, 67
   female, 45, 45f
   male, 44–45, 44f
Reproductive technologies, 433–446
   assisted. See Assisted reproductive technologies (ARTs)
   infertility, 435–437, 437f
   PGD. See Preimplantation genetic diagnosis (PGD)
   "savior siblings," 434, 434f, 445
Reprogramming, cell, 36
Rescue karyotyping, 418t, 419
Research studies. See specific studies and kinds of studies
Resistance genes, 297
Restriction enzymes, 330, 398–399, 399f, 408, 408f
RET gene, 223t
Retinal, 118cc
Retinitis pigmentosa (RP), 129
Retinoblastoma, 385, 385f, 394
Retroviruses, 212, 214
Rett syndrome, 17, 120, 186
Reverse transcriptase, 212, 214, 349cc
Reverse transcriptase inhibitor, 359t
Reverse transcription PCR, 176f
Reverse vaccinology, 367
Rh blood type
   elliptocytosis and, 104–105
   Rh incompatibility, 361–362, 362f
Rheumatic fever, 359t
Rheumatoid arthritis, 359t, 365
Rh incompatibility, 361–362, 362f
Rhino DNA Indexing System (RhODIS), 181
Ribonucleic acid. See RNA (ribonucleic acid)
Ribose, 167, 179
Ribosomal RNA (rRNA), 185–186, 185f, 199
Ribosomes, 18, 21f, 22f, 39, 185, 185f
Ribozymes, 407
Ring chromosomes, 247t, 259–260, 260f, 260t
RISC (RNA-induced silencing complex), 407, 407f
Risk
   of being carrier for recessive disorders, 281–282, 283f
   dietary patterns and cancer risk, 388
   empiric risk, 135–136, 135t, 136f, 143t, 157, 157t
   genetic and environmental risks, 131, 131f
   of inheriting disease, 282, 283f
   polygenic risk score, 394

RNA (ribonucleic acid), 4, 13. See also specific types of RNA
   DNA distinguished from, 168, 184–185, 184t, 185f
   processing, 188–189, 188f
   structure and types, 184–186, 184f–186f, 184t, 185t
   tracking RNA to pinpoint time of death, 202–203
RNAi (RNA interference), 407, 407f
RNA-induced silencing complex (RISC), 407, 407f
RNA interference (RNAi), 407, 407f
RNA polymerase, 184, 186, 187, 187f, 188f, 199
RNA primer, 174, 175, 175f
"RNA tie club," 191
Robertsonian translocations, 255–257, 256f, 257cc, 262
Roderick, T. H., 266
Romanov family, 101
Rosenberg, Noah, 343
Rothmund-Thomson syndrome, 65, 65t
Rough endoplasmic reticulum, 21f, 22f, 23
RP (retinitis pigmentosa), 129
RPE65 deficiency, 428cc
rRNA (ribosomal RNA), 185–186, 185f
RTTN gene, 228
Rubella (German measles), 64
Running, as human trait, 323f, 324
Run of homozygosity, 287, 288f
Russell, Karen, 164
Russell-Silver syndrome, 129

S

Sahelanthropus tchadensis, 307, 308f
Salmonella, 38, 226
Salmonella bacteria, 368
Salmonella typhi, 300
Sanfillippo syndrome, 414, 423
Sanger, Frederick, 177, 178
"Sanger sequencing," 177, 178
"San" people (Africa), 307
Saqqaq Eskimo people (Greenland), 321
Sargasso Sea, metagenomics of, 11
SARS (Severe acute respiratory syndrome), 349cc, 367
SARS-CoV-2, 349cc, 349f
Satellites, of chromosomes, 245
"Savior siblings," 434, 434f, 445
Scaffold proteins, 172, 173f
Schizophrenia, 162, 232
   abnormal imprinting in, 126
   effects of LSD similar to, 153
   prenatal nutrition and, 208, 209cc

   prevalence, 150t
   sibling pair study, 147
   signs and symptoms, 156–157, 157f
   synaptic pruning in, 157–158, 157t, 158f
SCIDs (severe combined immune deficiencies), 357, 357f, 357t, 428cc
   RAG1 gene mutation as cause, 292cc
   X-linked, 428cc
Scleroderma (systemic sclerosis), 223t, 361
SCN1A gene, 42
Scrapie, 197, 198f
Scrotum, 44, 44f
Secondary findings, 420
Secondary immune response, 353
Secondary oocytes, 50–51, 50f, 51f
Secondary sex characteristics, 114
Secondary sex ratio, 115
Secondary (2°) structure of proteins, 195, 195f, 199
Secondary transfer DNA, 337
Secretion, in organelles, 19, 21, 22f, 23
Segregation, law of, 73, 74–75, 75f, 75t, 87
Selective serotonin reuptake inhibitors (SSRIs), 155, 156f
Self-renewal, 18, 40
SELP (P-selectin) gene, 312t
Semiconservative replication, 174, 174f, 179
Seminal vesicles, 44–45, 44f
Sequence maps, 266, 267f
Sequence tagged site (STS), 269f
Sequencing by synthesis, 178
Sequential polar body analysis, 444, 444f, 444t
Serotonin, 155
Severe acute respiratory syndrome (SARS), 349cc, 367
Severe combined immune deficiencies (SCIDs), 357, 357f, 357t, 428cc
   RAG1 gene mutation as cause, 292cc
   X-linked, 428cc
Sex, 111–129
   China's one-child policy, 111–112
   homosexuality, 115, 115t
   inherited traits, 116–121, 117f, 127
      color vision, 117, 118cc–119cc
      sex-influenced traits, 122, 127
      sex-limited traits, 121–122, 127
      solving problems in, 120–121
      X-linked dominant, 120, 120f, 121t
      X-linked recessive, 117–120, 117f, 119f, 121t
   parent-of-origin effects, 124–126, 127

genomic imprinting, 124–125, 124f, 125f, 127
  imprinting diseases, 125–126, 126f
  sperm and egg production timetables, 126, 127
phenotype and, 113–115, 114f
sex chromosomes, 113, 113f
sex ratio, 115–116, 127
sexual development, 112–113, 112f
transgender identity, bioethics, 116
transgender identity, genetic basis of (bioethics), 116
X inactivation, 55, 122–124
  effect on phenotype, 123–124, 123f
  female as mosaic, 122–123, 122f
Sex chromosomes, 6, 13, 113, 113f. See also X chromosome; Y chromosome
  aneuploidy in, 252–254, 252t
    female, 252–253
    female infertility due to, 437, 437f
    male, 253–254
Sex determination, 61
Sex-influenced traits, 122, 127
Sex-limited traits, 121–122, 127
Sex ratio, 115–116, 127
Sexual development, 112–113, 112f
Sexual identity, 115, 115t
Sexual orientation, 115, 115t
Sheep, scrapie in, 197, 198f
Shigatoxigenic (STEC) E. coli, 348f, 368
Shiga toxin, 368
Short tandem repeats (STRs), 316, 321, 328, 329, 330–332, 330t, 331f, 332f, 333f, 334f, 342–343, 343f
Siamese (conjoined) twins, 60, 60f
"Sick cilia disease," 27
Sickle cell disease, 298–299, 298f
  carriers of, 80, 420
  conditional probability problems, 85, 86f
  as gene mutation, 220, 220f, 221, 238
  genetic testing for, 418t
  history of research on, 221
  IPS in treatment of, 429
Signal transduction, 26, 39, 351
Single-gene disorders
  characteristics of, 72
  cystic fibrosis as, 72, 78, 79cc
  environmental factors, 131, 131f
  mutations, 44, 65
  penetrance of, 72
  population genetics, 278, 278f
  X-linked, 116
Single-gene inheritance, 71–90, 88

complexity of (rare), 76–80
dominance and recessiveness defined, 78, 80
eye color, 76, 77f, 131
illustrative problem, 78
modes of inheritance, 77–78, 77f, 77t, 79cc, 87
disorders of. See Single-gene disorders
following, 72–76
  characteristics of diseases, 72
  Mendel's experiments, 72–74, 73f, 74f
  terminology of, 74–75, 75f, 75t, 76f
following multiple genes, 80–83
  illustrative problem in, 81–82, 82f, 83f
  law of independent assortment, 80–81, 81f, 82f
meiosis and, 52
pedigree analysis, 83–86, 88
  components of, 83, 83f
  conditional probability problem, 85, 86f
  exome and genome sequencing and, 85–86, 86f
  history of, 84–85, 84f
  Mendel's laws and, 85, 85f
Single-gene mutations, 44, 65
Single-gene traits, 72
Single nucleotide polymorphisms (SNPs), 328, 329, 339, 341
  of African Americans, 320f, 321
  copy number variants distinguished from, 230, 232
  genetic genealogy, 342–343, 343f
  as genetic markers, 317, 317f
  in GWAS studies, 139–141, 139t, 140f, 145
  in K. pneumoniae, 368
  schizophrenia and, 157
siRNAs (small interfering RNAs), 213t, 407, 407f
Site-directed mutagenesis, 226
Skeletal muscle fibers, 99
Skin
  collagen mutation and, 222, 222f
  human, 323–324, 323f
Skin color, 134–135, 134f
SLC6A4 gene, 164
SLC6A11 gene, 312t
SLC35F3 gene, 312t
SLCO1B1 gene, 423t
Sleep, 162
  disturbed, depression and, 155

familial advanced sleep phase syndrome, 151, 163
narcolepsy, 150–151, 151f
Sleep paralysis, 150
"Sloppy" DNA polymerase, 235, 236
SMA (spinal muscular atrophy), 94, 223t, 272, 406
Small interfering RNAs (siRNAs), 213t, 407, 407f
Small nuclear ribonucleoproteins (snRNPs), 188–189
Small nuclear RNAs (snRNAs), 213t
Small nucleolar RNA (snoRNA), 213t
Smallpox, 349cc, 362, 363, 363f
Smith-Lemli-Opitz syndrome, 298t
SMN1 gene, 406
SMN2 gene, 406
Smoking cessation drug, 155f
Smooth endoplasmic reticulum, 21f, 22f, 23
SMPD1 gene, 294t
snoRNA (small nucleolar RNA), 213t
SNPs. See Single nucleotide polymorphisms (SNPs)
snRNAs (small nuclear RNAs), 213t, 407, 407f
snRNPs (small nuclear ribonucleoproteins), 188–189
Somatic cell(s), 6, 13, 18, 36
Somatic gene therapy, 425
Somatic mosaicism, 218, 218f, 238
Somatic mutations, 219–220, 219f, 238
Space, gene expression in, 182–183
Spanish flu of 1918, 349cc
SPATA16 (spermatogenesis-associated protein 16), 436cc
Speech, as human trait, 323
Sperm, 44, 67, 438, 442
Spermatid, 49, 49f
Spermatocyte, 49, 49f
Spermatogenesis, 49–50, 49f, 52f, 67
  vs. oogenesis, 52f
Spermatogenesis-associated protein 16 (SPATA16), 436cc
Spermatogonium, 49, 49f, 67
Spermatozoa, 49, 49f, 67
Sperm selection, 417, 418t
S phase, 29–30, 29f, 31f, 39, 174, 184
Sphingomyelin lipidosis (Niemann-Pick disease type A), 432
Spinal cord injury
  gene expression response to, 404, 405f, 406t
  IVF with ICSI and, 440, 440f
  male infertility due to, 435

Index I-23

Spinal muscular atrophy (SMA), 94, 223t, 272, 406
Spindle, mitotic, 29–30, 40
Spindle assembly checkpoint, 31f, 32, 376f
Spleen, 351, 351f
Spliceosome, 188
Splice-site mutations, 227–229, 228f, 238
Spongiform encephalopathy, 298t
Spontaneous abortion
   "blighted ovum," 52
   due to atypical chromosomes, 252, 253, 255, 256, 256f, 259
   due to teratogens, 64
   maternal age and, 435
Spontaneous mutation, 224–225, 224f
   mutational hot spots, 224–225, 225f
   rates of, 224
Sporadic cancers, 377, 377f
SRY gene, 61, 112, 113f
SSRIs (selective serotonin reuptake inhibitors), 155, 156f
Stahl, Franklin, 174
Staphylococcus aureus, 297, 297f
Starvation. See Malnutrition
Stearoyl-CoA desaturase-1, 144t
STEC (Shigatoxigenic) *E. coli*, 348f, 368
Steel syndrome, 287, 288f, 293, 302
Stem cells, 6–7, 13, 18, 33f
   adult, 36, 36t
   applications, 36–37
   bioethics of use, 37
   cancer stem cells, 378–379, 379f
   cell lineages, 33–35, 33f, 34f
   embryonic stem (ES) cells, 35, 36, 36t, 40
   healing, 35, 35f
   iPS. See Induced pluripotent stem (iPS) cells
   PSCs. See Pluripotent stem cells (PSCs)
   sources, 35–36, 36t
   tourism, 37
   from umbilical cord, 37, 57
"Stemness," in cancer cells, 379, 380f
Stickler syndrome, 221t
"Sticky ends," 399, 399f
STIM1 gene, 312t
Streptococcus pneumoniae, 166–167, 367
STRs (short tandem repeats), 316, 321, 328, 329, 330–332, 330t, 331f, 332f, 333f, 334f, 342–343, 343f

Structural abnormalities, chromosomes, 254–260, 260t, 262. See also Chromosome(s)
STS (sequence tagged site), 269f
Sturtevant, Alfred, 103
Stuttering, twin study of, 142
Subclades, 317
Subfertility, 435
Submetacentric chromosomes, 244, 245f, 261
Substance use disorders, 152–153, 153f
Substrate reduction therapy, 424, 424t
Subtelomeres, 244, 244f
"Successful aging genes," 67
Sugar-phosphate backbone, 170–171, 171f, 172f, 175, 175f, 179
Sugar(s), in gene expression, 211, 212t
Suicide, schizophrenia and, 156
Sunburn peeling, 33, 33f
"Superovulation," 436
Surrogate motherhood, 438, 444t
Survivins, 31
Susceptibility testing, 418t
Sutton, Josiah, 328–329
Sweat, 324
Sweetie Pie (pig), 366
Swine flu of 2009, 349cc
Synapses, 149, 162
Synaptic pruning, in schizophrenia, 157–158, 157t, 158f
SYNGAP1 gene mutation, 420
Synonymous codons, 191, 199
Synteny, 316
Systemic lupus erythematosus, 359t, 360
Systemic sclerosis (scleroderma), 223t, 361

T

TAAD (thoracic aortic aneurysm and dissection), 427, 429
TALENs (transcription-activator-like effector nucleases), 408, 408t, 429
Tandem duplication, 229, 238
TATA binding program, 186, 187f
TATA box, 186, 187f, 207f
Tau protein, 65, 96cc
Tautomers, 224
Tay-Sachs disease, 23, 294t
   carriers of, 78, 283
   complete dominance in, 92
   as lethal allele disorder, 92
   near-disappearance of, 290
T cell receptors, 355, 356f
T cell(s), 356t
   CD4 T cells, 355, 357, 358f
   cytotoxic, 355, 356, 356f, 356t
   helper T cells, 351f, 355, 356t, 358, 358f
   in immunity, 351, 351f, 353, 353f, 355–356, 356f
Telocentric chromosomes, 245, 245f
Telomerase, 375–376
Telomerase RNA, 213t
Telomeres, 32, 32f, 40, 213t, 243, 243f, 375–376
Telophase, 31, 31f, 40
Telophase I, 47f, 48
Telophase II, 48, 48f
Template strand, of DNA, 184, 184f, 199
Teratogens, 62–64, 68
Teratoma, 125
TERT gene mutation, 376
Tertiary sex ratio, 115
Tertiary (3°) structure of proteins, 195–196, 195f, 199
Test cross, 75, 76f, 87, 103
Testes, 44, 44f, 67
Testosterone, 113, 114, 114f
Tetrachromats, 119cc
Tetrahydrocannabinol (THC), 153, 154cc
TGFβR (transforming growth factor β receptor), 97, 222t
Thalassemia, 220
Thalidomide, 62, 63, 97
Thalidomide babies, 63
Thanatophoric dysplasia, 53t
THC (tetrahydrocannabinol), 153, 154cc
Thiamine transport, 312t
Thoracic aortic aneurysm and dissection (TAAD), 427, 429
Three Identical Strangers (film), 138
"Three-parent" embryos, 99, 100
"Three strikes," 381, 382f, 382t
"Thrifty gene hypothesis," 144
Thymine (T), 4, 4f, 13, 170, 170f, 171, 172f, 184f, 184t
Thymus, 351, 351f, 357, 357f, 357t
Time of death, tracking, 202–203
Tissue(s), 6, 13
   types, 18, 18t
   uncontrolled tissue repair, 380, 380f
TLR 1, 6, 10 genes, 312t
TNF (tumor necrosis factor), 352, 352t, 365
Tobacco use disease, 312t
Tolerance, in drug addiction, 152, 153
Toll-like receptors, 351
Tomatoes, 10, 406
Torsion dystonia, 284
Tortoiseshell cats, 123

Total ridge count, 133, 133f
Totipotency, 35, 40
Touch DNA, 337
Toxic mutation, 218
TP53 tumor protein, 386f
Traits
  behavioral, 131
  dominant, 73
  inherited, 116–121, 117f, 127, 170
  Mendelian, 6, 7f
  multifactorial. *See* Multifactorial traits
  recessive, 73, 80
    as loss-of-function mutation, 80, 80f
    X-linked, 117–120, 117f, 119f, 121t
  sex-influenced, 122, 127
  sex-limited, 121–122, 127
  uniquely human, 323, 323f
  variability of, 136
  X-linked, 104
    dominant, 120, 120f, 121t
    recessive, 117–120, 117f, 119f, 121t
trans configuration, 103, 103f, 105, 271
Transcription, 4, 13, 184–189, 184f, 199
  defined, 186
  point mutations controlling, 227
  of proteins. *See* Protein(s)
  RNA processing, 188–189, 188f
  RNA structure and types, 184–186, 184f–186f, 184t, 185t
  steps in transcription, 186–187, 187f, 188f
  transcription factors, 186
Transcription-activator-like effector nucleases (TALENs), 408, 408t, 429
Transcription elongation, 186–187, 187f
Transcription factor binding, 186, 187f
Transcription factor(s), 113, 127, 186, 205, 206, 210
Transcription initiation, 186, 187f
Transcription termination, 187, 187f
Transcriptome, 202, 203, 205, 214
Transfer RNA (tRNA), 21, 39, 185, 185t, 186, 186f, 199
  mitochondrial diseases and, 99
  noncoding, 212, 213t
  in protein synthesis, 191–193, 192f, 193f
Transforming growth factor β receptor (TGFβR), 97, 222t
"Transforming principle," 167, 167f
Transgender identity, genetic basis of, 116
Transgenic organisms, 396, 396f, 400–404, 401t

Transitions, as point mutations, 227, 238
Translation, 4, 13, 189–194, 189f, 189t, 199
  building a protein, 191–193, 192f, 193f
  genetic code, 189–191, 189t, 190f, 191f
  of proteins. *See* Protein(s)
Translation elongation, 192, 193f
Translation initiation, 191–192, 192f
Translation termination, 193f
Translocation carrier, 256, 262
Translocation Down syndrome, 255–258, 260t
Translocation(s), 247t, 248, 255, 256, 258, 260f, 260t
Transmissible spongiform encephalopathy, 298t
"Transmitting males," 230, 231cc
Transplants, 365, 367
  altering immunity with, 365, 367
  rejection reactions, 365, 367
  types of, 365, 365f
  xenotransplants, bioethics of, 366, 366f
Transposons, 213, 213t, 214, 229–230, 229f, 297
Transversion, as point mutation, 227, 238
Traztuzumab (Herceptin), 365
T regs (regulatory cells), 361
TREM (triggering receptor expressed on myeloid cells), 97t
Treponema denticola, 38
Trichloroethylene exposure, 146
Trichothiodystrophy, 65, 65t, 235–236
Trichromats, 118cc
Triggering receptor expressed on myeloid cells *(TREM)*, 97t
Trio analysis, for autism, 160, 160f, 163
Triplet repeat diseases, 230. *See also* Fragile X syndrome
Triplet(s), of DNA bases, 229
Triplo-X, 252t, 253
Trisomy, 244, 244f, 246, 257cc
Trisomy 13, 249t, 251–252
Trisomy 18, 249t, 250–251, 251f
Trisomy 21, 249–250, 249t, 250f, 251f, 252, 262
  bioethics, 252
  maternal serum markers for, 252
  testing for, 244f, 246f
Trisomy 21 Down syndrome, 244f, 247t, 249, 251, 251f, 253
tRNA. *See* Transfer RNA (tRNA)

Trophoblast, 56
TRPV4 gene, 223t
Tubal (ectopic) pregnancy, 437, 437f
TUBB gene, 42
Tubulinopathy, 42
Tumor necrosis factor (TNF), 352, 352t, 365
Tumors, fibroid, 437, 437f
Tumor suppressor genes, 374, 375, 385–387, 394
  p53 gene mutation, 386, 386f
  retinoblastoma and, 385–386, 385f, 394
"Tune deafness," 138
Turner, Henry, 254
Turner (XO) syndrome, 252t, 254, 443
Tutu, Desmond, 307
21-hydroxylase deficiency, 114
Twins, 57, 60, 60f
  conjoined (Siamese), 60, 60f
  dizygotic (DZ), 57, 68, 138–139, 138t, 142
  monozygotic. *See* Monozygotic twins
Twin studies, 162
  altering gene expression in space, 182–183
  of complex traits, 138–139, 138t, 139f, 142
  Parkinson disease, 146
  sleep studies, 150
Typhoid fever, 298t, 300
Tyrosine kinase, 385
Tyrosine-related protein 1 *(TYRP1)* gene, 142, 142f
TYRP1 gene, 142, 142f

U

Ubiquitin, 196, 196f
Ulcerative colitis, 359t
Ullman, Emmerich, 366
Ullrich syndrome. *See* XO syndrome
Ultrasound, fetal, 245, 246f
Umbilical cord, 56, 57
Umbilical cord blood, 37
Uncombable hair syndrome, 110
Unfolded protein response, 196
Uniparental disomy (UPD), 260–261, 261f, 262
Uracil (U), 4, 4f, 13, 184f, 184t
"Urban microbiome," 11
Urinary disorders, 312t
Uterine (fallopian) tubes
  female infertility and, 437, 439
  GIFT and ZIFT technology, 440–441

V

Vaccines
  altering immunity with, 362–364, 363f
  history of vaccine technology, 363
  against measles, 359
  reverse vaccinology, 367
  using recombinant DNA technology, 400
Valine, 220
Valproic acid, 62
"Vanishing twin" phenomenon, 60
van Leeuwenhoek, Antonie, 38
Vaping, 63
Variable number of tandem repeats (VNTRs), 330
Variant of uncertain significance (VUS), 237, 387
Varicose veins, scrotal, 435
Vascular endothelial growth factor (VEGF), 378
Vas deferens. *See* Ductus deferens
Vectors
  cloning, 398, 399
  in gene therapy, 428cc–429cc
  in gene transfer therapy, 425
  mosquitoes as, destruction of, 409, 410f
  "self-inactivating," 428cc
VEGF (vascular endothelial growth factor), 378
veggie burger, 395–396
Venter, J. Craig, 266, 268, 270
Vesicles, 23, 39
Vetter, David, 357, 357f
Victim identification
  genetic testing, 418t
  genocide, 334–335
  natural disasters, 334
  reuniting families, 335–336, 336f
  World Trade Center, 333–334
Victoria, Queen of England, 84, 117
Vindija Cave (Croatia), 311, 313f
Viral DNA, 212
Viral infections
  birth defects caused by, 64
  use of CRISPRs against, 408
Viral teratogen, 64
Viruses, 349cc, 349f
  cancer-causing, 384
  from pigs, 366
  protein coats, 167, 168f
Visualization, in detecting chromosomes, 245, 246f

Vitamin A, 64
Vitamin C, 64
Vitamins, 20
VKORC1 gene, 422, 423t
VNTRs (variable number of tandem repeats), 330
VUS (variant of uncertain significance), 237, 387

W

Walking upright, 323f, 324
Wall, Kim, 336
Warfarin (Coumadin), 422, 423t
Watson, James, 166, 168–169, 169f, 169t, 171, 174, 184, 266, 268, 269
Wedgwood, Emma, 287
Weinberg, Wilhelm, 280
Werner syndrome, 65, 65t
WHIM syndrome, 242–243
Whole-genome duplication, 322–323
Whole genome sequencing, 274
Whole-genome shotgun sequencing, 268, 269f, 270f
Wild type phenotype, 74, 87
Wilkins, Maurice, 168, 169, 169t
Wilms' tumor, 385
Wilson disease, 20
Wing, Lorna, 161
Winkler, H., 266
Wnt4 gene, 113
"Wobble" position, 191
Wolffian ducts, 112, 112f
World Trade Center, identifying victims of, 333–334

X

Xanthomonas bacterium, 408
X chromosome, 104, 113
Xenograft, 365, 365f
Xenotransplants, 366, 366f
Xeroderma pigmentosum (XP), 236, 236f
X inactivation, 55, 122–124, 127
  in clones, 55
  effect on phenotype, 123–124, 123f
  female as mosaic, 122–123, 122f
  genome editing to treat 21 Down syndrome, 251f
  skewed, autoimmunity and, 360
XIST gene, 122
XIST long noncoding RNA, 407
XIST RNA sequence, 250, 251f

X-linked genes, 116, 127
X-linked hemophilia A, 229–230, 283
X-linked inheritance
  dominant, 120, 120f, 121t
  recessive, 117–120, 117f, 119f, 121t
X-linked lymphoproliferative disease, 357t
X-linked traits, 104
XO (Turner) syndrome, 252t, 254, 443
XP (xeroderma pigmentosum), 236, 236f
X rays, 226
XX male syndrome, 113
XXY (Klinefelter) syndrome, 252t, 253–254, 445
XXYY syndrome, 252t, 254
XY female syndrome, 113
XYY (Jacobs) syndrome, 252t, 254

Y

Y chromosome
  DNA sequencing of, 316–317, 317f
  gender and, 113, 113f
  haplogroups, 338, 339f, 339t, 340f, 340t
  male infertility due to deletions, 435
  in Neanderthals, 311
  pseudoautosomal regions, 113, 113f
Y-linked genes, 116, 127
Yolk sac, formation of, 56
Yoruba people (Nigeria), 306

Z

Zeta globin chains, 204
ZFNs (zinc finger nucleases), 408, 408t, 429
ZIFT (zygote intrafallopian transfer), 440–441, 444t
Zika virus, 372
  brain abnormalities due to, 64
  destruction of vectors for, 409
  teratogenicity of, 64
Zinc finger nucleases (ZFNs), 408, 408t, 429
Zinc fingers, 186
Zona pellucida, 53, 54f, 440, 441
Zygote(s), 68
  cleavage of, 53, 54f, 57t
  sex determination in, 117, 117f
Zygote intrafallopian transfer (ZIFT), 440–441, 444t